FATIGUE AND FRACTURE

UNDERSTANDING THE BASICS

Edited
by
F.C. Campbell

ASM International®
Materials Park, Ohio 44073-0002
www.asminternational.org

First printing, November 2012
Second printing, February 2022

Comments, criticisms, and suggestions are invited, and should be forwarded to ASM International.

Prepared under the direction of the ASM International Technical Book Committee (2011–2012), Bradley J. Diak, Chair.

ASM International staff who worked on this project include Scott Henry, Senior Manager, Content Development and Publishing; Karen Marken, Senior Managing Editor; Steve Lampman, Content Developer; Sue Sellers, Editorial Assistant; Bonnie Sanders, Manager of Production; Madrid Tramble, Senior Production Coordinator; and Diane Whitelaw, Production Coordinator.

Library of Congress Control Number: 2012940642

ISBN-13: 978-1-61503-976-0 (print)
ISBN-10: 1-61503-976-7 (print)
ISBN: 978-1-61503-976-0 (pdf)
ISBN: 978-1-62708-303-4 (electronic)

SAN: 204-7586

ASM International®
Materials Park, OH 44073-0002
www.asminternational.org

Printed in the United States of America

Contents

Preface

This book deals with the fatigue and fracture of engineering materials. Although modern fatigue analysis and fracture mechanics are mathematical disciplines, and mathematical rigor is normally found in texts on fatigue and fracture mechanics, I have endeavored to construct a more basic book that balances the major points of fatigue and fracture analysis without burdening the reader with too much complex mathematical development.

This book is ideal for the engineer with a basic knowledge of materials that is just starting to be involved with either failure analysis or fatigue and fracture analysis. It also provides a sound technical groundwork for further study of more advanced texts. This book is useful for the experienced failure analyst that needs a more thorough background in fatigue and fracture mechanics, or for the fatigue and fracture engineer that needs to know more about failure modes.

The first chapter gives a high level introduction to fatigue and fracture and describes some of the noteworthy failures that have occurred during the industrial age from 1900 to date. As a result of these failures, many of which were brittle, unexpected, and catastrophic in nature, there have been numerous changes to design philosophy, and after World War II, the evolution of fracture mechanics. A number of the life limiting factors are also introduced, including material defects, manufacturing defects, stress concentrations, elevated temperatures, and environmental degradation.

The second chapter covers the basics of the static mechanical properties of materials. Included are tension, compression, shear, torsion, and combined stress. Both the engineering and true stress-strain curves are covered in some detail. The importance of stress concentrations and residual stresses are also included.

The third chapter explains the difference between ductile and brittle fracture modes from both a macroscopic and microscopic level. The general characteristics, the macrostructural and microstructural aspects of both ductile and brittle fractures are explained. The ductile-to-brittle transition in steels is covered. Finally, intergranular failures are discussed along with embrittling causes.

The fourth chapter provides an introduction to fracture mechanics. The chapter starts with the pioneering work that Griffith conducted on brittle fracture that led to the development of linear elastic fracture mechanics. For metals that exhibit high levels of ductility, elastic plastic fracture mechanics is presented. Toughness tests discussed include: Charpy V-notch, drop weight tests, linear elastic fracture toughness, and nonlinear fracture mechanics testing (J-integral). The last section covers the important variables that affect fracture toughness; yield strength, loading rate, temperature, material thickness, and material orientation and anisotropy.

The fifth chapter covers the basics of fatigue. Topics covered include: high cycle fatigue, low cycle fatigue, fatigue life prediction, cumulative damage, fatigue crack nucleation and growth, the fracture mechanics approach to fatigue crack growth, crack closure, geometrical and manufacturing stress concentrations, temperature effects, fatigue life improvement methods, and fatigue design methodologies.

With the background provided by the previous chapters on fracture mechanics and fatigue, the sixth chapter covers the fatigue and fracture of three different classes of engineering alloys–steels, aluminum alloys, and titanium and titanium alloys.

The seventh chapter covers the important topic of structural joints–mechanically fastened and welded. It should be noted that many failures often start at joints. Mechanical fastener topics covered include threaded fasteners in tension, shear, and bearing and how preload affects joint performance. For welding, stress concentrations due to joint configurations, weld shape, and discontinuities are discussed along with methods to improve the fatigue and fracture resistance of welded joints.

The eighth chapter covers the basics of fracture control and damage tolerance analysis. Fracture control is the concerted effort to ensure safe operations without catastrophic failure by fracture, while damage tolerance is the property of a structure to sustain defects or cracks safely, until such time that action is (or can be) taken to eliminate the cracks by repair or by replacing the cracked structure or component. Important topics include residual strength, crack growth prediction, and nondestructive inspection.

The next two chapters cover the fatigue and fracture of ceramics and polymers in the ninth, and continuous fiber polymer matrix composites in the tenth. Ceramics are inherent brittle materials, while polymers may be either highly ductile or brittle. The fracture and fatigue behavior of ceramics is briefly covered. For polymers (i.e., plastics), important topics covered include: thermosets and thermoplastics, viscoelastic behavior, static strength, fatigue strength, fatigue crack propagation, impact strength, and the environmental performance of plastics.

The fatigue strength of continuous fiber polymer matrix composites, especially those with carbon fibers, is excellent. However, composites have a low resistance to impacts. Therefore, the thrust of Chapter 10 is to discuss impact events and the effects of delaminations and other defects on structural performance. The so-called "building block approach" to composite structural certification used in the aircraft industry is briefly covered.

The eleventh chapter covers high temperature failures with an emphasis on creep and stress rupture. Creep deformation mechanisms and failure modes are discussed with an emphasis on metallurgical instabilities and environmental effects. Creep life prediction, creep-fatigue interaction, and design for creep resistance are also included.

The topic of wear is covered in the twelfth chapter. Types of wear discussed include: abrasive wear, erosive wear, grinding wear, gouging wear, adhesive wear, and fretting wear. This is followed by wear in which fatigue is a major contributor: contact stress fatigue, subsurface-origin fatigue, surface-origin fatigue, subcase-origin fatigue and cavitation fatigue. Wear prevention methods are discussed.

In the thirteenth chapter, on environmentally-induced failures, first, some basic principles of electrochemical corrosion are covered and then some of the various types of corrosion, with an emphasis on stress corrosion cracking and corrosion fatigue. This is followed by a short section on corrosion control. The last section deals with high temperature oxidation which usually occurs in the absence of moisture.

The steps in the failure analysis process are outlined in the fourteenth chapter: collection of background data and samples, preliminary examination, nondestructive and mechanical testing, macroscopic and microscopic examination, determination of failure mechanism, chemical analysis, analysis by fracture mechanics, and testing under simulated service conditions.

I have included an appendix on defects that can lead to fatigue and fracture. It gives some degree of detail on defects that can lead to failure, including design deficiencies, material and manufacturing defects, and service life anomalies. Major topics include ingot related defects, forging imperfections, sheet forming imperfections, casting defects, heat treating defects, and weld discontinuities.

I would like to acknowledge the help and guidance of the staff at ASM International; Scott Henry, Steve Lampman, Karen Marken, Bonnie Sanders, Madrid Tramble, and Diane Whitelaw, for their valuable contributions.

F.C. Campbell

CHAPTER 1

Introduction to Fatigue and Fracture

IT IS OFTEN STATED that history repeats itself. Yet, when it comes to the failure of structural components and equipment, structural designers, manufacturers, and users do not want a repeat of history. The consequences and costs of fractured, cracked, corroded, and malfunctioned equipment are unwanted, dangerous, and expensive. Through the years, history has demonstrated that failures occur. History has also shown that the engineering communities have responded to prevent failure from occurring again. Some of the historic structural failures that have occurred in the 20th century are summarized in Table 1. These historic failures, as well as other failures, have revolutionized design philosophies, inspection techniques and practices, material development, and material processing and controls and have redefined the criteria for failure. Furthermore, the pursuit of understanding how and why these failures occurred has resulted in the development of structural integrity programs, enhanced analytical modeling and prediction techniques, accurate life-assessment methods, and a fortified commitment to avoid the recurrence of these failures through improved designs. The examples cited in Table 1 were serious and often tragic failures that had a great impact on structural designs and life-assessment developments. However, not all failures or malfunctions of equipment are as pivotal in history as those mentioned in Table 1. Yet, any failure, no matter how seemingly insignificant, should be investigated and the findings used to improve the design and increase the life and reliability of that component or equipment.

Industrial Significance of Fatigue

Fatigue is the process of progressive localized permanent structural change occurring in a material subjected to condi-

tions that produce fluctuating stresses and strains at some point (or points) and that may culminate in cracks or complete fracture after a sufficient number of fluctuations.

A simplistic view of the fatigue process is shown in Fig. 1. In this example (Fig. 1a), the component is first loaded from a zero load (stress) to some maximum positive value, and then the load starts reversing, falling back through zero to a maximum negative value and finally back to zero to complete one cycle. After a number of such cycles, a small crack will initiate, usually on or near the surface at a discontinuity such as a scratch or gouge. As more cycles accumulate, the crack grows until finally the remaining uncracked portion can no longer carry the load, and the component fractures. The fatigue lives of typical steel and aluminum alloys are shown in Fig. 1(b). If the stress is low enough for this steel alloy, it can be

Table 1 Historic failures and their impact on life-assessment concerns

Failure	Year	Reason for failure	Life-assessment developments
Titanic	1912	Ship hits iceberg and watertight compartments rupture.	Improvement in steel grades Safety procedures established for lifeboats Warning systems established for icebergs
Molasses tank failures	1919, 1973	Brittle fracture of the tank as a result of poor ductility and higher loads	Design codes for storage tanks developed Consideration given to causes for brittle fracture
Tacoma bridge failure	1940	Aerodynamic instability and failure caused by wind vortices and bridge design	Sophisticated analytical models developed for resonance Bridge design changed to account for aerodynamic conditions
World War II Liberty ships	1942–1952	1289 of the 4694 warships suffered brittle fracture or structure failure at the welded steel joints.	Selection of increased toughness material Improved fabrication practices Development of fracture mechanics
Liquefied natural gas (LNG) storage tank	1944	Failure and explosion of an LNG pressure vessel due to a possible welding defect and improperly heat treated material resulting in subsequent fatigue crack growth	Selection and development of materials with improved toughness at the service temperature of –160 °C (–250 °F)
Comet aircraft failures	1950s	Fatigue crack initiation in pressurized skins due to high gross stresses and stress-concentration effects from geometric features	Development of the fatigue "safe-life" approach Evaluation of the effects of geometry and notches on fatigue behavior Evaluation of the effects of stiffeners on stress distribution Establishment of aircraft structural integrity program (ASIP) in 1958
F-111 aircraft No. 94 wing pivot fitting	1969	Fatigue failure due to material defect in high-strength steel	Improved inspection techniques Change from fatigue "safe-life" to damage-tolerant design philosophy Development of materials with improved toughness
Seam-welded high-energy piping failures	1986–2000	Cavitation and creep voids in welds resulting in catastrophic high-energy rupture	Development of elevated-temperature life-assessment techniques for cavitation and creep failure
Aloha incident, Boeing 737	1988	Accelerated corrosion and multiple fatigue crack-initiation sites in riveted fuselage skin	Improved aircraft maintenance and inspection procedures Life-assessment methods developed for multiple-site damage (MSD)
Sioux City incident	1989	Hard alpha case present in titanium fan disk resulted in fatigue crack initiation and catastrophic failure.	Increased process controls on processing of titanium ingots Development of probabilistic design approach and analytical life assessment using dedicated computer programs for titanium disks
Earthquake in Kobe City, Japan, and Northridge, California	1994, 1995	Failure occurred in I-beams and columns due to joint configuration and welding practices that resulted in low ductility of the steel.	Development of earthquake-resistant structures Improved joint designs and welding practices for structural steels Improved controls on steel manufacture

Source: Ref 1

theoretically cycled forever; that is, it has a definite endurance limit. On the other hand, aluminum alloys do not have an endurance limit; if enough cycles are applied at even very low loads, they will eventually fail in fatigue.

The discovery of fatigue occurred in the 1800s when several investigators in Europe observed that bridge and railroad components were cracking when subjected to repeated loading. As the century progressed and the use of metals expanded with the increasing use of machines, more and more failures of components subjected to repeated loads were recorded. By the mid-1800s, A. Wohler proposed a method by which the failure of components from repeated loads could be mitigated and, in some cases, eliminated.

Undoubtedly, earlier failures from repeated loads had resulted in failures of components such as clay pipes, concrete structures, and wood structures, but the requirement for more machines made from metallic components in the late 1800s stimulated the need to develop design procedures that would prevent failures from repeated loads of all types of equipment. This activity was intensive from the mid-1800s and is still underway today. Even though much progress has been made, developing design procedures to prevent failure from the application of repeated loads is still a daunting task. It involves the interplay of several fields of technology, namely, materials engineering, manufacturing engineering, structural analysis (including loads, stress, strain, and fracture mechanics analysis), non-

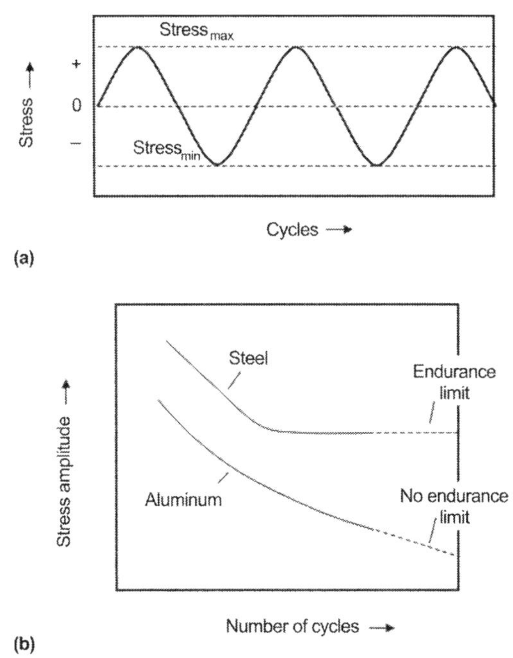

Fig. 1 The process of fatigue. (a) Cyclic loading. (b) Fatigue life of steel with an endurance limit and aluminum with no endurance limit. Source: Ref 2

destructive inspection and evaluation, reliability engineering, testing technology, field repair and maintenance, and holistic design procedures. All of these must be used in a consistent design activity that may be referred to as a fatigue design policy. Obviously, if other time-related failure modes occur concomitantly with repeated loads and interact synergistically, then the task becomes even more challenging. Inasmuch as humans always desire to use more goods and place more demands on the things we can design and produce, the challenge of fatigue is always going to be with us.

Until the early part of the 1900s, not a great deal was known about the physical basis of fatigue. However, with the advent of an increased understanding of materials, which accelerated in the early 1900s, a great deal of knowledge has been developed about repeated load effects on engineering materials. The fatigue process has proved to be very difficult to study. Nonetheless, extensive progress on understanding the phases of fatigue has been made in the last 100 years or so. It now is generally agreed that four distinct phases of fatigue may occur:

- Crack nucleation
- Structurally dependent crack propagation (often called the short crack or small crack phase)
- Crack propagation that is characterized by either linear elastic fracture mechanics, elastic-plastic fracture mechanics, or fully plastic fracture mechanics
- Final instability and failure

Each of these phases is an extremely complex process (or may involve several processes) in and of itself. For example, the nucleation of fatigue cracks is extremely difficult to study, and even pure fatigue mechanisms can be very dependent on the intrinsic makeup of the material. When extraneous influences are involved in nucleation, such as temperature effects (e.g., creep), corrosion of all types, or fretting, the problem of modeling the damage is formidable.

The Brittle Fracture Problem

> *Fracture is the separation of a solid body into two or more pieces under the action of stress.*

Fracture can be classified into two broad categories: ductile fracture and brittle fracture. As shown in the Fig. 2 comparison, ductile fractures are characterized by extensive plastic deformation prior to and during crack propagation. On the other hand, *brittle fracture takes place at stresses below the net section yield strength, with very little observable plastic deformation and a minimal absorption of energy. Such fracture occurs very abruptly with little or no warning and can take place in all classes of materials.* It is a major goal of structural engineering to develop methodologies to avoid such fractures, because they are associated with massive economic impacts and frequently involve loss of life.

It is difficult to identify exactly when the problems of failure of structural and mechanical equipment became of critical importance; however, it is clear that failures that cause loss of life have occurred for over 100 years. Throughout the 1800s, bridges fell and pressure vessels blew up, and in the late 1800s, railroad accidents in the United Kingdom were continually reported as "the most serious railroad accident of the week."

Those in the United States also have heard the hair-raising stories of the Liberty ships built during World War II. Of the 4694 ships considered in the final investigation, 24 sustained complete fracture of the strength deck, and 12 ships were either lost or broke in two. A spectacular example of this problem was the SS *Schenectady*, whose hull completely fractured while it was docked at its fitting-out pier. The fractured ship is shown in Fig. 3. In this case, the need for tougher structural steel was even more critical because welded construction was used in shipbuilding instead of riveted plate. In riveted plate construction, a running crack must reinitiate every time it runs out of a plate. In contrast, a continuous path is available for brittle cracking in a welded structure, which is why low notch toughness is a more critical factor for long brittle cracks in welded ships.

Similar long brittle cracks are less likely or rare in riveted ships, which were predominant prior to welded construction. Nonetheless, even riveted ships have provided historical examples of long brittle fracture due, in part, from low toughness. In early 1995, for example, the material world

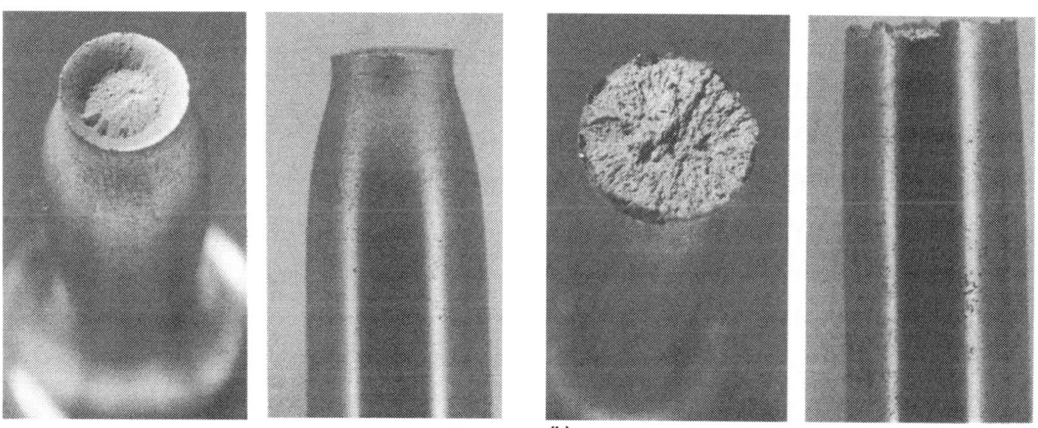

Fig. 2 Appearance of (a) ductile and (b) brittle tensile fractures in unnotched cylindrical specimens. Courtesy of G. Vander Voort. Source: Ref 3

was given the answer to an old question, "What was the ultimate cause of the sinking of the *Titanic*?" True, the ship hit an iceberg, but it now seems clear that because of brittle steel, "high in sulfur content even for its time," an impact that would clearly have caused damage perhaps would not have resulted in the ultimate separation of the *Titanic* into two pieces, where it was found in 1985. During the undersea survey of the sunken vessel with Soviet Mir submersibles, a small piece of plate was retrieved from 12,612 ft below the ocean surface. Examination by spectroscopy revealed a high sulfur content, and a Charpy impact test revealed the very brittle nature of the steel. However, there was some concern that the high sulfur content was, in some way, the result of 80 years on the ocean floor at 6000 psi pressures. Subsequently, the son of a 1911 shipyard worker remembered a rivet hole plug that his father had saved as a memento of his work on the *Titanic*. Analysis of the plug revealed the same level of sulfur exhibited by the plate from the ocean floor. In the years following the loss of the *Titanic*, metallurgists have become well aware of the detrimental effect of high sulfur content on the fracture of steel.

There are numerous other historical examples where material toughness was inadequate for design. The failure of cast iron rail steel for engine loads in the 1800s is one example. A large body of scientific folklore has arisen to explain structural material failures, almost certainly caused

Fig. 3 Brittle fracture of the SS *Schenectady*. Source: Ref 4

by a lack of tools to investigate the failures. An article on the building of the Saint Lawrence Seaway described the effect of temperature on equipment: "The crawler pads of shovels and bulldozers subject to stress cracked and crumbled. Drive chains flew apart, cables snapped and fuel lines iced up . . . *And anything made of metal, especially cast metal, was liable to crystallize and break into pieces*." It is difficult to realize that there still exists a concept of metal crystallization as a result of deformation that in turn leads to failure. Clearly, the development of fluorescence and diffraction x-ray analysis, transmission and scanning electron microscopes, high-quality optical microscopy, and numerous other analytical instruments in the last 75 years has allowed further development of dislocation theory and clarification of the mechanisms of deformation and fracture at the atomic level.

Brittle fracture has also plagued the aviation industry. In the 1950s, several Comets, the first commercial jet aircraft, produced in Britain, mysteriously exploded while in level flight. The cause was eventually traced to a design defect in which high stresses around the sharp corners on the windows caused small fatigue cracks to initiate, from which the fractures initiated. In the late 1960s and early 1970s, the U.S. fighter F-111 aircraft experienced catastrophic failure of the wing throughbox (the structure at which the wings join to the fuselage). Failures of the F-111 were related to the choice of a very brittle material (D6AC, a high-strength tool steel) and a heat treating procedure that produced nonuniform microstructures. In 1988, the upper fuselage of a Boeing 737 operated by Aloha Air fractured without warning during level flight over the Pacific Ocean. The reasons for this were related to corrosion of the aluminum alloy skin material and the frequent fuselage pressurization cycles resulting from many takeoffs and landings during short flights among the Hawaiian Islands.

In addition to the aforementioned, there are also numerous fracture examples of bridges, train wheels, and heavy equipment. In virtually every case, the reasons for brittle fracture can be found in inappropriate choice of materials, manufacturing defects, faulty design, and a lack of understanding of the effects of loading and environmental conditions. In all of the cases cited, there was severe economic loss and/or loss of life. For these reasons, it is an important engineering and ethical undertaking to reduce to an absolute minimum such accidents caused by brittle fracture.

In the previously mentioned examples, there are some common factors. Brittle fracture generally occurs in high-strength alloys (D6AC steel for the F-111 wing box; high-strength aluminum alloys for the Comets and 737), welded structures (Liberty ships, bridges), or cast structures (train wheels). It is significant that all failures started at small flaws which had escaped detection during prior inspections (in some cases, e.g., the F-111, many previous inspections). Subsequent analysis showed that in most instances, small flaws slowly grew as a result of repeated loads or a corrosive environment (or both) until they reached a critical size. After reaching critical size, rapid, catastrophic failure took place. The following sequence of events is usually associated with brittle fracture:

1. A small flaw forms either during fabrication (e.g., welding, riveting) or during operation (fatigue, corrosion).
2. The flaw then propagates in a stable mode due to repeated loads, corrosive environments, or both. The initial growth rate is slow and undetectable by all but the most sophisticated techniques. The crack growth rate accelerates with time, but the crack remains stable.
3. Sudden fracture occurs when the crack reaches a critical size for the prevailing load conditions. Final fracture is rapid, proceeding at almost the velocity of sound.

During the postwar period, predictive models for fracture control were pursued based on earlier work by Griffith, Orowan, and Irwin. Since the paper of Griffith in 1920 and the extensions of his basic theory by Irwin and others, we have come to realize that the design of structures and machines can no longer under all conditions be based on the elastic limit or yield strength. Griffith's basic theory is applicable to all fractures in which the energy required to make the new surfaces can be supplied from the store of energy available as potential energy, in the form of elastic strain energy. The elastic strain energy per unit of volume varies with the square of the stress and hence increases rapidly with increases in the stress level. One does not need to go to very high stress levels to store enough energy to drive a crack, even though this crack can be accompanied by considerable plastic deformation and hence consume considerable energy. Thus, self-sustaining cracks can propagate at fairly low stress levels.

Changes in Design Philosophy

Because of failures similar to those in Table 1, predicting performance and assessing the remaining life with greater confidence becomes increasingly important as costs for manufacturers and operators need to be reduced. Furthermore, the cost of failure is progressively greater as systems become more complex, downtime costs increase, and liability for failure increases. A brief discussion follows on the design process because it is important for failure investigators and life-assessment engineers to understand some of the design issues. Each structure has unique design requirements, but all structures are designed using some basic design principles. The relationship among the design phase, testing, systematic failure analysis, and life assessment of components is shown in Fig. 4.

One alternative for avoiding failures used in the past was to overdesign and to operate at very conservative loads. The economic penalties for both are increasingly significant; however, the economic penalties for failures are significant as well. It is necessary, then, to pay more attention to predicting and ensuring performance. Predicting and ensuring performance is fundamentally a part of the design process for buildings, power plants, aircraft, refineries, and ships. For any given design, the mission and the intended use are established. Predicting the performance and design life of a component depends on defining what life or performance is required in

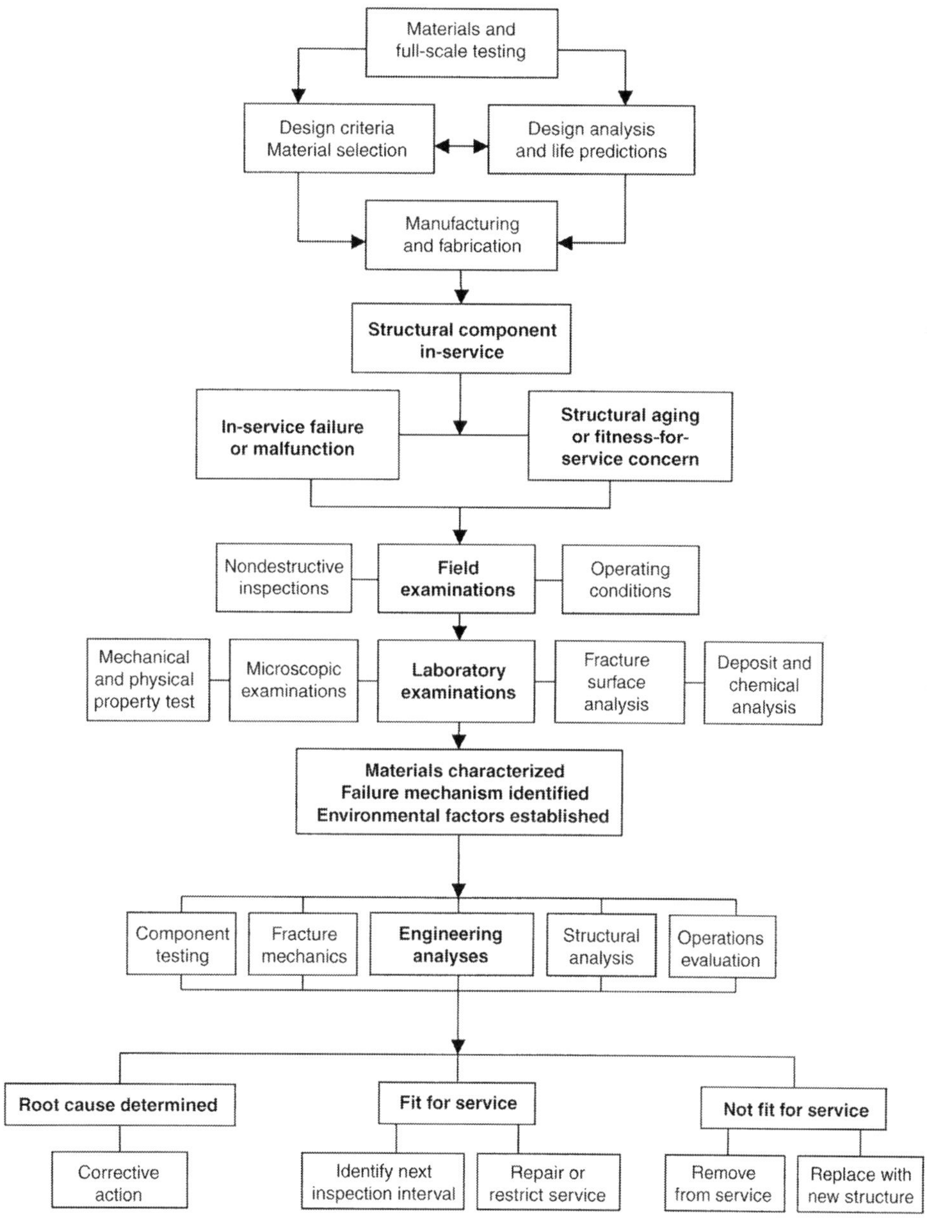

Fig. 4 Flow diagram showing the relationship between the design phase and the investigative tasks for in-service failure, structural aging, and fitness-for-service of structural components. Source: Ref 1

the anticipated combinations of mechanical and chemical environments. Defining performance may involve defining end points such as acceptable length of propagating cracks, maximum depth of propagating pits, acceptable remaining thickness of corroding pipes, maximum number of fatigue cycles or extent of cumulative damage, maximum number of plugged tubes, maximum number of failed circuits, maximum leakage, or appearance of a maximum area or number of rust spots. Defining such end points is a critical part of predicting life, because prediction defines when these end points will be reached and therefore when failure occurs.

Defining failure is also related to what is meant by the design life. For example, for the aerospace industry, a fighter aircraft may be designed for 8000 flight hours and analyzed for two lifetimes, or 16,000 flight hours. For the power industry, the design life of components is sometimes taken as 40 years. This means that the equipment is expected to perform satisfactorily at its rated output for 40 years. This is not to say that some maintenance is not necessary. However, to assert to a customer that a component has a 40 year design life, it is necessary to develop bases for such a claim. Such bases are usually provided by analyses, accelerated testing in the laboratory, and with prototype and model testing. As part of the life-assessment process, it is important to understand how a structural component—whether a pressure vessel, shaft, or structural member—is designed in order to understand how it may fail and to perform meaningful life assessment. For example, the first step in the design of any pressure vessel is to select the proper design code based on its intended use. A pressure vessel may be a power or heating boiler, a nuclear reactor chamber, a chemical process chamber, a hydrostatic test chamber used to test underwater equipment, or a pressure vessel for human occupancy. Once the intended use is identified, the appropriate design code can be selected. For example, pressure vessels use codes provided by many organizations and certifying agencies, such as the American Society of Mechanical Engineers (ASME), the American Bureau of Shipping, and European agencies that have similar pressure vessel design codes. Strict adherence to these codes for the design, fabrication, testing, and quality control and assurance allows the finished pressure vessel to be certified by the appropriate authorizing agency.

One of the first incentives to develop a pressure vessel code occurred after the Boston molasses tank incident in 1919, when the tank failed by overstress and consequently released more than 2 million gallons of molasses, resulting in the loss of life and property. Even after that catastrophic failure and understanding the nature of the failure, another molasses tank failure occurred in New Jersey in 1973. The destruction caused by this molasses tank incident is shown in Fig. 5. These molasses tank incidents demonstrate how important it is to prevent failures, and they underscore that good designs consider the operating conditions and limitations of materials of construction.

The next step in the design process is to identify the design parameters, such as configuration, design pressure, and others. An example of a design parameter list applicable for a chemical process chamber is presented in Table 2. These design parameters are the same parameters considered when conducting a pressure vessel failure investigation and life assessment.

Given the design parameters, the proper material(s) is selected for the structural component. Safety and economy are often the governing factors when selecting a material for pressure vessels. Materials selection is based

Fig. 5 Failed molasses tank, which fractured suddenly in New Jersey in March 1973. This catastrophic and sudden brittle fracture resulted in the release of the molasses in the tank, similar to the Boston molasses tank disaster in 1919. Source: Ref 1

Table 2 Pressure vessel design parameters

Required design code(a)	Penetration and location requirements
Basic chamber configuration. (Cylindrical or spherical; flat, spherical, or elliptical end details; etc.)	Contents and/or process within the pressure vessel
Internal volume capacity	Estimated operational pressure and temperature cycle history (number of cycles at what pressures and temperatures over the vessel's lifetime)
Minimum inside diameter	Piping, external and internal attachment requirements
Minimum inside length	Test chamber surroundings (enclosed in building or exposed to elements)
Chamber orientation (for cylindrical chambers, longitudinal axis vertical or horizontal)	Test chamber physical geographical location
Support configuration (saddle supports, bottom cylindrical skirt, legs, etc.)	Vessel special material requirements (vessel material other than carbon steel, internal cladding, etc.)
Maximum internal operating pressure	Vessel protection requirements (painted surfaces, stainless steel overlays at seals, cathodic protection, etc.)
Maximum external operating pressure (vacuum, etc.)	Fabrication requirements
Design operating temperature range	Material selection

(a) ASME Boiler and Pressure Vessel Code, Section VII, Div. 1, 2, or 3; American Bureau of Shipping; other. Source: Ref 1 on mechanical, corrosion, creep, toughness, and thermal properties as applicable. When welding is used, the appropriate weld material must be selected based on the chosen base material. The material is assigned an allowable stress value based on its ultimate and yield strengths and operating temperature range. The allowable stress value is then used in design equations or compared to results obtained from detailed analyses.

The design process then proceeds with the determination of the sizes and/or thickness of the various components. The design process is completed with the creation of the engineering and fabrication drawings. These drawings should include the dimensional information but also specify materials, weld identification, weld procedures, and required weld inspections. Other helpful information to include on the drawings is basic parameters such as design pressure, design temperature range, design code, and other information deemed necessary for the particular structural component.

Structural Design Approaches. The criteria of failure are determined by the strength of materials, fracture toughness, creep resistance, fatigue behavior, and the corrosion resistance of materials. In the strength-of-materials design approach, one typically has a specific structural geometry (assumed to be defect free) for which the load-carrying capacity must be determined. To accomplish this, a calculation is first made to determine the relation between the load and the maximum stress that exists in the structure. The maximum stress is then compared with the strength of the material. An acceptable design is achieved when the maximum stress is less than the strength of the material, suitably reduced by a factor of safety. It can be assumed that failure will not occur unless the maximum stress (σ_{max}) exceeds the yield strength of the material (σ_y). To ensure this, a factor of safety (S) can be introduced to account for material variability and/or unanticipated greater service loading. The strength-of-materials approach is a good approach for materials with no defects and simple structures. The strength-of-materials approach and the engineering design regime based on a factor of safety are illustrated in Fig. 6.

Linear Elastic Fracture Mechanics. Brittle fractures, similar to those mentioned in Table 1, are avoided by using a linear elastic fracture mechanics design approach. This approach considers that the structure, instead of being defect free, contains a crack. The governing structural mechanics parameter when a crack is present, at least in the linear approach, is an entity called the stress-intensity factor. This parameter, which is given the symbol K, can be determined from a mathematical analysis similar to that used to obtain the stresses in an uncracked component. For a relatively small crack in a simple structure, an analysis of the flawed structure beam would give to a reasonable approximation:

$$K = 1.12\sigma_{max}\sqrt{\pi a} \qquad \text{(Eq 1)}$$

where a is the crack length, and σ_{max} is the stress that would occur at the crack location in the absence of the crack. The basic relation in fracture mechanics is one that equates K to a critical value. This critical value is often taken as a property of the material called the plane-strain fracture toughness, conventionally denoted as K_{Ic}. When equality is achieved between K and K_{Ic}, the crack is presumed to grow in an uncontrollable manner. Hence, the structure can be designed to be safe from fracture by ensuring that K is less than K_{Ic}. The Liberty warship fractures are a classic example of a structural failure caused by K_{Ic} exceeding K and uncontrolled crack growth.

The essential difference from the strength-of-materials approach is that the fracture mechanics approach explicitly introduces a new physical parameter: the size of a (real or postulated) cracklike flaw. In fracture mechanics, the size of a crack is the dominant structural parameter. It is the specification of this parameter that distinguishes fracture mechanics from conventional failure analyses. The generalization of the basis for engineering structural integrity assessments that fracture mechanics provides is portrayed in terms of the failure boundary shown in Fig. 6. Clearly, fracture mechanics considerations do not eliminate the traditional approach. Structures using reasonably tough materials (high K_{Ic}) and having only small cracks (low K) will lie in the strength-of-materials regime. Conversely, if the material is brittle (low K_{Ic}) and strong (high yield strength), the presence of even a small crack is likely to trigger fracture. Thus, the fracture mechanics assessment is a crucial one.

The special circumstances that would be called into play in the upper right corner of Fig. 6 are worth noting. In this regime, a cracked structure

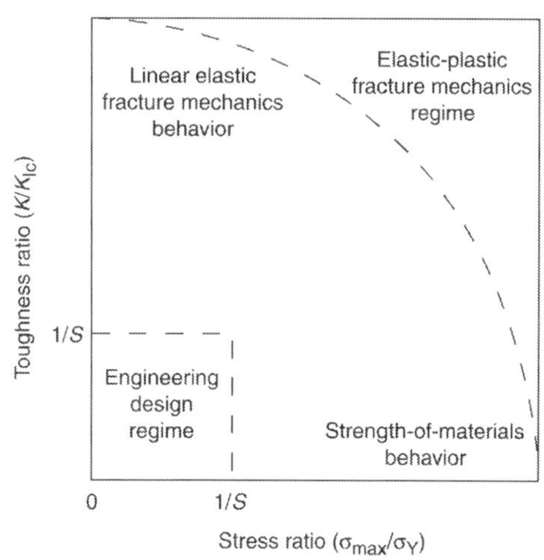

Fig. 6 General plot of the ratios of the toughness and stress showing the relationship between linear elastic fracture mechanics and strength of materials as it relates to fracture and structural integrity. Source: Ref 5

would experience large-scale plastic deformation prior to crack extension. Life assessment of aircraft and power plant equipment stems largely from the development of the damage tolerance philosophy based on fracture mechanics. Damage tolerance is the philosophy used for maintaining the structural safety of commercial transport, military aircraft, structures, and pressure vessels. The use of fracture mechanics and damage tolerance has evolved into the design program for structures that are damage tolerant, that is, designed to operate with manufacturing and in-service-induced defects.

Damage tolerance evaluation has been interpreted in the past as a means to allow continued safe operation in the presence of known cracking. This interpretation is incorrect. No regulations allow the strength of the structure to be knowingly degraded below ultimate strength (1.5× limit). The damage tolerance evaluation is merely a means of providing an inspection program for a structure that is not expected to crack under normal circumstances but may crack in service due to inadvertent circumstances. If cracks are found in primary structures, they must be repaired. The only allowable exception is through an engineering evaluation, which must show that the strength of the structure will never be degraded below ultimate strength operations or in-service conditions.

After many major fatigue failures in the 1950s on both military and commercial aircraft, the most notable of which were the DeHavilland Comet failures in early 1954, the U.S. Air Force (USAF) initiated the Aircraft Structural Integrity Program (ASIP) in 1958. The fatigue methodology adopted in the ASIP was the reliability approach, which became known as the safe-life method. This safe-life approach, used in the development of USAF aircraft in the 1960s, involved analysis and testing to 4 times the anticipated service life. On the commercial scene, another philosophy, fail safety, was introduced in the early 1960s, and a choice between safe-life and fail-safe methods was allowed by commercial airworthiness requirements. However, it was found that the safe-life method did not prevent fatigue cracking within the service life, even though the aircraft were tested to four lifetimes to support one service life (i.e., scatter factor of 4). One notable example is the F-111 aircraft No. 94, which crashed in 1969. The F-111 aircraft had a safe life of 4000 flight hours. However, there was a material defect in the F-111 aircraft, which used high-strength steel (ultimate tensile strength of 1655 to 1793 MPa, or 240 to 260 ksi; toughness of approximately 66 MPa$\sqrt{m}$, or 60 ksi$\sqrt{in.}$) for the wing box (Fig. 7). The defect was not observed during inspection, and a fatigue crack initiated and grew for only about 0.38 mm (0.015 in.). The aircraft was flown safely for 107 flights, at which time catastrophic failure occurred, causing destruction of the aircraft.

The leak-before-break design approach is prevalent in pressure-containing equipment such as pressure vessels and piping used in nuclear and fossil fuel power generation plants, refineries, and chemical plants. Failure analysis and life assessment of pressure-containing systems is essential. Although leak-before-break failures are not catastrophic, they are costly and can affect plant operations. Therefore, analyses are often performed to predict when the next internal or external inspection should be performed. Typical life-limiting mechanisms include stress-corrosion cracking, fatigue, and thermal fatigue. Welded structures that could initiate a crack are often susceptible to these mechanisms. The leak-before-break concept generally refers to a pressure-containment system failure in which a part-through-wall crack extends to become a through-the-wall crack, thus allowing fluid to escape. If no further crack extension occurs, then the loss of fluid is detected and no further crack growth occurs. Alternatively, when a through-the-wall crack propagates along the wall, a catastrophic event can occur. Power plant piping materials that are ductile, such as stainless steel and nickel-base alloys, often leak before break. Figure 8 shows small-bore, socket-welded piping that will initiate fatigue

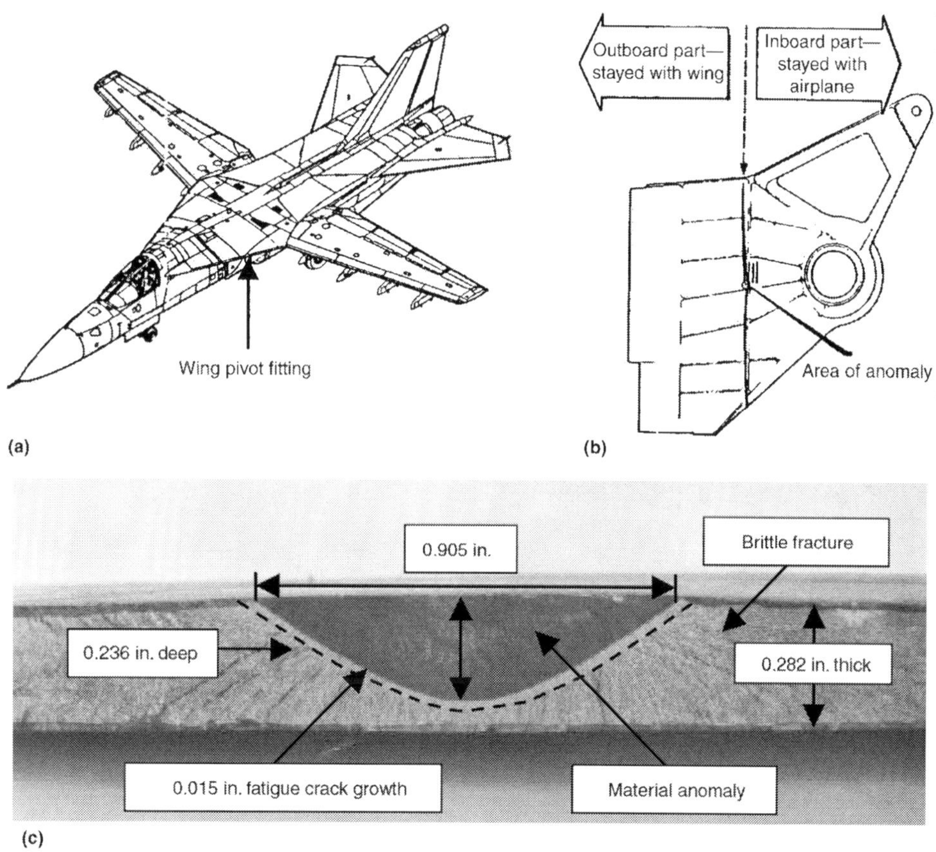

Fig. 7 Fatigue cracking in an aircraft wing fitting for the F-111 aircraft No. 94 that crashed in 1969. (a) and (b) Location of the left wing pivot box fitting. The 22 mm (0.91 in.) material defect was not observed during inspection, and a fatigue crack initiated and grew for only about 0.38 mm (0.015 in.) before unstable brittle fracture occurred. Source: Ref 1

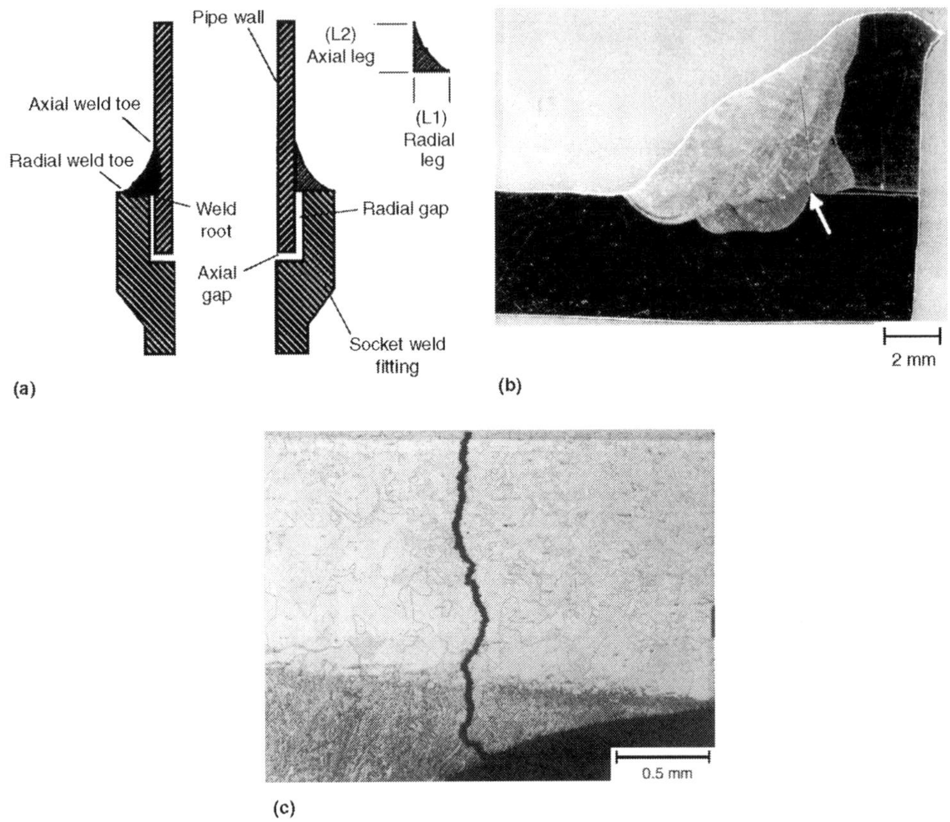

Fig. 8 Stainless steel piping such as small-bore piping is designed to leak before break. A fatigue crack either initiates at the toe or the root of the weld. (a) Typical socket fitting with a fillet weld. (b) Micrograph of a cross section through a socket-welded joint showing fatigue crack that initiated from the weld root and extended through the weld. (c) Micrograph of axial weld toe crack showing a fatigue-induced crack that extended through the pipe wall. Source: Ref 1

cracks at either the toe of the weld or the root of the weld. These ductile socket-welded pipes leak before catastrophic failure occurs.

Elevated-Temperature Concerns. For elevated-temperature equipment and structures subjected to steady-state or cyclic stresses, the principal design considerations are creep control, oxidation prevention through the use of oxidation-resistant materials or coatings, and selection of materials that have good stress-rupture and creep properties. The criteria for failure are to not go below a minimum stress-rupture strength for a given operating stress and temperature, and to not operate above a certain temperature that alters the microstructure or oxidizes the material.

Corrosion Allowances. Designs are configured such that the operating or loading stresses can be minimized for safe operations. It is necessary to consider the effects that an environment will have on the material; it is just as important as considering structural loads on a component. It is important that environments are known and controlled in such a way that corrosion is minimized on all surfaces. This means that designs consider effects of crevices, galvanic couples, flows, stresses, and temperatures to ensure that all the surfaces of materials will be minimally degraded within the design life.

A common design approach for pressure vessels and tanks to deal with corrosion is to provide a corrosion allowance, which takes the form of additional thickness based on available information on rates of general corrosion over the design life. For example, a carbon steel vessel designed for 25 years of service in sulfuric acid at a corrosion rate of 5 mils/yr (0.005 in./yr) would have a corrosion allowance of 125 mils (0.125 in.). However, such allowances cannot deal with stress-corrosion cracking, pitting, intergranular cracking, or effects of long-range cells. Use of a corrosion allowance can be disastrously misleading, because its use suggests that all corrosion problems have been solved. It should be pointed out that exceeding the corrosion allowance does not necessarily mean the vessel would fail or is unsuitable for service. It is an indication that the vessel should be evaluated for continued service.

Life-Limiting Factors

To assess the life of a structure, one must first understand the factors that limit life. It is the purpose of the following discussion to briefly discuss some of the aspects that control the durability and remaining life of structures. Each structure or structural system has unique parameters of life expectancy. For example, the life of an aircraft is dependent on a number of variables, such as flight profiles, usage rate, external environment, materials selection, and geometry. For fossil fuel plant equipment that is exposed to elevated temperatures, life-limiting factors involve pressure, temperature, environment, fatigue cycles, and time at temperature. The following life-limiting factors are common to most structures and should be considered in a failure analysis and a life assessment:

- Material defects
- Fabrication practices
- Stress, stress concentration, and stress intensity
- Temperature
- Thermal and mechanical fatigue cycles
- Corrosion concerns
- Improper maintenance

Material Defects. With the advent of strict quality-assurance controls and nondestructive inspection systems, manufacturing and material defects are not as common as in early years. Yet, they have and will be a potential concern in structural components. To determine if a defect is a cause for failure, it is important that the component or structure undergoes a comprehensive failure investigation to determine if such is the case. Defects greatly affect remaining life, because the time for crack initiation has been reduced.

Defects are either material defects or mechanical defects. The material defect may manifest itself because of a metallurgical anomaly. This anomaly or metallurgical defect may occur as a result of faulty furnace control during heat treating, improper plating that may result in unwanted hydrogen in high-strength steel, processing problems (e.g., forging laps, shuts, or undesirable grain structure), welding defects, and so forth. For example, a material defect, shown in Fig. 7, caused the historic failure of an F-111 aircraft in 1969. Another example of a preexisting material defect is the failure of a titanium disk on United Flight 232, which failed near Sioux City, Iowa, in July 1989. The Sioux City disaster was due to a manufacturing problem in a titanium fan disk, which had the presence of hard alpha case. As a result of the Sioux City incident, a probabilistic life-assessment approach was taken to evaluate titanium disks. Thus, defects occur as a result of both material processing and manufacturing assembly.

A mechanical defect is often induced after the manufacturing process when the structure is in service. A mechanical defect such as a dent, gouge, impact, or deformation of the material can reduce component life. The result is often yielding or cracking of the structural component. If the mechanical defect is located at a high-stress location, the reduction in life can be significant and possibly catastrophic. For example, damage at a fastener hole in a wing skin spar mechanically induced during the drilling of the hole is shown in Fig. 9. The failure analyst was able to identify the location of the damage and the extent of fatigue crack growth, which was used to establish inspection intervals.

Fabrication practices such as welding can affect the lifetime of a part. For example, in the fabrication of pressure vessels or pressure piping, unfavorable results from welding, such as brittle cracking in the heat-

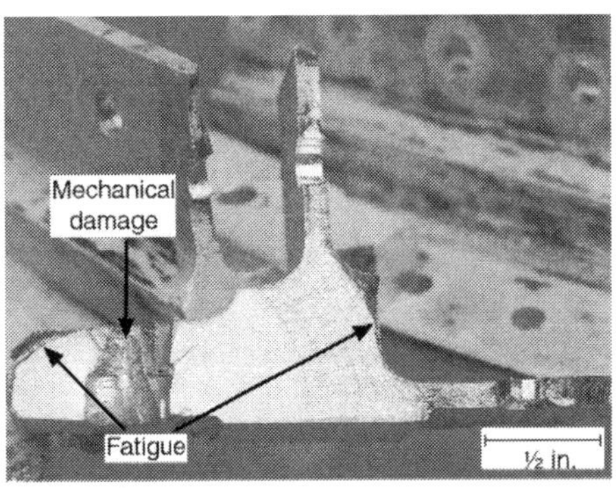

(a)

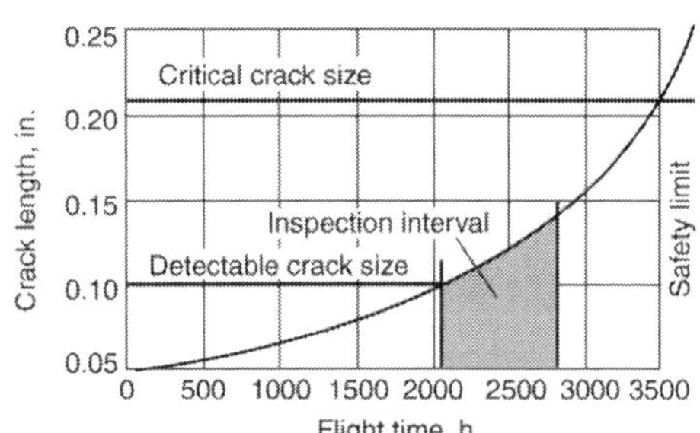

(b)

Fig. 9 Fractographic evaluation for life-assessment purposes. (a) Wing spar that had mechanically-induced damage at a fastener hole (indicated by an arrow). (b) Plot showing how fracture information is used to establish initial and recurring inspection requirements. Source: Ref 1 affected zone, often result from the use of steels containing excessive amounts of residual elements that increase hardenability and consequently their susceptibility to cracking. Therefore, complete control of composition is of the utmost importance when welding is involved. In producing welds without embrittlement, it is necessary to follow the proper welding procedure.

Sometimes imperfections are undetected and become evident during operations. For example, small cracks were not detected in a carbon steel ASTM A105 steam drum nozzle that was arc gouged. Consequently, during a cold startup, a 46 cm (18 in.) crack occurred, as shown in Fig. 10. Prior to the cold startup, the steam drum had been in service for five years with no cracking problems. Investigation of the cracking problem determined that crack initiation occurred from the root of an arc-gouged notch. A metallurgical cross section through the arc-gouged notch revealed microcracks and a very hard layer (54 HRC) that were not removed after arc gouging the nozzle. The carbon steel base metal was 72 to 88 HRB. The

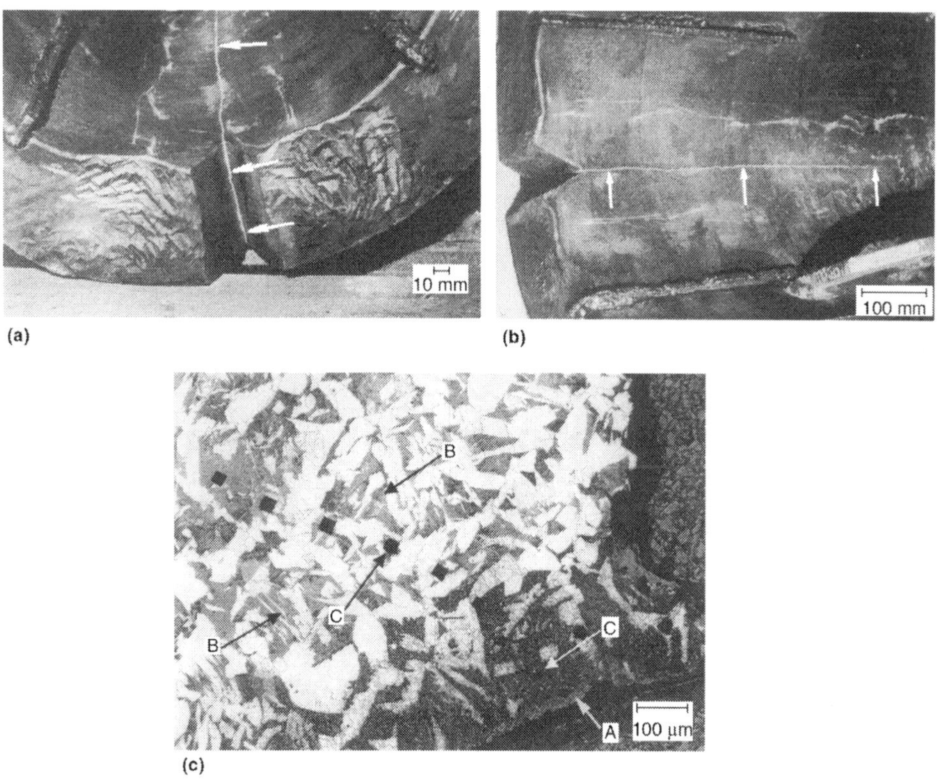

Fig. 10 Effect of welding on the life of a carbon steel structure. (a) and (b) 46 cm (18 in.) long crack found in a carbon steel as-forged nozzle that was arc gouged. Failure occurred after five years in service during cold startup procedure. (c) Micrograph showing a hardened layer (A) and the as-forged (acicular) microstructure (B). The black square shapes (C) are diamond pyramid hardness indentations. Source: Ref 1

crack extended during the cold startup because of preexisting microcracks and an as-forged nozzle that had poor fracture toughness.

Operating and Nonoperating Environmental Effects. Probably the single most significant element affecting structural service life and structural integrity is the actual usage or operating condition. Very early in the design and development phase, design usage is identified. The design usage for an aircraft system considers design capability, operating environments, intended mission goals, types of weapons, and power plant capabilities. Based on all considerations, the design usage is an attempt to truly identify the total environment as to how a structural component or system will be utilized in its expected service life. At that time, it is probably adequate in that it represents the intentions for how the structure is going to be tested and eventually operated. However, the actual usage may be different. An example of this is unanticipated cracking that occurred in an aircraft structure at one structural control point location because the operational spectrum was 11 times more severe than the actual design fatigue spectrum.

The influence and degradation of material properties and therefore the reduction of life cycle is often caused by the strong influence of the operating and nonoperating environments. Life-limiting factors occur both while the structure or equipment is in service and when not in use. For example, while the aircraft is operational, the dynamic factors (fatigue, flutter, and vibrations) are the responsible parameters for determining the life aspects of selected components, but while the aircraft is sitting on the ground, corrosion behavior is normally the dominant driver. In a similar manner, process equipment subjected to elevated temperatures can suffer reduced creep life, but equipment that is left stagnant with a corrosive medium in a pressure vessel is also detrimental. For pressure vessels and piping equipment, actual usage or operating conditions can even change from company to company or plant to plant. For example, the operation of a land-based turbine can vary from plant to plant within the same company.

Stress, Stress Concentration, and Stress Intensity. In the earlier design days, the principal design consideration was stress, that is, the load to cause failure at a specific wall thickness or cross-sectional area. The criterion for failure could be yielding resulting in plastic deformation or failure, or fracture due to exceeding the ultimate tensile strength of the material. The stress on structural members is still critical, yet, with the increased understanding of crack growth and failure mechanisms, stress-concentration effects and the stress intensity at a crack tip must also be considered.

High-stress regions in a structural component are typically located around geometric details, such as a hole, fillet radius, or notch. These high local stress regions are described in terms of the remote or far-field stress of a component by multiplying the remote stress by a stress-concentration factor. Stress-concentration factors for various geometric details are derived from the theory of elasticity and/or obtained experimentally. Stress-concentration curves are available for different design geometries. These curves are useful when investigating a design and doing a life assessment.

Stress intensity is the controlling factor in subcritical crack growth propagation rates and the identification of the critical crack size for the onset of rapid overload (critical fracture). Stress and crack geometry are parameters that determine the stress intensity. The stress-intensity factor describes the stress distribution at a crack tip. For a small edge crack in a thin, flat plate subject to a uniform tensile stress, the stress-intensity factor can be determined from:

$$K_I = Y\sigma\sqrt{\pi a} \quad \text{(Eq 2)}$$

where K_I is the stress-intensity factor, σ is the applied stress, and a is the crack length. The subscript on the stress-intensity factor refers to the mode of crack loading. Equation 2 is simplified; often each geometry and crack will have a correction factor (Y) to account for the crack shape, loading direction, and geometry. For example, an edge crack normal to the free edge that is loaded in tension would have a Y factor of 1.12 (Eq 3). For each crack and structure geometry, the approximate stress-intensity factor could be determined by:

$$K_I = 1.12\sigma\sqrt{\pi a} \quad \text{(Eq 3)}$$

Temperature Effects. Temperature affects structures because it can reduce the stress to failure, cause the material to be more brittle, or oxidize the surface. One of the first questions the failure investigator asks is, "What was the temperature when the failure occurred, or what is the operating temperature?" If the answer to this question is unknown, then the investigator deduces the temperature based on physical evidence and material properties.

Life of a metal component at elevated temperature, when subjected to either steady or fluctuating stress, is limited. In contrast, at ambient temperatures and in the absence of a corrosive environment, the life of a component in steady-state load conditions may be unlimited, provided operating loads do not exceed the yield strength of the specific metal and wear is not serious. Stress produced at elevated temperature produces a condition of continuous strain called creep. By definition, *creep is deformation as a function of time at constant load or stress*. Creep, after a period of time, may terminate in stress-rupture fracture, also known as creep rupture.

Most brittle fractures of steels occur because structural components are operating at temperatures below their ductile-to-brittle transition temperature or the nil-ductility transition temperature. Consequently, the structure is unable to maintain subcritical crack growth, and failure occurs in a catastrophic manner. The ductile-to-brittle transition temperature can be above room temperature, putting at risk structures that operate below room temperature. This occurs in tanks such as molasses tanks, pressure vessels, ships such as the Liberty warships, and high-strength alloy steel structures.

Metallic structures oxidize when subjected to elevated temperatures. For example, carbon steel boiler tube used above 540 °C (1000 °F) is prone to oxidation. An increase in oxidation resistance can be achieved with the addition of alloying elements such as chromium and molybdenum. A boiler tube of 9Cr-1Mo steel can have good oxidation resistance to 650 °C (1200 °F). Oxidation and high-temperature corrosion is also controlled through the use of protective coatings.

Thermal and Mechanical Fatigue Cycles. Fatigue fractures result from cyclic stressing, which progressively propagates a crack or cracks until the remaining section is no longer able to support the applied load. For example, pressure vessels and pressure piping are subject to high static stresses arising from the pressure of contained liquids or gases, to stresses resulting from misalignment of components, and to residual stresses induced during welding. The cyclical component may be added mechanically, by vibration of associated equipment or pulsation from a compressor. It may also be added thermally, resulting in thermal fatigue if the component is cycled through a temperature range in service. Fatigue cracks nucleate from a stress raiser such as a discontinuity or a notch, which produces triaxial stressing in the material. The stress raiser may be macroscopic in size, such as a notch or discontinuity, or microscopic and not visible.

Mechanical conditions are not the only sources of cyclic loading, which can contribute to fatigue failure. Transient thermal gradients within a vessel can induce plastic strains. If these thermal gradients are applied repeatedly, the resulting cyclic strain can induce failure. This process is known as thermal fatigue. Thermal fatigue is often considered a low-cycle failure mechanism that occurs due to operating conditions. *Thermal fatigue can be defined as the gradual deterioration and eventual cracking of a material by alternating heating and cooling during which free thermal expansion is partially or fully constrained.* The constraint of a part or material does not allow free expansion to occur. For example, piping attachments that are mechanically constrained will develop thermal fatigue cracks when thermally cycled.

The cyclic thermally-induced stresses associated with frequent startups and shutdowns are often more severe than stresses due to steady-state operations, and designers cannot always predict the use cycles of equipment that owner-operators impose. Equipment sometimes is operated only during peak demand periods; this imposes many startups and shutdowns, with attendant thermal stresses on pressure vessels and piping systems. When cracking occurs, it is in areas least able to accommodate severe cyclic loading; such areas are likely to be rigid or massive sections under conditions of high constraint or thermal stress.

Corrosion Concerns. In most cases, corrosion itself does not lead to structural failure, but it is the most expensive aspect of maintaining structural integrity. The effect of corrosion must be addressed by denoting the types of corrosion and where they can occur, inspection techniques to find the corrosion, and repair or replacement methods to remove the offending material. There are many corrosion mechanisms, such as stress-corrosion cracking, hydrogen embrittlement, and general corrosion, that reduce the life expectancy of components. These life-limiting corrosion mechanisms must be identified and considered in performing life assessments.

Improper Maintenance. The manufacturer of structures, equipment, and components often requires special maintenance, as identified in handbooks, structural repair manuals, and technical orders. It is imperative that these manuals be rigidly adhered to. It is easy to devise methods that may be shorter and easier than those discussed in the manual, but the manufacturer selected the particular procedures for sound reasons. It is dangerous to deviate from these methods, because the deviation may place undue stress or cause latent damage to components apart from those being examined. This high stress or latent damage can lead to premature failure of components, which may result in undesirable consequences. Proper maintenance and inspection of tanks, aircraft, and pressure vessels is mandatory. Extreme care during maintenance and the inspection following maintenance must be taken to ensure that proper procedures and techniques are rigorously followed.

ACKNOWLEDGMENTS

Portions of the material for this chapter came from "Industrial Significance of Fatigue Problems" by D.W. Hoeppner and "Fracture and Structure" by C.Q. Bowles, both articles in *Fatigue and Fracture,* Vol 19, *ASM Handbook,* ASM International, 1996.

REFERENCES

1. Failure Analysis and Life Assessment of Structural Components and Equipment, *Failure Analysis and Prevention,* Vol 11, *ASM Handbook,* ASM International, 2002
2. F.C. Campbell, Fatigue, *Elements of Metallurgy and Engineering Alloys,* ASM International, 2008
3. W.T. Becker and D. McGarry, Mechanisms and Appearances of Ductile and Brittle Fracture in Metals, *Failure Analysis and Prevention,* Vol 11, *ASM Handbook,* ASM International, 2002
4. S.D. Antolovich and B.F. Antolovich, An Introduction to Fracture Mechanics, *Fatigue and Fracture,* Vol 19, *ASM Handbook,* ASM International, 1996

5. M.F. Kanninen and C.H. Popelar, *Advanced Fracture Mechanics,* Oxford University Press, 1985

SELECTED REFERENCES

- S. Atallah, U.S. History's Worst LNG Disaster, *Firehouse,* Jan 1979, p 29
- D.J. Benac and D.N. Hopkins, Investigation of Fatigue-Induced Socket-Welded Joint Failures for Small-Bore Piping Used in Power Plants, *Pract. Fail. Anal.,* Vol 1 (No. 2), April 2001, p 71–82
- D.J. Benac and R.A. Page, Integrating Design, Maintenance, and Failure Analysis to Increase Structural Valve Integrity, *Pract. Fail. Anal.,* Vol 1 (No. 3), June 2001, p 31–43
- Y. Billah and R. Scanlan, Resonance, Tacoma Narrows Bridge Failure and Undergraduate Physics, *J. Phys.,*Vol 59 (No. 2), Feb 1991, p 118–123
- C.Q. Bowles, Fracture and Structure, *Fatigue and Fracture,* Vol 19, *ASM Handbook,* ASM International, 1996
- W.R. Hendricks, The Aloha Incident: A New Era for Aging Aircraft, *Structural Integrity of Aging Aircraft,* S.N. Altussi et al., Ed., Springer-Verlag, 1991, p 153
- V.A. Hindas, "F-111 Design Experience: Use of High Strength Steels," AIAA Second Aircraft Design and Operations Meeting
- D.W. Hoeppner, Industrial Significance of Fatigue Problems, *Fatigue and Fracture,* Vol 19, *ASM Handbook,* ASM International, 1996
- "Interim Guidelines: Evaluation, Repair, Modification and Design of Steel Moment Frames," FEMA-267, Federal Emergency Management Agency
- H.P. Leighly, Jr., B.L. Bramfitt, and S.J. Lawrence, RMS Titanic: A Metallurgical Problem, *Pract. Fail. Anal.,*Vol 1 (No. 2), April 2001, p 10–13, 33–37
- N1819U, NTSB-AAR-90-06, Aircraft Accident Report, United Airlines, Inc., Nov 1990
- S. Ross, Tacoma Narrows, 1940, *Construction Disasters,* 1st ed., McGraw-Hill, 1984, p 216–239

Fatigue and Fracture—Understanding the Basics
F.C. Campbell, editor

CHAPTER 2

Mechanical Behavior

MANY MATERIALS, both during initial fabrication and later when placed in service, are subjected to forces or loads. In order that excessive deformation or failure does not occur, it is important to know what effects these loads have on the part. The mechanical behavior of a material reflects the relationship between its response or deformation to an applied load or force. Important mechanical properties are strength, hardness, stiffness, and ductility.

There are three principal ways in which a load may be applied, namely, tension, compression, and shear, as illustrated in Fig. 1. A large number of mechanical property tests have been developed to determine a material response to applied loads or forces. Because the results of these tests are often used by regulatory agencies, it is important that these test methods are consistent and standardized. In the United States, ASTM International is the most widely recognized source of standardized testing procedures.

Tensile Properties

The tensile test is the most common test used to evaluate the mechanical properties of materials. Its chief use is to determine the properties related to the elastic design of structures. In addition, the tensile test gives information on the plasticity and fracture of a material. The main advantages of the tensile test are that the stress state is well established, the test has been carefully standardized, and the test is relatively easy and inexpensive to perform.

The tensile properties of a material are determined by applying a tension load to a specimen and measuring the elongation or extension in a load frame, such as the one shown in Fig. 2. The load can be converted to engineering stress (s) by dividing the load by the original cross-sectional area of the specimen:

$$s = \frac{P}{A_0} \text{ in pascals (Pa) or pounds per square inch (psi)} \quad \text{(Eq 1)}$$

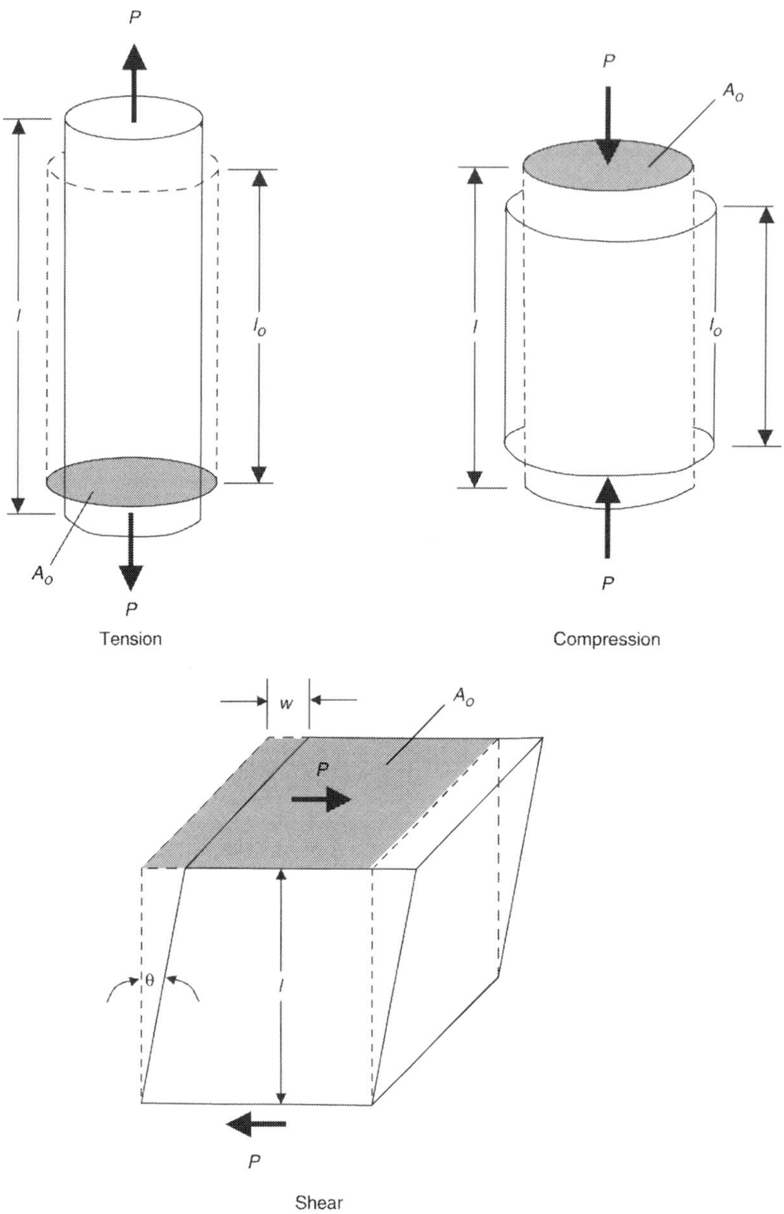

Fig. 1 States of stress. Source: Ref 1

where P is the load, and A_o is the original cross-sectional area.

The engineering strain (e) can be calculated by dividing the change in gage length by the original gage length:

$$e = \frac{l - l_o}{l_o} \qquad \text{(Eq 2)}$$

where l is the gage length, and l_o is the original gage length.

A typical stress-strain curve for a metal is shown in Fig. 3. The shape and magnitude of the stress-strain curve of a metal depends on its composition, heat treatment, prior history of plastic deformation, and the strain rate, temperature, and state of stress imposed during the testing. The different stress-strain behaviors for a brittle metal, one with limited ductility, and a fully ductile metal are shown in Fig. 4, and some typical stress-strain curves for a number of engineering alloys are shown in Fig. 5. The

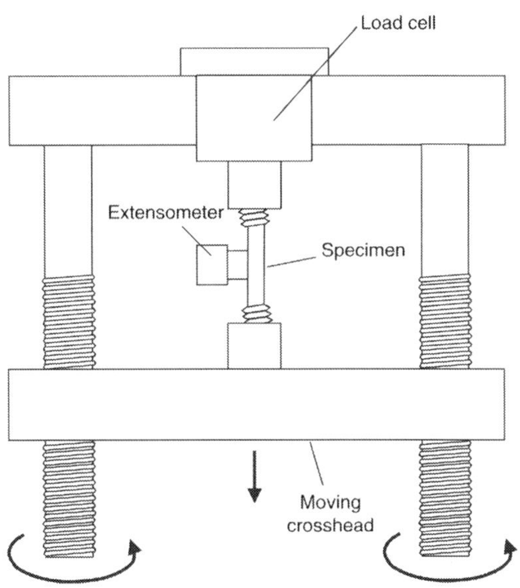

Fig. 2 Typical tensile test setup. Source: Ref 2

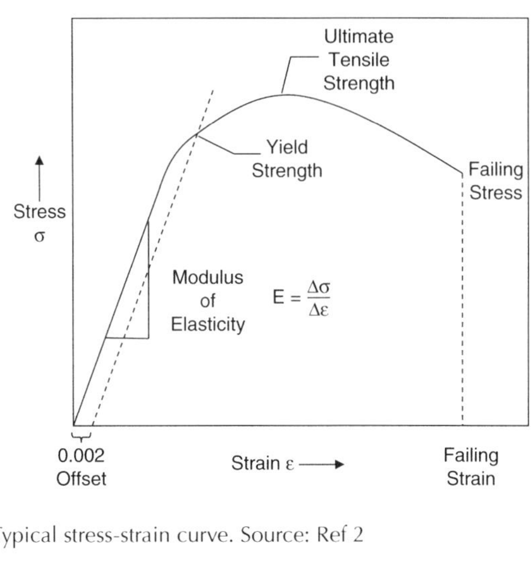

Fig. 3 Typical stress-strain curve. Source: Ref 2

parameters used to describe the stress-strain curve of a metal are the tensile strength, yield strength or yield point, percent elongation, and reduction in area. The first two are strength parameters, and the last two are indications of ductility.

As long as the specimen is loaded within the elastic region, the strain is totally recoverable and the specimen will return to its original length when the load is removed. However, when the load exceeds a value correspond-

Fig. 4 Different stress-strain responses. Source: Ref 2

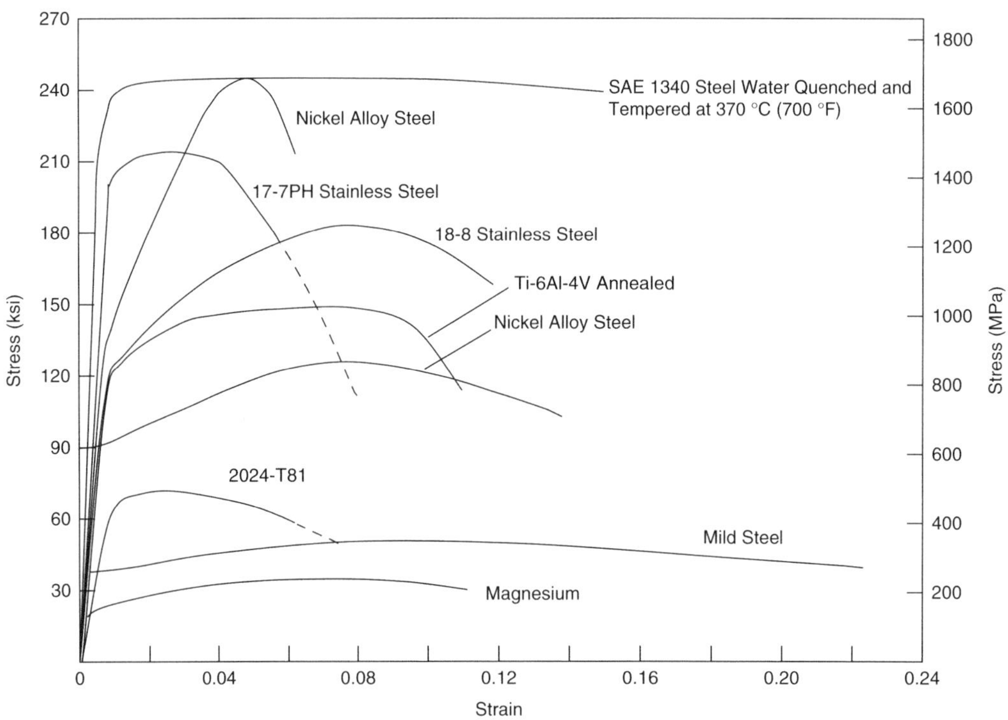

Fig. 5 Typical engineering stress-strain curves. Source: Ref 2

ing to the yield stress, the specimen undergoes plastic deformation and is permanently deformed when the load is removed. The stress to produce continued plastic deformation increases with increasing strain as the metal strains or work hardens. To a good engineering approximation, the volume remains constant during plastic deformation ($Al = A_0 l_0$), and as the specimen elongates, it decreases uniformly in cross-sectional area along its gage length. Initially, strain hardening more than compensates for this decrease in area, and the engineering stress continues to rise with increasing strain. However, eventually a point is reached where the decrease in area is greater than the increase in deformation load from strain hardening. This condition will be reached first at some point in the specimen that is slightly weaker than the rest. All further plastic deformation is then concentrated in this region, and the specimen begins to neck or thin down locally. Because the cross-sectional area is now decreasing far more rapidly than the deformation load is being increased by strain hardening, the engineering stress continues to decrease until fracture occurs.

As shown in Fig. 3, the ultimate tensile strength is the maximum stress that occurs during the test. Although the tensile strength is the value most often listed from the results of tensile testing, it is not generally the value that is used in design. Static design of ductile metals is usually based on the yield strength, because most designs do not allow any plastic deformation. However, for brittle metals that do not display any appreciable plastic deformation, tensile strength is a valid design criterion, reduced by an appropriate factor of safety.

With most metals, there is a gradual transition from elastic to plastic behavior, and the point at which plastic deformation actually begins is difficult to define with precision. The transition from elastic to plastic deformation is illustrated in Fig. 6. When a specimen is loaded into the plastic range and then unloaded, the elastic strain immediately disappears. With the passage of time, the anelastic strain recovers. The remaining strain is the plastic strain. In uniaxial loading, three criteria for the initiation of yielding have been used: the elastic limit, the proportional limit, and the yield strength.

The elastic limit shown in Fig. 7 is the greatest stress the material can withstand without any measurable permanent strain remaining after the complete release of load. However, the determination of the elastic limit requires a tedious incremental loading-unloading test procedure. For this reason, it is often replaced by the proportional limit. The proportional limit is the highest stress at which stress is directly proportional to strain. It is obtained by observing the deviation from the straight line portion of the stress-strain curve. Finally, the yield strength is the stress required to produce a small specified amount of plastic deformation. The usual definition of this property is the offset yield strength determined by the stress corresponding to the intersection of the stress-strain curve offset by a

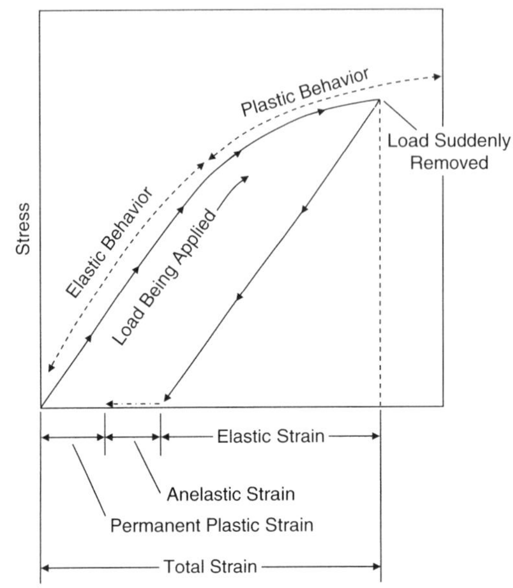

Fig. 6 Elastic and plastic behavior during tensile loading. Source: Ref 3

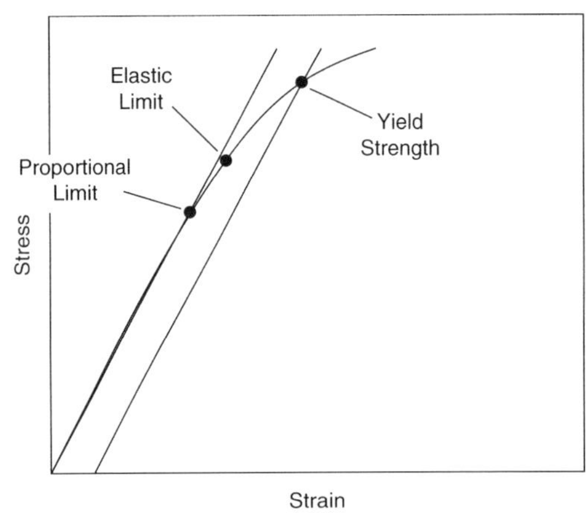

Fig. 7 Transition from elastic to plastic behavior. Source: Ref 2

specified strain. For metals without a definite yield point, the yield strength is determined by drawing a straight line parallel to the initial straight line portion of the stress-strain curve. The line is normally offset by a strain of 0.2% (0.002). The offset yield strength is often referred to in Great Britain as the proof stress, where offset values are either 0.1 or 0.5%. The yield strength obtained by an offset method is commonly used for design and specification purposes, because it avoids the practical difficulties of measuring the elastic or proportional limit. As previously stated, yield strength is generally a more important design parameter than tensile strength, because the possibility of plastic yielding is unacceptable for almost all structures.

Some materials have essentially no linear portion in their stress-strain curve, for example, soft copper and gray cast iron. For these materials, the offset method cannot be used, and the usual practice is to define the yield strength as the stress to produce some total strain, for example, $e = 0.005$. On the other hand, some metals, particularly annealed low-carbon steel, show a localized heterogeneous type of transition from elastic to plastic deformation that produces a definite yield point in the stress-strain curve. Rather than having a flow curve with a gradual transition from elastic to plastic behavior, metals with a yield point produce a flow curve or a load-elongation diagram similar to that shown in Fig. 8. The load increases steadily with elastic strain, drops suddenly, fluctuates about some approximately constant value of load, and then rises with further strain.

Ductility. Measures of ductility that are obtained from the tension test are the engineering strain at fracture (e_f) and the reduction of area at fracture (q). Both are usually expressed as percentages, with the engineering strain at failure often reported as the percent elongation. Both of these properties are obtained after fracture by putting the specimen back together and taking measurements of the final length (l_f) and final specimen cross section at fracture (A_f). Percent elongation can be determined by:

$$\%\ \text{elongation}\ (e_f) = \frac{l_f - l_o}{l_o} \times 100 \qquad \text{(Eq 3)}$$

where l_o is the original gage length, and l_f is the final length of the gage section.

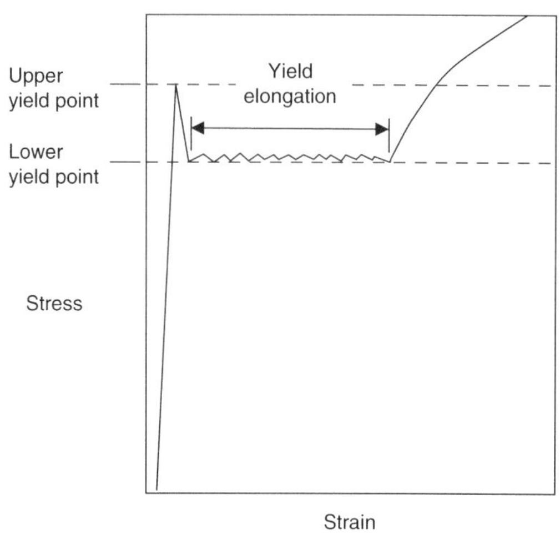

Fig. 8 Discontinuous yielding in plain carbon steels. Source: Ref 4

Likewise, reduction in area can be determined by:

$$\% \text{ reduction in area}(q) = \frac{A_o - A_f}{A_o} \times 100 \quad \text{(Eq 4)}$$

where A_o is the original area of the gage section, and A_f is the final area of the gage section at fracture.

Because an appreciable fraction of the deformation is concentrated in the necked region of the specimen, the value of e_f will depend on the gage length l_o over which the measurement was taken. The smaller the gage length of the specimen, the greater will be the contribution from the neck and the higher the value of e_f. Because percent elongation is sensitive to the gage length in the manner shown in Fig. 9, it is important to record the gage length used when reporting percent elongation. To eliminate this difficulty and to provide a measure of ductility that correlates with forming operations in which the gage length is very short, it is possible to determine the zero-gage-length elongation (e_o). From the constancy-of-volume relationship for plastic deformation ($Al = A_o l_o$):

$$\frac{l}{l_o} = \frac{A_o}{A} = \frac{1}{1-q} \quad \text{(Eq 5)}$$

$$e_o = \frac{l - l_o}{l} = \frac{A_o}{A} - 1 = \frac{1}{1-q} - 1 = \frac{q}{1-q} \quad \text{(Eq 6)}$$

This represents the elongation based on a very short gage length near the fracture.

When a tensile specimen is loaded, it elongates in the direction of the load and contracts in the transverse or lateral direction. The ratio of the

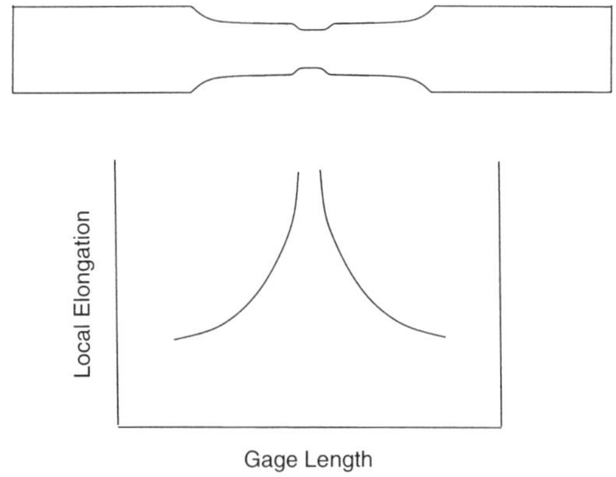

Fig. 9 Variation of local elongation along gage length. Source: Ref 5

lateral strain (e_t) to the tensile or longitudinal strain (e_l) is known as Poisson's ratio (ν):

$$\nu = -\frac{\text{Lateral strain}}{\text{Tensile strain}} = -\frac{e_t}{e_l} \qquad \text{(Eq 7)}$$

Values of Poisson's ratio for most materials range from 0.3 to 0.5.

The slope of the initial linear portion of the stress-strain curve is the modulus of elasticity, or Young's modulus, as shown in Fig. 3:

$$E = \frac{\Delta\sigma}{\Delta\varepsilon} \qquad \text{(Eq 8)}$$

The modulus of elasticity (E) is a measure of the stiffness of the material. The greater the modulus of elasticity, the smaller will be the elastic strain resulting from a given stress. In other words, metals with higher moduli of elasticity are stiffer than those with lower moduli. Because the modulus of elasticity is needed for computing deflections of beams and other members, it is another important design value.

The modulus of elasticity is determined by the binding forces between atoms. Because these forces cannot be changed without changing the basic nature of the material, the modulus of elasticity is one of the most microstructure insensitive of the mechanical properties. Generally, it is only slightly affected by alloying additions, heat treatment, or cold work.

Resilience and Toughness

Resilience is the ability of a metal to absorb energy when elastically deformed and then to return it when it is unloaded. It is usually measured by the modulus of resilience, which is the strain energy per unit volume required to stress the material from zero to the yield stress. In uniaxial tension, the modulus of resilience (U_R) is:

$$U_R = \frac{s_o^2}{2E} \qquad \text{(Eq 9)}$$

where s_o is the yield strength, and E is the modulus of elasticity. For applications where the metal must not undergo permanent plastic deformation but resist energy loads, the best materials will have a high yield stress and a low modulus of elasticity. An example is golf club driver heads that use heat treated titanium on the face to give a higher rebound than the older steel-faced heads.

Toughness is the ability of a metal to absorb energy in the plastic range. Although there are a number of approaches to defining toughness, one of the oldest is to consider it as the total area under the stress-strain curve.

This area is an indication of the amount of work per unit volume that can be done on a material without causing it to fail. As shown in Fig. 10, this definition of toughness implies that toughness is a function of both strength and ductility.

True Stress-Strain Curve

The engineering stress-strain curve, based on the original dimensions of the specimen, is used in engineering design because most structural designs do not permit yielding or plastic deformation. However, for analysis of metal forming operations, the true stress-strain curve, also known as the flow curve, should be used. The true stress-strain curve is based on the actual dimensions of the test specimen as it undergoes plastic deformation; it has the shape shown in Fig. 11. The true stress (σ) is defined as:

$$\sigma = \frac{P}{A} \text{ in pascals (Pa) or pounds per square inch (psi)} \qquad \text{(Eq 10)}$$

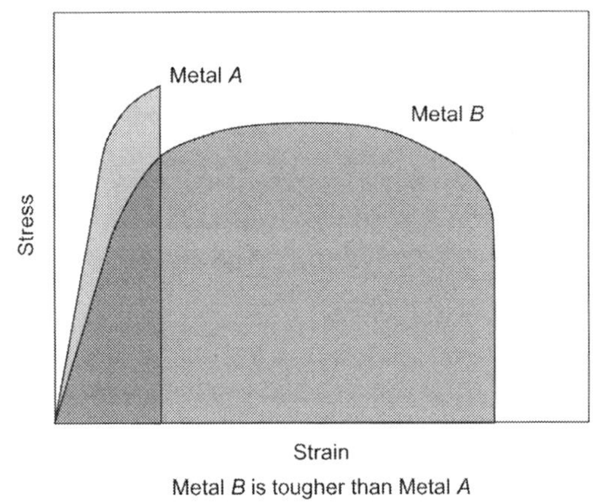

Fig. 10 Area under stress-strain curve as an indicator of toughness. Source: Ref 2

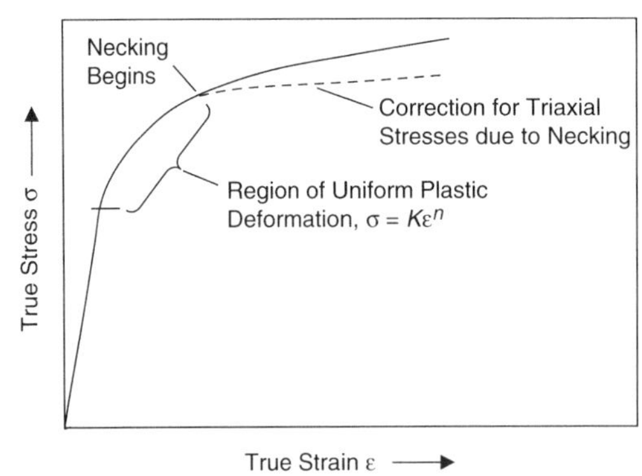

Fig. 11 True stress-true strain behavior. Source: Ref 2

where P is the load, and A is the instantaneous cross-sectional area.

The true stress is related to the engineering stress as follows:

$$\sigma = \frac{P}{A} = \frac{P}{A_o}\frac{A_o}{A} = s\frac{A_o}{A} \quad \text{(Eq 11)}$$

Using the constancy-of-volume relationship:

$$\frac{A_o}{A} = \frac{l}{l_o} = e + 1 \quad \text{(Eq 12)}$$

therefore:

$$\sigma = \frac{P}{A_o}(e+1) = s(e+1) \quad \text{(Eq 13)}$$

which is applicable up to the onset of necking.

The true strain (ε) is defined as:

$$\varepsilon = \int_{l_o}^{l} \frac{dl}{l} = \ln\frac{l}{l_o} \quad \text{(Eq 14)}$$

where l is the gage length, and l_o is the original gage length.

The relationship between the true strain and the engineering strain is:

$$\varepsilon = \ln(1-e) \quad \text{(Eq 15)}$$

The two values of strain give almost identical values for strains up to 0.1. This relationship is also only valid up to the onset of necking.

Beyond the onset of necking, the true strain is based on the actual area or diameter measurements. Because of the constancy of volume ($A_o l_o = Al$), the true strain can be written in terms of either length or area:

$$\varepsilon = \ln\frac{l}{l_o} = \ln\frac{A_o}{A} = \ln\frac{(\pi/4)D_o^2}{(\pi/4)D^2} = 2\ln\frac{D_o}{D} \quad \text{(Eq 16)}$$

For cylindrical test specimens, the true strain at failure (ε_f) can be determined from the reduction of area (q) by:

$$\varepsilon_f = \ln\frac{1}{1-q} \quad \text{(Eq 17)}$$

The flow curve for many metals in the region of uniform plastic deformation can be expressed by the Holloman power curve relation:

$$\sigma = K\varepsilon^n \quad \text{(Eq 18)}$$

where n is the strain-hardening exponent, and K is the strength coefficient. Equation 18 can be rewritten as:

$$\log \sigma = \log K - n \log \varepsilon \qquad \text{(Eq 19)}$$

As shown in Fig. 12, a log-log plot of true stress versus true strain up to the maximum load (onset of necking) will give a straight line if the Holloman relationship holds for the metal in question. The slope of the straight line is n, and K is equal to the true stress at a true strain $\varepsilon = 1.0$. The strain-hardening exponent can have values from $n = 0$ for a perfectly plastic solid up to $n = 1.0$ for a perfectly elastic solid. Metals typically have values of n between 0.10 and 0.50. It should be noted that n is equal to:

$$n = \frac{d(\log \sigma)}{d(\log \varepsilon)} = \frac{d(\ln \sigma)}{d(\ln \varepsilon)} = \frac{\varepsilon d\sigma}{\sigma d\varepsilon} \qquad \text{(Eq 20)}$$

Necking begins at the maximum load of the engineering stress-strain curve. The engineering stress-strain curve continues upward as the metal strain hardens, and this counteracts the reduction in cross-sectional area due to deformation. Eventually, a point is reached where the increase in strain hardening can no longer keep up with the reducing area, and the stress begins to decrease until fracture. On the other hand, because the true stress-strain curve is based on the instantaneous cross-sectional area, the curve continues to rise until failure.

The strain rate $\dot{\varepsilon} = d\varepsilon/dt$ can also influence the flow stress, particularly at elevated temperatures. The strain-rate sensitivity of metals is low at room temperature but increases with temperature, especially at temperatures above one-half of the absolute melting point. A series of curves for 6061-O aluminum are shown in Fig. 13. Note that the strain-rate sensitiv-

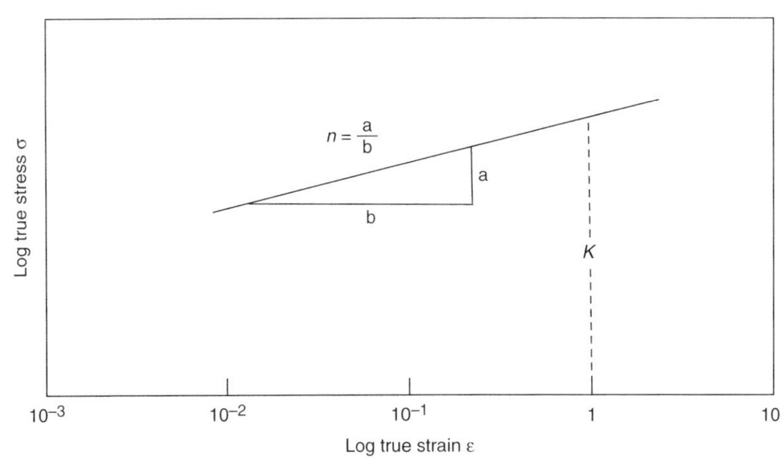

Fig. 12 Plot to determine strain-hardening exponent (n) and strength coefficient (K). Source: Ref 6

ity is much greater at the higher temperatures. For deformation occurring at constant volume, the relationship between flow stress and strain rate, at a constant strain and temperature, is given by:

$$\sigma = C\dot{\varepsilon}^m \qquad \text{(Eq 21)}$$

where m is the strain-rate sensitivity, which is the slope of a plot of log σ versus log $\dot{\varepsilon}$, as shown in Fig. 14, and C is the strain hardening coefficient.

The strain-rate sensitivity (m) can also be determined by measuring the change in flow stress (σ) due to a change in $\dot{\varepsilon}$ at a constant strain (ε) and temperature (T):

$$m = \frac{\Delta \log \sigma}{\Delta \log \dot{\varepsilon}} = \frac{\log(\sigma_2/\sigma_1)}{\log(\dot{\varepsilon}_2/\dot{\varepsilon}_1)} \qquad \text{(Eq 22)}$$

In hot working operations, m is typically in the range of 0.1 to 0.2.

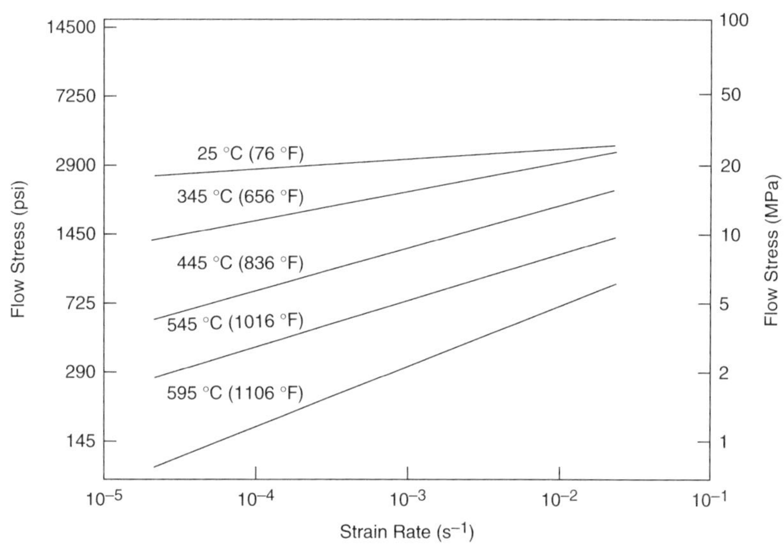

Fig. 13 Flow stress vs. strain rate for 6063-O aluminum. Source: Ref 7

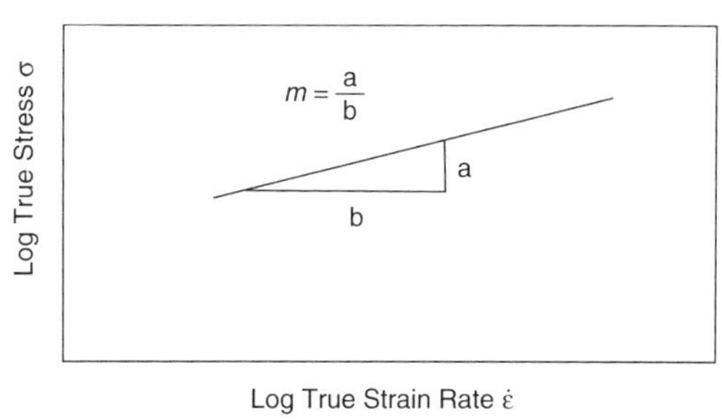

Fig. 14 Plot to determine strain-rate sensitivity (m). Source: Ref 2

The strain-rate sensitivity describes the ability of a material to resist plastic instability or necking. For superplasticity, *m* is usually greater than 0.5, with the majority of superplastic materials having an *m* value in the range of 0.5 to 0.8, where a value of 1.0 would indicate a perfectly superplastic material. The presence of a neck in a material undergoing a tensile strain results in a locally high strain rate and, for a high value of *m*, to a sharp increase in the flow stress within the necked region; that is, the neck undergoes strain hardening, which restricts its further development. Therefore, a high strain-rate sensitivity resists neck formation and leads to the high tensile elongations observed in superplastic materials. The flow stress decreases and the strain-rate sensitivity increases with increasing temperature and decreasing grain size. The elongation to failure tends to increase with increasing *m*.

Stress Concentrations

Geometrical features, such as holes, fillets, and radii, produce higher local stresses than encountered in the body of the material. The hole shown in Fig. 15 has 3× higher tensile stresses at the top and bottom of the hole than in the body of the material. These higher stresses are a result of the inability of the stresses to pass through the hole. As shown in Fig. 16, the fibers in the adjoining regions must carry this portion of the load in addition to their normal share. In essence, there is a concentration of force lines at the notch. Because the hole wants to elongate under the action of the load, compressive stresses are produced at the sides of the hole. Such stress increases are described by the stress-concentration factor (*K*):

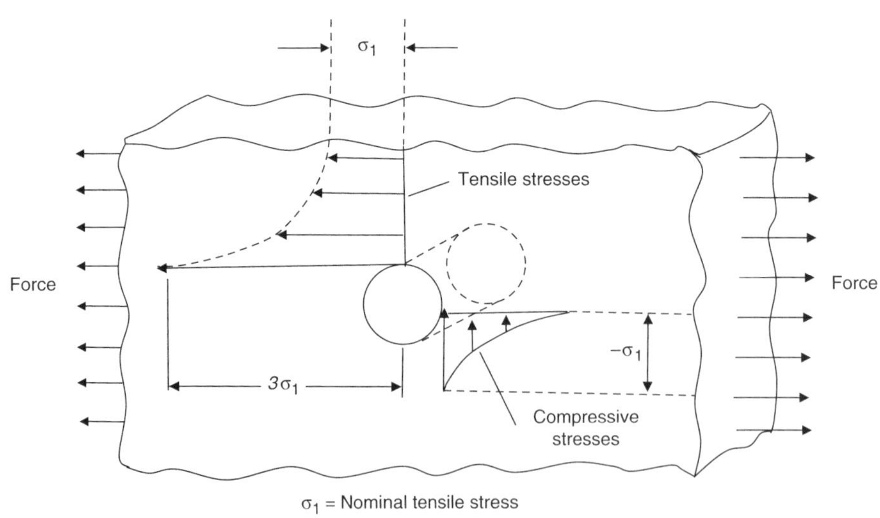

Fig. 15 Stress-concentration effect around a hole. Source: Ref 3

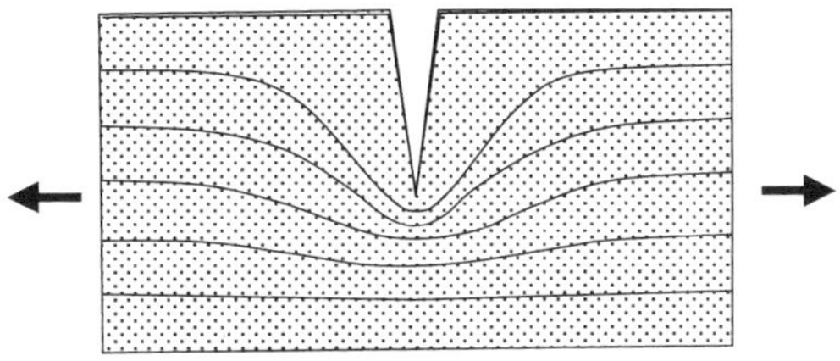

Fig. 16 Lines of force around notch. Source: Ref 8

$$K = \frac{\text{Maximum actual stress}}{\text{Nominal stress}} \quad \text{(Eq 23)}$$

While the stress-concentration factor (K) for a round hole is 3, much higher stress-concentration factors occur when there is a sharp notch in the material. The stress-concentration factors for three different geometries are shown in Fig. 17. Note that the value of the stress-concentration factor is strongly dependent on the actual geometry, and that values significantly higher than the value for a round hole are achieved for some geometries. Values for a wide range of geometries can be found in handbooks dealing with stress analysis.

Because most metals undergo plastic deformation at sufficiently high stresses, the stress values corresponding to the theoretical stress concentrations are often not developed. Ductile metals deform locally in the highly stressed regions and thus partially relieve the stress. Often, stress concentrations are not dangerous in ductile metals under static loads. However, in fatigue, even ductile metals are generally affected by stress concentrations. Most brittle metals are sensitive to stress-concentration effects under both static and dynamic loading.

Notched Tensile Test

Ductility measurements on standard smooth tensile specimens do not always reveal metallurgical or environmental changes that can lead to reduced local ductility. The tendency for reduced ductility in the presence of a triaxial stress field caused by a notch is called notch sensitivity. A common way of evaluating notch sensitivity is a tension test using a notched specimen. The notched tensile test has been used extensively for investigating the properties of high-strength steels, for studying hydrogen embrittlement in steels and titanium, and for investigating the notch sensitivity of high-temperature alloys.

The most common notched tensile specimen uses a 60° notch with a root radius of 0.025 mm (0.001 in.) or less introduced into a round (circumferential notch) or flat (double-edge notch) tensile specimen. Usually,

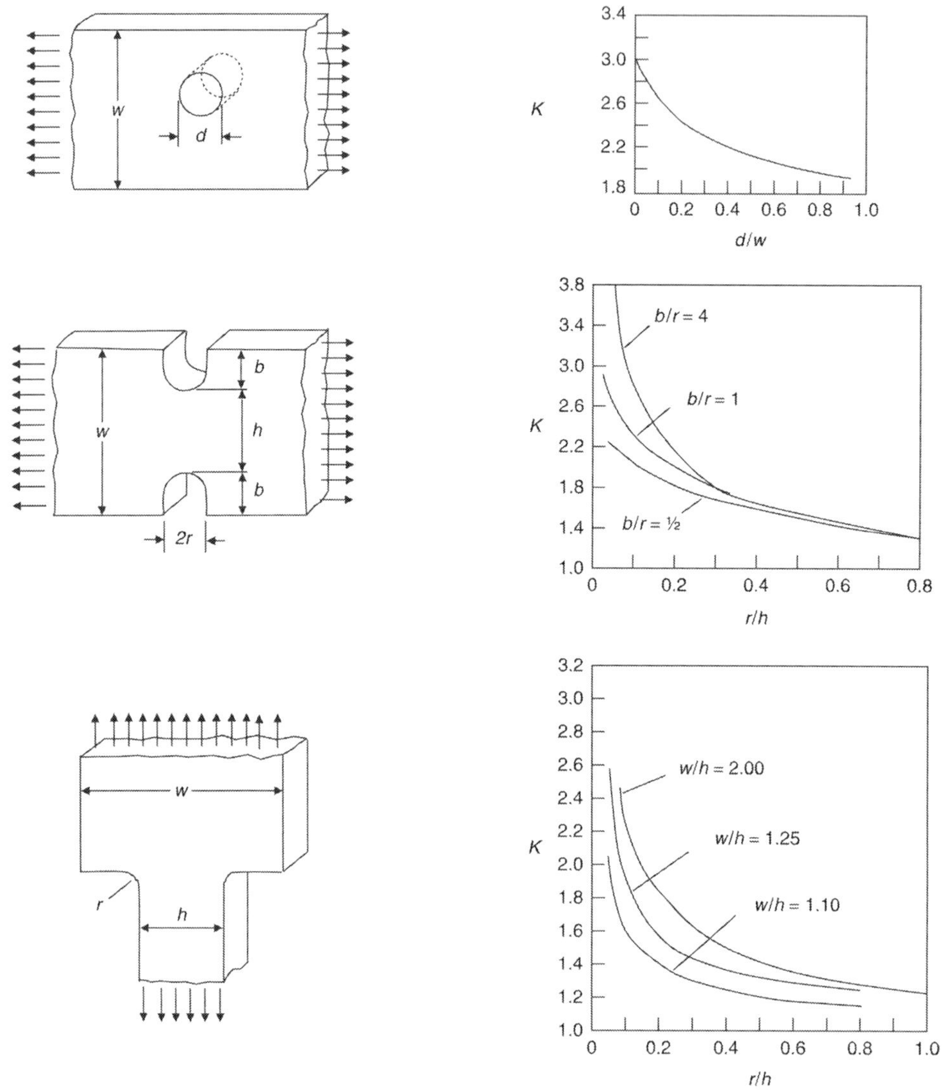

Fig. 17 Stress-concentration factors for three geometries. Source: Ref 9

the depth of the notch is such that the cross-sectional area at the root of the notch is one-half of the area in the unnotched section. The specimen is loaded in tension until fracture occurs. The notch strength is defined as the maximum load divided by the original cross-sectional area at the notch. Because of the plastic constraint at the notch, this value will be higher than the tensile strength of an unnotched specimen if the material possesses some ductility. Therefore, the common way of detecting notch brittleness, or high notch sensitivity, is by determining the notch strength ratio (NSR):

$$NSR = \frac{\text{Net tensile strength for notched specimen}}{\text{Tensile strength for unnotched specimen}} \quad \text{(Eq 24)}$$

If the NSR is less than 1, the material is considered to be notch brittle.

As strength, hardness, or some metallurgical variable restricting plastic flow increases, the metal at the root of the notch is less able to flow, and fracture becomes more likely. Notch brittleness can be considered to begin at the strength level where the notch strength begins to fall or, more conventionally, at the strength level where the NSR becomes less than unity. The sensitivity of notch strength for detecting metallurgical embrittlement in an alloy steel is illustrated in Fig. 18. Note that the elongation measured on a conventional smooth specimen was unable to detect the fall in notch strength produced by tempering in the 315 to 480 °C (600 to 900 °F) range.

Compression

Many metalworking operations, such as rolling, extrusion, and forging, are performed with the workpiece under compressive loads. A compression test is conducted in a manner similar to the tensile test, except that the force is compressive and the specimen contracts along the direction of the stress. However, the deformation shown in Fig. 1 is ideal because frictional forces develop at the two ends and cause barreling, as shown in Fig.

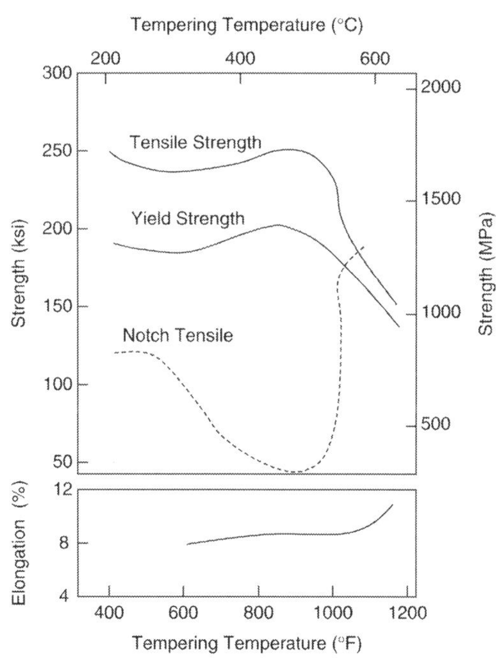

Fig. 18 Unnotched and notched properties of an alloy steel. Source: Ref 10

19. During deformation, as the metal spreads over the compression anvils to increase its diameter, frictional forces oppose the outward flow of metal. This frictional resistance occurs in the part of the specimen in contact with the anvils, while the metal at the specimen midheight can flow outward. This leads to the barreled shape, and internally a region of undeformed metal is created near the anvil surfaces. The addition of lubrication and/or synthetic fluorine-containing resin strips helps in reducing the frictional forces.

A solid under hydrostatic pressure (Fig. 20) undergoes a volume change called dilatation. If the volume change is ΔV, the dilatation (ε_{kk}), or volume strain, is defined as:

$$\varepsilon_{kk} = \frac{\Delta V}{V} \quad \text{(Eq 25)}$$

Because positive pressure causes a volume reduction, the negative of dilatation is proportional to the pressure, so that:

$$p = -K\varepsilon_{kk} \quad \text{(Eq 26)}$$

where K is known as the bulk modulus.

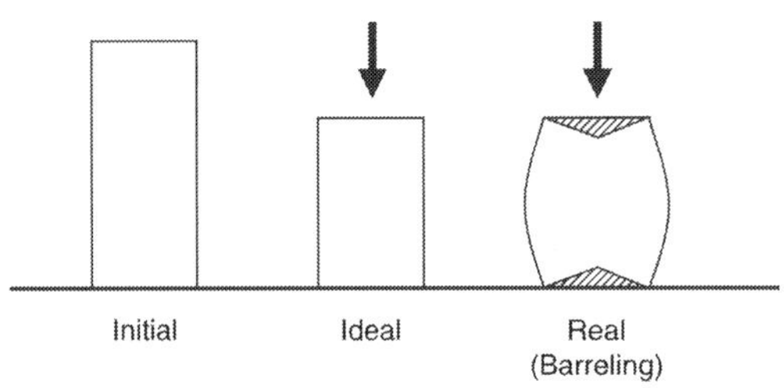

Fig. 19 Barreling during compression test. Source: Ref 2

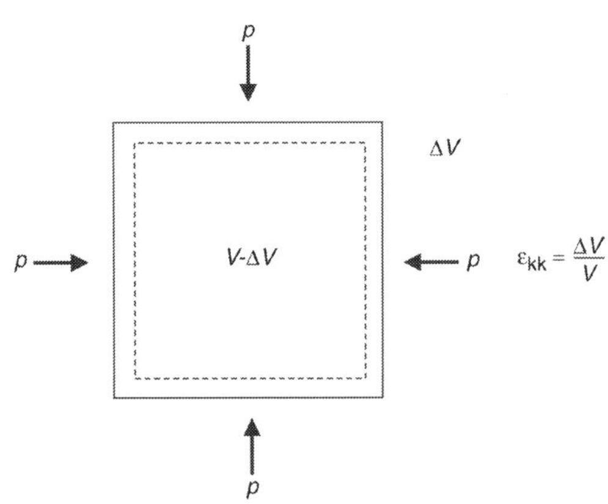

Fig. 20 Dilatation under hydrostatic pressure. Source: Ref 2

Annealed metals have approximately the same tensile and compressive yield strengths. When a metal is plastically deformed in tension, it displays the familiar work-hardening effect. However, the same tensile deformation that increases the tensile yield strength in tension reduces the compressive yield strength, a phenomenon known as the Bauschinger effect. In addition, plastic deformation in compression decreases the yield strength in tension. This directionality of strain hardening occurs in all polycrystalline metals but is especially pronounced in some metals, such as titanium and zirconium. For example, the titanium alloy Ti-6Al-4V develops more than a 50% difference between its tensile and compressive yield strengths after only a 2% tensile elongation. The Bauschinger effect is relieved by elevated temperature and can be totally eliminated by hot working or a stress-relief anneal after cold working.

When a tensile specimen is loaded to a plastic strain (point 2 in Fig. 21), it strain hardens because the yield point (point 1) has been exceeded. When the load is removed, the specimen contracts elastically along path 2-3. If the metal is then loaded in compression, its compressive yield strength (point 4) is considerably less than its tensile yield strength (point 2). The Bauschinger effect is caused by the tendency of dislocations to pile up and become entangled during the initial tensile loading, producing strain hardening. However, when the metal is loaded in compression, it is easier for the dislocations to reverse their path and travel backward in a direction that now contains fewer dislocations. Compressive loading from point 3 also produces Frank-Read dislocation loops of the opposite sign of

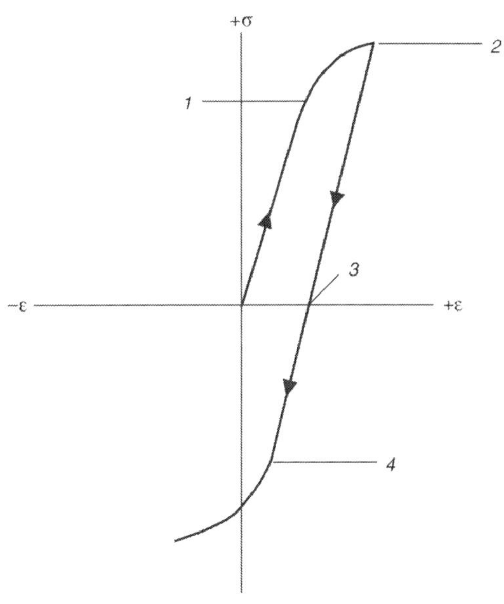

Fig. 21 Bauschinger effect. Source: Ref 2

those generated during tension loading, and these annihilate each other when they meet.

Shear and Torsion

For tests performed using a pure shear force, as shown in Fig. 1, the shear stress (τ) is computed according to:

$$\tau = \frac{P}{A_o} \tag{Eq 27}$$

where P is the load or force imposed parallel to the upper and lower faces, each of which has an area of A_o. The shear strain (γ) is defined as the tangent of the strain angle (θ):

$$\gamma = \frac{w}{l} = \tan\theta \tag{Eq 28}$$

but for small elastic strains, $\gamma = \theta$.

The units for shear stress and strain are the same as for their tensile counterparts. The shear modulus (G) is the slope of the linear portion of the shear stress-strain curve:

$$G = \frac{\tau}{\gamma} \tag{Eq 29}$$

Torsion is a variation of pure shear where a structural member is twisted. Torsional forces produce a rotational motion about the longitudinal axis of one end of the member relative to the other end. Torsional tests are normally performed on cylindrical solid shafts or tubes. In torsion, each element of the material deforms in pure shear, as shown in Fig. 22. The shear strain (γ) in an element is given by:

$$\gamma = r\theta/L \tag{Eq 30}$$

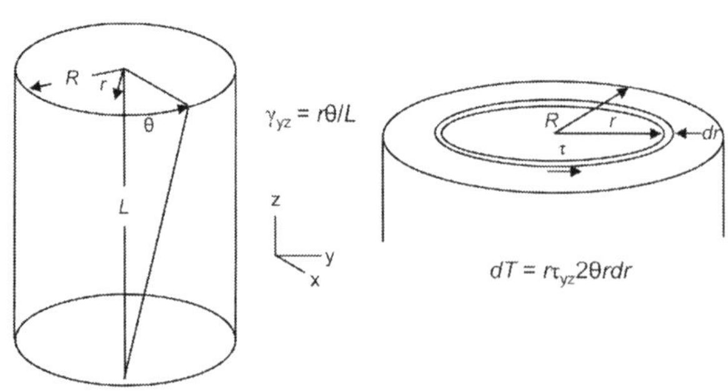

Fig. 22 Schematic of torsion test. Reprinted with the permission of Cambridge University Press. Source: Ref 11

where r is the radial position of the element, θ is the twist angle, and L is the specimen length. The shear stress (τ) cannot be measured directly or even determined unequivocally from the torque. This is because the shear stress (τ) depends on γ, which varies with radial position. Therefore τ depends on r. Consider an annular element of radius r and width dr having an area $2\pi r dr$. The contribution of this element to the total torque (T) is the product of the shear force on it ($\tau \cdot 2\pi r dr$) and the lever arm (r). Therefore, the torque is:

$$dT = 2\pi\tau r^2 dr$$

and

$$dT = 2\pi\tau_{yz}\int_0^R \tau r^2 dr \qquad \text{(Eq 31)}$$

Equation 31 cannot be integrated directly because τ depends on r. Integration requires substitution of the stress-strain (τ-γ) relation. Handbook equations for torque are usually based on assuming elasticity. In this case, $\tau = G\gamma = Gr\theta/L$. Substituting this and Eq 30 into Eq 31 gives:

$$T = 2\pi(\theta/L)G\int_0^R r^3 dr = (\pi/2)(\theta/L)GR^4 \qquad \text{(Eq 32)}$$

Substituting $GR\theta/L = \tau_s$ into Eq 32, $T = (\pi/2)R^3\tau_s$. Now, solving for the shear stress (τ_s) gives:

$$\tau_s = \frac{2T}{\pi R^3} \qquad \text{(Eq 33)}$$

If the bar is not elastic, Hooke's law cannot be assumed. The other extreme is when the entire bar is plastic and the material does not work harden. In this case, τ is a constant.

Stress-Strain Relationships

When the state of stress is multiaxial, it is necessary to calculate the elastic strains induced by the multiaxial state of stress in three mutually perpendicular directions by using the generalized Hooke's law:

$$e_x = \frac{1}{E}\left[\sigma_x - v\left(\sigma_y + \sigma_z\right)\right] \qquad \text{(Eq 34)}$$

$$e_y = \frac{1}{E}\left[\sigma_y - v\left(\sigma_x + \sigma_z\right)\right] \qquad \text{(Eq 35)}$$

$$e_z = \frac{1}{E}\left[\sigma_z - v\left(\sigma_x + \sigma_y\right)\right] \qquad \text{(Eq 36)}$$

where σ_x, σ_y, and σ_z are the normal stresses in three directions, and ε_x, ε_y, and ε_z are the elastic strains in the three coordinate directions.

The modulus of elasticity (E), the shear modulus (G), the bulk modulus (K), and Poisson's ratio (ν) are related to each other according to:

$$E = 2G(1-\nu) \quad G = \frac{E}{2(1+\nu)} \quad \nu = \frac{E-2G}{2G}$$

$$K = \frac{E}{3(1-2\nu)} \quad E = \frac{9KG}{3K-G} \quad \nu = \frac{3K-E}{6K}$$

and for many materials, $K \sim E$, $G \sim \frac{3}{8}E$, and $\nu \sim 0.33$.

Combined Stresses

For combined stress states that have a combination of tensile, compression, and shear stresses, the stress state can be resolved into principal stresses. Principal stresses act on axes that differ from the axes along which the stresses are acting and represent the maximum and minimum values of the normal stresses for the particular point under consideration. The principal stresses along the x and y axes are:

$$\sigma_1 = \frac{\sigma_x + \sigma_y}{2} + \sqrt{\left(\frac{\sigma_x - \sigma_y}{2}\right)^2 + \tau_{xy}^2} \quad \text{(Eq 37)}$$

$$\sigma_2 = \frac{\sigma_x + \sigma_y}{2} - \sqrt{\left(\frac{\sigma_x - \sigma_y}{2}\right)^2 + \tau_{xy}^2} \quad \text{(Eq 38)}$$

$$\tau_1 = \pm\sqrt{\left(\frac{\sigma_x - \sigma_y}{2}\right)^2 + \tau_{xy}^2} \quad \text{(Eq 39)}$$

where σ_1, σ_2, and τ_1 are the principal stress components, and σ_x, σ_y, and τ_{xy} are the calculated stress components shown in Fig. 23, all of which are determined at any particular point. Principal stresses can also be determined graphically, using the Mohr's circle construction shown in Fig. 24. More detailed descriptions of Mohr's circle can be found in almost any book dealing with mechanics or strength of materials.

Yield Criteria

In many instances, for example, metalworking operations, the material is not subjected to simple tensile and compression stresses; instead, a triaxial stress state will exist. For the case of simple uniaxial loading, the onset of yielding can accurately be predicted to occur at the yield strength

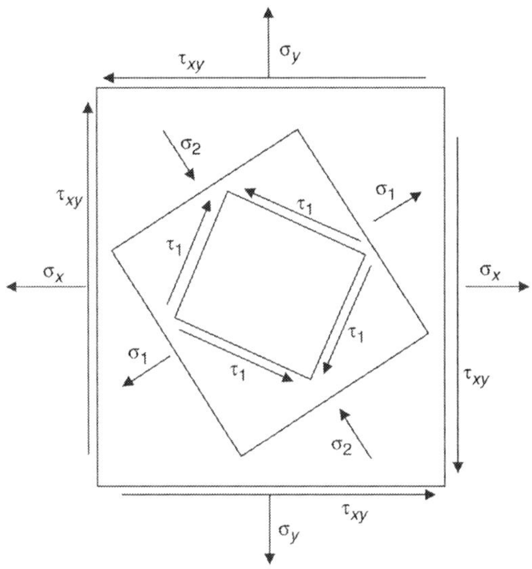

Fig. 23 Relative orientation of stresses. Source: Ref 12

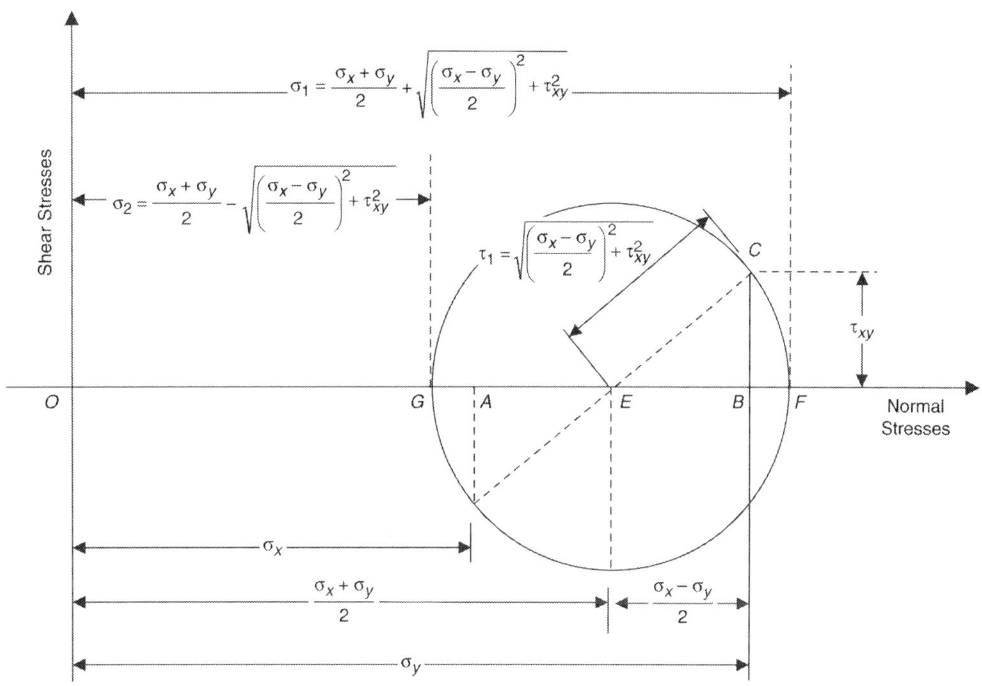

1. Let the axes *x* and *y* be chosen to represent the directions of the applied normal and shearing stresses, respectively.
2. Construct to suitable scale distances $OA = \sigma_x$, $OB = \sigma_x$, and $BC = AD = \tau_{xy}$.
3. With point *E* as a center, construct the circle DFC.
4. Then *OF* and *OG* are the principal stresses σ_1 and σ_2, respectively, and *EC* is the maximum shear stress τ_1.

Fig. 24 Mohr's circle for determination of principal stresses. Source: Ref 12 of the material during a simple tensile test. If the loading is more complicated, and a multiaxial state of stress is produced by the loads, the onset of yielding can no longer be predicted by comparing any one of the normal stress components with uniaxial material yield strength, not even the maximum principal normal stress. The onset of yielding for multiaxially stressed parts is more accurately predicted through the use of a combined stress theory of failure. The two most widely accepted theories for predicting the onset of yielding are the distortion energy theory (von Mises criterion) and the maximum shearing stress theory (Tresca criterion). The distortion energy theory is usually somewhat more accurate, while the maximum shearing stress theory may be slightly easier to use.

In the distortion energy theory, yielding is predicted to occur in a multiaxial state of stress when the distortion energy per unit volume becomes equal to or exceeds the distortion energy per unit volume at the time of yielding in a simple uniaxial stress test using a specimen of the same material. Therefore, yielding is predicted by the distortion energy theory if:

$$\frac{1}{\sqrt{2}}\left[(\sigma_1-\sigma_2)^2+(\sigma_2-\sigma_3)^2+(\sigma_3-\sigma_1)^2\right]^{1/2} \geq \sigma_y \qquad \text{(Eq 40)}$$

In the maximum shearing stress theory, yielding is predicted to occur in a multiaxial state of stress when the maximum shearing stress magnitude becomes equal to or exceeds the maximum shearing stress magnitude at the time of yielding in a simple uniaxial stress test using a specimen of the same material. In this instance, yielding is predicted by the maximum shearing stress theory if:

$$\sigma_1-\sigma_3 \geq \sigma_y \qquad \text{(Eq 41)}$$

where σ_1, σ_2, and σ_3 are the principal stresses at a point, ordered such that $\sigma_1 > \sigma_2 > \sigma_3$, and σ_y is the uniaxial yield strength in tension.

A comparison of the two criteria is shown in Fig. 25. In the simpler Tresca criterion shown in Fig. 25(a), stress combinations lying within the solid line envelope do not result in plastic flow, while those lying outside the envelope develop plastic flow. The situation is the same for the von Mises criterion shown in Fig. 25(b), which also shows the Tresca criterion for comparison and shows that the Tresca criterion is the more conservative of the two. However, experimental data shown in Fig. 25(c) fit the von Mises criterion somewhat better, but the difference between the two is not great.

Residual Stresses

Residual stresses are the stresses that remain in the material as a result of some manufacturing process. They may be mild and unnoticeable, moderate as in the slight warpage of a piece of sheet metal, or severe enough to cause cracking and splitting apart of a freshly quenched piece of steel.

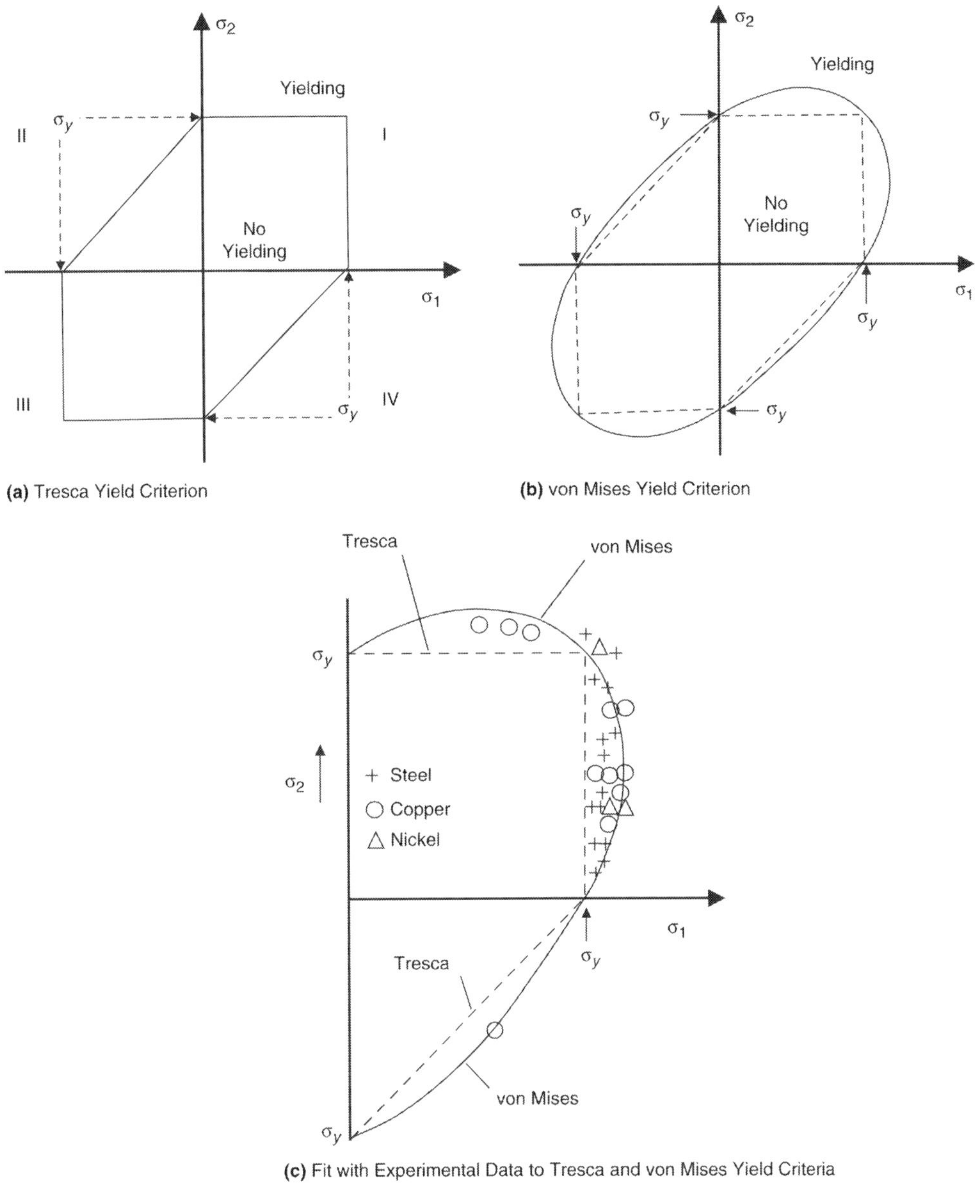

Fig. 25 Comparison of yield theories. Source: Ref 13

Plastic deformation due to cold working of a metal always produces residual stresses. However, the resulting stress pattern varies greatly, depending on the type of working employed. If the metal is only lightly worked, so that plastic deformation is limited to regions near the surface, the surface layers will be left in a state of residual compression. An example of this type of stress pattern is shown in Fig. 26(a), which may be produced by processes such as shot peening, light cold drawing, or surface rolling. In general, compressive stress states on the surface are desirable because in fatigue loading, the tensile fatigue stresses have to overcome the residual compressive stresses before the metal starts being loaded in tension. However, if the working is severe, so that deformation occurs throughout the cross section of the metal, there will be tensile residual stresses on the surface, as shown in Fig. 26(b). Using the same logic as for compression residual stresses, tensile residual stresses can drastically lower the fatigue strength because the residual stresses are additive to the applied tensile stresses. Therefore, heavily worked sections are frequently given a final light deformation to convert the residual surface tensile stresses to compressive stresses, as shown in Fig. 26(c), for example, in light surface rolling of heavily worked cold-drawn steel.

Another major source of residual stresses is the nonuniform change in volume because of thermal effects. In many cases, such as fusion welding, several thermal factors may act simultaneously to produce a final complex pattern of residual stresses. These factors include thermal expansion/contraction on heating and cooling, a change in volume, and a change from the liquid to the solid state. When a hot solid metal is quenched in water, the surface cools more rapidly than the center and thus shrinks faster. This nonuniform shrinkage causes plastic deformation of the hotter, softer interior. After the surface has cooled, the interior continues to contract as its temperature continues to fall. When both the surface and interior are able to support large stresses, the contraction of the interior produces compressive stresses at the surface, similar to the pattern shown in Fig. 26(a). These stresses can be quite large, approximately −207 MPa (−30 ksi) in steel and approximately −100 MPa (−15 ksi) in softer metals such as aluminum. The use of a milder quench will reduce the residual stress level.

In steels, the formation of martensite from austenite during quenching results in an expansion. This expansion during martensite formation can

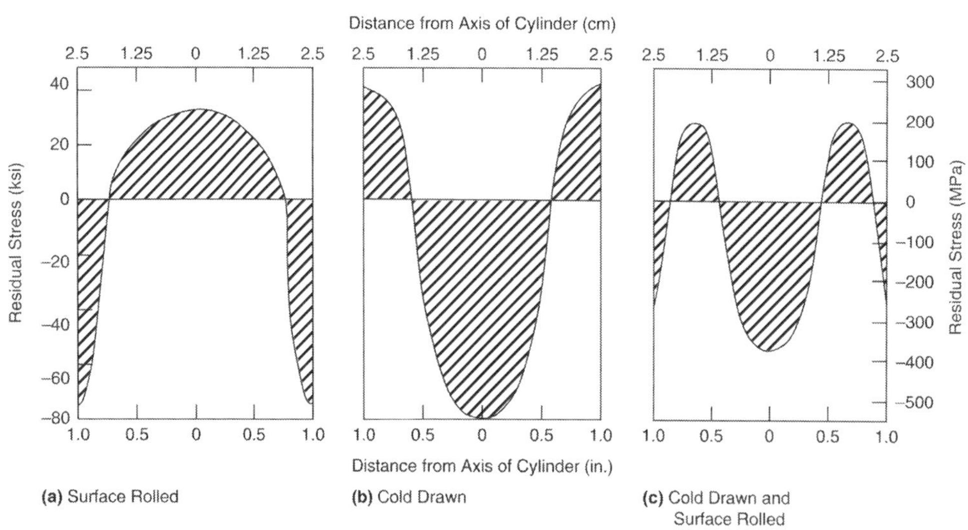

Fig. 26 Residual stress patterns in 5 cm (2 in.) diameter steel bars. Source: Ref 3

result in tensile residual stresses at the surface and compressive residual stresses in the interior, similar to that shown in Fig. 26(b), with maximum residual tensile stresses at the surface approaching +550 MPa (+80 ksi). In the actual hardening of steel, the stress patterns are usually much more complex than this because the final pattern is due to a combination of thermal contraction and martensitic expansion. However, this does help explain why freshly quenched steels can often crack and spall. Thus, it is common practice to initiate tempering as soon as possible after quenching. Combine these residual stresses with stress concentrations, such as sharp corners, and it is easy to see how problems can develop. Typical causes of distortion and cracking are sharp corners, re-entrant angles, and abrupt changes in section. Milder quenches and immediate tempering are helpful in reducing problems.

Stresses induced during casting can result from the shrinkage of the metal during solidification. These stresses can be great enough to cause the formation of cracks while the casting is solidifying, commonly known as hot tearing. After approximately 80% of solidification is complete, the dendrites become interlocked. Hot tearing can occur during the last 20% of solidification and is related to the length of the temperature interval over which the remaining solidification occurs.

Hardness

Hardness is the resistance to penetration, and the majority of hardness testers force a small sphere, pyramid, or cone into metal by means of an applied load. A number is obtained, and the hardness can often be correlated to the tensile strength of the metal. For example, the tensile strength of a steel in pounds per square inch is roughly equal to 500 times its Brinell hardness number. The correlations between tensile strength and hardness for steels, brass, and nodular cast iron are shown in Fig. 27.

A hardness test is easy to execute and provides information on the yield strength of a material. In brief, a hard material (the indentor) is placed above the sample surface. A load is then placed on the indentor that then penetrates the surface of the specimen. The size of the resulting indentation, a measure of the permanent deformation that the material undergoes during the test, relates to the material hardness. A small indentation indicates high hardness. When the load is divided by the indentation area (sometimes the projected area of the indentation), a hardness number (H) is defined. Hardness has units of stress and, because it relates to permanent deformation, H correlates with σ_y. The stress state arising during a hardness test differs from that in a tensile test, however. This results in H being greater than σ_y. To a good approximation, $H \sim (2.5 \text{ to } 3.0) \cdot \sigma_y$; the approximation $H = 3\sigma_y$ is often used.

A number of important hardness tests are summarized in Fig. 28. One of the most widely used is the Rockwell hardness tester, shown in Fig. 29.

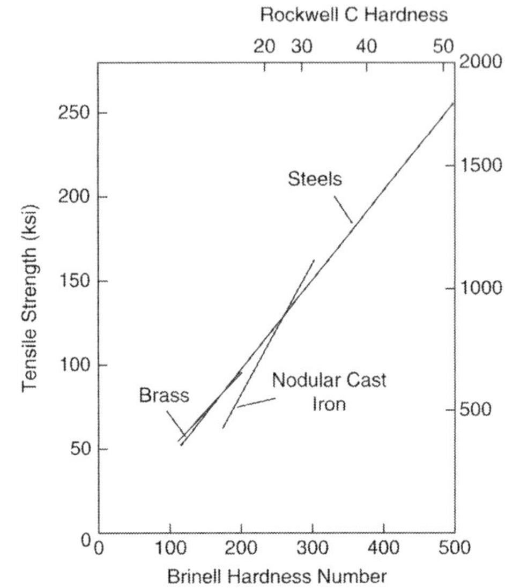

Fig. 27 Correlation of hardness with tensile strength. Source: Ref 2

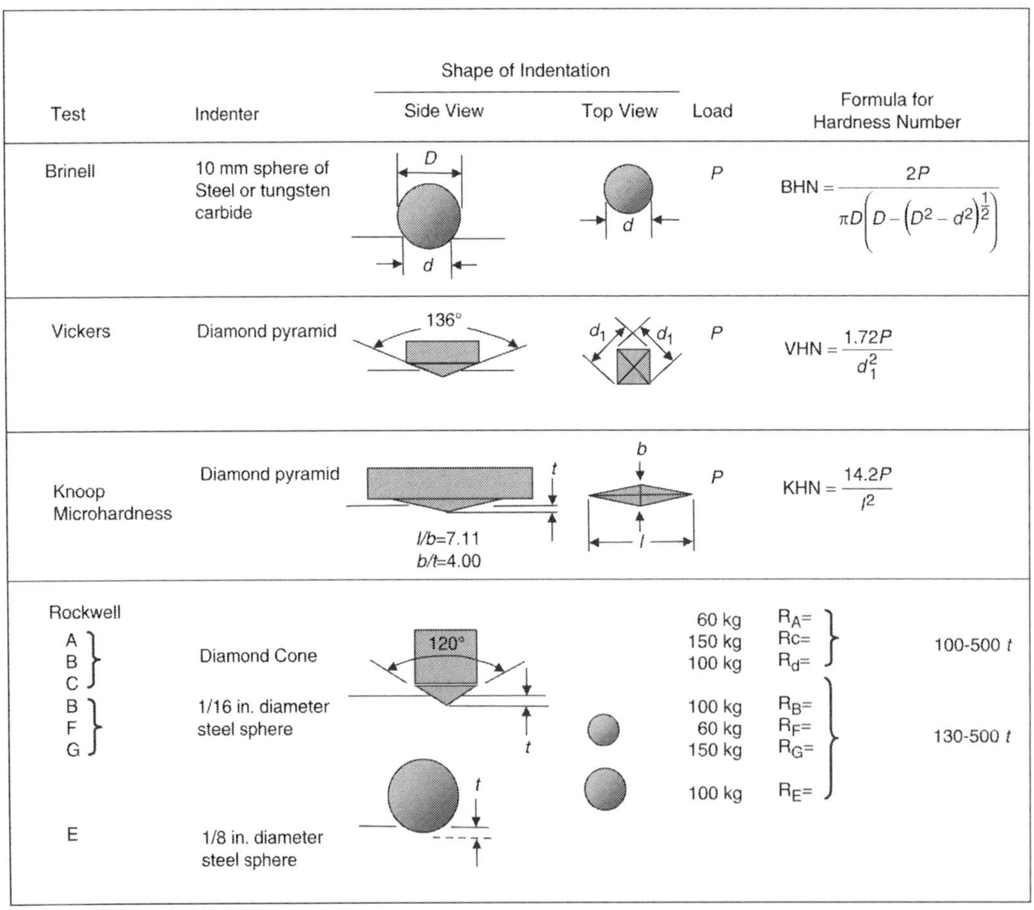

Fig. 28 Hardness tests. Source: Ref 14

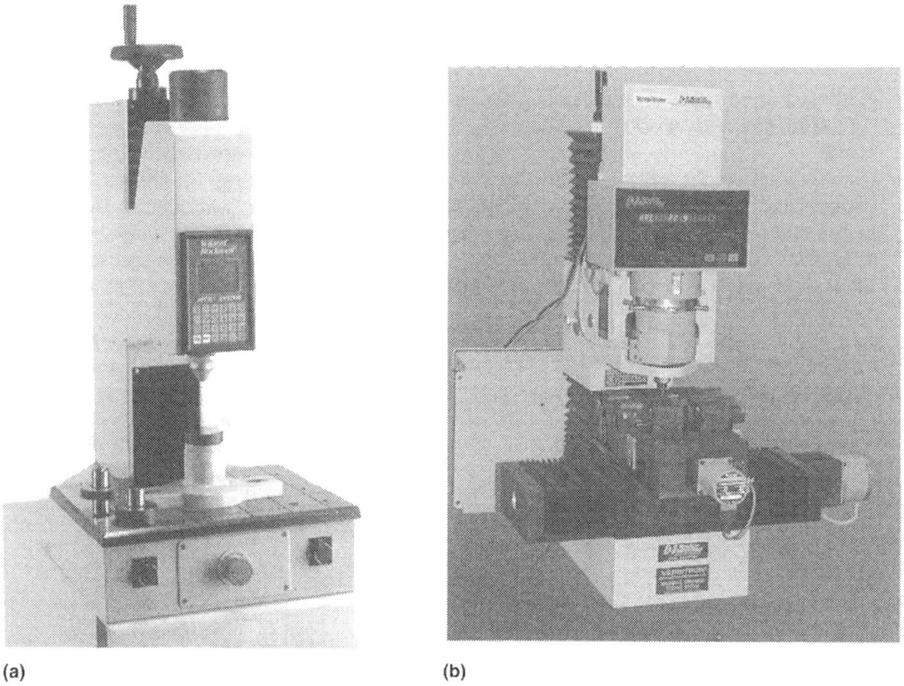

Fig. 29 (a) High-speed Rockwell tester. (b) Automated Rockwell tester for high-rate testing. Source: Ref 2

For determining the Rockwell hardness, a small load is initially applied, followed by a larger load. The penetrator is forced into the surface, and when the load is released, a dial indicator indicates the hardness, which is a function of the depth reached by the penetrator. The Brinell hardness tester is another widely used test for large steel parts, such as crankshafts, where a minimum of depth of penetration is needed. Microhardness tests can be made on individual constituents of a microstructure, with the hardness of particles as small as 0.18 mm (0.007 in.) being determined.

ACKNOWLEDGMENT

The material in this chapter came from "Mechanical Behavior" in *Elements of Metallurgy and Engineering Alloys* by F.C. Campbell, ASM International, 2008.

REFERENCES

1. W.D. Callister, *Fundamentals of Materials Science and Engineering,* 5th ed., John Wiley & Sons, Inc., 2001
2. F.C. Campbell, Mechanical Behavior, *Elements of Metallurgy and Engineering Alloys,* ASM International, 2008
3. A.G. Guy and J.J. Hren, *Elements of Physical Metallurgy,* 3rd ed., Addison-Wesley Publishing, 1974

4. B.L. Ferguson, Design for Deformation Processes, *Materials Selection and Design,* Vol 20, *ASM Handbook,* ASM International, 1997
5. G.E. Dieter, *Mechanical Metallurgy,* 3rd ed., McGraw-Hill Book Co., 1986
6. G.E. Dieter, Mechanical Behavior Under Tensile and Compressive Loads, *Mechanical Testing and Evaluation,* Vol 8, *ASM Handbook,* ASM International, 2000
7. R. D'Antonia, C. Caciag, and K. Mukherjee, *Metall. Trans.,* Vol 1, 1970, p 1771
8. M.A. Meyers and K.K Chawla, *Mechanical Metallurgy—Principles and Applications,* Prentice-Hall Inc., 1984
9. G.H. Neugebauer, *Prod. Eng.,* Vol 14, 1943, p 82–87
10. G.B. Espey, M.H. Jones, and W.F. Brown, *ASTM Proceedings,* Vol 59, American Society for Testing Materials, 1959, p 837
11. W.F. Hosford, *Mechanical Behavior of Materials,* Cambridge University Press, 2005
12. J. Marin, *Mechanical Properties of Materials and Design,* McGraw-Hill, 1942
13. T.H. Courtney, *Mechanical Behavior of Materials,* 2nd ed., McGraw-Hill Book Co., 2000
14. H.W. Hayden, W.G. Moffatt, and J. Wulff, *The Structure and Properties of Materials,* Vol III, Wiley, 1965

CHAPTER 3

Ductile and Brittle Fracture

THE TERMS *ductile* and *brittle* are opposites, like the colors black and white. For many parts, it is desirable to have metal properties that are predominantly hard, strong, and wear resistant, accepting the accompanying brittleness. Examples are gear teeth, antifriction bearings, and various types of cutting and forming tools. However, in many other parts, it is desirable to have metal properties that are predominantly ductile, accepting the relatively low hardness, strength, and wear resistance. Examples are household aluminum foil, flat metal that must be shaped, and wire for cold forming into coat hangers, nails, and wire for wrapping.

However, in a very large number of parts, the most desirable properties are shades of gray rather than the extremes of black or white. In these cases, neither extreme brittleness nor high ductility is entirely satisfactory. Rather, the properties must be balanced to a desirable blend of characteristics. Examples are springs, fasteners of various types, shafts, pressure vessels, blades, hand tools, and many other types of parts. These various conditions can be achieved by the proper choice of the metal composition as well as heat treatment or mechanical deformation to achieve the optimum combination of properties desired for a particular part. Many nonmetals lack ductility and normally are subject to brittle fracture; examples are chalk, rock, brick, and—under normal conditions of use—glass. However, as soon as glass is mentioned, one must be careful to qualify its brittleness as a function of temperature. Ordinary glass is brittle at ambient temperatures, but when heated in a flame or a furnace, it becomes extremely plastic or ductile. Then, it is readily moldable into an infinite variety of shapes. Indeed, heating glass is the major method of forming glass bottles, art objects, and the like.

Ductile and *brittle* are terms often used to describe the amount of macroscale plastic deformation that precedes fracture. The presence of brittle fracture is a concern, because catastrophic brittle fracture occurs due to the elastic stresses that are present and usually propagates at high speed, sometimes with little associated absorbed energy. Fracture occurring in a brittle manner cannot be anticipated by the onset of visible permanent distortion. It happens so fast that it does not allow time to safely shut down operating equipment. In addition, brittle fracture cannot be arrested by a removal of the load, except in very rare circumstances.

It is important to note that there is no universally accepted dividing line for ductile and brittle behavior in terms of strain at fracture or in terms of energy absorption. Relative ductility observed during tensile testing is an arbitrary basis for defining ductility. For example, it has been suggested that a material has adequate ductility when the reduction in area (RA) is between 26 and 18%, has limited ductility when the RA is between 18 and 2%, and is brittle when the RA is below 2%. Ductile behavior is also associated with high energy absorption at fracture, and adequate toughness or ductility is often defined by impact data, where criteria to determine whether the fracture is ductile or brittle involve some minimum level of absorbed energy at the service temperature of interest, say 14 or 20 J (10 or 15 ft · lbf).

Ductile Fracture

Ductile fracture results from the application of an excessive force to a metal that has the ability to deform permanently, or plastically, prior to fracture. Thus, the property of ductility is simply the ability of the material to flow or deform, which may or may not lead to fracture, depending on the magnitude of the force applied.

The property of ductility is somewhat related to the property of toughness, although the latter is usually measured in the presence of a notch or other stress concentration. The Charpy V-notch impact test is commonly used as a measure of toughness. The ability to absorb energy and deform plastically prior to fracture is characteristic of both ductility and toughness. This ability to absorb energy is a valuable property of ductile metals. For example, many modern automobiles are designed so that in the event of a front end collision, the car will not stop instantly but will fold like an accordion. This absorption of energy serves to slow down the rate of deceleration of the passenger compartment and lessen the impact on the occupants of the car.

Ductility also permits metal parts to be formed without fracture and to be adjusted in shape. A simple example is familiar to all persons who wear metal eyeglasses: periodically it is necessary to have the frames adjusted because of the various stresses that are applied to the frames during service. The adjustment is possible because of the ductility, or formability, of the metal in the frame. If the metal were brittle, adjustment would not be possible, because the frames would fracture. Actually, the ductility works both ways: it enables adjustment of the frames, but it also permits the distortion that makes adjustment necessary.

Characteristics of Ductile Fracture

Ductile fractures have characteristics that are different from those of brittle fractures. However, it must be recognized that many fractures contain some of the characteristics of both types. Ductile fractures have the following characteristics:

- There is considerable gross permanent or plastic deformation in the region of ductile fracture, as shown in Fig. 1. In many cases, this may be present only in the final rupture region of a fracture that may have originated with a fatigue or brittle fracture.
- Ductile fractures are those in which the shear stress exceeds the shear strength before any other mode of fracture can occur. Therefore, the micromechanism of fracture is in the shear direction, but this is not always obvious on macroexamination. The surface of a ductile fracture is not necessarily related to the direction of the principal tensile stress, as it is in a brittle fracture.
- The characteristic appearance of the surface of a ductile fracture is dull and fibrous. This is caused by deformation on the fracture surface.
- Ductile fractures proceed only as long as the material is being strained; that is, stop the deformation and the crack stops propagating. At the other extreme, once a brittle crack is initiated, it propagates through the material at velocities approaching the speed of sound, with no possibility of arresting it.

The classic example of a ductile fracture is a tensile specimen that has necked down prior to fracture. A typical fracture of this type, shown in Fig. 2, is the so-called cup-and-cone fracture characteristic of ductile metals pulled in tension. It is instructive to study this type of fracture in some detail:

- The narrowing, or necking, indicates that there has been extensive stretching, or elongation, of the grains of metal in the reduced area, particularly near the fracture itself.

Fig. 1 Ductile fracture in 1038 steel bolt. The bolt was annealed to a hardness of 95 HRB (equivalent to 15 HRC) and shows tremendous permanent deformation. Source: Ref 1

- Shear stress dominates deformation and ductile fracture. In most cases, the 45° plane of maximum shear stress components is not obvious or readily observed. However, Fig. 3 shows a low-carbon cast steel tensile specimen in which there were numerous casting discontinuities

Fig. 2 Type 302 stainless steel tensile specimen with the typical cup-and-cone fracture characteristic of ductile metals fractured in tension. In this case, the slant fracture at the surface of the test specimen was in both directions; in other instances, it may be in only one direction, forming a perfect cup and cone. Source: Ref 1

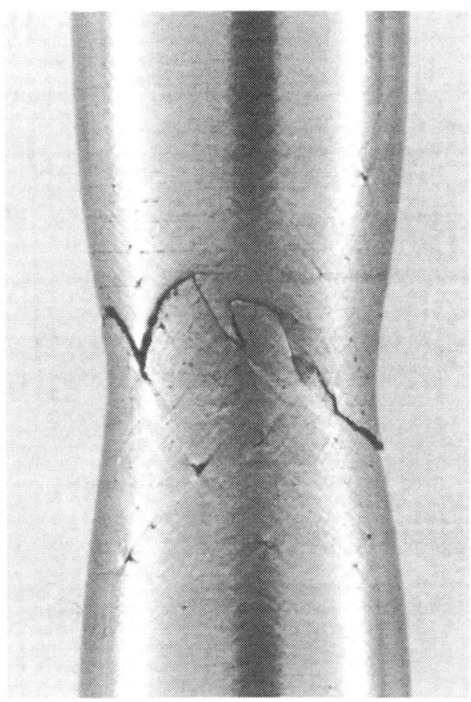

Fig. 3 Low-carbon cast steel test specimen emphasizing 45° shear aspect of tensile fracture of a ductile metal. Diagonal ridges are Lüders lines or stretcher strains; porosity in steel shows many localized fractures. Source: Ref 1

that tend to emphasize the 45° shear aspect of tensile fracture. The diagonal ridges, frequently called Lüders lines or stretcher strains, are easily visible in this photograph.

- A tensile cup-and-cone fracture originates with many tiny internal fractures called microvoids near the center of the reduced area. These voids occur after the tensile strength has been attained and as the stress (or load on the test machine) is dropping toward the fracture stress, as shown in Fig. 4, region (c).
- A ductile fracture starts near the center of the reduced section in tensile loading and then spreads outward toward the surface of the necked-down area. Before the fracture reaches the surface, however, it suddenly changes direction from generally transverse to approximately a 45° angle. It is this slant fracture—frequently called a shear lip—that forms the cup-and-cone shape characteristic of many tensile fractures of ductile metal. This slant fracture is useful for study of many fractures, because it represents the end of the fracture process at that location. Tensile fracture of a relatively thin section of a ductile metal may be entirely slant fracture. However, as the thickness increases, the percentage of slant fracture around the central origin area will decrease, sometimes resembling a picture frame, on a relatively thick rectangular section.

Shear is defined as that type of force that causes or tends to cause two contiguous parts of the same body to slide relative to each other in a direction parallel to their plane of contact. It should be noted that shear can be considered on both a macroscale and a microscale. Macroshear involves lateral cutting, or shearing as in direct, or transverse, shear. The cutting action of a pair of scissors (shears) is a common example. Microshear in-

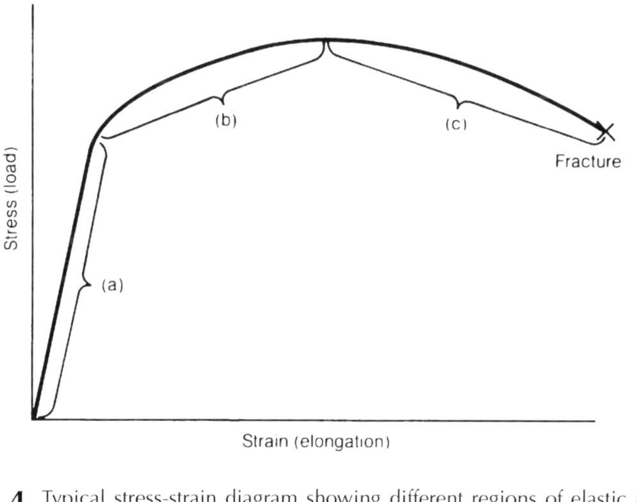

Fig. 4 Typical stress-strain diagram showing different regions of elastic and plastic behavior. (a) Elastic region in which original size and shape will be restored after release of load. (b) Region of permanent deformation but without localized necking. (c) Region of permanent deformation with localized necking prior to fracture at "X." Source: Ref 1

volves microscopic sliding on shear planes of the atomic crystals. Microshear causes plastic, or permanent, deformation.

Microstructural Aspects of Ductile Fracture

It is on a microscopic scale that the characteristics of ductile deformation and fracture really become unique. It is necessary to examine both the changes that are visible with a light microscope and those that can be examined only with an electron microscope at magnifications unattainable by a light microscope.

The behavior that occurs within a typical ductile metal immediately prior to fracture in a tensile test is shown in Fig. 5. Figure 5(a) is a photograph at low magnification (6×) of the necked-down portion of a tensile specimen made from hot rolled 1020 steel immediately prior to fracture. To obtain this photograph, it was necessary to stop the test after the maxi-

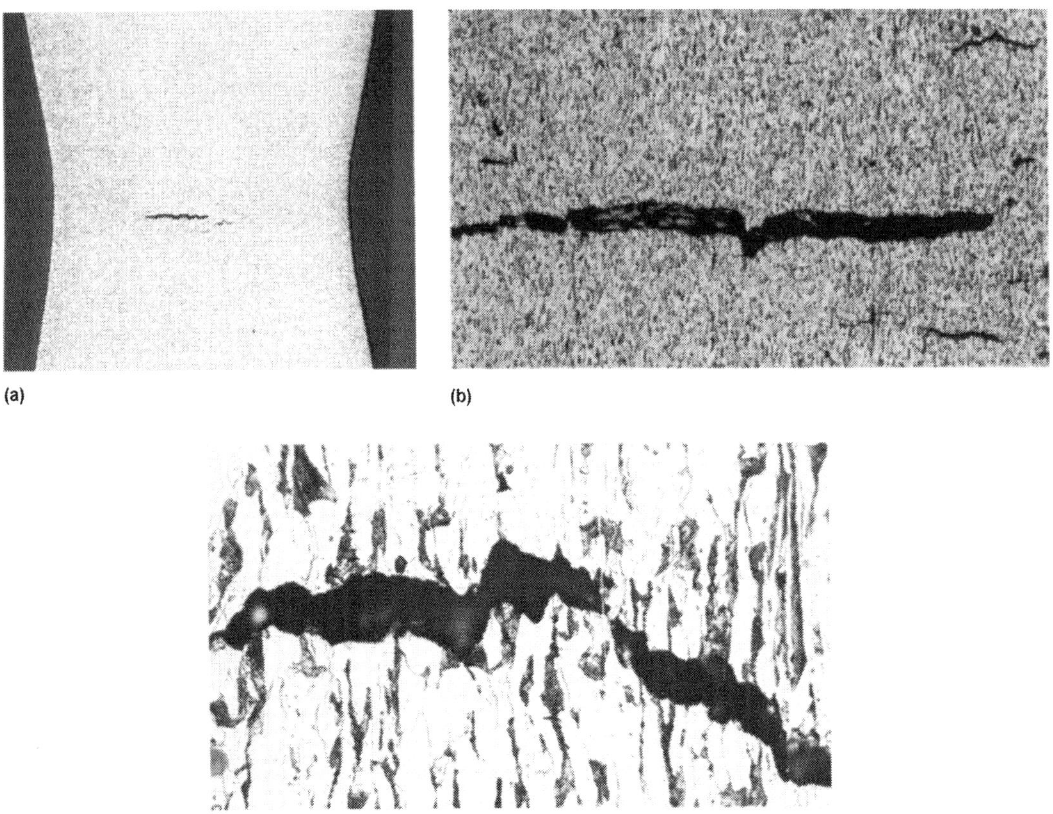

(a) (b) (c)

Fig. 5 Photographs at three different magnifications of longitudinal section of tensile test specimen of hot rolled 1020 steel. Test was stopped immediately prior to fracture (region c of Fig. 4). (a) At 6× magnification, shows internal cracking, reduced section, and severely deformed grain structure. (b) At 50× magnification, shows one large crack and several smaller cracks and elongated grains. (c) At 250× magnification, shows lack of fit due to distortion; also elongated ferrite (light) and pearlite (dark) areas. Source: Ref 1 mum load had been attained, in the region of the stress-strain curve just prior to the fracture at "X" in Fig. 4.

The specimen was sectioned longitudinally and metallographically prepared by polishing and etching to reveal the grain structure. Note that the major crack is near the center of the reduced section and that there is a smaller crack close to it nearer the center. Figure 5(b) is a higher-magnification (50×) view of the center area showing a large crack and many smaller cracks in the same general area. Many smaller cracks have joined together to form the major cracks. There is also tremendous vertical (axial) elongation of the original equiaxed (nondirectional) grains of the hot rolled steel. Grain distortion of this kind is characteristic of ductile deformation and fracture and is frequently observed in examination of damaged and failed parts, provided the plane examined is in the direction of deformation, as in this photograph.

Figure 5(c) is a view at still higher magnification (250×) of part of a crack that had not yet separated completely when the test was stopped. Note that the opposite sides of the crack do not fit each other because of the distortion. This is also characteristic of ductile fracture; the opposite surfaces cannot be fitted closely together as they can in a truly brittle fracture. Again, note the axial elongation of the grains; the white areas are soft ferrite, and the dark areas are harder pearlite.

The magnification of these three photographs is too low to reveal the details of the internal fractures or cracks that developed on the fracture surface during the fracturing process. The weakest areas, or at least those that contain some minute imperfections or inclusions, separate first into microvoids, or tiny cavities. As soon as the first internal fracturing occurs, the cross section at these locations is slightly reduced, thereby putting a higher local stress on the surrounding areas, which are then likely to form their own microvoids. These microvoids coalesce rapidly, and the process continues until fracture occurs. For this reason, the term frequently used for this process of formation of tiny voids and their growing together is *microvoid coalescence* (MVC).

The process of MVC continues to form larger cracks, which have approximately half of each cavity on each side of the fracture surface as a tiny cup or dimple. Therefore, the actual fracture surface of a ductile metal is essentially a mass of dimples, or half-voids, which usually can be seen only with the aid of an electron microscope; this is termed a dimpled rupture fracture surface, as shown in Fig. 6. Examination of these dimples is exceedingly useful in studying fractures because the dimples are extremely sensitive to the direction of the stresses that formed them.

The Process of Microvoid Coalescence. In MVC, voids nucleate (initiate), grow, and coalesce to develop the final fracture surface that is dimpled. The most common process is one of pore (incipient crack) formation at or in a particle, followed by either void growth and linkage or by localized shear band slip deformation in the intervoid ligaments.

Ductile Crack Nucleation. The dominant and common mechanism for creation of a ductile crack nucleus is a combination of debonding at particle interfaces and cracking of second-phase particles. There may be alternative crack nucleation mechanisms in high-purity single-phase materials that contain a minimum number of very small particles, but the most common site of ductile crack nucleation is at second phases and/or inclusions in the matrix, resulting from plastic deformation.

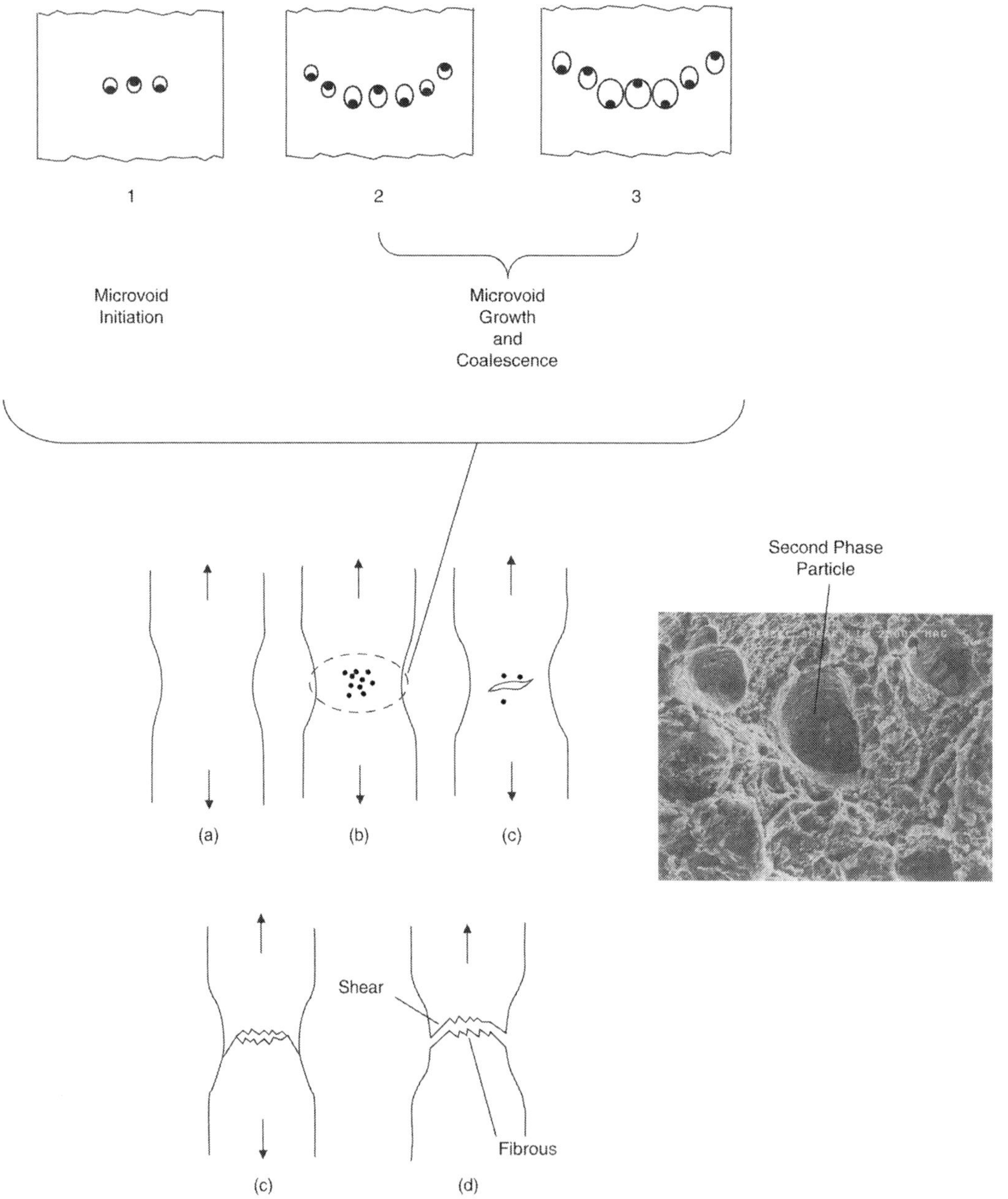

Fig. 6 Microvoid coalescence during ductile failure. Source: Ref 2

Two types of particles can be involved: inclusions and second phases. In polyphase alloys, the presence of eutectic or eutectoid structures also affects crack nucleation and propagation. In cast alloys, the presence of both microscale interdendritic porosity and macroscale shrinkage cavities provide holes for crack nucleation. There is little difference in void growth from shrinkage porosity and a hole containing an unbonded inclusion.

The incipient crack can form in four ways:

- Hole formation around an often weakly bonded particle
- Cracking of a particle
- Growth from a precracked particle, for example, from prior working operations
- Cracking in a matrix of limited ductility in which faceted particles cause stress concentration in the matrix-faceted nitride inclusions in a high-strength, low-alloy steel

Inclusions tend to have a low-strength interface bond (if any) with the matrix and have mechanical properties different from those of the matrix. In steels, these inclusions fall into three general groups:

- Stringer types that are sulfides and perhaps contain additional phases
- Spheroidal types (oxides and sulfides)
- Faceted types (nitrides)

The sulfide inclusions are ductile and low strength, while the oxides and nitrides are brittle. In addition, second-phase particles have mechanical properties different from those of the matrix, but experimental evidence suggests that there is often some bond strength (partial coherency) between these particles and the matrix (i.e., carbide particles in a low-carbon steel).

As a material is subjected to a load, debonding occurs in directions of maximum strain in the matrix, with the more weakly bonded materials debonding first. For unidirectional loading, debonding occurs at the interface generally normal to the applied load, as in Fig. 7(a) and (b). An example is shown in Fig. 8. When the applied loading creates a tensile hydrostatic stress (often referred to as a triaxial stress) in the material, debonding occurs all around the particle, as in Fig. 7(c). An example of volumetric debonding at a MnS inclusion is shown in Fig. 9. If there is sufficient load transfer to an elongated particle, cracking is also observed at approximately half the length of the particle, as in Fig. 7(b).

The common observation for steels is that fracture is initiated by (relatively large) sulfide stringers. If the material is quenched and tempered, debonding and fracture of carbide occurs at a greater strain. Particles that are totally or partially coherent with the matrix (e.g., carbides in steel) require substantial stress for decohesion to start at the particle-matrix interface. Careful work using steels containing small, spheroidized carbides and MnS inclusions shows that debonding occurs first around the larger and more weakly bonded MnS inclusions and then at a greater strain by debonding at the carbide particles. Fracture has also been associated with clusters (bead stringers) of oxide particles.

When material has been previously cold worked, it is often observed that not only is there debonding at matrix-particle interfaces, but also the particles themselves are cracked (Fig. 10). These cracked particles then provide additional crack nucleation sites. Elongated particles, such as MnS, crack at approximately half-length due to load transfer from the matrix to

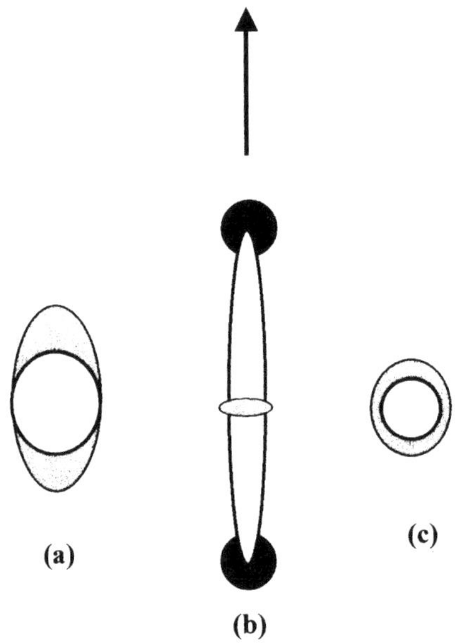

Fig. 7 Schematic of debonding at a matrix-particle interface with unidirectional stress. (a) Plane stress loading of an inclusion (no interfacial bond) causes debonding at the particle caps. (b) Debonding and fracture of high-aspect-ratio particles (elongated inclusion) due to shear transfer. Fracture is indicated at half-length. (c) More general (uniform) debonding from a large triaxial (i.e., tensile hydrostatic) stress. Source: Ref 3

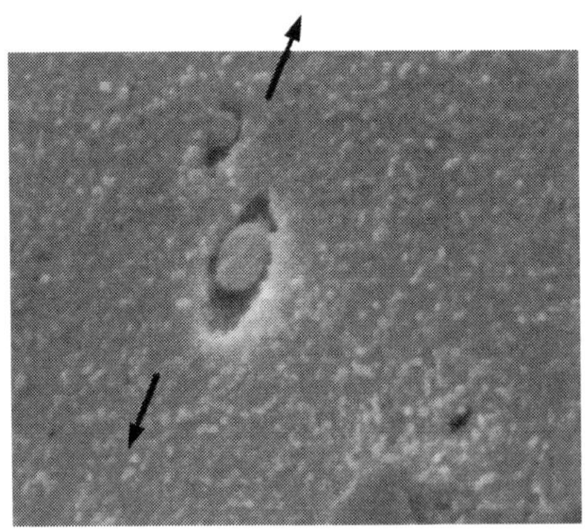

Fig. 8 Debonding associated with a MnS inclusion in a bainitic microstructure. Loading direction indicated. Source: Ref 4

the particle. Large second-phase brittle particles may crack by cleavage caused by dislocation pileup. Similarly, particle cracking is also observed metallographically when the material is initially in the annealed condition and the component is examined metallographically after fracture.

In materials containing a dual distribution of particle sizes, for example, MnS inclusions and spheroidal carbides in a quenched and tempered steel, a dual dimple size is often seen on the fracture surface. In this case, the MnS inclusions are larger than the carbide particles and have lower bond energy to the matrix than do the carbide particles. Fracture initiation is initially by debonding and void growth at the MnS-matrix interface (large dimples). After a larger strain, debonding then occurs at the smaller carbide particles, resulting in smaller dimples than the voids from MnS-matrix debonding, as shown in Fig. 11.

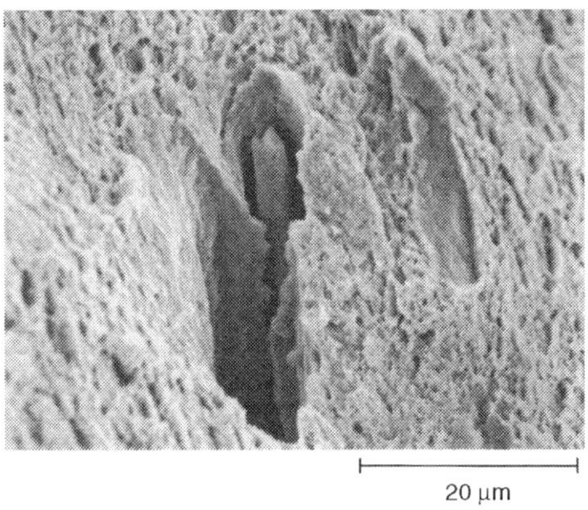

Fig. 9 Debonding of a MnS particle in a 4140 quenched and tempered steel due to a bending stress. The component was forged, and considerable directionality (banding and fibering) existed in the material. Crack propagation from bottom to top in the photograph. Courtesy of M. West, University of Tennessee. Source: Ref 3

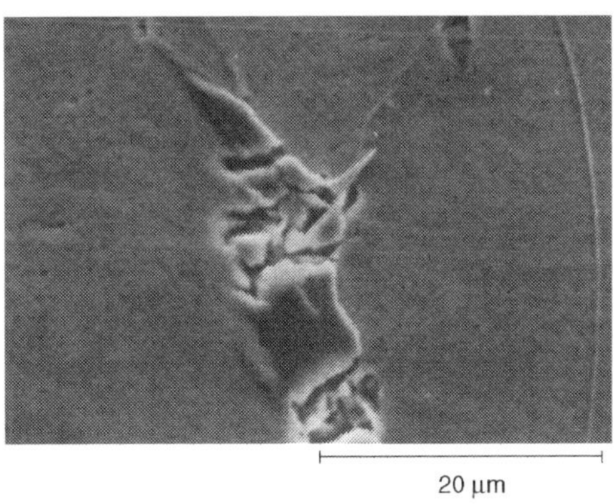

Fig. 10 Cracked cementite particle in a cold rolled low-carbon steel (approximately 0.1% C). High-magnification view of a cracked cementite particle showing multiple cracks and shattering. Courtesy of R. Holman, University of Tennessee. Source: Ref 3

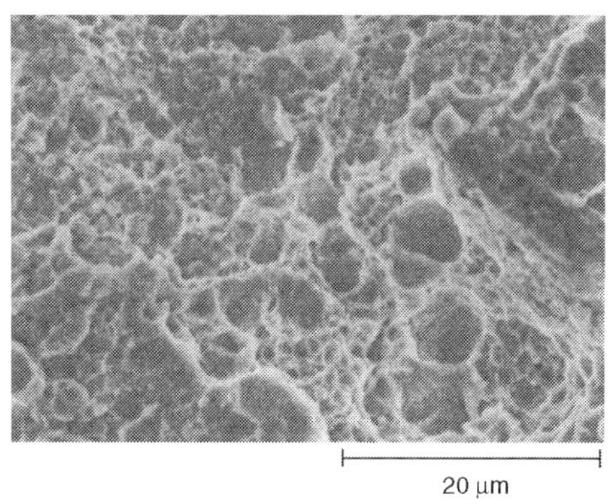

Fig. 11 Dual dimple size observed in a 4150 steel. Material was isothermally transformed at 190 °C (375 °F) and was not tempered. Tested as a Charpy V-notch specimen at 0 °C (30 °F). Source: Ref 4

Void Growth and Coalescence. The incipient cracks (voids) associated with the particles may grow by two mechanisms, depending on particle size and spacing:

- Growth of an incipient crack by enlargement and linkage of the voids
- Slip band deformation in ligaments between the voids

For large particles relatively closely spaced, and especially in the presence of constraint, void coalescence typically occurs on the plane of maximum normal stress. This mechanism of void coalescence is referred to as pure ductile tearing (Fig. 12a).

When the particles are more widely spaced, the process of void coalescence can involve slip deformation in the interligament region between the voids. Deformation of the ligaments may occur from localized shear bands or by necking of the ligaments between the voids. A good example of the former is the mechanism known as void sheet formation, which is a common fracture mode in steels containing dual-sized particles, that is, larger MnS inclusions and smaller spheroidal carbides. Fracture initiates at large MnS inclusions, followed by microscale strain localization on planes of high shear stress. These planes randomly contain smaller carbide particles, which subsequently nucleate voids. This is referred to as void sheet fracture and results in a characteristic zig-zag pattern of void growth, shown schematically in Fig. 12(b). An example of void sheet formation between MnS inclusions is shown in Fig. 13(a).

In addition, necking can occur in the ligaments between the particles. With sufficient plastic flow, the interparticle ligaments can show sufficient ductility to display necking with a large height, similar to that seen in single crystals. The example in Fig. 14 shows ductile MVC fracture in an aluminum-silicon cast alloy, which is quite different in appearance for MVC in steels. In the aluminum-silicon alloy shown, fracture in the matrix is initiated by cleavage in the silicon particles.

In summary, two types of microscale ductile crack propagation are recognized, depending on whether growth of incipient cracks occurs on a plane of maximum normal stress or on a plane of high shear stress:

- Pure ductile tearing by void coalescence, usually on the plane of maximum normal stress
- Zig-zag slip deformation on planes of high shear stress between voids

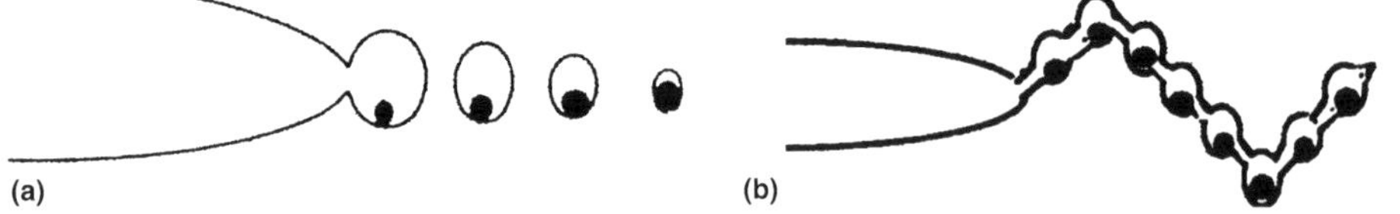

Fig. 12 Schematic of (a) ductile tearing along plane normal to normal stress and (b) zig-zag path of void sheet fracture along shear planes. Source: Ref 3

However, the relation between these two microscale mechanisms and the macroscale of crack growth is less clear. For example, the zig-zag path from void sheet formation in Fig. 13(a) does not necessarily result in macroscale shear bands. The macroscale fracture surface may be normal to the load, while the microscale paths of fracture may occur along the shear planes with a small step height.

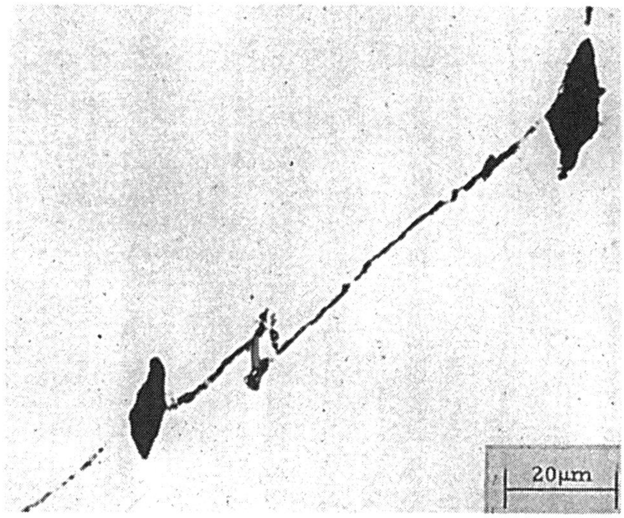

(a)

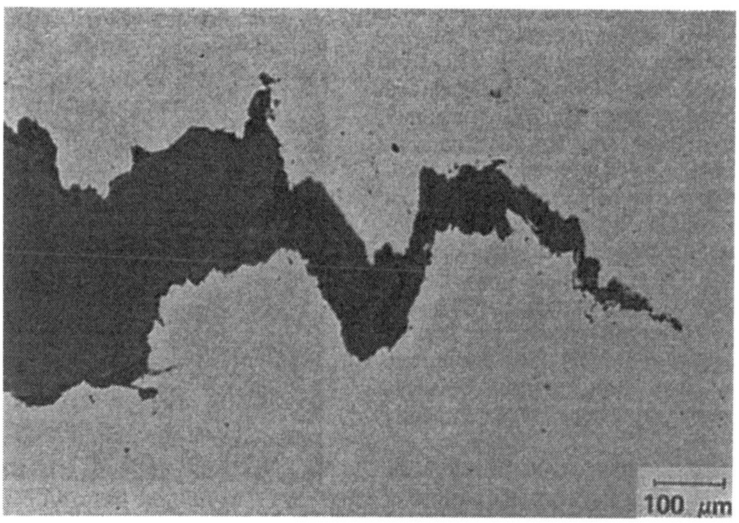

(b)

Fig. 13 Examples of ductile fracture on shear planes. (a) Void sheets from propagation of a crack between widely spaced inclusions within a shear band in a 4340 steel. Stress axis is vertical. Reprinted with permission from ASTM STP600, *Fractography—Microscopic Cracking Processes,* copyright ASTM International, 100 Barr Harbor Drive, West Conshohocken, PA, 19428. Source: Ref 5. (b) Ductile crack growth in high-strength, low-alloy steel (A710). Source: Ref 6

Nonetheless, the direction of crack growth at higher scales may also occur on either a plane of maximum normal stress or on a plane of maximum shear stress, depending on the degree of constraint (and thus component geometry), the size, and the spacing of particles that initiate incipient cracks. For example, Fig. 13(b) shows a zig-zag path of ductile fracture at a higher scale than the microstructural scale of void sheeting shown in Fig. 13(a). Ductile fracture along shear planes can also occur on the macroscale.

It is important to recognize that microstructural factors and/or localized deformation (such as microscale or macroscale shear bands) can influence macroscale direction. For example, even though the mechanism of void sheet formation does not normally result in a macroscale zig-zag path of fracture, macroscale shear bands can influence the direction of crack growth, as shown in Fig. 15. There is no apparent flat, central fibrous region, and there are five macroscale shear bands and additional small, un-

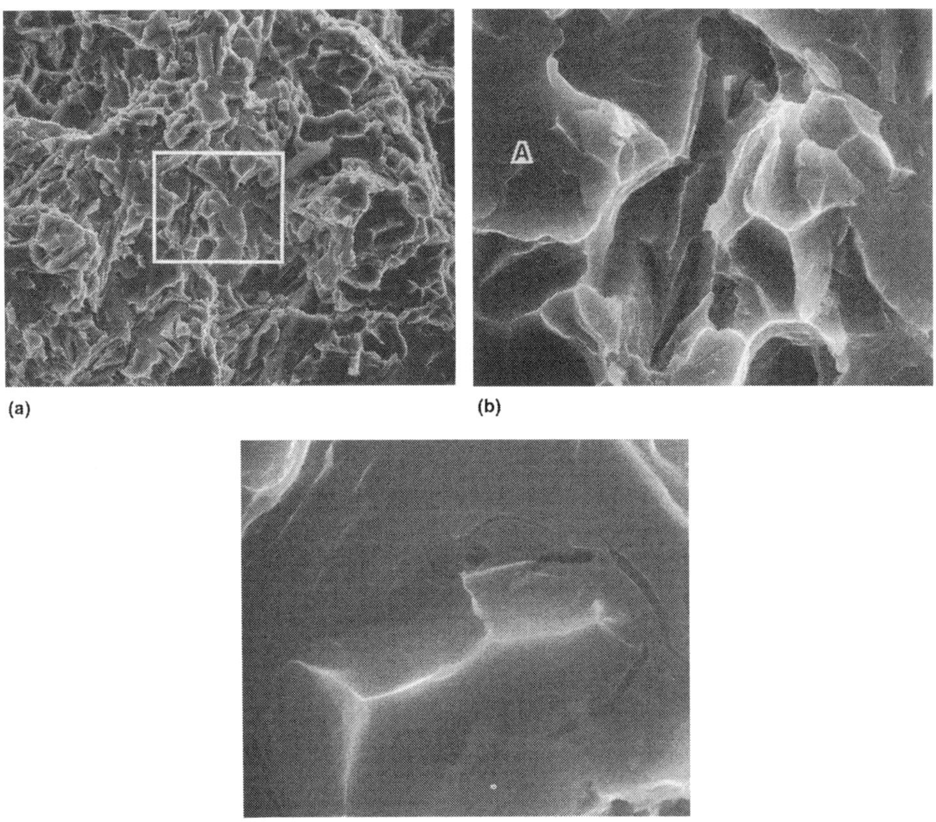

Fig. 14 Microvoid coalescence in an aluminum-silicon alloy (A380) loaded in tension. (a) Fracture surfaces consist of cleaved particles (i.e., silicon) and ridged fracture of the aluminum. Original magnification: 200×. (b) Higher-magnification (1440×) view of boxed region. (c) Fractured aluminum ligament surrounded by intermetallic particles at higher magnification (2880×). Source: Ref 7

resolved bands across the specimen. Fracture initiation occurred in the center of the specimen. The material in this case is slow-cooled HY-100 steel with a severely banded microstructure of alternating bands of soft ferrite and hard granular ferrite. The test specimen orientation was perpendicular to the banded microstructure, which more clearly shows the shear banding.

Dimple shapes have been studied extensively and are shown schematically in an easy-to-understand form in Fig. 16. The three basic modes of MVC shown in the figure are further described as follows:

- *Tension.* As illustrated schematically in Fig. 16(a), the tensile force at the center of the specimen causes microvoids to form first near the center and then to spread to nearby areas that must then carry more stress because the cross section is now smaller. Under an axial force, the microvoids are not skewed, or tilted, in any particular direction; thus, the fracture surface consists of equiaxed dimples when viewed perpendicular to the surface. The angle of view is particularly important when the dimples are studied in a scanning electron microscope. If they are viewed from an angle, they can appear foreshortened and not equiaxed.
- *Shear.* As shown in Fig. 16(b), the dimples become elongated at a pure shear fracture surface, such as that of a torsional fracture of a ductile metal. The dimples can become so elongated that they no longer closely resemble dimples except for their C-shaped ends. Also note that the C's are in opposite directions on the opposed fracture surfaces, because each represents one end of a dimple that was pulled in opposite directions. On torsional fractures, these dimples are frequently damaged by rubbing from the opposite sides, so that the surface observed is simply a mass of circular rub marks.

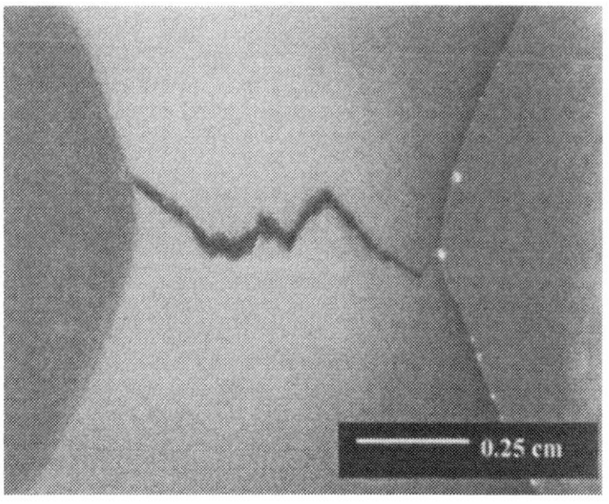

Fig. 15 Profile of fracture path in a notched bar specimen taken from the long transverse direction of HY-100 steel plate with a severely banded microstructure (alternating bands of ferrite and granular ferrite). Loading direction was perpendicular to the bands. Fracture is reported to initiate near the centerline. Note the apparent absence of a central, flat fibrous zone. Source: Ref 8

- *Tensile Tearing.* As shown in Fig. 16(c), this mode of fracture is somewhat similar to the pure tension mode shown in Fig. 16(a), except that the fracture actually originates at the edge of the metal rather than at the center. This is due to a bending force on the part that causes tensile fracture with equiaxed dimples at the region close to the origin, while the actual tensile tearing causes the C-shaped dimples to form near the end of the fracture, opposite the origin.

In addition to these three basic modes of MVC and fracture, there are several others, as shown in Fig. 17. The three modes just discussed appear at the upper left of this figure.

The effect of the direction of fracture can be seen dramatically on many bending fractures of ductile metals, even if the exterior has been case hardened. Close to the surface on the convex side of the fracture (the origin surface), the dimples are essentially round and equiaxed because of the tensile stress at that location, as shown in Fig. 18(a). In pure bending,

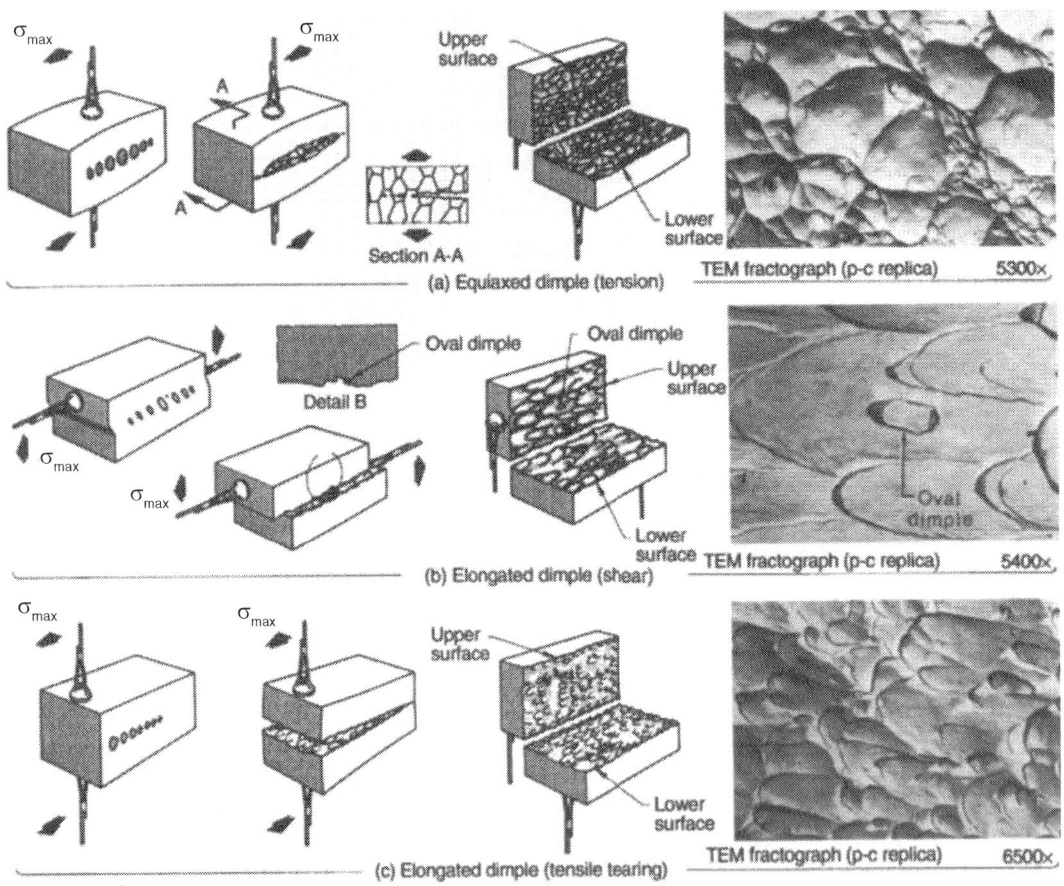

Fig. 16 Influence of direction of principal normal stress on the shape of dimples formed by microvoid coalescence. TEM, transmission electron microscopy. Source: Ref 9

the fracture starts on this convex side and progresses across the part to the opposite side, which acts as a hinge as the fracture opens up.

When the fracture approaches the hinge (concave) side of the fracture, the local stress is no longer pure tension but is in the tensile tearing mode, as shown in Fig. 18(c). At high rates of loading, the dimples can become

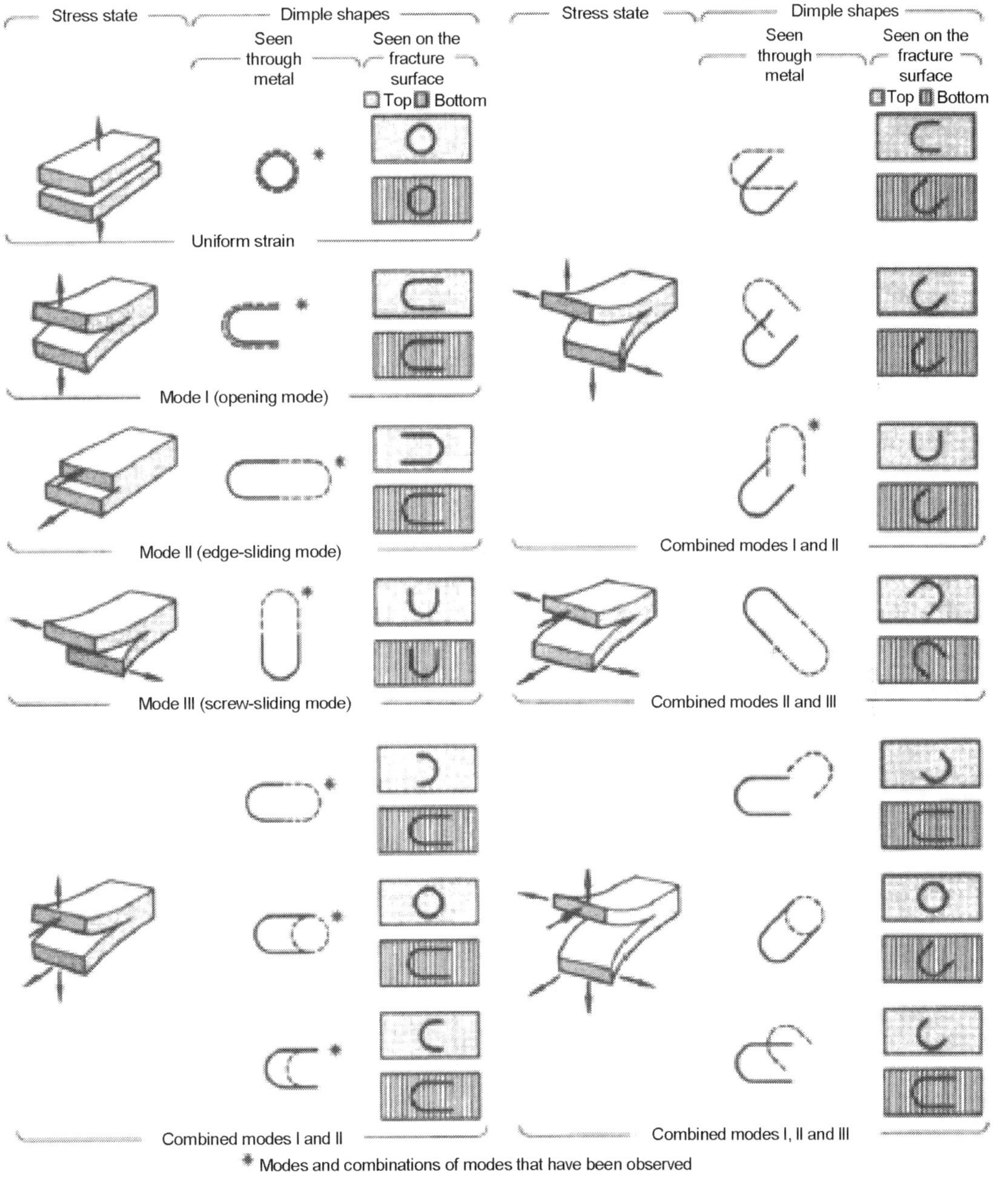

Fig. 17 Fourteen probable combinations of mating dimple shapes, resulting from different stress states that caused metal at the crack tip to deform by various modes. Source: Ref 9

extremely elongated, as in Fig. 18(c). Under lower rates of loading, the dimples are usually not extremely elongated, as in Fig. 18(b). Also, if there is a tensile force on the part in addition to the bending force, the dimple elongation will not be as dramatic as it is in this sequence of a rapid, pure bending fracture. Note that the closed end of the elongated dimples points back toward the origin.

The middle fractograph, Fig. 18(b), is a view of the fracture surface in the interior of the part where the transition between equiaxed and elongated dimples occurred. Note that on one side of the fractograph the dimples are equiaxed, while on the other side, they start to become elongated.

This series of three scanning electron microscope fractographs is from the same fracture surface resulting from an impact in a high-velocity automotive accident.

Cautions in Interpretation

At this point, it is prudent to insert some cautions about interpretation of the evidence visible with this type of microscopic examination. The

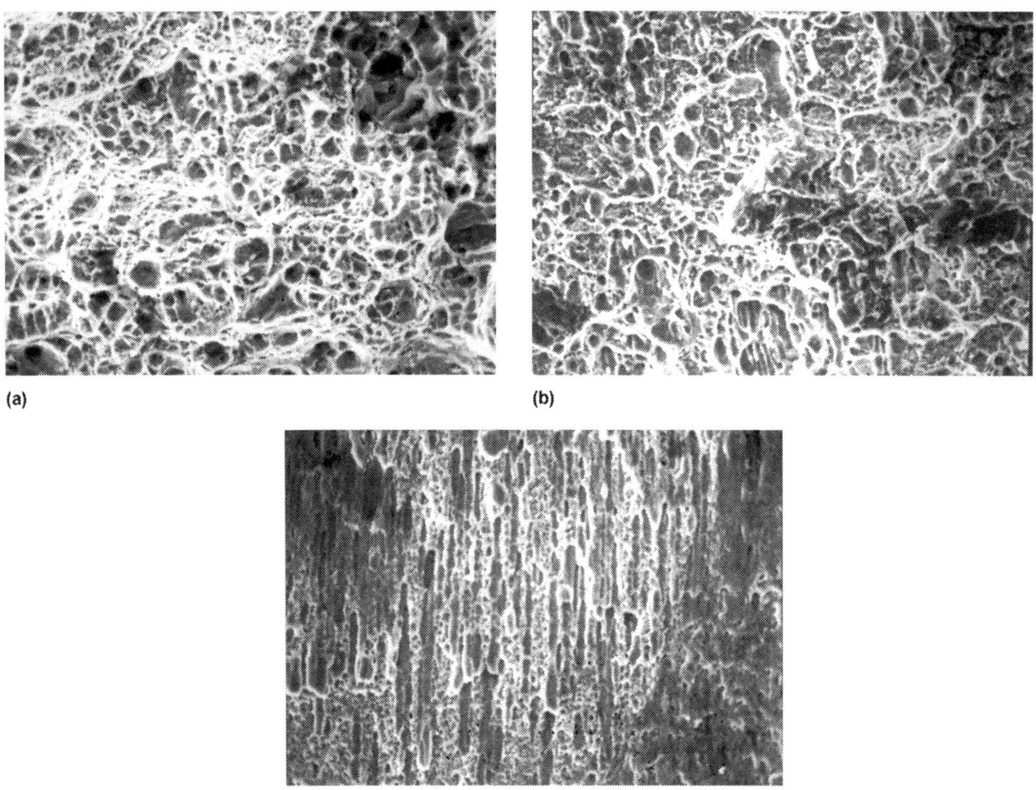

Fig. 18 Bending fracture. (a) Equiaxed, dimples near origin. Original magnification: 600×. (b) Transition. Original magnification: 600×. (c) Elongated, dimples near finish. Original magnification: 600×. Source: Ref 1

analyst must be aware of complicating factors extraneous to the fracture itself:

- *Mechanical Damage.* The fracture of a metal part is always a violent event because of the sudden release of energy. Fracture surfaces frequently are rubbed or pounded against each other, causing rub marks, abrasion, dents, smearing, or other scars of postfracture damage. It is not unusual for much of the fracture surfaces to be obliterated, so that microscopic examination as discussed previously is difficult or impossible. The microscopic features on fracture surfaces are fragile and are easily damaged. When this occurs, one must rely on the macrofeatures of the part or parts for useful evidence in analyzing the fracture.
- *Chemical Damage.* Surface pits resembling the dimples of ductile fracture may instead be the result of severe etching with corrosive acids (or bases, in some metals) or of cleaning techniques used to remove atmospheric corrosion. In other cases, a corrosive film, or layer, may have formed on the fracture surface and completely obliterated the underlying character of the surface.
- *Cavities may not be dimples.* In certain metals, cuplike cavities may be present on a fracture surface, but the cavities may not be the dimples characteristic of ductile fracture. In cast metals and weldments, particularly, this type of rounded cavity may actually be evidence of gas porosity. Gas porosity occurs because the inevitable gases in castings and weldments (which are actually localized castings) cannot escape to the atmosphere before the metal solidifies and traps the gas in tiny, smooth-walled bubbles within the metal. This is particularly true in welds in metals such as aluminum, which has a high rate of thermal conductivity. In effect, large, cold members having high thermal conductivity act as heat sinks, causing rapid solidification of the weld metal and entrapping gas bubbles before they can escape. In ductile fractures of metals with extensive gas porosity, it may be quite difficult to determine which cavities are true dimples and which are evidence of gas porosity.
- *Mixed Fracture Modes.* During study of a fracture surface, one should not be surprised to find different fracture modes. A predominantly ductile fracture surface may have certain locations with cleavage or intergranular fracture because of differences within the metal. These differences may result from crystalline orientation with respect to the fracture stress or from differences in the microstructure of the metal. For example, a steel with a partly pearlitic, partly ferritic microstructure may fracture with dimpled rupture in the ferrite regions but fracture by cleavage in the pearlite regions. A notched bar impact test specimen may fracture by cleavage in some areas and with dimpled rupture in others. A hydrogen embrittlement fracture may have certain areas with intergranular fracture and others with dimpled rupture. No true fatigue striations are possible in a single overload fracture; how-

ever, there may be somewhat similar-appearing parallel ridges resulting from fracture through pearlite or other lamellar structures or resulting from mechanical rubbing either during or after the fracture. These spurious marks can resemble striations that can confuse the analyst, who has a difficult task even without these complications. Of course, if cyclic loading has in fact occurred, true fatigue striations may also be present in addition to any or all of the other fracture modes.

Brittle Fracture

Brittle means little or no permanent deformation prior to fracture, usually accompanied by high hardness and strength, but with little tolerance for discontinuities. Brittle metals are in daily use as normal engineering materials, and, as long as they are properly handled, they are very satisfactory for many types of service. Hardened tool steels, gray cast iron, and many other metals are routinely used, within their limitations, with satisfactory results. However, if a tool such as a metal cutting file is bent, it will suddenly snap with a brittle fracture; there will be little or no permanent deformation, and the pieces may be placed back together in perfect alignment. In general, it is characteristic of very hard, strong, notch-sensitive metals to be brittle, although research work attempts to raise the useful strength of these metals without the danger of brittle fracture. Conversely, it is generally true that softer, weaker metals usually have ductile behavior. Gray cast iron is an exception. This metal is brittle because it contains a very large number of internal graphite flakes, which act as internal stress concentrations and limit the ability of the metal to flow or deform, which is necessary for ductile behavior.

These and certain other metals are known and expected to be brittle, but they are normally used satisfactorily in applications that are suitable for brittle metals, where there is little or no danger of fracture. Certain other common metals—particularly low-carbon and medium-carbon steels, which are widely used in industry—are normally considered to have ductile properties and are normally used in applications where the ability to adjust by plastic deformation is desired. However, under certain combinations of circumstances, these normally ductile steels can fracture in a totally brittle manner. This completely unexpected behavior has been the cause of many disasters in the past and can still cause disasters if the lessons of the past are not heeded. Also, it is not necessary to have high applied stresses on the part or structure. Brittle fracture can occur solely because of residual tensile stresses, with no applied load, or with any combination of applied and residual stresses.

Characteristics of Brittle Fracture

Brittle fractures have certain characteristics that permit them to be identified:

- There is no gross permanent or plastic deformation of the metal in the region of brittle fracture. Compare the brittle fracture shown in Fig. 19 with the ductile fracture previously shown in Fig. 1.
- The surface of a brittle fracture is perpendicular to the principal tensile stress. Thus, the direction of the tensile stress that caused the fracture to occur can be readily identified.
- Characteristic markings on the fracture surface frequently, but not always, point back to the location from which the fracture originated. In the case of flat steel, such as sheet, plate, or flat bars, and also case-hardened regions, there are characteristic V-shaped chevron or herringbone marks that point toward the origin of the fracture. In many instances, these marks are extremely fine and very difficult to recognize unless a strong light is positioned so that it just grazes the projections of the surface texture. A number of illustrations and photographs of these marks are shown in Fig. 20 through 25. Brittle fractures of some parts may have a pattern of radial lines, or ridges, emanating from the origin in a fanlike pattern. Again, it may be difficult to perceive the texture of the fracture surface unless the light is carefully controlled. Typical fractures are shown in Fig. 26 through 28. Brittle fractures of extremely hard, fine-grained metals usually have little or no visible fracture pattern. In these cases, it may be very difficult to locate the origin with certainty.

The preceding discussion of the ductile/brittle transition primarily concerns carbon and alloy steels as well as certain nonaustenitic stainless

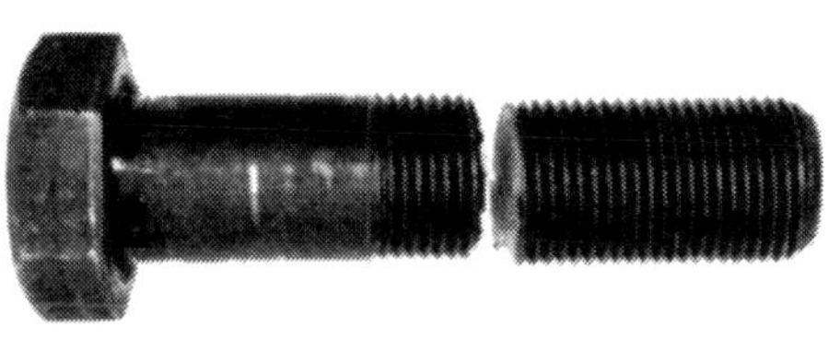

Fig. 19 This brittle bolt was water quenched with a hardness of 47 HRC but had no obvious deformation. This is the same 1038 steel shown in Fig. 1, except that the heat treatment was different. Source: Ref 1

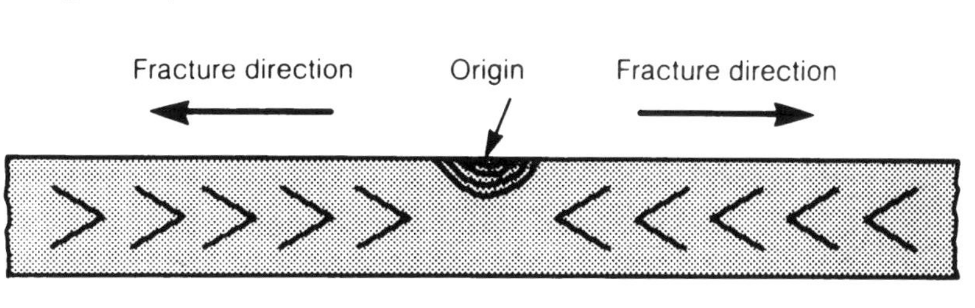

Fig. 20 Sketch of pattern of brittle fracture of a normally ductile steel plate, sheet, or flat bar. Note the classic chevron or herringbone marks that point toward the origin of the fracture, where there usually is some type of stress concentration, such as a welding defect, fatigue crack, or stress-corrosion crack. The plane of the fracture is always perpendicular to the principal tensile stress that caused the fracture at that location. Source: Ref 10

Fig. 21 Fragment of a thick-wall fractured drum. The fracture, which started at the right side of the photo, ran rapidly to the left, resulting in a well-defined chevron pattern. Source: Ref 11

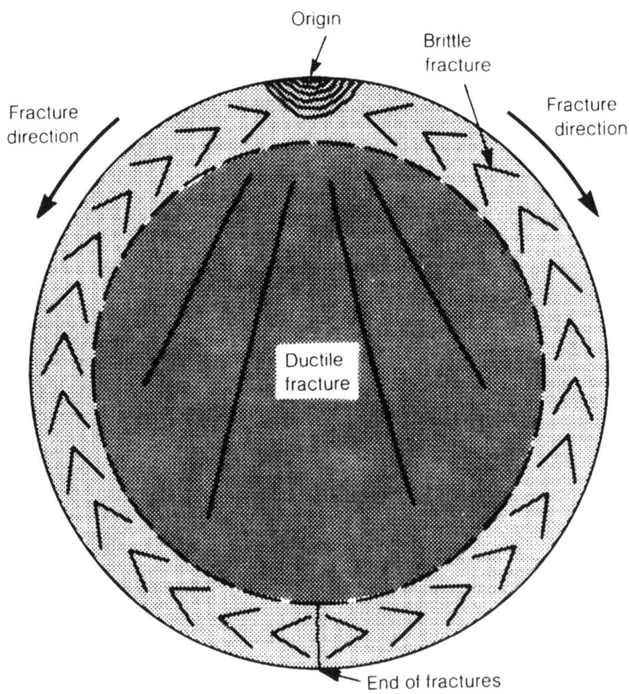

Fig. 22 Sketch of a fractured case-hardened shaft showing chevron marks pointing back toward the fracture origin. If brittle fracture continues completely around the part, the two separate fractures may form a step where they meet opposite the origin. The interior, or core, is likely to have a ductile fracture with dimpled rupture on a microscale. Source: Ref 10

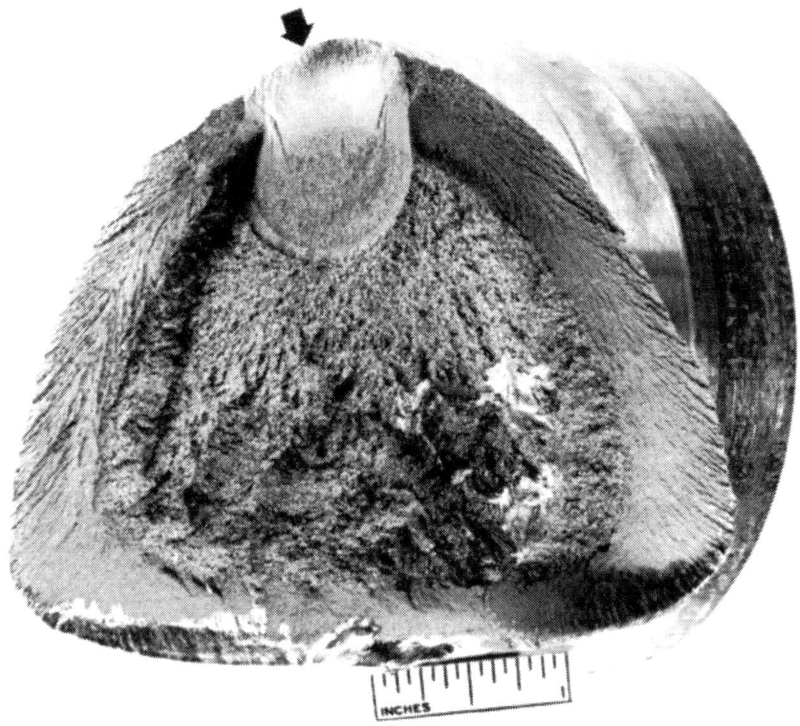

Fig. 23 Surface of a torsional fatigue fracture that caused brittle fracture of the case of an induction-hardened axle of 1541 steel. The fatigue crack originated (arrow) at a fillet (with a radius smaller than specified) at a change in shaft diameter near a keyway runout. Case hardness was approximately 46 HRC at the surface. Note the well-defined chevron marks in the two brittle fractures pointing toward the roughly circular fatigue portion of the fracture. Note also that this fracture is at a 45° angle to the shaft axis, as is typical of fatigue and brittle fracture of shafts in torsion. Source: Ref 10

steels. Other metals with body-centered cubic crystal structures behave similarly but are less common. Most nonferrous metals, such as alloys of aluminum and copper, and austenitic stainless steels have crystal structures that are not susceptible to the ductile/brittle transition characteristic of body-centered cubic metals.

Brittle fractures usually propagate by either or both of two fracture modes: cleavage or intergranular. In most cases, it is necessary to study the fracture surface with an electron microscope. Because very high magnifications are usually not necessary, a scanning electron microscope is usually preferred to a transmission electron microscope.

Microstructural Aspects of Brittle Fracture

Macroscopic brittle fracture may occur by cleavage fracture, intergranular fracture, or geometrically constrained ductile fracture (i.e., plane-strain MVC). Brittle fracture can occur in all materials, amorphous and crystalline, by cleavage. Additionally, in polycrystalline polyphase metal-

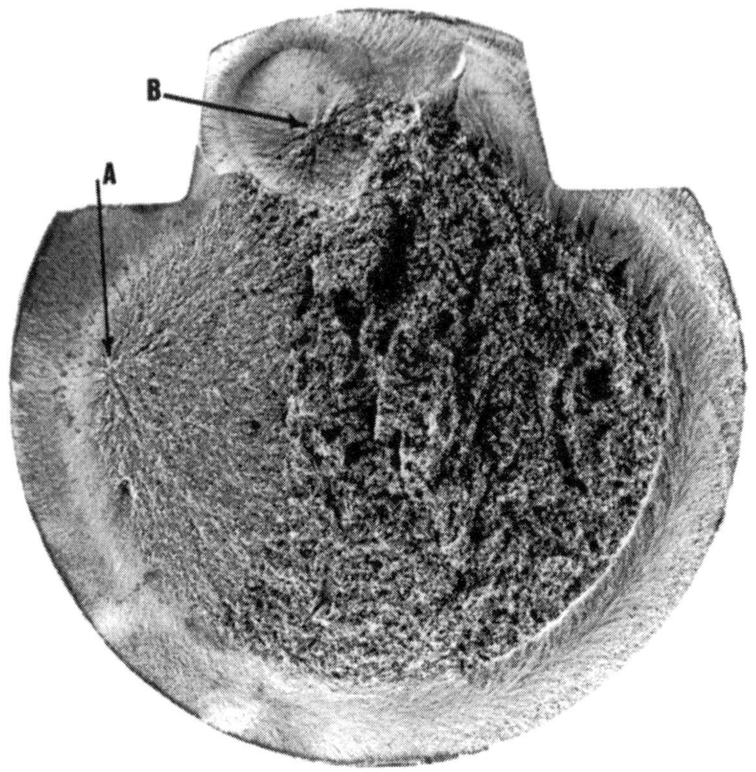

Fig. 24 Fatigue fracture of an 8.25 cm (3.25 in.) diameter induction-hardened shaft of 1541 steel after fatigue testing in rotary bending. Fatigue fracture origins A and B were subsurface due to the steep induction-hardened gradient and lack of an external stress concentration. Fatigue crack A, the larger of the two, propagated through the core and case regions on the left side; then the shaft suddenly fractured in a brittle manner around the right side of the case. Note the chevron marks on the lower right pointing in a clockwise direction toward fracture origin A. Source: Ref 10

lic materials, fracture can occur at or adjacent to grain boundaries, and cleavage fracture can occur in large second phases present in the microstructure, which then provide crack nuclei for ductile crack propagation in the matrix.

Cleavage fracture at a macroscale (low magnification) is characterized by high light reflectivity and a relatively flat surface. At the microscale, it is a series of flat regions or ledges that are often faceted. Higher magnification shows the ledges to be connected by ligaments described as river lines (Fig. 29) or fans (Fig. 30). The river lines coalesce as they propagate and have the appearance of water flowing downstream (Fig. 31). However, river lines cannot always propagate across grain boundaries in crystalline material, and at the microscale, crack initiation may occur in more than one location. Thus, the river lines only indicate the local direction of crack propagation, which can be opposite that of the crack propagation at the macroscale.

Fig. 25 Surface of a brittle fracture in a cold drawn, stress-relieved 1035 steel axle tube. Fracture originated at a weld defect (arrow) during testing in very cold weather. Note the well-defined chevron marks clockwise from the arrow pointing back toward the origin. Note also that the steel around the access hole below the grease fitting is actually necked down along the tube axis. The upper side of the tube is deformed and torn in a ductile manner. Source: Ref 10

The normal stress acting on a plane is of importance in propagating cleavage cracks. Consider the cracked plate shown in Fig. 32. When crack propagation from the preexisting imperfection (assumed through-the-thickness crack) occurs, the crack develops "wings" as it propagates and curves so as to reorient to propagate on the macroscale plane of maximum normal stress. At the microscale in crystalline material, a more complicated process is required to cause this macroscale appearance to be curved.

Totally brittle fracture in metals at the microscopic level (ideal cleavage or pure cleavage) occurs only under certain well-defined conditions, primarily when the component is in single-crystal form and has a limited number of slip systems, and is correctly described as cleavage fracture. More commonly in metals, the fracture surface contains varying fractions of transgranular cleavage and evidence of plastic deformation by slip. When both fracture processes operate intimately together, and especially in the fracture of quenched and tempered steels, the fracture process is

(a)

(b)

Fig. 26 Brittle fracture of solid rocket motor case during proof test. (a) Catastrophic brittle fracture of a 660 cm (260 in.) diameter solid propellant rocket motor case made of 18% Ni, grade 250, maraging steel. The case fractured at a repaired weld imperfection during a hydrostatic pressure test. Fracture occurred at approximately 57% of the intended proof stress. All welds of the case had been carefully inspected by x-ray and ultrasonic inspection. Arrows indicate origin of fracture. (b) Light fractograph of crack origin in the weld-related catastrophic fracture. Original magnification: 1%×. A crack was found beneath a gas tungsten arc repair weld on the inner surface of the case (surface at top in this fractograph) in the heat-affected zone of a longitudinal submerged arc assembly weld. The crack was approximately 3.6 cm (1.4 in.) long, parallel to and 1.20 cm (0.47 in.) beneath the outer surface of the case (surface at bottom). Radial marks are visible that confirm the fracture proceeded to the left and to the right. Source: Ref 12

termed quasi-cleavage. The dividing line between the terms *cleavage* and *quasi-cleavage* is somewhat arbitrary.

Metallic material fracture surfaces showing large fractions of cleavage may show cracking on more than one crystallographic plane within a given grain, leading to the most common feature associated with brittle

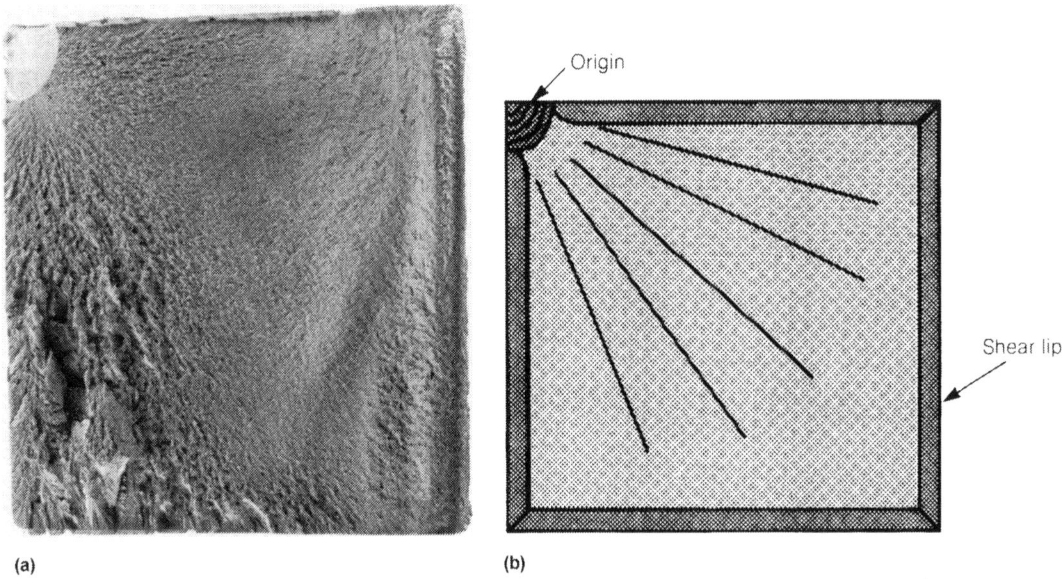

Fig. 27 Brittle fracture of D6B steel equalizer bar. (a) Fracture surface of a large (~13.3 × 15 cm, or 5.25 × 6 in.) equalizer bar made from D6B steel heat treated to a hardness of 45–47 HRC. This bar, which supports the front end of a large crawler tractor, was in service for approximately 200 h and was then returned to the laboratory, where it was flexed in high-stress, low-cycle fatigue, fracturing at 60,000 cycles. Note that the fatigue crack zone at upper left has a very fine texture, the result of continual testing. Note also the radial, fanlike ridges emanating from the fracture origin and the very small shear lips around the periphery of the brittle fracture. This fracture of a very large, high-strength part released a large amount of energy in a very short time. (b) Sketch of pattern of brittle fracture in a moderately hard, strong metal. Fracture originated at a sharp stress concentration that grew to the critical flaw size for that metal. The sharp stress concentration is frequently, although not always, a fatigue crack or a stress-corrosion crack. Note the fan-shaped pattern radiating from the origin region in the upper left corner. When viewed under the electron microscope, this type of fracture is likely to reveal a cleavage or quasi-cleavage fracture mode. Source: Ref 10

Fig. 28 Origin (at arrow) of a single-load brittle fracture that initiated at a small weld defect. Note also a fatigue fracture in the upper right corner. Radial ridges emanate from the origin in a fan-shaped pattern. The brittle part of the fracture is bright and sparkling, in contrast to the dull appearance of the fatigue zone and the thin shear lips at top and bottom surfaces. Source: Ref 10

faceted fracture, river lines, as previously noted. These lines may form by a ductile process, but the slip deformation that created them is not resolvable and such fracture is not described as quasi-cleavage. Alternatively, with less constraint, the connecting ligaments may become sufficiently large that MVC is observed in thin bands weaving through the general cleavage surface, as in Fig. 33.

Considerable evidence exists that cleavage fracture is facilitated by the presence of a normal stress on the cleavage plane. However, in crystalline materials there is convincing evidence that cleavage fracture occurs by prior dislocation motion to create a cleavage crack nucleus, and therefore, a shear stress is also required on one or more slip planes. If the material is amorphous and isotropic, the cleavage plane is simply the macroscale plane having the largest principal stress. However, if interatomic bond en-

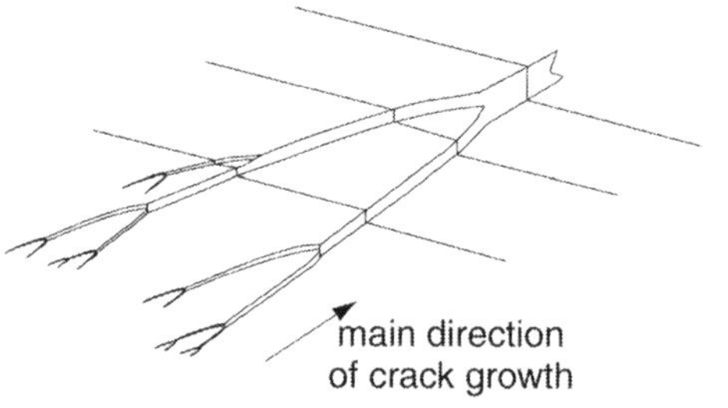

Fig. 29 Schematic of a river pattern. Crack growth is in the direction of crack coalescence. River patterns may be visible at the macroscale in organic glasses and brittle polymers but are visible only at the microscale in metallic materials. Source: Ref 13

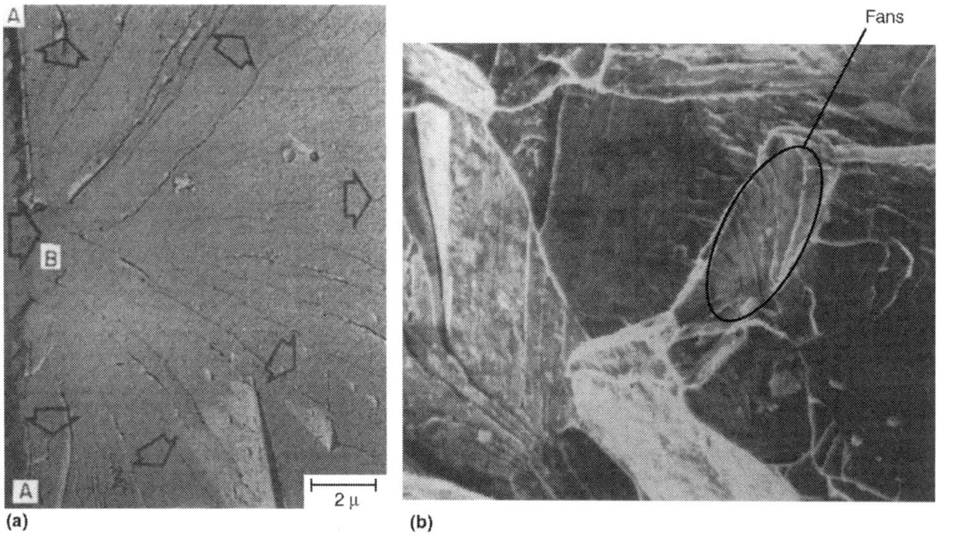

Fig. 30 Fans. (a) Examples of fans in a two-stage transmission electron microscopy replica of a cleavage fracture surface of iron. The river lines point back to the crack initiation site. (b) Fans on a scanning electron microscopy image. Source: Ref 14, 15

ergy is anisotropic, the resolved normal stress on the weakest bonded plane apparently controls cleavage fracture, for example, in a layered structure.

If the material is crystalline, bonding energies across planes of atoms are typically anisotropic so that cleavage fracture occurs on specific crys-

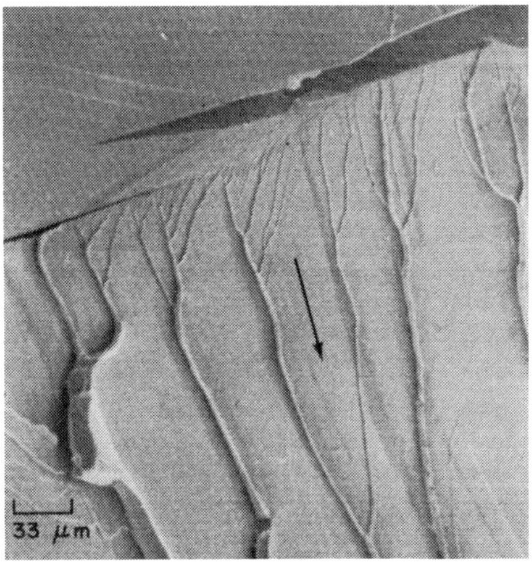

Fig. 31 River lines on a cleavage fracture surface. Direction of growth is parallel to the direction of crack coalescence, as indicated by the arrow. Cracks must reinitiate at a boundary containing a twist (mode III) deformation component. Source: Ref 16

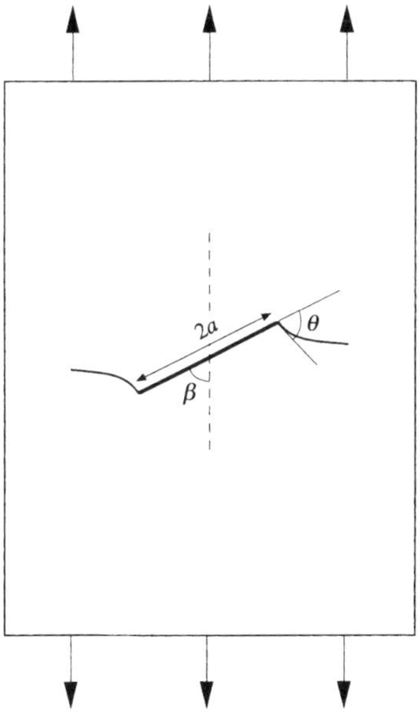

Fig. 32 Macroscale brittle crack propagation due to combined mode I and mode II loading. As cracks grow from the preexisting cracklike imperfection, crack curvature develops because of growth on a plane of maximum normal stress. Source: Ref 13

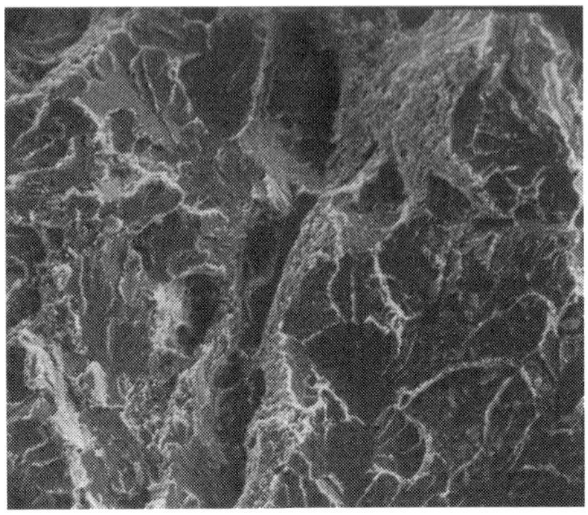

Fig. 33 Microscale quasi-cleavage fracture in an O1 tool steel tested at room temperature. Predominantly cleavage cracking with patches and ribbons of microvoid coalescence. Source: Ref 17

tallographic plane families in individual grains. In general, cleavage is observed in less symmetrical crystal structures (e.g., hexagonal close-packed, or hcp, and orthorhombic) and in the body-centered cubic (bcc) materials, but cleavage is unlikely in the face-centered cubic lattice.

The specific cleavage plane family and the cleavage plane multiplicity (number of nonparallel planes) vary with crystal structure, and there may be more than one cleavage plane family in a particular crystal structure. For example, bcc materials typically cleave on the family of basal planes (three planes, mutually perpendicular), but cleavage has also been reported on the {1,1,0} family of planes. Cleavage fracture has not been observed in the bcc alkali metals. The hcp materials are routinely considered to cleave on the basal plane (only one plane in the family), but this is not necessarily true. There may be multiple cleavage plane families. For example, zinc cleaves readily on the basal plane, but magnesium cleaves only with difficulty on the basal plane. Beryllium cleaves on both the basal plane and on the {1, 1, –2, 0} family of planes (3 planes, not mutually perpendicular but perpendicular to the basal planes).

Tilt and Twist Boundaries and Loading Mode. As shown schematically in Fig. 34, cleavage fractures frequently initiate on many parallel cleavage planes. As the fracture advances, however, the number of active planes decreases by a joining process that forms progressively higher cleavage steps. This network of cleavage steps is known as a river pattern. Because the branches of the river pattern join in the direction of crack propagation, these markings can be used to establish the local fracture direction.

A tilt boundary exists when principal cleavage planes form a small angle with respect to one another as a result of a slight rotation about a common axis parallel to the intersection (Fig. 34a). In the case of a tilt boundary, the cleavage fracture path is virtually uninterrupted, and the cleavage planes and steps propagate across the boundary. However, when the principal cleavage planes are rotated about an axis perpendicular to the boundary, a twist boundary results (Fig. 34b). Because of the significant misalignment of cleavage planes at the boundary, the propagating fracture reinitiates at the boundary as a series of parallel cleavage fractures connected by small (low) cleavage steps. As the fracture propagates away from the boundary, the numerous cleavage planes join, resulting in fewer individual cleavage planes and higher steps. Thus, when viewing a cleavage fracture that propagates across a twist boundary, the cleavage steps do not cross but initiate new steps at the boundary (Fig. 34b). Most boundaries, rather than being simple tilt or twist, are a combination of both types and are referred to as tilt-twist boundaries. Cleavage fractures exhibiting twist and tilt boundaries are shown in Fig. 35(a) and Fig. 36, respectively.

Feather markings are a fan-shaped array of very fine cleavage steps on a large cleavage facet (Fig. 37a). The apex of the fan points back to the fracture origin. Large cleavage steps are shown in Fig. 37(b). Tongues are occasionally observed on cleavage fracture (Fig. 35b). They are formed when a cleavage fracture deviates from the cleavage plane and propagates a short distance along a twin orientation.

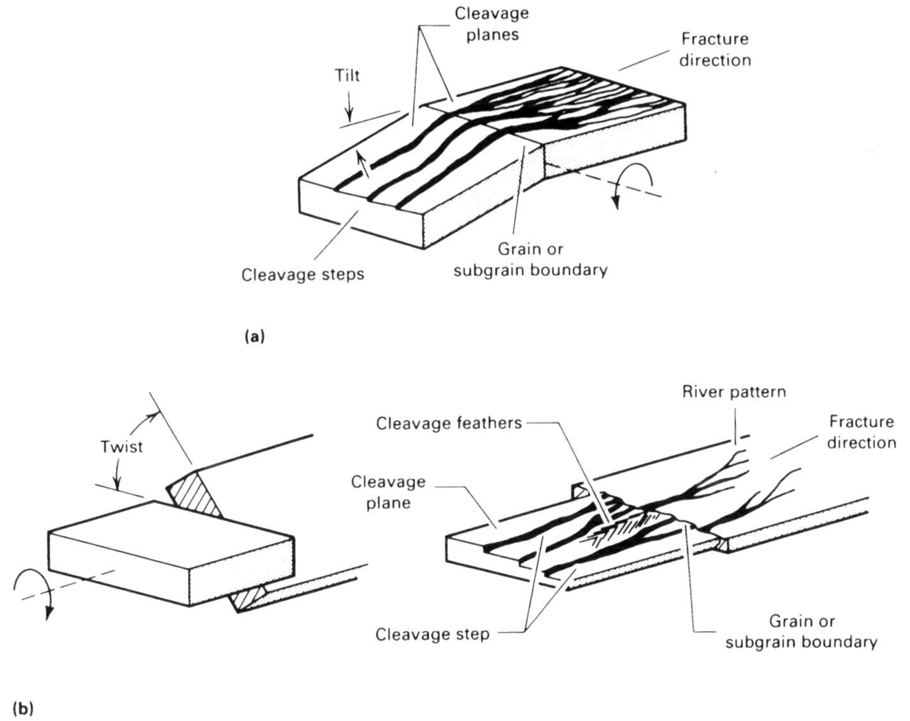

Fig. 34 Schematic of cleavage fracture formation showing the effect of subgrain boundaries. (a) Tilt boundary. (b) Twist boundary. Source: Ref 18

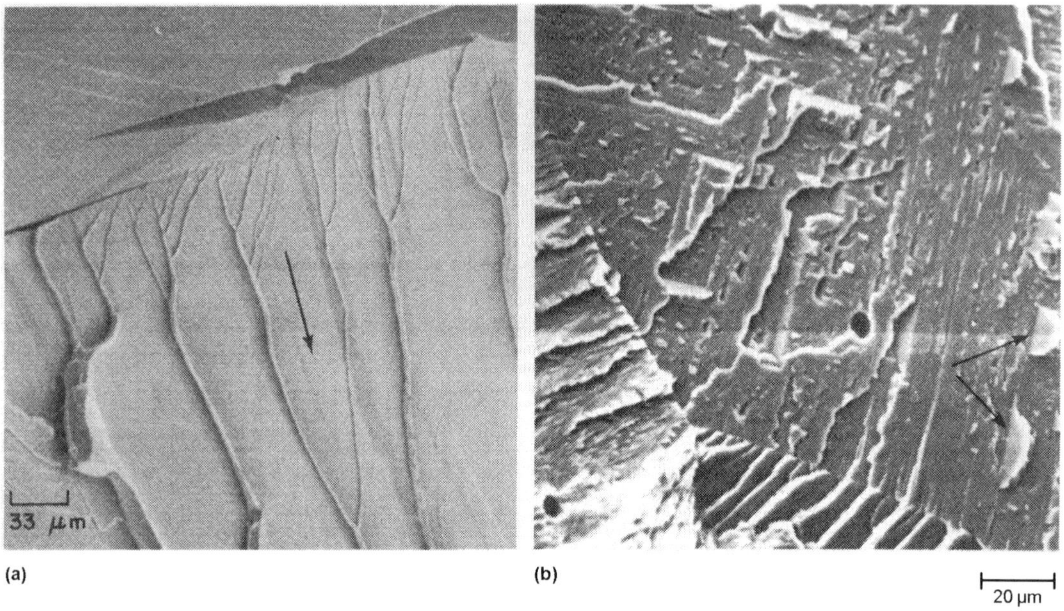

Fig. 35 Examples of cleavage fractures. (a) Twist boundary, cleavage steps, and river patterns in an Fe-0.01C-0.24Mn-0.02Si alloy that was fractured by impact. (b) Tongues (arrows) on the surface of a 30% Cr steel weld metal that fractured by cleavage. Source: Ref 18

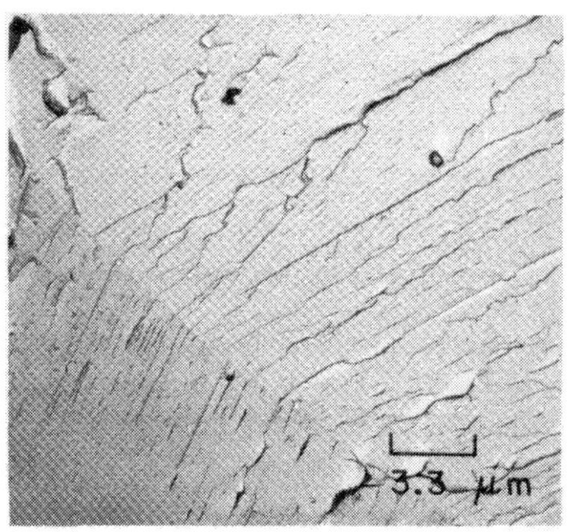

Fig. 36 Cleavage fracture in Armco iron showing a tilt boundary, cleavage steps, and river patterns. Transmission electron microscopy replica. Source: Ref 18

Macroscopic Ductile and Brittle Fracture Surfaces

A fracture is macroscopically brittle at the macroscale if it is on a plane normal to the maximum normal stress (condition 4 in Fig. 38). A fracture is considered to be macroscopically ductile when the fracture surfaces are inclined to an imposed load (slant fracture or plane stress). Toughness is

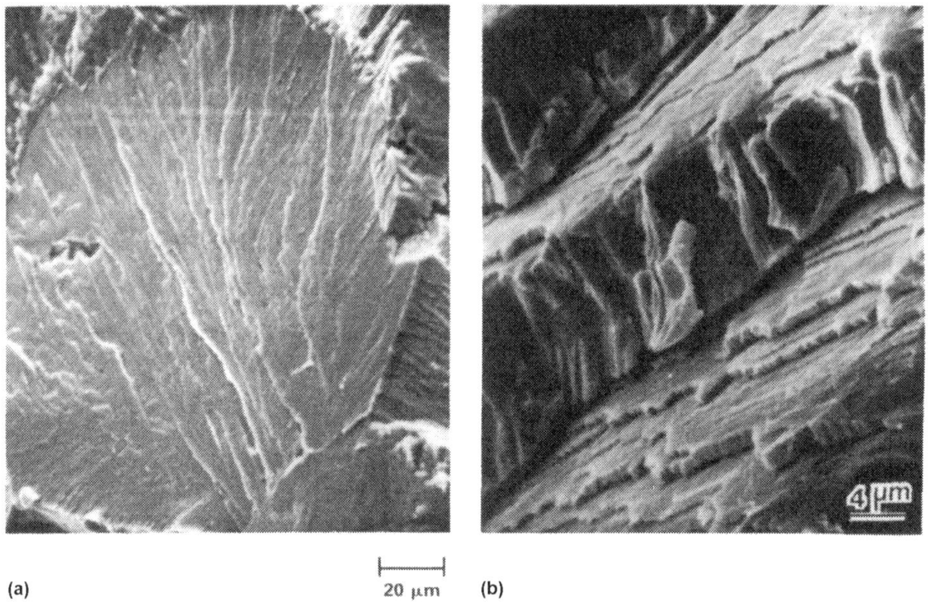

Fig. 37 Examples of cleavage fractures. (a) Feather pattern on a single grain of a chromium steel weld metal that failed by cleavage. (b) Cleavage steps in a Cu-25at.%Au alloy that failed by transgranular stress-corrosion cracking. Courtesy of B.D. Lichter, Vanderbilt University. Source: Ref 18

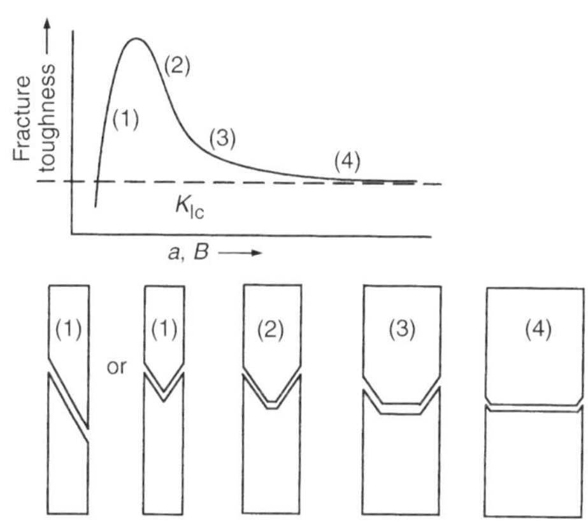

Fig. 38 Schematic of variation in fracture toughness and macroscale features of fracture surfaces for an inherently ductile material. As section thickness (B) or preexisting crack length (a) increases, plane-strain conditions develop first along the centerline and result in a flat fracture surface. With further increases in section thickness or crack size, the flat region spreads to the outside of the specimen, decreasing the widths of the shear lips. When the minimum value of plane-strain toughness (K_{Ic}) is reached, the shear lips have very small width. Source: Ref 16

higher under conditions of plane stress, because the additional work expended in work-hardening deformation contributes to fracture resistance under load. A fracture surface displaying both types of planes can be described as a mixed-mode fracture or, alternatively, by indicating the presence of shear lips on the fracture surface.

The local state of stress created by a load on a component geometry may cause crack propagation (i.e., critical fracture) that results in a fracture surface with a macroscale appearance, that is:

1. Totally ductile
2. Totally brittle
3. Initially brittle, then ductile
4. Initially ductile, then brittle
5. Mixed mode (ductile and brittle)

In the latter two cases (4 and 5), the ductile appearance may not be directly visible at the macroscale. Initially ductile fractures (case 4) are usually associated with rising-load ductile tearing, or the initial ductility may be inferred by transverse strain at the crack tip. The size of the plastic zone may be microscale in this case. Mixed-mode ductile and brittle cracking (case 5) would be inferred due to the presence of an intimate mixture of cleavage and MVC at the microscale or by the presence of shear lips at the macroscale.

The fracture appearance that occurs depends on the microstructure (strength and ductility) of the material and the degree of constraint associated with the presence of a cracklike imperfection. However, it also must be noted that some of the aforementioned criteria are based on macroscopic conditions or appearances and do not consider the microscopic mechanisms (i.e., slip, twinning, and cleavage) that cause fracture. A fracture may appear to be macroscopically brittle, but the cracking process may occur by a ductile mechanism. Examples in which the cracking mechanism is ductile but for which there is no or little visual macroscopic distortion include monotonic loading of a component containing a crack-like imperfection (plane-strain MVC fracture induced by part and crack geometry), long-life cyclic loading, and elevated-temperature failure (intergranular creep fracture). The major point is that the terms *ductile* and *brittle* should be used carefully with respect to the scale of observation or the description of fracture mechanisms. The distinction is important, because macroscopic brittle fractures can occur from the microscopic mechanism of ductile cracking.

Constraint is created by longer cracks, thicker sections, and a decreased crack tip radius. If the material is inherently brittle (say, a steel below the ductile-to-brittle transition temperature), crack initiation is expected at or near the preexisting cracklike imperfection, and the crack is expected to propagate in a microscale brittle manner. When the material has some inherent ductility, the fracture process is influenced by component and crack geometry, creating various fracture surface features.

Ductile cracking by MVC can result in a macroscale brittle fracture when the cracking is constrained by the geometry of the part and/or crack. With geometric constraint, plastic strain may be concentrated and lead to fracture without visible macroscale deformation. The microscale cracking mechanism is ductile, but geometric constraint limits macroscale distortion. This type of fracture may best be referred to as plane-strain MVC.

Consider first the effects of section thickness for an intermediate value of crack length and a sharp crack tip. For thin sections, there is little constraint imposed by a stress concentrator, so that the fracture process occurs essentially under conditions of plane stress, resulting in complete slant fracture (condition 1 in Fig. 38). As the section thickness increases, constraint, which is higher along the centerline than at the free surfaces, becomes sufficiently large to create plane-strain conditions and result in flat fracture (Fig. 38, condition 4). The slant fracture surfaces (Fig. 38, conditions 2 and 3) are described as shear lips, or, alternatively, the fracture can be described as mixed mode. Orientation of the shear lips may be used to identify the crack initiation location, which is helpful because chevrons or radial marks may not be present. The direction of crack propagation is parallel to the shear lips. Further increases in section thickness spread constraint toward the sides of the specimen, decreasing the width of the shear lips and ultimately resulting in a fracture that is essentially 100% flat (Fig. 38, condition 4).

Depending on the level of constraint and fracture toughness when fracture initiates, the stored elastic strain energy may or may not be sufficient to drive the crack completely across the specimen. A common situation in laboratory testing is the "pop-in" of a crack in the specimen; that is, a small crack suddenly forms under plane-strain conditions, with a concurrent drop in load. The load then rises, and crack propagation continues by ductile tearing. The process can repeat more than once, leaving telltale crack arrest marks on the fracture surface (Fig. 39).

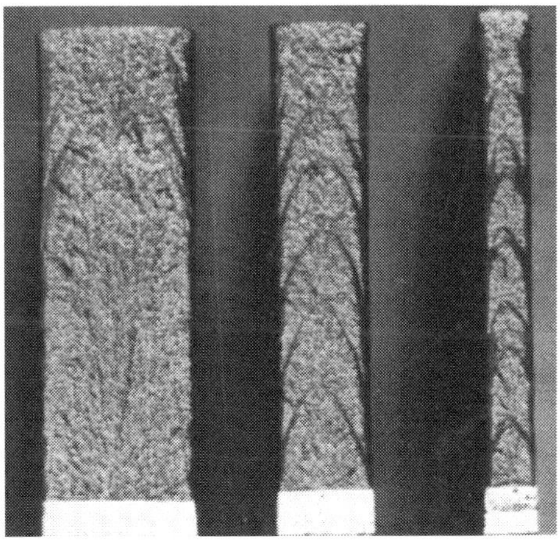

Fig. 39 Crack arrest lines on edge-notched tension specimens. Material thickness: 13 mm (1/2 in.), 10 mm (3/8 in.), and 6 mm (1/4 in.). Note the distance for first arrest, which increases with section thickness, and note that the arrest lines are not closed along the centerline in the 13 mm (1/2 in.) thick specimen, indicating full constraint at that location. Source: Ref 17

Note that the crack arrest marks indicate crack tunneling along the centerline and that they are matte in appearance compared to the generally shiny reflective surface. The curvature of the arrest lines delineates the crack front and indicates the direction of crack propagation. Chevrons created on a flat fracture surface point back to the crack initiation site, while arrest lines point in the direction of crack propagation. Microscale examination shows that the arrest lines are created by a change in fracture mechanism. Microscale examination of the fracture surface shows that the highly reflective regions of the fracture surface are created by cleavage or quasi-cleavage while the thin, arced arrest regions failed by MVC.

Ductile-to-Brittle Transition in Steels

As shown in Fig. 40, some bcc and hcp metals, and steels in particular, exhibit a ductile-to-brittle transition when loaded under impact. At high temperatures, the impact energy is high and the failure modes are ductile, while at low temperatures, the impact energy is low and the failure mode changes to a brittle fracture. The transition temperature is sensitive to both alloy composition and microstructure. For example, reducing the grain size of steels lowers the transition temperature. Not all metals display a ductile-to-brittle transition. Those having a face-centered cubic structure, such as aluminum, remain ductile down to even cryogenic temperatures.

Brittle fracture of normally ductile steels has occurred primarily in large, continuous, boxlike structures, such as box beams, pressure vessels, tanks, pipes, ships, bridges, and other restrained structures frequently joined with welded construction. The factors that must be present simultaneously to cause brittle fracture in a normally ductile steel are:

- *A stress concentration must be present.* This may be a weld defect, a fatigue crack, a stress-corrosion crack, or a designed notch, such as a sharp corner, thread, hole, or the like. The stress concentration must be

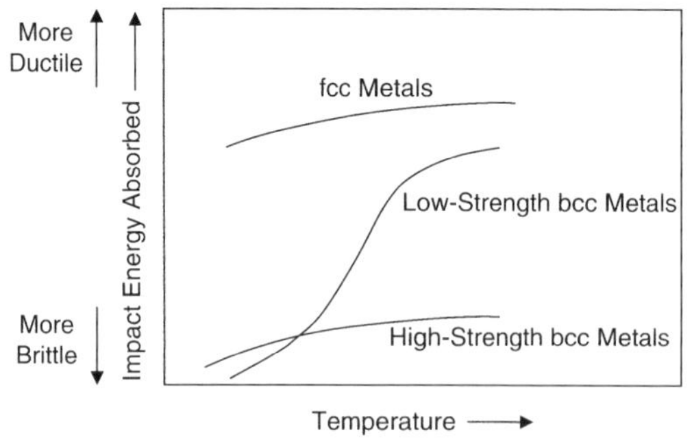

Fig. 40 Ductile-to-brittle temperature comparison for different metals. bcc, body-centered cubic; fcc, face-centered cubic. Source: Ref 2

large enough and sharp enough to be a critical flaw in terms of fracture mechanics.

- *A tensile stress must also be present.* This tensile stress must be high enough to provide microscopic plastic deformation at the tip of the stress concentration. One of the major complexities is that the tensile stress need not be an applied stress on the structure but may be a residual stress that is completely within the structure. In this case, the stress is not obvious or easily measured, as is the applied stress. The part or structure can be completely free of an external or applied load—just lying on a bench or floor, for example—and still experience instantaneous, sudden, catastrophic brittle fracture. This type of occurrence is within the experience of many persons who have worked with metals, particularly welded, torch-cut, or heat treated steels.
- *The temperature must be relatively low for the steel concerned.* The problem is that the definition of metal/temperature interrelationships is inexact, very much subject to the type of test used to try to understand whether or not a particular steel is actually subject to brittle fracture under certain conditions. However, regardless of the type of test used to try to establish the ductile-to-brittle transition temperature, the general results are the same: the lower the temperature for a given steel, the greater the possibility that brittle fracture will occur. For some steels, for example, the ductile-to-brittle transition temperature under certain conditions may be above room temperature.

As noted, the absence of any one of the factors—stress concentration, high tensile stress, relatively low temperature, susceptible steel—that contribute to brittle fracture of a ductile steel will prevent this problem from occurring.

For the following reasons, however, the choices are sometimes limited on what can be done:

- Stress concentrations are often present by design, such as in necessarily sharp corners, threads, holes, and grooves, or in unintentional stress concentrations, such as fatigue cracks, stress-corrosion cracks, weld defects, and arc strikes. Great care can be taken to minimize these stress concentrations, but they may occur despite the best safeguards.
- Tensile stresses are usually inevitable during service loading, depending on the type of part or structure. However, care can be taken to ensure that damaging tensile residual stresses are eliminated or minimized. This is particularly true when shrinkage stresses from welding are involved.
- Temperature can be controlled in certain applications but not in others. For example, certain processing equipment may operate continually at

elevated temperatures. In this case, brittle fracture may not be a consideration unless there is a damaging environmental factor, such as absorption of hydrogen or hydrogen sulfide.

In many other applications, exposure to relatively low temperature for the steel involved may be a real possibility and thus a real problem if the other contributing factors are likely to exist. Therefore, the steel itself may be the only factor that can be controlled in order to prevent brittle fracture of normally ductile steel. The metallurgical trends that tend to decrease the likelihood of brittle fracture of steel are low carbon content, moderate manganese content, high manganese-to-carbon ratio, inclusion of certain alloying elements, fine grain size, deoxidation of steel, and heat treatment to produce tempered martensitic or lower bainitic microstructures. Prevention of brittle fracture is entirely possible if advantage is taken of the recent technology that has led to the development of steels specifically for increased notch toughness.

Intergranular Failures

As shown in Fig. 41, failures can occur either transgranularly (i.e., through the grains) or intergranularly (i.e., along the grains). At room temperature, the grain boundaries are usually stronger than the grains themselves, and thus, fracture normally occurs in a transgranular manner. The occurrence of intergranular failures at room temperature often implies some embrittling behavior, such as the formation of brittle grain-boundary films, or the segregation of impurities, or inclusions that cluster at the grain boundaries. However, at temperatures high enough for creep to become dominant, the grain boundaries become weaker than the grains themselves, and intergranular failure modes are common.

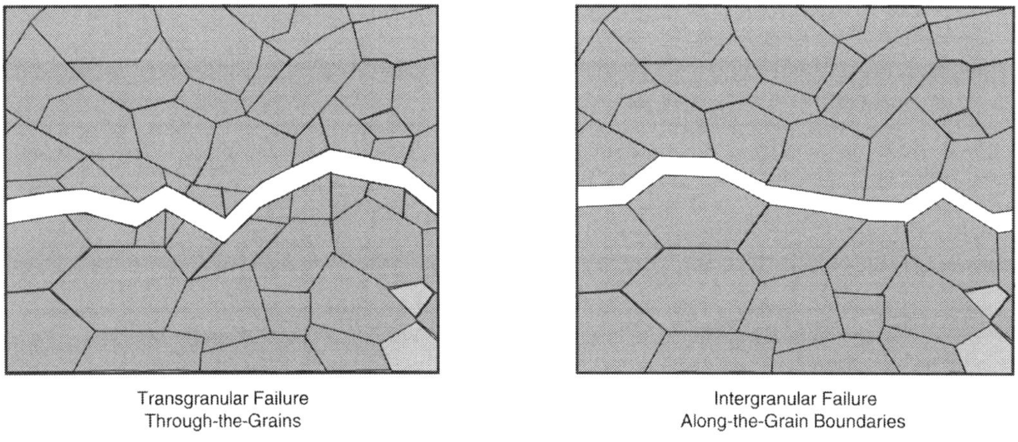

Fig. 41 Transgranular and intergranular failures. Source: Ref 19

Intergranular fractures are those that follow grain boundaries weakened for any of several reasons. An analogy may be made to a brick wall, which fractures through the mortar rather than through the bricks themselves. The mortar is analogous to the grain boundaries, while the bricks are analogous to the metal grains. A typical intergranular fracture is shown in Fig. 42. The reasons for weakened grain boundaries are frequently very subtle and poorly understood. Under certain conditions, some metals are subject to migration or diffusion of embrittling elements or compounds to the grain boundaries.

A fracture is referred to as decohesive rupture when it exhibits little or no bulk plastic deformation and does not occur by dimple rupture, cleavage, or fatigue. This type of fracture is generally the result of a reactive environment or a unique microstructure and is associated almost exclusively with rupture along grain boundaries. Grain boundaries contain the lowest-melting-point constituents of an alloy system. They are also easy paths for diffusion and sites for the segregation of such elements as hydrogen, sulfur, phosphorus, antimony, arsenic, and carbon; the halide ions, such as chlorides; as well as the routes of penetration by the low-melting-point metals, such as gallium, mercury, cadmium, and tin. The presence of these constituents at the boundaries can significantly reduce the cohesive strength of the material at the boundaries and promote an intergranular fracture, as shown in Fig. 43.

Intergranular fracture can be caused by several different mechanisms. The decohesive processes can involve the weakening of the atomic bonds,

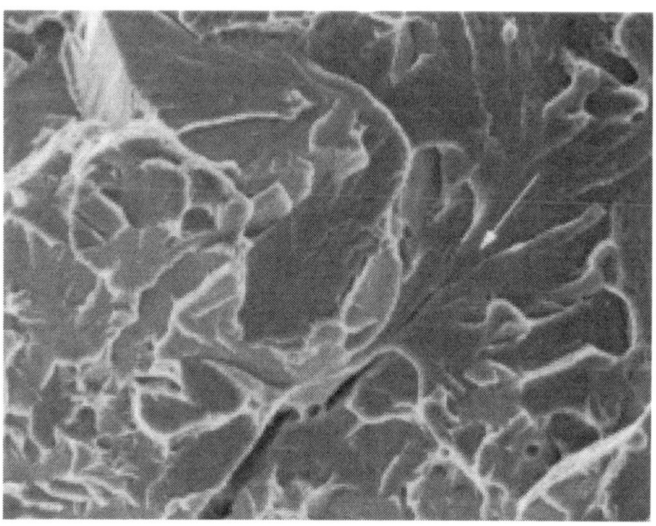

Fig. 42 Intergranular fracture in hardened steel, viewed under the scanning electron microscope. Note that fracture takes place between the grains; thus, the fracture surface has a "rock candy" appearance that reveals the shapes of part of the individual grains. Original magnification: 2000×. Source: Ref 10

the reduction in surface energy required for localized deformation, molecular gas pressure, the rupture of protective films, and anodic dissolution at active sites is associated with hydrogen embrittlement and stress-corrosion cracking (SCC). Decohesive rupture resulting from creep fracture mechanisms is discussed in greater detail in Chapter 11, "High-Temperature Failures," in this book.

The fracture of weak grain-boundary films, such as those resulting from grain-boundary penetration by low-melting-point metals, the rupture of melted and resolidified grain-boundary constituents (as in overheated aluminum alloys), or the separation of melted material in the boundaries before it solidifies (as in the cracking at the heat-affected zones of welds, a condition known as hot cracking), can produce a decohesive rupture.

Examples of intergranular fractures are shown in Fig. 44 through 46. An intergranular fracture resulting from hydrogen embrittlement is shown in Fig. 44, while Fig. 45 shows an intergranular fracture in a precipitation-hardenable stainless steel due to SCC. A fracture along a low-

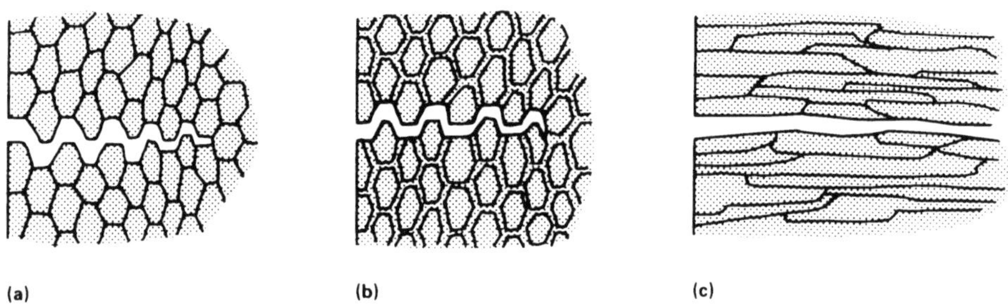

Fig. 43 Schematic illustrating intergranular fracture along grain boundaries. (a) Decohesion along grain boundaries of equiaxed grains. (b) Decohesion through a weak grain-boundary phase. (c) Decohesion along grain boundaries of elongated grains. Source: Ref 18

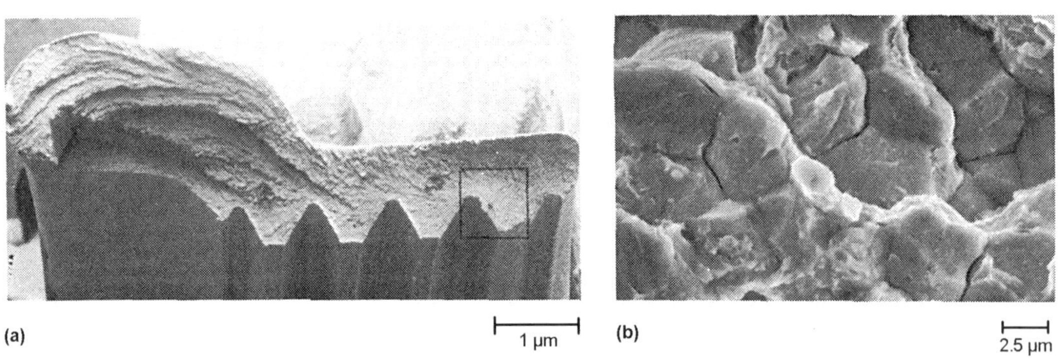

Fig. 44 Intergranular fracture in an AISI 8740 steel nut due to hydrogen embrittlement. Failure was due to inadequate baking following cadmium plating; thus, hydrogen, which was picked up during the plating process, was not released. (a) Macrograph of fracture surface. (b) Higher-magnification view of the boxed area in (a) showing typical intergranular fracture. Courtesy of W.L. Jensen, Lockheed Georgia Company. Source: Ref 18

strength grain-boundary film resulting from the diffusion of liquid mercury is shown in Fig. 46. When an intergranular fracture occurs along flattened, elongated grains that form nearly uninterrupted planes through the material, as in severely extruded alloys and along the parting planes of some forgings, a relatively smooth, featureless fracture results, as in Fig. 47.

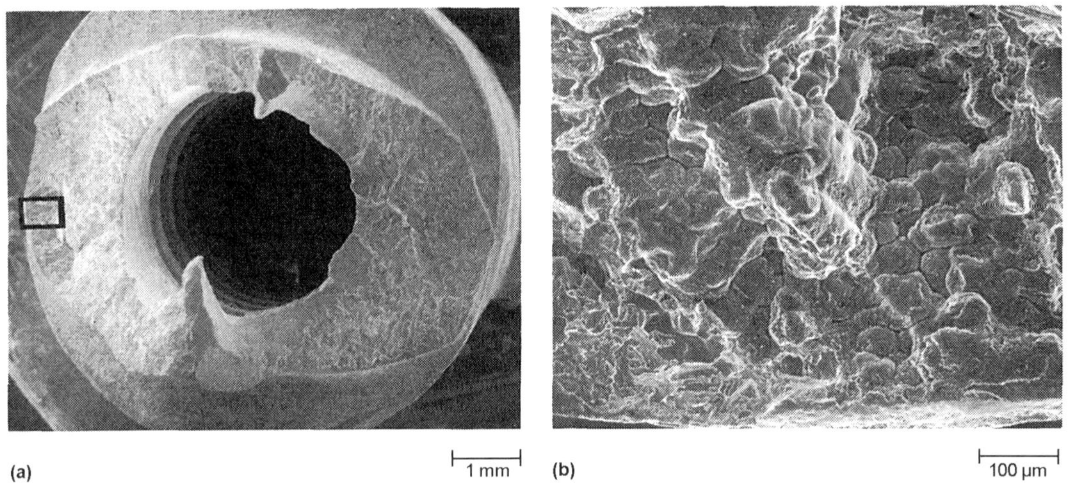

Fig. 45 17-4 PH stainless steel main landing gear deflection yoke that failed because of intergranular stress-corrosion cracking. (a) Macrograph of fracture surface. (b) Higher-magnification view of the boxed area in (a) showing area of intergranular attack. Courtesy of W.L. Jensen, Lockheed Georgia Company. Source: Ref 18

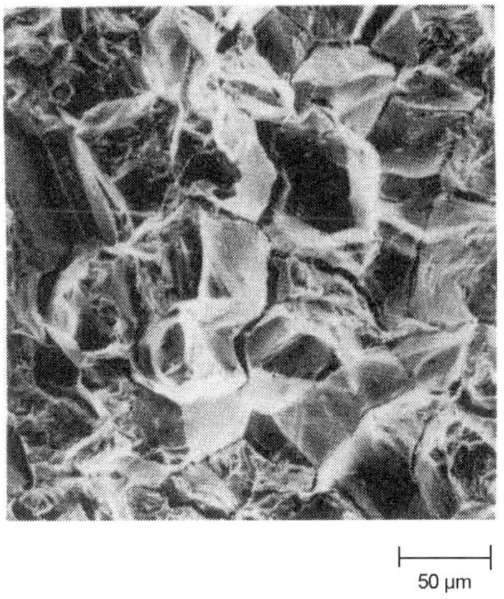

Fig. 46 Fracture surface of a Monel specimen that failed in liquid mercury. The fracture is predominantly intergranular with some transgranular contribution. Courtesy of C.E. Price, Oklahoma State University. Source: Ref 18

Fig. 47 Stress-corrosion fracture that occurred by decohesion along the parting plane of an aluminum alloy forging. Source: Ref 18

Causes of Steel Embrittlement

The major forms of embrittlement of steel are briefly discussed.

Strain Age Embrittlement. Most steels susceptible to the phenomenon of strain age embrittlement are low-carbon rimmed or capped steels that are severely cold worked during forming processes. Subsequent moderate heating during manufacture (as in galvanizing, enameling, or paint baking) or aging at ambient temperature during service may cause embrittlement.

Quench Age Embrittlement. Rapid cooling or quenching of low-carbon steels (0.04 to 0.12% C) from subcritical temperatures above approximately 560 °C (1040 °F) can precipitate carbides within the structure and also precipitation harden the metal. An aging period of several weeks at room temperature is required for maximum embrittlement.

Blue Brittleness. Bright steel surfaces oxidize to a blue-purple color when plain carbon and some alloy steels are heated between 230 and 370 °C (450 and 700 °F). After cooling, there is an increase in tensile strength and a marked decrease in ductility and impact strength caused by precipitation hardening within the critical temperature range.

Temper Embrittlement. Quenched steels containing appreciable amounts of manganese, silicon, nickel, or chromium are susceptible to temper embrittlement if they also contain one or more of the impurities antimony, tin, and arsenic. Embrittlement of susceptible steels can occur after heating in the range 370 to 575 °C (700 to 1070 °F) but occurs most rapidly at approximately 450 to 475 °C (840 to 885 °F).

500 °F Embrittlement. High-strength, low-alloy steels containing substantial amounts of chromium or manganese are susceptible to embrittlement if tempered in the range of 400 to 700 °F (200 to 370 °C) after hardening, resulting in tempered martensite. Steels with microstructures of tempered lower bainite also are subject to 500 °F embrittlement, but steels with pearlitic microstructures and other bainitic steels are not susceptible.

400 to 500 °C Embrittlement. Fine-grained, high-chromium ferritic stainless steels, normally ductile, will become embrittled if kept at 400 to 500 °C (750 to 930 °F) for long periods of time. Soaking at higher temperatures for several hours can restore normal ductility.

Sigma-Phase Embrittlement. Prolonged service at 560 to 980 °C (1040 to 1800 °F) can cause formation of the hard, brittle sigma phase in both ferritic and austenitic stainless steels and similar alloys. Impact strength is greatly reduced, particularly when the metal is cooled to approximately 260 °C (500 °F) or less.

Graphitization. Formation of graphite may occur in a narrow heat-affected zone of a weld in carbon and carbon-molybdenum steels held at temperatures over 425 °C (800 °F) for prolonged periods. The degree of embrittlement depends on the distribution, size, and shape of the graphite formed in the heat-affected zone.

Intermetallic Compound Embrittlement. Long exposure of galvanized steel to temperatures slightly below the melting point of zinc (420 °C, or 787 °F) causes zinc diffusion into the steel. This results in the formation of a brittle iron-zinc intermetallic compound in the grain boundaries.

Other types of embrittlement leading primarily to intergranular fracture are caused by environmental factors, including the following.

Neutron Embrittlement. Neutron irradiation of steel parts in nuclear reactors usually results in a significant rise in the ductile-to-brittle transition temperature of the steel. Metallurgical factors such as heat treating practice, microstructure, vacuum degassing, impurity control, and steel composition greatly affect susceptibility to this type of grain-boundary weakening.

Hydrogen Embrittlement. Hydrogen atoms diffuse readily into steel during processes such as acid pickling, electroplating, arc welding with moist or wet electrodes, and exposure to hydrogen sulfide. After stressing, delayed brittle fracture can occur, particularly in higher-strength steels.

Stress-Corrosion Cracking. Simultaneous exposure to a tensile stress (applied or residual) and to a relatively mild corrosive environment can cause brittle fracture in metal parts that may be either intergranular or transgranular, depending on conditions. If either factor is eliminated, SCC cannot occur.

Liquid Metal Embrittlement. Certain liquid metals can embrittle the solid metals with which they are in contact. A tensile stress is also required for brittle fracture to occur.

Each of the aforementioned types of embrittlement is the result of exposure to one or several environmental factors during manufacture, storage, or service. Each type is extremely complex and must be considered during any failure analysis investigation.

Combined Fracture Modes

It must not be assumed that brittle fracture always occurs solely by the cleavage or the intergranular fracture mode, as described previously. In most cases, one mode predominates but is not necessarily the only mode. For example, in a predominantly intergranular fracture, there probably will be regions, large or small, in the fracture surface that contain cleavage fracture as well. The reverse also is true. In other words, the mode of fracture that occurs at a particular location depends on the local composition, stress, environment, imperfections, and crystalline orientation of the grains. There may also be regions of tough, fibrous, dimple rupture fracture, particularly away from the origin of the major fracture.

It should be noted that cast metals generally tend to be less ductile (or more brittle) than wrought metals of the same composition under the same conditions. The reason is that various types of casting imperfections—particularly shrinkage porosity, gas porosity, and certain types of inclusions—act as internal stress concentrations in castings. In wrought metals, the hot working process refines the microstructure and changes the shape of many inclusions to long, thin stringers, which are usually less harmful.

ACKNOWLEDGMENTS

Portions of this chapter came from the chapters "Ductile Fracture" and "Brittle Fracture" by D.J. Wulpi in *Understanding How Components Fail*, 2nd ed., ASM International, 1999; "Mechanisms and Appearances of Ductile and Brittle Fracture in Metals" by W.T. Becker and D. McGarry and "Fracture Appearance and Mechanisms of Deformation and Fracture" by W.T. Becker and S. Lampman, both in *Failure Analysis and Prevention*, Vol 11, *ASM Handbook*, ASM International, 2002; and "Modes of Fracture" by V. Kerlins and A. Phillips in *Fractography*, Vol 12, *ASM Handbook*, ASM International, 1987.

REFERENCES

1. D.J. Wulpi, Ductile Fracture, *Understanding How Components Fail,* 2nd ed., ASM International, 1999
2. F.C. Campbell, Fracture, *Elements of Metallurgy and Engineering Alloys,* ASM International, 2008

3. W.T. Becker and D. McGarry, Mechanisms and Appearances of Ductile and Brittle Fracture in Metals, *Failure Analysis and Prevention,* Vol 11, *ASM Handbook,* ASM International, 2002
4. D.R. Johnson, "Toughness of Tempered Bainitic Microstructures in a 4150 Steel," Master's thesis, University of Tennessee, 1990
5. R.H. Van Stone and T.B. Cox, Use of Fractography and Sectioning Techniques to Study Fracture Mechanisms, *Fractography: Microscopic Cracking Mechanisms,* STP 600, C. Beachem and W. Warke, Ed., ASTM, 1976, p 18
6. T.A. Anderson, *Fracture Mechanics,* 2nd ed., CRC Press, 1995
7. G. Powell, S.-H. Cheng, and C.E. Mobley, Jr., *A Fractography Atlas of Casting Alloys,* Battelle Press, 1992
8. D. Chae, D.A. Koss, A.L. Wilson, and P.R. Howell, The Effect of Microstructural Banding on Failure Initiation of HY-100 Steel, *Metall. Trans. A,* Vol 31, 2000, p 995
9. *Fractography,* Vol 12, *ASM Handbook,* ASM International, 1987
10. D.J. Wulpi, Brittle Fracture, *Understanding How Components Fail,* 2nd ed., ASM International, 1999
11. M.E. Shank, Ed., *Control of Steel Construction to Avoid Brittle Failure,* Welding Research Council, 1957
12. J.E. Srawley and J.B. Esgar, "Investigation of Hydrotest Failure of Thiokol Chemical Corporation 260 Inch-Diameter SL-1 Motor Case," NASA TM X-1194, National Aeronautics and Space Administration, Washington, D.C.
13. D. Hull, *Fractography,* Cambridge University Press, 1999
14. G. Bodine, "The Effect of Strain Rate Upon the Morphology of High Purity 26% Chromium, 1% Molybdenum Ferritic Stainless Steel," Master's thesis, University of Tennessee, 1974
15. C.D. Beachem, Microscopic Fracture Processes, *Fracture,* Vol 1, Academic Press, 1968
16. W.T. Becker and S. Lampman, Fracture Appearance and Mechanisms of Deformation and Fracture, *Failure Analysis and Prevention,* Vol 11, *ASM Handbook,* ASM International, 2002
17. W.S. Lin, "The Ductile-Brittle Fracture Transformation: A Comparison of Macro and Microscopic Observation on Compact Tension Specimens," Master's thesis, University of Tennessee, 1974
18. V. Kerlins and A. Phillips, Modes of Fracture, *Fractography,* Vol 12, *ASM Handbook,* ASM International, 1987
19. F.C. Campbell., *Manufacturing Technology for Aerospace Structural Materials,* Elsevier Scientific, 2006

SELECTED REFERENCES

- T.L. Anderson, *Fracture Mechanics: Fundamentals and Applications,* 3rd ed., Taylor & Francis, 2005

- T.H. Courtney, *Mechanical Behavior of Materials,* 2nd ed., McGraw-Hill Book Co., 2000
- *Fractography,* Vol 12, *ASM Handbook,* ASM International, 1987
- R.W. Hertzberg, *Deformation and Fracture Mechanics of Engineering Materials,* 3rd ed., John Wiley & Sons, Inc., 1983
- W.F. Hosford, *Mechanical Behavior of Materials,* Cambridge University Press, 2005
- A.F. Liu, *Mechanics and Mechanisms of Fracture: An Introduction,* ASM International, 2005
- M. Myers and K. Chawla, *Mechanical Behavior of Materials,* 2nd ed., Cambridge University Press, 2009
- W. Soboyejo, *Mechanical Properties of Engineered Materials,* Marcel Dekker, 2002

CHAPTER 4

Fracture Mechanics

FRACTURE MECHANICS is the science of predicting the load-carrying capabilities of structures and components containing cracks. Fracture mechanics is based on a mathematical description of the stress field that surrounds a crack in a loaded body. Two categories of fracture mechanics are linear elastic fracture mechanics (LEFM) and elastic-plastic fracture mechanics (EPFM). LEFM is used when the crack tip is sharp and there is only a small amount of plastic deformation at or near the crack tip. LEFM is used for high-strength metals, such as high-strength steels, titanium, and aluminum alloys. EPFM is used when the crack tip is not sharp and there is some crack tip plasticity (blunting). EPFM is used in the design of materials, such as lower-strength, higher-toughness steels.

There are numerous methods and specimen configurations that have been developed to measure the toughness of various materials. In this chapter, five of the more common methods are covered: impact ductile-to-brittle transition temperature testing, drop-weight testing, plane-strain fracture toughness, J-integral, and crack tip opening displacement. Ductile-to-brittle transition temperature testing by methods such as the Charpy and Izod tests are the oldest and still most widely used.

Griffith's Theory of Brittle Fracture

The fracture strength of a solid should be a function of the cohesive forces that hold the atoms together. Using this criterion, the theoretical cohesive strength of a brittle elastic solid can be estimated to be in the range of $E/10$, where E is the modulus of elasticity. However, the true fracture strengths of real materials are much lower, normally 10 to as much as 1000 times below their theoretical values. In the 1920s, A.A. Griffith, while testing glass rods, observed that the longer the rod, the lower the strength. This led to the idea that the strength variation in the glass rods was due to defects, primarily surface defects. As the rods became longer, there was a higher probability of encountering a flaw large enough to cause failure. These flaws lower the fracture strength because they amplify the stress at the crack tip. This led to an instability criterion that considered the energy released in a solid at the time a flaw grows catastrophically under an applied stress.

Consider the internal crack shown in the Fig. 1 plate. As shown in the adjacent stress profile, the localized stress is high at the crack and then diminishes to the nominal applied stress (σ_o) at distances far removed from the crack. If it is assumed that the crack is similar to an elliptical hole through a plate and is oriented perpendicular to the applied stress, the maximum stress (σ_m) at the crack tip is given by:

$$\sigma_m = \sigma_o\left[1+\frac{2a}{b}\right] = \sigma_o\left[1+2\left(\frac{a}{\rho_t}\right)^{1/2}\right] \quad \text{(Eq 1)}$$

where $\rho_t = b^2/a$ is the radius tangential at the tip. Note that for a round hole ($a = b$), the quantity $\left[1+\frac{2a}{b}\right]$ reduces to 3, which is the conventional stress-concentration factor for a round hole, as described in Chapter 2, "Mechanical Behavior," in this book.

Here, σ_o is the magnitude of the applied nominal tensile stress, ρ_t is the radius of the crack tip, and a is the length of a surface crack, or one-half the length of an internal crack. For a relatively long microcrack that has

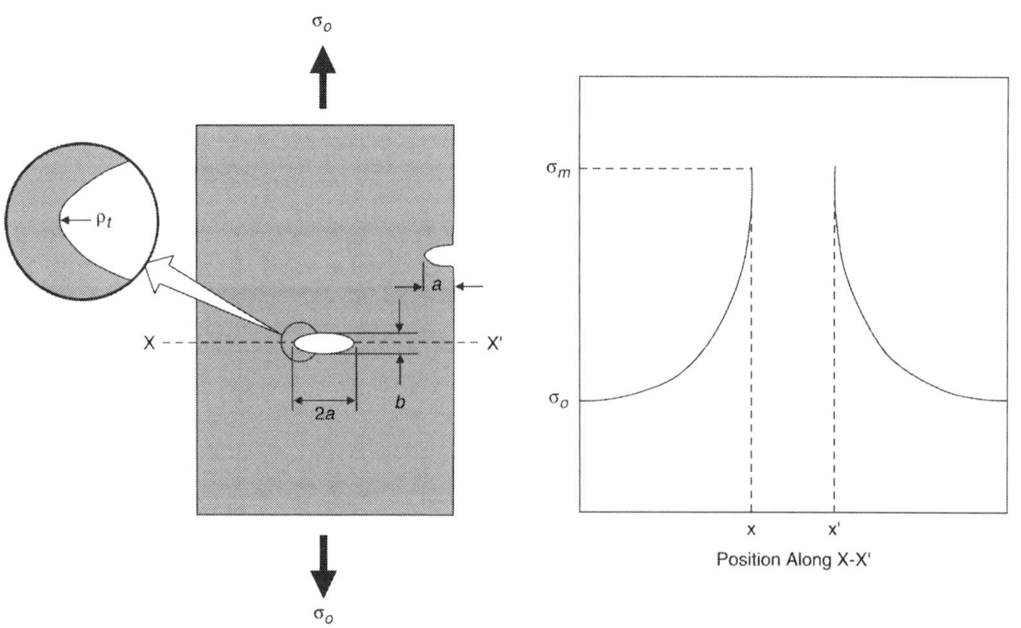

Fig. 1 Analysis of crack in wide plate. Source: Ref 1

small tip radius (sharp crack), the quantity $(a/\rho_t)^{1/2}$ is large and Eq 1 reduces to:

$$\sigma_m = 2\sigma_o\left(\frac{a}{\rho_t}\right)^{1/2} \quad \text{(Eq 2)}$$

The ratio σ_m/σ_o is the stress-concentration factor (K_t):

$$K_t = \frac{\sigma_m}{\sigma_o} = 2\left(\frac{a}{\rho_t}\right)^{1/2} \quad \text{(Eq 3)}$$

which is the amount the applied nominal stress is magnified at the crack tip. Note that the stress-concentration factor (K_t) increases with increasing crack length (a) and decreasing crack radius (ρ_t). Therefore, all cracks, if present, should be kept as small as possible. However, the effect of a stress raiser is much more significant in a brittle material than a ductile one. In a ductile material, when the concentrated stress exceeds the yield strength, the material will locally yield at the crack tip to relieve the stress. Because brittle materials do not have the ability to yield, the crack propagates through the material until it reaches a critical length, and the material fails. The theoretical strength of an elastic solid should be close to the cohesive strength of the material. This has been demonstrated with defect-free single-crystal whiskers that do indeed approach their theoretical strength.

Griffith developed a criterion for the elliptical crack in a plate using an energy balance approach. He equated the elastic strain energy that is stored in the material as it is elastically deformed to the surface energy created when two new free surfaces form during crack propagation. He concluded that the crack will propagate when the elastic energy released as a result of crack propagation exceeds the energy required to propagate the crack.

His analysis showed that the critical stress required to propagate a crack in a brittle material is:

$$\sigma_c = \sqrt{\frac{2E\gamma_s}{\pi a}} \quad \text{(Eq 4)}$$

where E is the modulus of elasticity, γ_s is the surface energy, and a is one-half the length of an internal crack.

Although this equation does not contain the crack tip radius ρ_t, Griffith's analysis was for brittle materials containing sharp cracks.

The Griffith equation is valid only for brittle materials that deform elastically, such as glasses and most ceramics. Because metals deform plastically, Orowan later modified the Griffith equation, replacing γ_s with γ_s +

γ_p, where γ_p is the plastic deformation associated with crack extension. Griffith's equation can then be rewritten as:

$$\sigma_c = \sqrt{\left[\frac{2E(\gamma_s + \gamma_p)}{\pi a}\right]} \quad \text{(Eq 5)}$$

If the material is highly ductile, then $\gamma_p > \gamma_s$ and:

$$\sigma_c = \sqrt{\frac{2E\gamma_p}{\pi a}} \quad \text{(Eq 6)}$$

In the 1950s, G.R. Irwin incorporated γ_s and γ_p into a single term, G_c, known as the critical strain energy release rate:

$$G_c = 2(\gamma_s - \gamma_p) \quad \text{(Eq 7)}$$

The strain energy release rate or the crack extension force (G) is the change in potential energy (U) of the system per unit increase in crack area ($G = dU/da$).

The original Griffith criterion can now be written for both brittle and ductile materials as:

$$G_c = \frac{\pi\sigma^2 a}{E} \quad \text{(Eq 8)}$$

Therefore, crack extension occurs when $\pi\sigma^2 a/E$ exceeds the value of G_c for the material in question.

Linear Elastic Fracture Mechanics

The fundamental ideas underlying the foundation of fracture mechanics stem from the work of Griffith, who demonstrated that the strain energy released upon crack extension is the driving force for fracture in a cracked material under linear elastic conditions. The elastic strain energy (U) is the work done by a load (P) causing a displacement (Δ):

$$U = \frac{P\Delta}{2} = \frac{CP^2}{2} \quad \text{(Eq 9)}$$

where C is equal to Δ/P, the elastic compliance.

The loss of elastic potential energy with crack extension of unit area A is defined as the strain energy release rate (G). For a crack extending at constant deflection or at constant load:

$$G = \frac{dU}{dA} = \left(\frac{P^2}{2}\right)\frac{dC}{dA} \quad \text{(Eq 10)}$$

This relationship characterizes the fracture resistance of structural materials by defining a critical strain energy release rate (G_c) at the critical load (P_c) when fracture occurs in a specimen with a known compliance function (dC/dA).

In LEFM, it is assumed that the part contains a crack or other flaw, the crack is a flat surface in a linear elastic stress field, and the energy released during rapid crack propagation is a basic material property that is not influenced by part size. As shown in Fig. 2, there are three modes of crack tip opening displacement. Mode I, tensile opening, is the most important. The other two, modes II and III, are sliding and tearing, respectively. It should be noted that mode I is by far the most important of the three, because it almost always turns out to be the limiting or critical case.

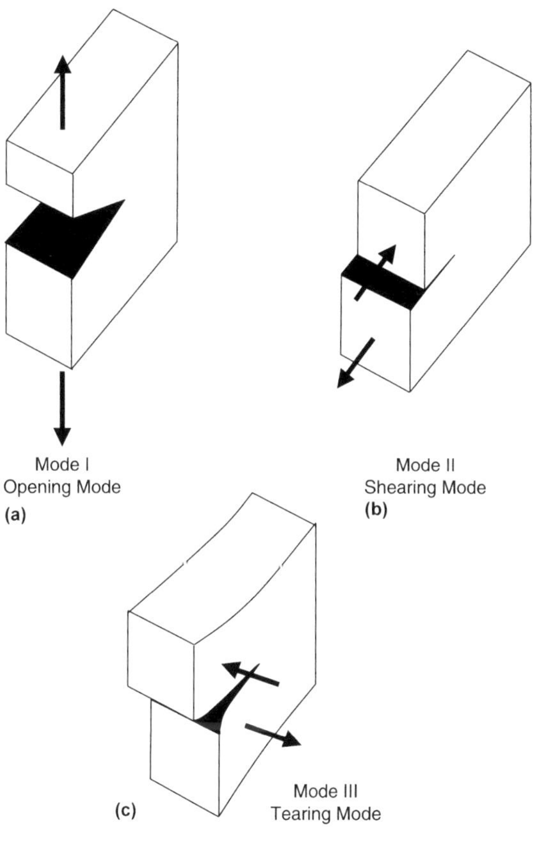

Fig. 2 Crack opening displacement modes. (a) Caused by stress normal to the crack face; considered most serious of loading modes, because $K_{Ic} < K_{IIc}, K_{IIIc}$; therefore, design usually based on K_I, K_{Ic}. (b) Shearing or forward shear; caused by in-plane shear stress; crack initiation during fatigue often results from shear; in mixed K_I-K_{II} mode, K_{II} causes crack to deflect until it reaches a mode I condition; mode II (and III) can produce friction as surfaces rub. (c) Tearing, or transverse shear; in mixed K_I-K_{III}, mode K_{III} sometimes causes crack to twist; mode III (and II) can produce friction as surfaces rub. Source: Ref 2

For mode I, the stresses acting on an element of material are shown in Fig. 3. The tensile (σ_x and σ_y) and shear (τ_{xy}) stresses are functions of both the radial distance (r) and the angle (θ), as follows:

$$\sigma_x = \frac{K}{\sqrt{2\pi r}}\cos\frac{\theta}{2}\left(1-\sin\frac{\theta}{2}\sin\frac{3\theta}{2}\right) \quad \text{(Eq 11)}$$

$$\sigma_y = \frac{K}{\sqrt{2\pi r}}\cos\frac{\theta}{2}\left(1+\sin\frac{\theta}{2}\sin\frac{3\theta}{2}\right) \quad \text{(Eq 12)}$$

$$\tau_{xy} = \frac{K}{\sqrt{2\pi r}}\sin\frac{\theta}{2}\cos\frac{\theta}{2}\cos\frac{3\theta}{2} \quad \text{(Eq 13)}$$

If the plate is thin, then $\sigma_z = 0$ and a condition of plane stress exists. On the other hand, if the plate is thick and $\varepsilon_z = 0$ and $\sigma_z = \nu(\sigma_x + \sigma_y)$, where ν is Poisson's ratio, then a condition of plane strain exists. In these equations, the parameter K is known as the stress-intensity factor. Note that the stress-intensity factor (K) and the stress-concentration factor (K_t), although similar, are not equivalent. In general, the stress-intensity factor is related to the applied stress and crack length by:

$$K = Y\sigma\sqrt{\pi a} \quad \text{(Eq 14)}$$

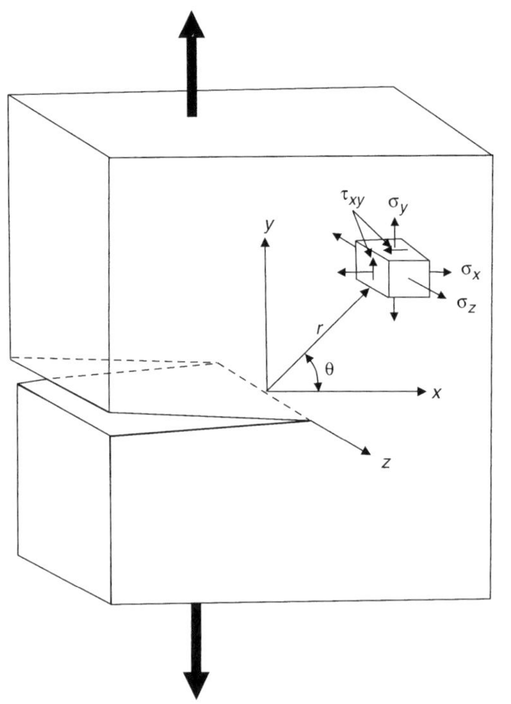

Fig. 3 Stresses at front of crack tip in mode I loading. Source: Ref 2

where Y, or $Y(a/w)$, is a dimensionless parameter that depends on the specific crack, specimen geometry, and type of loading. The stress-intensity factor (K) has the units of MPa$\sqrt{\text{m}}$ (ksi$\sqrt{\text{in.}}$). The stress-intensity factor (K) represents a single parameter that includes both the effect of the applied stress on a sample and the effect of a crack of a given size. The stress-intensity factor can have a simple relation to applied stress and crack length, or the relation can involve complex geometry factors for complex loading, various configurations of structural components, and variations in crack shapes. Examples of several stress-intensity factors for different geometries are shown in Fig. 4.

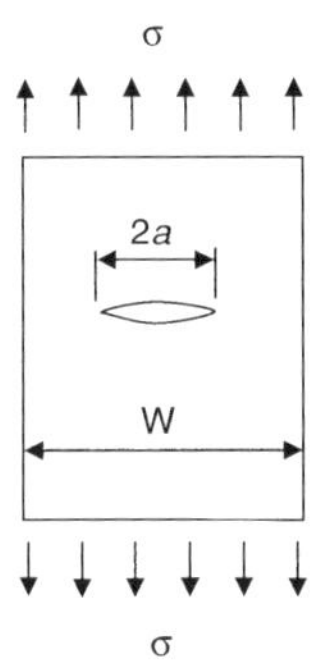

Center Crack: $K_I = Y\sigma\sqrt{\pi a}$

$$Y = 1 + 0.256\left(\frac{a}{W}\right) - 1.152\left(\frac{a}{W}\right)^2 + 12.200\left(\frac{a}{W}\right)^3$$

or $$Y = \sqrt{\sec\left(\frac{\pi a}{W}\right)}$$

or $$Y = \frac{1}{\sqrt{1-\left(\frac{2a}{W}\right)^2}}$$

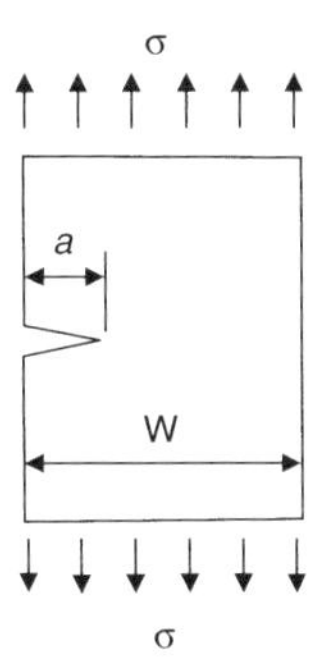

Single Edge Crack: $K_I = Y\sigma\sqrt{\pi a}$

$Y = 1.12$ for small cracks

or $$Y = 1.12 - 0.231\left(\frac{a}{W}\right) + 10.55\left(\frac{a}{W}\right)^2 - 21.72\left(\frac{a}{W}\right)^3 + 30.39\left(\frac{a}{W}\right)^4$$

up to a/W=0.6

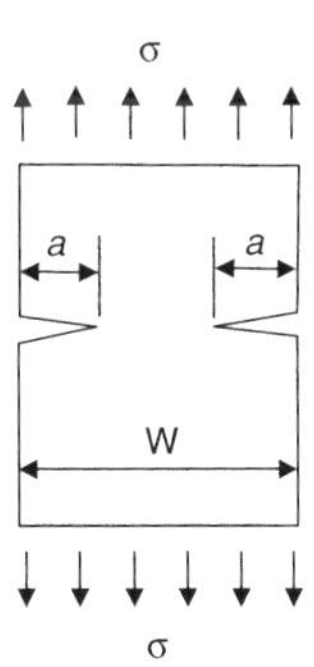

Double Edge Crack:

$Y = 1.12$ for small cracks

or $$Y = \frac{1.222 - 0.561\left(\frac{a}{W}\right) - 0.205\left(\frac{a}{W}\right)^2 + 0.471\left(\frac{a}{W}\right)^3 - 0.190\left(\frac{a}{W}\right)^4}{\sqrt{1-\frac{a}{W}}}$$

Fig. 4 Examples of crack configurations and stress-intensity factors. Source: Ref 2

When the applied stress (σ) exceeds some critical value (σ_c), the crack propagates and failure occurs. Therefore, there exists some critical value of K that corresponds to the critical value of applied stress (σ_c). This critical value of K is called the fracture toughness (K_c) and is simply:

$$K_c = Y\sigma_c\sqrt{\pi a} \qquad \text{(Eq 15)}$$

Fracture toughness (K_c) is a measure of the resistance of a material to brittle fracture when it contains a crack. As the ratio (a/w) approaches 0 for the case of a very wide plate with a short crack, the function $Y(a/w)$ approaches 1. The function $Y(a/w)$ for different geometries and different types of cracks can be found in handbooks on fracture mechanics.

The strain energy release rate and stress-intensity approaches are related:

$$K^2 = E'G \qquad \text{(Eq 16)}$$

where for plane stress, $E' = E$, the elastic modulus; for plane strain, $E' = E/(1 - \nu^2)$, where ν is Poisson's ratio. Thus, it is equivalent to attribute the driving force for fracture to the crack tip stress field, which is proportional to K or to the elastic strain energy release rate, G. The stress-intensity K is used more commonly than G, because K can be computed for different structural geometries using stress analysis techniques.

For relatively thin specimens (plane stress), the value of K_c will depend on the specimen thickness B, as indicated in Fig. 5. As the specimen becomes thicker and thicker, the plane-stress condition transitions into plane strain. When the fracture toughness (K_c) finally reaches a constant and lower value, this is known as the plane-strain fracture toughness (K_{Ic}).

$$K_{Ic} = Y\sigma_c\sqrt{\pi a} \qquad \text{(Eq 17)}$$

The subscript "I" refers to the mode I crack opening, because it is almost always the limiting value in fracture analysis. Because plastic deformation, which can blunt a sharp crack, is not possible in brittle elastic solid materials, these materials will have a low plane-strain fracture toughness and are subject to sudden and catastrophic failures. More-ductile materials, which can plastically deform and blunt cracks, will naturally have much higher values of K_{Ic} and will normally exhibit more graceful failure modes. It should be noted that the plane-strain fracture toughness (K_{Ic}) of a material is a material property, much as the yield strength or the elongation. This is not true for the plane-stress condition, where the fracture toughness (K_c) varies with the specimen thickness. The plane-strain fracture toughness (K_{Ic}) depends on a number of factors, such as temperature, strain rate, material processing, and the final microstructure. The value of K decreases with increasing strain rate, decreasing temperature, and increasing yield strength. It generally increases with finer grain sizes. The value of K_{Ic} determined for a given material is unaffected by specimen dimensions or the type of loading, provided that the specimen dimensions are large enough relative to the plastic zone to ensure plane-strain conditions at the crack tip. Therefore, plane-strain fracture toughness (K_{Ic}) is particularly pertinent in materials selection because, unlike other measures of toughness, it is independent of specimen configuration.

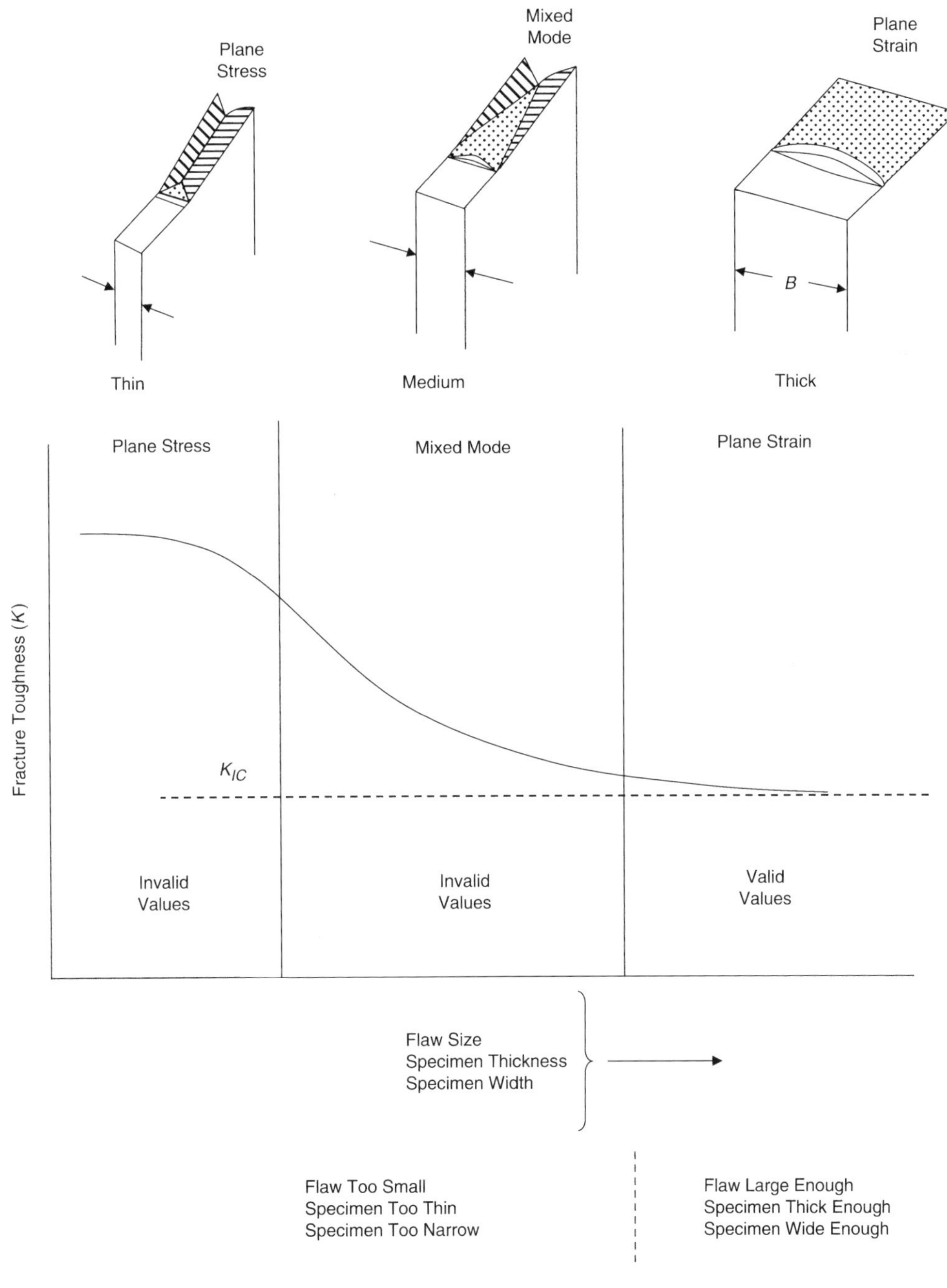

Fig. 5 Variation of K_c with specimen dimensions. Source: Ref 2

Plasticity Corrections

When the stress at the crack tip exceeds the yield strength, the material in the vicinity of the crack tip begins to deform plastically and absorb energy. As shown in Fig. 6, instead of rising to the tensile stress (σ_c) predicted by linear elastic analysis, the peak stress is reduced to the yield strength (σ_y). This explains why metals are so much tougher than brittle materials, such as ceramics, that do not yield.

Irwin proposed that the existence of a plastic zone makes the crack act as if it were longer than its actual size. In other words, as a result of crack tip plasticity, the displacements are longer and the stiffness is lower than predicted by elastic analysis. He proposed that the effective length of the crack is equal to the actual length plus the radius of the plastic zone:

$$a_{\text{eff}} = a - r_{\text{p}} \qquad \text{(Eq 18)}$$

where

$$r_{\text{p}} \approx \frac{1}{2\pi}\left(\frac{K}{\sigma_{\text{ys}}}\right)^2 \text{ for plane stress} \qquad \text{(Eq 19)}$$

and

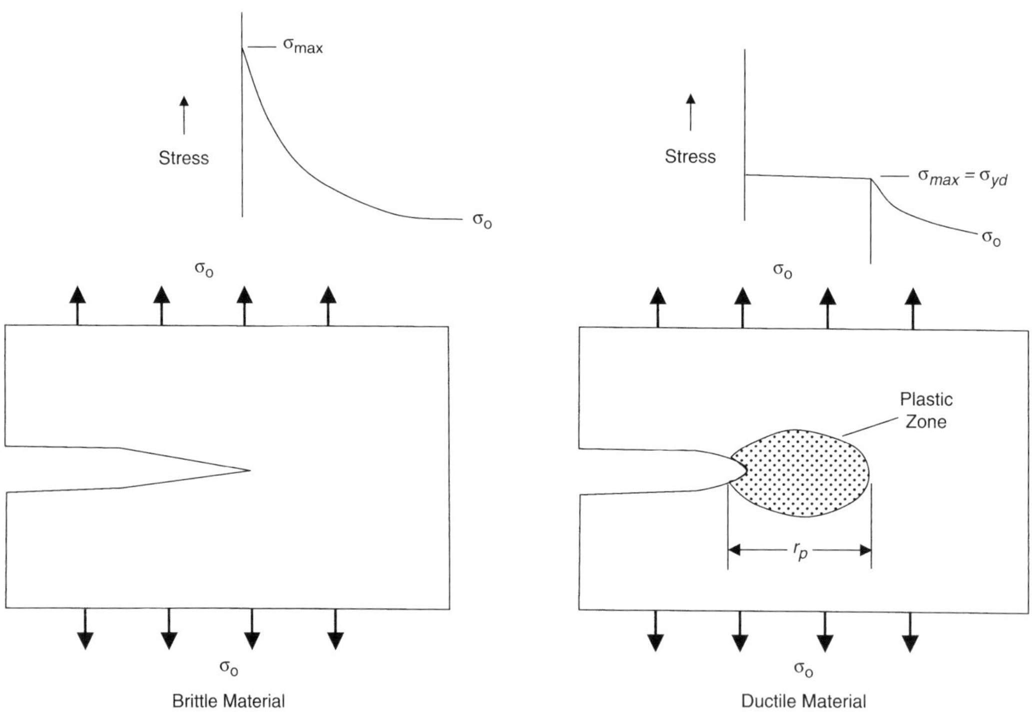

Fig. 6 Comparison of stress at crack tip for brittle and ductile material. Source: Ref 2

$$r_p \approx \frac{1}{6\pi}\left(\frac{K}{\sigma_{ys}}\right)^2 \text{ for plane strain} \qquad \text{(Eq 20)}$$

As result of the constraint imposed by the triaxial stress state in plane strain, the plastic zone (r_p) is less for plane strain than for plane stress. Thus, the plastic enclave for a thick specimen will have a larger plastic zone at the edges, due to plane-stress conditions, than in the middle, which is under pure plane-strain conditions (Fig. 7).

Size Effect. A two-dimensional stress state is assumed in a bulk material when one of the dimensions of the body is small relative to the others. A two-dimensional stress state called plane strain develops when plastic deformation at the crack tip is severely limited. This is promoted by thick sections, high strength, and limited ductility. In contrast, a two-dimensional stress state called plane stress develops when much more plastic deformation occurs around the crack tip. This is promoted by low-strength ductile materials and very thin sections of high-strength materials. The difference between plane strain and plane stress is based on the presence or absence, respectively, of transverse constraint in material deformation in the vicinity of the crack tip.

As specimen thickness (B) increases, σ_y increases from σ_{ys} (the engineering yield stress of the material at 0.2% strain) to $(3\sigma_{ys})^{1/2}$ because of a geometric constraint to plastic deformation associated with a transition from plane-stress to plane-strain conditions. The maximum value of σ_y is reached when the plastic zone size is limited to approximately 5% of the thickness. Thus, in a given material, the plastic zone size may vary with thickness by a factor of 3, leading to a strong dependence of K_c on thickness, as shown in Fig. 8. The inflection point of the curve in Fig. 8 occurs at approximately $(K_{Ic}/\sigma_{ys})^2$. Therefore, for maximum toughness:

$$B < \left(\frac{K_{Ic}}{\sigma_{ys}}\right)^2 \text{ or } K_{Ic} > \sigma_{ys}(B)^{1/2} \qquad \text{(Eq 21)}$$

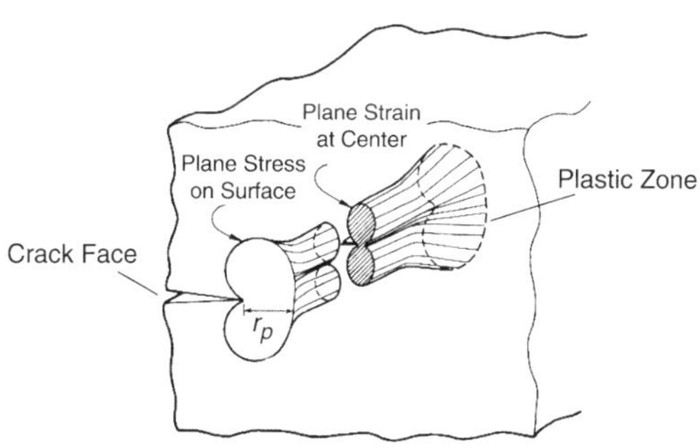

Fig. 7 Plastic zone formation at crack tip. Source: Ref 2

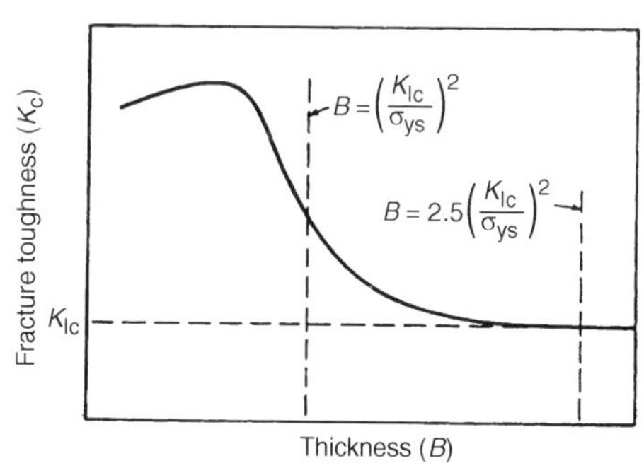

Fig. 8 Fracture toughness transition in structural alloys. Source: Ref 3

Equation 21 is useful in materials selection. However, of great importance is the fact that the curve in Fig. 8 approaches an asymptote at a thickness of $B \geq 2.5\ (K_{Ic}/\sigma_{ys})^2$. At this point, the fracture toughness value, K_{Ic}, is a material constant, independent of further increase of thickness.

Example: Maximum Stress to Fracture. Suppose that the fracture toughness of an aluminum alloy has been determined to be 30 MPa$\sqrt{\text{m}}$ (27 ksi$\sqrt{\text{in.}}$), and a penny-shaped crack of diameter 1.6 cm (0.63 in.) has been located in a thick plate that is to be used in uniaxial tension. Calculate the maximum allowable stress that can be imposed with fracture. Assume plane-strain conditions prevail. The yield stress of the material is 500 MPa (72.5 ksi).

Solution. The stress-intensity parameter formula for a penny-shaped crack is given by:

$$K = 2\sigma\left(\frac{a}{\pi}\right)^{1/2} \qquad \text{(Eq 22)}$$

where a is the crack radius, and σ is the applied stress.

At fracture, the applied stress intensity is equal to the plane-strain fracture toughness. In equation form, $K = K_{Ic}$. Rearranging Eq 22 and substituting appropriate values gives:

$$\sigma_f = K/2[\pi/a]^{1/2} = 30/2[\pi/0.008]^{1/2} = 297 \text{ MPa or } 43 \text{ ksi}$$

Thus, fracture will occur well below the material yield stress. This calculation shows that there is no guarantee that fracture will not occur simply because the nominal applied stresses are below the yield stress.

Example: Calculation of the Maximum Safe Flaw Size. Maraging steel (350 grade) has a yield of approximately 2450 MPa (355 ksi) and a toughness of 55 MPa$\sqrt{\text{m}}$ (50 ksi$\sqrt{\text{in.}}$). A landing gear is to be fabricated from this material, and the design stresses are 70% of yield. Assuming that flaws must be 0.25 cm (0.1 in.) to be detectable, is this a reasonable stress at which to operate? Assume that small edge cracks are present. The stress-intensity parameter for this geometry is:

$$K = 1.12\sigma\sqrt{\pi a} \qquad \text{(Eq 23)}$$

Solution. The flaw size at which fracture occurs is calculated by noting that at fracture $K = K_{Ic}$:

$$a_f = 0.797/\pi(K_{Ic}/\sigma_a)^2 = \frac{0.797}{\pi}\left[\frac{55}{1715}\right]^2 = 2.6\times10^{-4} \text{ m, or } 0.026 \text{ cm}$$

Thus, critical flaws may escape detection even though the design stresses for the part are below the yield stress. Consequently, the stress is too high to ensure safe operation of the landing gear. It must be reduced to the point at which the critical flaw size is greater than the minimum detectable crack size (0.25 cm).

Elastic-Plastic Fracture Mechanics

High-toughness structural materials undergo extensive plastic deformation prior to fracture. Therefore, the concepts of LEFM must account for elastic-plastic behavior. Three basic methods of EPFM include the crack tip opening displacement (CTOD), the *J*-integral, and the *R*-curve methods. These tests are intended to provide specialized measurements of fracture properties, as follows:

- *CTOD:* Full range of fracture toughness; for slow loading rates
- *J-integral:* Elastic-plastic fracture toughness; for slow loading rates
- *R-curve:* Resistance to fracture extension; for elastic-plastic fracture and slow loading rates

Crack Tip Opening Displacement

The concept of the CTOD and crack mouth opening displacement (CMOD) is shown schematically in Fig. 9, which shows a sample specimen before and after (hidden lines) deformation. Note that the CMOD is evaluated at the load line (centerline of the loading), and the CTOD is evaluated at the crack tip. Test methods used for evaluating the CTOD are British Standard 7448, Part 1 and ASTM E 1290.

The CTOD concept is a crack tip strain criterion for fracture. For a crack in an elastic body, the crack opening displacement (v) at a distance r from the crack tip is given by the displacement equation:

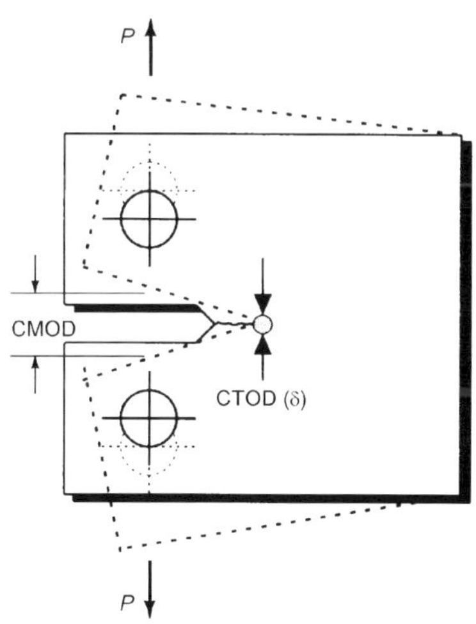

Fig. 9 Sample specimen showing the definition of crack mouth opening displacement (CMOD) and crack tip opening displacement (CTOD). CTOD is the diameter of the circular arc at the blunted crack tip and should not be confused with the plastic zone. Source: Ref 4

$$v = 2K/\pi E' (2\pi r)^{1/2} \quad \text{(Eq 24)}$$

where for plane stress, $E' = E$; for plane strain, $E' = E/(1 - \nu^2)$, where ν is Poisson's ratio.

Under conditions of small-scale yielding, the displacement at the crack tip (δ) can be calculated by assuming the effective crack tip ($a_{eff} = a + r_y$) is at a distance, r_y, from the actual crack tip:

$$\delta = 2v = 4K/\pi E' (2\pi r_y)^{1/2} = 4K^2/\pi E' \sigma_y \quad \text{(Eq 25)}$$

Theoretically, fracture occurs when $\delta = \delta_c$, the critical CTOD. In practice, a characteristic value for δ exists only for the crack initiation event; significantly more scatter exists for δ measured at maximum load or final fracture.

The CTOD approach is limited by the analytical and experimental uncertainties of the crack tip region. Analytically, δ is defined as the CTOD at the interface of the elastic-plastic boundary and the crack surface. Experimentally, δ is calculated from displacement measurements taken remotely from the crack tip, because direct physical measurements are not precise. Further uncertainty is introduced by the term σ_y in Eq 25, which may vary by 75%, depending on the degree of elastic constraint—a crack tip characteristic that cannot be measured directly.

The CTOD approach offers a significant improvement over linear elastic methods in the plastic range. An empirical correlation, known as a design curve, relates CTOD, crack size, and applied strain for a wide range of structural and material combinations. For many years, only the CTOD test measured toughness for a brittle, unstable fracture event using a nonlinear fracture parameter. In addition, the method allows the measurement of toughness after a "pop-in," which is described as a discontinuity in the load versus displacement record usually caused by a sudden, unstable advance of the crack that is subsequently arrested.

The J-Integral

The *J*-integral can be regarded as a change of potential energy of the body with an increment of crack extension. The general expression for *J*, as defined by Rice, is given by:

$$J = \int_{\Gamma} \left[W dy - \bar{T} \left(\frac{\partial \bar{u}}{\partial x} \right) ds \right] \quad \text{(Eq 26)}$$

where $\bar{T}$ is the traction vector acting on a segment ds of contour Γ, $\bar{u}$ is a displacement vector (the displacement on an element along arc s), and Γ is any contour in the x-y plane that encircles the crack tip. The parameter Γ is taken in a counterclockwise direction, starting from one crack face, ending on the opposite face, and closing the crack tip (Fig. 10). W is the strain energy density, given by:

$$W = \int_0^{\varepsilon_{ij}} \sigma \cdot d\left(\varepsilon_{ij}\right) \qquad \text{(Eq 27)}$$

That is:

$$W = \int [\sigma_x + d\varepsilon_x + \tau_{xy} + d\gamma_{xy} + \tau_{xz} + d\gamma_{yx} + \sigma_y + d\varepsilon_y + \tau_{yz} + d\gamma_{yz} + \sigma_z + d\varepsilon_z] \qquad \text{(Eq 28a)}$$

For generalized plane stress, Eq 28(a) is reduced to:

$$W = \int \left[\sigma_x + d\varepsilon_x + \tau_{xy} + d\gamma_{xy} + \sigma_y + d\varepsilon_y \right] \qquad \text{(Eq 28b)}$$

The total magnitude of J consists of two parts: an elastic part and a fully plastic part. For small-scale yielding or linear-elastic material behavior, J is equivalent to Irwin's strain energy release rate (G). Therefore, the portion for elastic-J is directly related to K, and the plastic part vanishes. However, unlike G or K, J can be used as a generalized fracture parameter for small- or large-scale yielding. In fact, J is used as an all-purpose fracture index in some literature. Circumstantially, the J reported could have been a J with large-scale yielding or just an elastic-J. The latter is actually a value of K^2 divided by E, or by $E/(1 - \nu^2)$, depending on the crack tip constraint (i.e., plane stress or plane strain).

General Procedure for Determining *J*. The finite-element method can be used to determine the stress and strain for points on the path of the selected contour Γ. The results are then substituted into Eq 26 and 27 for integration. To carry out the integration indicated in Eq 26 and 27, a relationship between stress and strain that realistically describes the plasti-

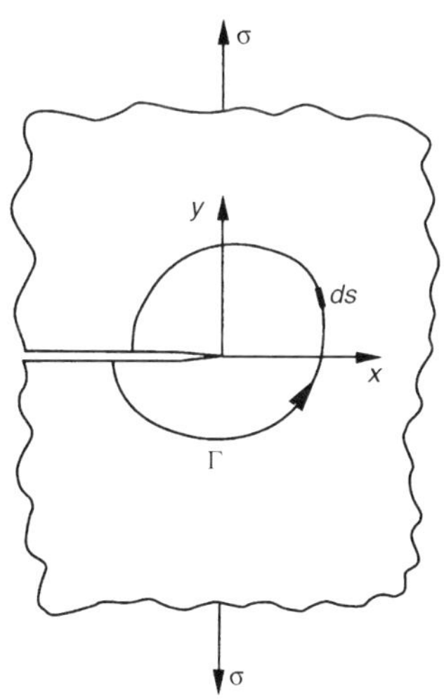

Fig. 10 Definition of the *J*-integral. Source: Ref 5

cally deforming material is needed. It has been shown that Eq 28(b) can be reduced to:

$$W = \frac{1}{2E}\left(\sigma_x + \sigma_y\right)^2 + \left(\frac{1+\nu}{E}\right)\left(\tau_{xy}^2 - \sigma_x \cdot \sigma_y\right) + \int_0^{\bar{\varepsilon}_p} \bar{\sigma} \cdot \left(d\bar{\varepsilon}_p\right) \qquad \text{(Eq 29)}$$

where $\bar{\sigma}$ and $\bar{\varepsilon}_p$ are equivalent stress and equivalent plastic strain, respectively, and:

$$\bar{\sigma} = \left(\sigma_x^2 - \sigma_x\sigma_y + \sigma_y^2 + 3\tau_{xy}^2\right)^{1/2} \qquad \text{(Eq 30)}$$

where σ_x, σ_y, and τ_{xy} are determined using a finite-element computer code. To execute the finite-element code, a stress-strain curve is entered into each element in the finite-element model (or at least into those elements along the selected contour Γ). The stress-strain curve can be represented by an empirical equation, or the *x*-*y* values of the curve are entered physically. For segments of Γ on which $\bar{\sigma}$ is greater than the material proportional limit and *dy* is nonzero, the value of $\bar{\varepsilon}_p$ that corresponds to an $\bar{\sigma}$ value is obtained from the stress-strain curve. Both $\bar{\sigma}$ and $\bar{\varepsilon}_p$ are then substituted into Eq 29 to compute the plastic component of *J*. Physically, the integral in Eq 29 may be interpreted as the area under the plastic stress-strain curve. Therefore, the plastic part of *J* is the integral in Eq 29, and the elastic part of *J* includes all the terms in Eq 26, excluding the integral term in Eq 29.

For very ductile materials, considerable care must be taken in evaluating the stress $\bar{\sigma}$, because a small error in $\bar{\sigma}$ will lead to large errors in estimating $\bar{\varepsilon}_p$ and thus may contribute large errors to the *J*-integral. In studying Eq 29, it is clear that *W* will have a unique value only if unloading is prohibited at every point in the structure. Monotonic loading conditions prevail throughout a cracked body under steadily increasing applied loads, provided that crack extension does not occur. Because in any calculation of *J*, the crack length is held constant, *W* will be unique, and a valid *J*-integral is obtained.

Rice also noted that *J* is path independent. That is, the values of *J* would be identical whether the contour Γ is taken far away or very close to the crack tip. In addition, there is no restriction as to what specific course Γ should follow. To confirm the path independence of *J*, another contour farther away from the crack tip (i.e., far from the plastic zone and primary elastic) was also evaluated. Although the analysis result for the second path is not shown here, the values of *J* calculated from the two paths were found to agree within 1%. Therefore, path independence had been verified.

Handbook Solutions for the Fully Plastic J_p. As mentioned earlier, the total *J* is a superimposition of an elastic part and a fully plastic part:

$$J = J_e + J_p \qquad \text{(Eq 31)}$$

For mode 1, the term J_e is directly related to K_I by way of:

$$J_e = \frac{K_I^2}{E'} \qquad \text{(Eq 32)}$$

where $E' = E$ for plane stress and $E' = E/(1 - \nu^2)$ for plane strain. The existing handbook solution for K can be used for a structural configuration under consideration. Close-form solutions for J_p are available.

Example: J-Integral Calculation for Center-Cracked Panel. For a center through-thickness crack subjected to far-field uniform tension having a total crack length $2a$ and width W, the fully plastic solution for J is:

$$J_p = \alpha\sigma_0\varepsilon_0 a \cdot (1 - 2a/W) \cdot h_1 \cdot \left(\frac{P}{P_0}\right)^{n+1} \qquad \text{(Eq 33)}$$

where σ_0 is the yield stress, ε_0 is the yield strain, and h_1 is a function of $2a/W$ and the strain-hardening exponent, n; its value is given in Fig. 11 (for plane stress) and Fig. 12 (for plane strain). In these figures, the dimension b is equal to one-half of the plate width. The material constants n and α are defined by the pure stress-strain law:

$$\frac{\varepsilon}{\varepsilon_0} = \alpha\left(\frac{\sigma}{\sigma_0}\right)^n \qquad \text{(Eq 34)}$$

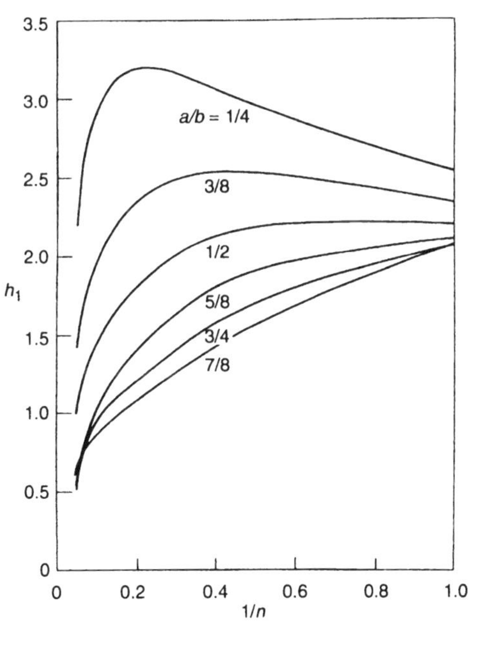

Fig. 11 h_1 versus $1/n$ for a center-cracked panel in tension under plane stress. Source: Ref 6

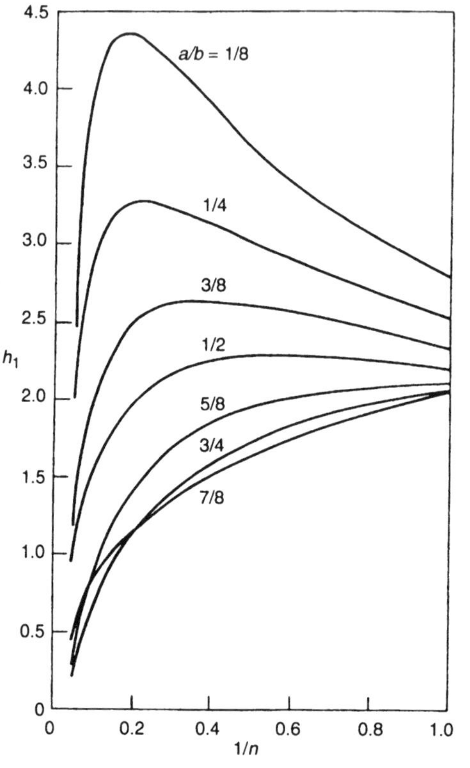

Fig. 12 h_1 versus $1/n$ for a center-cracked panel in tension under plane strain. Source: Ref 6

P is the applied load per unit thickness, and:

$$P_0 = \Psi \cdot (W - 2a) \cdot \sigma_0 \quad \text{(Eq 35)}$$

where $\Psi = 1$ for plane stress, and $\Psi = 2/\sqrt{3}$ for plane strain.

The R-Curve

The R-curve characterizes the resistance of a material to fracture during slow and stable propagation of a crack. An R-curve graphically represents this resistance to crack propagation of the material as a function of crack growth. With increasing load in a cracked structure, the crack extension force (G) at the crack tip increases. However, the material at the tip presents a resistance (R) to crack growth. According to Irwin, failure will occur when the rate of change of the crack extension force ($\partial G/\partial a$) equals the rate of change of this resistance to crack growth in the material ($\partial R/\partial a$). The resistance of the material to crack growth (R) increases with an increase in the size of the plastic zone.

Because the plastic zone size increases nonlinearly with a, R will also be expected to increase nonlinearly with a, while G increases linearly with a. Figure 13 shows the instability criterion: the point of tangency between the curves of G versus a and R versus a. The R-curve for a brittle material is shown in Fig. 13(a), and the R-curve for a ductile material is shown in Fig. 13(b). Crack extension occurs when $G > R$. Consider the G line for a stress σ', shown in Fig. 13(b). At the stress σ', the crack in the material will grow only from a_0 to a, because $G > R$ for $a < a'$. When $G < R$ for $a > a'$, the crack does not extend beyond a'. As the load is increased, the position of the G line changes, as indicated in the figure. When G becomes tangent to R, unstable fracture ensues. The R-curve for a brittle material (Fig. 13a) is a square curve, and the crack does not extend at all until point $G = G_c$ is reached and then unstable fracture follows.

The R-curve method is another version of the Griffith energy balance. One can conveniently make this kind of analysis if an analytical expression for the R-curve is available. However, experimental determination of R-curves is complicated and time-consuming.

Charpy and Izod Impact Testing

As discussed in Chapter 2, "Mechanical Behavior," in this book, the area under the stress-strain curve is an indication of material toughness. Thus, metals that have a good combination of strength and ductility should exhibit good toughness. For metals that do not contain flaws and are

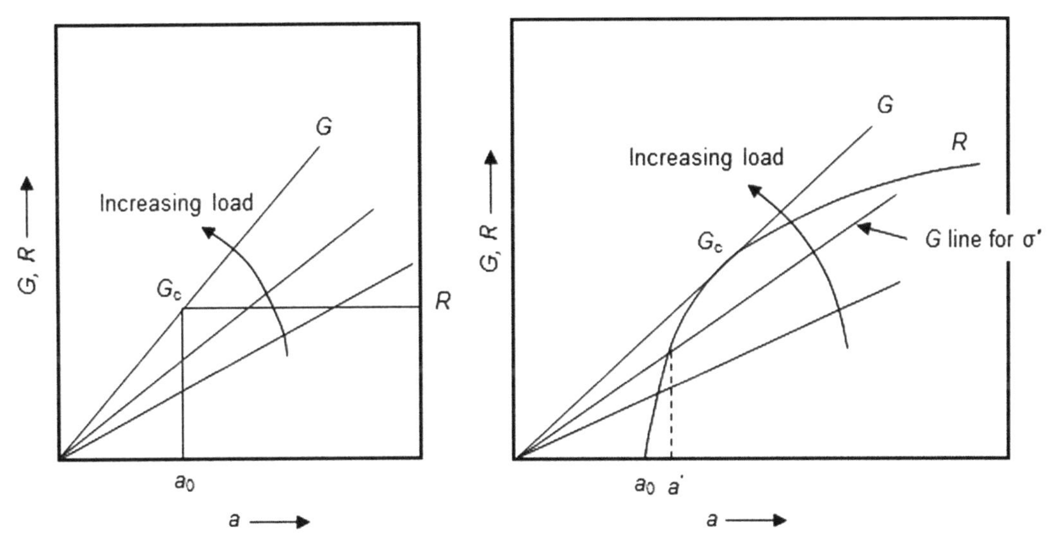

(a) (b)

Fig. 13 R-curves for (a) brittle material and (b) ductile material. Reprinted with the permission of Cambridge University Press. Source: Ref 7 loaded in uniaxial tension, this approach is valid. However, in many engineering applications, a notch is present that introduces a triaxial stress state. Because the triaxial stress state reduces the ability of the metal to deform plastically at the notch, toughness is reduced.

Prior to the formulation and maturity of LEFM, the notch toughness characteristics of low- and intermediate-strength steels were determined primarily by notched bar impact testing. The purpose of these tests is to determine the temperature at which a normally ductile failure transitions to a brittle failure. Therefore, tests are conducted at a series of different temperatures, and the impact loads (J or ft · lbf) are plotted versus test temperature. As the test temperature is lowered, there is a transition from ductile failures at high temperatures to brittle failures at lower temperatures (Fig. 14). This type of test is particularly applicable to steels. Because of their body-centered cubic (bcc) crystalline structure, steels undergo a change in their fracture behavior with changes in temperature. At high temperatures, fracture occurs in a ductile manner with large amounts of plastic deformation, while at low temperatures, fracture occurs in a brittle manner with little or no evidence of plastic deformation.

The most widely used specimen for characterizing the ductile-to-brittle transition behavior of steels is the Charpy V-notch impact specimen, which is described in ASTM E 23. The Charpy specimen (Fig. 15) is a three-point bend specimen containing a machined notch in the center of the face facing away from the impacting device. A cantilever bend test, the Izod test, is similar to the Charpy V-notch test and is preferred in Great Britain. The Izod specimen has a V-notch that is located toward one end of the specimen instead of in the middle, as in the Charpy V-notch specimen. The same testing machine can be used for both tests; however, the specimen geometry and computation of the impact toughness is different. The impact tester consists of a heavy pendulum that strikes the specimen at the bottom of its arc, where the velocity and kinetic energy are both at a maximum. As the specimen deforms and fractures, a portion of the kinetic energy of the pendulum is transferred to the specimen. The specimen is broken into two pieces as the pendulum continues its swing to a somewhat

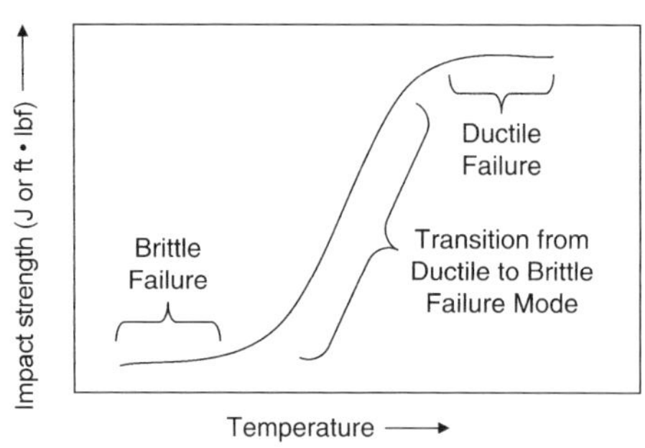

Fig. 14 Ductile-to-brittle transition typical of steels. Source: Ref 2

lower position than it was released from. The differences in these heights and the mass of the pendulum determine how much energy was absorbed by the specimen. Most impact testers have a gage that reports the energy level directly in J or ft · lbf. Impact tests such as the Charpy and Izod tests are severe tests because they occur at high strain rates and, due to their notches, have a high degree of triaxial loading that constrains plastic deformation at the notch.

Specimens are tested at different temperatures, and the impact notch toughness at each test temperature is determined from the energy absorbed during fracture. In addition, the percent shear (fibrous) fracture on the fracture surface and the change in the width of the specimen (lateral expansion) are frequently examined. An example of the ductile-to-brittle transition with temperature for each of these parameters is presented in

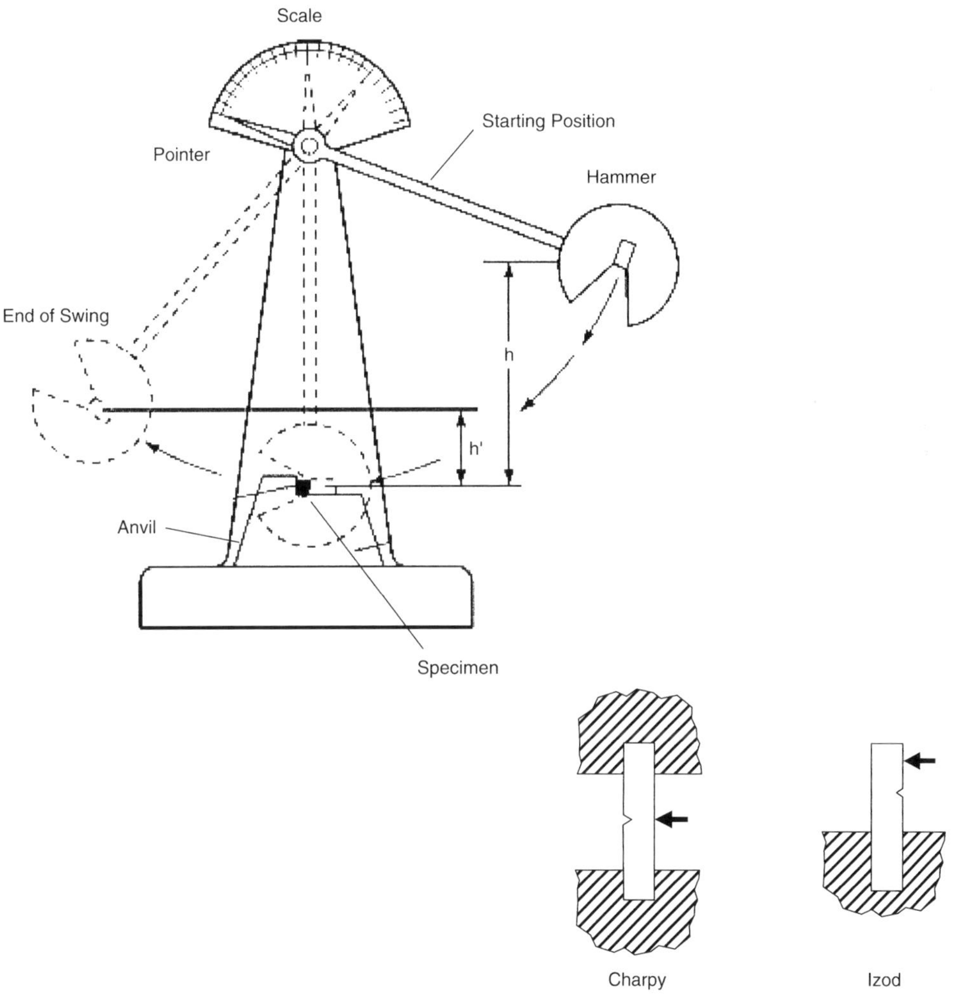

Fig. 15 Notched bar impact testing. Source: Ref 8

Fig. 16. The fracture surfaces of Charpy specimens tested at three different temperatures are shown in Fig. 17, exhibiting ductile, ductile-brittle, and brittle failure modes. The rate of change from ductile-to-brittle behavior depends on many factors, including the strength and composition of the material. Because the transition occurs over a range of temperatures, it is customary to define a single temperature within the transition range that reflects the behavior of the steel under consideration. Several equally useful definitions are in use, including the 15 ft · lbf temperature, the 15 mil lateral expansion temperature, and the 50% shear or ductile failure temperature.

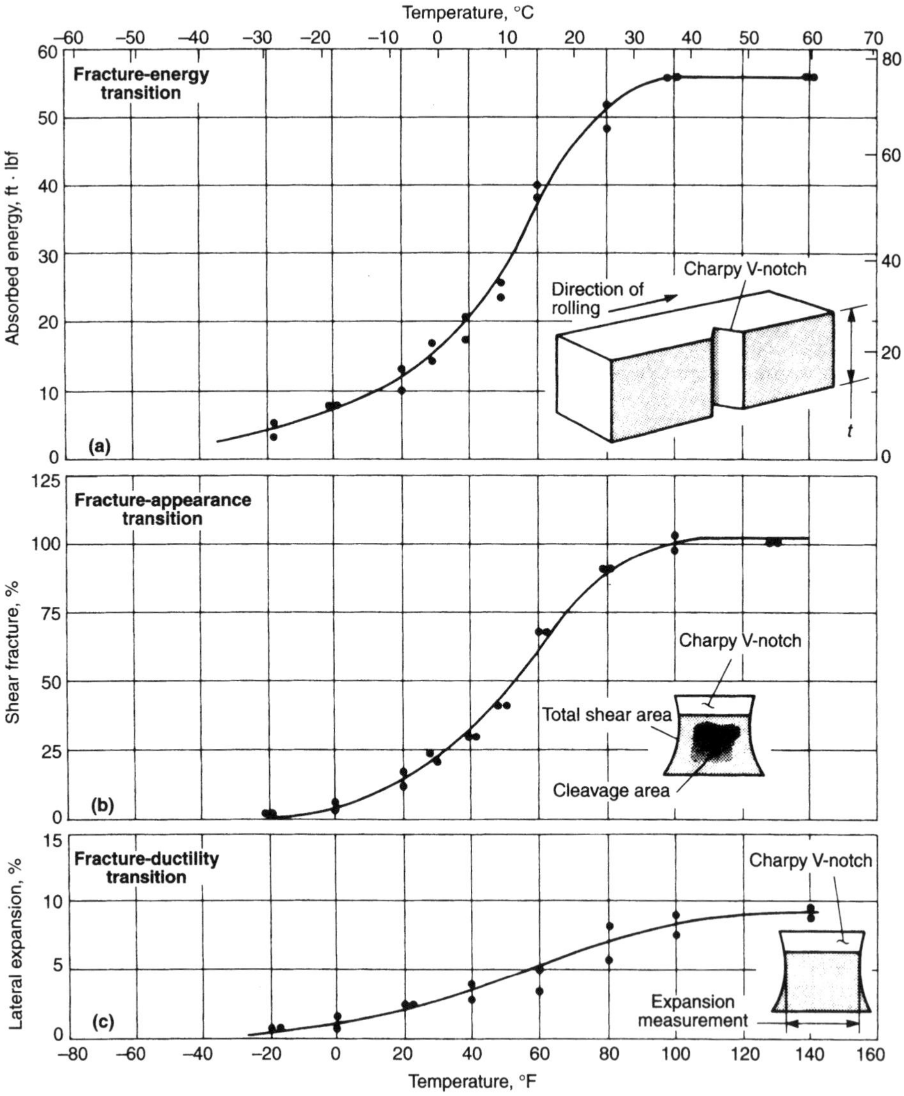

Fig. 16 Charpy impact ductile-to-brittle temperature transition criteria. Source: Ref 3

Several variations on the basic Charpy test specimen are shown in Fig. 18. The most common Charpy test uses a V-notched specimen. Three others are the Charpy keyhole test, Charpy U-notch test, and the precracked Charpy test. The Charpy keyhole and U-notch tests are similar to the Charpy V-notch test. Only the geometry of the machined notch is different. However, because the notch is not as severe, the keyhole and U-notch tests generally yield higher values of toughness and lower transition temperatures. They are more appropriate when testing materials that are less ductile or have a high notch sensitivity.

Applications of the notched bar impact test include comparisons of heat-to-heat variations in steels, evaluation of material behavior during

Fig. 17 Three Charpy V-notch impact test specimens of the same metallurgical conditions tested at three different temperatures. At the highest temperature (left), the fracture is virtually all shear. At intermediate temperature (center), the fracture is combined shear and cleavage. At the lowest temperature (right), the fracture is virtually complete cleavage. Note increased deformation with increased temperatures. Source: Ref 9

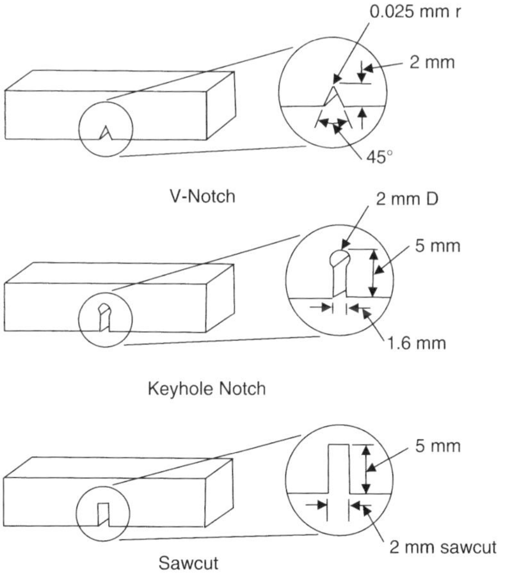

Fig. 18 Charpy impact specimen configurations. Source: Ref 2

either intentional or accidental high rates of loading, evaluation of the effect of irradiation on the embrittlement of steels, evaluation of the effects of microstructure and fabrication on toughness, and studies of the fundamental aspects of deformation in bcc materials. It should be noted that the notched bar impact tests do not provide a direct correlation to how an alloy will perform in service. However, they are useful indicators of how the material may behave in service. In spite of the limitations of notched bar impact testing, these tests are relatively simple, quick, and inexpensive methods for testing the dynamic fracture behavior of materials.

One on the major drawbacks of the Charpy test is that it does not provide much information about the fracture process itself. Therefore, instrumented Charpy tests have been developed. A strain gage is mounted on the arm of the pendulum, and a fast-triggered data-acquisition system records the impact. The data provide load-time profiles (Fig. 19) that show the different stages of deformation and fracture: general yield, maximum load, fast fracture, and arrest load after fast fracture. In addition, the actual energy absorbed can be obtained by accounting for the decrease in velocity of the pendulum as it fractures the specimen. The capability to separate the total energy absorbed into its different components makes the instrumented Charpy test a more effective analytical tool.

Drop-Weight Testing

Specimen size always presents a problem in any fracture testing. When conducting a series of notched bar impact tests using specimens of different sizes, at some temperature the largest specimens will be completely brittle while the smallest specimens will be completely ductile. The fractures for in-between specimens will vary from almost fully brittle to almost fully ductile. The problem of differing transition temperatures for full-sized parts and test specimens was discovered when a series of full-sized parts were tested using a giant pendulum-type impact machine. The results were compared with those determined using small standard test

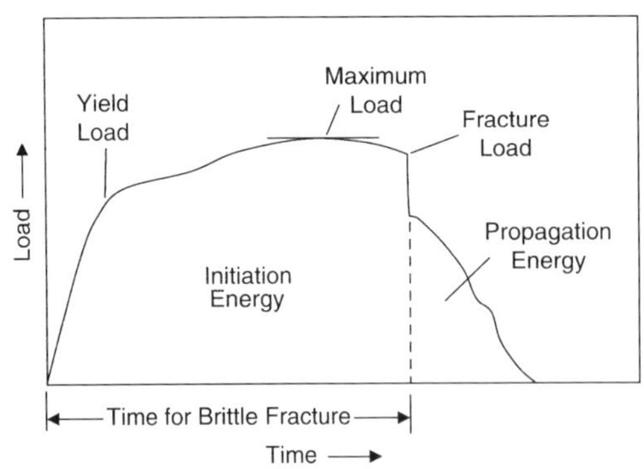

Fig. 19 Load-time curve for instrumented Charpy impact test. Source: Ref 10

bars made from the same material. A partial solution to this problem was the development of the drop-weight test (DWT) and the drop-weight tear test (DWTT). A third type, which actually is a derivative of the DWTT, is called the dynamic tear test (DT). These tests produce transition temperatures similar to those found when testing full-sized parts. However, such tests are adaptable only for plate specimens of limited sizes and have not become widely used. The essential features of these tests are discussed as follows.

Drop-Weight Test. The DWT specimen and setup are shown in Fig. 20 and described in ASTM Standard E 208. The crack inducer is a bead of hardfacing metal. The weld bead, deposited on one side of the specimen at the center, is purposely hard and brittle. A notch is made in the weld bead but not in the specimen itself. After cooling to the desired temperature, the specimen is placed, weld down, on rounded end supports and struck by a falling weight with sufficient energy to bend the specimen approximately 5°; the amount of weight and height depend on the strength of the material being tested. A cleavage crack forms in the bead as soon as incipient yield occurs (at approximately 3° deflection), thus forming the sharpest possible notch. When the crack reaches the specimen material, it will either propagate or stop. A series of tests are conducted over a range of temperatures to find the nil-ductility transition temperature (NDTT).

The specimen is then examined to see whether or not it has fractured. A specimen is considered broken if the crack extends to one or both sides of the surface. If the crack does not propagate to the edge, it is considered a "no break." If the weld notch is not visibly cracked, or if complete deflection does not occur, it is considered a "no test." Thus, this is a "go, no-go" type of test in that the specimen will either break or fail to break. The NDTT is the maximum temperature at which the specimen breaks. When minimum temperatures are set in material specifications, at least two specimens must be tested at the specified temperatures. All specimens tested must show a "no break" performance.

The drop-weight tear test can be used to determine the appearance of fractures in plain carbon or low-alloy pipe steels, with yield strengths less

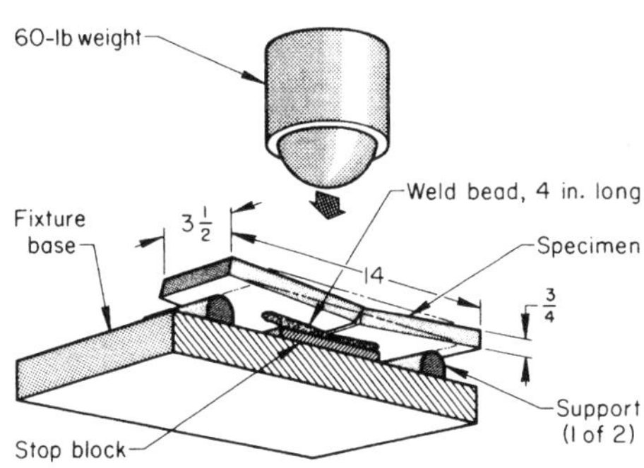

Fig. 20 Drop-weight test setup. Source: Ref 5

than 825 MPa (120 ksi) over the temperature range where the fracture mode changes from brittle to ductile. It has been used to evaluate line-pipe material for natural gas transmission lines. Specimens made from curved pipe may be flattened prior to testing. This test method is similar in some ways to the DWT. The transition fracture appearance occurs at the same temperature as for full-sized parts. It has the same sudden change from shear fracture to flat fracture as that observed in full-scale equipment. The test is relatively simple in terms of specimen preparation and lack of sensitivity to specimen preparation techniques. The results vary with specimen thickness in the same manner as actual parts.

A major shortcoming, as in the DWT, is that testing is confined to plate material thicknesses between 3 and 20 mm (0.125 and 0.75 in.). The test specimen is even larger than the DWT specimen: 75 mm (3 in.) wide by 300 mm (12 in.) long. Tests are made with the same apparatus used for the DWT, but the test fixture for holding the specimen is altogether different. A large pendulum-type machine can also be used, but the vertical weight dropping apparatus is more commonly used. Up to 2700 J (2000 ft · lbf) of energy may be required. The test specimen configuration and setup is shown in Fig. 21. The test is covered in ASTM Standard E 436.

The test result is evaluated by examining the broken pieces. The idea is to determine the percentages of the apparently flat and shear regions on the fracture surface, which are very different in appearance. The transition from one to the other is abrupt. There are two methods for making this evaluation: one is for percentages of shear from 45 to 100%, and the other for percentages from 0 to 45%. The acceptance criterion is percentage of shear at a specific temperature. The temperature at which 50% shear occurs is sometimes considered the ductile-brittle transition temperature. The procedure for measuring the DWTT percent shear area is detailed in ASTM Standard E 436.

The dynamic tear test belongs to the same family as the DWTT but uses a different specimen size. Like the DWTT, the use of either a drop-weight or a pendulum-type machine is allowed. Applicable thicknesses for the DT are 5 to 16 mm (0.2 to 0.6 in.), compared to 3 to 20 mm (0.125

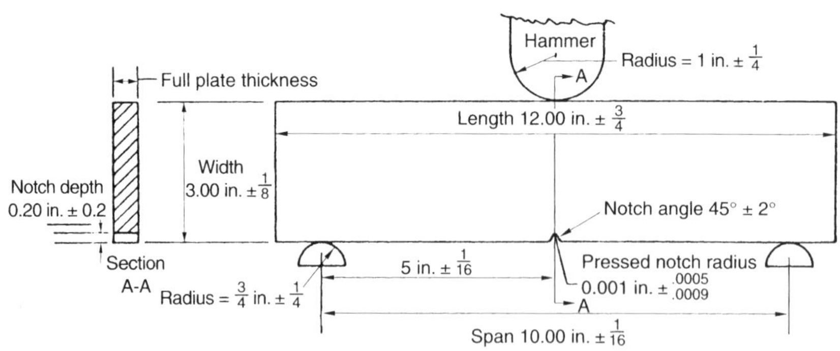

Fig. 21 Drop-weight tear test specimen and support dimensions. Source: Ref 5

to 0.75 in.) for the DWTT. The material to be tested should have a hardness not higher than 36 HRC. The hardness restriction is due to using a press-knife procedure for sharpening the specimen notch. Although test setups for both types of test are similar, the DT specimen is somewhat smaller than the DWTT specimen. Evaluation of the DT result is also based on fracture appearance; however, the method used to estimate the fracture appearance is not the same as for the DWTT. Details about the DT are given in ASTM Standard E 604.

From consideration of fracture mechanics and plastic-flow properties, and from numerous measurements, a ratio-analysis diagram (RAD) can be constructed to represent many of the conditions relating to materials selection for fracture control. Simplified versions of RADs for steels, aluminum alloys, and titanium alloys are shown in Fig. 22, 23, and 24, respectively. On the ordinate scales are values of DT energy and values of K_{Ic} that correspond to the DT values. The K_{Ic} scale is extended to the point corresponding to a 1-to-1 ratio of K_{Ic} to σ_{ys} (yield strength) for the best alloy of those to be considered. Sloping lines representing various ratios of K_{Ic} to σ_{ys} are drawn. By the equations of fracture mechanics, these ratios can be related to the critical flaw (crack) depth expected to initiate fracture in a plate of a particular thickness. Each sloping line is applicable to a certain critical thickness. Such a line can be constructed by plotting

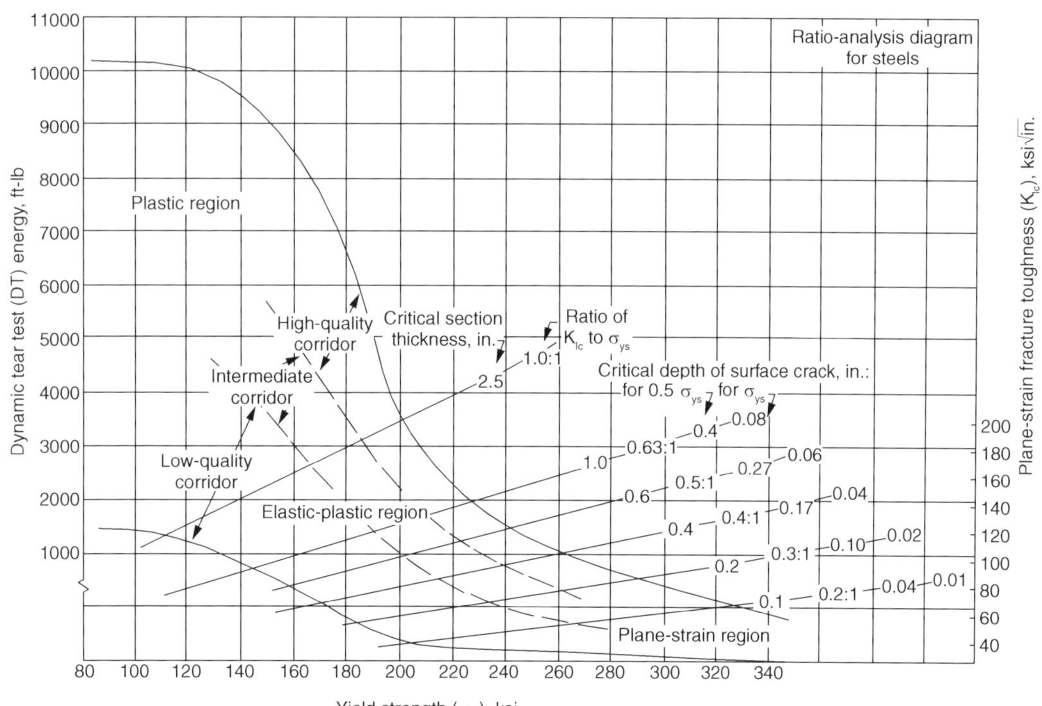

Fig. 22 Ratio-analysis diagram for steels, as prepared for trade-off analyses for components 25 mm (1.0 in.) thick. Data were determined using specimens 12.5 to 75 mm (0.5 to 3.0 in.) thick. Source: Ref 11

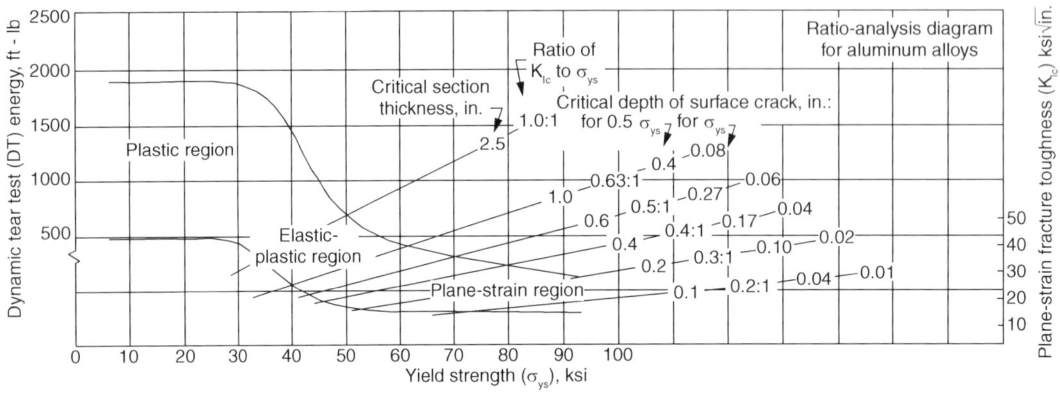

Fig. 23 Ratio-analysis diagram for aluminum alloys, as prepared for trade-off analyses for components 25 mm (1.0 in.) thick. Data were determined using specimens 25 to 100 mm (1.0 to 4.0 in.) thick. Source: Ref 11

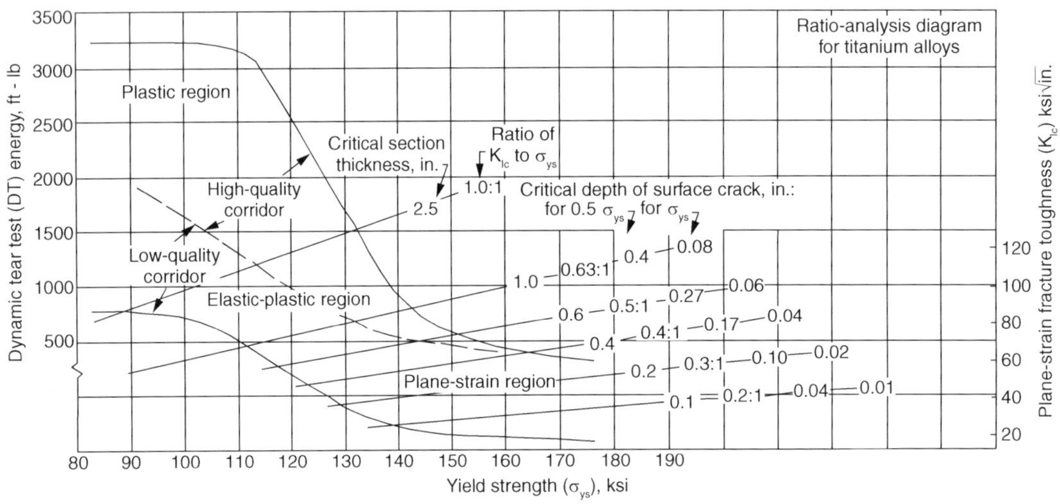

Fig. 24 Ratio-analysis diagram for titanium alloys, as prepared for trade-off analyses for components 25 mm (1.0 in.) thick. Data were determined using specimens 25 to 100 mm (1.0 to 4.0 in.) thick. Source: Ref 11

values of toughness versus yield strength for each alloy being considered for a design. If the thickness in the design is less than the critical thickness denoted by the sloping line, crack tips in the part will not be stressed under plane-strain conditions, and either elastic-plastic or gross yielding behavior can be expected. If design thickness is greater than critical thickness, the part can sustain even greater crack tip stress intensity before crack extension will begin.

Because the modulus of elasticity is implicit in the relations between K_{Ic} and crack size, different diagrams are required for steel, aluminum alloys, and titanium alloys. Furthermore, at a given yield strength, different alloys have different fracture toughness levels. Consequently, the diagrams may be divided into quality corridors based on test results for alloys with high, intermediate, and low toughness. With a RAD, it is then possible to guide the selection of alloys and working yield strengths so as to place the metal in either the ductile regime or the fracture mechanics regime, as may be appropriate for the cost or risk involved. Results obtained in DTs extend fracture analysis beyond the LEFM range.

Fracture Toughness Testing

Fracture toughness is an empirical material property that is determined by one or more of a number of standard fracture toughness test methods. In the United States, the standard test methods for fracture toughness testing are developed by ASTM International. Other industrial countries have equivalent standards-writing organizations that develop fracture toughness test standards.

Whether the fracture is ductile or brittle does not directly influence the deformation process that a component or specimen may undergo during the measurement of toughness. The deformation process is generally described as being linear elastic or nonlinear. This determines which parameter is used in the fracture toughness test characterization. All loading begins as linear elastic. For this, the primary fracture parameter is the crack tip intensity factor (K). If the toughness is relatively high, the loading may progress from linear elastic to nonlinear during the toughness measurement, and a nonlinear parameter is needed. The nonlinear parameters that are most often used in toughness testing are the J-integral, labeled J, and the crack tip opening displacement (CTOD), labeled δ. Because all loading starts as linear elastic, the nonlinear parameters are written as a sum of a linear component incorporating K and a nonlinear component.

Load displacement curves showing linear elastic, elastic-plastic, and fully plastic behavior are shown in Fig. 25. Although the J_{Ic} value can be

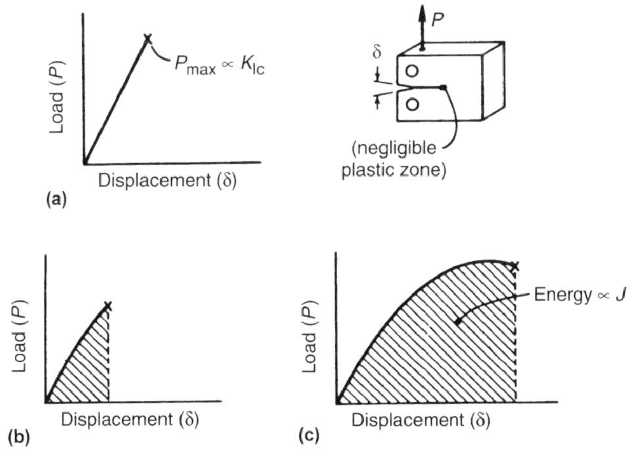

Fig. 25 Load displacement behavior observed in fracture toughness tests. (a) Linear elastic. (b) Elastic-plastic (failure before limit load). (c) Fully plastic (exhibits a limit load). (a) shows brittle behavior (K_{Ic} is measured). (b) and (c) show ductile behavior (J_{Ic} is measured). Source: Ref 3

used to obtain an estimate of K_{Ic}, denoted $K_{Ic}(J)$, it should be understood that K_{Ic} and $K_{Ic}(J)$ may represent significantly different fracture behavior. K_{Ic} is the critical K level for the linear elastic case, at which significant measurable extension of the crack occurs, often triggering unstable fracture. For valid measurements of K_{Ic}, the critical K level must be reached prior to significant plastic deformation; that is, the plastic zone size is negligible, less than 2% of the crack length and the thickness. In contrast, $K_{Ic}(J)$ for elastic-plastic or fully plastic fracture indicates an estimate K_{Ic} calculated from J_{Ic}, the J-integral value at which the first measurable extension of the crack occurs. Because plastic deformation prior to the onset of cracking does not invalidate the J_{Ic} measurement, smaller specimens can be used. Either (a) significant amount of stable crack extension under rising load is displayed before final fracture occurs, or (b) the elastic-plastic behavior culminates in plastic instability.

The fracture toughness test is generally conducted on a test specimen containing a preexisting defect; usually the defect is a sharp crack introduced by fatigue loading and is called the precrack. The test is conducted on a machine that loads the specimen at a prescribed rate. Measurements of load and a displacement value are taken during the test. The data resulting from this are subjected to an analysis procedure to evaluate the desired toughness parameters. These toughness results are then subjected to qualification procedures (or validity criteria) to see if they meet the conditions for which the toughness parameters are accepted. Values meeting these qualification conditions are labeled as acceptable standard measures of fracture toughness. The standard fracture toughness test then has the following ingredients:

- Test specimens, types, and preparation
- Loading and instrumentation requirements
- Measurements
- Data analysis
- Qualification of results

The fracture toughness value of a material is determined in the laboratory, with the most generally accepted test methods following standards published by ASTM International. Currently, ASTM International has standards E 23 for notch toughness testing, E 399 for plane-strain fracture toughness testing, and E 813, E 1737, and E 1290 for elastic-plastic fracture toughness testing. Other standard fracture test methods have been developed, such as the British Standard 7448, Part 1.

Fracture toughness is defined as resistance to the propagation of a crack. This propagation is often thought to be unstable, resulting in a complete separation of the component into two or more pieces. Actually, the fracture event can be stable or unstable. With unstable crack extension, often associated with a brittle fracture event, the fracture occurs at a well-defined point, and the fracture characterization can be given by a single value of the fracture parameter. With stable crack extension, often associated with a ductile fracture process, the fracture is an ongoing process that cannot be readily described by a point. This fracture process is characterized by a crack growth resistance curve, or *R*-curve. This is a plot of a fracture parameter versus the ductile crack extension, Δa. An example *K*-based *R*-curve is shown in Fig. 26. Sometimes, a single point is chosen on the *R*-curve to describe the entire process; however, this is mostly done for convenience and does not give a complete quantitative description of the fracture behavior.

Plane-Strain Fracture Toughness (K_{Ic}) Test (ASTM E 399)

Fracture mechanics and fracture toughness testing began with a strictly linear elastic methodology using the crack tip stress-intensity factor, *K*. Later, nonlinear parameters were developed. However, the first test methods developed used the linear elastic parameters and were based on *K*.

The first fracture toughness test that was written as a standard was the K_{Ic} test method, ASTM E 399. This test measures fracture toughness that develops under predominantly linear elastic loading with the crack tip region subjected to near plane-strain constraint conditions through the thickness. The test was developed for essentially ductile fracture conditions but can also be used for brittle fracture. As a ductile fracture test, a single point to define the fracture toughness is desired. To accomplish this, a point where the ductile crack extension equals 2% of the original crack length is identified. This criterion is illustrated schematically with a *K*-*R*

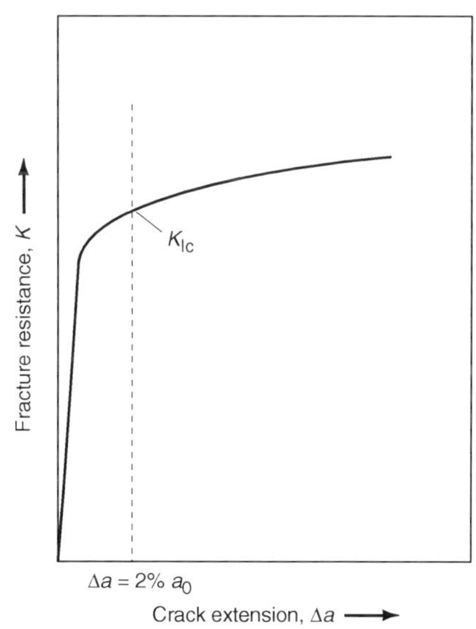

Fig. 26 Schematic of *K*-based crack resistance *R*-curve with definition of K_{Ic}. Source: Ref 12

curve in Fig. 26. This criterion gives a somewhat size-dependent measurement, and so validity criteria are chosen to minimize the size effects as well as restrict the loading to essentially linear elastic regime.

Five different specimen geometries are allowed. These are the single-edge notched bend specimen, SE(B), compact specimen, C(T), arc-shaped tension specimen, A(T), disk-shaped compact specimen, DC(T), and the arc-shaped bend specimen, A(B). Many of these specimen geometries are used in the other standards as well. They are shown schematically in Fig. 27(a) through (e). The bend and compact specimens (Fig. 27a and b, re-

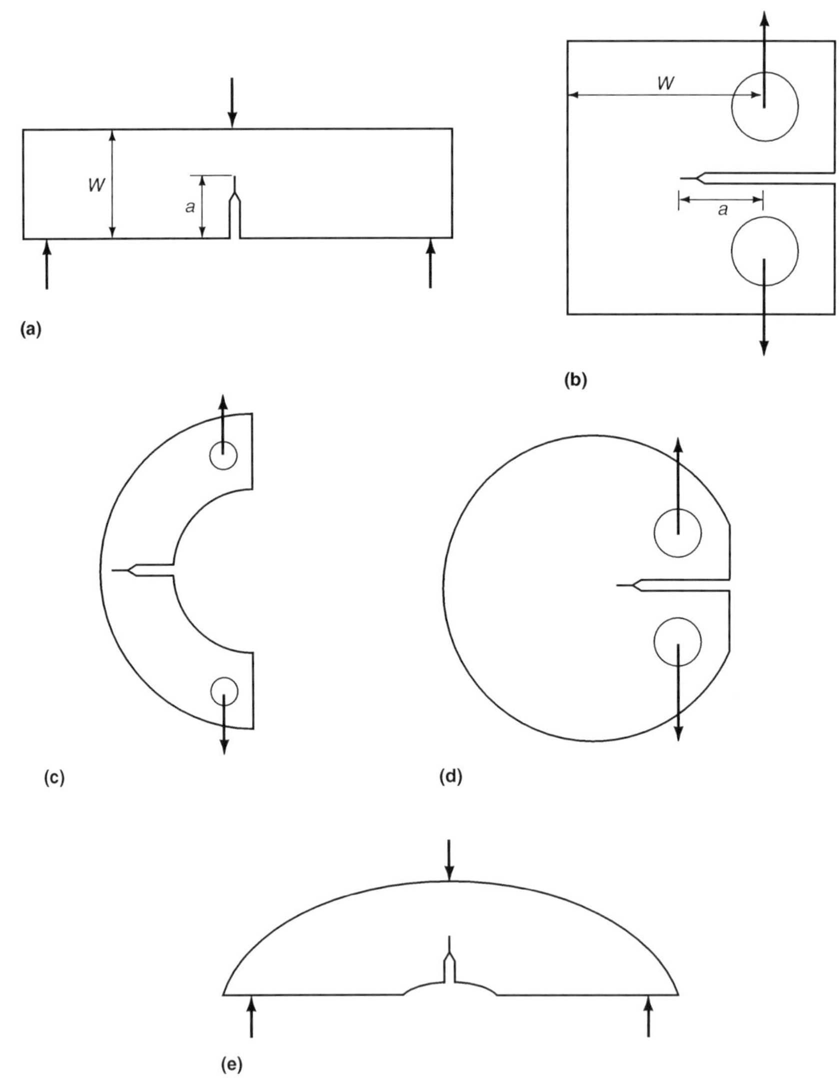

Fig. 27 Specimen types used in the K_{Ic} test (ASTM E 399). (a) Single-edge notched bend, SE(B). (b) Compact specimen, C(T). (c) Arc-shaped tension specimen, A(T). (d) Disk-shaped compact specimen, DC(T). (e) Arc-shaped bend specimen, A(B). Source: Ref 12

spectively) are traditional fracture toughness specimens used in nearly every fracture toughness test method. The other three are special geometries that represent special component structural forms. Therefore, most fracture toughness tests are conducted with either the edge notched bend or compact specimens. The choice between the bend and compact specimen is based on:

- The amount of material available (the bend takes more)
- Machining capabilities (the compact has more detail and costs more to machine)
- The loading equipment available for testing

All of the specimens for the K_{Ic} test must be precracked in fatigue before testing. The standard E 399 gives the details for precracking.

The choice of the specimen also requires a choice of the size. Because the validity criteria depend on the size of the specimen, it is important to select a sufficient specimen size before conducting the test. However, the validity criteria cannot be evaluated until the test is completed; therefore, choosing the correct size is a guess that may turn out to be wrong. There are guidelines for choosing a correct size, but there is no guarantee that the chosen size will pass the validity requirement. The test specimens must also be chosen so that the proper material is sampled. This means that the location in the material source and the orientation of the sample must be correct and accounted for. As the specimens are being prepared, requirements for tolerances on such things as locations of surfaces, size and location of the notch and pin holes, and surface finishes must be followed.

Most tests are conducted on either closed-loop servo-hydraulic machines or constant-rate crosshead drive machines. Special loading fixtures must be used for the test. Two types can be used; the choice of loading fixture depends on the test specimen chosen. The bend specimens SE(B) and A(B) use a bend fixture. The tension specimens C(T), DC(T), and A(T) require a pin-and-clevis loading. These two types of fixtures are illustrated in Fig. 28(a) and (b). Note in Fig. 28(a) that the bend loading is three-point, which is the case for all bend-loaded specimens. Also note in Fig. 28(b) that the clevis has a loading flat at the bottom of the pin hole. This allows free rotation of the specimen arms during the test and is essential for obtaining valid results.

For the K_{Ic} test, a continuous measurement of load and displacement is required during the test. The load is measured by a load cell, which should be on all loading machines. The measurement of displacement is usually done with a strain-gaged clip gage that is positioned over the mouth of the crack in the specimen, as shown for the bend specimen in Fig. 28(a). The standards give guidelines for the working requirements of the load and displacement gages used in the tests.

The loading of the specimen is done at a prescribed rate. It must be done fast enough so that any environmental or temperature interactions

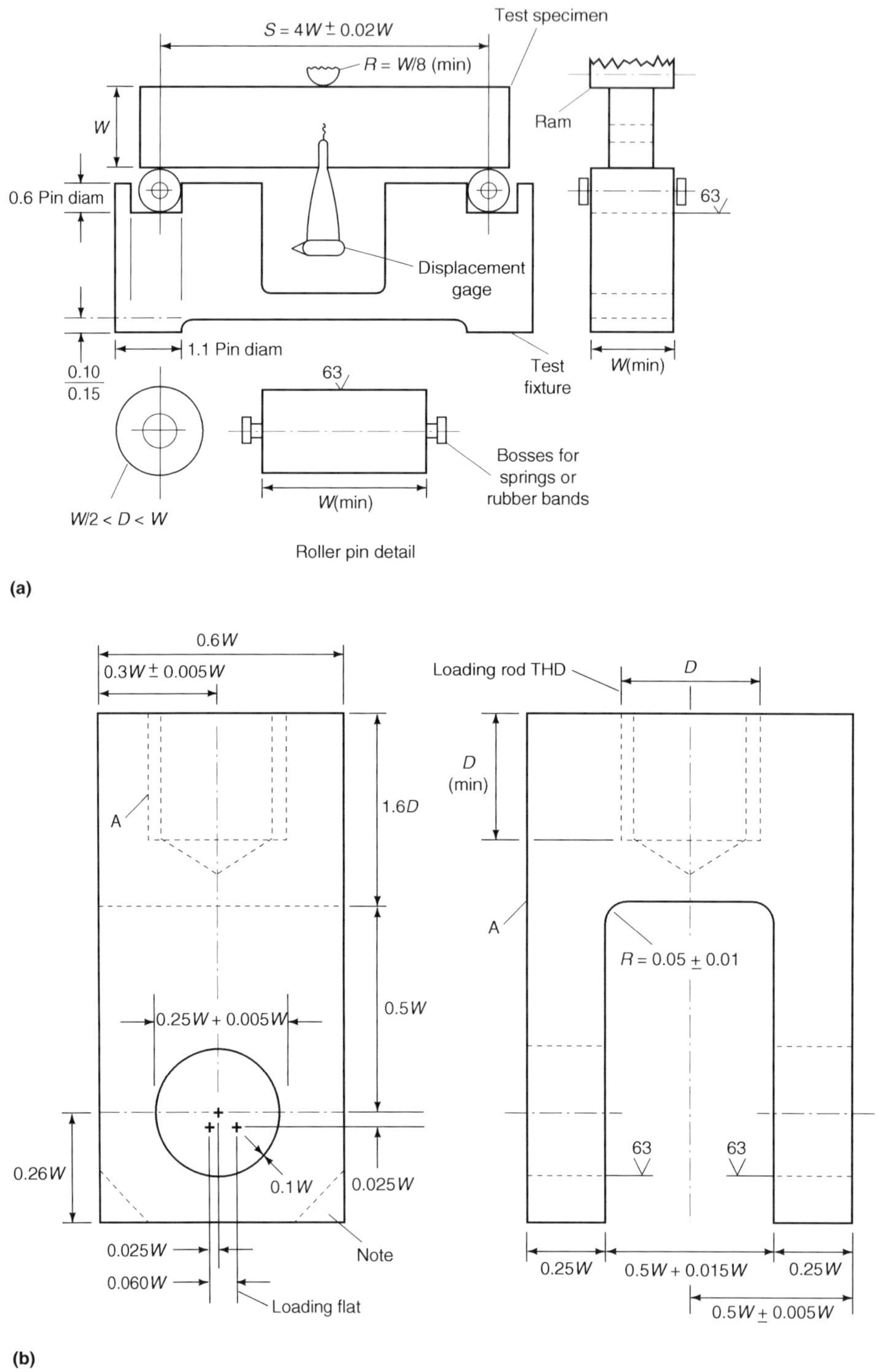

Fig. 28 Test fixtures for the K_{Ic} test specimens. (a) Fixtures for the bend test. (b) Clevises for the compact specimen. Source: Ref 12

are not a problem. On the other hand, it must be done slowly enough so that it is not considered a dynamically loaded test. For the K_{Ic} test, the load must be applied at a rate so that the increase in K is given by the range 0.55 to 2.75 MPa$\sqrt{m}$/s. The loading is done in displacement control, which usually means test machine crosshead control. During the loading, the load and displacement are measured continuously. This can be done autographically or digitally.

The load-versus-displacement record provides the basic data of the test. The data are then analyzed to determine a provisional K_{Ic} value, labeled K_Q. This provisional value is determined from a provisional load, P_Q, and the crack length. The P_Q value is determined with a secant slope on the load-versus-displacement record, as shown in Fig. 29. The P_Q value is determined by drawing the original loading slope of the load-versus-displacement record. A slope of 5% less than the original slope is then drawn from the origin. For a monotonically increasing load, the P_Q is taken where the 5% secant intersects the load-versus-displacement curve; this is illustrated as type I in Fig. 29. For other records in which an instability or other maximum load is reached before the 5% secant, the maximum load reached up to and including the possible intersection of the 5% secant is the P_Q. Type II illustrated in Fig. 29 is an example of one of the other types of load-versus-displacement records. The 5% secant corresponds to approximately 2% ductile crack extension; this may be physical crack extension or effective crack extension related to plastic zone development. Unstable failure before reaching the 5% offset also marks a measurement point for P_Q at the maximum load reached at the point of instability.

The P_Q value is used to determine the corresponding K_Q value. This is calculated from the equation:

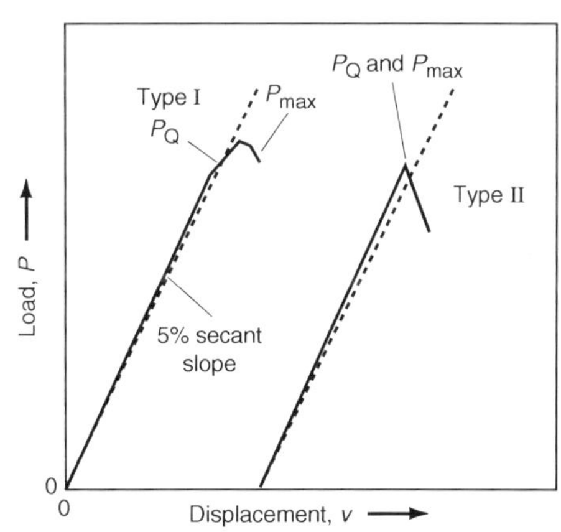

Fig. 29 Typical load-versus-displacement records for the two types of K_{Ic} testing. Source: Ref 12

$$K = Pf\left(a/W\right)/B\sqrt{W} \qquad \text{(Eq 36)}$$

where P is load, B and W are specimen thickness and width, and $f(a/W)$ is a calibration function that depends on the ratio of crack length to specimen width (a/W) and is given in the standard. For the calculation of K, a crack length value (a) is required. This comes from a physical measurement on the fracture surface of a broken specimen half. The specimen must be fractured into halves if it is not already that way after the test. The crack length is measured to the tip of the precrack using an averaging formula in the test standard. This value of crack length normalized with width (W) is used in the calibration function $f(a/W)$ to determine the K_Q value.

The K_Q is a provisional K value that may be the K_{Ic} if it passes the validity requirements. The two major validity requirements are to ensure that crack resistance does not increase significantly with crack growth and that linear elastic loading and plane-strain thickness are achieved. The first of these two requirements is quantified as:

$$\frac{P_{max}}{P_Q} \leq 1.10 \qquad \text{(Eq 37)}$$

which limits the R-curve behavior to an essentially flat trend and ensures some physical crack extension. The second requirement is:

$$a, B \geq 2.5\left(\frac{K_Q}{\sigma_{ys}}\right) \qquad \text{(Eq 38)}$$

which guarantees linear elastic loading and plane-strain thickness. P_{max} is the maximum value of load reached during the test. An example of P_{max} is shown in Fig. 29; σ_{ys} is the 0.2% offset yield strength. Values of K_Q that pass these validity requirements are labeled as valid K_{Ic} and are reported as such in a standard prescribed reporting of test results.

Correlations of K_{Id}, K_{Ic}, and Charpy V-Notch Impact Energy Absorption. It is possible to measure dynamic toughness in a quasi-static fracture test, which basically involves applying a fast loading rate to conduct an ordinary K_{Ic} test. Procedures for this rapid-load $K_{Ic}(t)$ test are given in a special annex to ASTM Standard E 399 for plane-strain fracture testing. For conventional K_{Ic}, the maximum loading rate is limited to 2.75 $MPa\sqrt{m}$/s (2.5 $ksi\sqrt{in.}$/s). Any loading rate faster than that is considered rapid-load fracture toughness. Usually, these tests are done in the range of 10 to 10^4 $MPa\sqrt{m}$/s (9 to 9×10^3 $ksi\sqrt{in.}$/s). The term K_{Id} is also used, where the subscript "d" stands for dynamic. Alternatively, the value of K_{Id} can be obtained by applying an impact load to a conventional K_{Ic} test coupon. Actually, the use of the term *dynamic fracture toughness* is quite confusing. ASTM International does not clearly define this quantity. Yet the term K_{Id} appears in the literature, which may mean $K_{Ic}(t)$ or values that were converted from data developed from notched bar impact tests. A nearly true K_{Ic} value probably comes from impacting the precracked Charpy specimen.

The Charpy V-notch impact specimen is the most widely used specimen for material development, specifications, and quality control. Moreover, because the Charpy V-notch (CVN) impact energy absorption curve for constructional steels undergoes a transition in the same temperature zone as the impact plane-strain fracture toughness (K_{Id}), a correlation among these test results has been developed for the transition region and is given by:

$$\frac{(K_{Id})^2}{E} = 5(CVN) \qquad \text{(Eq 39)}$$

where K_{Id} is in ksi$\sqrt{in.}$, E is in ksi, and CVN is in ft · lbf.

The validity of this correlation is apparent from the data presented in Fig. 30 for various grades of steel ranging in yield strength from approximately 36 to approximately 140 ksi and in Fig. 31 for eight heats of SA 533B, class 1, steel. Consequently, a given value of CVN impact energy absorption corresponds to a given K_{Id} value, which in turn corresponds to a given toughness behavior at lower rates of loading.

A relationship between K_{Ic} and upper-shelf Charpy V-notch impact energy absorption, shown in Fig. 32, was developed empirically for steels having room-temperature yield strength (σ_{ys}) higher than approximately 110 ksi and is given by:

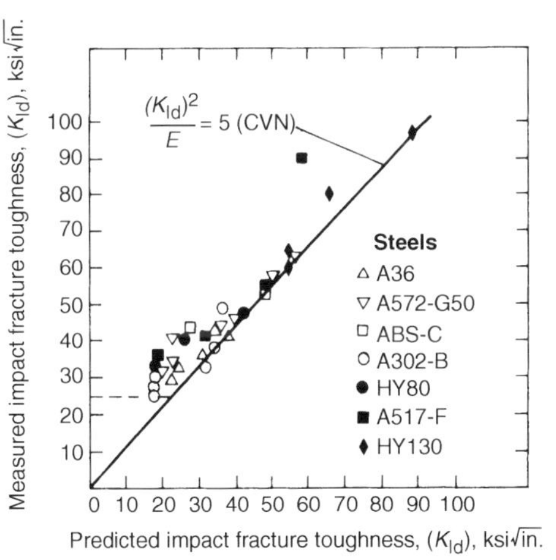

Fig. 30 Correlation of plane-strain impact fracture toughness and impact Charpy V-notch energy absorption for various grades of steel. Source: Ref 3

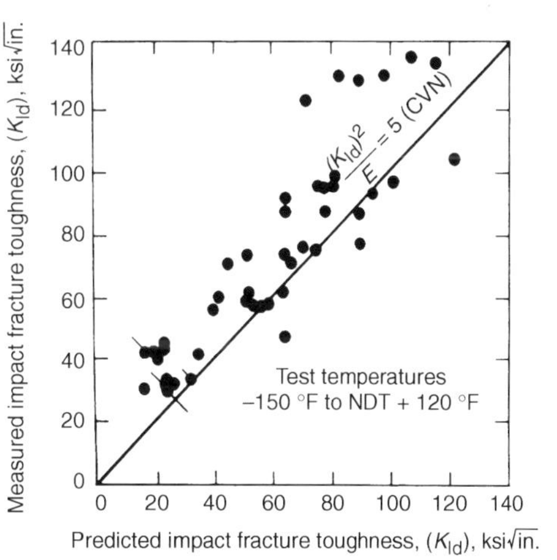

Fig. 31 Correlation of plane-strain impact fracture toughness and impact Charpy V-notch energy absorption for SA 533B, class 1, steel. NDT, nil-ductility transition. Source: Ref 3

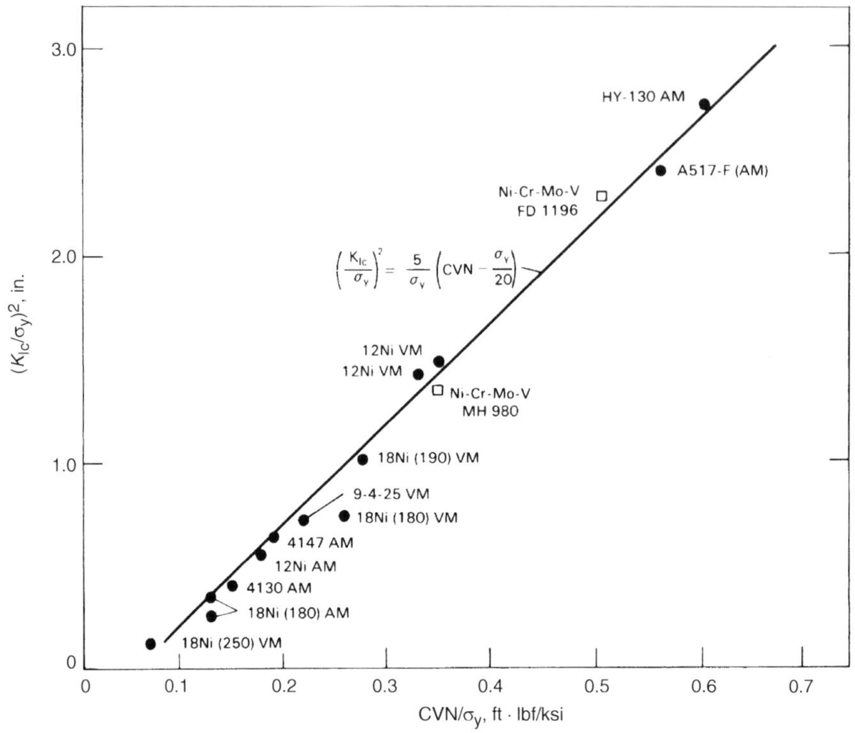

Fig. 32 Relation between plane-strain fracture toughness (K_{Ic}) and Charpy V-notch (CVN) impact energy. Tests conducted at 27 °C (80 °F). VM, vacuum melted; AM, air melted. Source: Ref 3

$$\left(\frac{K_{Ic}}{\sigma_{ys}}\right)^2 = \frac{5}{\sigma_{ys}}\left(CVN - \frac{\sigma_{ys}}{20}\right) \qquad \text{(Eq 40)}$$

where K_{Ic} is in ksi$\sqrt{\text{in.}}$, σ_{ys} is the yield strength in ksi, and CVN is energy absorption in ft · lbf for a Charpy V-notch impact specimen tested in the upper-shelf (100% shear fracture) region.

At the upper shelf, the effects of loading rate and notch acuity are not as critical as in the transition region. The effect of loading rate is to elevate the yield strength by approximately 25 ksi. Thus, Eq 40 may be used to calculate K_{Id} values by replacing σ_{ys} with the dynamic yield strength (σ_{yd}), where $\sigma_{yd} \approx \sigma_{ys} + 25$ ksi. This use of Eq 40 to calculate K_{Id} is consistent with the observation that, in the upper-shelf region, the dynamic fracture toughness of steels is higher than the static fracture toughness.

Nonlinear Fracture Toughness Testing

For many materials used in structures, it is desirable to have high toughness, a value at least high enough so that the structure would not reach K_{Ic} before significant yielding occurs. For these materials, it is necessary to use the nonlinear fracture parameters to measure fracture toughness properties. The two leading nonlinear fracture parameters are J and δ. For high-toughness materials, fracture is often by a ductile mechanism, but this is not necessarily the case for all materials.

J-Integral Fracture Toughness. The J_{Ic} measurement is shown schematically in Fig. 33. Typically, a series of deeply notched compact specimens are precracked to $a/W \approx 0.6$, and each specimen is loaded to a J-level in the region where crack extension is anticipated. The load-versus-displacement curve is recorded on an x-y recorder, with displacement being measured at the load line. Then, the specimen is unloaded and heated to tint the region of crack extension. The specimen is fractured, and the crack extension (Δa) is measured from the exposed fracture surface. J is calculated from the load-displacement record and specimen dimensions using:

$$J = \left(\frac{A}{Bb}\right) f\left(\frac{a_0}{W}\right) \quad \text{(Eq 41)}$$

where A is the area under the load-versus-displacement curve, $b = (W - a)$ is the uncracked ligament, and $f(a_0/W)$ is a function of crack length.

The results of a test series are plotted as J versus Δa. On the same graph, the line defined by Eq 42 is drawn:

$$J = 2\bar{\sigma}\Delta a \quad \text{(Eq 42)}$$

where $\bar{\sigma}$ is the flow stress (the average of the yield and ultimate strengths). The intersection of the J-Δa plot and the $J/2\bar{\sigma}$ line is defined as J_{Ic}, the value of J at the onset of crack extension. Apparent crack extension at Δa values of less than $J/2\bar{\sigma}$ is attributed to deformation at the crack tip instead

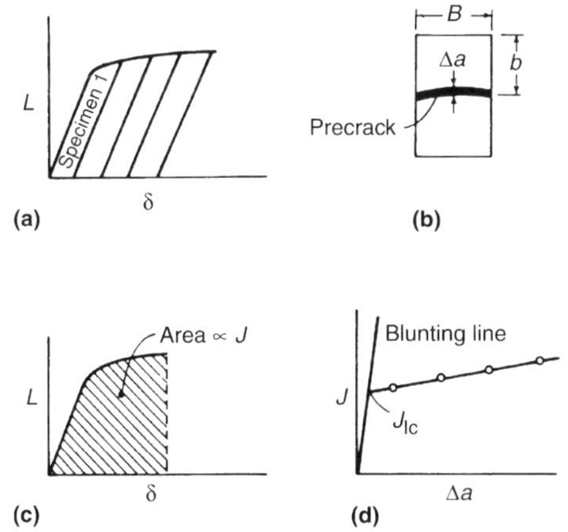

Fig. 33 J-resistance curve test method for J_{Ic} determination. (a) Test records. (b) Heat-tinted fracture surface. (c) J calculation. (d) Resistance curve.
Source: Ref 3

of material separation. The linear elastic plane-strain fracture toughness can be estimated from J_{Ic} as follows:

$$K_{Ic}(J) = \left[\frac{J_{Ic}E}{1-\nu^2}\right]^{1/2} \quad \text{(Eq 43)}$$

where E is Young's modulus, and ν is Poisson's ratio.

Crack Tip Opening Displacement (CTOD) (ASTM E 1290). The CTOD method of fracture toughness measurement was the first one that used a nonlinear fracture parameter to evaluate toughness. The first CTOD standard was written by the British Standards Institution. Subsequently, ASTM E 1290 was written as the U.S. version of this test method. The basic idea of the test method is to evaluate a fracture toughness point for brittle fracture or to evaluate a safe point for the case of ductile fracture. The primary measurements of toughness are at unstable fracture before significant ductile crack extension, labeled δ_c, unstable fracture after significant crack extension, δ_u, or the point of maximum load in the test, δ_m.

The CTOD standard uses the same bend and compact specimens that are used in the J_{Ic} test; thus, the same loading fixtures are used. For many years, the CTOD test was the only one that measured toughness for a brittle, unstable fracture event using a nonlinear fracture parameter. In addition, the method allows the measurement of toughness after a pop-in, which is described as a discontinuity in the load-versus-displacement record usually caused by a sudden, unstable advance of the crack that is subsequently arrested.

Variables Affecting Fracture Toughness

Fracture toughness is highly dependent on many variables, many of which can dramatically alter the fracture resistance characteristics of a structure, mechanical component, or test specimen. During analysis and design, consideration of yield strength, loading rate, temperature, thickness, and orientation should all be addressed. When fracture toughness data are generated, laboratory conditions for testing should model actual service conditions as practically as possible. Although not a complete list, some of the more important variables that affect fracture toughness are presented in this section.

Yield Strength. A general phenomenon with many (but not all) steels and high-strength alloys is that as the yield strength increases, the facture toughness decreases. This consideration of yield strength, in tandem with fracture toughness, is especially important in design applications where the material will be taken to the upper limits of stress. A compilation of yield strength and corresponding fracture toughness values for various aircraft structural materials is presented in Table 1.

Table 1 Yield strength and fracture toughness (K_{Ic}) of various plate materials tested at room temperature

Material identification	Process description	Yield strength MPa	Yield strength ksi	K_{Ic} MPa · m$^{1/2}$	K_{Ic} ksi · in.$^{1/2}$
Steel					
4340	260 °C temper	1495–1640	217–238	50–63	45–57
D6AC	540 °C temper	1495	217	102	93
HP 9-4-20	550 °C temper	1280–1310	186–190	132–154	120–140
HP 9-4-30	540 °C temper	1320–1420	192–206	90–115	82–105
10 Ni (vim)	510 °C temper	1770	257	54–56	49–51
18 Ni (200)	Marage	1450	210	110	100
18 Ni (250)	Marage	1785	259	88–97	80–88
18 Ni (300)	Marage	1905	277	50–64	45–58
Aluminum					
2014-T651	...	435–470	63–68	23–27	21–25
2020-T651	...	525–540	76–78	22–27	20–25
2024-T351	...	370–385	54–56	31–44	28–40
2024-T851	...	450	65	23–28	21–25
2124-T851	...	440–460	64–67	27–36	25–33
2219-T851	...	345–360	50–52	36–41	33–37
7050-T73651	...	460–510	67–74	33–41	30–37
7050-T73651	...	515–560	75–81	27–31	25–28
7050-T73651	...	400–455	58–66	31–35	28–32
7050-T73651	...	525–540	76–78	29–33	26–30
7050-T73651	...	560	81	26–30	24–27
Titanium					
T1-6Al-4V	Mill annealed	875	127	123	112
T1-6Al-4V	Recrystallized annealed	815–835	118–121	85–107	77–90

Note: These values are dependent on other factors, such as the actual plate thickness that specimens are cut from. Source: Ref 13

The general relationship between yield strength and toughness may not be the case for structural steels used in various other applications. Modern steelmaking techniques can routinely increase the yield strength of the material, with little or no sacrifice in toughness. In some cases, an increase in toughness may be achieved in tandem with an increase in yield strength. The important point is that the user of a material should be careful not to enhance one material quality (such as yield strength) at the expense of another material quality (such as toughness).

Loading Rate. There is a tendency for many materials, such as steels, to exhibit different values of fracture toughness at different loading rates. In terms of stress-intensification rates ($\dot{K}$), impact tests are done at relatively fast rates in the range of 10 to 10^4 MPa$\sqrt{m}$/s (9.1 to 9.1 × 10^3 ksi$\sqrt{in.}$/s), as opposed to the quasi-static fracture test rates of less than 2.75 MPa$\sqrt{m}$/s (2.5 ksi$\sqrt{in.}$/s). In general, impact test fracture toughness results tend to be lower (more conservative) than fracture toughness results determined with a test done according to ASTM E 399 or E 813.

It should be noted that impact loading and rapid loading rate are not the same thing. In fact, a rapid loading rate can have strain rates an order of magnitude less than those of an impact test. Some examples of rapid loading are dynamic bridge loading and ship slamming, although they are often thought to be impact loading situations. This misunderstanding arose from early observations of steam locomotives crossing bridges and feeling the "hammering" of the counterweights on the driving wheels.

Temperature. As shown in Fig. 34, fracture toughness may increase or decrease as temperature is lowered, depending on metallurgical factors. Because fracture toughness is highly dependent on temperature, an effort should be made early in the design process to establish characteristic temperatures that model the service applications of the structure or mechanical component under consideration. Many design applications, such as aircraft, pressure vessels, and turbines, run at varying temperatures, and it may be necessary to generate a spectrum of fracture toughness information regarding temperature.

The greatest challenge in assessing fracture toughness at a range of temperatures is determining what test method is appropriate. Modern structural steels used in land and marine vehicles and structures, such as high-strength, low-alloy steel, quenched and tempered steel, and carbon steel, exhibit significant plasticity to invalidate plane-strain toughness testing at any temperature. For other materials, such as the higher-strength materials used in aerospace applications, plane-strain fracture toughness testing may be appropriate at lower to moderate temperatures but not for higher service temperatures. Methods such as J-integral, CTOD, and even CVN testing are available for higher-temperature testing, but the test method should be carefully chosen.

Material Thickness (Plane Stress and Plane Strain). Material thickness is an especially important consideration in the design for fracture, because thicker sections can lead to a decrease in the fracture toughness of

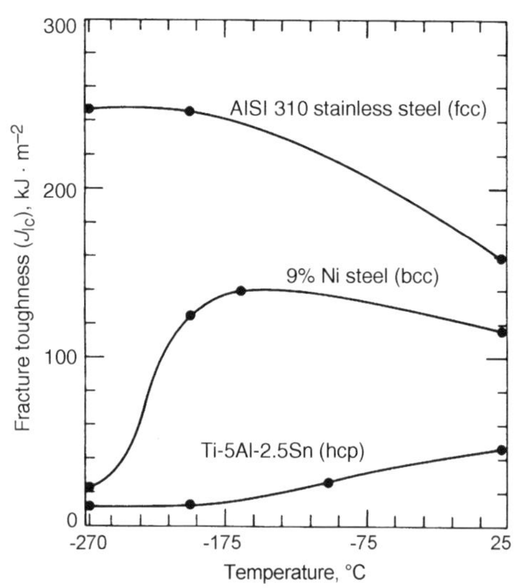

Fig. 34 Temperature dependence of fracture toughness for alloys, illustrating characteristic behavior for three different crystal structures. fcc, face-centered cubic; bcc, body-centered cubic; hcp, hexagonal close-packed. Source: Ref 3

the structure or component part under consideration. What this means is that as the thickness of a plate (or section) is increased, the fracture toughness value can decrease, at least to a limiting value. This is important for the designer to note, because a design "fix" to a highly stressed component may be to increase the cross-sectional area of the part, thus reducing the nominal stress. However, the fracture toughness of the component may be decreased by the change in geometry, and it should be reevaluated in terms of adequacy for design.

Material thickness is linked with plane-stress and plane-strain behavior in fracture mechanics due to the dependency of the stress state at the crack tip on the thickness. When a component is thin, there is essentially no lateral constraint in the direction of the thickness, and plane-stress conditions will dominate. Conversely, when a component is very thick, there is a great deal of lateral constraint, and plane-strain conditions can prevail. In general, increasing the specimen thickness will decrease the fracture toughness value to a limiting (critical) value.

Material Orientation and Anisotropy. When a material has mechanical properties that are independent of direction, it is said to be isotropic. Conversely, if the material has mechanical properties that depend on a particular or preferred orientation(s), it is said to be anisotropic.

Processing of engineering alloys usually involves mechanical working at intermediate and high temperatures. Hot working can produce highly directional grain structures (e.g., aluminum alloys) and significant degrees of texturing (e.g., titanium alloys). When the crack plane is parallel to the rolling direction, segregated impurities and intermetallics that lie in these planes represent easy fracture paths, and the toughness is low. However, when the crack plane is perpendicular to these weak planes, crack tip blunting and stress reduction take place, thus effectively toughening the material.

To provide a common scheme for describing material anisotropy, ASTM International has standardized the following six orientations: L-S, L-T, S-L, S-T, T-L, and T-S. As shown in Fig. 35, the first letter denotes the direction of the applied load; the second letter denotes the direction of crack growth. In designing for fracture toughness, consideration of anisotropy is very important, because different orientations can result in widely differing fracture toughness values. Table 2 shows how fracture toughness can vary among materials, as well as orientation.

ACKNOWLEDGMENTS

The material for this chapter came from "Fracture Toughness and Fracture Mechanics" and "Fracture Toughness Testing" by J.D. Landes, both in *Mechanical Testing and Evaluation*, Vol 8, *ASM Handbook*, ASM International, 2000; "Design for Fracture Toughness" by M.P. Blinn and R.A. Williams in *Materials Selection and Design*, Vol 20, *ASM Handbook*,

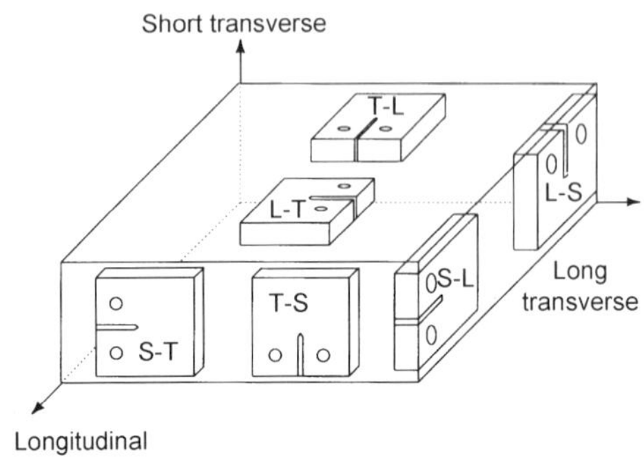

Fig. 35 Specimen orientation scheme showing the longitudinal, long transverse, and short transverse directions. There are six possible specimen designations: L-S, L-T, S-L, S-T, T-L, and T-S. The first letter denotes the direction of the applied load; the second letter denotes the direction of crack growth. Source: Ref 14

Table 2 Fracture toughness of some high-strength alloys

Alloy	Yield strength MPa	Yield strength ksi	Orientation(a)	Temperature °C	Temperature °F	Plane-strain fracture toughness(b) MPa√m	Plane-strain fracture toughness(b) ksi√in.
2042-T351	385	56	L-T	29	84	31	28
2024-T351	292	42	S-L	32	90	21	19
7075-T651	530	77	L-T	28	82	32	29
7075-T651	446	64.5	S-L	29	84	21	19
4140	1379	200	L-T	24	75	65	59
4140	1586	230	L-T	24	75	55	50
4340	1455	211	L-T	21	70	83	75.5
D6AC	1496	217	L-T	21	70	102	93
HP9-4-20	1282	186	L-T	26	79	151	137
HP9-4-20	1310	190	T-L	26	79	138	125.5
250 Maraging	1607	233	L-T	24	75	86	78
250 Maraging	1600	232	T-L	24	75	86	78
Ti-6Al-4V	889	129	L-T	24	75	64	58
Ti-6Al-4V	910	132	T-L	24	75	68	62
Ti-6Al-4V	883	128	S-L	24	75	75	68
Inconel 718	1041	151	T-L	24	75	87	79
Inconel 718	986	143	S-L	24	75	73	66

(a)The first letter gives the direction normal to the crack plane, while the second letter gives the direction of crack propagation. In the L-T orientation, the crack is normal to the rolling direction (L) and propagates in the transverse direction (T). (b) All numbers are rounded off so as not to imply a greater precision than can be justified by the experimental procedure. Source: Ref 15

ASM International, 1997; "An Introduction to Fracture Mechanics" by S.D. Antolovich and B.F. Antolovich and "Fracture Resistance of Structural Alloys" by K.S. Ravichandran and A.K. Vasudevan, both in *Fatigue and Fracture*, Vol 19, *ASM Handbook*, ASM International, 1996; and "Fracture" in *Elements of Metallurgy and Engineering Alloys* by F.C. Campbell, ASM International, 2008.

REFERENCES

1. W.D. Callister, *Fundamentals of Materials Science and Engineering,* 5th ed., John Wiley & Sons, Inc., 2001
2. F.C. Campbell, Fracture, *Elements of Metallurgy and Engineering Alloys,* ASM International, 2008

3. Fracture Toughness and Fracture Mechanics, *Mechanical Testing and Evaluation,* Vol 8, *ASM Handbook,* ASM International, 2000
4. R.D. Venter and D.W. Hoeppner, "Crack and Fracture Behavior in Tough Ductile Materials," report submitted to the Atomic Energy Control Board of Canada, Oct 1985
5. A.F. Liu, *Mechanics and Mechanisms of Fracture: An Introduction,* ASM International, 2005
6. V. Kumar, M.D. German, and C.F. Shih, "An Engineering Approach for Elastic-Plastic Fracture Analysis," Report NP-1931, Electric Power Research Institute, July 1981
7. M. Myers and K. Chawla, *Mechanical Behavior of Materials,* 2nd ed., Cambridge University Press, 2009
8. H.W. Hayden, W.G. Moffatt, and J. Wulff, *The Structure and Properties of Materials,* Vol III, John Wiley & Sons, 1965
9. D.J. Wulpi, Brittle Fracture, *Understanding How Components Fail,* 2nd ed., ASM International, 1999
10. G.E. Dieter, *Mechanical Metallurgy,* 3rd ed., McGraw-Hill Book Co., 1986
11. W.S. Pellini, "Criteria for Fracture Control Plans," NRL 7406, U.S. Naval Research Laboratory, Washington, May 1970
12. J.D. Landes, Fracture Toughness Testing, *Fatigue and Fracture,* Vol 19, *ASM Handbook,* ASM International, 1996
13. H.O. Fuchs and R.I. Stephens, *Metal Fatigue in Engineering,* John Wiley & Sons, Inc., 1980
14. "Test Method for Plane-Strain Fracture Toughness of Metallic Materials," E 399-90, *Annual Book of ASTM Standards,* ASTM
15. *Damage Tolerant Design Handbook,* MCIC-HB-01, Metals and Ceramics Information Center, Battelle Columbus Laboratories, including 1st and 2nd supplements, Jan 1975

SELECTED REFERENCES

- T.L. Anderson, *Fracture Mechanics: Fundamentals and Applications,* 3rd ed., Taylor & Francis, 2005
- D. Broek, *Elementary Engineering Fracture Mechanics,* 4th ed., Kluwer Academic Publishers, 1986
- D. Broek, *The Practical Use of Fracture Mechanics,* Kluwer Academic Publishers, 1988
- T.H. Courtney, *Mechanical Behavior of Materials,* 2nd ed., McGraw-Hill Book Co., 2000
- R.W. Hertzberg, *Deformation and Fracture Mechanics of Engineering Materials,* 3rd ed., John Wiley & Sons, Inc., 1983
- W.F. Hosford, *Mechanical Behavior of Materials,* Cambridge University Press, 2005
- M. Myers and K. Chawla, *Mechanical Behavior of Materials,* 2nd ed., Cambridge University Press, 2009

CHAPTER 5

Fatigue of Metals

FATIGUE FAILURES occur due to the application of fluctuating stresses that are much lower than the stress required to cause failure during a single application of stress. It has been estimated that fatigue contributes to approximately 90% of all service failures due to mechanical causes. Fatigue is a problem that can affect any part or component that moves. Automobiles on roads, aircraft wings and fuselages, ships at sea, nuclear reactors, jet engines, and land-based turbines are all subject to fatigue failures. Fatigue was initially recognized as a problem in the early 1800s when investigators in Europe observed that bridge and railroad components were cracking when subjected to repeated loading. As the century progressed and the use of metals expanded with the increasing use of machines, more and more failures of components subjected to repeated loads were recorded. Today, structural fatigue has assumed an even greater importance as a result of the ever-increasing use of high-strength materials and the desire for higher performance from these materials.

Stress Cycles

There are three basic factors necessary to cause fatigue: a maximum tensile stress of sufficiently high value, a large enough variation or fluctuation in the applied stress, and a sufficiently large number of cycles of the applied stress. There are many types of fluctuating stresses. Several of the more common types encountered are shown in Fig. 1. A completely reversed stress cycle is commonly used in testing where the maximum and minimum stresses are equal. This is the type of stress produced by the R.R. Moore rotating beam fatigue machine shown in Fig. 2, which is similar to what a shaft may encounter during service. Because this was the original type of machine used to generate fatigue data, quite a bit of the data in the literature is for fully reversed bending with no mean stress applied on top of it. Another common stress cycle is the repeated stress cycle

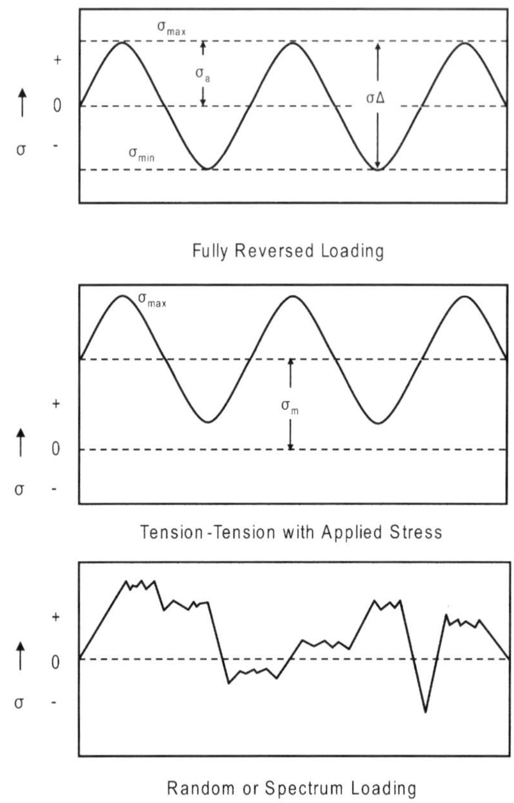

Cyclic Stress Range: $\sigma\Delta = \sigma_{max} - \sigma_{min}$

Cyclic Stress Amplitude: $\sigma_a = \dfrac{\sigma_{max} - \sigma_{min}}{2}$

Mean Stress: $\sigma_m = \dfrac{\sigma_{max} + \sigma_{min}}{2}$

Stress Ratio: $R = \dfrac{\sigma_{min}}{\sigma_{max}}$

Fig. 1 Typical loading cycles. Source: Ref 1

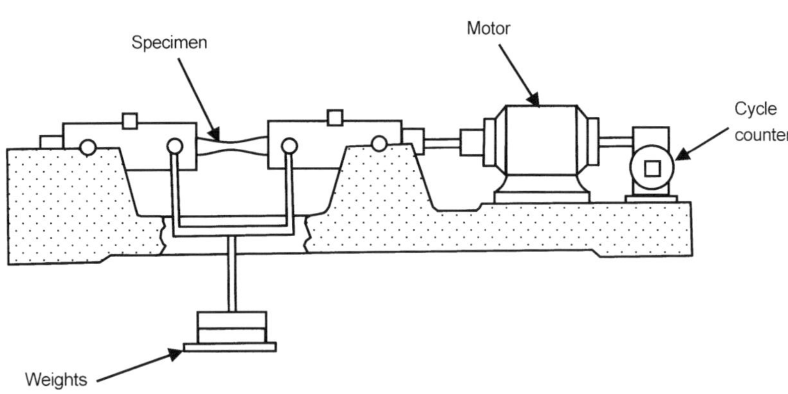

Fig. 2 Schematic of R.R. Moore reversed bending fatigue machine. Source: Ref 2

in which there is a mean stress (σ_m) applied on top of the maximum and minimum stresses. The maximum and minimum stresses do not necessarily have to be equal in value. The last type shown is the random or irregular stress cycle in which the part is subjected to random loads during service. Although a majority of the fatigue data in the literature is for fully reversed bending, there are also axial test machines (Fig. 3) that are capable of tension and compression loading in both the high- and low-cycle fatigue ranges. These modern test frames are closed-loop servo-hydraulically controlled and can be programmed with almost any desired fatigue spectrum.

A fluctuating stress is made up of two components: a mean or steady stress (σ_m) and an alternating or variable stress (σ_a). The stress range (σ_r) is the difference between the maximum and minimum stress in a cycle:

$$\sigma_r = \sigma_{max} - \sigma_{min} \quad \text{(Eq 1)}$$

The alternating stress is one-half the stress range:

$$\sigma_a = \frac{\sigma_r}{2} = \frac{\sigma_{max} - \sigma_{min}}{2} \quad \text{(Eq 2)}$$

The mean stress is the algebraic sum of the maximum and minimum stress in the cycle:

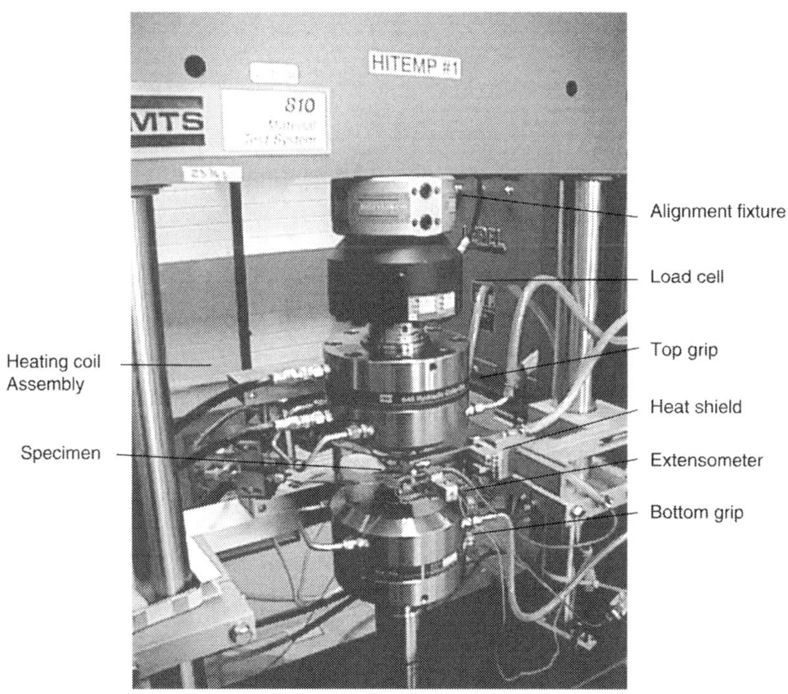

Fig. 3 Modern fatigue test frame. Source: Ref 1

$$\sigma_m = \frac{\sigma_{max} + \sigma_{min}}{2} \quad \text{(Eq 3)}$$

Two ratios frequently used in presenting fatigue data are:

$$\text{Stress ratio:} \quad R = \frac{\sigma_{min}}{\sigma_{max}} \quad \text{(Eq 4)}$$

$$\text{Amplitude ratio:} \quad A = \frac{\sigma_a}{\sigma_m} = \frac{1-R}{1+R} \quad \text{(Eq 5)}$$

High-Cycle Fatigue

High-cycle fatigue involves a large number of cycles ($N > 10^5$ cycles) and an elastically applied stress. High-cycle fatigue tests are usually carried out for 10^7 cycles and sometimes 5×10^8 cycles for nonferrous metals. Although the applied stress is low enough to be elastic, plastic deformation can take place at the crack tip. High-cycle fatigue data are usually presented as a plot of stress (S) versus the number of cycles to failure (N). A log scale is normally used for the number of cycles. The value of the stress (S) can be the maximum stress (σ_{max}), the minimum stress (σ_{min}), or the stress amplitude (σ_a). The S-N relationship is usually determined for a specified value of the mean stress (σ_m) or one of the two ratios R or A.

The fatigue life is the number of cycles to failure at a specified stress level, while the fatigue strength is the stress at which failure does not occur at a predetermined number of cycles. As the applied stress level is decreased, the number of cycles to failure increases. Normally, the fatigue strength increases as the static tensile strength increases. For example, high-strength steels heat treated to over 1380 MPa (200 ksi) yield strengths have much higher fatigue strengths than aluminum alloys with 480 MPa (70 ksi) yield strengths. A comparison of the S-N curves for steel and aluminum are shown in Fig. 4. Note that steel not only has a higher fatigue strength than aluminum, it also has an endurance limit. In other words, below a certain stress level, the steel alloy will never fail. On the other hand, aluminum does not have a true endurance limit. It will always fail if tested to a sufficient number of cycles. Therefore, the fatigue strength of aluminum is usually reported as the stress level it can survive at a very large total number of cycles, usually 5×10^8 cycles. It should be noted that there is a considerable amount of scatter in fatigue test results. It is therefore important to test a sufficient number of specimens to obtain statistically meaningful results.

A family of S-N curves for a material tested at various stress ratios is shown schematically in Fig. 5. If the stresses are fully reversed, the stress ratio (R) is –1. If the stresses are partially reversed, R is a negative number less than 1. If the stress is cycled between a maximum stress and no load,

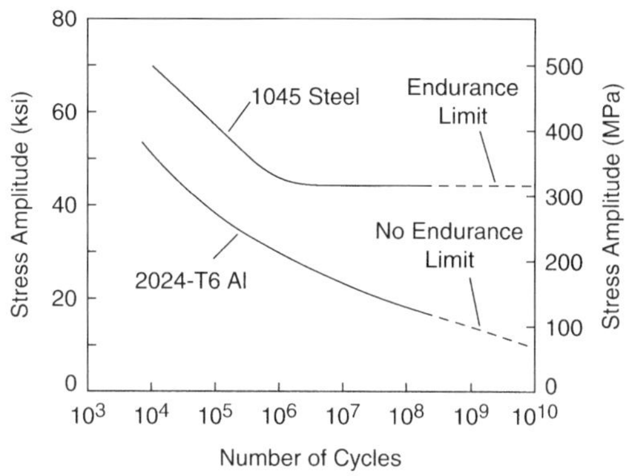

Fig. 4 Comparison of steel and aluminum fatigue behavior. Source: Ref 3

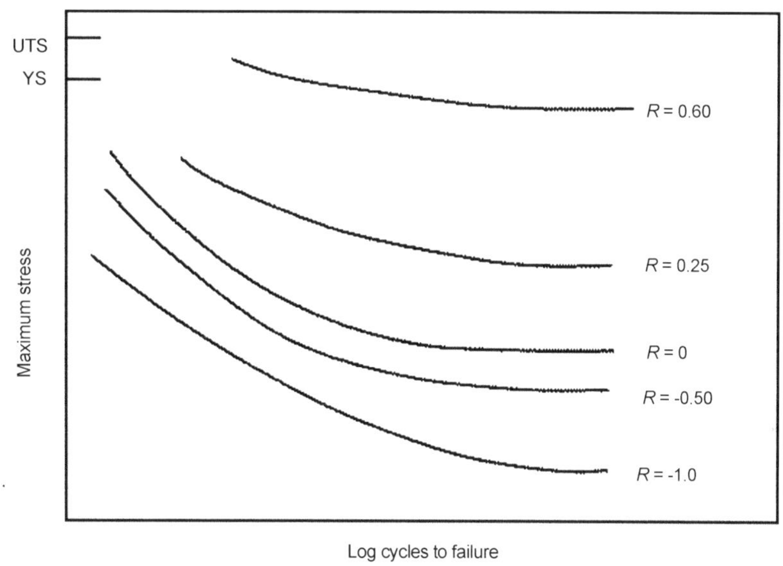

Fig. 5 Influence of stress ratio *R* on fatigue strength. UTS, ultimate tensile strength; YS, yield strength. Source: Adapted from Ref 4

stress ratio *R* is zero. If the stress is cycled between two tensile stresses, stress ratio *R* is a positive number less than 1. A stress ratio *R* of 1 indicates no variation in stress, and the test would become a sustained-load creep test rather than a fatigue test. For carbon and low-alloy steels, *S-N* curves typically have a fairly straight slanting portion at low cycles, changing into a straight horizontal line at higher cycles, with a sharp transition between the two.

For design purposes, data must often be presented by methods other than *S-N* curves. Certain values of *N* may require determination of safe combinations of stress amplitude and mean stress, or maximum stress and mean stress, during the design development of a structural member. This is accomplished by constructing a graph that shows the relationship between the stress components in question at a constant number of cycles. Such graphs are called constant-life diagrams. A constant-life diagram is constructed using actual test data originally presented as a family of *S-N* curves, as shown in the conversion procedure illustrated in Fig. 6. Figure 7 shows constant fatigue-life diagrams from 10^4 to 10^7 cycles for smooth specimens (solid lines) and for notched specimens (dashed lines) for an age-hardened aluminum alloy. Both maximum/minimum stress coordinates and alternating/mean stress coordinates are used for plotting. In addition, lines of constant *A* and *R* ratios are also entered. The lines for the

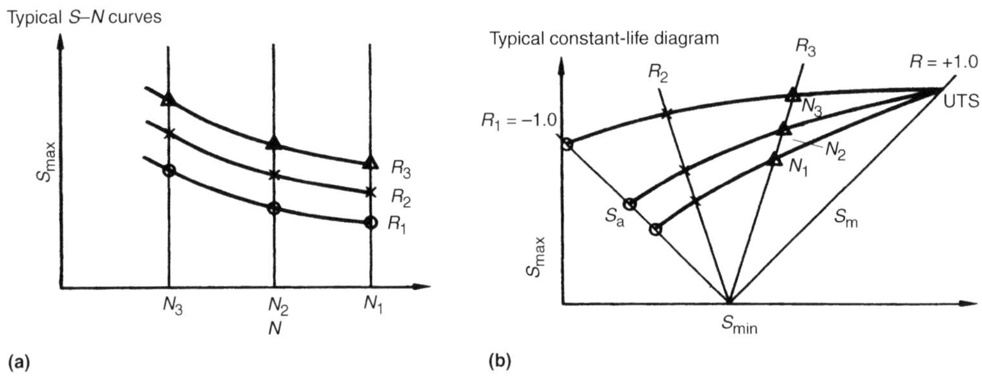

Fig. 6 Constructing a constant-life diagram from a set of *S-N* curves containing three *R* ratios, where $R = -1$ is the fully reversed cyclic loading condition and $R = +1$ is the monotonic ultimate strength of the material (no cyclic load). (a) Typical *S-N* curves. (b) Typical constant-life diagram. Source: Ref 5

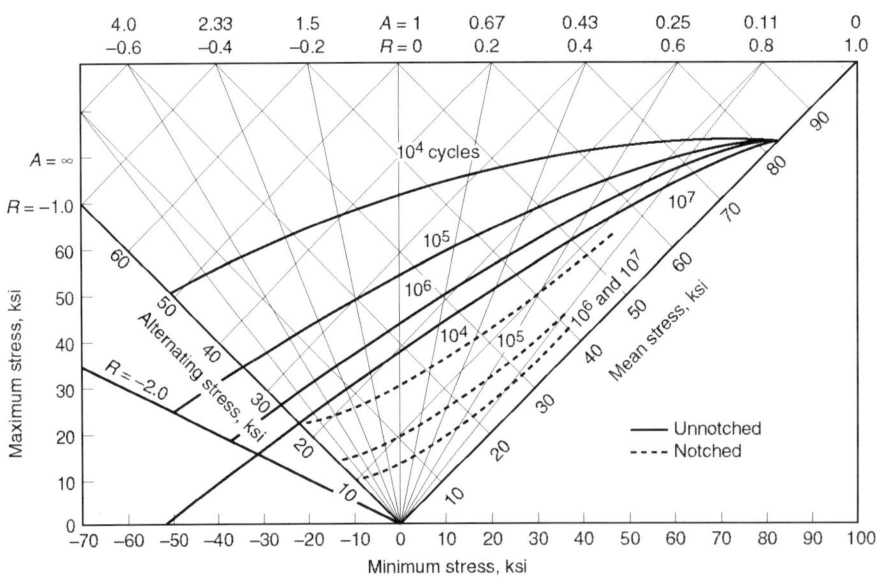

Fig. 7 Constant-life fatigue diagrams for unnotched (solid lines) and notched (dashed lines) 7075-T6 specimens. Source: Ref 6

smooth specimens end at a mean stress of 565 MPa (82 ksi), which corresponds to the ultimate strength of the aluminum alloy. Although the diagram is very busy, it shows, in addition to the fatigue information, how the same data can be plotted differently.

For a large number of steels, there is a direct correlation between tensile strength and fatigue strength; that is, higher-tensile-strength steels have higher endurance limits. The endurance limit is normally in the range of 0.35 to 0.60 of the tensile strength. This relationship holds up to a hardness of approximately 40 HRC (~1240 MPa, or 180 ksi, tensile strength), and then the scatter becomes too great to be reliable (Fig. 8). This does not necessarily mean it is wise to use as high a strength steel as possible to maximize fatigue life because, as the tensile strength increases, the fracture toughness decreases and environmental sensitivity increases. In addition, the endurance limit of high-strength steels is extremely sensitive to surface condition, residual stress state, and the presence of inclusions that act as stress concentrations.

Fatigue cracking can occur quite early in the service life of the member by the formation of a small crack, generally at some point on the external surface. The crack then propagates slowly through the material in a direction roughly perpendicular to the main tensile axis, as illustrated in Fig. 9.

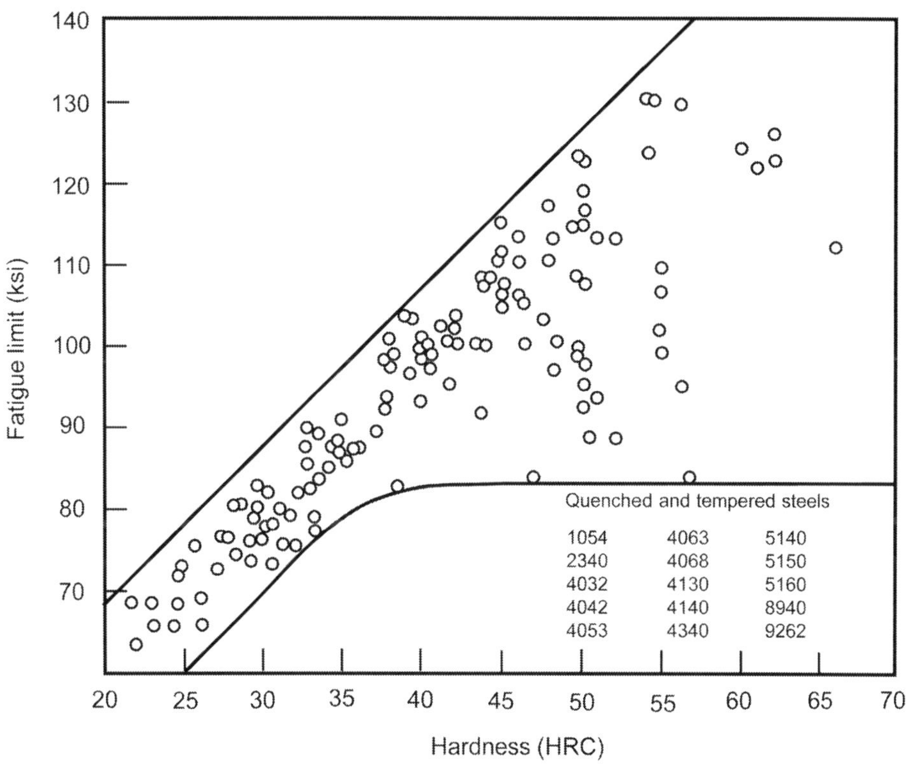

Fig. 8 Endurance limit versus hardness for steels. Source: Ref 7

Ultimately, the cross-sectional area of the member is reduced to the point that it can no longer carry the load, and the member fails in tension. The fracture surface of a fatigued high-strength part is shown in Fig. 10. The portion of the fracture surface due to fatigue crack growth and the portion finally cracked due to overload are clearly evident.

As previously mentioned, much of the fatigue data in the literature has been determined for completely reversed bending with $\sigma_m = 0$. However, the effects of mean stress are important, and an increase in mean stress will always cause a reduction in the fatigue life (Fig. 11). A number of relationships have been developed that allow the effects of a mean stress to be predicted from fully reversed bending data. Goodman developed a

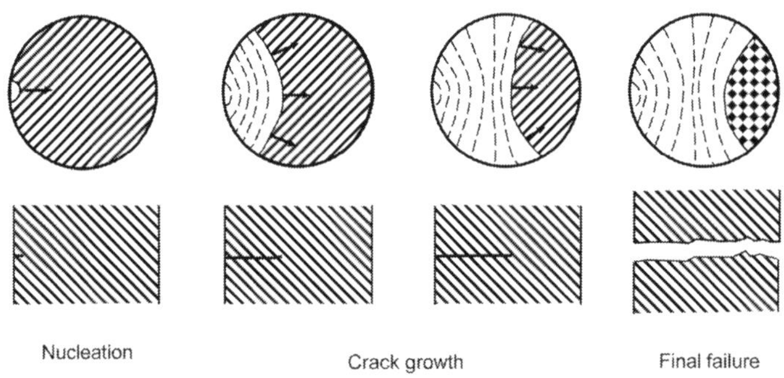

Fig. 9 Typical propagation of fatigue crack. Source: Ref 8

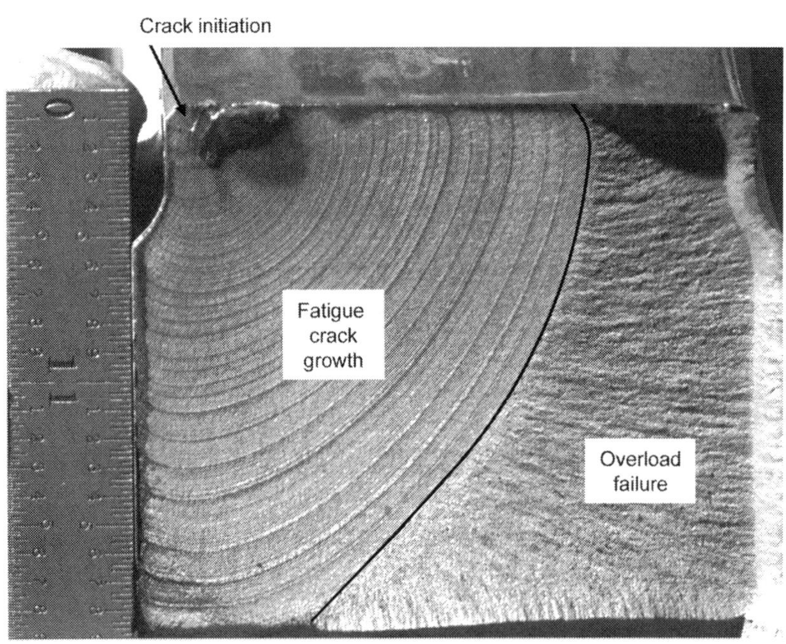

Fig. 10 Fatigue crack growth in high-strength steel part. Source: Ref 2

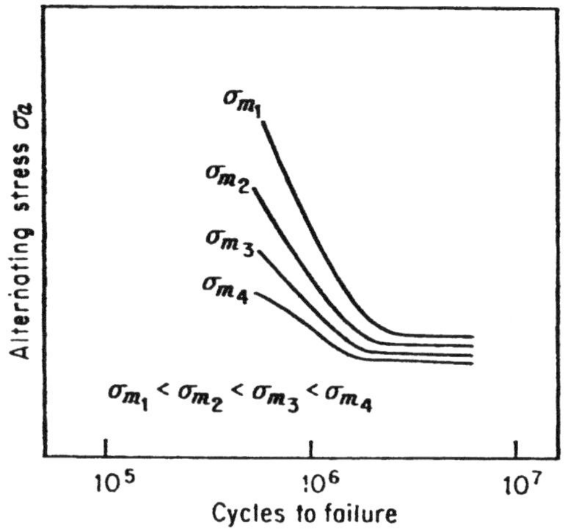

Fig. 11 Effect of mean stress on fatigue life. Source: Ref 5

straight-line relationship, while Gerber used a parabolic relationship (Fig. 12). Test data for ductile metals usually fall closer to the Gerber parabolic curve; however, because of the scatter in fatigue data and the fact that notched data fall closer to the Goodman line, the more conservative Goodman relationship is often used in practice. If the design is based on yield rather ultimate strength, as most are, then the even more conservative Soderberg relationship can be used. Mathematically, the three relationships can be expressed by:

$$\sigma_a = \sigma_e \left[1 - \left(\frac{\sigma_m}{\sigma_u} \right)^x \right] \quad \text{(Eq 6)}$$

where

$x = 1$ for the Goodman line
$x = 2$ for the Gerber curve
σ_e = fatigue limit for completely reversed bending

If the design is based on the yield strength, then the yield strength (σ_y) can be used instead of the ultimate strength (σ_u).

Data Scatter. During *S-N* curve development, if more than one specimen is tested at the same stress level, so that the loading and testing conditions are duplicated within experimental accuracy, a different number of cycles to failure will generally result. It is helpful to plot a histogram of number of failed specimens at a certain cycle interval, as shown in Fig. 13. Normalization of the histogram is accomplished by dividing the ordinate by the total number of specimens, 57 in this case. Then the ordinate gives the percentage of specimens that fail at a given number of cycles to failure. This method of testing and plotting reveals that fatigue contains considerable scatter. The scatter has been found in carefully performed tests

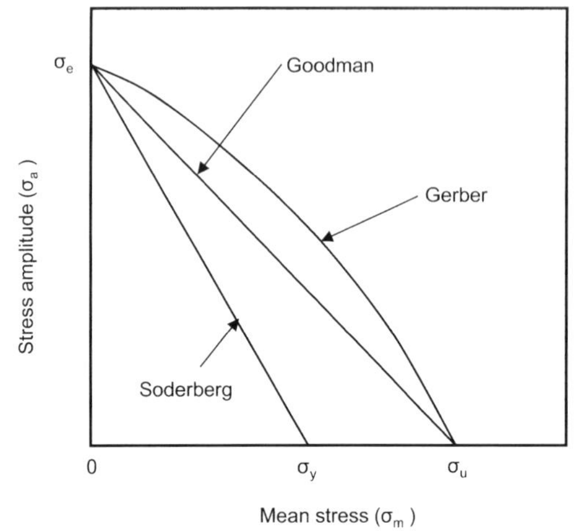

Goodman $\sigma_a = \sigma_e\left[1-\left(\frac{\sigma_m}{\sigma_u}\right)\right]$

Soderberg $\sigma_a = \sigma_e\left[1-\left(\frac{\sigma_m}{\sigma_y}\right)\right]$

Gerber $\sigma_a = \sigma_e\left[1-\left(\frac{\sigma_m}{\sigma_u}\right)^2\right]$

where
σ_e = Fatigue strength for *N* cycles under zero mean stress
σ_a = Fatigue strength for *N* cycles under mean stress of σ_m
σ_u = Ultimate tensile strength
σ_y = Yield strength

Fig. 12 Comparison of Goodman, Gerber, and Soderberg relationships. Source: Ref 2

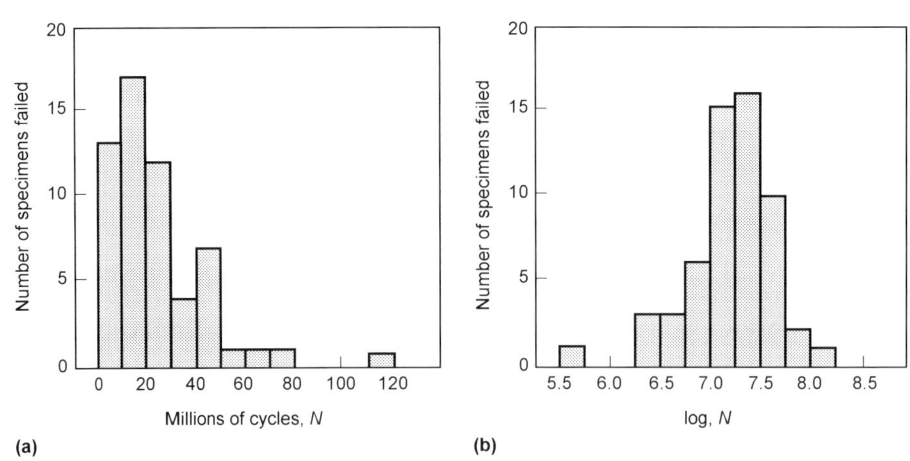

Fig. 13 Histograms showing fatigue-life distribution for 57 specimens of a 75S-T6 aluminum alloy tested at 207 MPa (30 ksi). Note the influence of a (a) linear or (b) logarithmic plot of cycles to failure *N* on the shape of the histogram. Source: Ref 9

and has been established as a basic property of the fatigue strength. Note also the difference in the histograms when plotted on a linear and a log scale.

To appreciate the effects of scatter, it is advantageous to keep the relation shown in the schematic of Fig. 14 in mind. In this figure, a normalized histogram, as shown in Fig. 13, is superposed on the fatigue curve at three different stress levels. The ordinates of the bell-shaped curves give the percentage of specimens expected to fail at a given number of cycles. The peaks of the distributions can be connected by a fatigue curve. This curve would indicate the number of cycles to failure at which the largest percentage of specimens are expected to fail. If only a few tests are run, the results are expected to be close to this curve. The width of the band in Fig. 14 indicates that all specimens are expected to fail in this interval. The distributions become broader as the stress levels decrease, indicating an increase in scatter.

It is possible to treat fatigue as a probabilistic process and to use methods of probability theory for design. A probabilistic design of components is often used if a large number of components are involved and if it is part of a reliability analysis. In the majority of cases, a probabilistic fatigue design is not performed because of the expense and time involved in generating the data. However, it is always important to keep the statistical nature of fatigue strength in mind.

Low-Cycle Fatigue

During cyclic loading within the elastic regime, stress and strain are directly related through the elastic modulus. However, for cyclic loading that produces plastic strains, the responses are more complex and form a hysteresis loop, as shown in Fig. 15. From point *O* up to point *A*, there is tension. Unloading from *A* and entering into compression, we arrive at

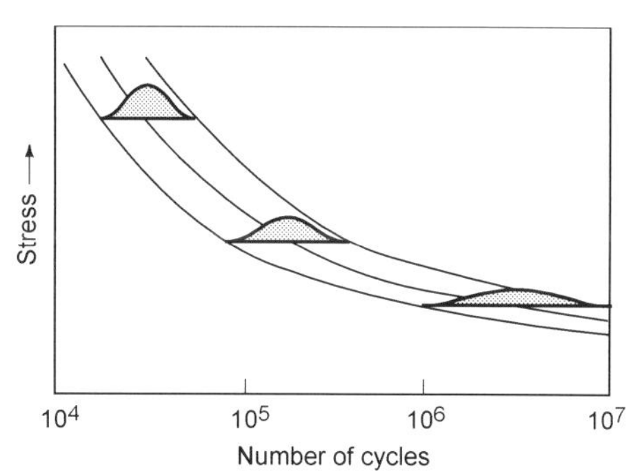

Fig. 14 Schematic showing the fatigue curve with the distribution of lives at three different stress levels. Note that the distribution widens as the stress level decreases. Source: Ref 10

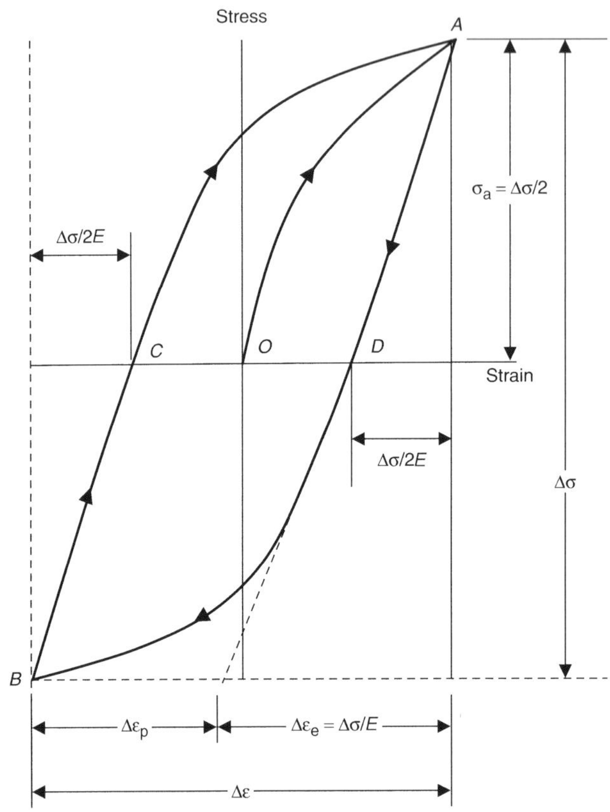

Fig. 15 Hysteresis loop for cyclic loading. Source: Ref 1

point *B*. Reloading from *B* and reversing the stress direction, we go back to point *A*. Points *A* and *B* represent the cyclic stress and strain limits. The total strain ($\Delta\varepsilon$) consists of both elastic and plastic components:

$$\Delta\varepsilon = \Delta\varepsilon_e + \Delta\varepsilon_p \tag{Eq 7}$$

where $\Delta\varepsilon_e$ is the elastic strain ($\Delta\varepsilon_e = \Delta\sigma/E$), and $\Delta\varepsilon_p$ is the plastic strain.

The plastic strain ($\Delta\varepsilon_p$) is the width of the loop at its center, that is, the distance *CD* in Fig. 15. The area of the hysteresis loop is equal to the work done or the energy loss per cycle.

In cyclic strain-controlled fatigue, the strain amplitude is held constant during cycling. Because plastic deformation is not completely reversible, the stress-strain response during cycling can change, largely depending on the initial condition of the metal. The metal can either undergo cyclic strain hardening, cyclic strain softening, or remain stable. Cyclic strain hardening and softening are illustrated schematically in Fig. 16. In both cases, the hysteresis loops change with successive cycles. Cyclic hardening leads to an increasing peak strain with increasing cycles. Strong metals tend to cyclically soften, and low-strength metals tend to cyclically harden. However, the hysteresis loop tends to stabilize after a few hundred cycles, when the material attains an equilibrium condition for the imposed strain level.

Typically, metals harden if $\sigma_u/\sigma_y \geq 1.4$ and soften if $\sigma_u/\sigma_y \leq 1.2$. The reason for hardening or softening is related to the dislocation microstructure of the metal. When the metal is highly work hardened and the dislocation density is high, cyclic strain allows the rearrangement of dislocations into more stable networks, thereby reducing the stress at which plastic deformation occurs. Conversely, when the initial dislocation density is low, the cyclic strain increases the dislocation density, increasing the amount of elastic strain and stress on the material.

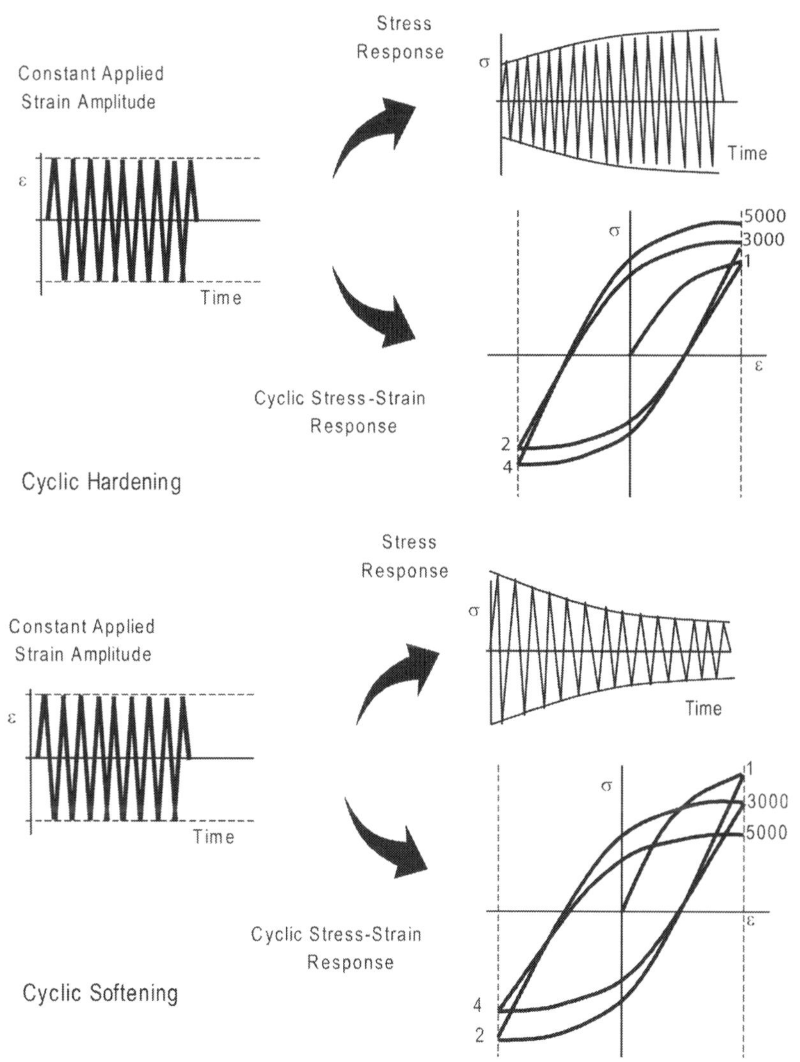

Fig. 16 Cyclic hardening and cyclic softening. Source: Ref 1

Fatigue-Life Prediction

Ever since Wöhler's work on railroad axles subjected to rotating-bending stresses, fatigue data have been presented in the form of an S_a-log N_f curve, where S_a is the stress amplitude and N_f is cycles to failure, as shown in Fig. 17(a). Although an endurance limit is generally observed for many steels under constant stress amplitude testing, such a limit does not exist for high-strength steels or nonferrous metals such as aluminum alloys. As a matter of fact, a single large overload that is common in many air, sea, ground vehicle, and electronic applications can unpin dislocations, thereby causing the endurance limit to be eradicated.

In approximately 1900, Basquin showed that the S_a-log N_f plot could be linearized with full log coordinates (Fig. 17b) and thereby established the exponential law of fatigue. In axial tests using engineering stress, the curve "bends over" at short lives and extrapolates to the ultimate tensile strength (S_u) at ¼ cycle. Further, in comparing axial test results to rotating-bending test results, it is observed that rotating bending gives significantly longer lives, particularly in the low-cycle region (Fig. 18). The reason for the deviation is the method of calculation of the fiber stress in a

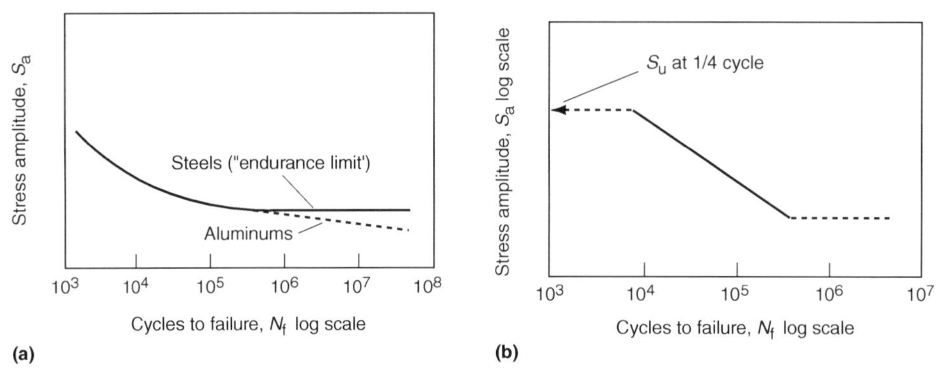

Fig. 17 (a) Stress versus log cycles to failure curve. (b) Log stress versus log cycles to failure curve. Source: Ref 11

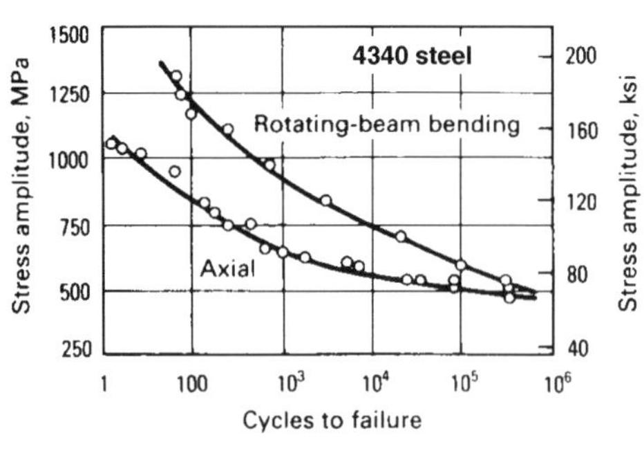

Fig. 18 Stress versus log cycles to failure curves for bending and axial loading tests of 4340 steel. Source: Ref 11

bending type of test, which is an elasticity equation, whereas fatigue is caused by plastic deformation (to-and-fro slip). Thus, the assumption of elastic response in a cyclic environment can be and is often erroneous. This fact is certainly true in the presence of a notch or other geometric or metallurgical discontinuity. Such possibilities exist in most common engineering materials.

If true stress amplitudes are used instead of engineering stress, the entire stress life plot may be linearized, as illustrated in Fig. 19. Thus, stress amplitude can be related to life by a power-law relationship:

$$\sigma_a = \sigma_f'(2N_f)^b \quad \text{(Eq 8)}$$

where σ_a is equal to $\Delta\sigma/2$ in a zero mean constant amplitude test, and σ_a is the true stress amplitude; $2N_f$ is the number of reversals to failure (1 cycle = 2 reversals); σ'_f is the fatigue strength coefficient; and b is the fatigue strength exponent (Basquin's exponent). The parameters σ'_f and b are fatigue properties of the metal. The fatigue strength coefficient (σ'_f) is approximately equal to σ_f for many metals. The fatigue strength exponent (b) varies between approximately –0.05 and –0.12.

In approximately 1955, Coffin and Manson, who were working independently on the thermal fatigue problem, established that plastic strain-life data could also be linearized with log-log coordinates (Fig. 20). As with the true stress life data, the plastic strain-life data can be related by the power-law function:

$$\frac{\Delta\varepsilon_p}{2} = \varepsilon_f'(2N_f)^c \quad \text{(Eq 9)}$$

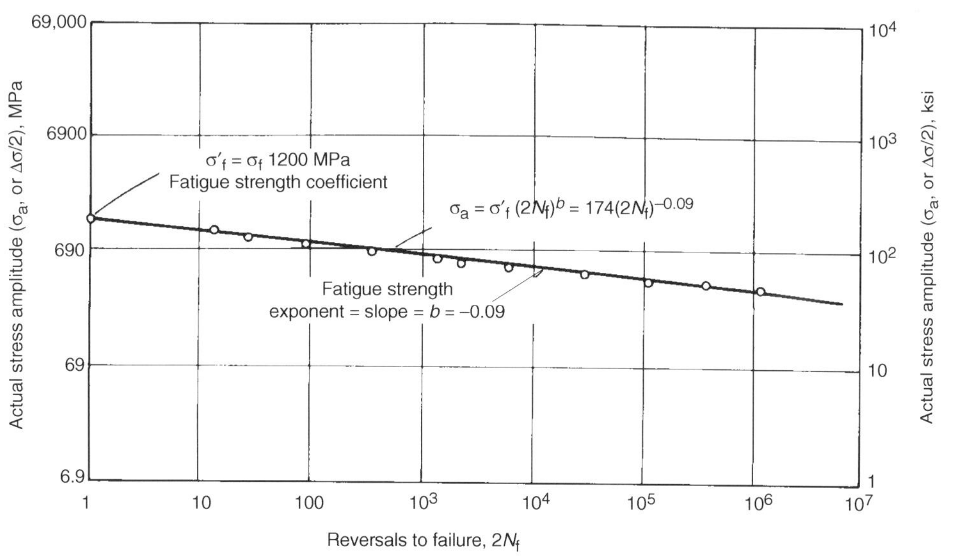

Fig. 19 Log true stress versus log reversals to failure of 4340 steel. From *Fatigue Design Handbook*, SAE. Source: Ref 11

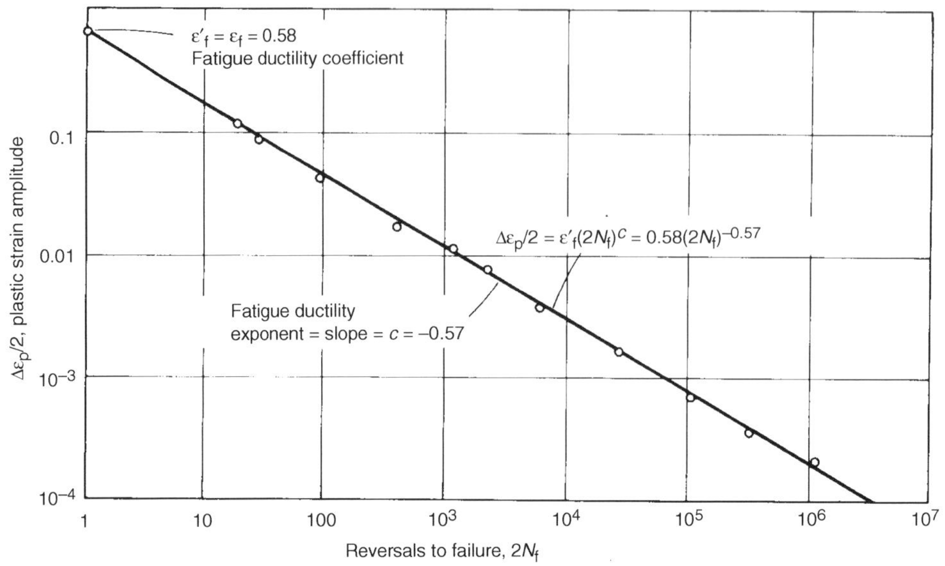

Fig. 20 Log plastic strain versus log reversals to failure of 4340 steel. From *Fatigue Design Handbook*, SAE. Source: Ref 11

where $\Delta\varepsilon_p/2$ is the plastic strain amplitude, ε'_f is the fatigue ductility coefficient, and c is the fatigue ductility exponent. The parameters ε'_f and c are also fatigue properties, where ε'_f is approximately equal to ε_f for many metals, and c varies between approximately –0.5 and –0.7 for many metals.

Total strain has two components: elastic and plastic, or $\varepsilon = \varepsilon_e + \varepsilon_p$. Expressed as the strain amplitudes from a constant amplitude, zero mean strain-controlled test:

$$\frac{\Delta\varepsilon}{2} = \frac{\Delta\varepsilon_e}{2} + \frac{\Delta\varepsilon_p}{2} \qquad \text{(Eq 10)}$$

Because $\sigma_a = \sigma'_f(2N_f)^b$ and:

$$\frac{\Delta\varepsilon_e}{2} = \frac{\sigma_a}{E} \qquad \text{(Eq 11)}$$

one can divide Eq 8 by E, the modulus of elasticity, to obtain:

$$\frac{\Delta\varepsilon_e}{2} = \frac{\sigma'_f}{E}(2N_f)^b \qquad \text{(Eq 12)}$$

Combining Eq 9, 10, and 12:

$$\frac{\Delta\varepsilon}{2} = \underbrace{\frac{\sigma'_f}{E}(2N_f)^b}_{\text{Elastic}} + \underbrace{\varepsilon(2N_f)^c}_{\text{Plastic}} \qquad \text{(Eq 13)}$$

Equation 13 is the foundation for the strain-based approach to fatigue and is called the strain-life relationship. Further, the two straight lines, one for the elastic strain, and one for the plastic strain, can be plotted, as has been done in Fig. 21.

Several conclusions may be drawn from the total strain-life curve in Fig. 21. At short lives, less than $2N_t$ (the transition fatigue life, where $\Delta\varepsilon_p/2 = \Delta\varepsilon_e/2$), plastic strain predominates and the metal ductility will control performance. At longer lives, greater than $2N_t$, the elastic strain is more dominant than the plastic, and strength will control performance. An ideal material would be one with both high ductility and high strength. Unfortunately, strength and ductility are usually a trade-off. The optimum compromise must be tailored to the expected load or strain environment being considered in a real history for a fatigue analysis.

By equating the elastic and plastic components of total strain, the transition fatigue life can be calculated as:

$$2N_f = \left(\frac{\varepsilon_f' E}{\sigma_f'}\right)^{1/(b-c)} \quad \text{(Eq 14)}$$

This is the point on the plot of strain life where the elastic and plastic strain-life lines intersect.

Four fatigue properties have been introduced:

- σ'_f, fatigue strength coefficient
- ε'_f, fatigue ductility coefficient
- b, fatigue strength exponent
- c, fatigue ductility exponent

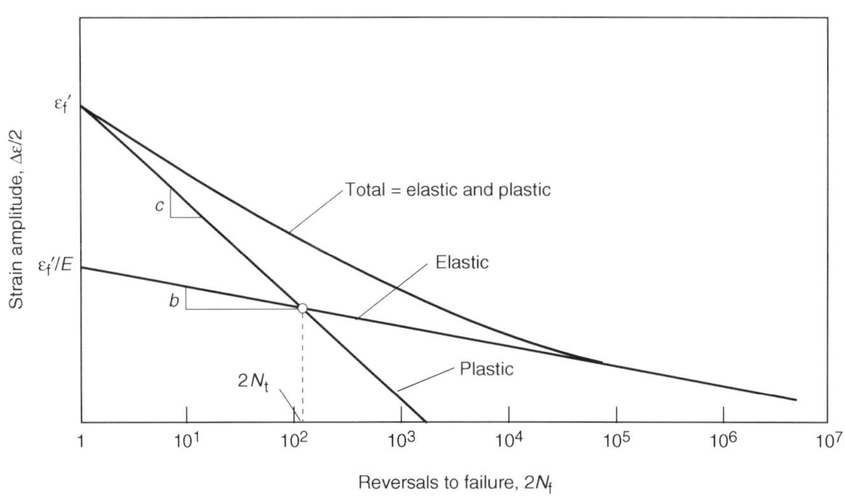

Fig. 21 Log strain versus log reversals to failure. Source: Ref 11

A functional relationship between strain and life has been introduced. A means of accounting for plastic strain, which causes fatigue, is therefore available (Eq 13):

$$\frac{\Delta\varepsilon}{2} = \frac{\sigma_f'}{E}(2N_f)^b + \varepsilon(2N_f)^c$$

These relationships apply to wrought metals only. When internal defects govern life, as is the case with cast metals, higher-hardness wrought steels, and weldments, these principles are not directly applicable.

Cyclic stress-strain material properties may be related in the following manner:

$$K' = \frac{\sigma_f'}{\varepsilon_f'^{n'}} \quad \text{(Eq 15)}$$

Through energy arguments:

$$b = -n' / (1 + 5n') \quad \text{(Eq 16)}$$

and:

$$c = 1/(1 + 5n') \quad \text{(Eq 17)}$$

Thus:

$$n' = b/c \quad \text{(Eq 18)}$$

which allows a relationship between fatigue properties and cyclic stress-strain properties. If average values of b and c (–0.09 and –0.6, respectively) are inserted into Eq 18, $n' \sim 0.15$ results. This is in agreement with the observation that, in general, the average value of n' for most metals is close to 0.15.

However, it must be pointed out that the log-log linear, two straight lines, elastic-plastic approach does not always describe the results of strain-life testing. Several alloys, including the important alloys SAE 4340, 2024-T4 Al, 7075-T6 Al, and Ti-8Al-1Mo-1V, do not exhibit a linear relationship for either elastic or plastic strain life. The heterogeneous deformation in aluminum alloys also causes deviation from a singular straight line description for elastic and plastic strain-life lines. In the absence of adequate data on constant strain amplitude, it is often necessary to approximate the strain-life curve from monotonic tensile properties.

Example—Estimating Fatigue of Hardened Steel. The following examples illustrate the fatigue estimation of hardened steel.

Fatigue Strength Limit (S_{fl}). For many steels with a hardness less than approximately 500 HB, the fatigue limit S_{fl} at 2×10^6 reversals is approximated by:

$$S_n\,(\text{ksi}) \cong \frac{S_u}{2} \cong 0.25(\text{HB}) \qquad \text{(Eq 19)}$$

where HB is the Brinell hardness number. For example, for a steel of 200 HB:

$$\frac{S_u}{2} \approx \text{HB}/2 \approx 100 \text{ ksi} \qquad \text{(Eq 20a)}$$

$$S_{fl} \approx 50 \text{ ksi} \qquad \text{(Eq 20b)}$$

Often, the 0.1% offset yield stress from the cyclic stress-strain curve may be used to approximate S_{fl}. For high-strength steels and nonferrous metals, it is more appropriate to use $\frac{1}{3}S_u \approx S_{fl}$ at 10^8 cycles. In general, however, it is probably more conservative to use $\frac{1}{2}S_u$ at 10^6 cycles for all metals.

Fatigue Strength Coefficient (σ'_f). A reasonably good approximation for the fatigue strength coefficient is:

$$\sigma'_f \approx \sigma_f \text{ (corrected for necking)} \qquad \text{(Eq 21)}$$

or for steels to approximately 500 HB:

$$\sigma_f\,(\text{ksi}) \approx (S_u + 50) \qquad \text{(Eq 22)}$$

For example, a steel of 200 HB:

$$\sigma'_f \approx \sigma_f \approx 150 \text{ ksi} \qquad \text{(Eq 23)}$$

Thus, the intercept at one reversal of the elastic strain-life line is:

$$\frac{\Delta\varepsilon_e}{2} \cong \frac{\sigma_f}{E} \cong \frac{\sigma_f}{30\times10^3} = 0.005 \qquad \text{(Eq 24)}$$

Fatigue Strength Exponent (b). As mentioned previously, b varies from –0.05 to –0.12 and for most metals has an average of –0.085. In approximating the fatigue strength at 2×10^6 reversals with $\frac{1}{2}S_u$, it may be shown that:

$$b \cong -1/6 \log\left(\frac{300}{100}\right) \cong -1/6(0.477) \cong -0.0795 \qquad \text{(Eq 25)}$$

One may now construct the elastic strain-life line, as illustrated in Fig. 22, by either the slope and intercept or the intercept and the fatigue limit at 2×10^6 reversals.

Fatigue Ductility Coefficient (ε'_f). It is a common approximation to set the fatigue ductility coefficient equal to the true fracture ductility ($\varepsilon'_f \approx \varepsilon_f$).

For the 200 HB steel, which is very ductile, the percent reduction in area (%RA) is approximately 65%. Therefore:

$$\varepsilon \cong \ln\left(\frac{1}{1-\mathrm{RA}}\right) \cong \ln\left(\frac{1}{1-0.65}\right) \cong 1 \qquad \text{(Eq 26)}$$

Fatigue Ductility Exponent (c). The fatigue ductility exponent (c) is not as well defined as the other fatigue properties. According to Coffin, c is approximately –0.5, whereas according to Manson, c is approximately –0.6. Morrow has shown that for many metals, c varies between –0.5 and –0.7, or an average of –0.6.

Plotting of Strain-Life Curve. Instead of using a slope c to construct the plastic strain-life line, it is advantageous to note the empirical representation of the hardness and transition fatigue life shown in Fig. 23. For the 200 HB steel in this example, the transition fatigue life is $2N_t \approx 6 \times 10^4$ reversals. By connecting the intercept of $\varepsilon'_f \approx \varepsilon_f = 1$ and the point on the elastic strain-life line at the value of $2N_t$, the plastic strain-life line is constructed. One may now plot the plastic strain-life line and algebraically add to it the elastic strain-life line to obtain the total strain-life curve, as illustrated in Fig. 24.

It should be clear after the examples given that the manner in which metals resist cyclic straining is dependent on both strength and ductility. An idealized situation is depicted in Fig. 25. Consider the steel at 600 HB (a strong metal that resists strain elasticity on the basis of its high strength) compared to the steel at 300 HB (a ductile metal that resists strain plastically on the basis of its superior ductility). The tough steel at 400 HB resists strain by a combination of both its strength and ductility. However, this does not mean that the 400 HB steel is the best material for a specific

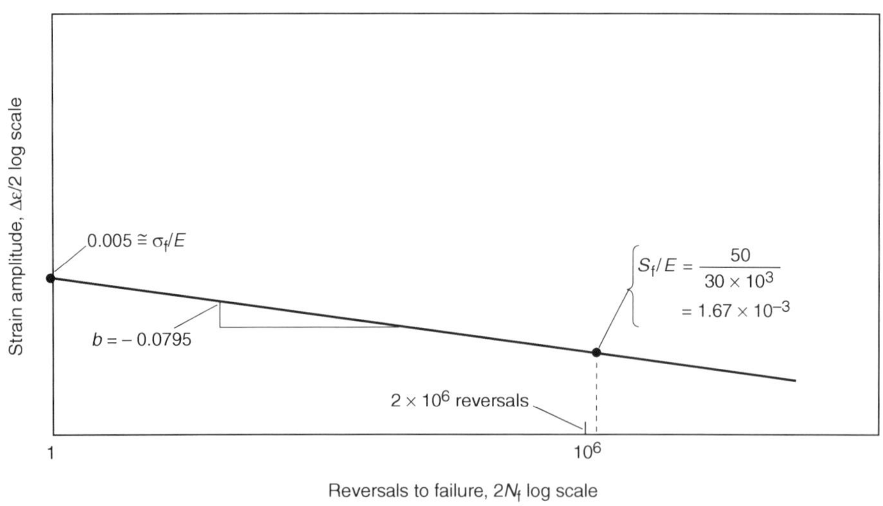

Fig. 22 Log elastic strain versus log reversals to failure. Source: Ref 11

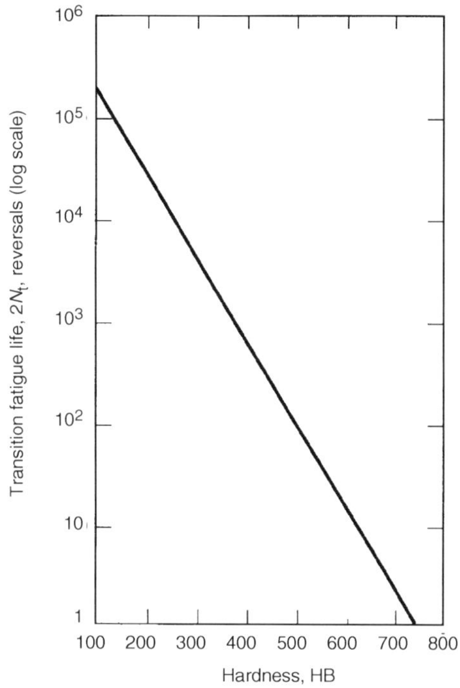

Fig. 23 Log transition failure life versus Brinell hardness for steels. Source: Ref 12

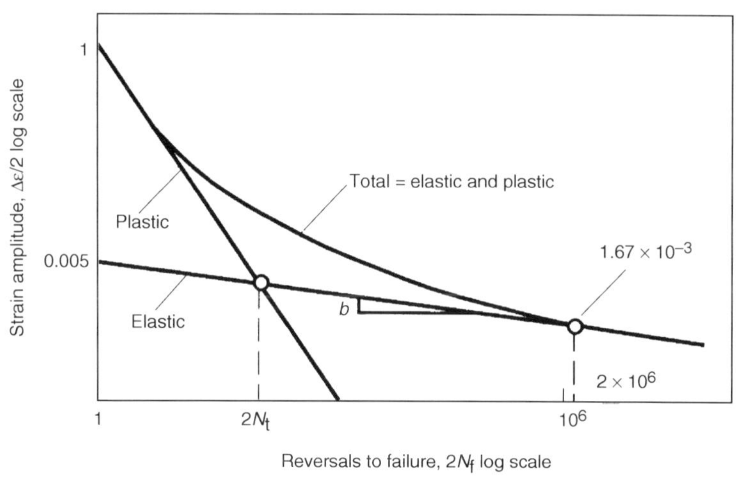

Fig. 24 Estimated curve of log strain versus log reversals to failure for a steel (200 HB). Source: Ref 11

duty cycle that must be resisted in actual design application. The best material must be tailored to the application.

All of the strain-life curves in Fig. 25 intersect at a strain of 0.01, with life to failure of approximately 2×10^3 reversals (1000 cycles). The real trend for a variety of steels of varying hardness and microstructures is shown in Fig. 26. Note that the SAE 1010 (a low-carbon, low-hardness steel used in many ground vehicle components) has a transition fatigue life of approximately 10^5 reversals. Therefore, even at 10^6 reversals, there will be a certain portion of plastic strain present that would not be accounted for in the stress-based approach to fatigue. However, the advent

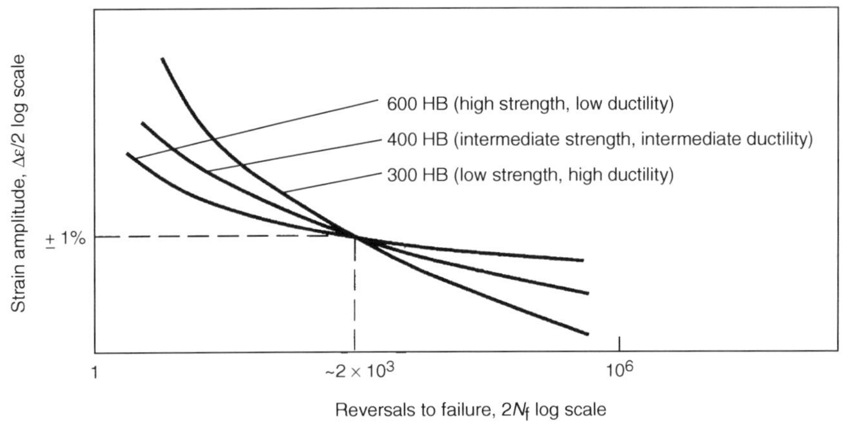

Fig. 25 Strain-life curves for a steel at three different hardness levels (approximation). Source: Ref 11

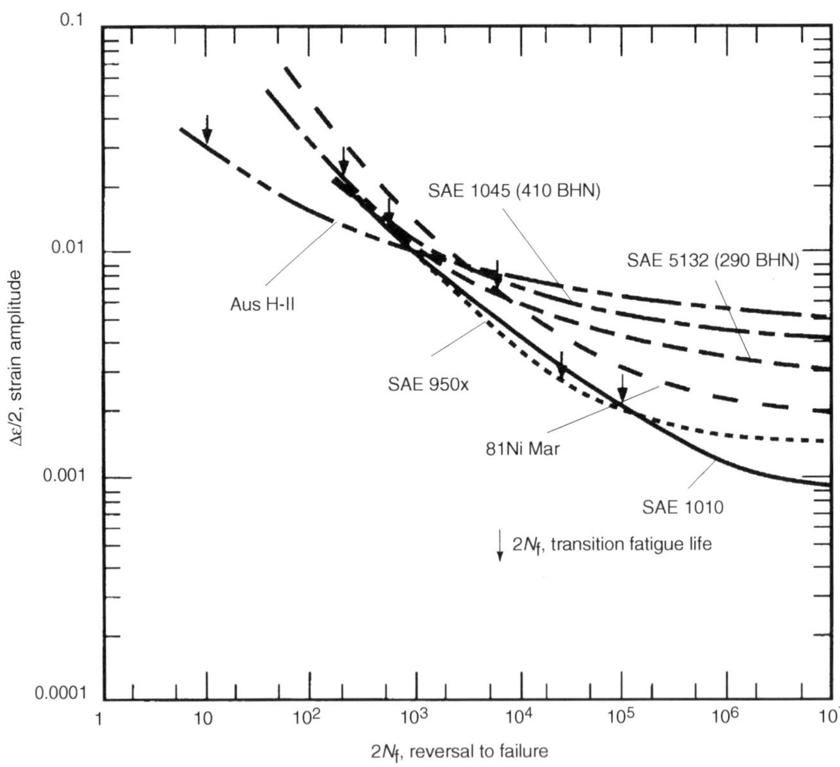

Fig. 26 Strain-life curves for steels with varying microstructures and hardness. Source: Ref 13

of modern, closed-loop electrohydraulic testing machines has made the strain-based test procedure and data presentation fairly commonplace.

Cumulative Damage

Fatigue tests are often conducted under simple conditions, such as constant amplitude and constant frequency. However, in real structures the loading conditions are rarely simple. Many structures are subjected to a range of fluctuating loads, mean stress levels, and variable frequencies. Thus, it is important to be able to predict the life of a component subjected to variable amplitude loading using data generated in constant amplitude laboratory tests. Cumulative damage theories consider the fatigue process to be one of damage accumulation until the life of the component is exhausted.

Consider the fatigue-life diagram in Fig. 27. At a constant stress σ_1, the life is 200 cycles, while at σ_2 the life is 400 cycles. According to the cumulative damage theory, in going from *A* to *B* or *C* to *D*, the fatigue life is gradually exhausted. That is, at points *A* and *C*, 100% of the life is available, while at points *B* and *D*, the respective lives are completely exhausted. If fatigue damage accumulates in a linear manner, each cycle contributes the same amount of damage at a given stress level. For example, at σ_1, on cycling the metal from *A* to *E*, one-fourth of the available fatigue life is exhausted. If the stress level is now lowered to σ_2, then the percentage of life exhausted at σ_1 is equivalent to the percentage of life exhausted at σ_2. That is, one-fourth of the fatigue life at σ_2 is equivalent to one-fourth of the fatigue life at σ_1. Thus, in descending from *E* to *F*, we go

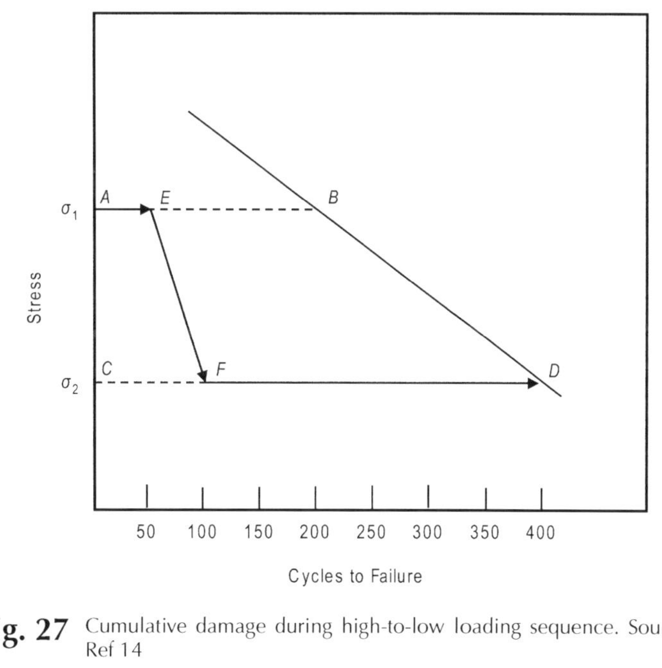

Fig. 27 Cumulative damage during high-to-low loading sequence. Source: Ref 14

from 50 to 100 cycles, and because only one-fourth of the fatigue life was exhausted at σ_1, three-fourths of the fatigue life is still available at σ_2 (i.e., 300 cycles). The same kind of change can be described for a low-to-high stress traverse.

Cumulative damage during fatigue is often determined using the Palmgren-Miner rule, which assumes that the total life of a part can be estimated by adding up the percentage of life consumed by each stress level:

$$\frac{n_1}{N_1}+\frac{n_2}{N_2}+\ldots\ldots+\frac{n_k}{N_k}=1 \text{ or } \sum_{j=1}^{j=k}\frac{n_j}{N_j}=1 \quad \text{(Eq 27)}$$

where $n_1, n_2 \ldots n_k$ represent the number of cycles at a specific stress level, and $N_1, N_2 \ldots N_k$ represent the fatigue life in cycles at the same stress level.

This rule should be used with caution. It is a linear relationship that does not consider the effects of understressing or overstressing. Understressing, in which relatively low stress levels are initially applied, can result in longer fatigue lives at a higher stress level due to localized strain hardening. On the other hand, overstressing, in which high stresses are initially applied, can result in shorter fatigue lives at a low stress level, due to damage induced by the higher stress levels.

Fatigue Crack Nucleation and Growth

Fatigue cracks generally initiate in a highly stressed region of a component subjected to cyclic stresses of sufficient magnitude. The crack propagates under the applied stress through the material until complete fracture results. On the microscopic scale, the most important feature of the fatigue process is the nucleation of one or more cracks under the influence of reversed stresses that exceed the flow stress, followed by the development of cracks at persistent slip bands or at grain boundaries. Subsequently, fatigue cracks propagate by a series of opening and closing motions at the tip of the crack that produce, within the grains, striations that are parallel to the crack front.

The process of fatigue consists of three stages:

- *Stage I:* Initial fatigue damage leading to crack nucleation and crack initiation
- *Stage II:* Progressive cyclic growth of a crack (crack propagation) until the remaining uncracked cross section of a part becomes too weak to sustain the loads imposed
- *Stage III:* Final, sudden fracture of the remaining cross section

Stage III occurs during the last stress cycle when the cross section cannot sustain the applied load. The final fracture is the result of a single overload, which can be a brittle, ductile, or mixed-mode fracture

Crack Initiation

In general, fatigue cracks initiate and propagate in regions where the strain is most severe. Because most engineering materials contain defects, and thus regions of stress concentration that intensify strain, most fatigue cracks initiate and grow from structural defects. Under the action of cyclic loading, a plastic zone develops at the defect tip. This zone of high deformation becomes an initiation site for a fatigue crack. The crack propagates under the applied stress through the material until complete fracture results.

Fatigue cracks form at the point(s) of maximum local stress and minimum local strength, usually at or near the surface of the part. The local stress pattern is determined by the shape of the part, including such local features as surface and metallurgical imperfections that concentrate macroscopic stress, and by the type and magnitude of the loading. Strength is determined by the material itself, including all discontinuities, anisotropies, and inhomogeneities present. Local surface imperfections, such as scratches, mars, burrs, and other fabrication flaws, are the most obvious flaws at which fatigue cracks start. Surface and subsurface material discontinuities in critical locations are likely initiation sites. Inclusions of foreign material, hard precipitate particles, and crystal discontinuities, such as grain boundaries and twin boundaries, are examples of microscopic stress concentrators in the material matrix.

Although crack initiation usually starts at a notch or other surface discontinuity, even in the absence of a surface defect, crack initiation will eventually occur due to the formation of persistent slip bands (PSB), so called because traces of the bands persist even when the surface damage is polished away. Slip bands are a result of the systematic buildup of fine slip movements, on the order of only 1 nm. However, the plastic strain within the PSB can be as much as 100 times greater than that in the surrounding material. The back-and-forth movement of the slip bands leads to the formation of intrusions and extrusions at the surface, eventually leading to the formation of a crack, as illustrated in Fig. 28. The initial crack propa-

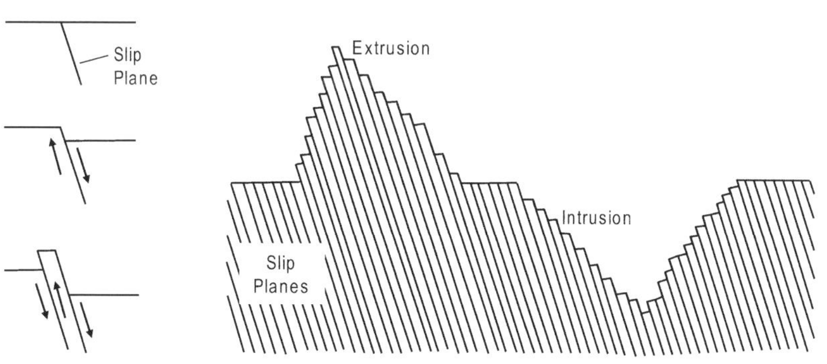

Fig. 28 Development of extrusions and intrusions during fatigue crack initiation. Source: Ref 8

gates parallel to the slip bands. The crack propagation rate during stage I is very low, on the order of 1 nm per cycle, and produces a practically featureless fracture surface. The crack initially follows the slip bands at approximately 45° to the principal stress direction. When the crack length becomes sufficient for the stress field at the tip to become dominant, the overall crack plane changes and becomes perpendicular to the principal stress, and the crack enters stage II.

Stage I may be difficult to envision because of its submicroscopic nature. Even with electron microscopes, it is very difficult to examine and may not be discernible at all, depending on geometry, metal, environment, and stress level. A stage I fracture never extends over more than approximately two to five grains around the origin. In each grain, the fracture surface is along a well-defined crystallographic plane, which should not be confused with a cleavage plane, although it has the same brittle appearance. There are usually no fatigue striations associated with a stage I fracture surface. In some cases, depending on the material, environment, and stress level, a stage I fracture may not be discernible.

On a microscopic level, cracks develop at persistent slip bands or at grain boundaries. The local directions of propagation are controlled to some extent by crystallographic planes and may form on or may be parallel to slip bands in grains near the surface. In interior grains, cleavage cracks often form at the intersection of slip bands with grain boundaries. A variety of crystallographic features has been observed to nucleate fatigue cracks. In pure metals, tubular holes that develop in persistent slip bands, extrusion-intrusion pairs at free surfaces, and twin boundaries are common sites for crack initiation.

Cracking may also initiate at grain boundaries in polycrystalline materials, even in the absence of inherent grain-boundary weakness. At high strain rates, the grain boundary seems to be the preferred site for crack nucleation. Nucleation at a grain boundary appears to be purely a geometrical effect related to plastic incompatibility, whereas nucleation at a twin boundary is associated with active slip on crystallographic planes immediately adjacent and parallel to the twin boundary.

The aforementioned processes also occur in alloys and heterogeneous materials. However, alloying and commercial production practices introduce segregation, inclusions, second-phase particles, and other features that disturb the structure, and these have a dominant effect on the crack nucleation process. In general, alloying that enhances cross slip or twinning that increases the rate of work hardening will retard crack nucleation. On the other hand, alloying usually raises the flow stress of a metal, which speeds up crack propagation.

In high-strength materials containing spheroidal second-phase particles, secondary crack nucleation ahead of the main crack front (slipless fatigue) occurs around such particles. In slipless fatigue, cracks form along lattice planes that are unfavorably located relative to the maximum tensile stress.

In observing locations of crack nucleation, the possibility of environment-related mechanisms, including pitting corrosion, stress-corrosion cracking, and other effects of a hostile environment, must be considered. For example, a great number of fatigue failures in otherwise lightly loaded structures originate in fretted areas. In any structure having mechanically fastened joints with some relative motion, fretting provides a possible failure origin. Once a fatigue crack has been nucleated, its rate and direction of growth are controlled by localized stresses and by the structure of the material at the tip of the crack.

Fatigue Crack Propagation

Stage II crack growth occurs when the stage I crack changes direction and propagates in a direction normal to the applied stress. The transition is from one or two shear planes in stage I to many parallel plateaus separated by longitudinal ridges in stage II. The plateaus are usually normal to the direction of maximum tensile stress (Fig. 29a and b). A transition from stage I to stage II in a coarse-grained specimen of aluminum alloy 2024-T3 is shown in Fig. 29(c). The presence of inclusions rich in iron and silicon did not affect the fracture path markedly. The inclusions, which were fractured, ranged from 5 to 25 μm in diameter. In Fig. 29(c), the stage II area shows a large number of approximately parallel fatigue patches containing very fine fatigue striations that are not resolved at the magnification used. Fine striations are typical in stage II but are usually seen only under high magnification.

Crack growth proceeds by a continual process of crack sharpening followed by blunting, as shown in the Fig. 30 illustration. Crack propagation during crack growth often produces a pattern of fatigue striations (Fig. 31), with each striation representing one cycle of fatigue. Although striations are indicative of fatigue, fatigue failures can occur without the formation of striations. Striations are microstructural details that are best examined with a scanning electron microscope and are not visible to the naked eye.

In the roots of small fatigue crack notches, the local stress state is triaxial (plane strain). This reduces the local apparent ductility of the material and helps control the orientation of the crack as long as the crack is small. After a crack has nucleated and propagated to a larger size, it becomes a macroscopic stress raiser and can be more influential than any stress raiser already in the part. At this point, the crack tip will take over control of the fracture direction. Subsequently, the orientation of the crack surface will depend on the stress field at the crack tip and will often follow a series of void coalescences in advance of the crack front. Early crack extension occurs under plane-strain conditions and gives a typical fine-grained, flat-faced surface that, when produced under random loading or sequences of high and low stress amplitudes, exhibits characteristic beach marks.

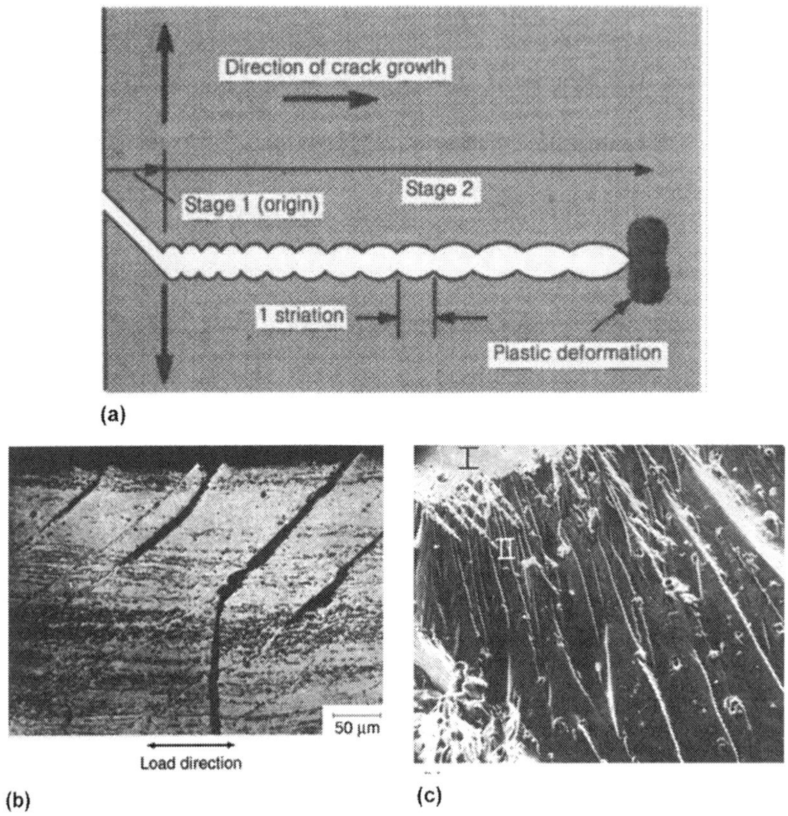

Fig. 29 Transition from stage I to stage II fatigue. (a) Highly enlarged schematic cross-sectional sketch. The edge of the metal is at left. When tensile forces repeatedly act on the surface, the microstructural changes of stage I cause a submicroscopic crack to form. With each repetitive opening, the crack jumps a small distance (one striation). Note that the spacing of each striation increases with the distance from the origin, assuming the same opening stress. The metal at the tip of the fatigue crack (right) is plastically deformed on a submicroscopic scale. (b) Change in fracture path from stage I (top of photo) to stage II (bottom). (c) Transition of a fatigue fracture in a coarse-grained specimen of aluminum alloy 2024-T3. Source: Ref 15, 16

As a general rule, low-stress, high-cycle fatigue produces flat-faced (plane-strain) fractures. The fracture surface appears fine grained and lightly polished near the crack nucleation site, where the stress intensification is least. The surface becomes progressively rougher and more fibrous as the crack grows and the intensity of stress increases. On high-stress, low-cycle fatigue surfaces, found in certain areas of all complete fatigue fractures, the surface is fibrous, rough, and more typical of plane-stress loading conditions, where the general fracture direction is at 45° to the main tensile load.

In large structural components, the existence of a crack does not necessarily imply imminent failure of the part. Significant structural life may remain in the cyclic growth of the crack to a size at which a critical failure

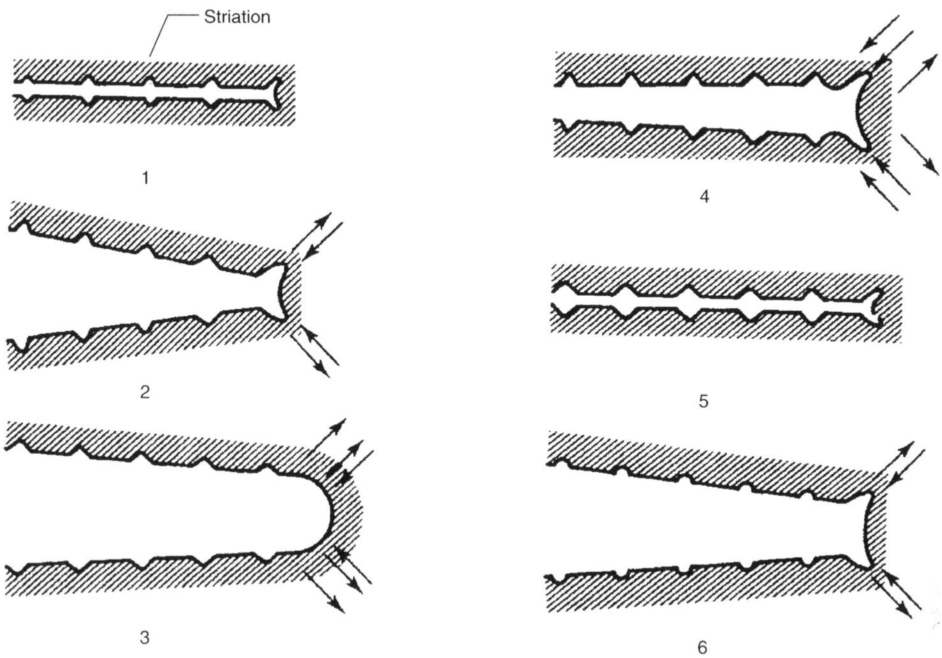

Fig. 30 Fatigue crack propagation. Source: Ref 17

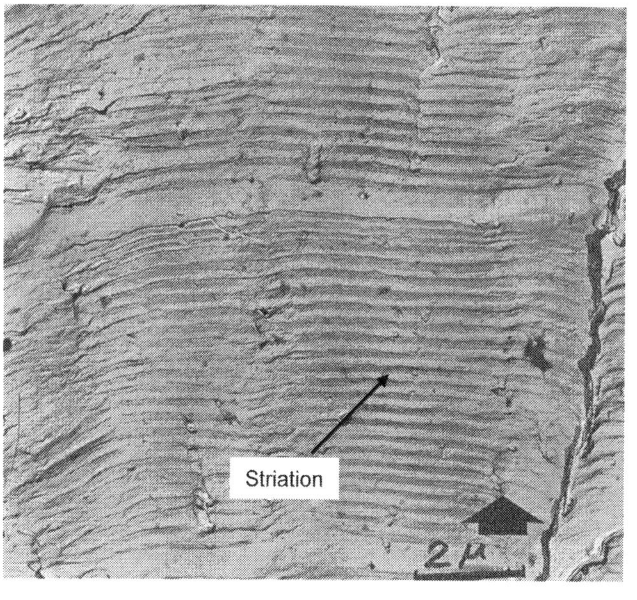

Fig. 31 Scanning electron micrograph showing fatigue striations. Source: Ref 2

occurs. The growth or extension of a fatigue crack under cyclic loading is principally controlled by maximum load and stress ratio. However, as in crack initiation, a number of additional factors may exert a strong influence, including environment, frequency, temperature, grain direction, and other microstructural factors.

Beach Marks. The most characteristic feature usually found on fatigue fracture surfaces is beach marks, which are centered around a common point that corresponds to the fatigue crack origin. Also called clamshell, conchoidal, and arrest marks, beach marks are perhaps the most important macroscopic characteristic feature in identifying fatigue failures. Beach marks can occur as a result of changes in loading or frequency or by oxidation of the fracture surface during periods of crack arrest from intermittent service of the part or component.

An example of beach marks on the fracture surface of a 1040 steel shaft is shown in Fig. 32. The beach marks are produced by changes in crack

Fig. 32 Rotating-bending fatigue fracture of a keyed shaft of 1040 steel, approximately 30 HRC. The fatigue crack originated at the lower left corner of the keyway and extended almost through the entire cross section before final rupture occurred. A prominent beach mark pattern is visible, appearing to swing counterclockwise on this section because the rotation was in a clockwise direction. Source: Ref 15

growth rates from the constantly changing stress intensity at the crack tip. The initial crack growth rate is slow but increases with the increasing stress intensity at the crack tip as the crack extends further. Beach marks may also be produced by load variations during service, spectrum load fatigue testing, or changes in stress intensity from geometric features of the part. However, many fatigue fractures produced under conditions of uninterrupted crack growth and without load variations do not exhibit beach marks. Moreover, sometimes a fracture surface may have features that look like fatigue beach marks. For example, a surface from a stress-corrosion cracking fracture that reveals features similar to fatigue beach marks is shown in Fig. 33.

Ratchet marks are macroscopic features that may be seen on fatigue fractures in shafts and flat leaf springs. They may also occur in ductile fractures of overtorqued fasteners. In fatigue fractures, ratchet marks are the result of multiple fatigue crack origins, each producing a separate fatigue crack zone. In shafts, the separately initiated cracks normally are propagated on planes slightly inclined to the plane of the shaft diameter (Fig. 34). As two approaching cracks meet, a small step is formed. The small steps around the periphery of a shaft are the ratchet marks. Although ratchet marks are most apparent on the periphery of fractures in shafts, the stepped appearance is characteristic whenever fatigue cracks emanate

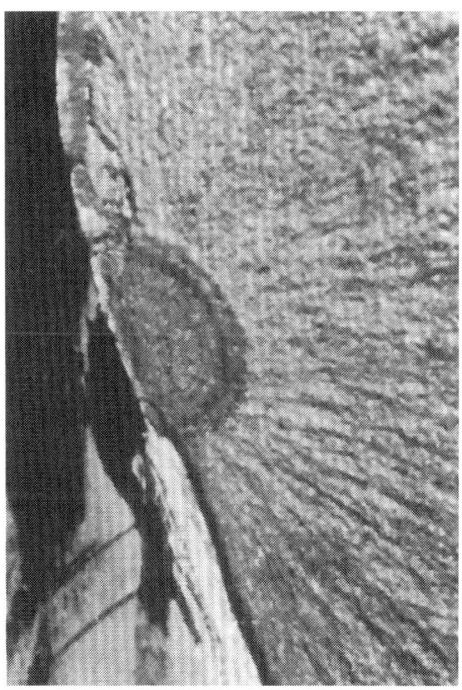

Fig. 33 Beach marks on a 4340 steel part caused by stress-corrosion cracking. Tensile strength of the steel was approximately 1780 to 1900 MPa (260 to 280 ksi). The beach marks are a result of differences in the rate of penetration of corrosion on the surface. They are in no way related to fatigue marks. Original magnification: 4×. Source: Ref 18

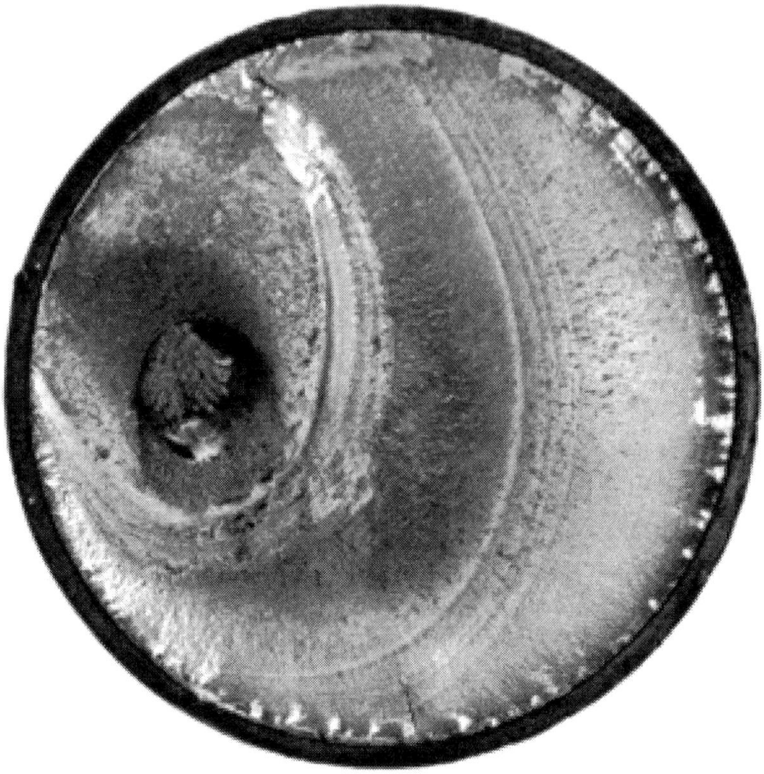

Fig. 34 Surface of a fatigue fracture in a 1050 steel shaft, with hardness of approximately 35 HRC, that was subjected to rotating bending. Presence of numerous ratchet marks (small shiny areas at surface) indicates that fatigue cracks were initiated at many locations along a sharp snap ring groove. The eccentric pattern of oval beach marks indicates that the load on the shaft was not balanced; note final rupture area near left side. Source: Ref 15

from several origins and subsequently meet to form one principal crack front. The occurrence of ratchet marks requires almost simultaneous initiation of several fatigue cracks, a situation that is favored by high stress and, to a lesser extent, by the presence of high stress-concentration factors, such as corrosion pits or grinding burns.

Striations. In the electron microscope examination of fatigue fracture surfaces, the most prominent features found are patches of finely spaced parallel marks, known as fatigue striations. Fatigue striations are oriented perpendicular to the microscopic direction of crack propagation, as in Fig. 35(a). In general, two types of striations have been recognized: ductile striations (Fig. 35b) and brittle striations (Fig. 35c). The brittle striations are connected by what seems to be cleavage fracture along sharply defined facets. Numerous river markings separating these facets run normal to the striations (Fig. 35c). Brittle striations in aluminum alloys usually are an indication of a corrosive environment. Ductile striations are more common.

With uniform loading, fatigue striations generally increase in spacing as they progress from the origin of fatigue. Each striation is the result of a single cycle of stress, but every stress cycle does not necessarily produce a striation. Striation spacing depends strongly on the level of applied loading. The clarity of the striations depends on the ductility of the material. Striations are more readily visible in ductile materials. Thus, patches of fatigue striations in high-strength steel are less visible than in an aluminum alloy.

At high rates of crack growth (0.0025 mm, or 10^{-4} in., per cycle or more), the striations become wavy and develop a rough front. A large plastic zone exists in front of the crack, which may cause extensive secondary cracking. Each secondary crack propagates as a fatigue crack, cre-

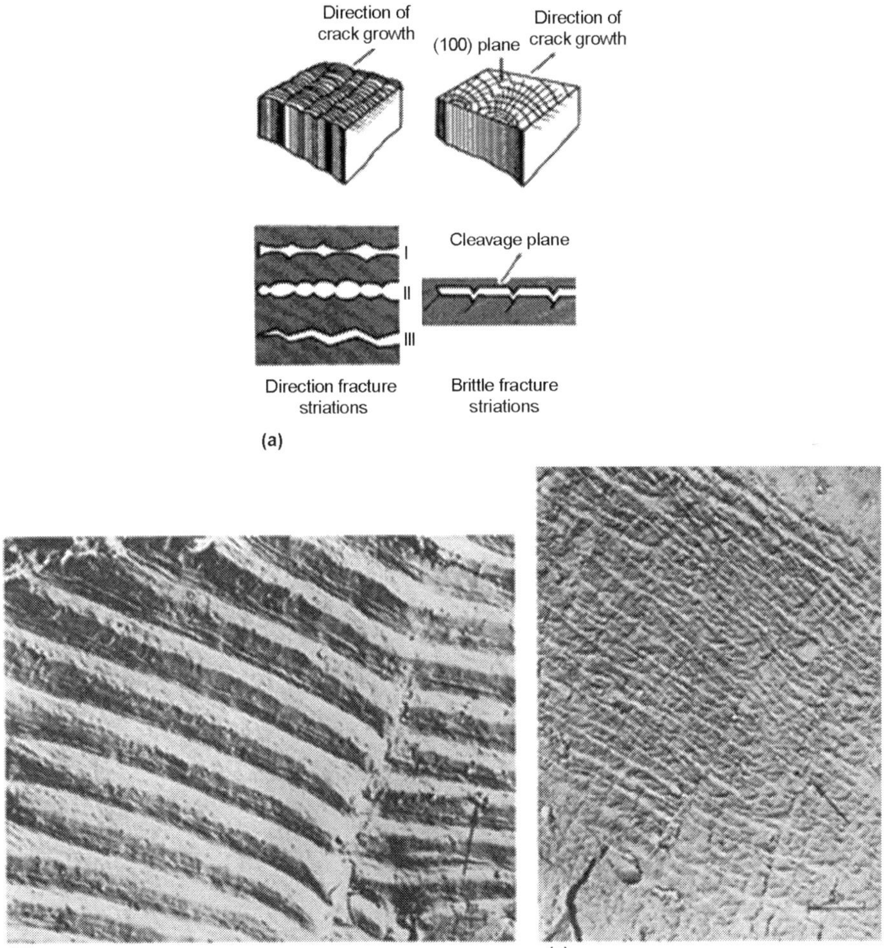

Fig. 35 Ductile and brittle striations. (a) Schematic of different types of ductile and brittle striations. (b) Ductile striations in 718 aluminum alloy. (c) Brittle fatigue striations of 2014 aluminum alloy. Note cleavage facets running parallel to direction of crack propagation and normal to striations. Source: Ref 19

ating a network of secondary striations. The local crack direction may differ markedly from the overall direction of crack propagation because of the many changes in direction of the local fracture path.

In steels, fatigue striations that are formed at ordinary crack growth rates are not always as well defined as they are in aluminum alloys. The striations that are formed in aluminum alloys at very low crack growth rates (less than 1.3×10^{-4} mm, or 5×10^{-6} in. per cycle) are difficult to resolve and often cannot be distinguished from the network of slip lines and slip bands associated with plastic deformation at and near the crack front as it propagates through the section.

Final Fracture (Stage III)

Ultimate failure occurs when the fatigue crack becomes long enough that the remaining cross section can no longer support the applied load.

Final fracture occurs when the crack has grown to the critical size for overload failure. The size of the final fracture zone depends on the magnitude of the loads, and its shape depends on the shape, size, and direction of loading of the fractured part. The final fracture zone of a fatigue fracture surface is often fibrous, resembling the fracture surfaces of impact or fracture toughness test specimens of the same material.

Two features of the final fracture zone aid in determining the origin of fracture. First, fatigue usually originates at the surface, and therefore, the fatigue origin is not included in the shear lip zones of the overload region. In tough materials, with thick or round sections, the final fracture zone will consist of a fracture by two distinct modes: tensile fracture (plane-strain mode) extending from the fatigue zone and in the same plane, and shear fracture (plane-stress mode) at 45° to the surface of the part bordering the tensile fracture.

In thin sheet metal pieces having sufficient toughness, final fracture occurs somewhat differently. As the crack propagates from the fatigue zone, the fracture plane rotates around an axis in the direction of crack propagation until it forms an angle of approximately 45° with the loading direction and the surface of the sheet. The fracture plane, inclined 45° to the load direction, can occur on either a single-shear or a double-shear plane, as in Fig. 36.

The second characteristic of a fast fracture zone is chevron marks that point back to the origin of fracture. This is shown in Fig. 37 for a torsional fatigue fracture of a steel shaft. Beach marks fan out from the origin at a sharp corner in a keyway at the top. Beyond the beach marks, cracking progressed by fast fracture along the hardened perimeter, producing two sets of chevron marks, pointing back to the crack origin.

The macroscopic fracture appearance will depend on part geometry, type of loading, and stress level. Typical fracture appearances are shown in Fig. 38 and 39.

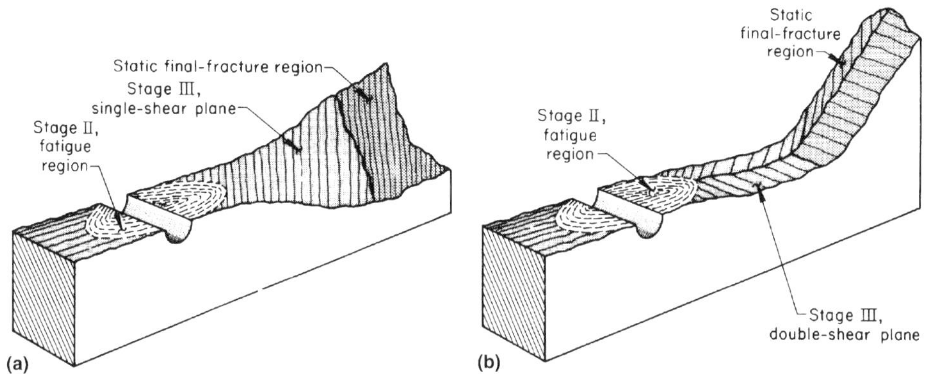

Fig. 36 Fracture planes that are 45° to the direction of loading. (a) Single-shear plane. (b) Double-shear plane. Source: Ref 18

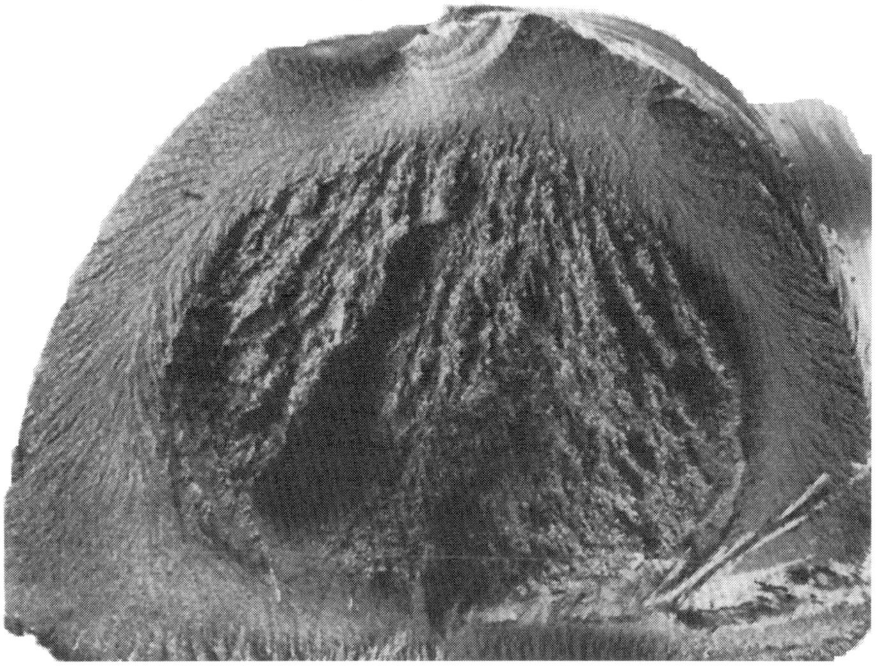

Fig. 37 Surface of a torsional fatigue fracture in an induction-hardened 1041 (1541) steel shaft. The shaft fractured after 450 h of endurance testing. Original magnification: 1.25×. Source: Ref 18

Fracture Mechanics Approach to Fatigue Crack Propagation

Linear elastic fracture mechanics assumes that all structures contain flaws. Cracks grow from an initial size (a_0) to a critical size (a_c) corresponding to failure as a function of the number of load cycles (Fig. 40).

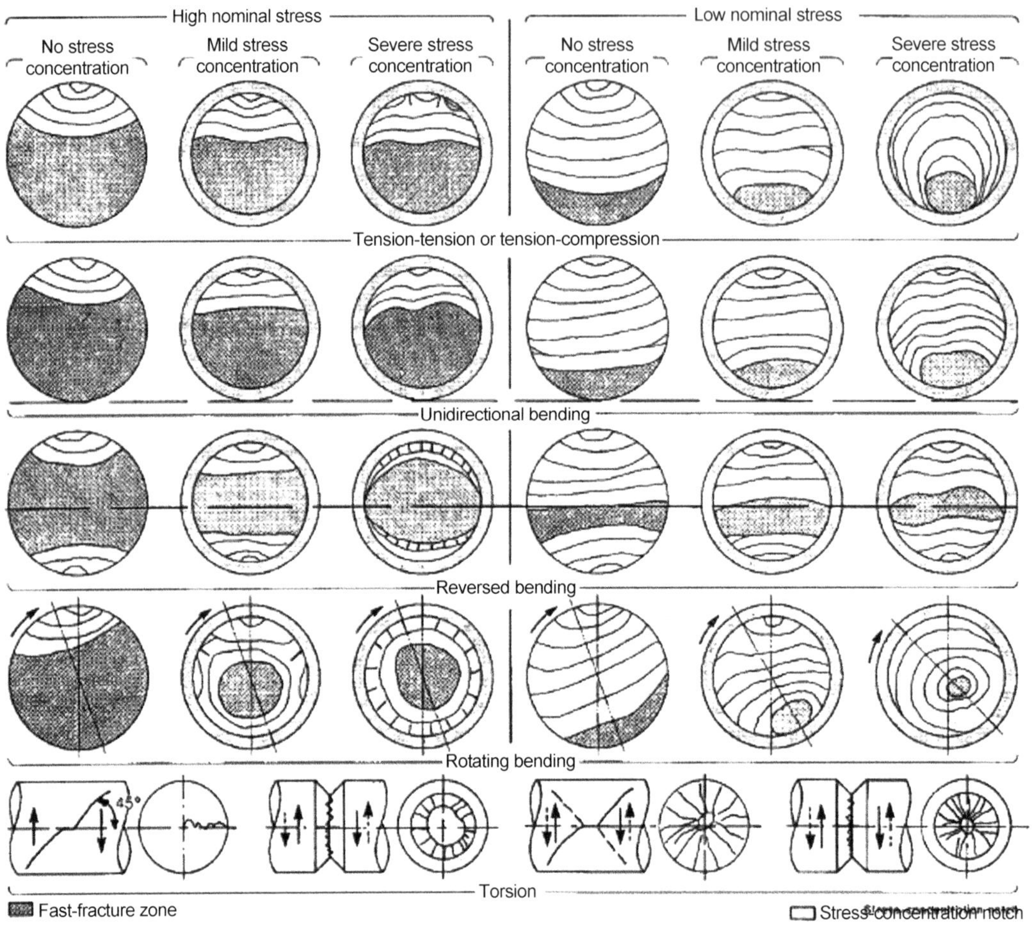

Fig. 38 Schematic representation of fatigue fracture surface marks produced in smooth and notched cylindrical components under various loading conditions. Note that the final rupture zones (fast fracture zones) on the left half of the figure, which had a high nominal stress, are considerably larger than the corresponding final rupture zones on the right half, which had low nominal stresses. The individual lines in the sketches represent the general configuration of the progression of the fatigue crack, as recorded by possible submicroscopic striations and macroscopic beach marks. The long dashed lines in reverse and unidirectional bending indicate the bending axes. Also note the radial ratchet marks between origins of the high nominal stress fractures. In the torsional fatigue fractures (bottom row), note that unidirectional fatigue (left) is at an approximate 45° angle to the shaft axis, while reversed torsional fatigue of a cylindrical shaft (second from right) has an X-shaped pattern, with the origin in either the longitudinal or the transverse direction. Torsional fatigue cracks in stress concentrations tend to be very rough and jagged because of many 45° cracks. Source: Modified from Ref 20

The crack growth rate (*da/dn*) can be determined from the slope of the curve. Initially, the crack growth rate is slow but increases with increasing crack length. Of course, the crack growth rate is also higher for higher applied stresses. If one can characterize the crack growth, it is then possible to estimate the service life or inspection intervals required under specific loading conditions and service environment. In the fracture mechanics approach to fatigue crack growth, the crack growth rate, or the amount of crack extension per loading cycle, is correlated with the stress-intensity

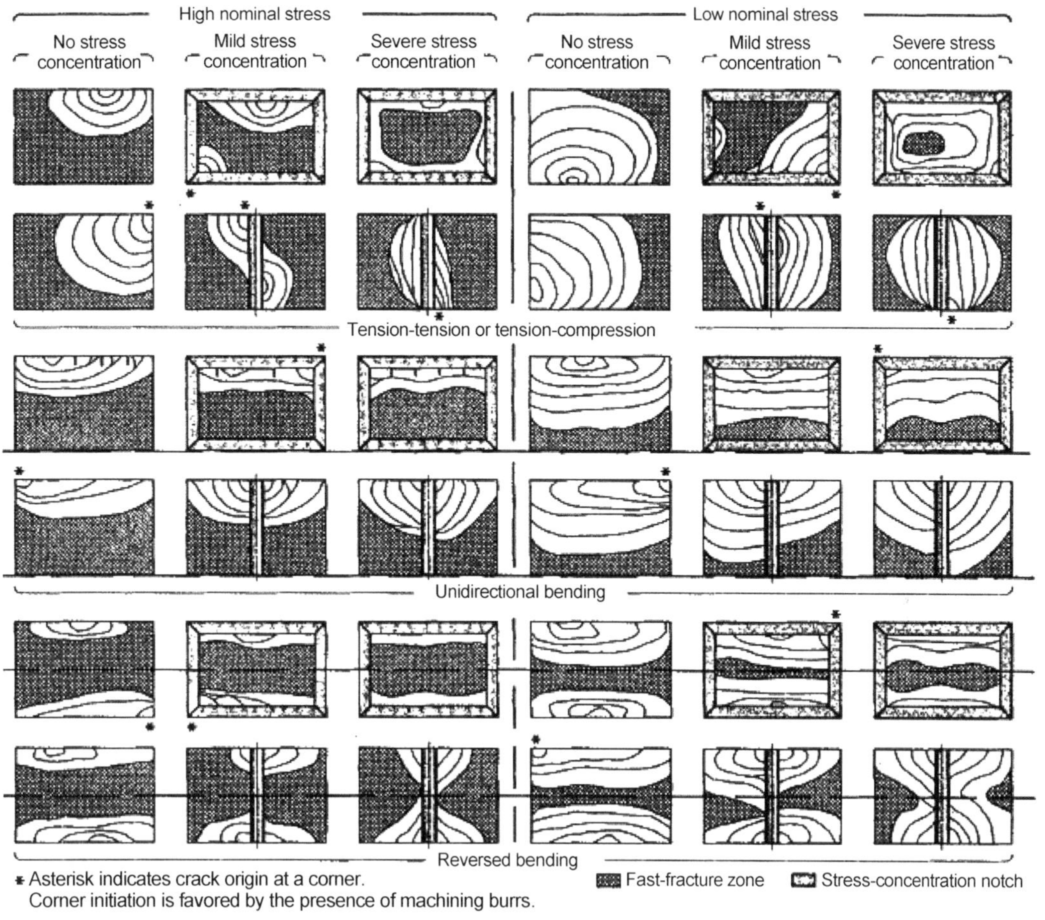

Fig. 39 Schematic representation of fatigue fracture surface marks produced in components with square and rectangular cross sections and in thick plates under various loading conditions. Note that the final rupture zones (fast fracture zones) on the left half of the figure, which had a high nominal stress, are considerably larger than the corresponding final rupture zones on the right half, which had low nominal stresses. The individual lines in the sketches represent the general configuration of the progression of the fatigue crack, as recorded by possible submicroscopic striations and macroscopic beach marks. The long dashed lines in reverse and unidirectional bending indicate the bending axes. Note that fatigue origins at corners of rectangular shapes or at ends of drilled holes are quite common in this type of part. Also, multiple fatigue origins are quite common with parts under high nominal stress; ratchet marks perpendicular to the surface separate adjacent fatigue areas. Source: Modified from Ref 20

parameter (K). This approach makes it possible to estimate the useful safe life and inspection intervals. An idealized da/dn versus ΔK curve is shown in Fig. 41. In region I, ΔK_{th} is the fatigue crack growth threshold, which is at the lower end of the ΔK range, where crack growth rates approach zero. In region II, the crack growth rate is stable and essentially linear and can be modeled by power-law equations, such as the Paris equation:

$$da/dn = C(\Delta K)^m \quad \text{(Eq 28)}$$

where a is the flaw or crack size; n is the number of cycles; C and m are constants related to material variables, environment, temperature, and fa-

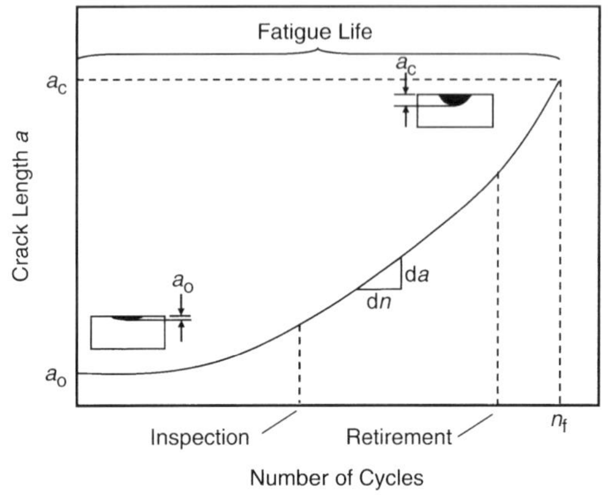

Fig. 40 Crack length as a function of cycles. Source: Ref 1

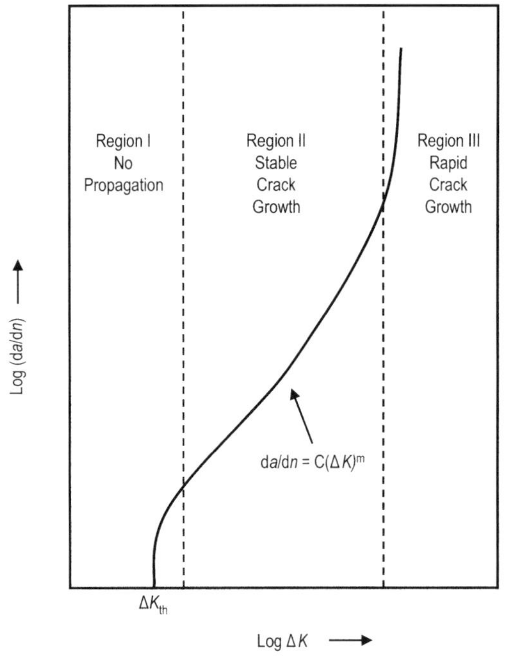

Fig. 41 Crack propagation curve for fatigue loading. Source: Ref 1

tigue stress conditions; and $\Delta K = \Delta K_{max} - \Delta K_{min}$ is the stress-intensity parameter range. The constants C and m are material parameters that must be determined experimentally. Typically, m is in the range of 2 to 4 for metals and 4 to 100 for ceramics and polymers.

Because the zone ahead of the crack that experiences cyclic plasticity (i.e., the fatigue plastic zone) is very small, plane-strain conditions can develop even for small thicknesses. This is an important conclusion because it means that data can be obtained from thin specimens and applied quite generally.

During region II growth in the linear crack growth region, the Paris law can be used to determine the number of cycles to failure; ΔK can be expressed in terms of $\Delta\sigma$:

$$\Delta K = \Delta K_{max} - \Delta K_{min} = Y\sigma_{max}\sqrt{\pi a} - Y\sigma_{min}\sqrt{\pi a} = Y\Delta\sigma\sqrt{\pi a} \quad \text{(Eq 29)}$$

where Y depends on the specific specimen geometry. Thus, the Paris law becomes:

$$\frac{da}{dn} = C\left(Y\Delta\sigma\sqrt{\pi a}\right)^m \quad \text{(Eq 30)}$$

One of the goals of fatigue analysis is to be able to predict the fatigue life of structures. Therefore, rearranging Eq 28, one can solve for the fatigue life (n):

$$dn = \frac{da}{C(\Delta K)^m} \quad \text{(Eq 31)}$$

which may then be integrated to give:

$$n_f = \int_0^{n_f} dn = \int_{a_o}^{a_c} \frac{da}{C(\Delta K)^m} \quad \text{(Eq 32)}$$

Substitution of the expression for ΔK (Eq 32) gives:

$$n_f = \int_{a_o}^{a_c} \frac{da}{C\left(Y\Delta\sigma\sqrt{\pi a}\right)^m} = \frac{1}{C\pi^{\frac{m}{2}}(\Delta\sigma)^m} \int_{a_o}^{a_c} \frac{da}{Y^m a^{\frac{m}{2}}} \quad \text{(Eq 33)}$$

It is assumed that $\Delta\sigma$ (or $\sigma_{max} - \sigma_{min}$) is constant and that Y depends on the crack length and therefore cannot be removed from within the integral. For cases where Y depends on crack length, these integrations will generally be performed using numerical techniques. During region II crack growth, some metals are sensitive to the load ratio (R), as shown in Fig. 42 for 7075-T651 aluminum plate. The Paris equation does not account for the load ratio; however, there are other expressions that do account for the sensitivity of crack growth rate to the load ratio. One of the most utilized is the one developed by Foreman and his associates:

$$\frac{da}{dn} = \frac{C(\Delta K)^m}{(1-R)K_c - \Delta K} \quad \text{(Eq 34)}$$

In these expressions, the initial crack length (a_o) and the final or critical crack length (a_c) need to be determined. Crack lengths can be detected

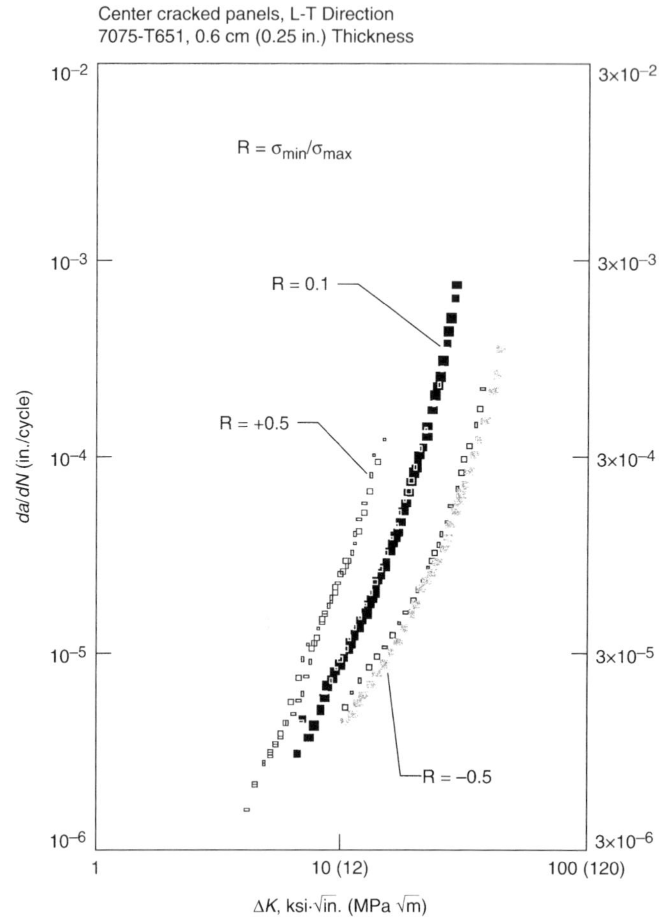

Fig. 42 Stress ratio effects on an aluminum alloy. Source: Ref 1

using a number of nondestructive testing techniques. If no cracks are detectable, then one must assume that a crack exists at the lower end of the resolution of the detection system being used. To determine the critical crack length (a_c), the crack will eventually grow to a length at which the metal immediately fails, that is:

$$K_{max} \rightarrow K_c \quad \text{(Eq 35)}$$

or

$$Y\sigma_{max}\sqrt{\pi a_f} \rightarrow K_c \quad \text{(Eq 36)}$$

Solving for a_c gives:

$$a_c = \frac{K_c^2}{\pi Y^2 \sigma_{max}^2} \quad \text{(Eq 37)}$$

Note that even if a component contains a detectable crack, it can remain in service, provided that it is periodically inspected. This philosophy forms the basis for what is known as the damage tolerance design approach.

Finally, in region III, the crack growth rate accelerates because the fracture toughness of the material is approached, and there is a local tensile overload failure.

Example Problem—Calculation of Fatigue Lives. The crack growth rate of ferritic/pearlitic steels is given by the equation:

$$\frac{da}{dn} = 3.6 \times 10^{-10}(\Delta K)^{3.0} \tag{Eq 38}$$

where ΔK is given in units of ksi$\sqrt{\text{in.}}$, and da/dn is in units of in./cycle. Assume that a part contains an edge crack that is 0.05 in. long. The stresses vary from 0 to 50 ksi, and the fracture toughness is 100 ksi$\sqrt{\text{in.}}$ Compute the life of the part.

Solution. For this geometry, the stress-intensity parameter is given by:

$$\Delta K = 1.12\Delta\sigma\sqrt{\pi a}$$

Rigorously, the expression should be $\Delta K = \Delta\sigma\sqrt{\pi a} + \sigma\sqrt{a} \cdot (1/2) \cdot a^{-1/2} \cdot \Delta a$. However, because $\Delta\sigma$ in a given cycle is much larger than Δa, to a very good approximation it may be written $\Delta K = \Delta\sigma\sqrt{\pi a}$.

Using the information given in the problem statement and the aforementioned expression, the crack growth rate is given by:

$$\frac{da}{dn} = 3.52 \times 10^{-4}(a)^{3/2}$$

This equation can be integrated from the initial condition of $n = 0$ and $a = a_o = 0.05$ in. to the final condition of $n = n_f$ and $a = a_f$. The final crack length (a_f) occurs when $K_{max} = K_{Ic}$. Integrating yields:

$$n_f = 2.84 \times 10^3 \int_{a_o}^{a_f} \frac{da}{a^{3/2}}$$

Integration of the right side of the equation gives:

$$n_f = 5.68 \times 10^3 \left[\frac{1}{\sqrt{a_{a_o}}} - \frac{1}{\sqrt{a_f}}\right]$$

As mentioned previously, a_f depends on the fracture toughness and maximum stress. For the geometry being considered, this is given by:

$$a_f = \frac{1}{\pi}\left(\frac{K_{Ic}}{\sigma}\right)^2 = \left[\frac{100}{1.12 \times 50}\right]^2 = 1.02 \text{ in.}$$

Substituting into the expression for life gives:

$$n_f = 5.68 \times 10^3 \left[1/\sqrt{0.05} - 1/\sqrt{1.02} \right] = 1.98 \times 10^4$$

Thus, the part is expected to last almost 20,000 cycles.

The representation of the crack growth rate in Eq 28 is essentially correct for a reasonably wide domain of stress-intensity ranges. However, additional factors must be considered at both low and high stress-intensity ranges. When the stress-intensity range is low, a point will be reached where the average crack growth rate approaches the interatomic spacing. Physically, a crack cannot propagate a fraction of an interatomic spacing; however, it must be remembered that the measured crack growth rate in a typical experiment represents an average across an entire crack front. In some regions, the crack may be stopped, while in others it may be moving. For practical purposes, the threshold stress-intensity range (ΔK_{th}) occurs when the average crack growth rate is less than 10^{-12} m/cycle. For values of ΔK less than ΔK_{th}, the crack growth rate is effectively zero. Equation 28 cannot be used to describe the crack growth behavior in the threshold regime. Also, when the stress-intensity parameter becomes so large that it approaches the fracture toughness (K_{Ic}), the rate of crack growth becomes much more rapid than that predicted by Eq 28, because, in addition to the fatigue process, a considerable amount of local tensile failure occurs during each cycle. Again, Eq 28 no longer provides a correct representation of the crack growth behavior. Over the broad range of stress-intensity parameters, fatigue crack growth rates may be represented by an equation of the form:

$$\log \frac{da}{dn} = C_1 \sinh\left[C_2 \log(\Delta K) + C_3 \right] + C_4 \qquad \text{(Eq 39)}$$

where C_1 is a material constant, and C_2, C_3, C_4 are functions of load ratio, frequency, and temperature, respectively.

Numerous other empirical expressions have also been proposed to describe crack growth behavior from ΔK_{th} to values of ΔK when K_{max} approaches K_{Ic}. Regardless of the representation, the life may always be computed by integrating the crack growth curve, as was done in the preceding example problem. It is worth noting that use of Eq 28 is usually conservative for the threshold-to-intermediate crack growth rate regime. While Eq 28 underestimates the growth rate for large ΔK values near the upper limit, the underestimate is small. This is because as the stress intensity approaches the fracture toughness of the material, the crack is growing so rapidly that very few cycles are accumulated in this regime. Use of threshold data must be done very judiciously due to the fact that a small error in threshold can lead to significant overestimates of the fatigue life.

The life of components containing preexisting flaws can be computed in principle using fracture mechanics concepts. For complex load/crack geometries and stress patterns, the equations for ΔK are complex. In addition, the da/dn versus ΔK relationships may be quite complicated or perhaps not even available in functional form. Such conditions do not lead to easily evaluated integrals for the fatigue life, as was the case in the example problem.

However, the life of a component can always be computed through stepwise numerical integration techniques similar in principle to the technique that was used in the example problem. In this scheme, an increment of crack growth (Δa) is selected. The number of cycles associated with a growth increment Δa is obtained from $\Delta N = \Delta a(dn/da)$. All of the Δns are added up from the initial crack length to the final crack length. In such cases, it is most efficient to use computers to do the actual computations.

Crack Closure

During fatigue cycling, the crack growth rate can slow due to crack closure. For example, if the structure is subjected to a series of high tensile overloads that are followed by loads at lower stress levels, the crack may temporally stop propagating, leading to a decrease in the crack growth rate da/dn and consequently have a longer fatigue life than would normally be expected. Crack closure mechanisms effectively reduce the ΔK during fatigue cycling, thus retarding the crack growth. Because ΔK is equal to $K_{max} - K_{min}$ and the actual ΔK is reduced because crack closure is hindered, ΔK is effectively smaller and the crack growth rate is then reduced. Four mechanisms that can cause crack closure are shown in Fig. 43. Plasticity-induced closure results from compressive residual stresses developing in the plastic wake. Roughness-induced closure is due to crack meandering, in which the crack deviates from mode I displacements and is subject to mode II shear displacements. These displacements cause a mismatch between the upper and lower crack faces, which results in contact between the crack faces during cycling. Oxide debris-induced closure results from corrosion products becoming wedged in the crack interface. Finally, in fluid-induced closure, fluid pressure can act as a wedge preventing total crack closure.

The Short Crack Problem

For long cracks, under conditions of applicability of linear elastic fracture mechanics (LEFM), there exists a threshold stress-intensity range (ΔK_{th}), below which no fatigue crack growth occurs. The value of the cyclic threshold stress intensity depends on a variety of factors: the microstructure of the material; the test environment; the load ratio (R); various

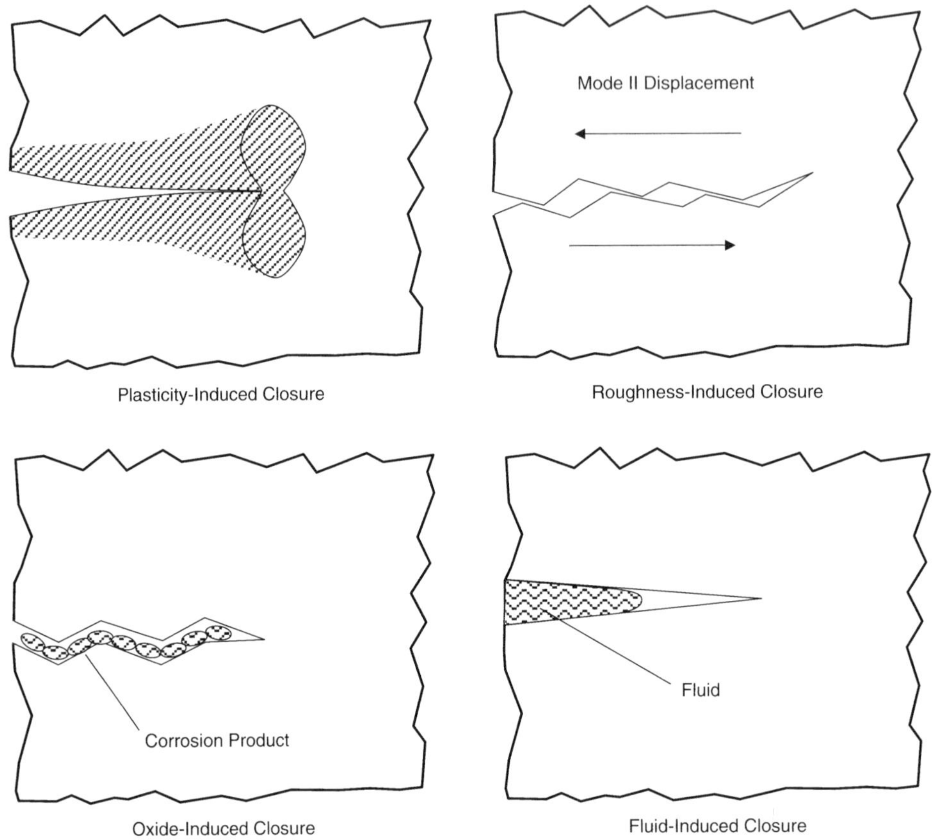

Fig. 43 Fatigue crack closure mechanisms in metals. Source: Ref 21

crack tip factors, such as the amount of overload and cold work; experimental techniques; and the geometry of the specimen. It has been observed that short fatigue cracks, in metals and polymers, can propagate at rates different from those of the corresponding long fatigue cracks under the influence of the same driving force. Generally, for a given ΔK, the growth rates of small cracks are higher than those of long cracks.

A short crack is a crack that is smaller than the microstructural unit of the materials; for instance, a crack of length equal to grain or precipitate size is a short crack. In practice, one finds that long cracks can be between 1 and 20 mm (0.04 and 0.8 in.), while short cracks are smaller than 0.1 mm (0.004 in.). The anomalous growth of short cracks is shown in Fig. 44, a plot of log da/dn versus ΔK. Long cracks do not grow below a constant threshold (ΔK_{th}). However, in the short crack regimen, cracks grow below this threshold value, as indicated by the deviation from the dashed line in the figure. Short cracks propagate below the long crack threshold (ΔK_{th}). The fatigue crack growth rate of short cracks decreases progressively, until a minimum in crack velocity occurs at a crack length on the

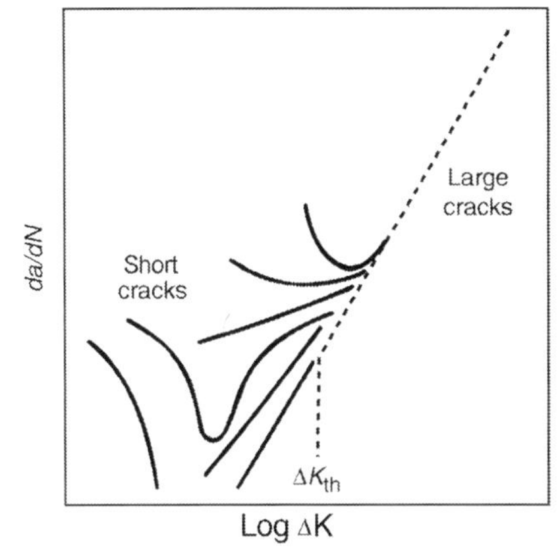

Fig. 44 Typical short crack growth rate behaviors in comparison to typical large crack behavior. Source: Adapted from Ref 22

order of the grain size; that is, $a \approx d$, where d is the grain size. This so-called short crack anomaly arises when the crack size approaches the dimension of the microstructural feature (e.g., grain size, inclusions, etc.). Under such circumstances, homogeneity is lost, which is implicit in the LEFM treatment of long cracks. The short crack is different from a long crack in terms of the internal stresses that it encounters. All short cracks grow in internal stress fields that accentuate the applied stress to the level at which the crack propagates. The fundamental problem is that the customary LEFM parameter ΔK is, strictly speaking, an invalid representation of the crack driving force in the presence of enhanced near-tip plasticity and microstructural inhomogeneity.

When the crack tip cyclic plastic zone size, and sometimes the crack itself, is embedded within the predominant microstructural unit (e.g., a single grain), the crack tip plastic strain range is determined by the properties of individual grains and not by the continuum aggregate. The growth rate acceleration of small cracks embedded within a single surface grain is primarily due to enhancement of the local plastic strain range that results from a lower yield stress for optimum slip in the surface grains. As a small crack approaches a grain boundary, the fatigue crack may accelerate, decelerate, or even arrest, depending on whether or not slip propagates into the contiguous grain. In turn, the transmission of slip across a grain boundary depends on the grain orientation, the activities of secondary and cross slip, and the plane of slip. The transition of the small crack from one grain to another may require a change in the crack path, which may also influence crack closure. Thus, the resulting crack growth behavior is very sensitive to the crystallographic orientation and properties of individual grains located within the cyclic plastic zone. As the crack grows, the number of grains interrogated by the crack tip plastic zone increases, and the statistically averaged material properties become smoother.

This short crack behavior is very troublesome and has caused numerous inaccuracies in life predictions. Unfortunately, while some proposed solutions to modeling short crack behavior have been found to be satisfactory for specific circumstances, there is no current model that can be applied universally. Short crack behavior was first documented in the mid-1970s, extensively investigated in the 1980s, and remains an active research topic. The problem is now well-enough understood to facilitate some standardization of concepts, test methods, and analysis techniques, but short crack technology is not yet routinely applied in industrial practice.

Geometrical Stress Concentrations

In most structures, fatigue cracking usually initiates at a stress concentration. The stress concentration may be inherent in the design, such as a filet, hole, thread, or other geometrical feature, or the stress concentration can result from a manufacturing process, such as a rough surface finish or residual tensile stresses introduced by heat treatment.

The effect of geometrical stress concentrations on fatigue is often studied by testing notched specimens. When a notch is present in a specimen under uniaxial loading, three effects are present:

- There is an increase or concentration of stress at the root of the notch.
- There is a stress gradient from the notch toward the center of the specimen.
- A triaxial state of stress exists.

The dramatic reduction in fatigue life of normalized 4130 steel sheet containing different types of stress concentrations is shown in Fig. 45.

The effect of notches on fatigue strength is determined by comparing the *S-N* curves of notched and unnotched specimens. The data for the

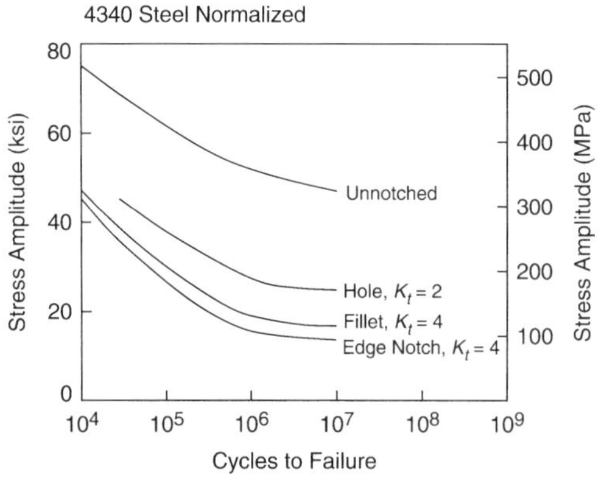

Fig. 45 Effect of geometrical stress concentrations on fatigue life. Source: Ref 23

notched specimens are usually plotted in terms of nominal stress based on the net section of the specimen. The effect of the notch in decreasing the fatigue strength is reported as the fatigue strength reduction factor, or the fatigue notch factor (K_f):

$$K_f = \frac{\text{Fatigue limit unnotched}}{\text{Fatigue limit notched}} \qquad \text{(Eq 40)}$$

For metals that do not have a definite fatigue limit, the fatigue notch factor is based on the fatigue strength at a specified number of cycles. Values of the fatigue notch factor vary with the severity of the notch, the type of notch, the material, the type of loading, and the applied stress level.

Notched fatigue data are also reported using a notch sensitivity factor (q):

$$q = \frac{K_f - 1}{K_t - 1} \qquad \text{(Eq 41)}$$

This relationship compares the theoretical stress-concentration factor (K_t) to the fatigue notch factor (K_f). In this relationship, a material that experiences no reduction in fatigue due to a notch will have a notch sensitivity factor of $q = 0$, while one that experiences a reduction in fatigue up to the full theoretical value will have a notch sensitivity factor of $q = 1$. The value of q is dependent on the material and the radius of the notch root, as illustrated by the plots shown in Fig. 46 for steels. The notch sensitivity factor q significantly decreases with smaller notch radii. Although this seems counterintuitive, it occurs because K_f increases more slowly than K_t with decreasing notch radius. For notches with large radii, K_f is almost equal to K_t. For materials with small notches, K_f is less than K_t. In addition, lower-strength metals are less affected than high-strength metals by geometric discontinuities that reduce the fatigue resistance, because high-strength metals have a limited capacity for deformation and crack tip blunting.

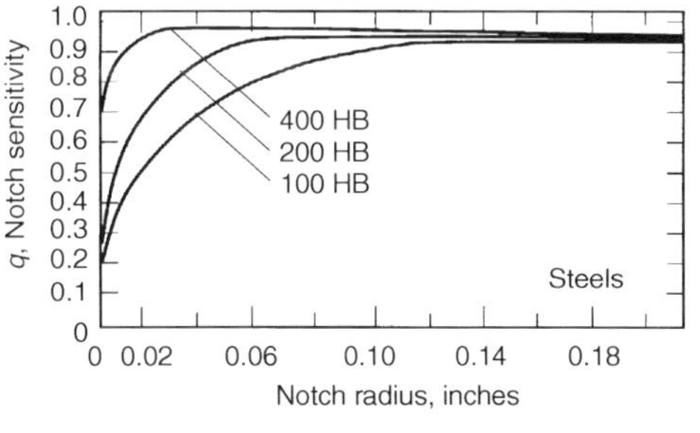

Fig. 46 Notch sensitivity versus notch radius for steels. Source: Ref 11

Manufacturing Stress Concentrations

Because almost all fatigue failures start at or near a surface, the surface finish and residual-stress state near the surface can have a profound effect on the fatigue strength. Since the early days of fatigue investigations, it was recognized that the fatigue life of a component is very dependent on the surface finish, as produced by machining or grinding operations. As shown in Fig. 47, highly polished steel specimens perform much better in fatigue than even carefully machined surfaces. Further degradation in fatigue strength occurs for hot rolled and forged surfaces. Finally, when corrosion is introduced, the fatigue life is seriously degraded.

Parts that are formed at room temperature will contain residual stresses. For example, a surface of a part that was formed in tension will contain residual compressive stresses, and a surface that was in compression during forming will contain residual tension stresses. Because fatigue always occurs under tensile loading, the surface with the residual tensile stresses will be the more prone to fatigue cracking. Similar to forming, some quenching operations during heat treatment can result in a tensile residual-stress pattern on the surface that will adversely affect fatigue strength.

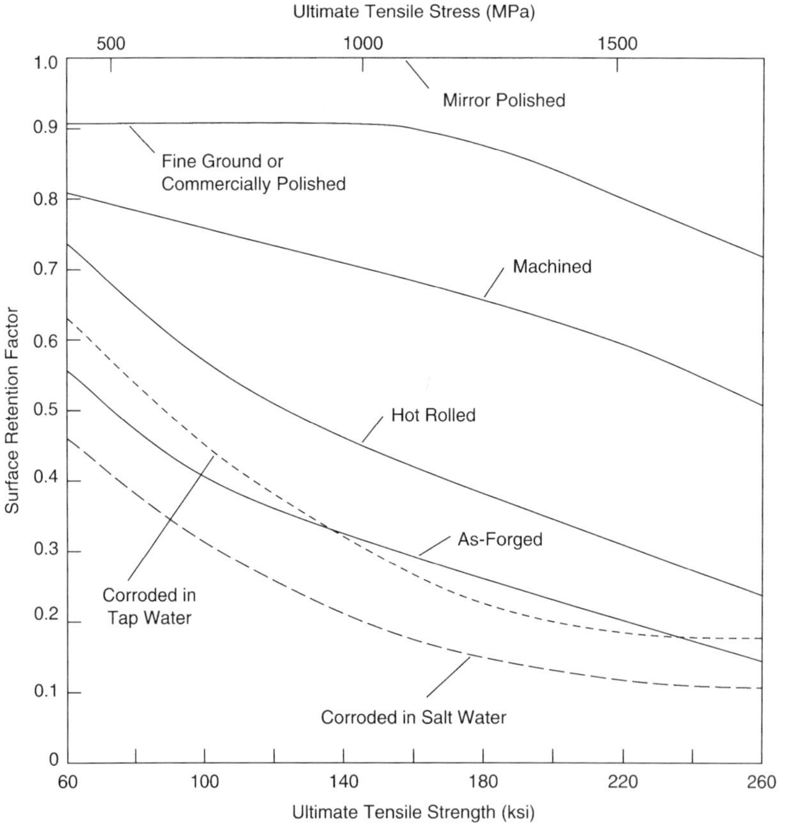

Fig. 47 Effect of surface finish on fatigue of steels. Source: Ref 24

The fatigue data shown in Table 1 for 4340 steel forgings tested in fully reversed bending illustrate two important points. First, vacuum melting, which removes inclusions, improves both the longitudinal and transverse endurance limits. Second, because forging tends to elongate inclusions in the longitudinal or working direction, the fatigue endurance limit is higher in the longitudinal direction. The lower transverse endurance limit in steels containing inclusions is attributed to stress concentrations at the inclusions, which can be quite high when an elongated stringer is oriented transverse to the applied tensile stress. For the case of the vacuum-melted steel, the transverse limit is still somewhat lower than for the longitudinal direction, indicating the extreme sensitivity of transverse fatigue properties to microstructure.

Traditional machining methods, which are the principal means of metal removal, include chip producing, such as milling, drilling, turning, broaching, reaming, and tapping, and abrasive machining methods, such as grinding, sanding, and polishing. These machining operations can produce surface layer alterations when abusive machining conditions are used. In general, abusive machining promotes higher temperatures and excessive plastic deformation. Gentle machining operations occur when a sharp tool is employed and when machining conditions result in reduced machining forces.

The surface characteristics produced by the low-stress grinding of AISI 4340 steel quenched and tempered to 50 HRC are shown in Fig. 48(a). The low-stress condition produced no visible surface alterations, while the abusive grinding condition (Fig. 48b) produced an untempered martensite layer 0.03 to 0.13 mm (0.001 to 0.005 in.) deep with a hardness of 65 HRC. Below this white layer there was an overtempered martensitic zone with a hardness of only 46 HRC. The hardness returned to its normal value at a depth of 0.30 mm (0.012 in.) below the surface (Fig. 48c). Abusive grinding also tends to produce a residual stress within the altered surface layer. Abusive grinding produced high tensile stresses in the altered zone, while low-stress grinding produced a surface with small compressive stresses (Fig. 48d). Fatigue tests conducted on flat specimens indicate that abusive grinding may seriously reduce the fatigue strength, as shown in Fig. 48(e). In this example, abusive grinding depressed the endurance limit from 760 MPa (110 ksi) for low-stress grinding to 520 MPa (75 ksi).

Other traditional machining processes also impart residual stresses in the surface layers. In milling, the residual stress tends to be compressive

Table 1 Effect of inclusions in 4340 forged steel

Property	Electric furnace melted	Vacuum melted
Longitudinal fatigue limit, MPa (ksi)	800 (116)	959 (139)
Transverse fatigue limit, MPa (ksi)	545 (79)	827 (120)
Ratio transverse/longitudinal	0.68	0.86
Hardness, HRC	27	29

Source: Ref 25

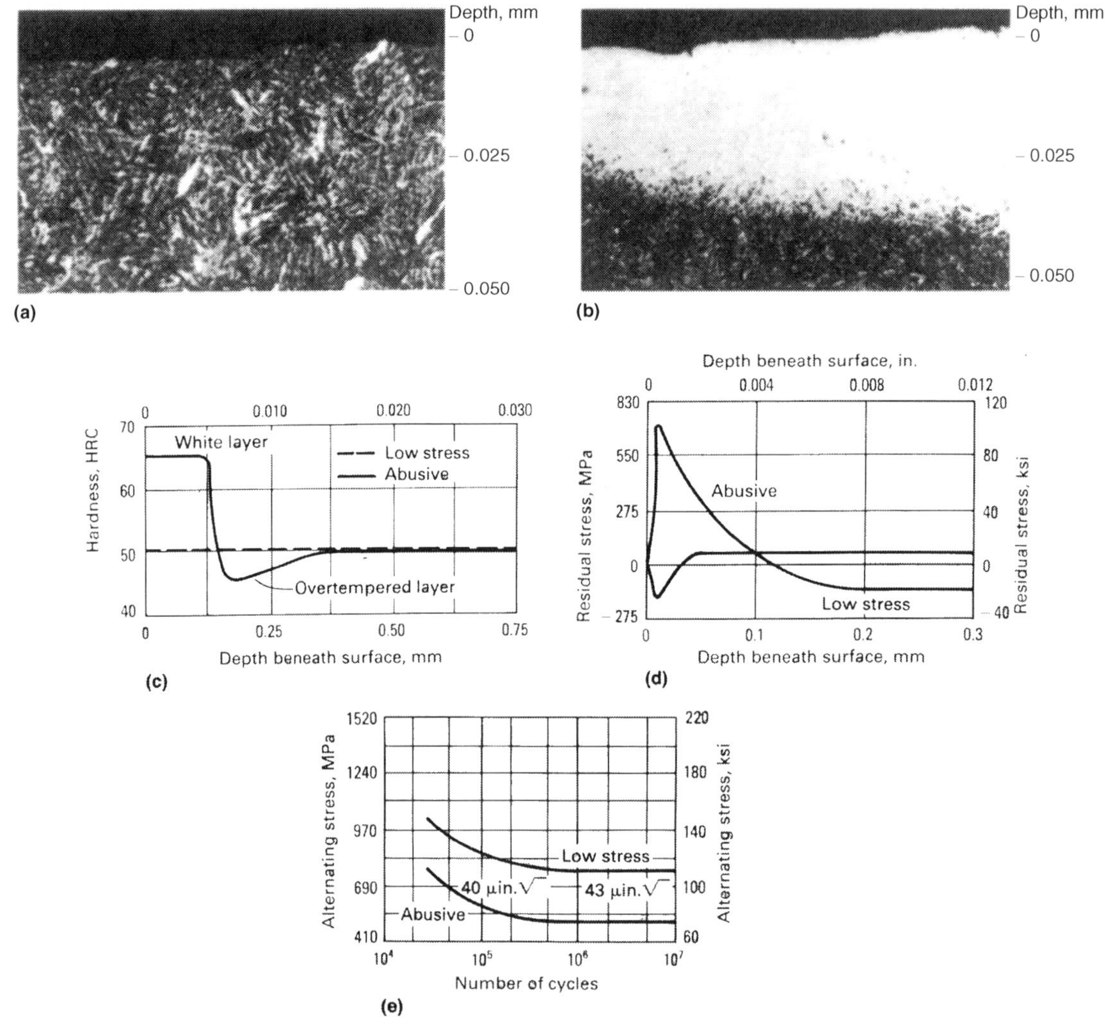

Fig. 48 Surface characteristics produced by low-stress and abusive grinding of AISI 4340 steel. (a) Low-stress grinding produced no visible surface alterations. (b) The untempered martensitic white layer shown from abusive conditions has a hardness of 65 HRC and is approximately 0.025 to 0.050 mm (0.001 to 0.002 in.) deep. (c) Plot of microhardness alterations showing a total heat-affected zone of 0.33 mm (0.013 in.) from abusive conditions. (d) Plot of residual stress. (e) Effect on fatigue strength. Source: Ref 26

(Fig. 49). In surface milling 4340 steel, the stresses are tensile at the surface but go into compression below the surface.

Nontraditional machining includes a variety of methods for removing and finishing materials. Examples of nontraditional operations are electrical discharge machining, laser beam machining, electrochemical machining, electropolishing, and chemical machining.

- *Electrical discharge machining (EDM):* Tends to produce a surface with a layer of recast spattered metal that is usually hard, cracked, and porous to some degree. Under roughing conditions, globs of recast metal are spattered onto a white layer of rehardened martensite. An overtempered zone of down to 46 HRC was also found beneath the surface. The surface produced under finishing conditions also contained discontinuous patches of recast metal plus a thin layer of rehardened martensite 0.003 mm (0.0001 in.) deep.

- *Laser beam machining (LBM):* Tends to produce the same types of surface alterations as EDM. The intense heat generated by the laser beam results in a recast surface layer.
- *Electrochemical machining (ECM):* Is capable of producing a surface that is essentially free of metallurgical surface layer alterations. However, abusive ECM conditions tend to degrade surface roughness and adversely affect fatigue strength.
- *Electropolishing (ELP) and chemical machining (CM):* Surface softening is produced on most materials by electrochemical machining, as well as by electropolishing and chemical machining, also referred to as chemical milling. This softening may be severe enough and deep enough to affect the fatigue strength and other mechanical properties of metals and may necessitate postprocessing.

The surface alterations produced by abusive metal-removal conditions generally have a minor effect on the static mechanical properties. How-

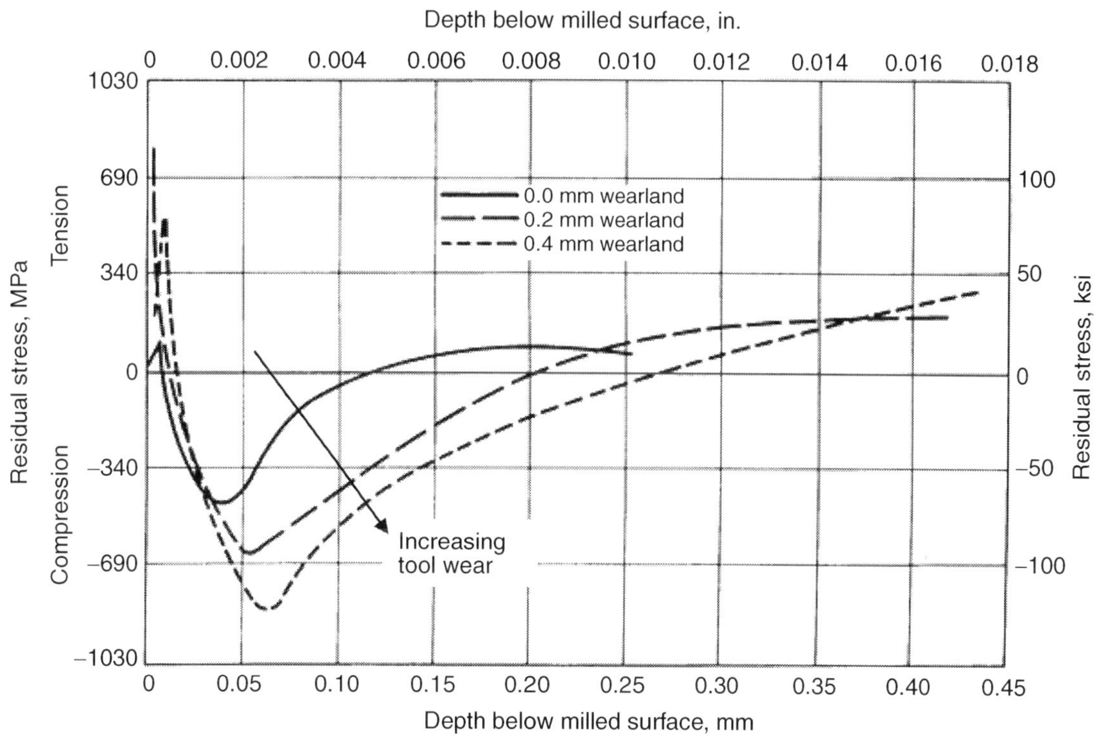

Fig. 49 Residual stress from surface milling 4340 steel quenched and tempered to 52 HRC. Note that while increased tool wear produced higher compressive residual stresses below the surface, it also increased the tensile residual stresses at the surface. Source: Adapted from Ref 27

ever, a major exception to this is a marked decrease in ductility and tensile strength of materials that have been processed using EDM, followed by stress-relief heat treatment.

Surface alterations produced in machining are known to affect the fatigue and stress-corrosion cracking properties of many materials. Extensive investigations have been performed on high-strength alloys, and data illustrating the effect of some machining methods on fatigue strength are given in Table 2. Electropolishing of 4340 steel resulted in a 12% decrease in fatigue strength compared to gentle grinding. Chemical machining of Ti-6Al-4V resulted in an 18% drop in fatigue strength compared to gentle grinding, while the electrochemical machining and electrical discharge machining of Inconel 718 produced a 35% and a 63% drop, respectively, in endurance limit compared to gentle grinding. Generally, ECM, ELP, and CM produce a stress-free surface. These surfaces exhibit reduced fatigue strength when compared to a gently ground surface, because the fatigue enhancement by the compressive stress associated with low-stress grinding is not present.

Endurance limits vary with selected changes in grinding parameters. When abusively grinding, there is a tendency to form patches or streaks of untempered martensite or overtempered martensite on the surface. It has been found that the presence of either one of these two microconstituents is usually associated with a significant drop in fatigue strength. For example, the presence of a depth of untempered martensite as little as 0.01 mm (0.0005 in.) or as large as 0.09 mm (0.0035 in.) produces a decrease in endurance limit from 760 MPa (110 ksi) down to 480 to 520 MPa (70 to 75 ksi), as shown in Fig. 50. Typically, residual tensile stresses are involved in this condition. Retempering of the workpiece does not correct the problem. Although it tempers the untempered martensite and reduces its brittleness, it does not restore the softening of the overtempered martensite to its prior condition. In addition, the tempering cycle does not reduce the residual tensile stresses formed in the abusive grinding operation.

Table 2 Effect of method of machining on fatigue strength

		Endurance limit in bending, 10^7 cycles		
Alloy	Machining operation	MPa	ksi	Gentle grinding, %
4340 steel, 50 HRC	Gentle grinding	703	102	100
	Electropolishing	620	90	88
	Abusive grinding	430	62	61
Ti-6Al-4V, 32 HRC	Gentle milling	480	70	113
	Gentle grinding	430	62	100
	Chemical milling	350	51	82
	Abusive milling	220	32	52
	Abusive grinding	90	13	21
Inconel 718, aged, 44 HRC	Gentle grinding	410	60	100
	Electrochemical machining	270	39	65
	Conventional grinding	165	24	40
	Electrical discharge machining	150	22	37

Source: Ref 26

The effect on fatigue strength from EDM and grinding of aged Inconel 718 is illustrated in Fig. 51. After gentle grinding, Inconel 718 had an endurance limit of 410 MPa (60 ksi). With either gentle or abusive EDM, the endurance limit dropped to 150 MPa (22 ksi).

Fatigue-Life Improvement

One of the most effective methods of improving fatigue life is to induce residual compressive stresses on the surfaces of the part. This is often ac-

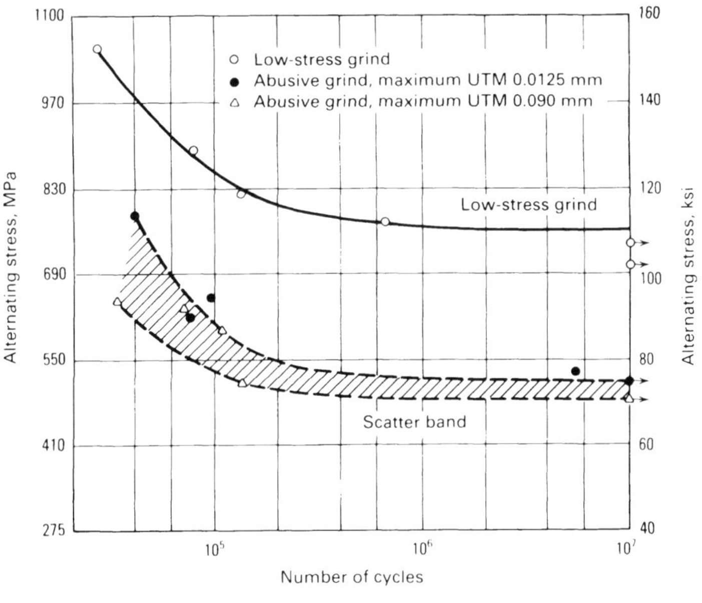

Fig. 50 Loss of fatigue strength from the abusive grinding of 4340 steel quenched and tempered to 50 HRC. UTM, untempered martensite. Source: Ref 27

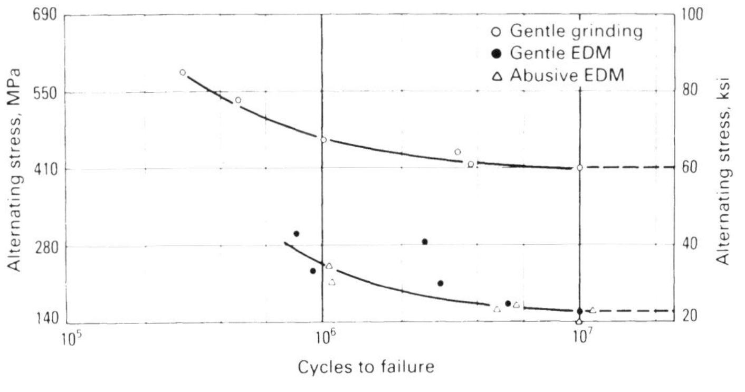

Fig. 51 Effect of electrical discharge machining (EDM) and grinding on the fatigue strength of Inconel 718. Fatigue tests involved cantilever bending at room temperature and zero mean stress. Source: Ref 26

complished by shot peening or by surface rolling with contoured rollers. The advantage of shot peening is that it can be applied to large surfaces and to contoured or irregular surfaces. In shot peening, fine steel or cast iron shot is propelled into the surface at high velocities. The severity of the stress produced by shot peening is controlled by measuring the residual deformation of shot-peened specimens called Almen strips. In addition to inducing compressive stress on the surface, the surface layers are also strengthened by cold working. Shot peening not only enhances the properties of gently machined materials but also improves the properties of metals that have been processed by machining techniques that tend to abuse or degrade fatigue strength and other mechanical properties.

Important variables in shot peening include the hardness of the shot, the size, the shape, and the velocity. Shot peening must be carefully controlled so as to not introduce surface damage to the part. The surfaces produced by three different peening processes on 7050 aluminum are shown in Fig. 52. Because aluminum alloys are rather soft, glass beads are often used instead

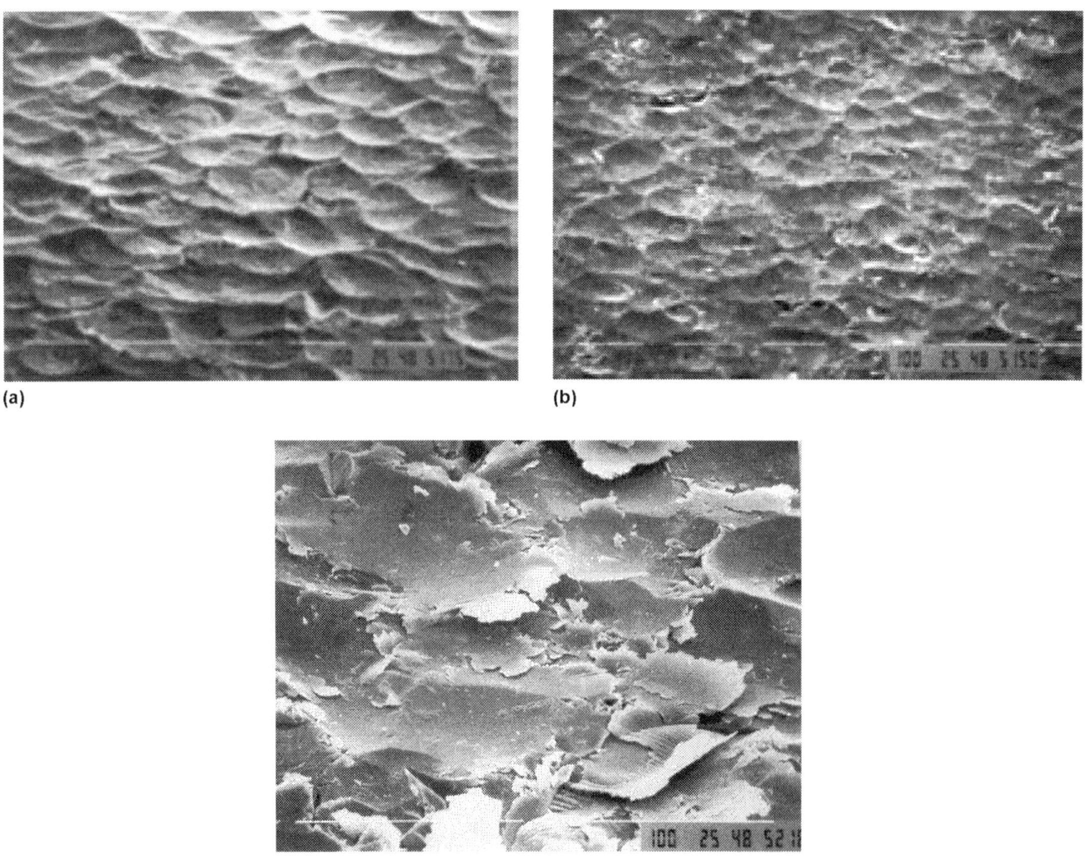

(a) (b) (c)

Fig. 52 Peened surfaces of 7050 aluminum; (a) steel shot, (b) glass beads, (c) surface tears due to overpeening. Source: Ref 28 of steel shot. Overpeening (too high an intensity or too long a dwell time) can produce surface tears (Fig. 52c) that are detrimental to fatigue life.

Shot peening can improve the fatigue strength of machined metals (Table 3). For example, on 4340 steel at 50 HRC, shot peening improved the fatigue strength of a gently ground surface by 10%. For Inconel 718 heat treated to 44 HRC, shot peening after roughing EDM increased the fatigue strength to 130% of the gently ground level. Similar improvements in fatigue strength by shot peening occur on surfaces produced by ECM and ELP. Shot peening is more effective in high-cycle fatigue than in higher-stress low-cycle fatigue. In low-cycle fatigue, large stresses in the plastic range cause "fading" of the residual-stress pattern. It has also been shown that the compressive stresses introduced by shot peening help retard stress-corrosion cracking.

Steel parts subjected to wear conditions are often carburized or nitrided to harden the surface for greater wear resistance. Because both of these processes also introduce residual compressive stresses on the surface, there is also an improvement in fatigue life. The greatest improvement occurs where a high stress gradient occurs, as in bending or torsion, with less improvement in axial loading. The effect of nitriding on the fatigue strength of steel crankshafts is shown in Fig. 53. Nitriding is extremely effective in improving the fatigue strength of notched specimens. Other surface-hardening methods, such as flame or induction hardening, produce similar effects. Case hardening is more effective in components that have a stress gradient, as in bending or where there is a notch. For axially loaded components, fatigue cracks can form under the case in the softer, weaker core at the boundary between the two.

Castings can be hot isostatic pressed to help reduce internal porosity. Hot isostatic pressing (HIP) of aluminum castings is usually conducted using argon pressure at 100 MPa (15 ksi) and temperatures in the range of

Table 3 Effect of method of shot peening on fatigue strength

Alloy	Machining operation	Endurance limit in bending, 10^7 cycles, MPa	ksi	Gentle grinding, %
4340 steel, 50 HRC	Gentle surface ground	703	102	100
	Gentle surface ground and shot peened	772	112	110
	Abusive surface ground	430	62	61
	Abusive surface ground and shot peened	630	92	90
	Electropolished	620	90	88
	Electropolished and shot peened	660	96	94
Inconel 718, solution treated and aged, 44 HRC	Gentle surface ground	410	60	100
	EDM roughing	170	25	42
	EDM roughing and shot peened	540	78	130
	EDM finishing	170	25	42
	EDM finishing and shot peened	480	70	117
	ECM	285	41	68
	ECM and shot peened	560	81	135
	ELP	290	42	70
	ELP and shot peened	540	79	132

Note: Shot size: S110; shot hardness: 50–55 HRC; coverage: 300%. EDM, electrical discharge machine; ECM, electrochemical machining; ELP, electropolishing. Source: Ref 29

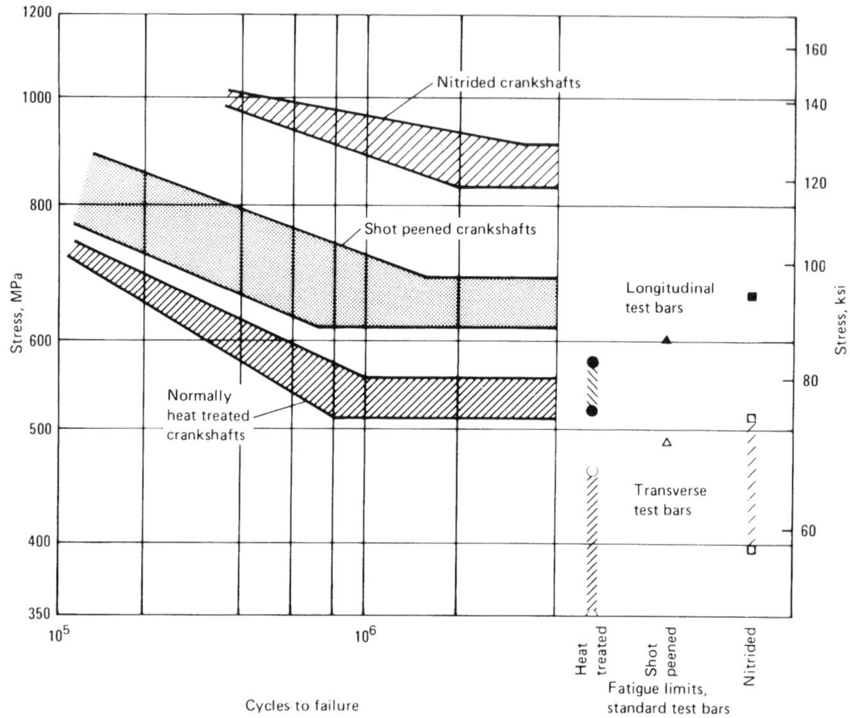

Fig. 53 Effect of nitriding and shot peening on fatigue behavior. Comparison between fatigue limits of crankshafts (*S-N* bands) and fatigue limits of separate test bars, which are indicated by plotted points at right. Steel was 4340. Source: Ref 30

480 to 525 °C (900 to 980 °F). Hot isostatic pressing usually results in improved mechanical properties, especially fatigue strength, but of course, it adds to the cost and cycle time. The improvement in fatigue life for the aluminum casting alloy A201.0-T7 as a result of HIP is shown in Fig. 54.

Methods for fatigue improvement of mechanically fastened joints and welded structures are given in Chapter 7, "Metallic Joints—Mechanically Fastened and Welded," in this book.

Fatigue Design Methodologies

Since the 1800s, a number of design philosophies or methodologies have evolved to deal with design against fatigue failures. To be usable in anything other than a comparative sense, fatigue properties must be consistent with one of three general fatigue design philosophies. Each of these has a concomitant design methodology and one or more means of representing testing data that provide the "properties" of interest. These are:

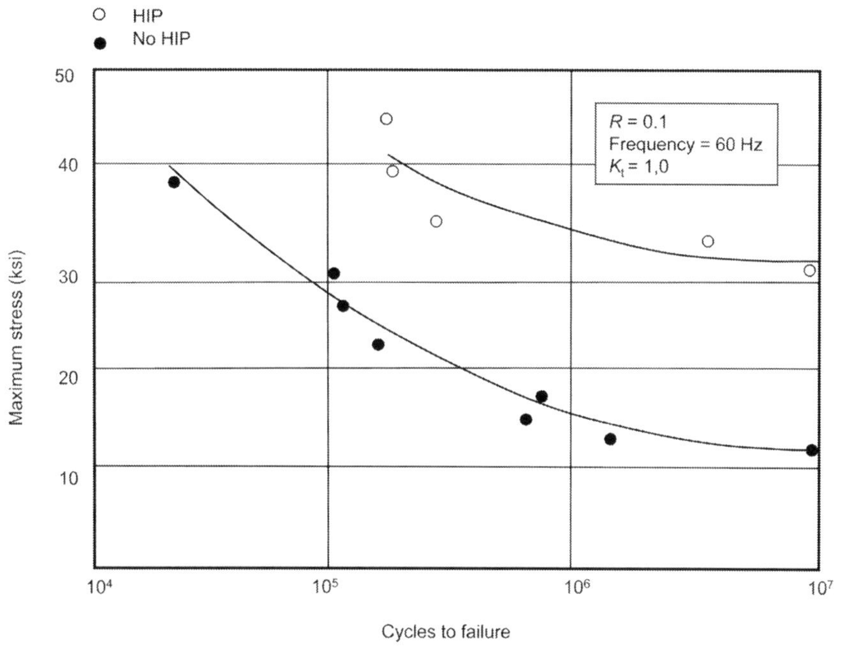

Fig. 54 Effect of hot isostatic pressing (HIP) on fatigue life of A201.0-T7 aluminum casting. Source: Ref 31

Design philosophy	Design methodology	Principal testing data description
Safe life, infinite life	Stress life	*S-N*
Safe life, finite life	Strain life	ε-N
Damage tolerant	Fracture mechanics	da/dn-ΔK

The safe-life, infinite-life philosophy is the oldest of the approaches to fatigue. Examples of attempts to understand fatigue by means of properties, determinations, and representations that relate to this method include August Wöhler's work on railroad axles in Germany in the mid-1800s. The design method is stress life, and a general property representation would be *S-N* (stress versus log number of cycles to failure). Failure in *S-N* testing is typically defined by total separation of the sample.

General applicability of the stress-life method is restricted to circumstances where continuum, "no cracks" assumptions can be applied. However, some design guidelines for weldments, which inherently contain discontinuities, offer what amount to residual-life and runout determinations for a variety of process and joint types that generally follow the safe-life, infinite-life approach. The advantages of this method are simplicity and ease of application, and it can offer some initial perspective on a given situation. It is best applied in or near the elastic range, addressing constant amplitude loading situations in what has been called the long-life, hence infinite-life regime.

The stress-life approach seems best applied to components that look like the test samples and are approximately the same size; this satisfies the similitude associated with the use of total separation as a failure criterion. Much of the technology in application of this approach is based on ferrous metals, especially steels. Other materials may not respond in a similar manner. Given the extensive history of the stress-life method, substantial property data are available, but beware of the testing conditions employed in producing older data.

Through the 1940s and 1950s, mechanical designs pushed to further extremes in advanced machinery, resulting in higher loads and stresses and thus moving into the plastic regime of material behavior and a more explicit consideration of finite-lived components. For these conditions, the description of local events in terms of strain made more sense and resulted in the development of assessment techniques that used strain as a determining quantity. The general data (property) presentation is in terms of ε-N (log strain versus log number of cycles or number of reversals to failure). The failure criterion for samples is usually the detection of a small crack in the sample or some equivalent measure related to a substantive change in load-deflection response, although failure may also be defined by separation.

Employment of strain is a consistent extension of the stress-life approach. As with the safe-life, infinite-life approach, the strain-based safe-life, finite-life philosophy relies on the "no cracks" restriction of continuous media. While considerably more complicated, this technique offers advantages: it includes plastic response, addresses finite-life situations on a sounder technical basis, can be more readily generalized to different geometries, has greater adaptability to variable amplitude situations, and can account for a variety of other effects. The strain-life method is better suited to handling a greater diversity of materials; for example, it is independent of assuming steellike response for modification factors. Because it does not necessarily attempt to relate to total failure (separation) of the part but can rely on what has become known as "initiation" for defining failure, it has a substantial advantage over the stress-life method. Difficulties in applying the method arise because it is more complex, is more computationally intensive, and has more complicated property descriptions. In addition, because this method does not have as extensive a history, properties may not be as readily available.

From a design standpoint, there are some circumstances where inspection is not a regularly employed practice, impractical, unfeasible, or occasionally physically impossible. These situations are prime candidates for the application of the safe-life techniques when coupled with the appropriate technologies to demonstrate the likelihood of failure to be sufficiently remote.

The notable connection between the two techniques described previously is the necessary assumption of continuity (i.e., "no cracks"). However, many components, assemblies, and structures have cracklike discontinuities induced during service or repair or as a result of primary or secondary processing, fabrication, or manufacturing. It is abundantly clear that in many instances, parts containing such discontinuities do continue to bear load and can operate safely for extended periods of time. Developments from the 1960s and before have produced the third design philosophy: damage tolerant. It is intended expressly to address the issue of cracked components.

In the case where a crack is present, an alternative controlling quantity is employed. Typically this is the mode I stress-intensity range at the crack tip (ΔK_I), determined as a function of crack location, orientation, and size within the geometry of the part. This fracture mechanics parameter is then related to the potential for crack extension under the imposed cyclic loads for either subcritical growth or the initiation of unstable fracture of the part. It is markedly different from the other two approaches. Property descriptions for the crack extension under cyclic loading are typically da/dn-ΔK_I curves (log crack growth rate versus log stress-intensity range).

The advantage of the damage-tolerant design philosophy is obviously the ability to treat cracked objects in a direct and appropriate fashion. The previous methods only allow for the immediate removal of cracked structure. Use of the stress-intensity values and appropriate data (properties) allows the number of cycles of crack growth over a range of crack sizes to be estimated and fracture to be predicted. The clear tie of crack size, orientation, and geometry to nondestructive evaluation (NDE) is also a plus. Disadvantages are possibly computationally intensive stress-intensity factor determinations, greater complexity in development and modeling of property data, and the necessity to perform numerical integration to determine crack growth. In addition, the predicted lives are influenced considerably by the initial crack size used in the calculation, requiring quantitative development of probability of detection for each type of NDE technique employed. Related to the initial crack size consideration is the inability of this approach to model effectively that the component was actually suitable for modeling as a continuum, which eliminates the so-called initiation portion of the part life. The damage tolerance approach is covered in more detail in Chapter 8, "Fracture Control and Damage Tolerance Analysis," in this book.

Although much progress has been made in the design against fatigue failures, they still occur with disturbing frequency. For any design, it is imperative that part or component testing be conducted prior to placing the part in service. In addition, it is important that the test conditions are representative of the actual environment that the part will experience in service.

ACKNOWLEDGMENTS

Portions of the material for this chapter came from "Fatigue Fracture" by D.J. Wulpi in *Understanding How Components Fail*, 2nd ed., ASM International, 1999; "Fatigue Failures" in *Failure Analysis and Prevention*, Vol 11, *ASM Handbook*, ASM International, 2002; "Design for Fatigue Resistance" by E. Krempl and "Fundamentals of Modern Fatigue Analysis for Design" by M.R. Mitchell, both in *Fatigue and Fracture*, Vol 19, *ASM Handbook*, ASM International, 1996; and "Fatigue" in *Elements of Metallurgy and Engineering Alloys* by F.C. Campbell, ASM International, 2008.

REFERENCES

1. F.C. Campbell, Fatigue, *Elements of Metallurgy and Engineering Alloys,* ASM International, 2008
2. H.W. Hayden, W.G. Moffatt, and J. Wulf, *The Structure and Properties of Materials,* Vol III, John Wiley, 1965
3. S. Kalpakjian, *Manufacturing Engineering and Technology,* 3rd ed., Addison-Wesley Publishing Co., 1995
4. H.E. Boyer, *Atlas of Fatigue Curves,* ASM International, 1986
5. A.F. Liu, *Mechanics and Mechanisms of Fracture: An Introduction,* ASM International, 2005
6. E. Krempl, Design for Fatigue Resistance, *Materials Selection and Design,* Vol 20, *ASM Handbook,* ASM International, 1997
7. *Properties and Selection of Metals,* Vol 1, *Metals Handbook,* 8th ed., American Society for Metals, 1961
8. R.A. Higgins, *Engineering Metallurgy—Applied Physical Metallurgy,* 6th ed., Arnold, 1993
9. P.G. Forrest, *Fatigue of Metals,* Pergamon Press, 1962, p 47
10. P.E.K. Frost, J. Marsh, and L.P. Cook, *Metal Fatigue,* Clarendon Press, 1974, p 241
11. M.R. Mitchell, Fundamentals of Modern Fatigue Analysis for Design, *Fatigue and Fracture,* Vol 19, *ASM Handbook,* ASM International, 1996
12. Y. Higashida and F.V. Lawrence, "Strain Controlled Fatigue Behavior of Weld Metal and Heat-Affected Base Metal in A36 and A514 Steel Welds," Fracture Control Program Report No. 22, University of Illinois, College of Engineering, Aug 1976
13. R.W. Landgraf, Cycle Deformation Behavior of Engineering Alloys, *Proceedings of Fatigue—Fundamental and Applied Aspects Seminar,* Aug 15–18, 1977 (Remforsa, Sweden)
14. M.A. Meyers and K.K. Chawla, *Mechanical Metallurgy—Principles and Applications,* Prentice-Hall Inc., 1984
15. D.J. Wulpi, Fatigue Fracture, *Understanding How Components Fail,* 2nd ed., ASM International, 1999

16. D.A. Ryder, "The Elements of Fractography," AGARDograph No. AGARD-AG-155-71, Advisory Group for Aerospace Research and Development of NATO, 1971
17. C.M. Laird, STP 415, American Society for Testing and Materials, 1966, p 130
18. Fatigue Failures, *Failure Analysis and Prevention,* Vol 11, *ASM Handbook,* ASM International, 2002
19. R.M.N. Pelloux, *Met. Q.,* Nov 1965, p 34
20. *Fractography and Atlas of Fractographs,* Vol 9, *Metals Handbook,* 8th ed., American Society for Metals, 1974, p 68–70, 85–89
21. T.L. Anderson, *Fracture Mechanics: Fundamentals and Applications,* 3rd ed., Taylor & Francis, 2005
22. R.C. McClung, K.S. Chan, S.J. Hudak, Jr., and D.L. Davidson, Behavior of Small Fatigue Cracks, *Fatigue and Fracture,* Vol 19, *ASM Handbook,* ASM International, 1996
23. J. Collins and S. Daniewicz, Failure Considerations, *Mechanical Engineers Handbook,* John Wiley & Sons, Inc., 1998
24. R.C. Juvinall, *Stress, Strain, and Strength,* McGraw-Hill Book Co., 1967
25. G.E. Dieter, *Mechanical Metallurgy,* 3rd ed., McGraw-Hill Book Co., 1986, p 375–431
26. M. Field, J.F. Kahles, and J.T. Cammett, A Review of Measuring Methods for Surface Integrity, *Ann. CIRP,* Vol 2 (No. 1), 1972
27. M. Field, J.F. Kahles, and W.P. Koster, Surface Finish and Surface Integrity, *Machining,* Vol 16, *ASM Handbook,* ASM International, 1989
28. P.K. Sharp and G. Clark, "The Effect of Peening on the Fatigue Life of 7050 Aluminium Alloy," DSTO-RR-0208, Defence Science and Technology Organisation, 2001
29. W.P. Koster, L.R. Gatto, and J.T.Cammett, Influence of Shot Peening on Surface Integrity of Some Machined Aerospace Materials, *Proceedings of the First International Conference on Shot Peening,* Sept 1981 (Paris, France), Pergamon Press, p 287–293
30. B. Boardman, Fatigue Resistance of Steels, *Properties and Selection: Irons, Steels, and High-Performance Alloys,* Vol 1, *ASM Handbook,* ASM International, 1990
31. J.R. Davis, *ASM Specialty Handbook: Aluminum and Aluminum Alloys,* ASM International, 1993

CHAPTER 6

Fatigue and Fracture of Engineering Alloys

ALTHOUGH VIRTUALLY ALL design and standard specifications require the definition of the tensile properties for a material, these data are only partly indicative of mechanical resistance to failure in service. Except for those situations where gross yielding or highly ductile fracture represents limiting failure conditions, tensile strength and yield strength are usually insufficient requirements for design of fracture-resistant structures. Strength by itself may not be sufficient if toughness, resistance to corrosion, stress corrosion, or fatigue are reduced too much in achieving high strength. This chapter covers some of the aspects of fatigue and fracture of three important engineering classes of structural alloys: steels, aluminum alloys, and titanium alloys.

Fracture Toughness of Steels

A relative comparison of the fracture toughness (K_{Ic}) of steel, titanium, and aluminum alloys is shown in Fig. 1. This figure shows that as the strength of each alloy class increases, the fracture toughness decreases. In addition, it shows that the alloys with the highest fracture toughness are steels, followed by titanium, and then aluminum.

The fracture toughness of high-strength materials is almost always inversely proportional to their yield strength, as shown in the data for steel alloys in Fig. 2. The high-fracture-toughness steels normally have more ductile low-carbon martensite and retained metastable austenite as dominant phases in their microstructures, while steels that contain large amounts of ferrite and pearlite have lower fracture toughness. Impurities, inclusions, and coarse microconstituents also lower the toughness. Decreases in grain size produce greater toughness and strength in virtually all metals and is the only method of simultaneously increasing both strength and toughness. A summary of the effects of microstructure on the toughness of steels is shown in Table 1.

For the fracture of a mild steel at low temperatures, it is reasonable to assume that fracture will occur when the conditions at the crack tip are such that the plastic constraint factor (pcf) elevates the yield stress (σ_{ys}) to a level equivalent to the fracture stress (σ_f). This idea may be expressed by the condition:

$$\text{pcf} \cdot \sigma_{ys} = \sigma_f \qquad \text{(Eq 1)}$$

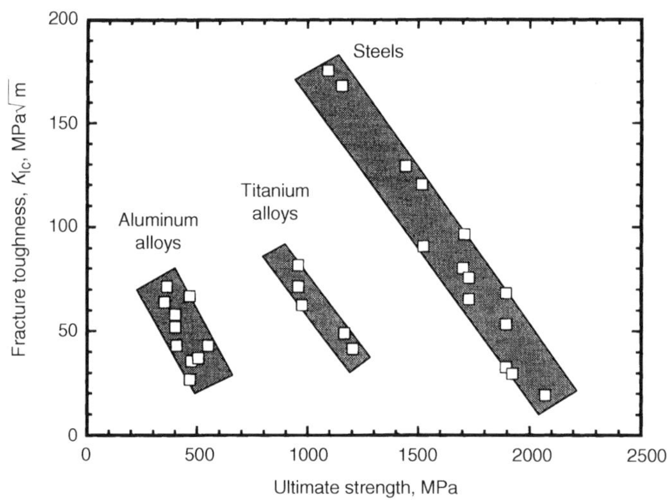

Fig. 1 Fracture toughness as a function of strength for high-strength structural alloys. Source: Ref 1

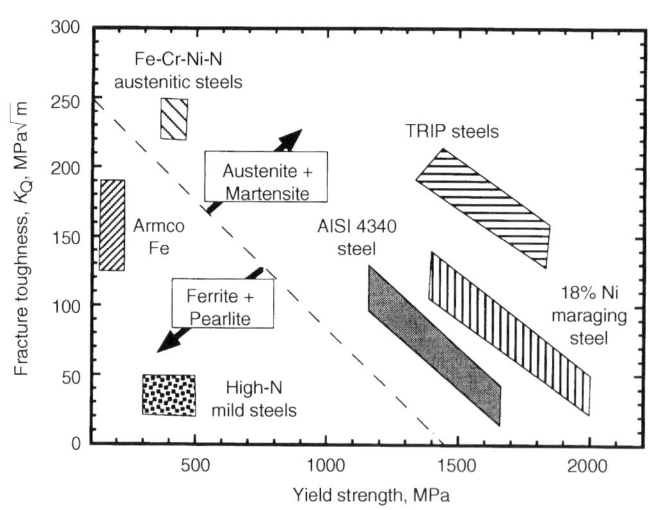

Fig. 2 Fracture toughness as a function of yield strength for structural steels. Source: Ref 1

Both the yield strength and the fracture stress increase with decreasing grain size, and the pcf increases with K_{Ic} and decreases with σ_{ys}. This can be understood by considering the case in which the fracture toughness is low and the yield strength is high. In this case, the low fracture toughness does not allow crack tip stresses to build up to sufficiently high values to exert significant constraint. Graphs showing the effects of the austenite grain size on fracture toughness are shown in Fig. 3.

Steels frequently contain inclusions, and these inclusions, in conjunction with other precipitates that may be present, have a major influence on the fracture toughness of steels. For example, MnS is frequently present, and for a given size of inclusion, it is known that the toughness decreases with an increasing volume fraction of MnS. This is equivalent to saying that the toughness increases with increasing spacing of the inclusions for a given inclusion size. Physically, this is quite understandable, because as the crack is loaded, the crack tip opening displacement will not be able to reach its maximum value due to cracking/decohesion of the inclusion and the linkup with the crack tip. In fact, K_{Ic} would be expected to vary di-

Table 1 Effects of microstructural variables on fracture toughness of steels

Microstructural parameter	Effect on toughness
Grain size	Increase in grain size increases K_{Ic} in austenitic and ferritic steels
Unalloyed retained austenite	Marginal increase in K_{Ic} by crack blunting
Alloyed retained austenite	Significant increase in K_{Ic} by transformation-induced toughening
Interlath and intralath carbides	Decrease K_{Ic} by increasing the tendency to cleave
Impurities (P, S, As, Sn)	Decrease K_{Ic} by temper embrittlement
Sulfide inclusions and coarse carbides	Decrease K_{Ic} by promoting crack or void nucleation
High carbon content (>0.25%)	Decrease K_{Ic} by easily nucleating cleavage
Twinned martensite	Decrease K_{Ic} due to brittleness
Martensite content in quenched steels	Increase K_{Ic}
Ferrite and pearlite in quenched steels	Decrease K_{Ic} of martensitic steels

Source: Ref 1

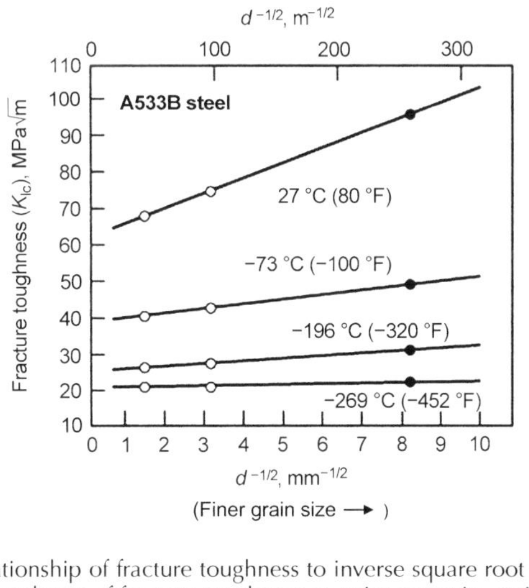

Fig. 3 Relationship of fracture toughness to inverse square root of grain size. Dependence of fracture toughness on prior-austenite grain size at four temperatures. Source: Ref 2

rectly with the square root of the inclusion spacing. It would also be expected that at a given inclusion spacing, the fracture toughness would increase with increasing strength of the inclusion/matrix interface above the yield strength. This variation is shown for a 0.45C-Ni-Cr-Mo-V steel (similar to 4340) in Fig. 4 for a wide range of conditions.

If the volume fraction is held constant and the size of the inclusions is increased, it is easier to nucleate voids at the inclusion/matrix interface, and the toughness will be lowered at a given spacing. It is clear from the data in Fig. 4 that it is very important to control the inclusion size and volume fraction in order to increase the fracture toughness. It should also be noted that there should be a cutoff to the behavior seen in Fig. 4, because if all inclusions were removed, the fracture toughness would reach some limiting value that is characteristic of the material being considered.

Other inclusions also play a role. For example, in quenched and tempered steels, the incoherent carbide particles are rather large, and it has been suggested that the toughness of these steels is limited by the formation of a void sheet between the inclusion voids associated with decohesion between the carbides and the matrix. Void sheet nucleation is shown in Fig. 5. Nucleation of this void sheet will have the effect of limiting the fracture toughness, because the energy absorption process associated with growth of the larger voids to the point of impingement will be interrupted, and early linkup will occur. When such mechanisms operate, it is usual to see both large and small voids on the fracture surface, such as shown in Fig. 6 for a 300-grade maraging steel. The nucleation of the void sheet is

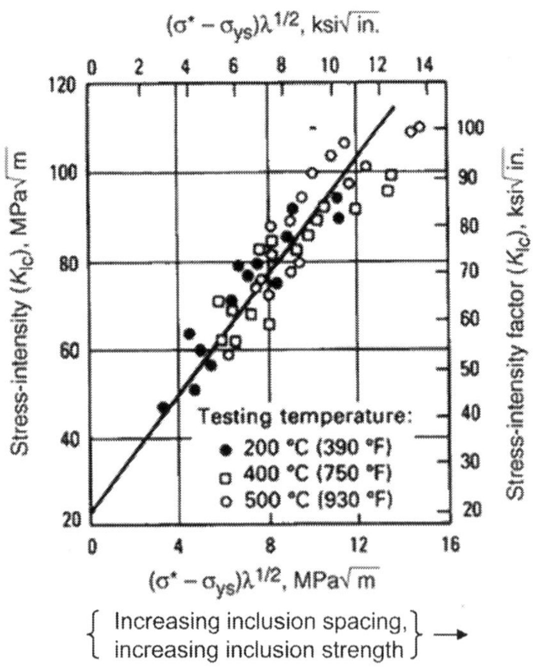

Fig. 4 K_{Ic} for a martensitic 0.45C-Ni-Cr-Mo-V steel as a function of inclusion spacing and yield strength. Source: Ref 3

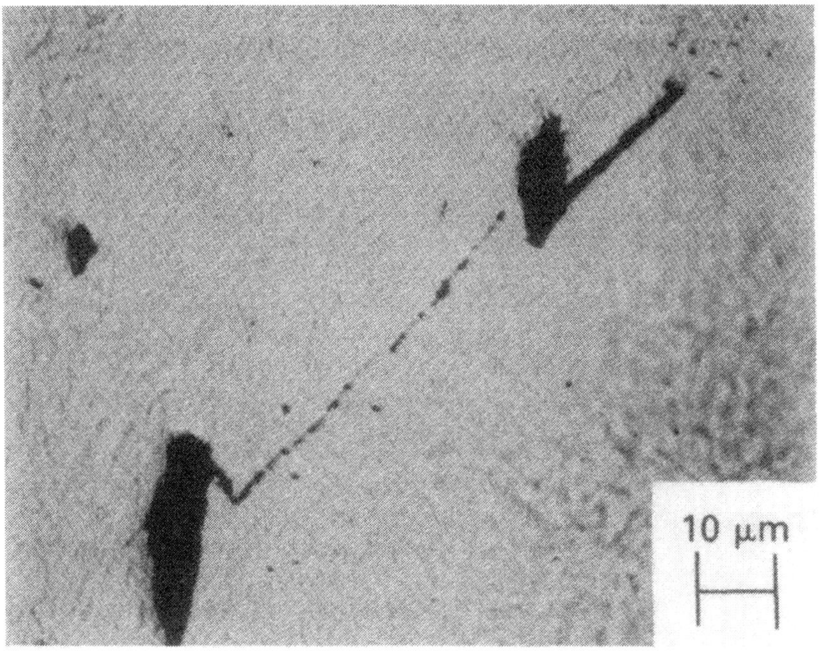

Fig. 5 Void sheet formation in steel. Source: Ref 4

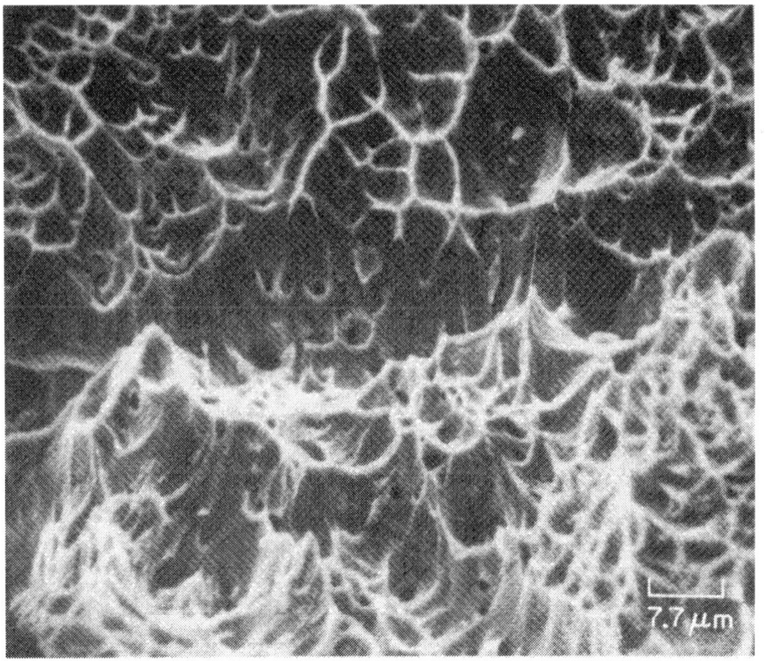

Fig. 6 Scanning electron microscopy view of the surface of the tensile test fracture in 18% Ni, grade 300 maraging steel, showing a portion of the central zone of the fracture, close to the origin. The surface here is composed of equiaxed dimples of two sizes. The large dimples probably formed at Ti(C,N) particles; all the dimples were caused by particles of some sort. Source: Ref 5

easier with large particle sizes. This mechanism explains the relatively high fracture toughness of maraging steels, which contain extremely fine strengthening precipitates, compared to the carbides in common quenched and tempered steels.

Maraging (martensite aging) steels, based on the Fe-Ni-Ti system, have very high strength and high fracture toughness and are designed with very low levels of carbon (<0.05%) and high levels of alloying elements (typically 18% Ni, 3% Ti, and 1 to 2% Co). The benefit of low carbon is twofold: crack-initiating carbides are absent, and the martensite matrix is more ductile. In addition, maraging steels, due to the high nickel content, contain retained austenite that causes extensive plastic deformation and strain-induced martensitic transformation at the crack tip during fracture, thus increasing the fracture resistance. The strengthening lost by the elimination carbides is replaced by precipitation of fine $(Fe,Ni)_3Ti$ phases from martensite obtained by aging. However, under certain conditions, precipitation of phases at grain boundaries leads to deterioration in toughness. Hence, chemistry, processing, and heat treatment conditions must be designed to avoid the grain-boundary precipitation. Typical heat treatment of maraging steels involves austenitization and oil quenching or air cooling, followed by aging at high temperature in the austenite + ferrite field.

Metastable Austenite-Based Steels. Transformation-induced plasticity (TRIP) steels with high contents of nickel and manganese retain the high-temperature face-centered cubic austenite at room temperature on quenching. By a judicious choice of composition, this austenite is designed to be metastable after quenching and transformable during deformation. The deformation and the volume change accompanying the austenite-to-martensite transformation increase the energy required for extension of a moving crack, resulting in high fracture toughness. The higher the tensile strain or crack opening displacement at the tip, the larger the volume fraction of transformed martensite and the higher the TRIP effect on toughness, as illustrated in Fig. 7. These steels possess the highest fracture toughness levels attainable in steels and hence are used in mining, drilling, and other applications requiring wear and erosion resistance.

Quenched and Tempered Steels. Fracture resistance in quenched and tempered steels is achieved by eliminating coarse alloy carbides, increasing hardenability to minimize ferrite formation, and alloying to retain austenite at room temperature.

An increase in the austenitization temperature, besides coarsening the grain size, dissolves carbides and nitrides present in steels. This eliminates the crack nucleation from carbides, thereby increasing the fracture toughness, as illustrated for the two steels in Fig. 8. The austenitization time is also critical, and both time and temperature are chosen according to the amount of carbon and alloying elements in the steel. The effect of tempering temperature on fracture toughness depends on the type of prior austenitization treatment, as shown in Fig. 9. For example, the heat treatment

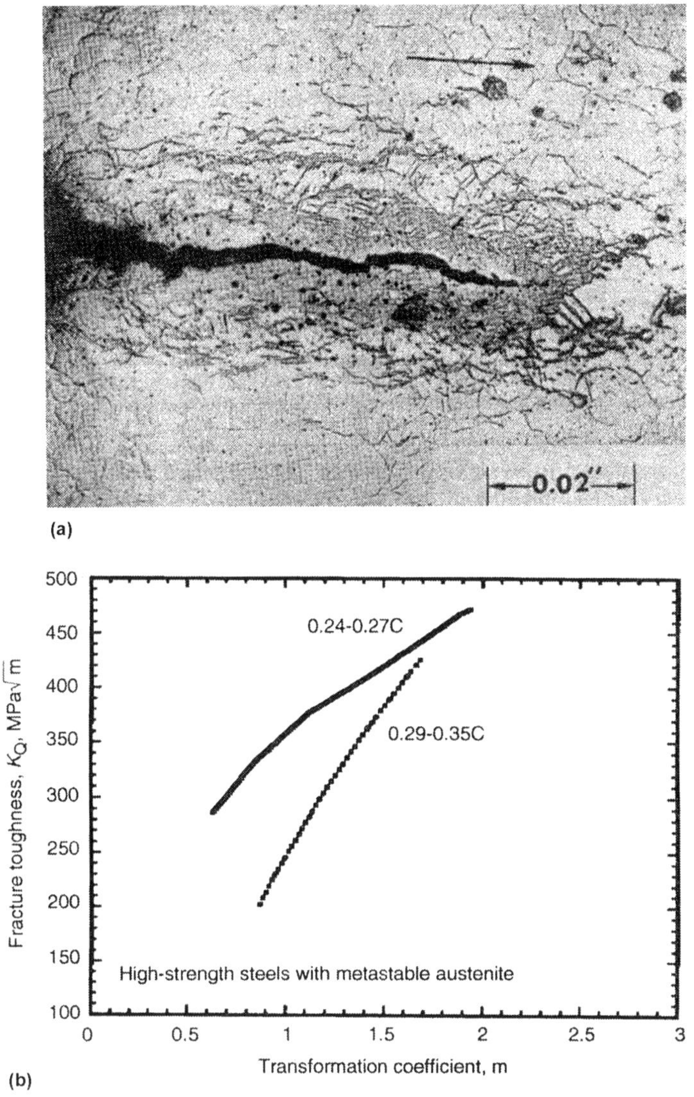

Fig. 7 Transformation-induced plasticity (TRIP) steel crack and toughness. (a) Formation of martensite around a crack in a TRIP steel. (b) Effect of austenite transformation on the fracture toughness of metastable austenitic steels. Source: Ref 1

involving austenitization at 1200 °C (2190 °F), followed by an ice brine quench and refrigeration in liquid nitrogen, results in high fracture toughness levels at all tempering temperatures, compared to austenitizing at 870 °C (1600 °F) and quenching in oil. The higher fracture toughness levels of the former are attributed to the formation of 100% refined martensite upon quenching. Heat treatment at 870 °C (1600 °F) results in a mixture of blocky ferrite and upper bainite having continuous films of carbides at the lath boundaries. This leads to low fracture toughness due to easy crack

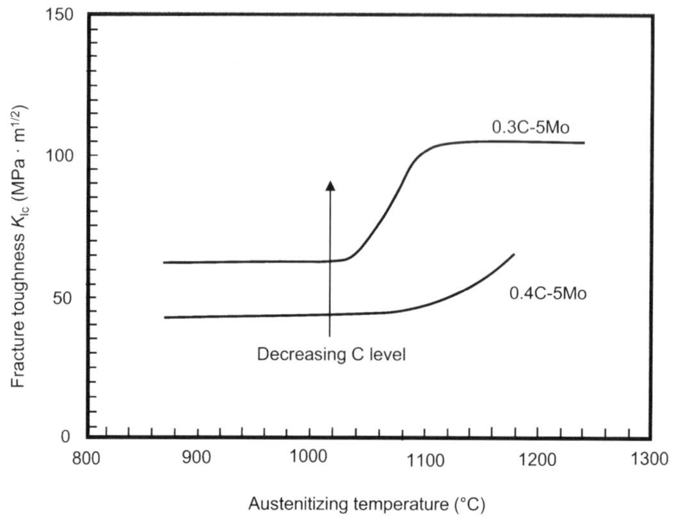

Fig. 8 Effect of austenitization temperature on the fracture toughness of two quenched and tempered steels. Source: Ref 1

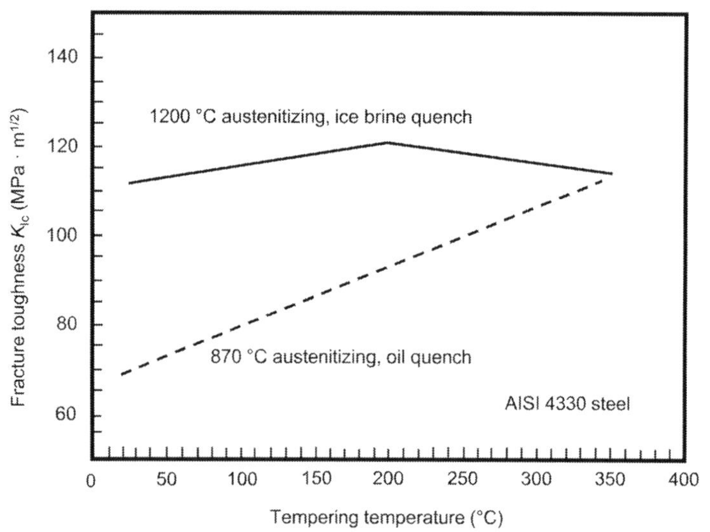

Fig. 9 Effect of tempering temperature on fracture toughness. Source: Ref 1

propagation along the martensitic lath boundaries at low tempering temperatures. However, at higher tempering temperatures, elimination of the continuous carbide films by spheroidization increases the fracture toughness. In addition, in some alloy steels, retained austenite contributes to further increases in fracture toughness, by either crack tip blunting or strain-induced transformation.

Fracture toughness in steels also depends on the nature of martensite—low-carbon ductile martensite or high-carbon twinned martensite. Supersaturated carbon in martensite increases the twinning to accommodate the strains in iron lattice due to carbon. The low fracture toughness levels of some martensitic steels at low tempering temperatures are due to the presence of brittle twinned martensite. An increase in the twin density of martensite results in low fracture toughness, as illustrated in Fig. 10.

In quenched and tempered steels, inclusions also decrease fracture toughness (Fig. 11) by promoting crack nucleation by inclusion fracture,

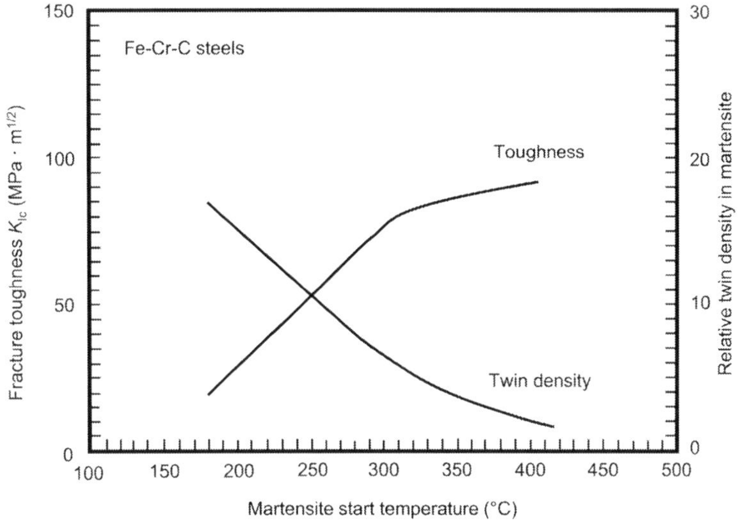

Fig. 10 Fracture toughness and martensite twin density as a function of martensite start temperature for an Fe-Cr-C steel. Source: Ref 1

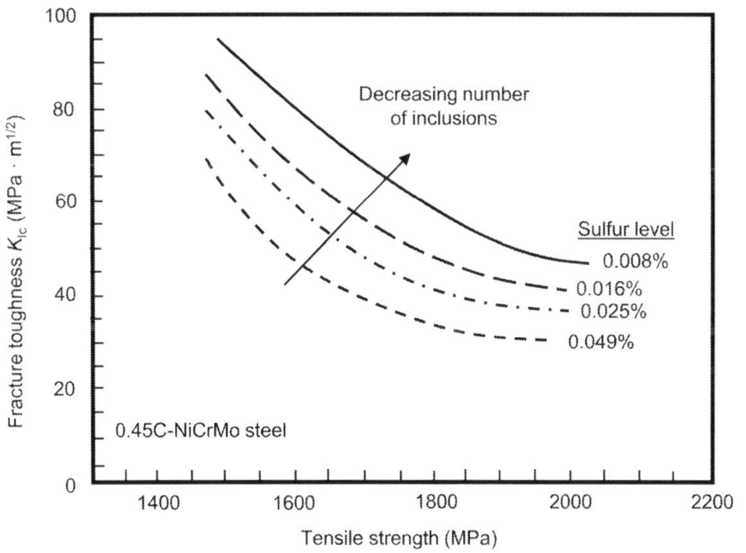

Fig. 11 Effect of sulfur on the fracture toughness vs. strength relationship. Source: Ref 1

void nucleation at the particle-matrix interface, and early coalescence. This reduces the extent of void growth before fracture, limiting the plastic energy absorption in the process zone at the crack tip. A decrease in the sulfur level in steel increases the spacing between the inclusions, thereby increasing the size of the plastically deformed process zone.

Fatigue of Steels

The *S-N* plot shown in Fig. 12 presents data for AISI-SAE 4340 steel, heat treated to a tensile strength of 860 MPa (125 ksi) in the notched and unnotched condition. Figure 13 shows the combinations of cyclic stresses that can be tolerated by the same steel when the specimens are heat treated to different tensile strengths ranging from 860 to 1836 MPa (125 to 265 ksi).

The effect of elevated temperature on the fatigue behavior of 4340 steel heat treated to 1090 MPa (158 ksi) is shown in Fig. 14. An increase in temperature reduces the fatigue strength of the steel and is most deleterious for those applications in which the stress ratio (*R*) lies between 0.4 and 1.0. A decrease in temperature may increase the fatigue limit of steel; however, parts with preexisting cracks may also show decreased total life as temperature is lowered, because of accompanying reductions in critical crack size and fracture toughness.

The effect of notches on the fatigue behavior of the ultrahigh-strength 300M steel is shown in Fig. 15. A K_t value of 2 is obtained in a specimen

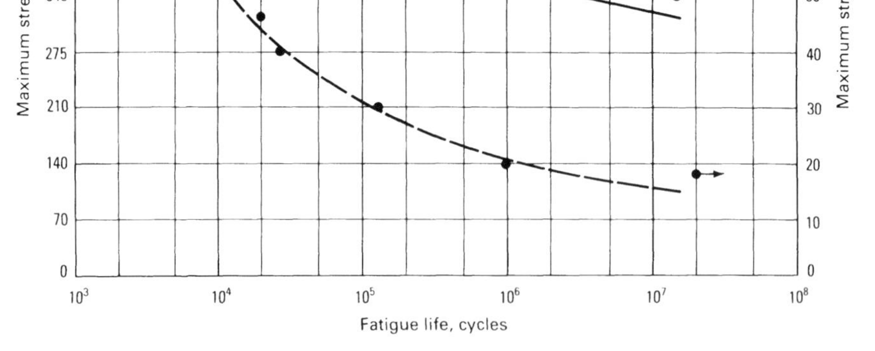

Fig. 12 Room-temperature *S-N* curves for notched and unnotched AISI 4340 alloy steel with a tensile strength of 860 MPa (125 ksi). Stress ratio *R* equals −1.0. Source: Ref 6

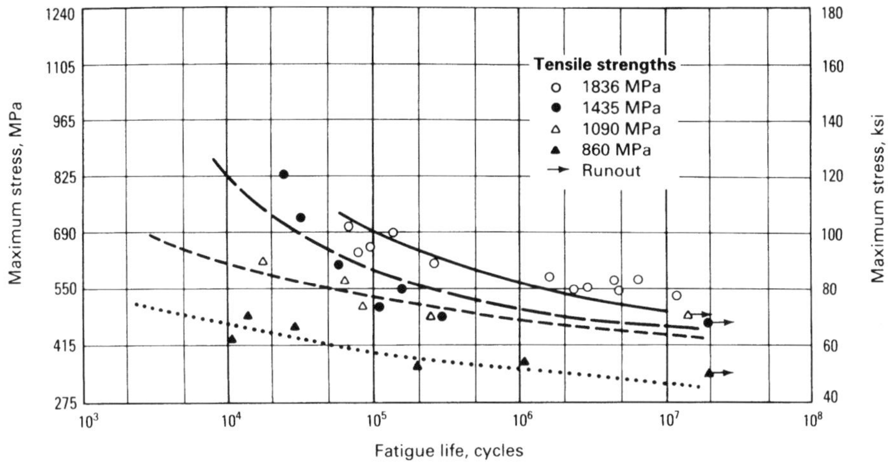

Fig. 13 Room-temperature *S-N* curves for AISI 4340 alloy steel with various ultimate tensile strengths and with $R = -1.0$. Source: Ref 6

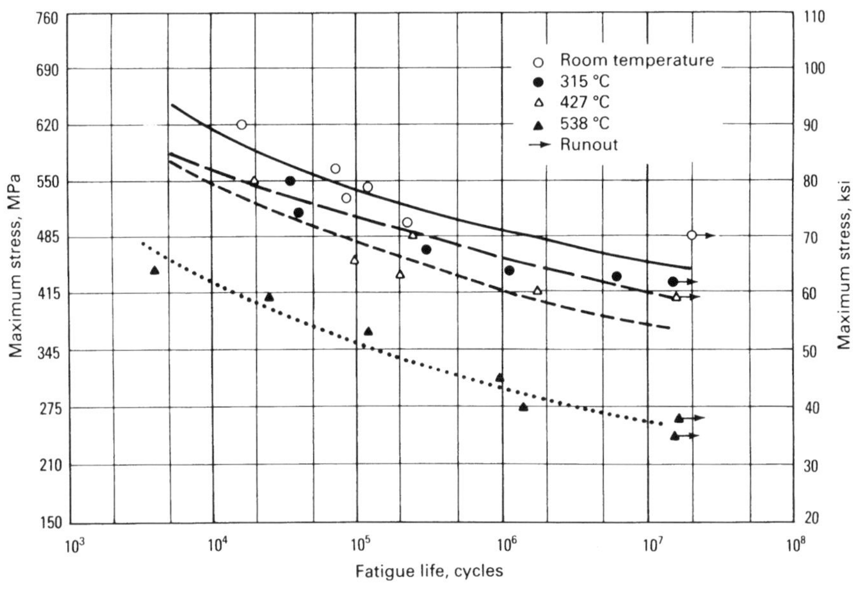

Fig. 14 *S-N* curves at various temperatures for AISI 4340 alloy steel with an ultimate tensile strength of 1090 MPa (158 ksi). Stress ratio *R* equals −1.0. Source: Ref 6

having a notch radius of approximately 1 mm (0.040 in.). For small parts, such a radius is often considered large enough to negate the stress concentration associated with any change in section. However, the significant effect of notches, even those with low stress-concentration factors, on the fatigue behavior of this steel is apparent.

Data such as those presented in Fig. 12 to 15 may not be directly applicable to the design of structures, because these graphs do not take into

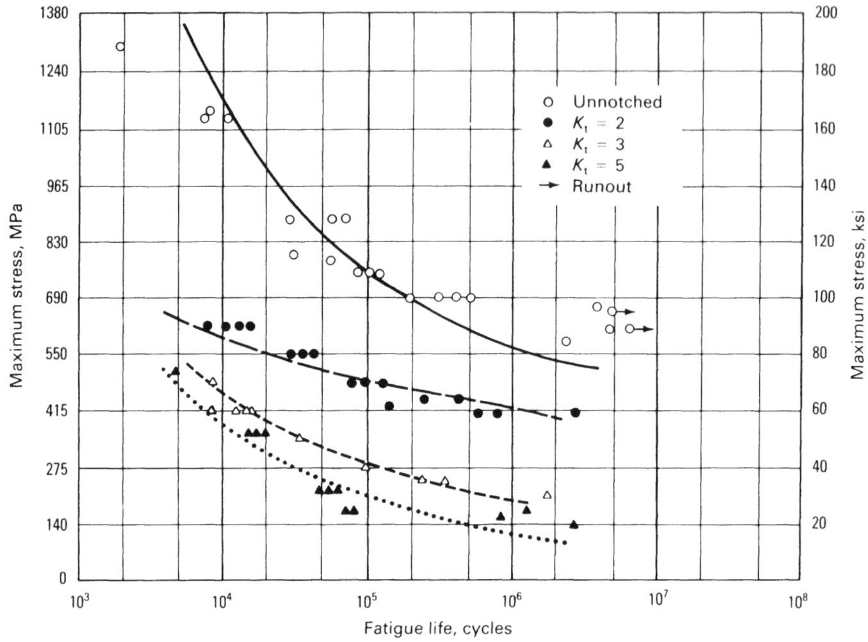

Fig. 15 Room-temperature *S-N* curves for a 300M steel with an ultimate tensile strength of 2000 MPa (290 ksi) having various notch severities. Stress ratio *R* equals 1.0. Source: Ref 6

account the effect of the specific stress concentrations associated with reentrant corners, notches, holes, joints, rough surfaces, and other similar conditions present in fabricated parts. The localized high stresses induced in fabricated parts by stress raisers are of much greater importance for cyclic loading than for static loading.

Stress raisers reduce the fatigue life significantly below those predicted by the direct comparison of the smooth specimen fatigue strength with the nominal calculated stresses for the parts in question. Fabricated parts in simulated service have been found to fail at less than 50,000 repetitions of load, even though the nominal stress was far below that which could be repeated many millions of times on a smooth, machined specimen.

Correction Factors for Test Data. The available fatigue data normally are for a specific type of loading, specimen size, and surface roughness. For instance, the R.R. Moore rotating beam fatigue test machine uses a 7.5 mm (0.3 in.) diameter specimen that is free of any stress concentrations (because of specimen shape and a surface that has been polished to a mirror finish) and that is subjected to completely reversed bending stresses. For the fatigue limits used in design calculations, the correction of fatigue life data can be obtained by multiplying the fatigue limit from testing (N_i) by three factors that take into account the variation in the type of loading, part diameter, and surface roughness:

$$\text{Design fatigue limit} = K_l \cdot K_d \cdot K_s \cdot N_i \quad \text{(Eq 2)}$$

where K_l is the correction factor for the type of loading, K_d for the part diameter, and K_s for the surface roughness. Values of these factors are given in Table 2 and Fig. 16.

Metallurgical Variables Affecting the Fatigue Life of Steels

The metallurgical variables having the most pronounced effects on the fatigue behavior of carbon and low-alloy steels are strength level, ductility, cleanliness of the steel, residual stresses, surface conditions, and aggressive environments.

Table 2 Correlation factors for surface roughness (K_s), type of loading (K_l), and part diameter (K_d) for fatigue life of steel parts

	Value for loading in		
Factor	Bending	Torsion	Tension
K_l	1.0	0.58	0.9(a)
K_d			
where $d \leq 10$ mm (0.4 in.)	1.0	1.0	1.0
where 10 mm (0.4 in.) $< d \leq 50$ mm (2 in.)	0.9	0.9	1.0
K_s	See Fig. 16		

(a) A lower value (0.06 to 0.85) may be used to take into account known or suspected undetermined bending because of load eccentricity. Source: Ref 7

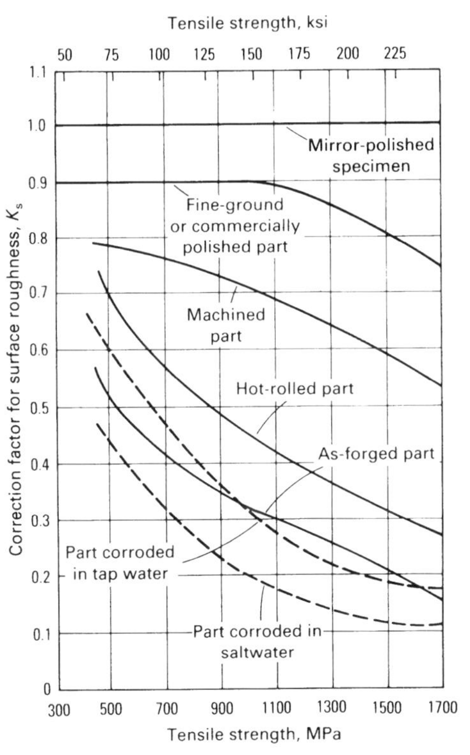

Fig. 16 Surface roughness correction factors for standard rotating beam fatigue life testing of steel parts. See Table 2 for correction factors from part diameter and type of loading. Source: Ref 7

Strength Level. For most steels with a hardness below 400 HB (not including precipitation hardening steels), the fatigue limit is approximately one-half of the ultimate tensile strength. Thus, any heat treatment or alloying addition that increases the strength (or hardness) of a steel can be expected to increase its fatigue limit, as shown in Fig. 17 for various low-alloy steels as a function of hardness. However, for medium-carbon steels, a higher hardness (or strength) may not be associated with improved fatigue behavior in the low-cycle regime (<10^3 cycles) due to the more limited ductility of the higher carbon contents.

Ductility is generally important to fatigue life only under low-cycle fatigue conditions. Exceptions to this include spectrum loading where there is an occasional overload with millions of smaller cycles, or in extremely brittle materials where crack propagation dominates.

Cleanliness of a steel refers to its relative freedom from nonmetallic inclusions. As previously discussed, these inclusions generally have a deleterious effect on the fatigue behavior of steels, particularly for long-life applications. The type, number, size, and distribution of nonmetallic inclusions may have a greater effect on the fatigue life of carbon and alloy steel than differences in composition, microstructure, or stress gradients. However, nonmetallic inclusions are rarely the prime cause of the fatigue failure of production parts; if the design fatigue properties were determined using specimens containing inclusions representative of those in the parts, any effects of these inclusions would already be incorporated in the test results.

Points on the lower curve in Fig. 18 represent the cycles to failure for a few specimens from one bar selected from a lot consisting of several bars

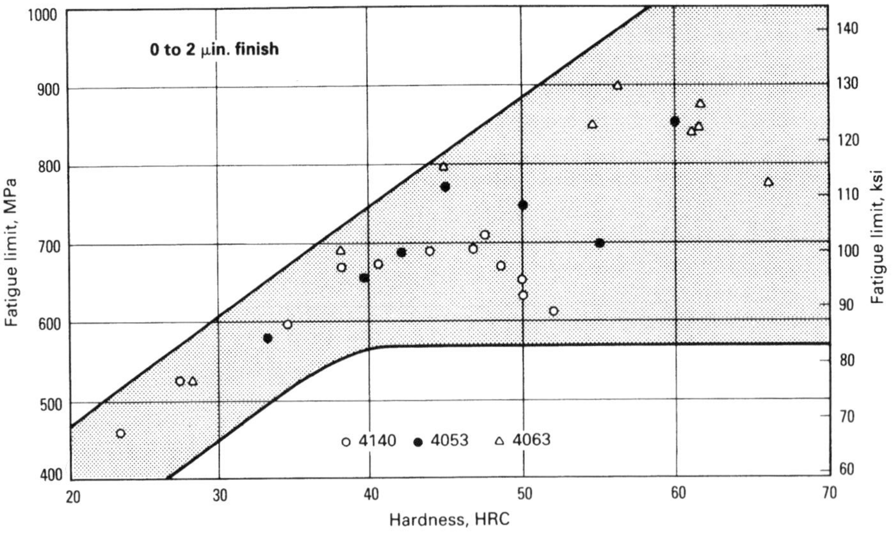

Fig. 17 Effect of carbon content and hardness on fatigue limit of through-hardened and tempered 4140, 4053, and 4063 steels. Source: Ref 8

of 4340H steel. Large spherical inclusions, approximately 0.13 mm (0.005 in.) in diameter, were observed in the fracture surfaces of these specimens. The inclusions were identified as silicate particles. No spherical inclusions larger than 0.02 mm (0.00075 in.) were detected in the other specimens. Large nonmetallic inclusions can often be detected by nondestructive inspection, and steels can be selected on the basis of such inspection. Vacuum melting, which reduces the number and size of nonmetalllic inclusions, is also used to increase the fatigue limits of premium-quality steels.

Surface conditions of a metal part, particularly surface imperfections and roughness, can reduce the fatigue limit of the part. This effect is most apparent for high-strength steels. The interrelationship between surface roughness, method of producing the surface finish, strength level, and fatigue limit is shown in Fig. 16, in which the ordinate represents the fraction of fatigue limit relative to a polished test specimen that could be anticipated for the combination of strength level and surface finish.

Fretting is a wear phenomenon that occurs between two mating surfaces. It is adhesive in nature and is primarily caused by vibration. Usually, fretting is accompanied by oxidation. Fretting usually occurs between two tight-fitting surfaces that are subjected to a cyclic, relative motion of extremely small amplitude. Fretted regions are highly sensitive to fatigue cracking. Under fretting conditions, fatigue cracks are initiated at very low stresses, well below the fatigue limit of nonfretted specimens.

Decarburization is the depletion of carbon from the surface of a steel part. As indicated in Fig. 19, it significantly reduces the fatigue limits of steel. Decarburization from 0.08 to 0.75 mm (0.003 to 0.030 in.) on AISI-SAE 4340 notched specimens that have been heat treated to a strength level of 1860 MPa (270 ksi) reduces the fatigue limit almost as much as a notch with $K_t = 3$. When subjected to the same heat treatment as the core

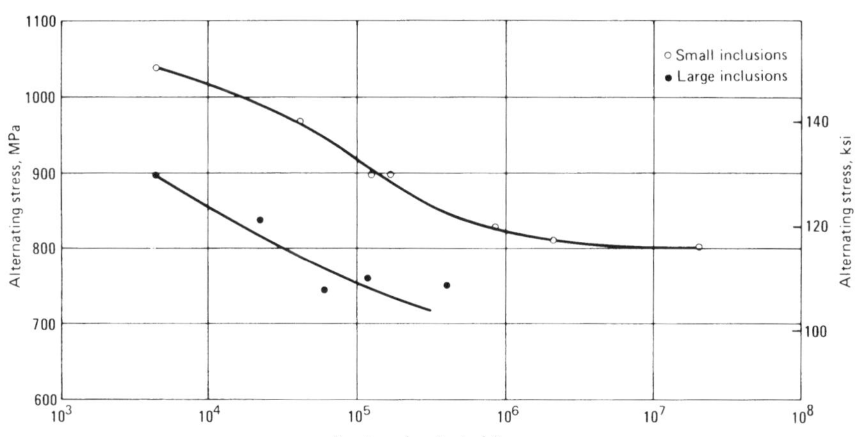

Fig. 18 Effect of nonmetallic inclusion size on fatigue. Steels were two lots of AISI-SAE 4340H; one lot (lower curve) contained abnormally large inclusions; the other lot (upper curve) contained small inclusions. Source: Ref 8

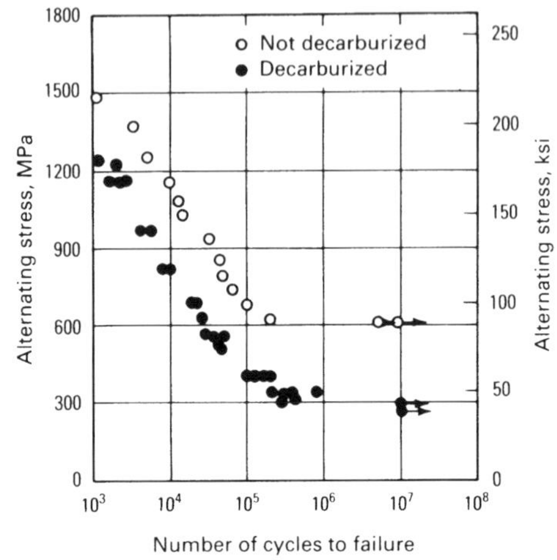

Fig. 19 Effect of decarburization on the fatigue behavior of a steel. Source: Ref 8

of the part, the decarburized surface layer is weaker and therefore less resistant to fatigue than the core. Hardening a part with a decarburized surface can also introduce residual tensile stresses, which reduces the fatigue limit. Results of research studies have indicated that fatigue properties lost through decarburization can be at least partially regained by recarburization (carbon restoration in the surfaces).

Surface Alloying. Carburizing, carbonitriding, and nitriding are three processes for surface alloying. In these processes, carbon, nitrogen, or both elements are diffused into the surface layer of the steel part. The solute atoms strengthen the surface layer of the steel and increase its bulk relative to the metal below the surface. The case and core of a carburized steel part respond differently to the same heat treatment. Because of its higher carbon content, the case is harder after quenching and harder after tempering. To achieve maximum effectiveness of surface alloying, the surface layer must be much thinner than the thickness of the part to maximize the effect of the residual stresses; however, the surface layer must be thick enough to prevent operating stresses from affecting the material just below the surface layer. A particular advantage of surface alloying in the resistance to fatigue is that the alloyed layer closely follows the contours of the part.

Surface Hardening. Induction, flame, laser, and electron beam hardening selectively harden the surface of a steel part; however, the steel must contain sufficient carbon to permit hardening. In each operation, the surface of the part is rapidly heated, and the part is quenched either by externally applied quenchant or by internal mass effect. This treatment forms a surface layer of martensite that is harder than the steel beneath it. Induction, flame, laser, and electron beam hardening can produce beneficial surface residual stresses that are compressive. By comparison, surface residual stresses resulting from through hardening are often tensile. The fatigue lives of through-hardened, carburized, and induction-hardened transmission shafts are compared in Fig. 20.

The importance of a proper case depth on fatigue life is shown in Fig. 21. The hardened case must be deep enough to prevent operating stresses from affecting the steel beneath the case. However, it should be thin enough to maximize the effectiveness of the residual stresses. Three advantages of induction, flame, laser, or electron beam hardening are:

- The core may be heat treated to any appropriate condition.
- The processes produce relatively little distortion.
- The part may be machined before heat treatment.

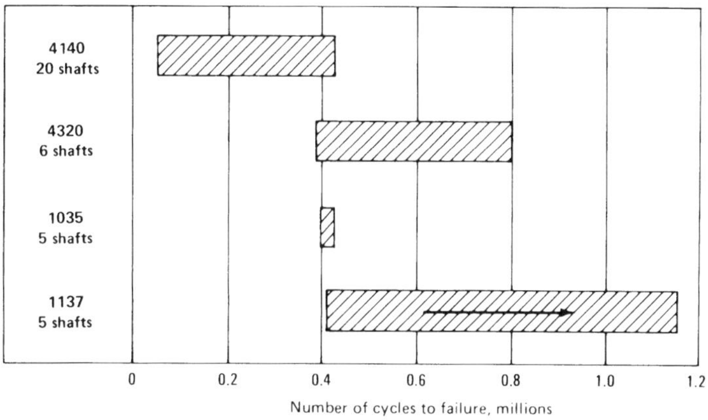

Steel	Surface hardness, HRC	Hardening process
4140	36–42	Through hardened
4320	40–46	Carburized to 1.0–1.3 mm (0.040–0.050 in.)
1035	42–48	Induction hardened to 3 mm (0.120 in.) min effective depth (40 HRC)
1137	42–48	Induction hardened to 3 mm (0.120 in.) min effective depth (40 HRC)

Fig. 20 Effect of carburizing and surface hardening on fatigue life. Comparison of carburized, through-hardened, and induction-hardened transmission shafts tested in torsion. Arrow in lower bar on chart indicates that one shaft had not failed after the test was stopped at the number of cycles shown. Source: Ref 8

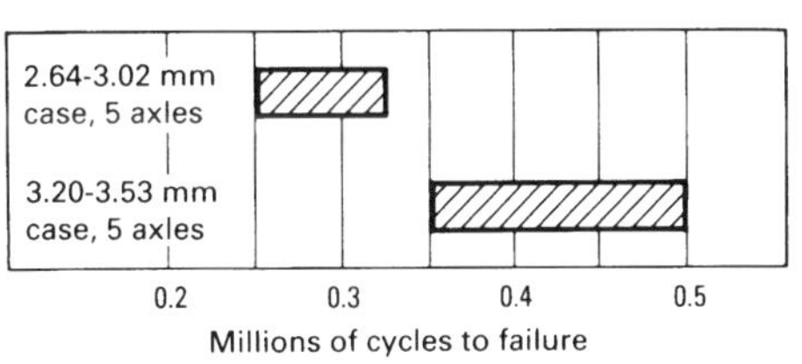

Fig. 21 Effect of case depth on fatigue life. Fatigue tests on induction-hardened 1038 steel automobile axle shafts 32 mm (1.25 in.) in diameter. Case depth ranges given on the chart are depths to 40 HRC. Shafts with lower fatigue life had a total case depth to 20 HRC of 4.5 to 5.2 mm (0.176 to 0.206 in.), and shafts with higher fatigue life, 6.4 to 7.0 mm (0.253 to 0.274 in.). Load in torsion fatigue was 2030 N · m (1500 ft · lbf), and surface hardness was 58 to 60 HRC after hardening. Source: Ref 8

Mechanical working of the surface of a steel part effectively increases the resistance to fatigue. Shot peening and skin rolling are two methods for developing compressive residual stresses at the surface of the part. Shot peening is also useful in recovering the fatigue resistance lost through decarburization of the surface. Decarburized specimens that were shot peened raised the fatigue limit from 275 MPa (40 ksi) after decarburizing to 655 MPa (95 ksi) after shot peening.

Tensile residual stresses at the surface of a steel part can severely reduce its fatigue limit. Such residual stresses can be produced by through hardening, cold drawing, welding, or abusive grinding. For applications involving cyclic loading, parts containing these residual stresses should be given a stress-relief anneal if feasible.

Aggressive environments can substantially reduce the fatigue life of steels. In the absence of the medium causing corrosion, a previously corroded surface can substantially reduce the fatigue life of the steel, as shown in Fig. 16.

Grain size of steel influences fatigue behavior indirectly through its effect on the strength and fracture toughness of the steel. Fine-grained steels have greater fatigue strength than coarse-grained steels.

Composition. An increase in carbon content can increase the fatigue limit of steels, particularly when the steels are hardened to 45 HRC or higher (Fig. 17). Other alloying elements may be required to attain the desired hardenability, but they generally have little effect on fatigue behavior.

Microstructure. For specimens having comparable strength levels, resistance to fatigue depends somewhat on microstructure. A tempered martensite structure provides the highest fatigue limit. However, if the structure as-quenched is not fully martensitic, the fatigue limit will be lower. Pearlitic structures, particularly those with coarse pearlite, have poor resistance to fatigue.

Steel Fatigue Crack Growth

The fatigue crack propagation of structural steels is often considered relatively insensitive to changes among grades. However, steel cleanliness has been shown to have a significant influence on fatigue crack growth rate behavior within a steel. For example, significant anisotropy or directionality of behavior between conventional and calcium-treated A588A steels is shown in Fig. 22. It has been demonstrated that nonmetallic inclusions can accelerate fatigue crack growth, particularly in the through-the-thickness (short transverse-longitudinal, or SL, and short transverse-long transverse, or ST) and transverse (long transverse-longitudinal, or TL) testing orientations, and particularly at higher ΔK levels. This is demonstrated for three other steel grades in Fig. 23. These results indicate that cleanliness and loading orientation must be considered when using fatigue

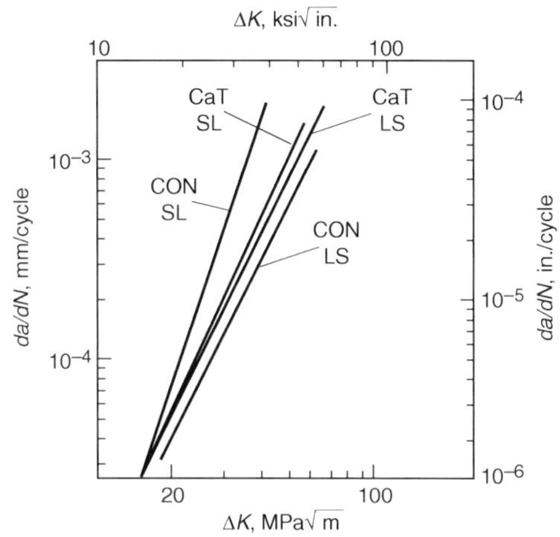

Fig. 22 Fatigue crack growth rate results for two A588A steels showing comparison of LS and SL testing orientations. CON, conventional; CaT, calcium treatment. Improved isotropy of the calcium-treated steel is noted. Source: Ref 9

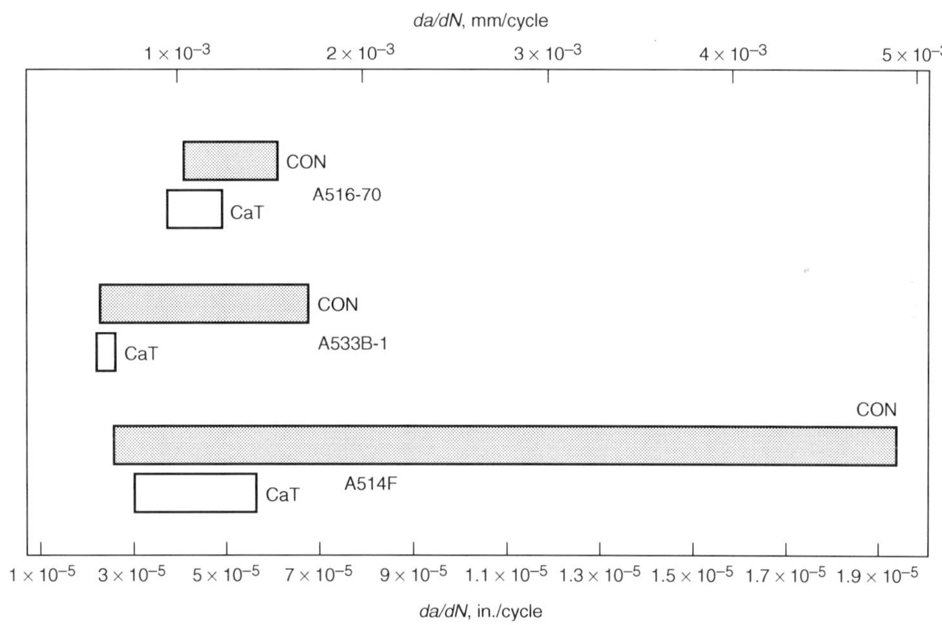

Fig. 23 Range of fatigue crack growth rates at $\Delta K = 55$ MPa · (m)$^{1/2}$ [50 ksi · (in.)$^{1/2}$] in six testing orientations for conventional (CON) and calcium-treated (CaT) quality plates of A516-70, A533B-1, and A514F. Improved isotropy of quality levels is demonstrated for calcium-treated plate. Source: Ref 10

crack growth rate data. If the environment of the application is other than benign air, there can be a further acceleration of fatigue crack growth. Salt water, for example, can have a strong effect, particularly at slower testing frequencies, as illustrated in Fig. 24.

Fracture Toughness of Aluminum Alloys

The fracture resistance of aluminum alloys is strongly sensitive to purity, aging, the presence of intermetallic compounds, thermomechanical treatment, grain size, and orientation or texture. A list of variables and the nature of their effect on the fracture toughness of aluminum alloys is given in Table 3.

Commercial aluminum alloy structural components are selected from the 2*xxx*, 5*xxx*, 6*xxx*, and 7*xxx* alloy groups, which offer medium to high strengths. Of these, the 5*xxx* and 6*xxx* alloys offer medium to relatively high strength, good corrosion resistance, and are generally so tough that fracture toughness is rarely a design consideration. In addition, the 5*xxx* alloys provide good resistance to stress corrosion in marine atmospheres

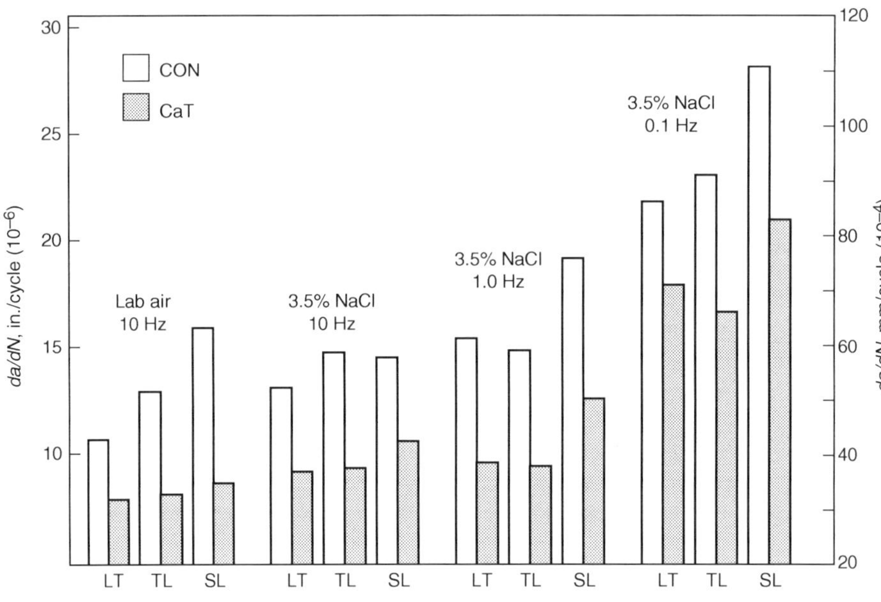

Fig. 24 Bar graphs comparing fatigue crack growth rates at $\Delta K = 55$ MPa · (m)$^{1/2}$ [50 ksi · (in.)$^{1/2}$] for conventional and calcium-treated A633C steels. CON, conventional; CaT, calcium treatment. The effect of testing frequency in salt water and steel quality is demonstrated. Source: Ref 9

Table 3 Effects of microstructural variables on fracture toughness of aluminum alloys

Variable	Effect on fracture toughness
Quench rate	Decrease in K_{Ic} at low quench rates
Impurities (Fe, Si, Mn, Cr)	Decrease in K_{Ic} with high levels of these elements
Grain size	Decrease in K_{Ic} at large grain sizes due to coarse grain-boundary precipitation
Grain-boundary precipitates	Increase in size and area fraction decrease K_{Ic}
Underaging	Increases toughness
Peak aging	Increases fracture toughness
Overaging	Decreases fracture toughness
Grain boundary segregates (Na, K, S, H)	Lower fracture toughness in Al-Li alloys

Source: Ref 1

and good welding characteristics. Notably, this class of alloys has been widely used in low-temperature applications that satisfy the most severe requirements of liquefied fuel storage and transportation at cryogenic temperatures. Alloys of the 6*xxx* class, with good formability and weldability at medium strengths, are widely used in conventional structural applications. The 2*xxx* and 7*xxx* alloys are generally used in applications involving highly stressed parts. Within the 2*xxx* and 7*xxx* alloys, certain alloys and tempers are available with combinations of high toughness and high strength. The stress-corrosion cracking resistance of 2*xxx* and 7*xxx* alloys is generally not as good as in the other aluminum alloy groups; however, service failures are avoided by good engineering practices and proper selection of alloy and temper, or a suitable protective system. The 2*xxx* and 7*xxx* alloys are widely used in aerospace applications.

Alloys of the 1*xxx* class are used primarily in applications where electrical conductivity, formability, ductility, and resistance to stress corrosion are more important than strength. The 3*xxx* alloys, widely used in piping applications, are characterized by relatively low strengths and very good toughness, ductility, formability, brazing, and welding characteristics.

Tough aluminum alloys such as those from the 1*xxx*, 3*xxx*, 5*xxx*, and most 6*xxx* series do not normally exhibit elastic unstable fracture; that is, these alloys are so tough that fracture toughness is rarely a design criterion.

Alloys for which fracture toughness is a meaningful design-related parameter (i.e., the 2*xxx* and 7*xxx* alloys) fall into two categories:

- Controlled-toughness, high-strength alloys, that is, those alloys developed primarily for their high fracture toughness at high strength
- Conventional high-strength alloys, tempers, and products for which fracture toughness is a meaningful design parameter but which are not normally used for fracture-critical components

Evaluations have shown that the fracture toughness of high-strength, precipitation-hardened 2*xxx* and 7*xxx* alloys is not adversely affected by high strain rate or moderate temperature reductions.

To develop high toughness, the microstructure must accommodate significant plastic deformation, and yet a microstructure that resists plastic deformation is needed for high strength. As indicated in Fig. 25, the 7*xxx* alloys have the highest combination of strength and toughness of any family of aluminum alloys. In the 7*xxx* alloys, the highest strength is associated with the T6 peak-aged temper. However, decreasing strength to acceptable levels by overaging provides a way to increase toughness (Fig. 26) as well as resistance to exfoliation, stress-corrosion cracking (SCC), and fatigue crack growth in some 7*xxx* alloys. Alloys of the 2*xxx* class are used in both the naturally aged and artificially aged conditions. Commercial naturally aged 2*xxx* alloys (i.e., T3 and T4 tempers) provide good combinations of toughness and strength. Artificial aging to precipitation-hardened T8 tempers produces higher strength with some reduction in toughness but in addition offers greater stability of mechanical properties at high temperatures and higher resistance to exfoliation and SCC.

The strength-fracture toughness interaction has been postulated to be the consequence of void linkup created by slip-induced breakdown of submicron strengthening particles, which occurs more readily at high strength levels. If the strengthening (matrix) precipitates are shearable, they may promote strain localization, which leads to premature crack nucleation

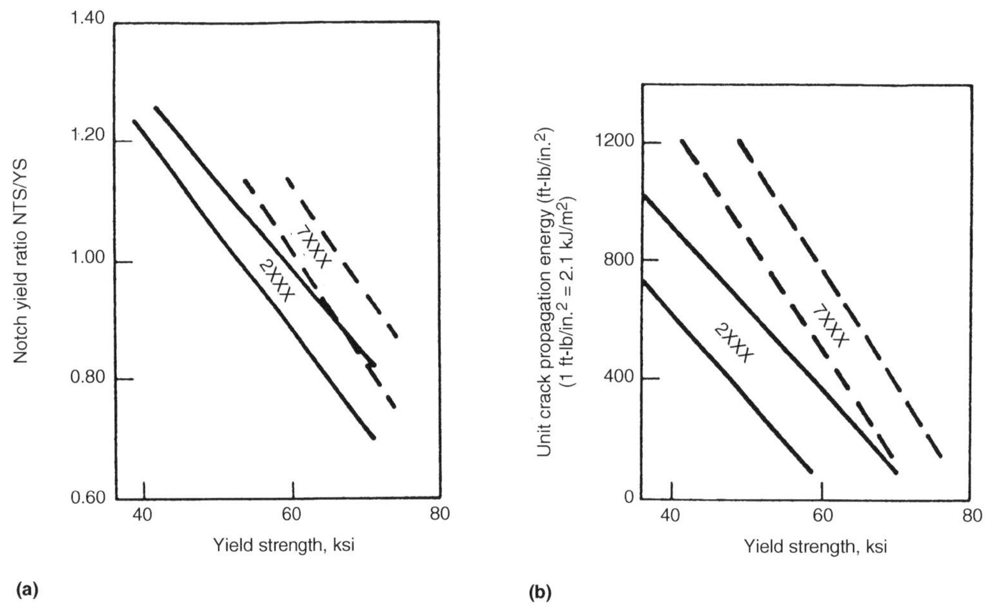

Fig. 25 Comparison of 2*xxx* and 7*xxx* commercial aluminum alloys. (a) Notch toughness vs. yield strength. (b) Unit propagation energy vs. yield strength. Source: Ref 11

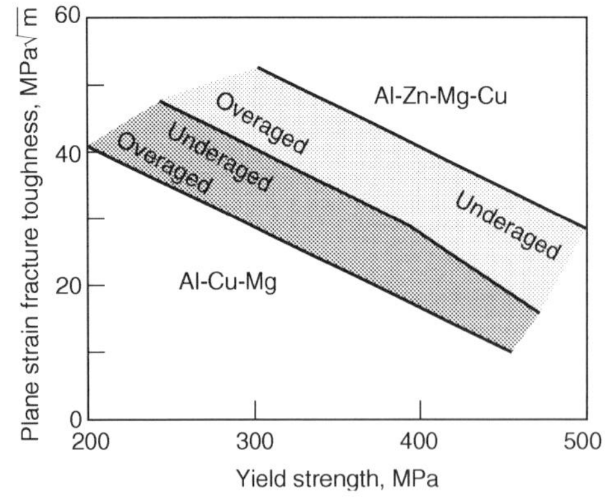

Fig. 26 Relationships of plane-strain fracture toughness to yield strength for the 2*xxx* and 7*xxx* series of aluminum alloys. Source: Ref 12

and low fracture toughness. Whether or not the strengthening precipitates are sheared or looped and bypassed by dislocations depends on alloy composition and aging treatment. During aging, heterogeneous precipitation usually occurs at grain and subgrain boundaries, resulting in soft, solute-denuded precipitation-free zones (PFZs) in the matrix adjacent to the boundaries. The combination of these soft zones, which can localize strain, and grain-boundary precipitates, which can aid in microvoid nucleation, also has an adverse effect on fracture toughness. It has been clearly demonstrated that the amounts, distribution, and morphology of alloy phases and second-phase particles in alloy microstructure have a large influence on toughness. An understanding of the interrelationships of alloy microstructure and fracture mechanisms has led to design of new commercial aluminum alloys offering optimum high strength and high toughness. Primarily, the alloy improvements have evolved through microstructural control obtained by increased purity, modified compositions, and better homogenization, fabrication, and heat treatment practices.

The balance between strength and toughness is greatly affected by a variety of processing parameters, including solution heat treatment, quenching efficiency, deformation prior to aging (for 2*xxx* alloys), and aging treatment. The solution heat treatment determines the amount of solute in solid solution and the vacancy content, which affects subsequent aging kinetics. Quenching affects both the microstructure and properties by determining the amount of solute that precipitates during cooling and that which is available for subsequent age hardening. It also affects the level of residual stresses that can influence manufacturing costs, fatigue, and corrosion behavior. After quenching, methods to obtain a balance of properties include cold working before aging, when practical (T8 temper), and selecting aging times and temperatures to minimize grain-boundary precipitates and PFZs. Deformation prior to aging aids in the nucleation and growth of the matrix precipitates, which decreases the time to reach peak strength. This, along with low-temperature aging, minimizes the amount of grain-boundary precipitates and PFZs (which adversely affect fracture toughness) at the desired strength level.

Alloy 2124, a higher-purity version of 2024, was the first 2*xxx* alloy developed to have high fracture toughness. The principal contribution to high toughness was increased purity (low iron and silicon), which minimizes formation of relatively large insoluble constituents (>1 μm). The detrimental effect of large constituent phases on the fracture toughness of aluminum alloys has been documented by many investigators. Constituent particles participate in the fracture process through void formation at particle/matrix interfaces or by fracturing during primary processing. Their volume fraction can be minimized by reducing impurity elements, for example, iron and silicon, and excess solute. The detrimental effect of dispersoids also depends on their size and the details of their interface with the matrix. Because zirconium particles are small and coherent with the matrix (strong interface), they are usually not involved in the fracture process.

The resultant improvement for production materials is shown in Fig. 27. Minimization of insoluble constituents by process control was used to develop 2124 and 7475 as higher-toughness versions of 2024 and 7075, respectively. The biggest gains in fracture toughness of 2*xxx* alloys by process control have been for the precipitation-hardened T8 tempers, which are widely used in applications requiring good resistance to exfoliation corrosion and SCC. The effect of purity on toughness of other alloys is shown in Table 4.

Grain size and degree of recrystallization can have a significant effect on fracture toughness. The desired degree of recrystallization depends on product thickness, that is, whether the part is under plane stress or plane

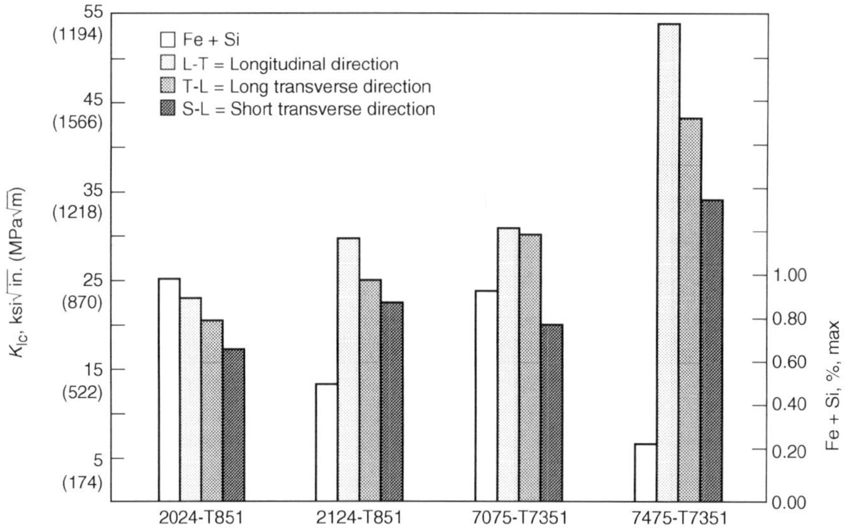

Fig. 27 Aluminum alloys 2124 and 7475 are tougher versions of alloys 2024 and 7075. High-purity metal (low iron and silicon) and special processing techniques are needed to optimize toughness in these materials. Source: Ref 13

Table 4 Effect of purity on the fracture toughness of some high-strength wrought aluminum alloys

Alloy and temper	Max Fe, %	Max Si, %	0.2% proof stress, MPa (ksi)	Tensile strength, MPa (ksi)	Fracture toughness, MPa · (m)$^{1/2}$ [ksi · (in.)$^{1/2}$] Longitudinal	Short transverse
2024-T8	0.50	0.50	450 (65)	480 (70)	22–27 (20–25)	18–22 (17–20)
2124-T8	0.30	0.20	440 (64)	490 (71)	31 (28)	25 (23)
2048-T8	0.20	0.15	420 (61)	460 (67)	37 (34)	28 (26)
7075-T6	0.50	0.40	500 (73)	570 (83)	26–29 (24–27)	17–22 (16–20)
7075-T73	0.50	0.40	430 (62)	500 (73)	31–33 (28–30)	20–23 (18–21)
7175-T736	0.20	0.15	470 (68)	540 (78)	33–38 (30–35)	21–29 (19–27)
7050-T736	0.15	0.12	510 (74)	550 (80)	33–39 (30–36)	21–29 (19–27)

Source: Adapted from Ref 14

strain. In thin products under plane stress, fracture is controlled by plasticity, and a small recrystallized grain size is preferable. If the grain size is small enough, plasticity will be enhanced without detrimental, low-energy, intergranular fracture. However, for thick products under plane strain, fracture is usually controlled by coarse particles, and an unrecrystallized grain structure is preferable.

Fatigue of Aluminum Alloys

Early work at Alcoa on large numbers of smooth and notched specimens demonstrated that wide variations in commercial aluminum alloys caused little or no detectable difference in fatigue strengths. When early fatigue crack growth experiments categorized fatigue crack growth rates of aluminum alloys into one band, for example, Fig. 28, it was generally thought that the fatigue resistance of all aluminum alloys was alike. As a consequence of these early beliefs, further efforts to develop fatigue-resistant aluminum alloys were minimized, and although several conceptual improvements have been advanced in laboratory experiments, none to date have reached commercial levels. For alloys developed to provide improved combinations of properties such as strength, corrosion resistance, and fracture toughness, fatigue resistance was determined as a last step before products were offered for sale, only to ensure that fatigue resistance was not degraded.

Despite early conclusions from laboratory data, users discovered that certain aluminum alloys performed decidedly better than others in service

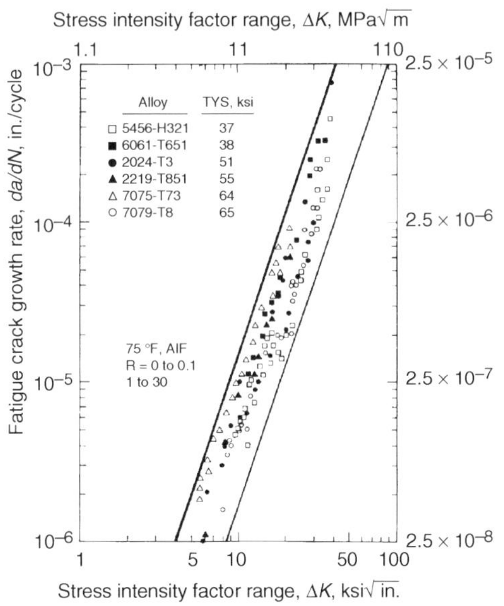

Fig. 28 Scatter band limits for fatigue crack growth rate behavior for a range of aluminum alloys. Source: Ref 15

when fluctuating loads were encountered, and therefore, any generalization that all aluminum alloys are alike in fatigue is not wholly appropriate. For example, alloy 2024-T3 has long been recognized as a better fatigue performer in service than alloy 7075-T6. In part, this may be explained by the higher design stresses used for 7075-T6. However, the results of Fig. 29(b) show alloy 7075-T6 to have a broader scatter for smooth specimens and a lower bound of performance for severely notched specimens that is below that for alloy 2024-T4 (Fig. 29a). Broader scatter is also evident for 7079 compared to 2014 (Fig. 29c and d).

S-N Fatigue

High-cycle fatigue characteristics commonly are examined on the basis of cyclic *S-N* plots of rotating beam-, axial-, or flexure-type sheet tests. Many thousands of tests have been performed, and a collection of aluminum alloy *S-N* data is contained in the publication *Fatigue Data Book: Light Structural Alloys* (Ref 17). Early work on rotating beam tests is summarized in Fig. 30. There seems to be greater spread in fatigue strengths for unnotched specimens than for notched specimens. This ap-

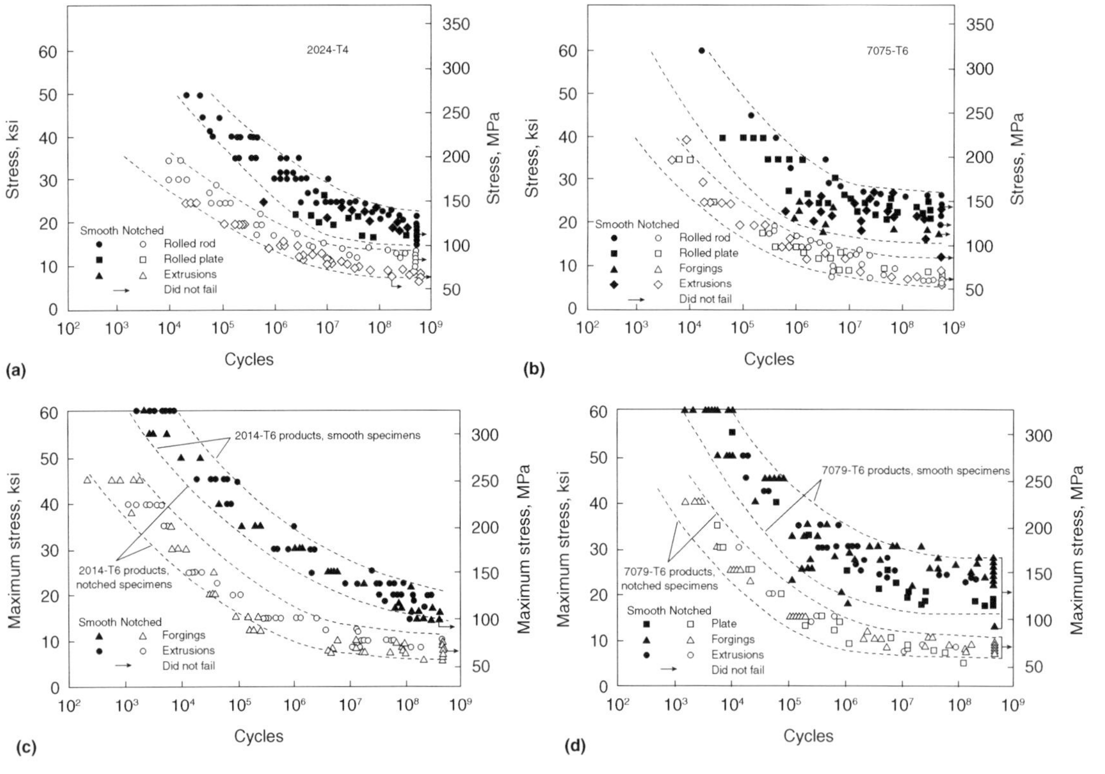

Fig. 29 Variation in rotating beam fatigue for (a) 2024-T4, (b) 7075-T6, (c) 2014-T6, and (d) 7079-T6 alloys. Notches (60°) were very sharp ($K_t > 12$), with a radius of approximately 0.005 mm (0.0002 in.). Results are from over a thousand rotating beam tests performed in the 1940s. Sources: Ref 16

pears to be evidence that the presence of a notch minimizes differences, thus suggesting similar crack propagation after crack initiation with a sharp notch. In this context, the spread in smooth fatigue life is partly associated with variations in crack initiation sources at surface imperfections or strain localizations. In general, however, the *S-N* approach does not provide clear distinctions in characterizing the crack initiation and crack propagation stages of fatigue.

The *S-N* response curves for rotating beam fatigue strength of unnotched aluminum alloys tend to level out as the number of applied cycles approaches 500 million. This allows some rating of fatigue endurance, and estimated fatigue limits from rotating beam tests have been tabulated for many commercial aluminum alloys. Fatigue limits should not be expected in aggressive environments, because *S-N* response curves do not tend to level out when corrosion fatigue occurs. Rotating beam strengths determined in the transverse direction are not significantly different from test results in the longitudinal direction.

Rotating beam data have also been analyzed to determine whether fatigue strength can be correlated with static strength. From a plot of average endurance limits (at 5×10^8 cycles) plotted against various tensile properties, there does not appear to be any well-defined quantitative relation between fatigue limit and static strength. This is consistent with results for most nonferrous alloys. It should be noted that proportionate increases in fatigue strength from tensile strengths do appear lower for age-hardened aluminum alloys than for strain-hardened alloys (Fig. 31). A similar trend appears evident for fatigue strength at 5×10^7 cycles (Fig. 32).

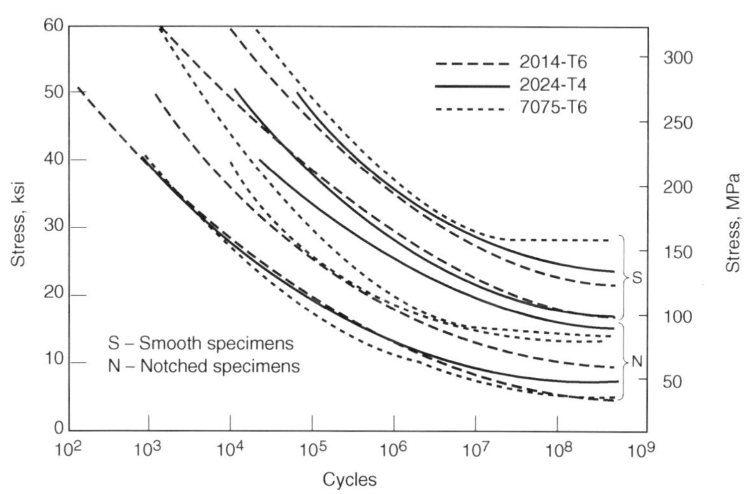

Fig. 30 Comparison of fatigue strength bands for 2014-T6, 2024-T4, and 7075-T6 aluminum alloys for rotating beam tests. Source: Ref 16

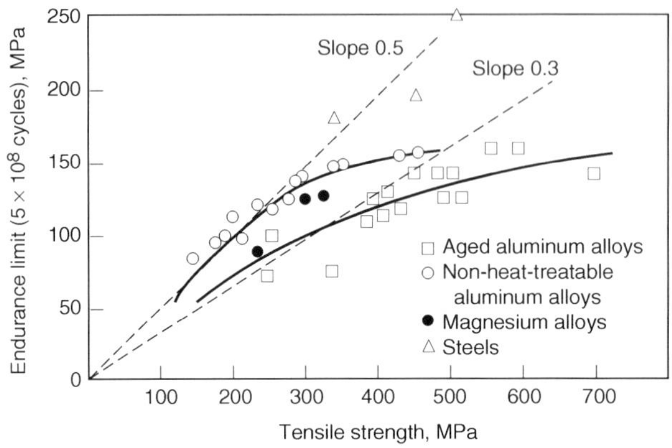

Fig. 31 Fatigue ratios (endurance limit/tensile strength) for aluminum alloys and other materials. Source: Ref 18

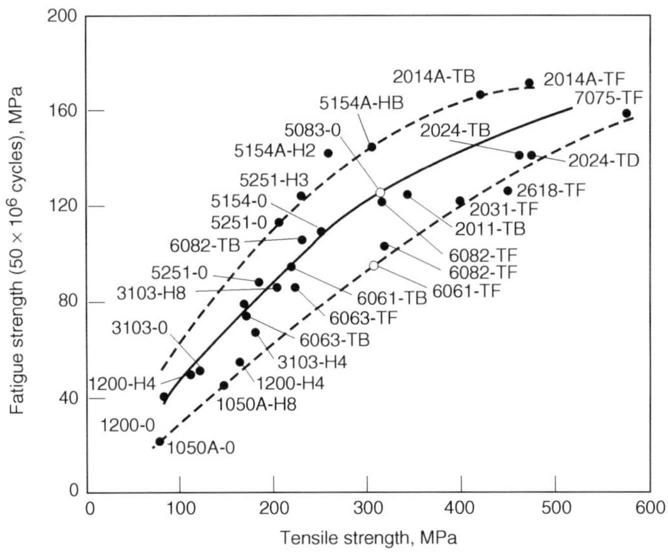

Fig. 32 Relationships between the fatigue strength and tensile strength of some wrought aluminum alloys. Source: Ref 11

Most aluminum alloys experience some reduction of fatigue strength in corrosive environments such as seawater, especially in low-stress, long-life tests (e.g., Fig. 33). Unlike sustained-load SCC, fatigue degradation by the environment may occur even when the direction of principal loading with respect to grain flow is other than short transverse. The fatigue response to the environment varies with the specific alloy, so final alloy selection for design should address this important interaction. When accumulating data for this purpose, it is recommended that any testing be conducted in a controlled environment, and preferably the environment of the intended application. However, an environment known to be more severe than that encountered in service is often used to conservatively establish baseline data and design guidelines. Because environmental interaction with fatigue is a rate-controlled process, interaction of time-dependent fatigue parameters such as frequency, waveform, and load history should be factored into the fatigue analysis.

Typically, the fatigue strengths of the more corrosion-resistant 5*xxx* and 6*xxx* aluminum alloys and tempers are less affected by corrosive environments than are the higher-strength 2*xxx* and 7*xxx* alloys, as indicated by Fig. 34. Corrosion fatigue performance of the 7*xxx* alloys may, in general, be improved by overaging to the more corrosion-resistant T7 tempers. For the 2*xxx* alloys, the more corrosion-resistant, precipitation-hardened T8-type tempers provide a better combination of strength and fatigue resistance at high endurances than the naturally aged T3 and T4 tempers.

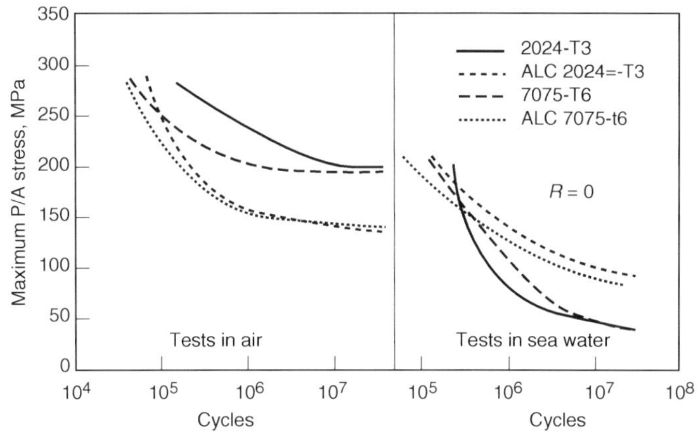

Fig. 33 Axial stress fatigue strength of 0.8 mm (0.030 in.) 2024, 7075, and clad sheet in air and seawater, $R = 0$. Source: Ref 19

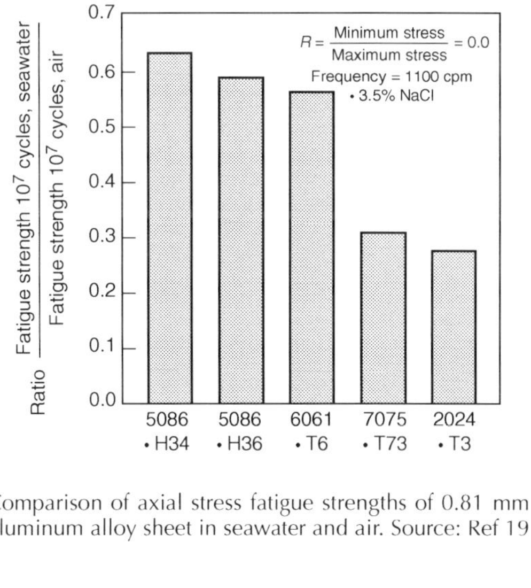

Fig. 34 Comparison of axial stress fatigue strengths of 0.81 mm (0.032 in.) aluminum alloy sheet in seawater and air. Source: Ref 19

However, artificial aging of the 2*xxx* alloys is accompanied by a loss in toughness with a resultant decrease in fatigue crack growth resistance at intermediate and high stress intensities. Interaction of a clad protective system with the fatigue strength of alloys 2024-T3 and 7075-T6 in air and seawater environments is shown in Fig. 33. In air, the cladding appreciably lowers the fatigue resistance. However, in seawater, the benefits of the cladding are readily apparent.

Many alloys have large inclusions, which may concentrate strain during cyclic deformation and lead to early crack nucleation. This detrimental effect can be reduced substantially by lowering the impurity levels. This is illustrated in Fig. 35, which shows that a significant improvement in the high-cycle fatigue life of 7075 alloy is obtained by lowering the iron and silicon content (7475 alloy).

Fatigue Crack Growth of Aluminum Alloys

In examining fatigue crack growth rate curves for many materials exhibiting very large differences in microstructure, the striking feature is the similarities between these curves, not the differences. This point is illustrated by Fig. 36, a compilation of data for 2*xxx*- and 7*xxx*-series aluminum alloys. The differences in crack growth rate between these alloys are important from the viewpoint of integrating along any one of them to obtain the lifetime of a structure, but from a mechanistic point of view, these differences are small. A larger range of metals can be represented by a single curve if the driving force (ΔK) is normalized by modulus. These data exclude the effect of environment (mainly water vapor), which is a major factor affecting fatigue crack growth rates. In general, fatigue crack

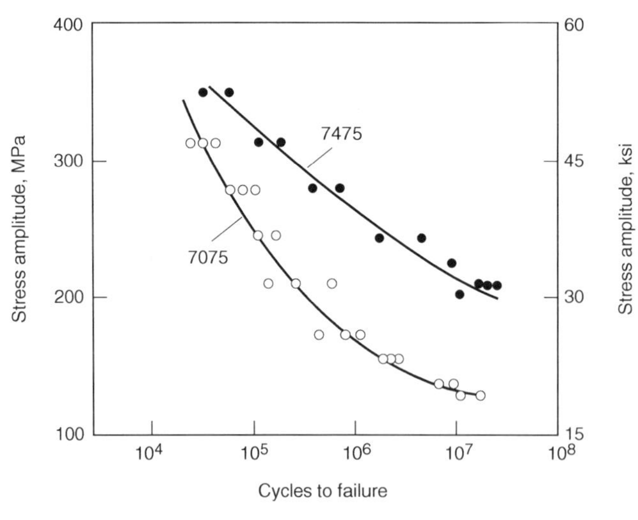

Fig. 35 Effect of inclusion density on the stress-life behavior of two 7*xxx* alloys: high-inclusion-density alloy 7075 and low-inclusion-density alloy 7475. Source: Ref 11

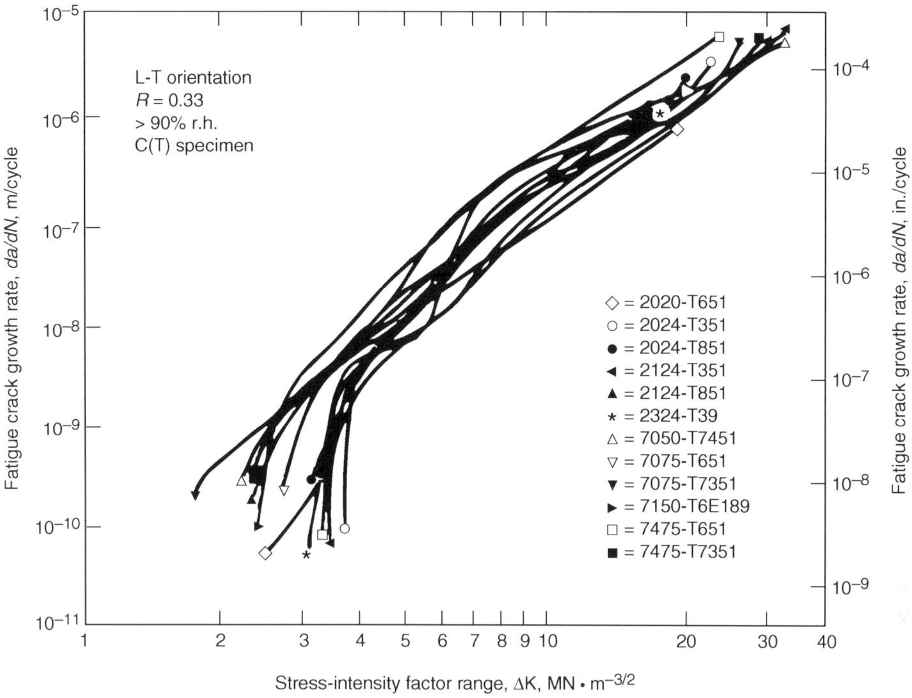

Fig. 36 Minor influences of differing microstructures on fatigue crack growth rate curves: data from twelve 2*xxx* and 7*xxx* aluminum alloys with different heat treatments. Source: Ref 20

growth rates in nonhostile environments fall within a relatively narrow scatter band, with only small systematic effects of composition, fabricating practice, or strength.

Those metallurgical factors that contribute to increased fracture toughness also generally contribute to increased resistance to fatigue crack propagation at relatively high ΔK levels. For example, as illustrated in Fig. 37, at low stress intensities the fatigue crack growth rates for 7475 are approximately the same as those for 7075. However, the factors that contribute to the higher fracture toughness of 7475 also contribute to the retardation of fatigue crack growth, resulting in two or more times slower growth for 7475 than for 7075 at ΔK levels equal to or greater than approximately 16 MPa $\cdot \sqrt{\text{m}}$ (15 ksi $\cdot \sqrt{\text{in.}}$). A similar trend has been observed for 2124-T851, which exhibits slower growth than 2024-T851. Smooth specimens of alloys 2024 and 2124 exhibit quite similar fatigue behavior. Because fatigue in smooth specimens is dominated by initiation, this suggests that the large insoluble particles may not be significant contributors to fatigue crack initiation. However, once the crack is initiated, crack propagation is slower in a material with relatively few large particles (2124) than in a material with a greater number of large particles (2024).

The extensive use of age-hardenable aluminum alloys at high strength levels, that is, greater than 520 MPa (75 ksi), has been hampered by poor

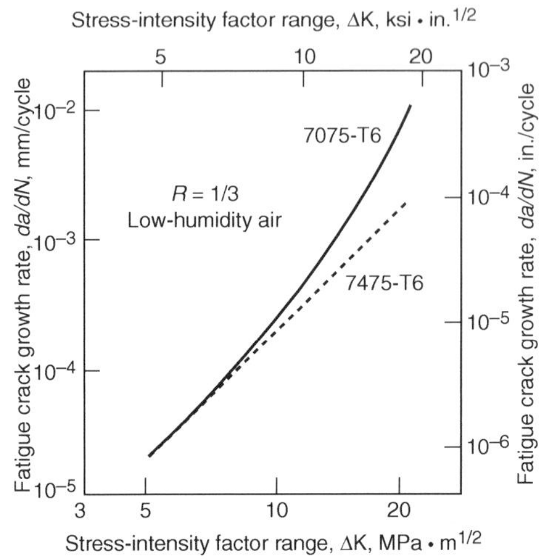

Fig. 37 Benefit of high-toughness alloy 7475 at intermediate and high stress intensity. Source: Ref 21

secondary properties of toughness, stress-corrosion resistance, and fatigue resistance, particularly in the short transverse direction. Some secondary property improvements have been obtained by employing slight changes in alloy chemistry, different grain-refining elements, or removal of the impurity elements iron and silicon. Such research has led to the development of alloys with improved fracture toughness and stress-corrosion resistance compared to the extensively used 7075. However, significant improvements in fatigue resistance have not been realized with these methods.

Fracture Toughness of Titanium Alloys

Due to their good combination of specific strength, ductility, and fracture toughness, titanium alloys are primarily used in aerospace applications. As in steels and aluminum alloys, this combination is achieved by the careful control of two-phase microstructures. Among the two phases (alpha, or α, and beta, or β), beta is more ductile and is preferable in increasing the fracture toughness of titanium alloys. The three broad classes of titanium alloys are near-alpha, alpha + beta, and beta alloys, grouped according to the levels of alpha or beta stabilizing elements. Typically, beta content by volume is near-α, <10%; α + β, 10 to 25%; and β, >25%. The fracture toughness/strength relationship maps for different titanium alloys are shown in Fig. 38. Metastable beta alloys possess the highest combination of strength and toughness. This arises from a large volume fraction of beta-phase and fine aged-alpha precipitates.

Unlike steels and aluminum alloys, titanium alloys are generally free from inclusions and intermetallics that form during solidification. In addition, there is no precipitation and coarsening of brittle phases, so the control of microstructure for fracture toughness is less difficult. However, microstructure still plays a major role in controlling the fracture toughness of titanium alloys. Microstructure control is primarily achieved by mechanical processes, such as hot/cold working and heat treatment involving solution treatment followed by quenching and aging or slow cooling. The fracture toughness values of a typical titanium alloy with different microstructures are listed in Table 5. In general, for a given beta-phase content, fracture toughness increases with an increase in the amount of lamellar α as well as an increase in the aspect ratio of alpha phase.

The dominant variables that influence fracture toughness in titanium alloys are the interstitial elements, grain size, microstructural morphology, and relative proportions of alpha and beta phases. These variables and the nature of their effect on fracture toughness are listed in Table 6.

Increases in oxygen and hydrogen levels in Ti-6Al-4V alloy decrease fracture toughness, as shown in Fig. 39 and 40, respectively. This is caused by an increase in the planarity slip, promoted by the ordering of the Ti_3Al phase, which causes easy crack nucleation at grain and phase boundaries. This tendency to ordering is also increased at high aluminum contents,

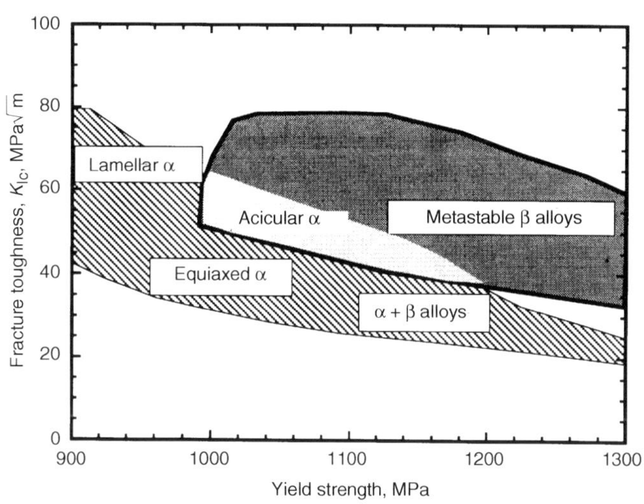

Fig. 38 Relationship between fracture toughness and strength for different classes of titanium alloys and microstructures. Source: Ref 1

Table 5 Fracture toughness levels of Ti-6Al-4V alloy in different microstructural conditions

Microstructure	0.2% yield strength, MPa (ksi)	Elongation, %	K_{Ic}, MPa · (m)$^{1/2}$ [ksi · (in.)$^{1/2}$]
Beta processed (aligned lamellar alpha)	903 (131)	12	78 (72)
Alpha + beta processed (equiaxed alpha in aged beta matrix)	917 (133)	16	53 (49)
Recrystallized (fully equiaxed alpha)	925 (134)	19	47 (43)

Source: Adapted from Ref 1

Table 6 Effect of microstructural variables on the fracture toughness of titanium alloys

Variable	Effect on fracture toughness
Interstitials (O, H, C, N)	Decrease in K_{Ic}
Grain size	Increase in grain size decreases K_{Ic}
Lamellar colony size	Increase in colony size increases K_{Ic}
Beta phase	Increases in beta volume fraction and continuity increase K_{Ic}
Grain-boundary alpha phase	Increases in thickness and continuity increase K_{Ic}
Shape of alpha phase	Increase in aspect ratio of alpha phase increases K_{Ic}
Orientation	Crack oriented for easy cleavage along basal planes gives low K_{Ic}

Source: Ref 1

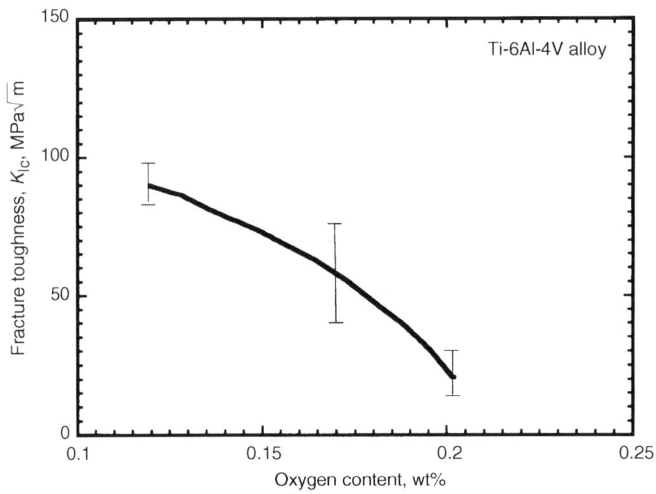

Fig. 39 Effect of oxygen level on the fracture toughness of Ti-6Al-4V alloy. Source: Ref 1

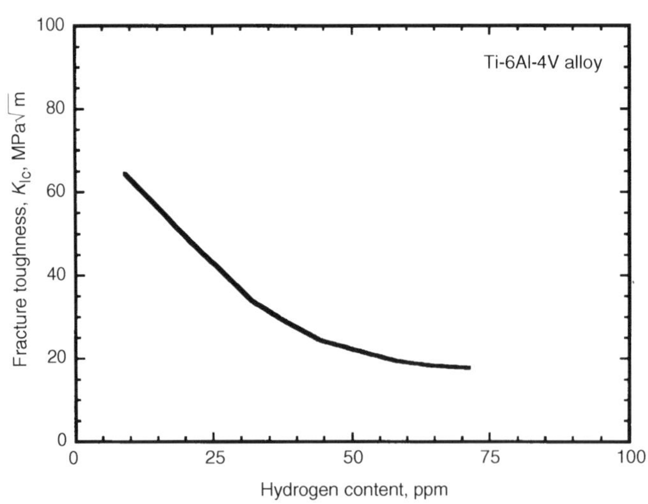

Fig. 40 Effect of hydrogen level on the fracture toughness of Ti-6Al-4V. Source: Ref 1

and hence, compositions of commercial alloys rarely exceed 6% Al. The presence of hydrogen in metallic alloys is almost always detrimental. In titanium, hydrogen causes cleavage and interface cracking due to the formation of hydrides (TiH_2). Alloys with high levels of beta phase can dissolve more hydrogen, thereby preventing the decrease in fracture toughness due to hydrogen. The other interstitial elements, carbon and nitrogen, have low solid solubility in titanium and form fine TiC and TiN dispersions when the solubility level is exceeded. These particles drastically decrease the ductility as well as the fracture toughness of titanium alloys and hence must be eliminated.

The effect of beta grain size on fracture toughness is illustrated in Fig. 41 for Ti-5.2Al-5.5V-1Fe-0.5Cu alloy. There is an inverse relationship of fracture toughness to beta grain size. As the grain size increases, the tendency to intergranular fracture increases, due to the weakening effect of fine 0.2 μm thick particles at the β-β grain boundary. This is primarily due to the increased density of grain-boundary precipitates as a result of the reduction in available grain-boundary area. The same alloy was heat treated differently to produce thick continuous alpha phase at the grain boundary, which increased fracture toughness, as shown in Fig. 42. However, for this to occur, the grain interior (aged beta matrix) should be stronger than alpha.

The orientation of crack plane in a fracture toughness test with respect to the rolling direction of the titanium alloy plate has a significant effect, due to the preferred orientation of hexagonal close-packed crystal grains having limited slip systems relative to the body-centered cubic and face-centered cubic crystals. The effect of orientation on the fracture toughness/strength relationship is illustrated in Fig. 43. While a strong inverse relationship between fracture toughness and yield strength exists for the longitudinal orientation, it is less prevalent in the transverse orientation.

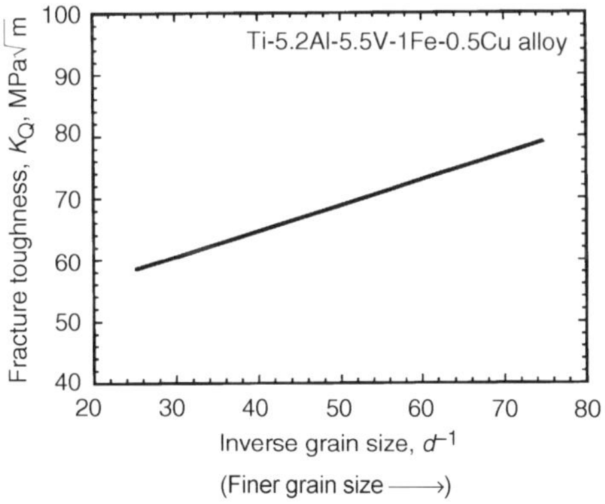

Fig. 41 Effect of grain size on the fracture toughness of a titanium alloy. Source: Ref 1

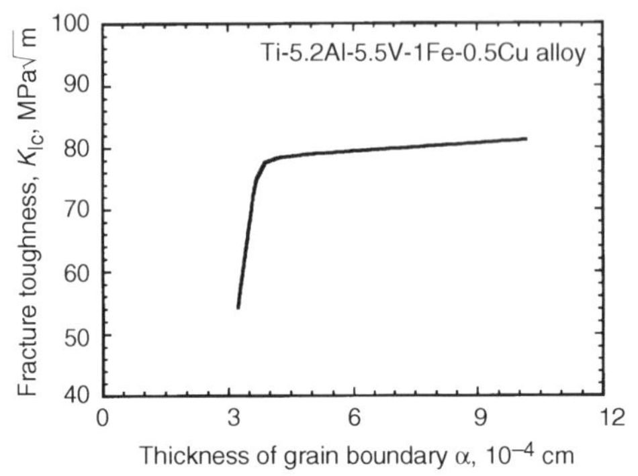

Fig. 42 Effect of the thickness of grain-boundary alpha phase on the fracture toughness of a titanium alloy. Source: Ref 1

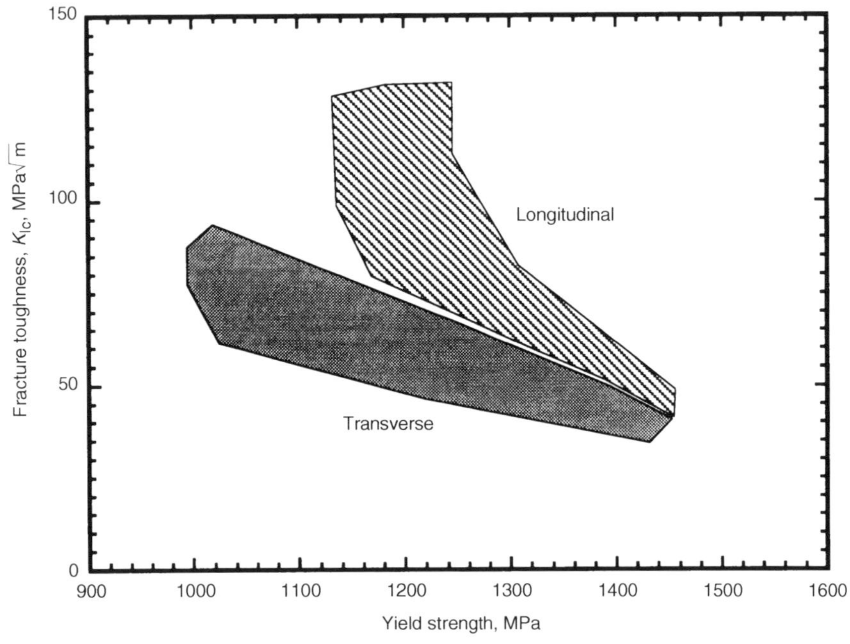

Fig. 43 Effect of crack plane orientation on the fracture toughness of Ti-6Al-4V alloy. Source: Ref 1

The effect of orientation on fracture toughness is due to the relative orientation of slip systems with respect to the crack plane.

The key to improving the combination of strength and toughness in titanium alloys is to increase the beta-phase content, increase the lamellar alpha volume fraction, increase the aspect ratio of alpha phase, and reduce planarity slip and interface embrittlement by reducing the levels of oxygen, hydrogen, carbon, and nitrogen.

Fatigue of Titanium Alloys

The fatigue life (N_F) of structural parts is the sum of crack nucleation life (N_I) and crack propagation life to final fracture (N_P): $N_F = N_I + N_P$. Fatigue tests on titanium alloys have shown that at high stress or strain amplitudes, the ratio N_I/N_F can be as small as 0.01. Therefore, the resistance to crack propagation determines fatigue life in the low-cycle fatigue regime. With regard to crack propagation in structural parts, it is useful to distinguish between small surface cracks (microcracks) and long through-cracks (macrocracks), because the dependence of crack growth on microstructural parameters such as grain size or phase dimensions can be contradictory. As a rule of thumb, the fatigue life for small, highly stressed components (small critical flaw size) is mainly controlled by microcrack growth, while for large components operating at low stress levels (large critical flaw size), macrocrack growth behavior is more important. As opposed to steels or aluminum alloys, titanium alloys are generally free of defects such as inclusions or pores. However, fatigue behavior in titanium alloys is very sensitive to surface preparation.

Fatigue life in unalloyed and commercially pure (CP) titanium depends on grain size, interstitial contents, and degree of cold work, as illustrated in Fig. 44. Decreasing the grain size from 110 to 6 μm improves the fa-

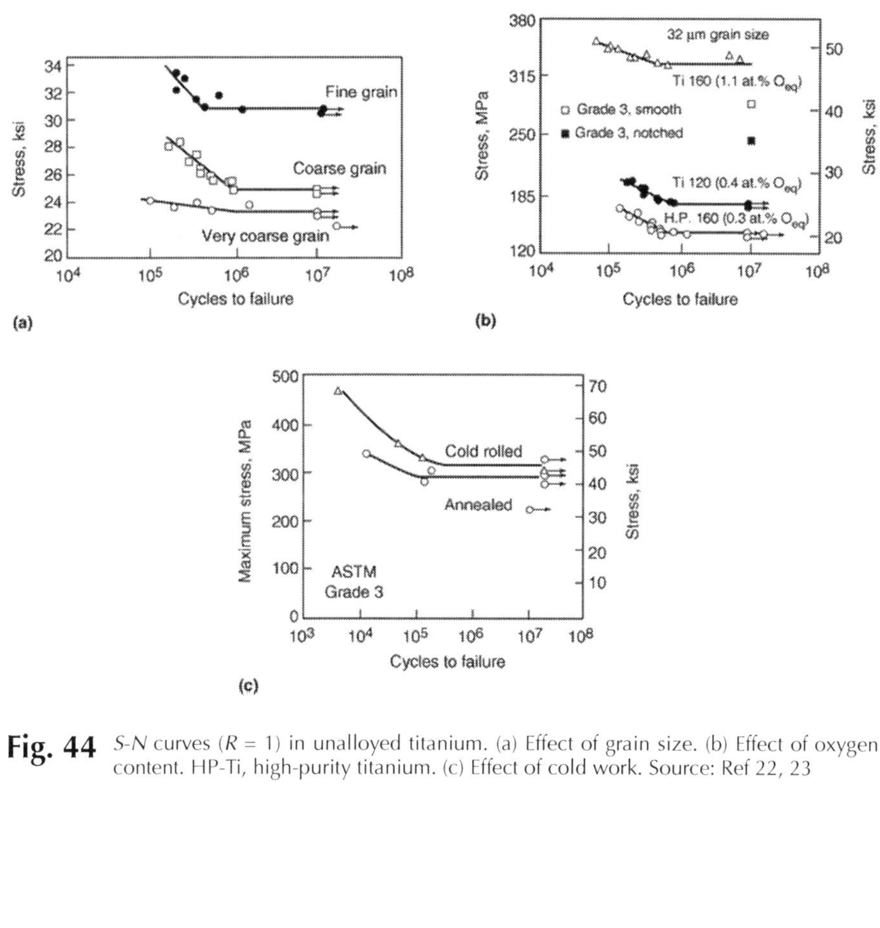

Fig. 44 *S-N* curves ($R = 1$) in unalloyed titanium. (a) Effect of grain size. (b) Effect of oxygen content. HP-Ti, high-purity titanium. (c) Effect of cold work. Source: Ref 22, 23

tigue limit (10^7 cycles) in CP titanium from 180 to 240 MPa (26 to 35 ksi) (Fig. 44a). The effect of oxygen content on the *S-N* curves of high-purity titanium, shown in Fig. 44(b), correlates with the increase in yield strength with higher oxygen contents. Cold work also increases the yield strength and thus improves the fatigue performance, as shown in Fig. 44(c).

In addition to alpha grain size, degree of age hardening, and oxygen content, the fatigue properties of the two-phase near-alpha and alpha-beta alloys are strongly affected by the morphology and arrangement of both the alpha and beta phases. Basically, fully lamellar, fully equiaxed, and duplex (primary alpha phase in a lamellar matrix) microstructures can be developed in near-alpha and alpha-beta alloys. Important microstructural parameters are the prior-beta grain size or colony size of the alpha and beta lamellae and the width of the alpha lamellae in fully lamellar microstructures. Additional parameters for duplex structures are grain size and volume fraction of the primary alpha phase. Tensile properties of the fully lamellar, fully equiaxed, and duplex microstructures of Ti-6Al-4V are compared in Table 7.

Typical crack nucleation sites are shown in Fig. 45. Fatigue cracks in fully lamellar microstructures nucleate at slip bands within the alpha lamellae (Fig. 45a) or at alpha zones along prior-beta grain boundaries. Because both resistance to dislocation motion and resistance to fatigue crack nucleation depends on the width of the alpha lamellae, there is a direct correlation between fatigue strength and yield stress. Fatigue cracks in the fully equiaxed microstructure nucleate at slip bands within the alpha grains (Fig. 45b). Thus, the fatigue strength correlates to the alpha grain size dependence of the yield stress. Fatigue cracks in duplex microstructures can nucleate in the lamellar matrix, at the interface between lamellar matrix and the primary alpha phase, or within the primary alpha phase. The exact nucleation site depends on cooling rate, volume fraction, and size of the primary alpha phase. An example of crack nucleation within the lamellar regions is shown in Fig. 45(c).

The fatigue life of these microstructures is shown in Fig. 46. A decrease in the width of the alpha lamellae from 10 to 0.5 μm in fully lamellar microstructures improves the fatigue strength from 480 to 650 MPa (70 to 94

Table 7 Tensile properties in Ti-6Al-4V (aged 24 h at 500 °C)

Microstructure	Width of α lamellae, μ m	0.2% yield strength, MPa (ksi)	RA, %
Fully lamellar	0.5	1040 (151)	16
	1	980 (141	18
	10	930 (135)	14
Fully equiaxed	2	1120 (162)	46
	6	1070 (155)	44
	12	1060 (154)	43
Duplex	0.5	1045 (152)	39
(10 μ m primary alpha size)	1	975 (141)	34

RA, reduction in area. Source: Adapted from Ref 24–26

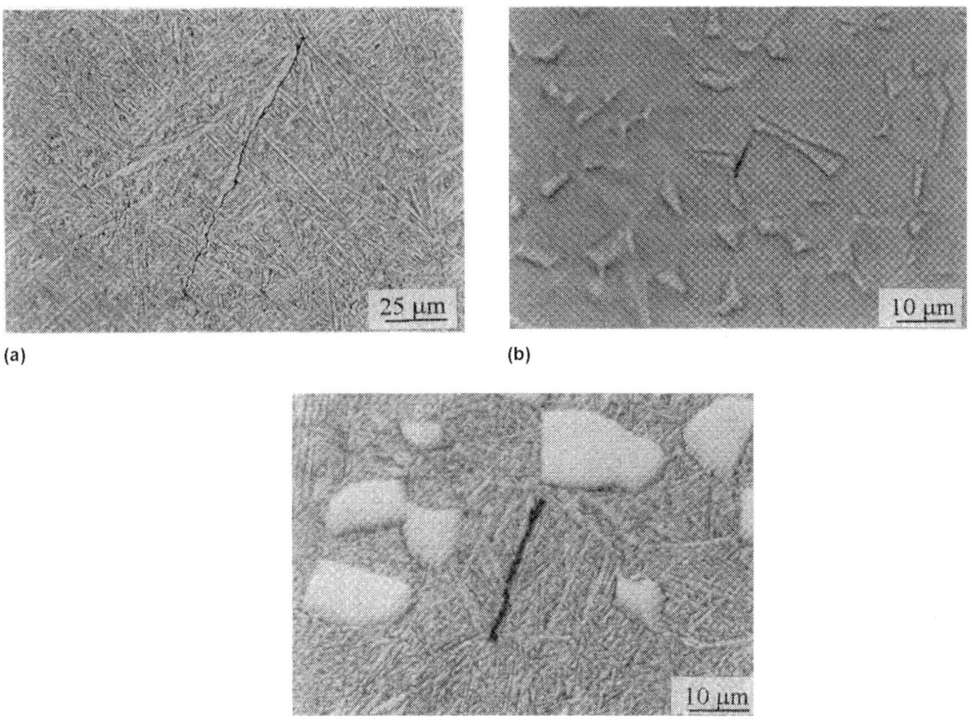

(a) (b) (c)

Fig. 45 Fatigue crack nucleation sites in Ti-6Al-4V. (a) Fully lamellar microstructure. (b) Fully equiaxed microstructure. (c) Duplex microstructure. Source: Ref 27

ksi), as shown in Fig. 46(a). Similarly, a decrease in the alpha grain size from 12 to 2 μm in fully equiaxed structures increases the fatigue strength from 550 to 720 MPa (80 to 104 ksi), as shown in Fig. 46(b). A decrease in the width of the alpha lamellae from 0.5 to 1 μm in duplex microstructures increases the fatigue strength from 480 to 575 MPa (70 to 83 ksi), as shown in Fig. 46(c). Reducing the prior-beta grain size in fully lamellar microstructures, as well as decreasing the primary alpha volume fraction in duplex structures, improves fatigue life in both low-cycle fatigue (LCF) and high-cycle fatigue (HCF).

Plots of da/dn versus ΔK for two extreme microstructures of Ti-6Al-4V are shown in Fig. 47. The crack growth resistance in the coarse lamellar microstructure is inferior to the equiaxed structure. Crack growth studies on fine lamellar and duplex structures in Ti-6Al-4V have shown that their da/dn-versus-ΔK curves lie in between the coarse lamellar and equiaxed structures.

For beta alloys, depending on the alloy class (solute rich or solute lean), the following microstructural parameters are important in determining fatigue life: beta grain size, degree of age hardening, and precipitate-free regions in the solute-rich alloys such as Beta C. In addition, grain-bound-

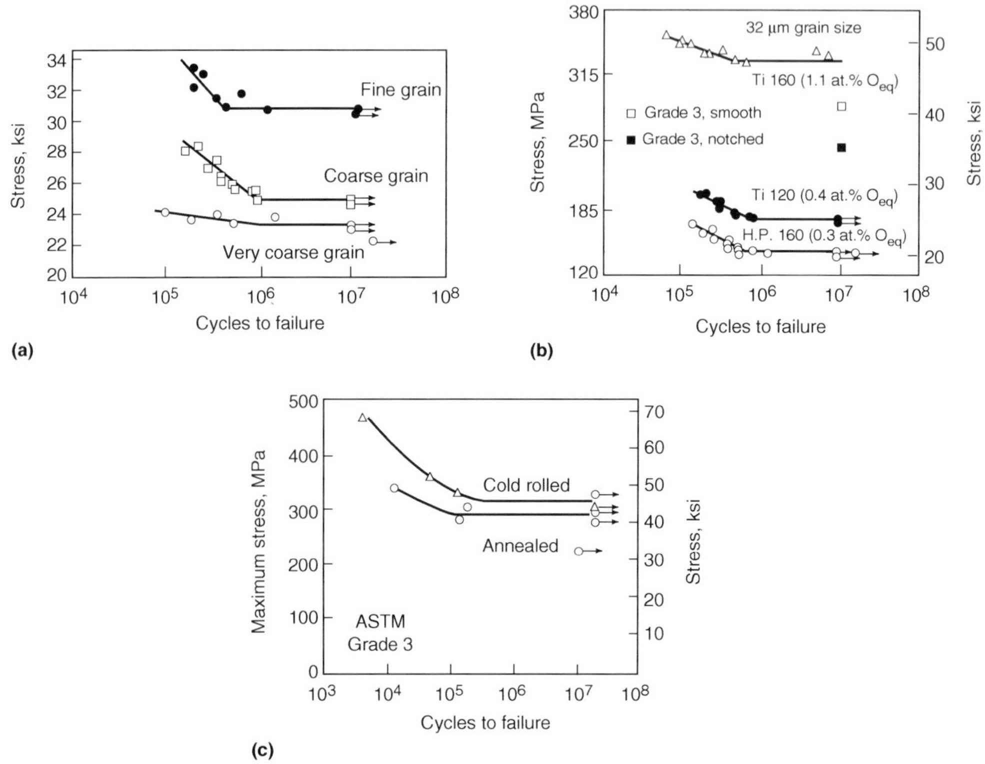

Fig. 46 *S-N* curves ($R = -1$) in Ti-6Al-4V. (a) Fully lamellar microstructure. Effect of width of alpha lamellae. (b) Fully equiaxed microstructure. Effect of alpha grain size. B/T-RD, basal/transverse texture, rolling direction; WQ, water quench. (c) Duplex microstructure. Effect of width of alpha lamellae. Source: Ref 24, 26

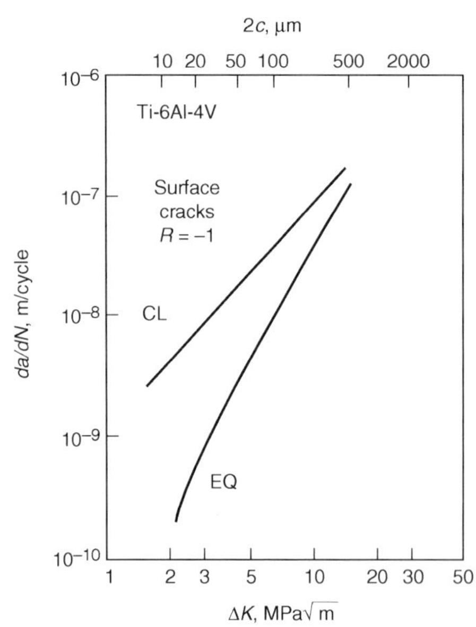

Fig. 47 *da/dn*-Δ*K* curves of microcracks in Ti-6Al-4V. CL, coarse lamellar; EQ, equiaxed. Source: Ref 28

ary alpha, primary alpha size, and volume fraction are important in solute-lean alloys such as Ti-10V-2Fe-3Al.

Effect of Texture on Fatigue Life. While beta alloys are considered fairly isotropic due to their body-centered cubic structure, the mechanical properties of alpha and alpha-beta alloys can be quite anisotropic. Due to the anisotropy of the hexagonal alpha lattice structure and the resulting directionality of mechanical properties in single crystals, polycrystals in textured material with high volume fractions of the alpha phase can be quite anisotropic with respect to fatigue performance. Because the texture in alpha-beta alloys can be varied to a larger extent than in alpha alloys, the effect of crystallographic texture on fatigue behavior has been mainly studied in Ti-6Al-4V. In contrast to beta-annealed (fully lamellar) microstructures, which normally have a nearly random texture, various types of textures can be developed in the alpha phase of fully equiaxed and duplex microstructures.

The four basic types of textures are basal (B), transverse (T), mixed basal/transverse (B/T), and weakly textured (WT). These textures are achieved by appropriate thermomechanical processing. The *S-N* curves of fully equiaxed microstructures of Ti-6Al-4V with various types of sharp textures are shown in Fig. 48. The highest HCF strength, approximately 725 MPa (105 ksi), was obtained by testing the B/T-type texture parallel to the rolling direction (RD). The lowest HCF strength, 580 MPa (84 ksi), was measured for a T-type texture tested perpendicular to the RD in the rolling plane (transverse direction, or TD).

High-Cycle Fatigue Strength. Fatigue limits (or endurance limits) represent the value of stress below which a material can presumably endure an infinite number of cycles. For many conditions, fatigue limits may be observed at 10^7 cycles or more. In other cases, however, fatigue limits are not observed. These cases are generally attributed to periodic overstrains or the absence of hardening, as with very low oxygen levels (Fig. 49). Also, the absence of a fatigue limit in Ti-6Al-4V alloy can be associated with subsurface initiations, especially at cryogenic temperatures.

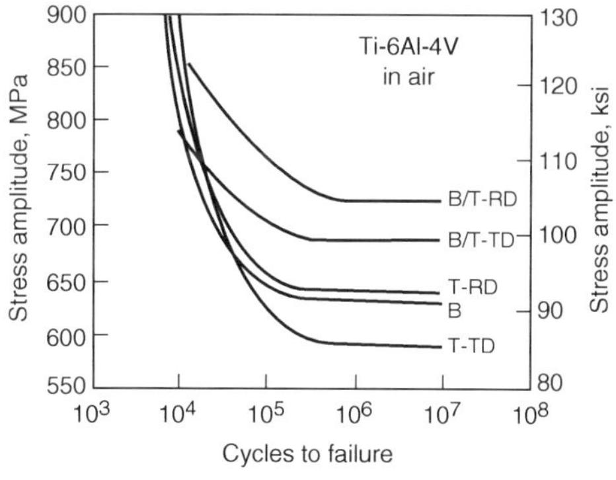

Fig. 48 *S-N* curves in Ti-6Al-4V (air). Effect of texture and loading direction. B, basal; T, transverse; RD, rolling direction; TD, transverse direction. Source: Ref 24

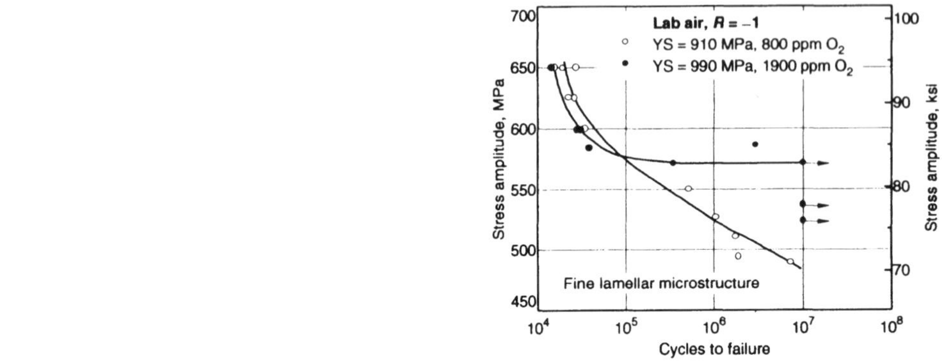

Fig. 49 Effect of low oxygen and yield strength on Ti-6Al-4V fatigue limits. Source: Ref 29

Approximately 80 to 90% of HCF life involves the nucleation of surface cracks, which can be greatly influenced by stress concentrations (such as corrosion pits or notches) or surface residual stress.

Notch Effects. The notched HCF strength of wrought Ti-6Al-4V at 10^7 cycles is typically between approximately 135 to 275 MPa (20 to 40 ksi) ranges for notched ($K_t = 3$) specimens (Fig. 50). The static tensile strength has even less influence on fatigue strength for notched specimens than for unnotched specimens (Fig. 51). It should also be noted that fatigue limits of unnotched Ti-6Al-4V and other titanium alloys reveal more scatter than quenched and tempered low-alloy steels. For example, extensive tensile and smooth bar fatigue testing on Ti-6Al-4V and regression analysis has been performed to see if a correlation exists between 10^7 cycle fatigue strength and yield or tensile strength. In both cases, the coefficient of correlation was smaller than 0.1, indicating that essentially no correlation exists. This tends to point out an important difference between titanium alloys and steels: that the effect of microstructures on fatigue is a more complex variable, especially for the different variations of microstructures associated with dual-phase materials such as alpha-beta alloys or the solute-lean beta alloys such as Ti-10V-2Fe-3Al.

Effects of Surface Treatments

Mechanical surface treatments such as shot peening, polishing, or surface rolling can be used to improve the fatigue life in titanium alloys. Because fatigue failure represents the sum of both crack nucleation and crack propagation life, the changes induced by such treatments can have contradictory influences on the fatigue strength. Surface roughness determines whether fatigue strength is primarily crack nucleation controlled (smooth) or crack propagation controlled (rough). For smooth surfaces, a work-hardened surface layer can delay crack nucleation due to the increase in strength. In rough surfaces, the crack nucleation phase can be absent, and

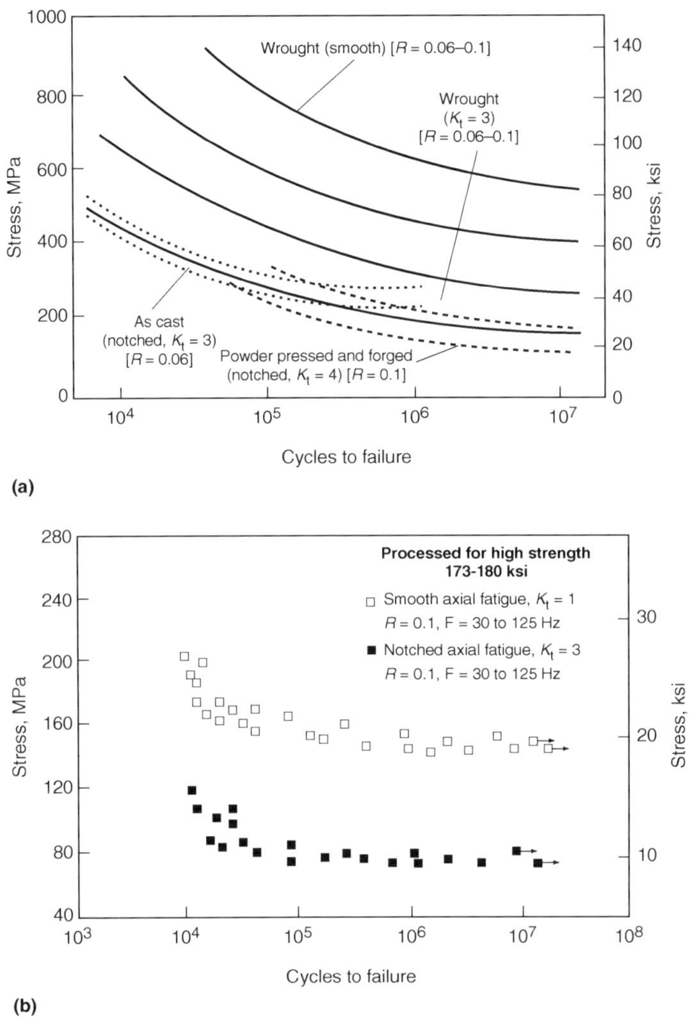

Fig. 50 Notch effects on (a) Ti-6Al-4V and (b) Ti-10V-2Fe-3Al. Source: Ref 30, 31

a work-hardened surface layer is detrimental to crack propagation due to the reduced ductility. Near-surface residual compressive stresses are clearly beneficial, because they can significantly retard microcrack growth once cracks are present (Table 8).

It must be kept in mind that the changes induced by mechanical surface treatment are not necessarily stable. In particular:

- Residual stresses can be reduced or eliminated by a stress-relief treatment.
- Degree of cold work can be removed by recrystallization.
- Surface roughness can be reduced by an additional surface treatment (e.g., polishing).

- Beneficial residual compressive stresses can be reduced by cyclic plastic deformation (i.e., during fatigue loading in service).

These factors can be used to design surface treatments for titanium alloys that are appropriate to the application. For example, Fig. 52 shows a series of *S-N* curves for Ti-6Al-4V at room temperature and at elevated temperature. As a reference condition, an electropolished (EP) surface is considered to be free of residual stresses and cold work and has a mirror finish. Compared with electropolishing, shot peening (SP) significantly improves the HCF strength at room temperature. However, at elevated temperature, shot peening lowers HCF strength to values below that of the reference condition. This can be explained by considering the individual contributions of the surface properties to fatigue life. For example, stress relieving

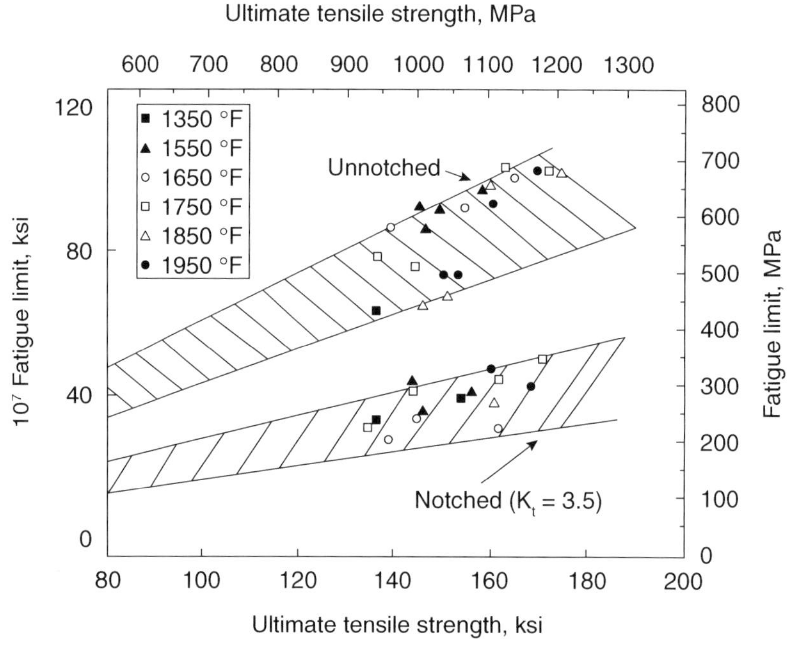

Fig. 51 Notched and unnotched fatigue limit of Ti-6Al-4V. 6.35 mm (0.25 in.) specimens cut from as rolled bar, solution treated at indicated temperatures, and cooled at various rates (furnace, air, water quench). Rotating beam fatigue at 8000 rpm. Fatigue limits at 10^7 cycles determined by highest stress amplitude at which specimens ran 10^7 cycles without failure. Source: Ref 32

Table 8 Effects of surface properties on fatigue life of titanium

Surface property	Crack nucleation effect	Microcrack propagation effect
Surface roughness	Accelerates	No effect
Degree of cold work or dislocation density	Retards	Accelerates
Residual compressive stresses	Minor or no effect	Retards

Source: Ref 33

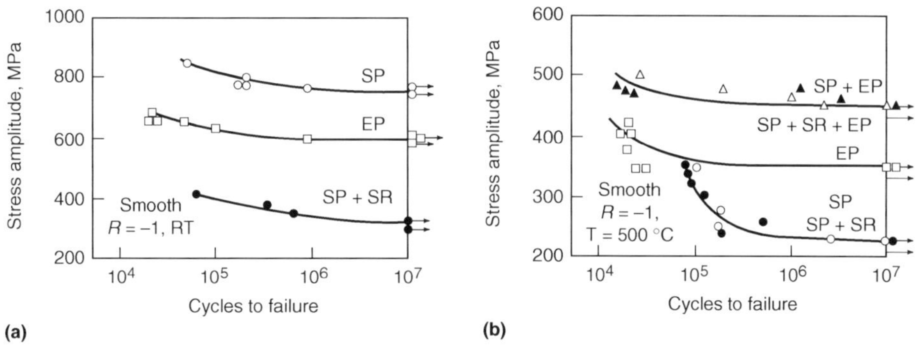

Fig. 52 *S-N* curves ($R = -1$) for Ti-6Al-4V with a fine lamellar microstructure. (a) Room temperature. (b) 500 °C (930 °F). SP, shot peened; EP, electropolished; SR, stress relieved. Source: Ref 34

(SR) 1 h at 600 °C (1110 °F) after shot peening (SP + SR) decreases the endurance limit of the shot-peened condition at room temperature but does not alter the HCF strength at 500 °C (930 °F). In effect, stress relieving is redundant when cyclic loading occurs at high temperature. If the compressive residual stresses were the only mechanism operating, one could conclude that shot peening cannot be used to improve fatigue performance at elevated temperatures. However, an additional surface treatment that reduces the surface roughness, in the present case, shot peening and electropolishing (SP + EP), demonstrates that work hardening the surface layer can also be exploited to improve HCF strength, irrespective of whether a stress-relief treatment is applied (SP + SR + EP) or not. The degree of near-surface cold work is not significantly altered by stress relieving and thus raises the HCF strength.

Microcrack growth rates in the conditions shot peening and shot peening + stress relieving are compared to the reference electropolishing in Fig. 53. At room temperature (Fig. 53a), crack growth is drastically retarded in shot peening, but it is accelerated in shot peening + stress relieving as compared to electropolishing. The difference in growth rate between curves shot peening and shot peening + stress relieving is caused by the residual compressive stresses in shot peening, while the difference between curves shot peening + stress relieving and electropolishing is related to the high dislocation density in the shot peening + stress relieving condition. At elevated temperature (Fig. 53b), no difference in microcrack growth was measured between shot peening and shot peening + stress relieving due to residual-stress relaxation at that temperature. The inferior performance of shot peening and shot peening + stress relieving compared to the electropolished reference is due to the negative effect of high dislocation densities (low residual ductility) on crack growth resistance.

Mechanical surface treatments such as shot peening and surface rolling can be applied alone or in combination with heat treatments to obtain op-

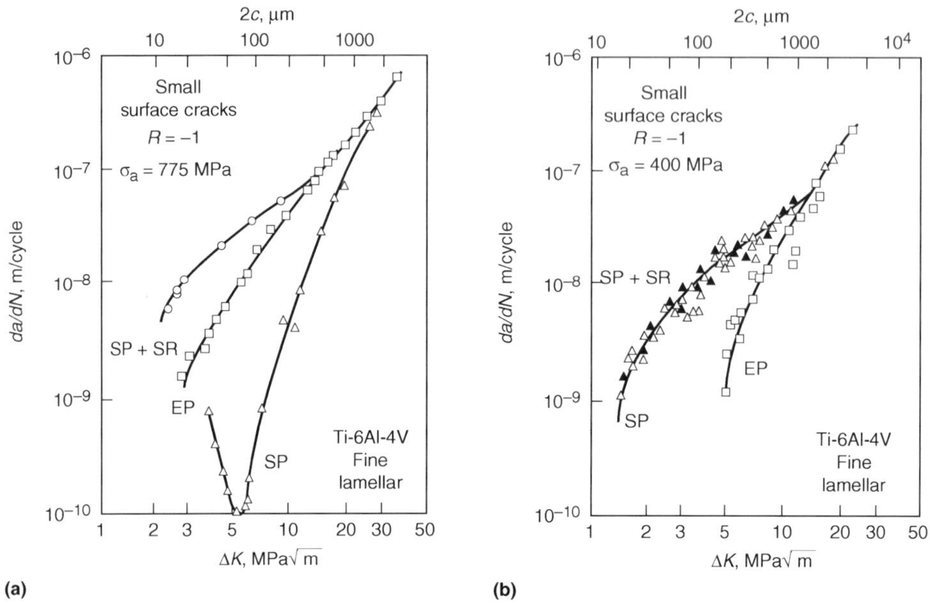

Fig. 53 Microcrack growth in Ti-6Al-4V. (a) Room temperature. (b) 500 °C (930 °F). SP, shot peened; EP, electropolished; SR, stress relieved. Source: Ref 34

timum properties in mechanically loaded titanium parts. The particular treatment applied should reflect the type of alloy (alpha, alpha-beta, or beta) to make use of its characteristic response to heat treatment and/or thermomechanical processing.

To improve the fatigue properties, all aerospace-grade titanium castings are processed by hot isostatic pressing (HIP) to close off any internal porosity. Typical HIP processing is done under argon pressure of 103 MPa (15 ksi) at 954 °C (1750 °F) for 2 h. It should be noted that HIP will collapse internal porosity but will not close off surface-connected porosity, hence the need to repair those areas by welding prior to HIP. The fatigue improvement of Ti-6Al-4V investment castings as a result of HIPing is shown in Fig. 54.

Fatigue Crack Growth of Titanium Alloys

When plotted as *da*/*dn* versus Δ*K*, fatigue crack growth (FCG) rates of titanium alloys generally lie between those of steels and aluminum alloys. This results from the use of Δ*K* to characterize the stress field at the crack tip. Δ*K* is not strictly valid for titanium alloys with a significant volume fraction of alpha grains, because the alpha phase is anisotropic, and the derivation of *K* assumes isotropic elastic behavior. Nonetheless, *K* is widely used to describe crack growth behavior in titanium alloys. Typical FCG behavior for titanium as compared to steel and aluminum at a low load ratio is shown in Fig. 55. More so than in steel and aluminum, micro-

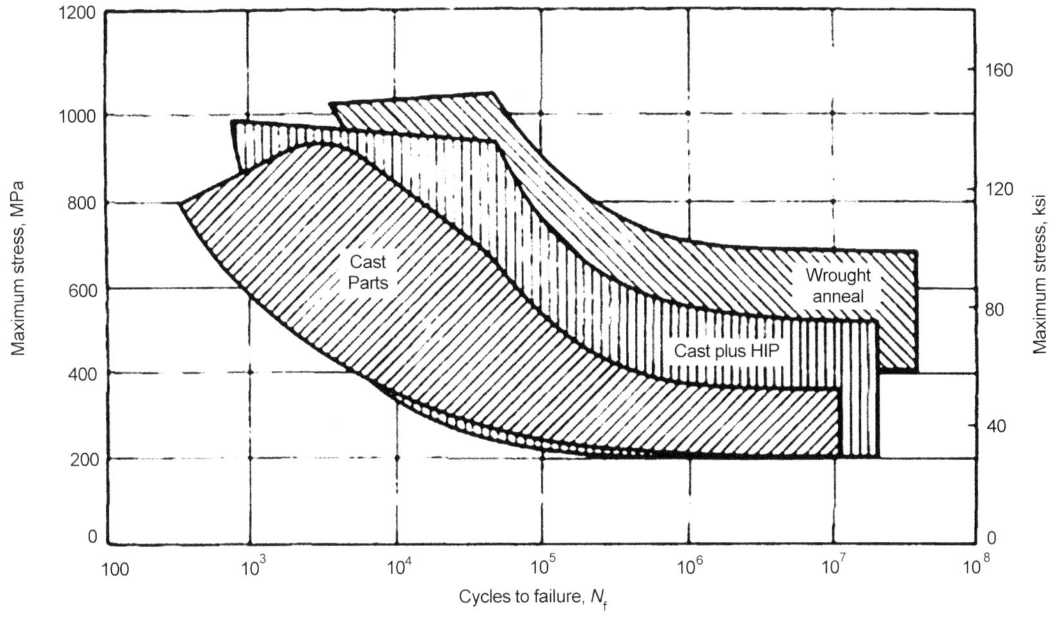

Fig. 54 Effect of hot isostatic pressing (HIP) on fatigue properties of Ti-6Al-4V investment castings. Room-temperature smooth bar, tension-tension fatigue, $R = +0.1$. Source: Ref 35

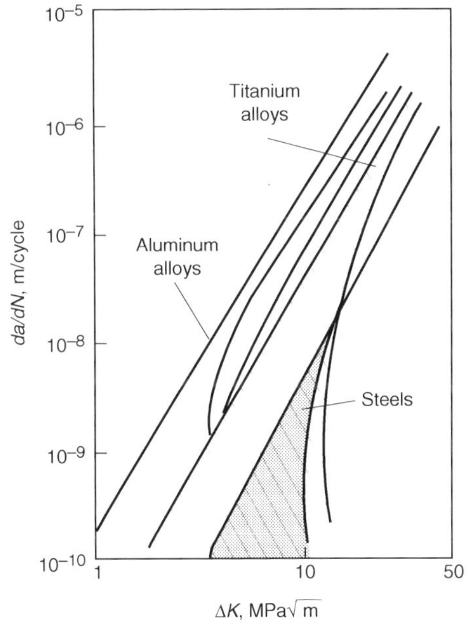

Fig. 55 Range of fatigue crack growth rates in titanium alloys, which lie between those of steels and aluminum alloys. Source: Ref 36, 37

structural variations can significantly influence FCG rates and thresholds, but they are more or less pronounced according to specific loading and environmental conditions. Fatigue crack growth is commonly held to be slower in coarse microstructures than in fine microstructures. However, this is only true when cracks are long and the stress ratio is low (i.e., less than roughly 0.3).

Metallurgical Effects on Titanium FCG

The possible influences of microstructure are specific to the alloy type. For example, in CP titanium or alpha alloys, grain size, interstitial (oxygen) content, and cold work can be varied. The influence of grain size on FCG at $R = 0.07$ for different grades of CP titanium is shown in Fig. 56. In general, the higher the oxygen content, the greater the influence of grain size, and vice versa. Fatigue crack growth rates in fine-grained material (i.e., 20 to 35 μm, or ASTM 7 to 8) are relatively insensitive to oxygen content. When grain size is increased to roughly 220 μm (ASTM 1.5), FCG rates not only are significantly lower than in fine-grained material, but a higher oxygen content can further lower FCG rates. Cold work would presumably cause slightly increased FCG rates by reducing residual ductility.

In near-alpha and alpha-beta alloys, the phase morphology (lamellar versus equiaxed) is the most important microstructural feature. Fatigue crack growth rates at $R = 0.1$ for Ti-6Al-4V in both mill-annealed and coarse lamellar (beta-annealed) conditions are shown in Fig. 57. The fine, mill-annealed microstructure exhibits much higher FCG rates than the

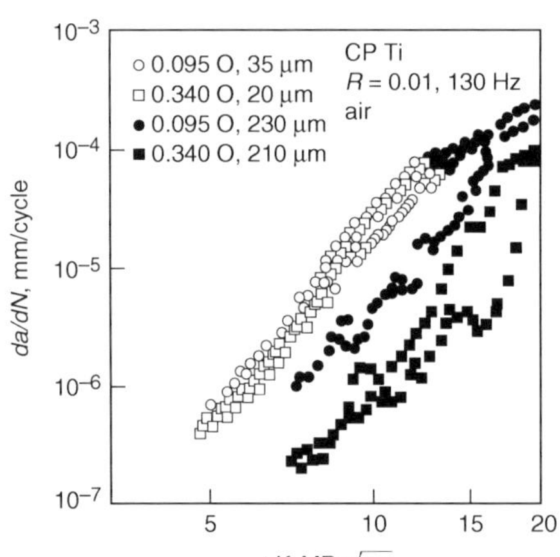

Fig. 56 Influence of grain size on fatigue crack growth at $R = 0.07$ for commercially pure titanium (CP Ti) with varying oxygen contents. Under these loading conditions, fatigue crack growth is lowered by coarse grains and increased oxygen content. Source: Ref 38

beta-annealed lamellar microstructure, particularly at lower ΔK values. Although this is commonly observed for long crack data measured on standard fracture mechanics specimens, when cracks are small or when specimen dimensions are small compared to the length of the crack front, coarse lamellar microstructures exhibit much higher FCG rates than fine microstructures.

When both alpha grains and lamellar regions are present, as in duplex microstructures, FCG rates are intermediate between those of fine-grained equiaxed and fully lamellar microstructures. Oxygen content also influences FCG rates in two-phase alloys in the same manner as in CP titanium; however, the effect is so minor as to be easily masked by other microstructural changes, presumably because the grain sizes in these alloys are always less than 20 μm. As was shown for CP titanium, FCG in fine-grained material is not very sensitive to interstitial content. The difference between FCG rates in mill-annealed Ti-6Al-4V in both the standard and extra-low interstitial grades is roughly a factor of 2.

In near-beta and metastable beta alloys, grain size, degree of age hardening, and amount of cold work are possible microstructural parameters to

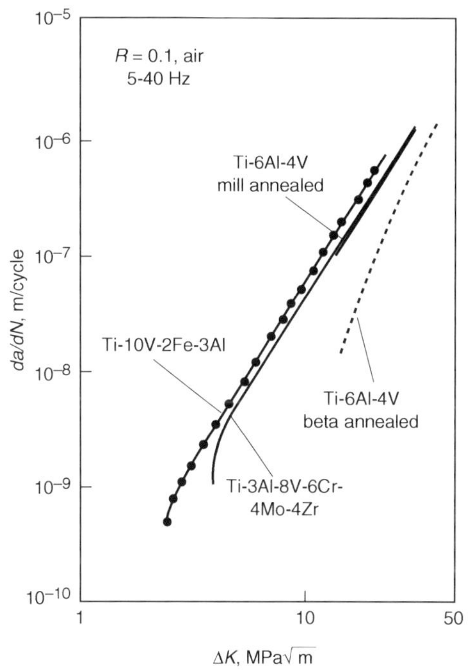

Fig. 57 Typical fatigue crack growth rates for Ti-6Al-4V in a lamellar and fine equiaxed condition and for Ti-10V-2Fe-3Al and Ti-3Al-8V-6Cr-4Mo-4Zr. Fatigue crack growth rates tend to be lower in lamellar microstructures than in fine equiaxed microstructures in Ti-6Al-4V. Fatigue crack growth in near-beta and metastable beta alloys is usually slightly higher than that of fine equiaxed alpha + beta alloys. Source: Ref 39–41

be considered. The available data suggest that FCG in these alloys is insensitive to grain size and degree of cold work (or residual deformation from hot working). Age hardening has only a minor influence on FCG. Typical FCG rates for the near-beta and metastable beta alloys Ti-10V-2Fe-3Al and Ti-3Al-8V-6Cr-4Mo-4Zr, respectively, are shown in Fig. 57. Fatigue crack growth rates are slightly higher than those of fine-grained near-alpha and alpha-beta alloys. This is partially due to the slightly lower elastic modulus of the beta matrix, which tends to shift FCG curves to lower ΔK values.

In CP titanium and in near-alpha and alpha-beta alloys, a preferred crystallographic orientation, or texture in the alpha phase, can affect FCG. A preferred orientation in the alpha phase can be indirectly assessed by measuring the elastic modulus. In commercial alloys, values of roughly 100 GPa (14.5×10^6 psi) indicate that few basal planes are loaded in tension, while values as high as 130 GPa (18.9×10^6 psi) can be found when many basal planes are loaded in tension. Normally, a higher elastic modulus is expected to lower FCG rates at a given ΔK. However, a high proportion of basal planes in the crack plane leads to more rapid crack growth in CP titanium and alpha-beta alloys, because cleavage is the preferred mode of crack advance on these planes. Figure 58 shows FCG rates in air for Ti-6Al-4V in a fine equiaxed condition (alpha grain size of 1 to 2 μm) loaded parallel and perpendicular to the basal planes. This texture effect is absent in vacuum, and the influence of texture becomes more pronounced in a salt water environment, suggesting that the acceleration mechanism is related to the environment. For near-beta and metastable beta alloys, few data are available regarding the effect of texture on FCG. Where different sample orientations in rolled plate have been investigated, virtually no difference in FCG behavior has been found, suggesting that the influence of texture is negligible.

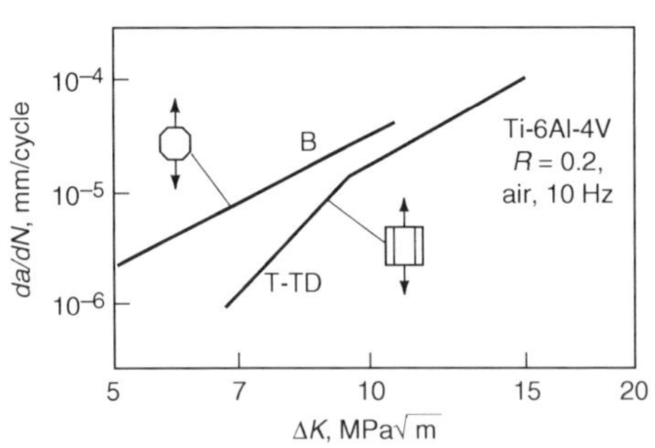

Fig. 58 Influence of texture on fatigue crack growth in Ti-6Al-4V. Fatigue crack growth rates are higher when basal planes are loaded in tension. The elastic modulus in tension for the basal texture (B) is 109 GPa (15.8×10^6 psi); for the transverse texture (T), 126 GPa (18.3×10^6 psi). The yield stress is roughly 1150 MPa (167 ksi) for both. TD, transverse texture tested perpendicular to the rolling direction. Source: Ref 42

ACKNOWLEDGMENTS

The material in this chapter came from "Fracture Resistance of Structural Alloys" by K.S. Ravichandran and A.K. Vasudevan, "Alloy Design for Fatigue and Fracture" by S.D. Antolovich, "Fracture and Fatigue Properties of Structural Steels" by A.D. Wilson, "Selecting Aluminum Alloys to Resist Failure by Fracture Mechanisms" by R.J. Bucci, G. Nordmark, and E.A. Starke, Jr., and "Fatigue and Fracture Properties of Titanium Alloys," all in *Fatigue and Fracture*, Vol 19, *ASM Handbook*, ASM International, 1996; and "Fatigue Resistance of Steels," by B. Boardman, *Properties and Selection: Irons, Steels, and High-Performance Alloys*, Vol 1, *ASM Handbook*, ASM International, 1990.

REFERENCES

1. K.S. Ravichandran and A.K. Vasudevan, Fracture Resistance of Structural Alloys, *Fatigue and Fracture,* Vol 19, *ASM Handbook,* ASM International, 1996
2. F.R. Stonsifer and R.W. Armstrong, *Fracture 1977,* Vol 2A, D.M.R. Taplin, Ed., Pergamon Press, 1977, p 1–6
3. A.H. Priest, in *Effect of Second-Phase Particles on the Mechanical Properties of Steel,* The Iron and Steel Institute, London, 1971
4. K.H. Schwalbe, *Eng. Fract. Mech.,* Vol 9, 1977, p 795–832
5. *Fractography,* Vol 12, *Metals Handbook,* 9th ed., ASM International, 1987, p 384
6. *Metallic Materials and Elements for Aerospace Vehicle Structures,* MIL-HDBK-5B, *Military Standardization Handbook,* U.S. Department of Defense, 1987
7. R.C. Juvinall, *Engineering Considerations of Stress, Strain and Strength,* McGraw-Hill, 1967
8. B. Boardman, Fatigue Resistance of Steels, *Properties and Selection: Irons, Steels, and High-Performance Alloys,* Vol 1, *ASM Handbook,* ASM International, 1990
9. A.D. Wilson, Fracture and Fatigue Properties of Structural Steels, *Fatigue and Fracture,* Vol 19, *ASM Handbook,* ASM International, 1996
10. A.D. Wilson, Fatigue Crack Propagation in Steels: The Role of Inclusions, *Fracture: Interaction of Microstructure, Mechanisms and Mechanics,* TMS-AIME, 1984, p 235–254
11. R.J. Bucci, G. Nordmark, and E.A. Starke, Jr., Selecting Aluminum Alloys to Resist Failure by Fracture Mechanisms, *Fatigue and Fracture,* Vol 19, *ASM Handbook,* ASM International, 1996
12. R. Develay, *Metals and Materials,* Vol 6, 1972, p 404
13. R.R. Senz and E.H. Spuhler, Fracture Mechanics Impact on Specifications and Supply, *Met. Prog.,* 1975, p 64–66
14. M.O. Speidel, *Metall. Trans. A,* Vol 6, 1975, p 631

15. "Standard Test Method for Determining Susceptibility to Stress-Corrosion Cracking of High Strength Aluminum Alloy Products," G 47, *Annual Book of ASTM Standards,* Section 3, Vol 03.02, ASTM
16. R. Templin, F. Howell, and E. Hartmann, "Effect of Grain-Direction on Fatigue Properties of Aluminum Alloys," Alcoa, 1950, and *ASTM Proc.,* Vol 64, p 581–593
17. *Fatigue Data Book: Light Structural Alloys,* ASM International, 1995
18. P.C. Varley, *The Technology of Aluminum and Its Alloys,* Newnes-Butterworths, London, 1970
19. B. Sarker, M. Marek, and E.A. Stacke, Jr., *Metall. Trans. A,* Vol 12, 1981, p 1939
20. W.H. Reimann and A.W. Brisbane, *Eng. Fract. Mech.,* Vol 5, 1973, p 67
21. J.T. Staley, "How Microstructure Affects Fatigue and Fracture of Aluminum Alloys," presented at International Symposium of Fracture Mechanics (Washington, D.C.), 1978
22. N.G. Turner and W.T. Roberts, *Trans. TMS-AIME,* Vol 242, 1968, p 1223
23. *Properties and Selection: Stainless Steels, Tool Materials, and Special-Purpose Metals,* Vol 3, *Metals Handbook,* 9th ed., American Society for Metals, 1980, p 376
24. M. Peters and G. Lütjering, Report CS-2933, Electric Power Research Institute
25. W. Trojahn, thesis, Ruhr University, Bochum, Germany, 1980
26. L. Wagner, G. Lütjering, and R.I. Jaffee, in *Microstructure/Property Relationships in Titanium Aluminides and Alloys,* TMS-AIME, 1991, p 521
27. L. Wagner and G. Lütjering, *Z. Metallkd.,* Vol 87, 1987, p 369
28. L. Wagner and G. Lütjering, in *Titanium '88: Science and Technology,* Les Editions de Physique, 1988, p 345
29. *Fatigue and Microstructure,* American Society for Metals, 1979, p 237
30. *Military Handbook: Titanium and Titanium Alloys,* MIL-HDBK-697A, U.S. Department of Defense, 1974
31. J.C. Williams and E. Starke, *Deformation, Processing and Structure,* American Society for Metals, 1984, p 326
32. L. Bartlo, STP 459, ASTM, 1969
33. M. Bauccio, Technical Note 9: Descaling and Special Surface Treatments, *Materials Properties Handbook: Titanium Alloys,* ASM International, 1994, p 1145–1158
34. H. Gray, L. Wagner, and G. Lütjering, in *Shot Peening,* H. Wohlfahrt, R. Kopp, and O. Vöhringer, Ed., DGM, 1987, p 467
35. R.R. Boyer, Processing of Titanium and Titanium Alloys, *Metals Handbook Desk Edition,* 2nd ed., ASM International, 1998

36. M.O. Speidel, Stress Corrosion Cracking and Corrosion Fatigue Fracture Mechanics, *Corrosion in Power Generating Equipment,* M.O. Speidel and A. Atrens, Ed., Plenum Press, 1983, p 85–132
37. P.K. Liaw, T.R. Leax, and W.A. Logsdon, Near-Threshold Fatigue Crack Growth Behavior in Metals, *Acta Metall.,* Vol 31, 1983, p 1581–1587
38. J.L. Robinson and C.J. Beevers, The Effects of Load Ratio, Interstitial Content and Grain Size on Low-Stress Fatigue Crack Propagation in Titanium, *Met. Sci. J.,* Vol 7, 1973, p 153–159
39. G.R. Yoder, L.A. Cooley, and T.W. Crooker, Observations on Microstructurally Sensitive Fatigue Crack Growth in a Widmanstätten Ti-6Al-4V Alloy, *Metall. Trans. A,* Vol 8, 1977, p 1737–1743
40. T.W. Duerig, J.E. Allison, and J.C. Williams, Microstructural Influences on Fatigue Crack Propagation in Ti-10V-2Fe-3Al, *Metall. Trans. A,* Vol 16, 1985, p 739–751
41. C.G. Rhodes and N.E. Paton, The Influence of Microstructure on Mechanical Properties in Ti-3Al-6Cr-4Mo-4Zr (Beta-C), *Metall. Trans. A,* Vol 8, 1977, p 1749–1761
42. M. Peters, A. Gysler, and G. Lütjering, Influence of Texture on Fatigue Properties of Ti-6Al-4V, *Metall. Trans. A,* Vol 15, 1984, p 1597–1605

CHAPTER 7

Metallic Joints—Mechanically Fastened and Welded

FAILURE IN ENGINEERING STRUCTURES are still common today, despite the fact that modern tools for designing structures are very sophisticated and readily available. Computer-aided design, finite-element stress analysis, computerized material property databases, and an array of process simulation tools are among the many design aids accessible to the engineer. In addition, over the past 20 to 30 years, national and international codes and standards have been developed for many industries, and these provide explicit guidelines on such issues as materials selection, design methods, standardized load histories, and safety factors.

In general, most of today's fabricated components and structures are either mechanically fastened or welded, and invariably, the joint is the most critical area from the performance perspective. An examination of structural and component failures documented in open literature over the past 50 years or so clearly indicates that failures predominantly start at joints.

Mechanically Fastened Joints

Mechanically fastened structural joints, such as riveted joints and high-strength bolted joints, are used in many fatigue-critical applications, and a considerable amount of fatigue testing of mechanically fastened structural joints has been performed to support the development of fatigue provisions in structural design specifications. In general, the important factors affecting the fatigue strength of mechanically fastened joints are clamping force, grip length, type of fastener, fastener material, fastener pattern, tension-shear ratio, tension-bearing ratio, and the type of materials being joined.

Typically, the fatigue resistance of high-strength friction-bolted joints is superior to that of riveted joints. For example, the fatigue strength of carbon steel joints fastened with A325 bolts approaches the yield strength of the connected material (Fig. 1). On the other hand, the fatigue resistance of carbon steel joints fastened with hot-driven rivets is practically identical to the fatigue resistance of carbon steel plates containing open circular holes.

Joint design and loading are probably the most critical factors. The static strength of the material may not always be a factor. In some codes, fatigue resistance of bolt joints may be considered independent of static strength. For example, in a Swedish design recommendation for bolted connections, the fatigue strength is actually reduced for higher-strength bolts, from a stress range of 100 to 120 MPa (14.5 to 17.4 ksi) at $N = 10^7$ cycles for grade 8.8 bolts to 80 MPa (11.6 ksi) for grade 12.9 and higher-strength bolts. In the European Convention for Constructional Steelworks' recommendation, a single *S-N* curve is given for all steel bolt materials up to grade 10.9 (maximum ultimate tensile strength of 1040 MPa, or 150 ksi), where the stress range at $N = 0.5 \times 10^6$ cycles is only 26.5 MPa (3.8 ksi). However, as shown in Fig. 2, the static strength may have a significant effect if the fastener contains rolled threads.

A summary of many tests on riveted plain carbon and alloy steels also demonstrates a joint fatigue strength at approximately 180 MPa (26 ksi) for 2×10^6 cycles, despite a range of static tensile strengths of 415 to 690

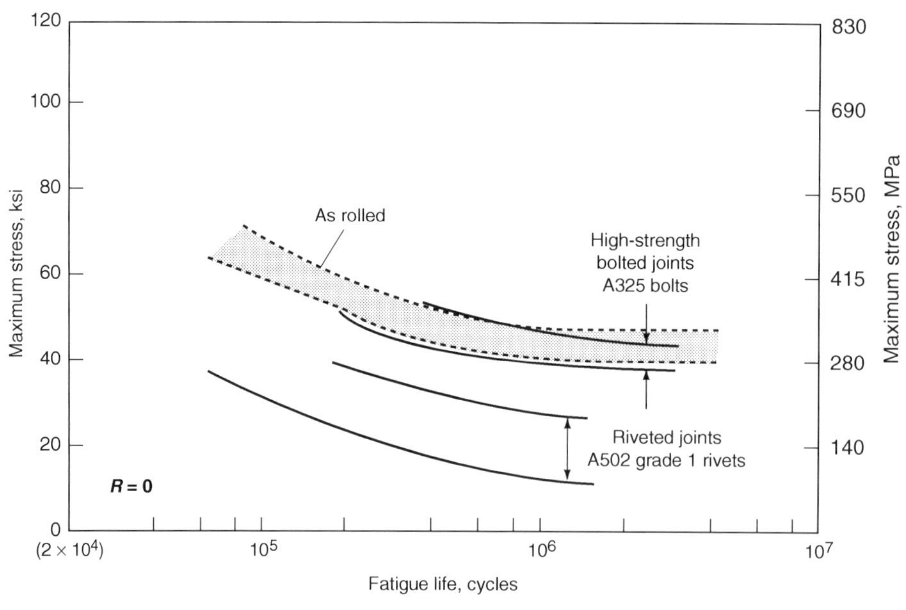

Fig. 1 Fatigue strength of carbon steel structural joints. Source: Ref 1

MPa (60 to 100 ksi). This illustrates the importance of joint design and loading over that of material strength. For example, as shown in Fig. 3, the tension/shear/bearing ratio is very important in the fatigue life of riveted joints. In particular, the stress concentrations at the hole surface degrades allowable bearing strength, which promotes the use of high-strength bolts instead of rivets. In a properly designed bolted joint, it is possible to carry the shear loads by friction.

Threaded Fasteners in Tension

Thread design and clamping forces are two key factors affecting the fatigue resistance of bolted joints. However, fastener strength can be an important variable when threads are rolled (Fig. 2). This is in direct contrast with machined threads, which demonstrate little or no static strength effect on fatigue endurance. This demonstrates the importance of surface

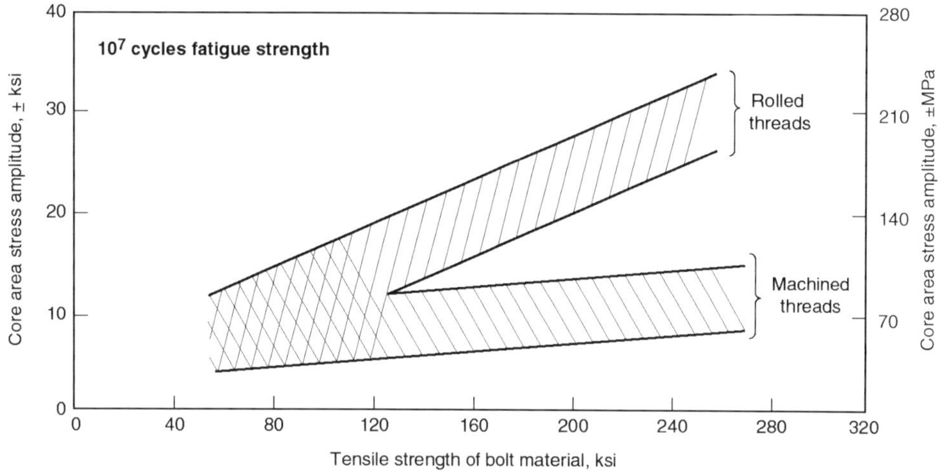

Fig. 2 Axial fatigue strength at 10^7 cycles of bolt-nut assemblies with rolled threads and machined threads ($R = -1$). Source: Ref 2

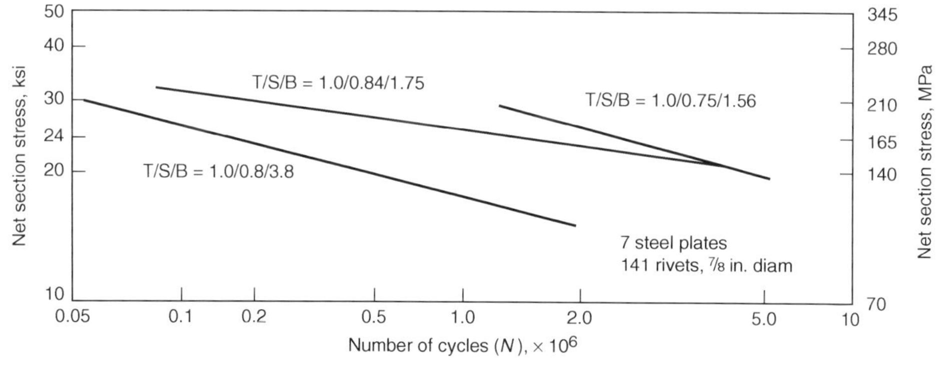

Fig. 3 Fatigue life of riveted joints with different tension (T)/shear (S)/bearing (B) ratios. Source: Ref 3

condition on fatigue crack initiation. Rolled threads have a superior fatigue resistance due to the combination of compressive residual stresses and smoother surface at the thread root.

The form of the threads, plus any mechanical or metallurgical surface condition, is much more important than nominal steel composition in determining the fatigue strength of a particular lot of bolts. However, surface composition is also a factor. Surface decarburization can cause a significant reduction in the fatigue resistance of steel bolts. Fatigue tests on 6.4 mm (0.25 in.) diameter steel bolts with a tensile strength of 917 MPa (133 ksi) showed that:

Decarburized layer, in.	Fatigue limit at 10^7 cycles, ksi
None	40
0.001	27
0.002	25

The weakest point of an axially loaded standard bolt and nut combination in fatigue is normally in the bolt threads at approximately one turn in from the loaded, or bearing, face of the nut, where the load transfer from nut to bolt is at maximum value. This area of stress concentration occurs because the bolt elongates as the nut is tightened, thus producing increased loads on the threads nearest the bearing face of the nut, which add to normal service stresses. This condition is alleviated to some extent by using nuts of a softer material that will yield and distribute the load more uniformly over the engaged threads.

Fatigue failures occur with much less frequency under the bolt head or at the thread runout and are usually machining errors. For example, failures beneath the bolt head can be caused by poor forging practice or grinding burns. Head failures have also been caused by improper upsetting procedures that result in broken flow lines after machining in the critical head-to-shank fillet region. The stress concentration at the runout of the thread can be reduced by reducing the bolt shank diameter to less than the thread root diameter or by a stress-relieving groove. The nut should never be run up to the end of the thread unless the shank is undercut below the thread.

Shape and size of the head-to-shank fillet are important, as is a generous radius from the thread runout to the shank. In general, the radius of this fillet should be as large as possible while, at the same time, permitting adequate head-bearing area. This requires a design trade-off between the head-to-shank radius and the head-bearing area to achieve optimum results. Cold working of the head fillet is another common method of preventing fatigue failure, because it induces a residual compressive stress and increases the material strength.

Effect of Thread Design. The principal design feature of a bolt is the threaded section, which establishes a notch pattern in design. Any measures that decrease stress concentration can lead to improved fatigue life. Typical examples of such measures are the use of increased root radius threads and internal thread designs that distribute the load uniformly over a large number of bolt threads.

The effect of thread form on the fatigue resistance of an axially loaded bolt is shown in Table 1. The Whitworth or British Standard thread (with rounded thread root and crest) is superior in fatigue when compared to the American Standard thread form with a flat crest and root. Axial load fatigue tests on "unified" screw threads for use in the United States of America, Canada, and Britain show a somewhat higher fatigue strength than the British Standard Whitworth (Table 2). The unified thread differs from the American Standard thread in that the former has a radius at the thread root while the latter has a flat thread root.

As previously noted, the use of softer nuts can reduce the stress concentration at the bearing face of the nut-bolt, where maximum stress concentration occurs. For example, the stress-concentration factor at the bearing face of the nut can be reduced from 3.4 to 2.5 by using an aluminum, rather than a steel, nut on a steel bolt. However, galvanic corrosion may occur in corrosive environments with such a combination of two different metals. Various nut design modifications can also improve fatigue resistance by improving the load distribution.

A general estimate of the notch effect on fatigue strength can be approximated using Peterson's relation for the fatigue notch factor (K_f = unnotched/notched fatigue strength):

Table 1 Effect of thread form on fatigue strength of bolt steel

Steel	Tensile strength, ksi	Thread form	Fatigue limit(a), ksi	Notch factor (K_f)(b)	Stress-concentration factor (K_t)
0.3% C	57.4	None	37	...	...
		Whitworth	21	1.76	3.86
		American Std.	13	2.84	5.62
SAE2320	109	None	73	...	...
		Whitworth	22	3.32	3.86
		American Std.	19	3.85	5.82

(a) Repeated tension (R = 0). (b) K_f = ratio of unnotched fatigue/notched fatigue limit. Source: Ref 4

Table 2 Fatigue strength of bolt steels with unified threads

Diameter, in.	Threads per in.	Tensile strength, ksi	Unified thread	Fatigue limit at 10^7 cycles, stress range, ksi
3.8	20	134	Ground	27.5
			Rolled	48.0
		87.5	Lath-cut	17.0
			Rolled	28.5
¾	10	141	Ground	29.5 (28.0)(a)
			Rolled	62.5 (50.0)(a)
		87.5	Lath-cut	18.0 (16.8)(a)
			Ground	16.5
			Rolled	34.5 (28.0)(a)
2½	10	76.0	Ground	12.5
	16		Ground	11.2

(a) Whitworth thread in parentheses. Source: Ref 5

$$K_f = 1 + \frac{K_t - 1}{1 + a/r} \quad \text{(Eq 1)}$$

where r is the notch root radius, and a is a material constant that is related to the ultimate tensile strength (UTS) such that $a \approx 110/\text{UTS}^2$ (a is in meters, and UTS is in MPa units). The stress-concentration factor (K_t) depends on the geometry. For each groove in a threaded bolt, the adjacent threads act as relief notches so that the stress-concentration factor is much lower than that of a single groove of the same geometry. The stress-concentration factor (K_t) of a single notch of thread shape has been estimated to be 4.3, while for similarly shaped parallel grooves, as in a threaded bolt, it was only 2.52. The effect of bolt diameter on K_t is shown in Fig. 4. This method allows an estimate of fastener fatigue limits from fatigue strength of smooth specimens. Rough estimates of a fatigue limit may agree with the fatigue limit given in some codes, while there are large discrepancies with other codes and rules.

Effect of Preload. The primary force on an unloaded but installed bolt is tension, which is set up by stretching the bolt during tightening, whereas the most important stress in the nut is the shear stress in the threads. The bolt behaves like a spring (Fig. 5a). When the bolt is preloaded or when the spring is stretched, a stress is induced in the bolt and a clamping force in the structure before any working load is encountered. As the working load is applied, the preloaded bolt does not encounter an additional load until the working load equals the preload in the bolt (Fig. 5b). At this point, the force between members is zero. As more load is applied, the bolt must stretch. Only beyond this point will any cyclic working load be transmitted to the bolt. When the bolt encounters an increase in load, being elastic, it stretches. If this stretch (strain) exceeds the elastic limit, the bolt yields plastically, taking a permanent set. The result is a loss in preload or clamping force (dashed curve, Fig. 5c). With a fluctuating load, this situation can cycle progressively, with continued loss of preload and possibly rapid fatigue failure.

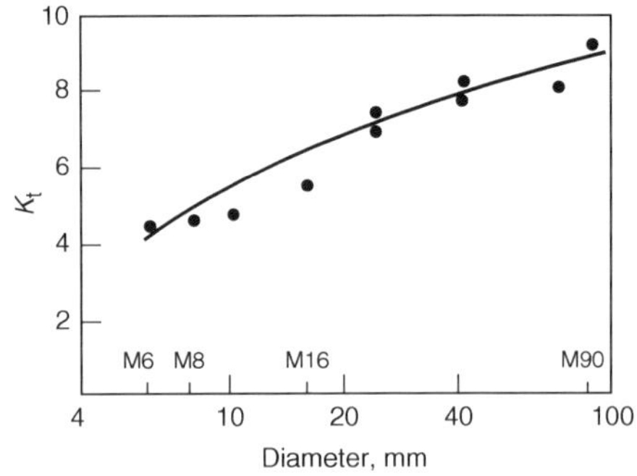

Fig. 4 Stress concentration factor (K_t) versus bolt diameter for bolts with standard metric threads. Source: Ref 6

To eliminate fatigue problems that occur at room temperature, the designer should specify as high an initial preload as practical. The optimum fastener torque values for applying specific loads to the joint have been determined for many high-strength fasteners. However, these values should be used with caution, because the tension produced by a selected torque value depends directly on the friction between the contacting threads. With the proper selection of materials, proper design of bolt-and-nut bearing surfaces, and the use of locking devices, the assumption is that the initial clamping force will be sustained during the life of the fastened joint. This assumption cannot be made for elevated-temperature design.

At elevated temperature, the induced bolt load will decrease with time as a result of creep, even if the elastic limits of the materials are not exceeded, and this can adversely affect fastener performance. Therefore, it is necessary to compensate for high-temperature conditions in advance when assembling the joint at room temperature. Failures that occur at elevated temperatures necessitate evaluation of the properties of the fastener mate-

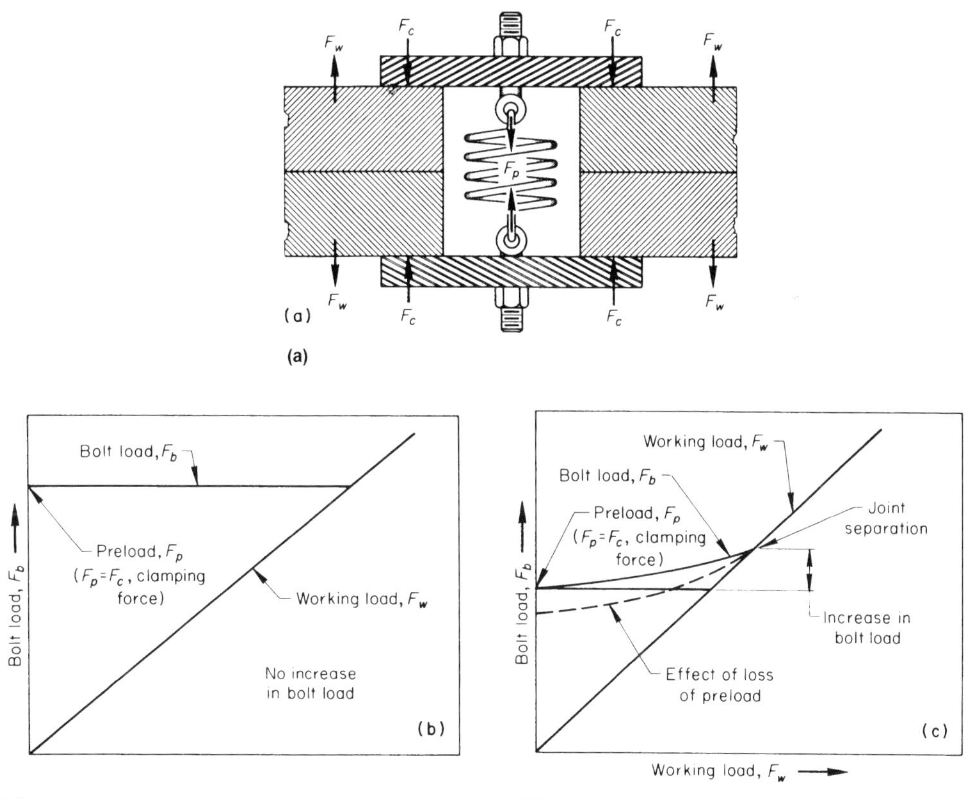

Fig. 5 Schematic showing the springlike effect of loading conditions on bolted joints. (a) Theoretical load condition for an elastic fastener and a rigid structure. (b) Ideal relationship of bolt load to working load with an elastic fastener and a rigid structure. (c) Actual relationship (both fastener and structure elastic) of bolt load to working load. Source: Ref 7

rial relative to the applied loads, operating temperature, and time under load at temperature.

The amount of clamping force that the fastener must provide to hold the assembly together must be sufficient to both maintain preloading and prevent slipping of the parts or opening of the joint when service loads are applied. The factors that primarily establish the preload requirement are the stiffness of the materials in the joint and the loads that are placed on the assembly.

Preloading, in effect, increases the area carrying the external load from the net area of the bolt to the sum of the clamped faying surfaces plus the bolt net area. The relationship between the bolt load (P_B), the bolt pretension (Q), and the externally applied load (P) is expressed by:

$$P_B = Q + \frac{P}{1 + \frac{K_p}{K_b}} \quad \text{(Eq 2)}$$

where K_b and K_p are the stiffnesses of the bolt and clamped parts, respectively. The cyclic load in a pretensioned bolt is a function of the external cyclic load and the stiffness ratio K_p/K_b. The magnitude of the pretension (Q) establishes the maximum working load, that is, the load at which the faying surfaces separate. The load carried by the bolt (P_B) may be decreased by decreasing the bolt stiffness (K_b), thus increasing the fatigue resistance of the bolt. For maximum fatigue resistance, soft or flexible inserts between the clamped parts should be avoided. The clamped parts should be flat and tight; that is, K_p should be high relative to K_b.

Pretension is gradually reduced during cyclic loading. It appears that loss of pretension is more rapid with machine-cut threads than with rolled threads and also when there are several substrates being joined. The interaction of elevated temperature and cyclic loading on bolt preload is also complicated. It appears that fatigue resistance may increase with temperature. However, at long lives, creep, rather than fatigue, may be the controlling failure mode.

Tightening of Bolts. The determination of the actual preload in a given bolt-nut combination is difficult. Tightening with a torque wrench is the simplest but is likely to be inaccurate unless calibrated. Various devices exist that allow the clamping force-torque relation to be established for each bolt-nut combination. Bolts with built-in indicators of elongation, crushable washers, and bolts with break-off driving elements are also used to control preload.

For estimating purposes, torque may be expressed approximately by:

$$T = K \cdot D \cdot P_B \quad \text{(Eq 3)}$$

where T is the torque, K is the torque coefficient, D is the nominal bolt size, and P_B is the bolt tension.

The unitless torque coefficient, K, may be taken as 0.2 but is, in reality, a function of thread conditions, geometry, friction between thread contact surfaces, friction between bearing face of turning element (nut or bolt head), and so on. A more exact torque-bolt tension relation for estimating purposes is shown in Fig. 6. However, in order for tension to be developed, the torque applied to a fastener must overcome friction under the head of the fastener and in the threads, and the fastener or nut must turn. Because friction can absorb as much as 90 to 95% of the energy applied to the fastener, as little as 5 to 10% of the energy is left for generating fastener tension, as shown in Fig. 7. If the amount of friction varies greatly, wide variations in clamping force are produced, which can mean loose or broken bolts, leading to assembly failures.

Therefore, if torque alone is measured, it can never be known with certainty whether the desired tension has been achieved. Unfortunately, it must be concluded that torque is a highly unreliable, totally inaccurate measurement for evaluation of the preload on a threaded fastener. For many noncritical fasteners, where safety or the functional performance of an assembly is not compromised, it may be acceptable to specify and

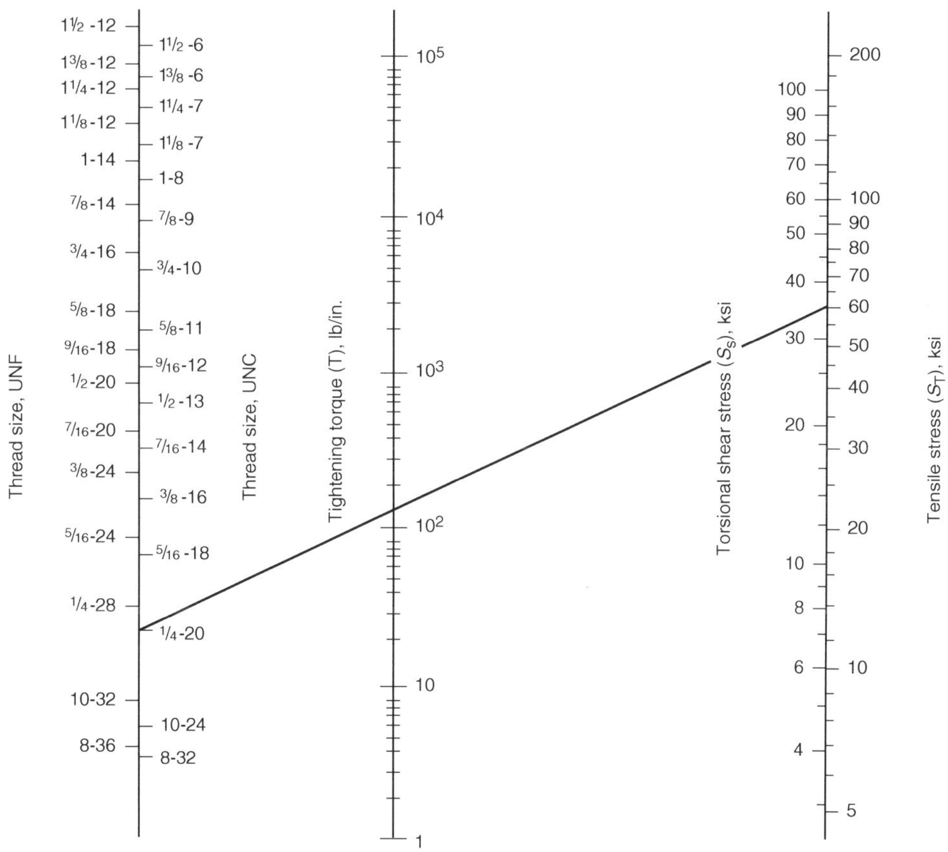

Fig. 6 Nomograph for torque on bolts. Source: Ref 8

monitor torque alone. The most common measurement tools are hand torque wrenches that are used for installation and torque audit measurements and rotary torque sensors that are used to measure installation torque dynamically. To ensure proper assembly of critical fasteners, more than torque must be measured.

Fastener tension can be measured using different devices, such as strain-gaged bolts or fastener force washers, or by using special techniques, such as ultrasonic bolt measurement. Although these devices and methods are useful for research, they are often impractical or costly for evaluating fastener tension in production quality-control efforts. The most common method to indirectly estimate fastener tension is to take torque measurements as the fastener is tightened or by using a fastener that has a nut that breaks away at a predetermined torque during installation. For example, the fastener shown in Fig. 8 contains a nut with a breakaway groove, which controls the amount of torque and preload on the pin. The nut is tightened down until a predetermined torque level is achieved, and then the top portion of the nut fractures.

Preload Control. Insufficient preload, caused by an inaccurate tightening method, is a frequent cause of bolted joint failure. There are six main methods used to control the preload of a threaded fastener, summarized as follows. However, whatever method is used to tighten a bolt, a degree of bolt preload scatter is to be expected:

- *Torque-controlled tightening:* Controlling the torque that a fastener is tightened to is the most popular means of controlling preload. The nominal torque necessary to tighten the bolt to a given preload can be determined either from tables or by calculation, using a relationship between torque and the resulting bolt tension. A fundamental problem with torque tightening is that because the majority of the torque is used

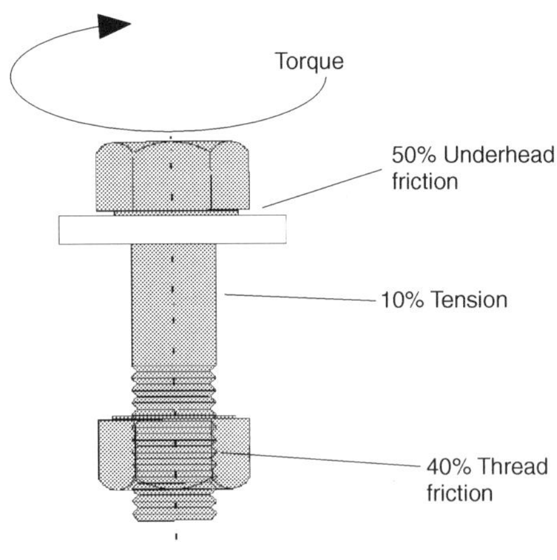

Fig. 7 Typical distribution of energy from torque applied to a bolted assembly. Source: Ref 9

to overcome friction, slight variations in the frictional conditions can lead to large changes in the bolt preload.

- *Angle-controlled tightening:* This method, also known as turn-of-the-nut method, has been applied for use with power wrenches. The bolt is tightened to a predetermined angle beyond the elastic range and results in a small variation in the preload due, in part, to the yield stress tolerance. The main disadvantages of this method lie in the necessity for precise experimental determination of the angle. In addition, the fastener can only sustain a limited number of re-torques before it fails.
- *Yield-controlled tightening:* An electronic control system is used that is sensitive to the torque gradient of the bolt being tightened. Rapid detection of the change in slope of this gradient indicates the yield point has been reached and stops the tightening process. This is achieved by incorporating sensors to read torque and angle during the tightening process. Because the angle of rotation and torque are both measured by the control system, permissible values can be used to detect fasteners that lie outside specification (having too low a yield, for example). A small degree of preload scatter still results from this method, due to the influence of friction. The method detects the yield point of the fastener under the action of combined tension and torsion. The higher the thread friction, the higher the torsional stress, which, for a given yield value, results in a lower preload due to a lower direct stress. Because of the cost of the tools necessary to use this method, widespread adoption of this method is unlikely.

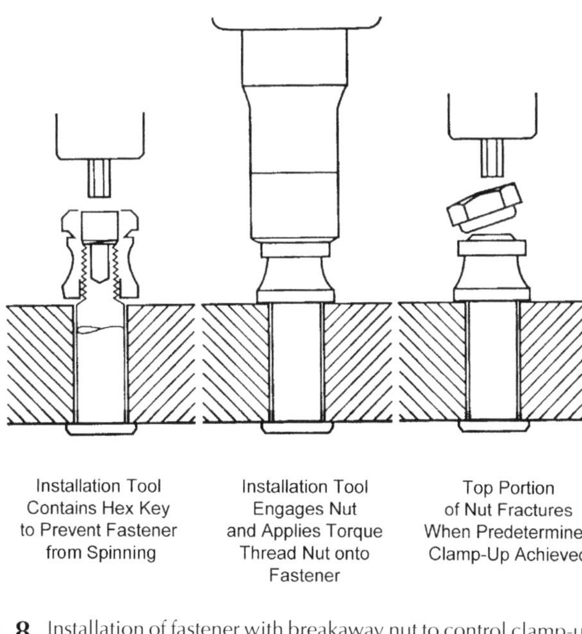

Fig. 8 Installation of fastener with breakaway nut to control clamp-up. Source: Ref 10

- *Bolt stretch method:* This method uses a small hydraulic ram that fits over the nut; the threaded portion of the bolt/stud protrudes well past the nut, and a threaded puller is attached. Hydraulic oil from a small pump acts upon the hydraulic ram, which in turn acts upon the puller. This is transmitted to the bolt, resulting in extension. The nut can then be rotated by hand. Control of the hydraulic pressure effectively controls the preload in the bolt.
- *Heat tightening:* This method uses the thermal expansion characteristics of the bolt. The bolt is heated and expands, the nut is indexed using the angle-of-turn method, and the system is then allowed to cool. As the bolt attempts to contract, it is constrained longitudinally by the clamped material, and a preload results. The process is slow and not widely used. It is generally used only on very large bolts.
- *Tension-indicating methods:* This category includes the use of special load-indicating bolts, load-indicating washers, and the use of methods that determine the length change of the fastener. There are a wide number of ways bolt tension can be indirectly measured. Special bolts have been designed that will give an indication of the force in the bolt.

The key to preventing self-loosening of fasteners is to ensure that there is sufficient clamp force present on the joint interface to prevent relative motion between the bolt head or nut and the joint; the joint is designed to allow for the effects of any plastic deformation and stress relaxation, and proven thread-locking devices are specified. There are a multitude of thread-locking devices available. Through the efforts of the American National Standards Subcommittee B18:2 on locking fasteners, three basic locking fastener categories have been established:

- *Free spinning:* The free-spinning type is a plain bolt with a circumferential row of teeth under the washer head. It is ramped, allowing the bolt to rotate in the clamping direction but lock into the bearing surface when rotated in the loosening direction.
- *Friction locking:* This category can be subdivided into two groups: metallic and nonmetallic. The metallic friction-locking fastener usually has a distorted thread that provides the desired torque. Nonmetallic friction-locking devices have plastic inserts (e.g., nylon) that provide a thread-locking function.
- *Chemical locking:* These are adhesives that fill the gaps between the male and female threads and bond them together. Such adhesives are now available in microencapsulated form and can be preapplied to the thread. When the fastener is engaged with its mating thread, the capsules are crushed and the shearing action of the rotating fastener mixes the epoxy and hardener, initiating the adhesive cure.

Effect of Mean Stress. Axially loaded mild steel bolts are insensitive to mean stress. A similar insensitivity to mean stress has been shown for axially loaded high-strength structural bolts. However, in the latter case, it was shown that the variability in observed lives increased with decreasing mean stress.

Studs. The fatigue resistance of studs is normally greater than that of bolts, because the loading at the first turn of thread engagement is less severe.

Bolts and Rivets in Bearing and Shear

Many fasteners transfer load through shear by bearing on the sides of holes in the joined substrates. Although proportioned in design by assuming all fasteners carry an equal share of the transferred load, the fasteners in the outermost rows carry more load than those in the inner rows.

Bolted Joints. In bolted shear joints, fatigue failures are initiated at:

- Bolt holes in the outermost row, where shear transfer is maximum
- A region of fretting
- Some geometric stress concentration other than a bolt hole

Failures at the first two sites are more likely than at some stress concentration other than the bolt hole. Fretting caused by relative movement of the plies is more common at long lives and relatively low stresses. In aircraft design, elaborate joints (e.g., scarfed joints with tapered substrates) produce a more uniform shear transfer, and therefore, fretting predominates. The influence of mean stress, bolt patterns, and joint dimensional parameters is similar to those of riveted joints.

Cold-Driven Rivet Joints in Sheet. At short lives (less than 10^5 cycles), rivets may fail in shear, while at longer lives, sheet failure occurs through the outermost line of rivets, with cracks initiated at the rivet holes. Fatigue life is insensitive to mean stress and is primarily a function of stress range.

In single-lap joints, the fatigue resistance increases with the ratio of rivet diameter (d) to sheet thickness (t), with the ratio of pitch (p) to d, and with the number of rows of rivets. Recommended proportions of riveted, single-lap joints for high fatigue resistance are listed in Table 3.

Double-lap joints are superior in fatigue resistance to single-lap joints due to offset loading in the latter. The number of rows of rivets in a double-lap joint appears to have less influence on fatigue resistance than in a single-lap joint, but again, a high d/t ratio is desirable. Also, the outer

Table 3 Riveted single-lap joint proportions for high fatigue resistance

				Edge distance, in.	
Joint	d/t	g/d	P/d	From row	From line
Single row	3	3	...	2	1½ to 2
Double row:					
In-line	4.5	4	3½	3	3
Staggered	4.5	4	2⅞ (min)	...	...

d, rivet diameter; *g*, gage; *t*, sheet thickness; *p*, pitch. Source: Ref 11 substrates should be at least 0.6 times the thickness of the inner substrate. Optimum proportions for double-lap joints are similar to those listed in Table 3.

Joints with flush-head rivets generally have poorer fatigue strength than those with protruding heads. Rivet hole preparation also has a significant influence on fatigue resistance. In order of decreasing fatigue strength, fabrication techniques are coin-dimpling, spin-dimpling, drilling, and machine countersinking.

Cold Working and Interference Fits. Many large structures are assembled with mechanical fasteners. In the aerospace industry, fatigue life improvement of aluminum structures is often accomplished with cold working of fastener holes and/or installing interference-fit fasteners. Because fatigue cracks often initiate at fastener holes in metallic structure, methods such as cold working fastener holes and interference-fit fasteners have been developed to improve fatigue life. Both cold working and interference-fit fasteners set up residual compressive stress fields in the metal immediately adjacent to the hole (Fig. 9). The applied tension stress dur-

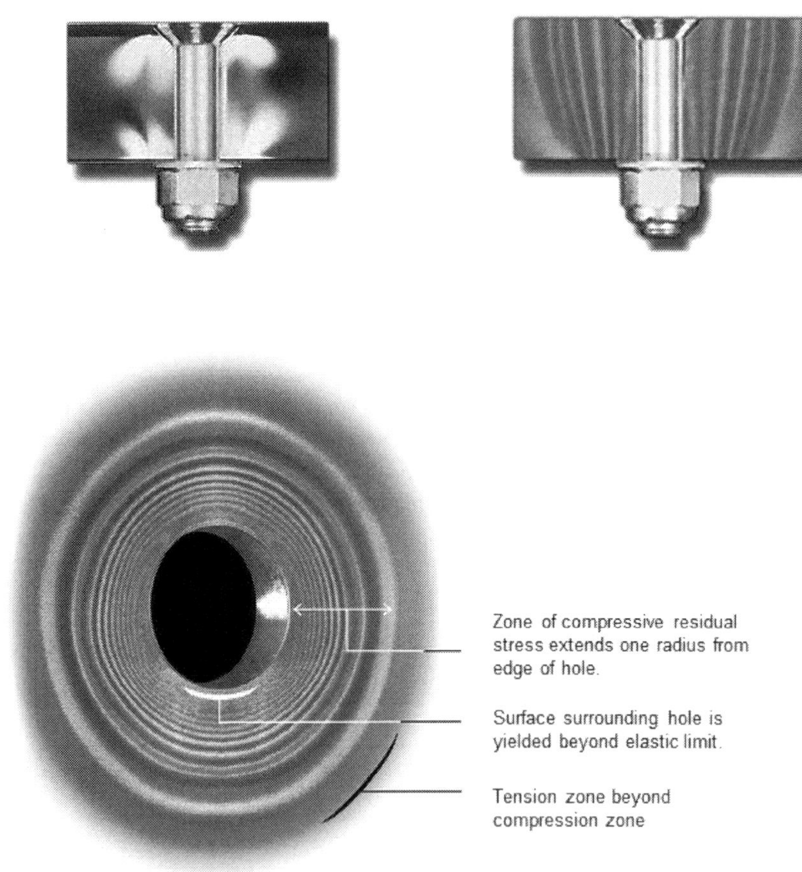

Fig. 9 Residual stress state around cold-worked hole. Courtesy of Fatigue Technology, Inc. Source: Ref 12

ing fatigue loading must then overcome the residual compressive stress field before the hole becomes loaded in tension. The fatigue improvement due to cold working in 2024-T81 aluminum is shown in Fig. 10.

Cold working of holes is usually conducted using either the split-sleeve or split-mandrel method. Both methods involve pulling a mandrel through the hole that expands the hole diameter, creating plastic deformation of material around the hole and a resulting residual compressive stress field. The residual stress field, depending on the material and the amount of expansion, will extend approximately one radius from the edge of the hole. In the split-sleeve process (Fig. 11), a stainless steel split sleeve is placed over a tapered mandrel and inserted into the hole. The hole is cold worked when the largest part of the mandrel is drawn back through the sleeve. After cold working, the sleeve is removed and discarded. In the split-mandrel process, a collapsible mandrel is placed in the hole, and as the mandrel is withdrawn, it expands to cold work the hole.

Interference-fit fasteners can also be used in metallic structures to improve fatigue life. When an interference-fit fastener is installed in metal, it also plastically deforms a small zone around the hole, setting up a compressive stress field, which again is beneficial when fatigue loading is primarily in tension. The amount of interference can vary, depending on structural requirements, but it is usually in the range of 0.08 to 0.1 mm (0.003 to 0.004 in.). In some highly loaded holes, both cold working and interference-fit fasteners are used. While both cold working and the use of interference-fit fasteners are proven methods of improving fatigue resis-

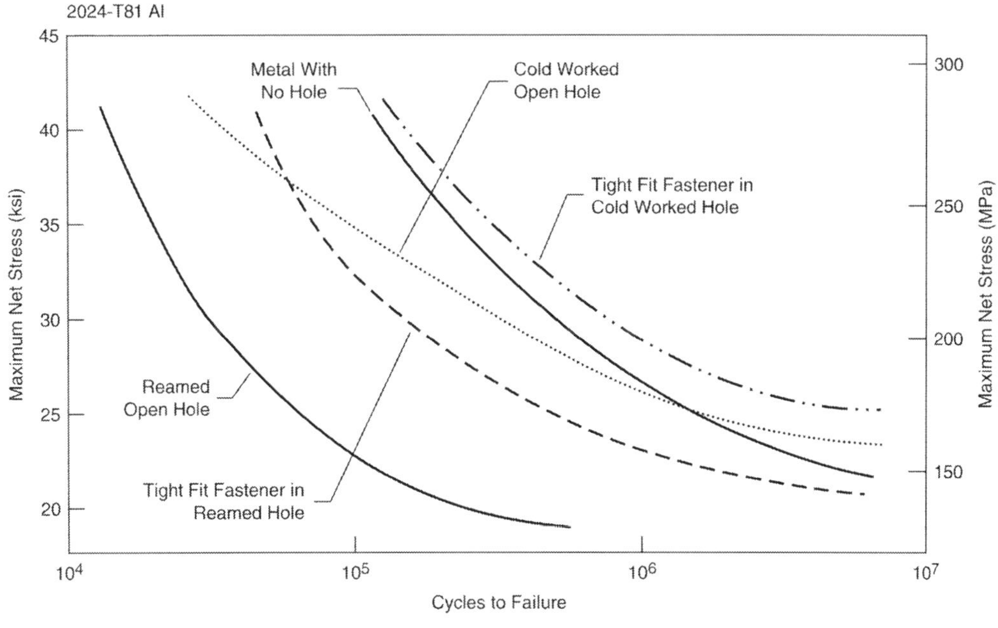

Fig. 10 Fatigue life improvement with cold working. Source: Ref 13

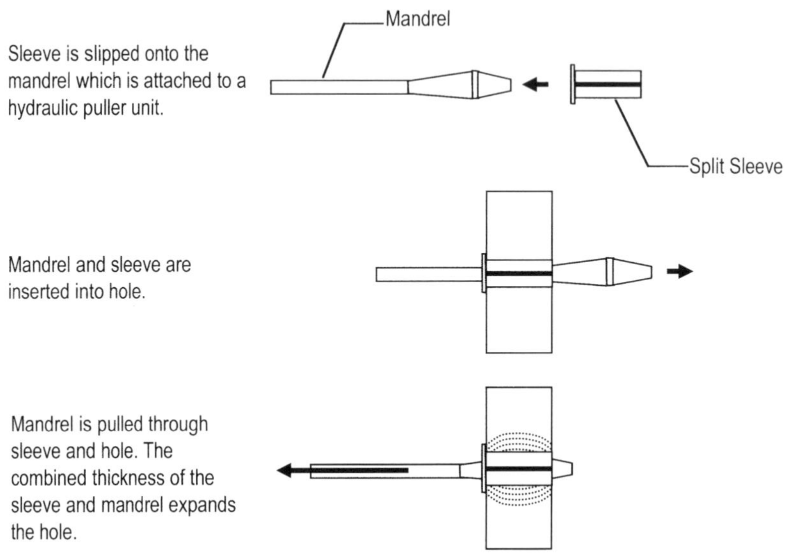

Fig. 11 Split-sleeve cold working process. Source: Ref 14

tance, both increase assembly costs and should only be specified when they are really needed.

Friction Joints. The most effective means of increasing the fatigue strength of joints is by tensioning the fasteners to clamp the plies together. The load is then transferred from one substrate to another by friction of the faying surfaces rather than by shear in the bolts bearing on the plies. The maximum, or slip, load that can be transferred through friction is:

$$P_S = K_S \cdot m \cdot n \cdot P_B \quad \text{(Eq 4)}$$

where K_S is the coefficient of slip (friction), m is the number of slip planes (number of substrates − 1), n is the number of bolts, and P_B is the bolt preload or clamping force.

Coefficients of slip for various structural steels are listed as follows (dry mill scale):

Steel	K_S
A36 (carbon steel)	0.35
A440 (high-strength low-alloy)	0.18
A514 (quenched and tempered)	0.20

An increase in fatigue strength with clamped joints is only achieved if relative slipping is eliminated, otherwise fretting failures are likely. Slipping is best prevented by careful positioning of the bolts over the entire contact area, using smallest possible gage and pitch (allowing for wrench clearance), tightening to a high bolt clamping force, and reducing the contact, or faying, surfaces to a minimum. Unnecessary overhang of the substrates beyond the first bolt row must be eliminated to prevent fretting.

Clamping causes the substrates to behave integrally as if welded together and to more uniformly distribute the total load over the fasteners. The stress concentrations at the holes are reduced by the compressive clamping forces, and fatigue failure is generally initiated away from the holes.

Again, it is essential in assembling bolted friction joints that a consistent, high bolt preload be maintained. This is accomplished through the calibration of torque or pneumatic impact cut-off wrenches. The bolt-tension calibrator is a hydraulic instrument that measures the bolt tension developed by tightening the bolt or nut. The test is made on at least three bolts of each lot (size-grip combination), and the torque or cut-off to develop the required tension is averaged.

As previously mentioned, an alternate criterion for tightening bolts is the turn-of-nut method (Fig. 12) developed in the railroad and automotive industries. Bolts are installed to a specified rotation from the "snug-tight" condition. Snug-tight condition is achieved when the bolt tension or preload equals 70% of the specified minimum breaking strength of the bolt (Table 4). Nut rotations when both faces are normal to the bolt axis are:

Bolt length, 1	Nut rotation from snug-tight
1 < 4 diam	⅓ turn
4*d* < 1 < 8*d*	½ turn
8*d* < 1 < 12*d*	⅔ turn

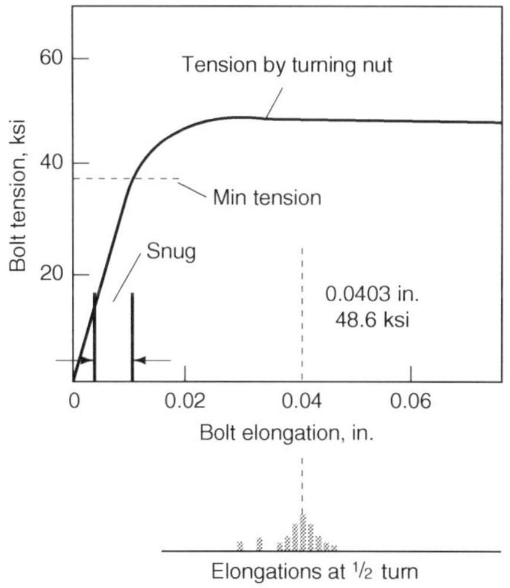

Fig. 12 Measurement of bolt elongation in a typical test joint. Source: Ref 2

Table 4 Minimum bolt breaking strengths

	Breaking strength, ksi	
Bolt diameter, in.	A325	A490
½	17.0	21.3
¾	40.1	50.1
1	72.7	90.9
1¼	102	145
1½	147	211

Source: Ref 15

The button-head, interference-body bolt, called, variously, interrupted-rib, structural-rib, or Dardelet-rivet bolt, combines the clamping action of a high-strength fastener with that of an interference fastener. The body fit of this fastener is an advantage in the assembly of galvanized materials with low interface friction. The body-bound fit metal-to-metal contact also ensures better electrical grounding of structures.

Hot-Driven Riveted Joints of Structural Plate. Important factors affecting the fatigue strength of riveted (hot-driven) and friction-bolted joints are clamping force, grip length, type of fastener, fastener pattern, tension-shear ratio, tension-bearing ratio, and type of steel in fastener or main material. Failure through the net section occurs in both riveted and bolted joints, but an increase in clamping force of the bolts can cause failure in the gross section. In a riveted joint, fatigue cracks are initiated at the rivet holes, while, in friction-bolted joints, they are initiated at the edges of the clamped substrates. It has been shown that the fatigue resistance of high-strength friction-bolted joints is superior to that of riveted joints and that the fatigue strength of carbon steel joints fastened with A325 bolts approaches the yield strength of the connected material (Fig. 1).

Some research indicates that no steel shows a superiority over carbon steel as plate material in riveted joints subjected to cyclic loading. It has been suggested that the effect of variation in clamping force within a joint or from one joint to another subordinates the relative superiority of one steel over another and produces the great experimental scatter. In addition, the fatigue resistance of riveted joints is very sensitive to the tension/shear/bearing ratio (Fig. 3), which masks variations in static strength. As previously noted, high-strength bolted joints transfer the loads through friction and therefore avoid the sensitivity.

Field Repair and Rehabilitation of Riveted Structures. The generally accepted method of repair or rehabilitation of elderly riveted structures to extend their service life has been the replacement of entire members that either contain fatigue damage or are suspected to be overstressed in cyclic loading. One program studied the effectiveness of rehabilitation by replacement of rivets with high-strength structural bolts in critical or fatigue-damaged locations as a more economical alternative to replacement of entire members. Constant-amplitude fatigue tests and variable-amplitude service simulation tests of full-scale model ore bridge joints and constant-amplitude tests of joints removed from an ore bridge showed that rehabilitation of fatigue-damaged members by fastener replacement increases fatigue life 2 to 6 times. The effect on fatigue life extension of increasing bolt tension above the minimum value accepted in present specifications and/or the effect of crack length (up to 25 mm, or 1 in., long) at rehabilitation were found to be of secondary importance.

Drilling holes at crack tips is not always effective in arresting fatigue cracks in structural members. Tests on welded crane runway girders have shown that, should the drilled hole miss the crack tip, crack growth could be reinitiated. It is often very difficult to ensure that the hole includes the crack tip. Therefore, it was concluded that this technique is neither sufficiently reliable nor applicable to be recommended as a universal repair for cracks. Instead, a drilled hole with a properly torqued high-strength bolt (70% or more of bolt breaking strength) could be considered as an expedient crack arrestor until a more reliable repair can be implemented. The combination of drilled hole and high-strength bolt has also been used as a temporary crack arrestor on railway cars.

Pin Joints. The fatigue strength of a single pin joint is extremely low in comparison to that of the base material. The fatigue strength is influenced by the stress-concentration factor (K_t), which, in turn, is a function of many parameters—geometric and material. The relationship between mean stress and stress range are further complicated by fretting.

Stress Concentrations in Pin Joints. The stress-concentration factor in the lug of a pin joint is defined as:

$$K_t = \frac{\text{Maxium stress at hole}}{\text{Net section stress}} \quad \text{(Eq 5)}$$

where the net section (on the transverse diameter of the hole) is $(D - d)/t$. D, d, and t are, respectively, lug width, hole diameter, and lug thickness.

As d/D approaches unity, the lug tends to act as a flexible strap, and K_t goes to unity. K_t is also a function of the edge distance (H), and for $H/D = 1$:

$$K_t = 0.6 + 0.95\,(D/d) \quad \text{(Eq 6)}$$

and for $H/D = 0.5$:

$$K_t = 0.85 + 0.95\,(D/d) \quad \text{(Eq 7)}$$

for values of D/d from 0.2 to 0.8.

In addition to the hole size relative to lug size affecting K_t, the stress-concentration factor is also influenced by:

- Lug head shape
- Lug "waisting"
- Clearance and interference of pin

- Pin material and lubrication
- Pin bending

A square-shaped lug head has a lower K_t than a rounded lug head. However, head shape has little effect at large values of D/d. Reducing the lug width on its loaded side (i.e., waisting) increases K_t. As the clearance between the pin and hole increases, K_t also increases. Interference fit reduces K_t, but this reduction varies inversely with load. It has been suggested that $K_t = 3$ is near optimum for fatigue strength design. The proportions for this K_t would be $D/d = 2$, $H/d = 1$, and $t/d = 1$.

Fatigue in Welded Joints

As shown in Fig. 13, attaching a weld to a load-carrying member cannot only reduce the fatigue strength substantially but also lower the fatigue limit. In this example, the fatigue limit of the welded component is one-tenth that of the plain component. As a consequence of this phenomenon, it is frequently found that in cyclically loaded welded components, the design stresses are limited by the fatigue strength of the welded joints.

There are several reasons why a weld reduces the fatigue strength of a component. These reasons fall into the following categories:

- Stress concentration due to weld shape and joint geometry
- Stress concentration due to weld imperfections
- Welding residual stresses

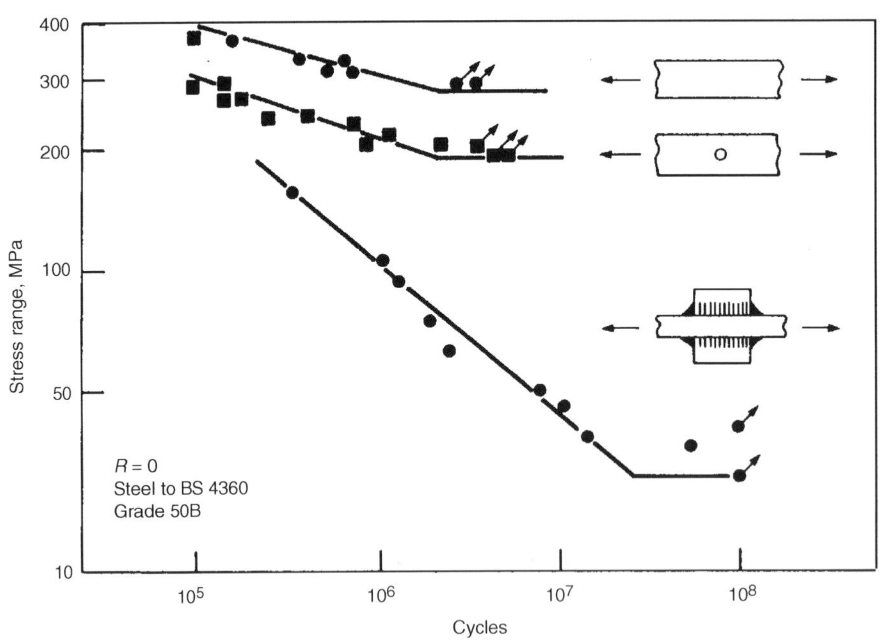

Fig. 13 Comparison of fatigue behavior of a welded joint and parent metal. Source: Ref 16

To be able to design against fatigue in welded structures, it is important to understand the influence of these factors on the performance of welded joints.

Stress Concentrations due to Weld Shape and Joint Geometry

Making a welded joint either increases or decreases the local cross-sectional dimensions where the parent metal meets the weld. Generally, any change in cross-sectional dimensions in a loaded member will lead to a concentration of stress. Thus, it is inevitable that the introduction of a weld will produce an increase in the local stress. The precise location and the magnitude of stress concentration in welded joints depends on the design of the joint and on the direction of the load.

Some examples of stress concentrations in butt welds and fillet welds are given in Fig. 14. The weld toe is often the primary location for fatigue cracking in joints that have good root penetration. In situations where the root penetration is poor or the root gap is excessive, or in load-carrying fillet welds where the weld throat is insufficient, the root area can become the region of highest stress concentration. Fatigue cracks in these situations start from the root of the weld and generally propagate through the weld (Fig. 15).

The geometry/shape parameters that influence fatigue of welded joints by affecting the local stress concentration include plate thickness (T), attachment toe-to-toe length (L), attachment thickness (t), weld toe radius (r), weld angle (θ), and the profile of the weld surface (convex versus concave) (Fig. 16). It is generally found that the fatigue strength of a welded joint decreases with increasing plate thickness (Fig.17a), increas-

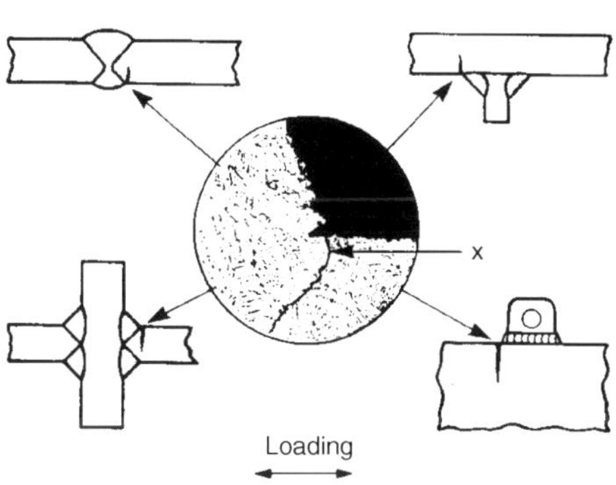

Fig. 14 Examples of stress concentrations in welded joints. Source: Ref 16

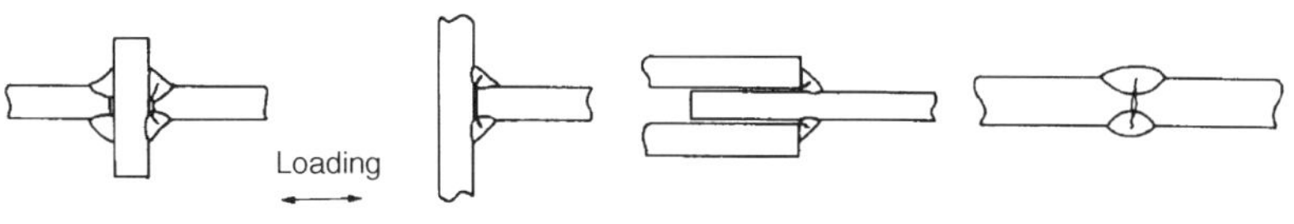

Fig. 15 Fatigue cracking from the weld root. Source: Ref 16

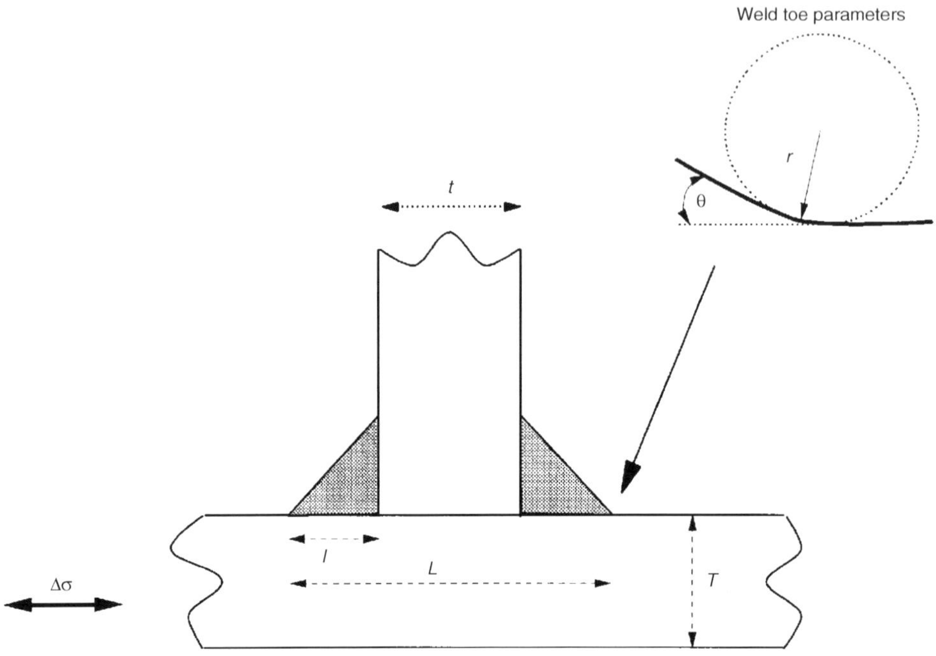

Fig. 16 Geometry parameters that affect weld toe stress concentration. *r*, weld toe root radius; θ, weld angle; *L*, toe-to-toe weld length. Source: Ref 16

ing attachment length (Fig. 17b), increasing weld angle (Fig. 17e), decreasing toe radius (Fig. 17f), and misalignment (Fig. 17c). The results given in Fig. 17 are experimental but have been confirmed by finite-element stress analysis of the local weld area. Decreasing the ratio of attachment length to thickness (L/T) from 2.0 to 0.375 will increase the relative fatigue strength by a factor of approximately 1.2. Similar results are obtained for changing the weld angle from 90° to 22.5°. By far the largest influencing parameter is the weld toe radius, which, for example, can increase the fatigue strength over the range $r = 1.0$ to 8.0 mm (0.04 to 0.3 in.) by a factor of 1.3, effectively doubling the life. The weld toe stress-concentration factor (K_t) is a function of all these geometry variables, and a number of formulas are available for butt welds and cruciform joints subjected to bending or tensile load. As an example, the following formula is for a T-joint subjected to bending load:

$$K_t = 1 + \left[\frac{1-\exp\left(-0.98\sqrt{\frac{T}{2l}+1}\right)}{1-\exp\left(-0.45\pi\sqrt{\frac{T}{2l}+1}\right)}\right]\cdot\left[\frac{0.13+0.65\left(1-\frac{r}{T}\right)^4}{\left(\frac{r}{T}\right)^{1/3}}\right]\tanh\left[\frac{\left(\frac{2l}{T}\right)^{1/4}}{1-\frac{r}{T}}\right] \quad \text{(Eq 8)}$$

with the validity boundaries $r/T = 0.02$ to 0.2, $l/T = 0.5$ to 1.2, and $\theta = 30$ to 80°.

Misalignment is another geometry-related parameter that can influence the fatigue performance of a welded joint. Misalignment can manifest itself in several ways, with angular and axial misalignments being the most common. Generally, fatigue strength decreases with increasing misalign-

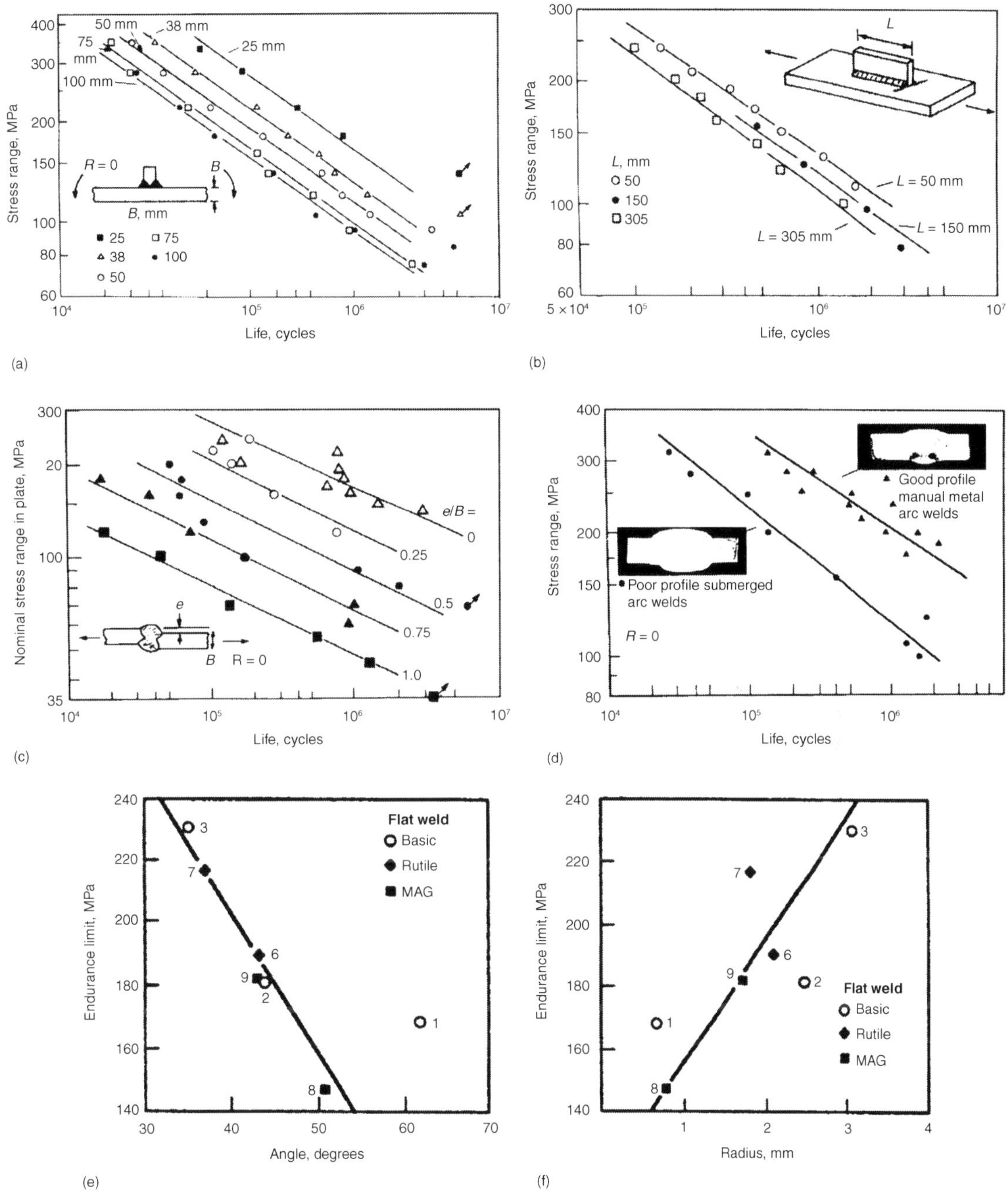

Fig. 17 Geometry factors affecting fatigue in welded joints. (a) Effect of plate thickness. (b) Effect of attachment length. (c) Effect of misalignment. (d) Effect of weld profile. (e) Effect of weld toe angle. Source: Ref 17. (f) Effect of weld toe radius. Source: Ref 16, 17

ment, as shown in Fig. 17(c). Formulas to calculate the local weld stress-concentration factor due to misalignment are now well established.

Stress Concentrations due to Weld Discontinuities

Stress analysis of an idealized model of a fillet weld loaded in the transverse direction shows that the stress-concentration factor (K_t) at the weld toe is approximately 3. This is comparable to K_t for a hole in a plate. In view of this, it would be reasonable to expect that the fatigue behavior of a fillet weld is similar to that of a plate with a hole. However, as shown in Fig. 13, the fatigue performance of the welded joint is substantially lower, implying that other factors come into play for welded joints.

As stated previously, weld imperfections are, to some extent, controllable and can be avoided during fabrication, or their effects can be included in the design. The difference observed in Fig. 13 has been attributed to the presence of microscopic features at the weld toe. These features are small, sharp, nonmetallic intrusions and are present in most, if not all, welds. The extent and distribution of these features varies with the welding process and also possibly with the quality of the steel plate and its surface condition. The exact source of these intrusions is not precisely known, but it is believed that slag, surface scale, and nonmetallic stringers from a dirty steel are the primary causes. An example of a weld toe intrusion is given in Fig. 18. The combined effect of these sharp, cracklike features and concentration of stress due to the weld geometry is that fatigue cracks initiate very early on, and most of the life is spent in crack propagation.

Planar weld imperfections (e.g., hydrogen cracks, lack of side-wall fusion) are clearly to be avoided because they will substantially reduce the fatigue life. Volumetric imperfections such as slag inclusions and porosity can be tolerated to some extent, because the notch effect of these imperfections is generally lower than that of the weld toe (Fig. 19).

Welding Residual Stresses

In welded structures, it is normally found that residual stresses are present in the weldment area, and these can be high enough to approach the yield strength of the material. These stresses occur as a result of the thermal expansion and contraction during welding, due to the constraint provided by the fabrication or by the fixtures, and due to the distortion in the structure during fabrication (often referred to as reaction stresses). These stresses are localized to the weld zone and are self-balancing (i.e., both tensile and compressive stresses are present). Transverse to the weld toe, the residual stress is typically tensile and can approach yield point. When a load cycle is applied to the structure, it is superimposed onto the residual-stress field, and the effective stresses acting at the weld joint can fluctuate down from yield level (Fig. 20). The range of each cycle remains

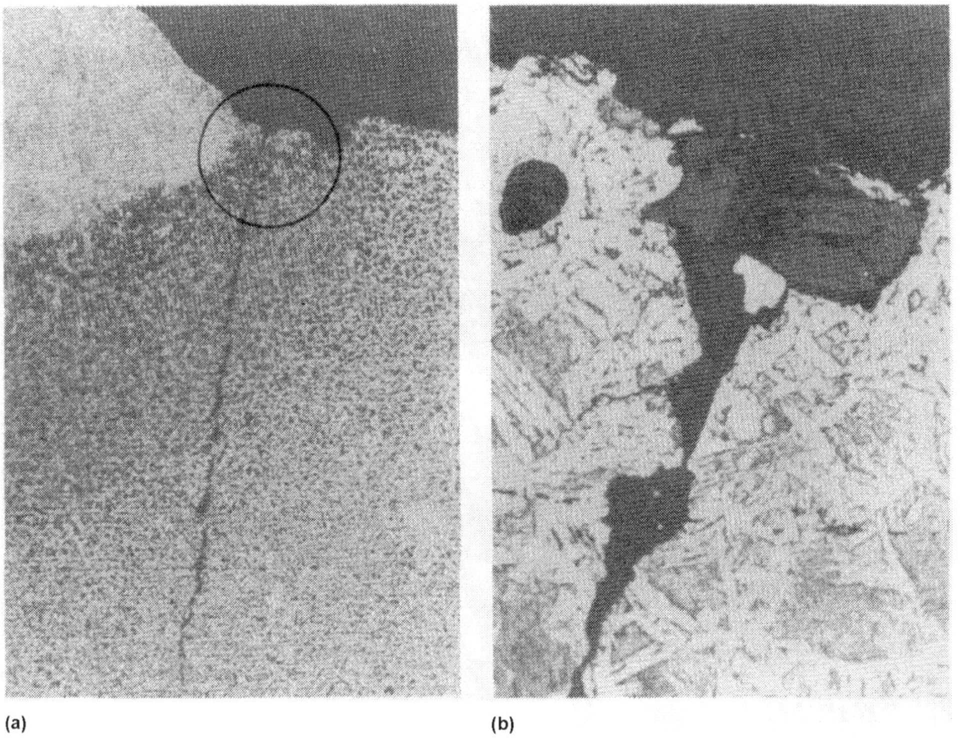

Fig. 18 Example of weld toe intrusions. Source: Ref 16

unchanged, but the effective mean stress can be significantly different from the applied mean stress. Because of the dominance of crack propagation in welded joints and the presence of high residual tensile stresses, the mean stress effect is negligible, and fatigue life is controlled by the stress range.

Reducing residual stresses using postweld heat treatment can improve fatigue life, but only if the applied load cycles are partially or fully compressive. For fully applied tensile loads, postweld heat treatment does not improve the fatigue life (Fig. 21b). Thus, it is important to know the exact nature of applied loads before a decision is made on the need to heat treat a welded structure. Indeed, it should be noted that stress relief of welded joints is never fully effective. Residual stresses up to yield point have been measured in stress-relieved pressure vessels and up to 75% of yield in stress-relieved nodes in offshore structures. Due consideration must be given to the complexity of the overall structure when stress relieving and to the time and temperature of the process.

Effect of Material Properties

Because crack propagation dominates the fatigue life of welded joints, material properties have no effect on fatigue strength. This is illustrated in

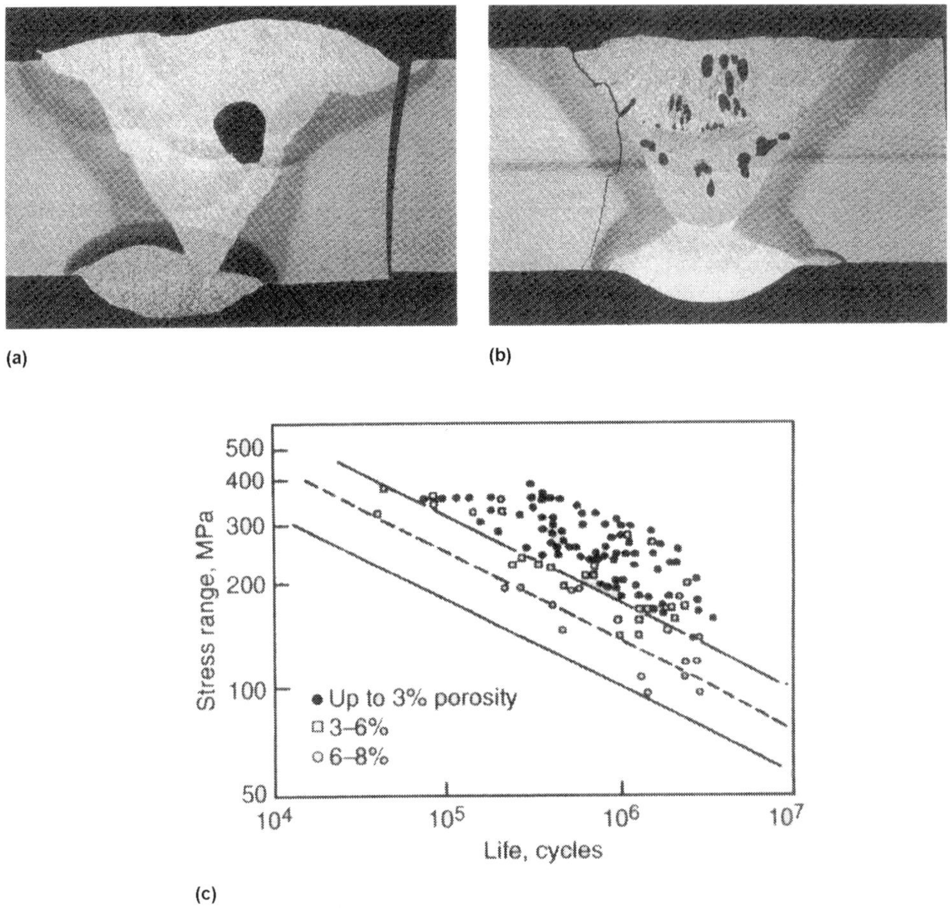

Fig. 19 Effect of volumetric defects on fatigue. (a) Slag inclusion in butt weld. Cracking from weld toe. (b) Porosity in butt weld. Cracking from weld toe. (c) Transverse groove welds containing porosity. Source: Ref 16

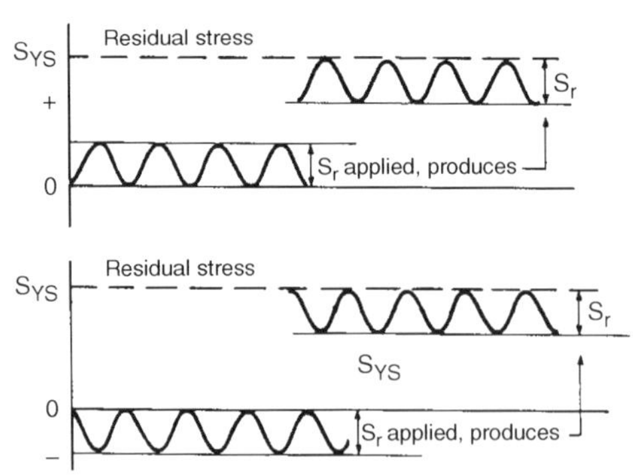

Fig. 20 Superposition effect of applied and local welding residual stresses. Source: Ref 16

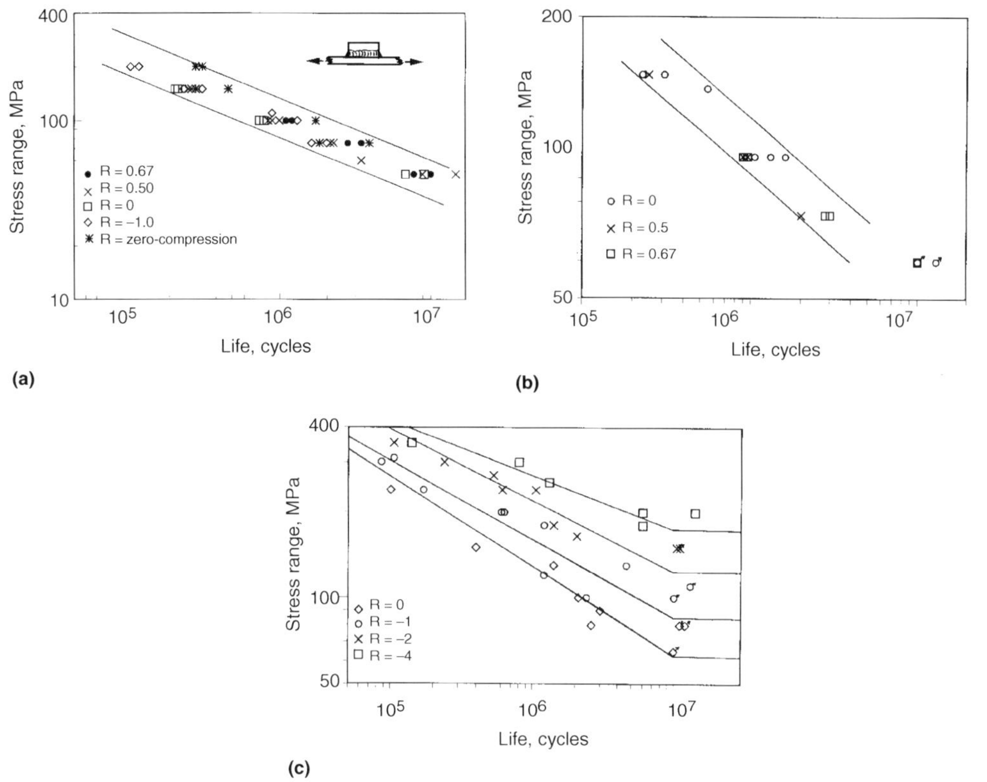

Fig. 21 Influence of welding residual stress on fatigue. (a) Effect of applied stress ratio on as-welded joints. (b) Effect of stress relief as a function of tensile load cycle. All specimens stress relieved. (c) Effect of stress relief as a function of tension-compression load cycle. All specimens stress relieved. Source: Ref 16

Fig. 22, where it can be seen that data points for steels with different strengths fall within the same scatter band. Thus, using a high-strength steel to improve fatigue life will not be beneficial for welded structures.

Microstructure. A fatigue crack starting at the weld toe will immediately grow into the heat-affected zone (HAZ) and then into the base metal. During this period, it will propagate through a variety of microstructures, and, as shown in Fig. 23, its growth rate will not be influenced in any way. Thus, the variety of HAZ microstructures (and hardness levels) in the weldment area have little or no effect on the rate at which the crack grows.

Fracture Toughness. As with tensile strength, the fracture toughness of the weld metal, the HAZ, or the base metal does not influence the crack growth rate. This can be deduced from Fig. 23, which represents a variety of materials with different strength levels and toughness values. The only influence of fracture toughness is limiting the size the fatigue crack could reach before the material fails in an unstable manner, because tougher materials are able to tolerate bigger fatigue cracks.

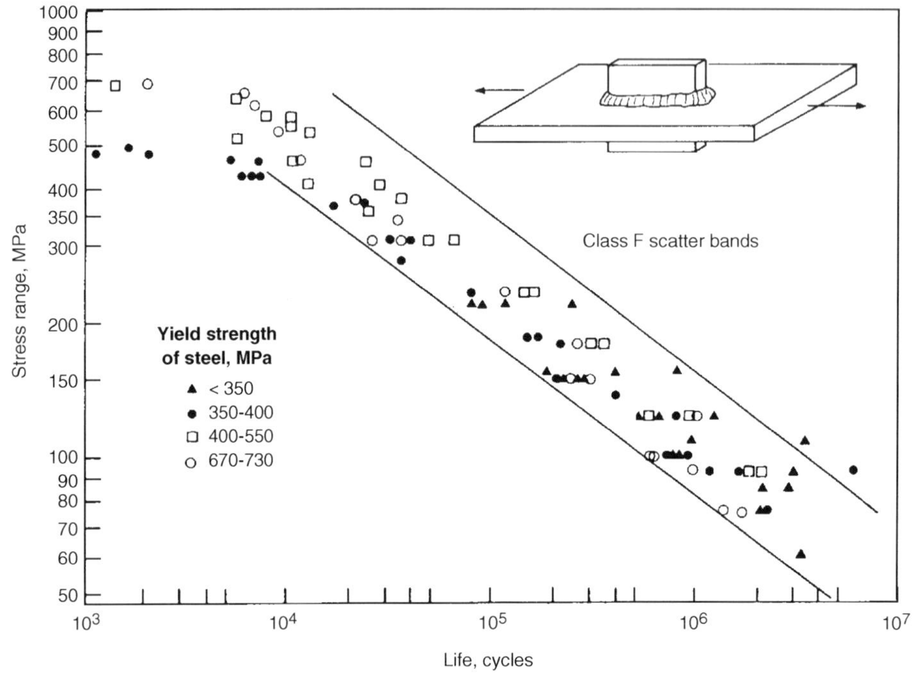

Fig. 22 Fatigue test results from fillet welds in various strengths of steel. Source: Ref 16

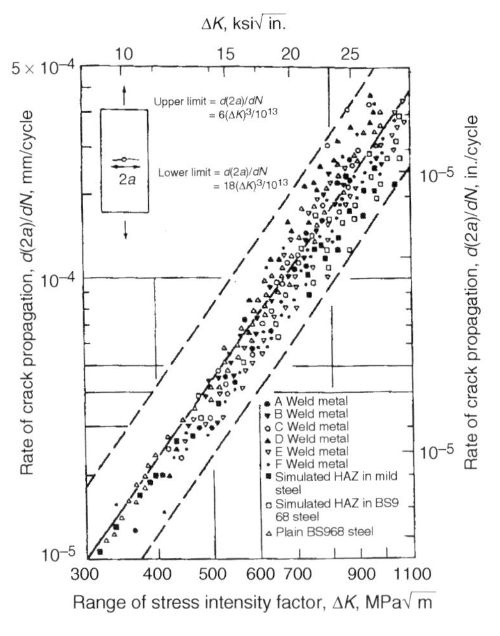

Fig. 23 Crack growth rate data showing no influence of weld metal, heat-affected zone (HAZ), or base metal. Source: Ref 16

Methods for Improving the Fatigue Life of Welded Joints

Postweld fatigue life improvement techniques can be used for increasing the fatigue life of welded joints. From the preceding sections, it can be seen that three factors influence fatigue of welded joints: stress concentrations due to joint and weld geometry, stress concentrations due to localized defects, and welding residual stresses. Improvement in fatigue life can be obtained by reducing the effects of one or more of these parameters. This is particularly true for cracks starting from the weld toe, which is by far the most common failure site. However, in load-carrying fillet welds, cracks can start at the weld root and propagate through the weld throat until failure occurs. In this case, additional weld metal to increase the weld throat dimension will result in reduction in shear stress and hence a corresponding increase in fatigue strength.

Because the most common failure site is the weld toe, many postweld treatments for this region have been developed to improve the fatigue life. These techniques rely largely on removing the detrimental intrusions at the weld toe, reducing the joint stress concentration, or modifying the residual-stress distribution. There are primarily two broad categories of postweld techniques: modification of the weld geometry and modification of the residual-stress distribution. A complete list is given in Fig. 24.

Methods that reduce the severity of the stress concentration or removal of weld toe intrusions include grinding, machining, or remelting. These techniques essentially focus on altering the local weld geometry by removing the intrusions and, at the same time, achieving a smooth transition between the weld and the plate. Typical profiles of burr-ground and gas tungsten arc welding (GTAW)-dressed welds are illustrated in Fig. 25. Because weld toe intrusions can be up to 0.4 mm (0.016 in.) in depth, the general guideline is that the grinding or machining operation must penetrate at least 0.5 mm (0.02 in.) into the parent plate. In these procedures, the depth, the diameter of the groove, and the direction of the grinding marks become important issues. In GTAW and plasma dressing, the weld toe is remelted in order to improve the local profile and also to "burn" away or move the intrusions. With these techniques, the position of the arc with respect to the weld toe and the depth of the remelt zone are two critical variables. Controlling the shape of the overall weld profile has also been recognized as a method of improving the fatigue life. The American Welding Society weld profile procedure using the "dime" test reduces the weld geometry K_t and hence increases the fatigue life.

Methods that modify the residual-stress field include heat treatment, hammer and shot peening, and overloading. Postweld heat treatment is known to reduce tensile residual stresses but does not eliminate them completely. The benefits of postweld heat treatment can be realized only if the applied loads are either partially or fully compressive. Overloading techniques rely on reducing the tensile residual-stress field and/or introducing compressive stresses at the weld toe. Exact loading conditions for a complex structure are difficult to establish, so this technique is rarely used.

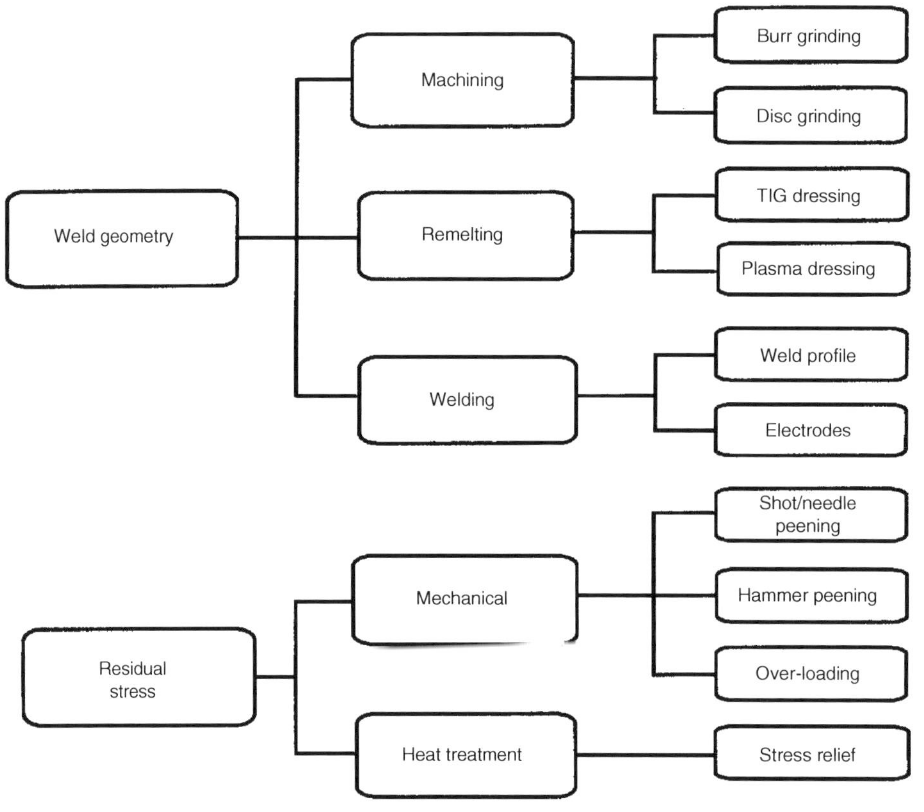

Fig. 24 Range of postweld fatigue life improvement techniques. TIG, tungsten inert gas. Source: Ref 16

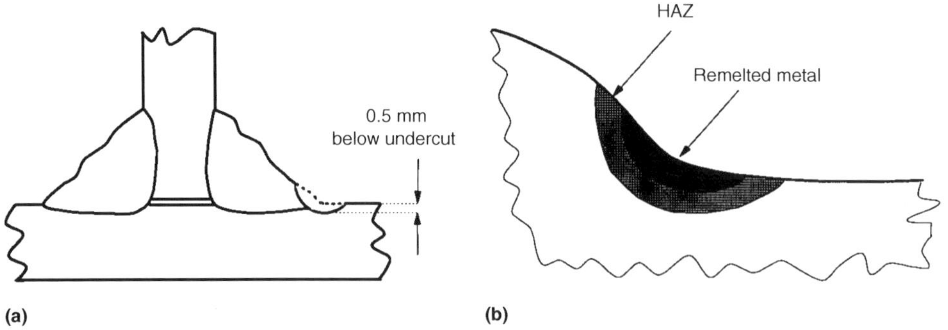

Fig. 25 Typical profiles for (a) burr grinding and (b) gas tungsten arc weld dressing of weld toe. HAZ, heat-affected zone. Source: Ref 16

To obtain significant improvements in fatigue strength, it is necessary to introduce compressive stresses in the local area in a consistent and repeatable manner. The three peening techniques (shot, needle, and hammer peening) aim to achieve this by cold working the surface of the weld toe. As with the grinding methods, it is necessary to penetrate the parent plate and deform to a depth of at least 0.5 mm (0.02 in.). Hammer peening is perhaps the best technique to do this, and it has the advantage of removing the weld toe intrusions by cold working them out. Peening techniques require special equipment and can present safety challenges for noise control. They are very difficult to automate and control in production.

A comparison of the improvement in fatigue strength obtained by some of these techniques is shown in Fig. 26. It can be seen that hammer peening is perhaps the best technique, perhaps because it reduces or eliminates intrusions as well as introduces compressive stresses. Burr grinding, which is easier to implement, also produces significant improvement in fatigue life. Disc grinding and burr grinding are probably the most frequently used methods for improving the fatigue life of welded joints.

Fracture Control in Welded Structures

The brittle fracture of the welded ships during World War II stimulated intensive research into avoiding catastrophic failures in fabricated structures. These early failures led to many studies to identify the parameters that must be controlled to avoid brittle fractures. Toughness of the mate-

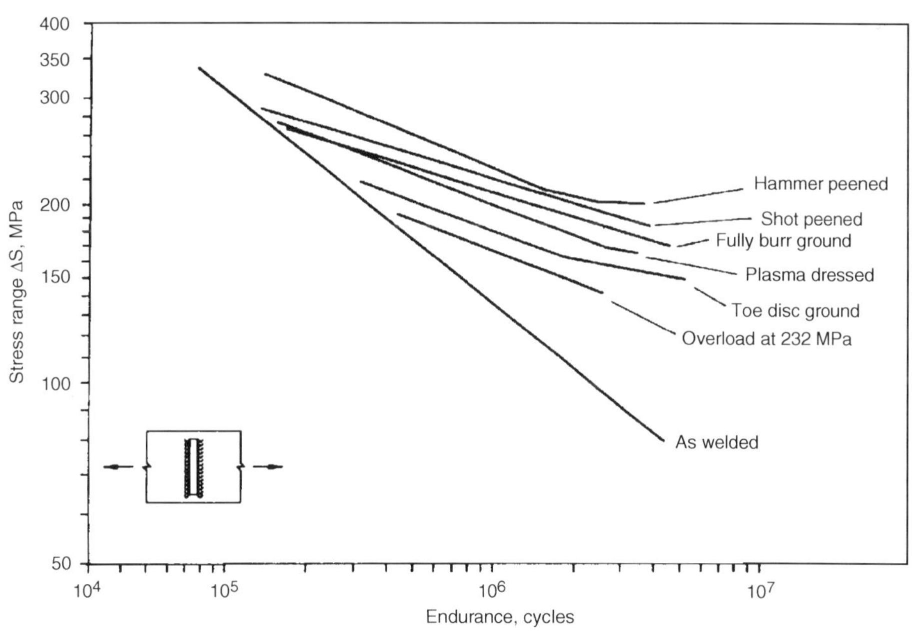

Fig. 26 Comparison of different improvement techniques. Source: Ref 16

rial was one parameter that emerged from these studies. Much of the early work focused on measuring the toughness of steels used in ship structures using notched-bar impact specimens. The most widely used impact specimen is the Charpy V-notch impact test, and today the test has become one of the necessary means of ensuring that the weld metal, the base metal, and the HAZ have sufficient toughness to avoid unstable fractures.

The Charpy test evaluates both the initiation of brittle fracture as well as its subsequent propagation. The significance attached to controlling fracture initiation or propagation has influenced much of the fracture research work, and it has resulted in two philosophies for fracture control.

U.S. researchers concluded that because of the highly heterogeneous nature of welds, it would be impractical and risky to try to control brittle fracture initiation in welded structures. It was argued that "pop-in" of fractures from brittle regions was likely and that the steel toughness should be sufficient to prevent a pop-in from propagating into a full-scale fracture. Pellini et al. used this concept to propose the first practical design tool linking the required fracture toughness to temperature and design stress. The approach was presented in the form of a fracture analysis diagram (Fig. 27) that has a constant shape, where the position of the curve along the temperature axis is indexed to the nil-ductility temperature (NDT), which is obtained from the drop weight test. This approach has been widely used for fracture control of many types of structures, and, as far as is known, there have been no brittle fractures in steels that meet the NDT criteria.

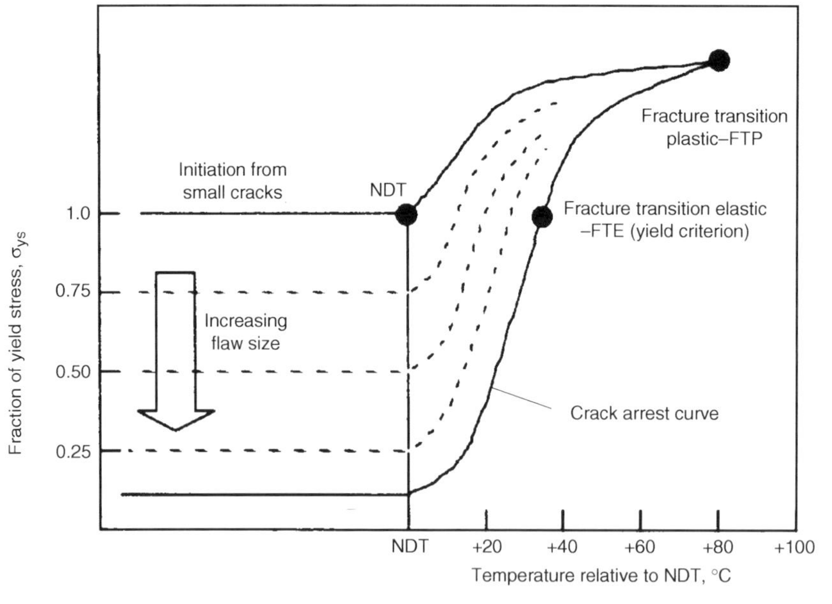

Fig. 27 Pellini's fracture analysis diagram. NDT, nil-ductility temperature. Source: Ref 16

The competing approach is to try to control initiation of the fracture. This has been advocated mostly by the Europeans and now more recently in North America. The initiation approach is intimately linked to fracture mechanics concepts, and in many codes and standards, it has been used to specify toughness requirements for weld and base metals and has been valuable in establishing flaw acceptance criteria for inspection. For welded joints, it was recognized that initiation of fracture took place under the influence of high residual tensile stresses and stress-concentration effects, which led to the use and advancement of postyield fracture criteria, primarily through the development of fracture mechanics approaches for crack tip opening displacement (CTOD). The onset of brittle fracture requires the presence of a flaw of critical dimensions, low toughness, and stress (primary and secondary) of a critical magnitude.

Factors Affecting Fracture Toughness

The toughness of a material can be measured using a standard Charpy V-notch specimen or by using a fracture mechanics test specimen. The commonly referenced results from Charpy impact tests include the impact energy required to fracture the specimen, the lateral expansion, and the percent shear fracture appearance at the specified temperature. The latter two are intended as measures of ductility. In contrast to the Charpy test, fracture mechanics toughness tests fall into two groups: linear elastic fracture mechanics tests (K_{Ic}) and elastic-plastic fracture mechanics tests (CTOD, J-curves, and R-curves). The major advantage of using fracture mechanics specimens is that the data from these tests can be used directly in fracture mechanics fitness-for-service procedures to assess the tolerance of the structure to the presence of defects and cracks. The parameters that affect the toughness of a material include:

- Material grade and chemistry
- Strength and hardness
- Microstructure and level of impurities
- Grain orientation
- Size of structure or specimen (constraint)
- Notch/crack acuity
- Loading rate
- Temperature

Some of these are discussed subsequently in the context of welded regions.

In conventional structural steels that have a ferritic microstructure, both Charpy and fracture mechanics toughness tests show a transition from ductile to brittle behavior with decreasing temperature. Charpy energy transition curves for a variety of steels and for a number of other materials are shown in Fig. 28. At higher temperatures, in the upper-shelf regime, a ferritic material behaves in a fully ductile manner and is capable of absorbing considerable amounts of energy when deformed. At these temperatures, ferritic materials are generally considered tough. As the temperature is reduced, a transition to lower-shelf behavior occurs. In this regime, ferritic materials are brittle, and fracture occurs in a cleavage mode. The decrease in toughness (energy absorbed) from upper-shelf behavior can be substantial, and for some materials the transition temperature range is very small.

From Fig. 28, it is clearly evident that toughness is a function of the material type and the yield strength. Furthermore, the transition behavior is affected by the size and type of the specimen, the loading rate, and the notch acuity. Generally, the toughness is reduced with increasing loading rate, increasing specimen size, increasing notch acuity, increasing thickness, and increasing crack depth (Fig. 29). Thus, neither the fracture strength nor the toughness is a sole predictor or criterion for fracture control. This can present challenges to a design engineer who would like to be able to predict conditions for the onset of brittle fracture from some simple specifiable material properties. Also, this indicates that it is important to match the test specimen size and test conditions closely with the local region in the actual structure.

For welded joints, the challenge of generating toughness information is increased further by the nonhomogeneous nature of the region. Variations in notch location and orientation can result in significantly different tough-

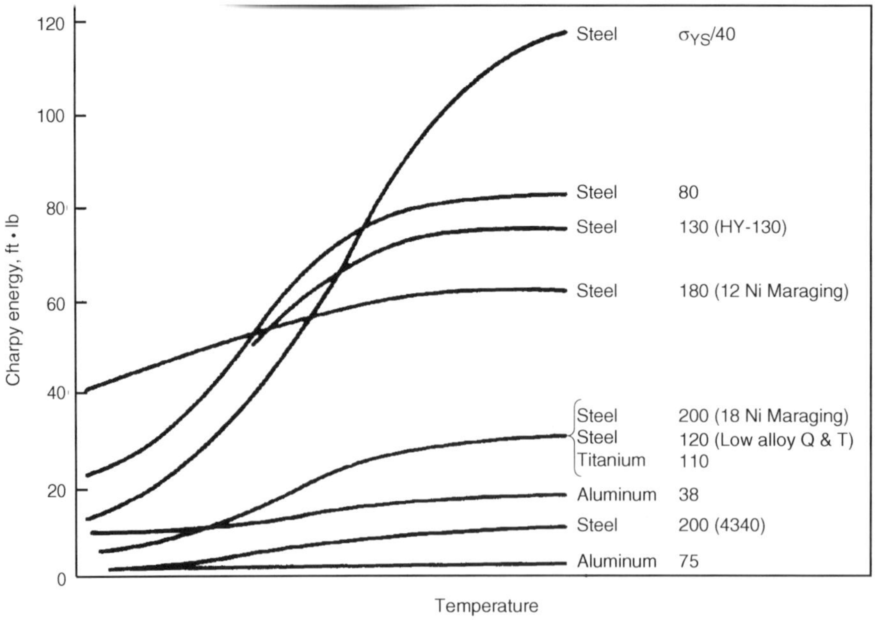

Fig. 28 Ductile-to-brittle transition curves for a variety of materials. Source: Ref 16

ness values. The weld HAZ presents particular difficulties in toughness measurements due to its narrow width. For example, in structural steels welded using typical arc energies, the transformed HAZ is usually no more than 3 to 4 mm (0.12 to 0.16 in.) wide. As shown in Fig. 30, within this distance there are wide ranges of microstructures and properties, and the critical region of interest may be 1 mm (0.04 in.) or less. Despite their small size, brittle regions within the HAZ can have a significant influence on the integrity of a structure with respect to failure by brittle fracture. A number of catastrophic brittle fractures of engineering structures, including pressure vessels, storage tanks, and bridges, in which the fracture started in the HAZ testify to this.

Some examples of notch locations and orientations in the weld metal and the HAZ for measurement of fracture toughness are illustrated in Fig. 31. While placement of the notch in the weld metal is simple and straightforward, HAZ notch location can be difficult. Some industries have developed specific procedures that simplify Charpy testing of HAZ and ask for through-the-thickness notches located at the fusion boundary (FB), FB + 2 mm (0.08 in.), and FB + 5 mm (0.2 in.). Although a certain amount of information on the variation in HAZ toughness can be obtained from these tests, their primary aim is to ensure that a certain level of quality in workmanship is maintained. Results obtained from Charpy testing of the HAZ must be treated with caution, because it has been shown that the blunt

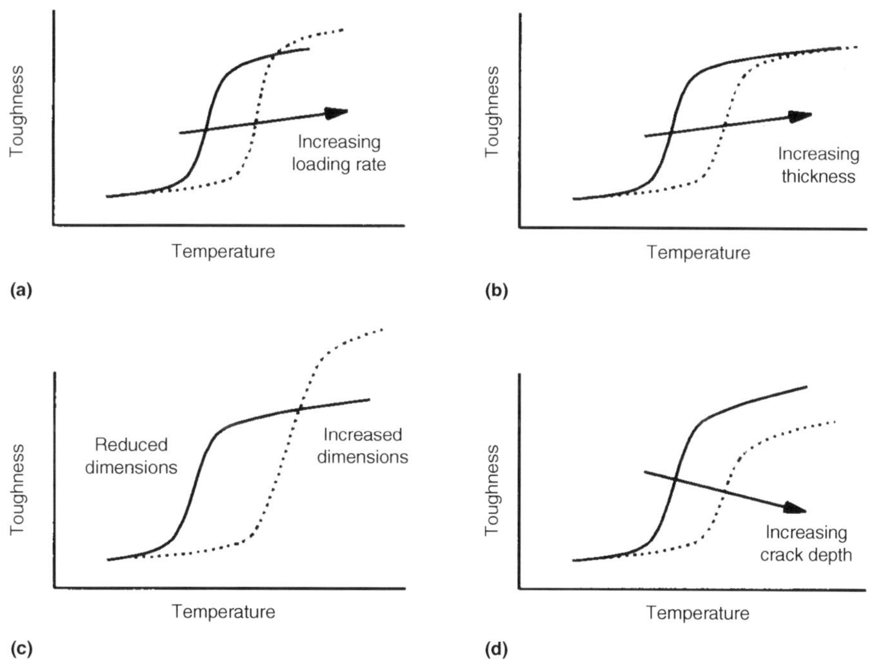

Fig. 29 Effects of (a) loading rate, (b) thickness, (c) specimen dimensions, and (d) crack depth on the ductile-to-brittle transition behavior. Source: Ref 16

notch of this type of test specimen can miss sampling local regions of low toughness. This caution also applies to Charpy testing of multipass welds, because it is well known that in the transition regime, depending on the orientation of the Charpy notch, bimodal fracture behavior can be obtained at the same test temperature. This is because the notch of one test specimen may sample a microstructure exhibiting upper-shelf behavior, and a specimen from the adjacent location may sample a low-toughness microstructure. Thus, weldment toughness data can exhibit significant scatter, particularly in the transition regime, and care must be taken in interpreting the results.

Although both Charpy and fracture mechanics toughness test standards are now well established for parent material, the complex nature of the weld joint has hindered the development of standardized procedures for measuring the toughness. An essential requirement for tests on welded joints is that the test welds should be fully representative of the service structure of interest. This requirement is based on the knowledge that the fracture toughness of weld metals and weld HAZs may be critically dependent on such factors as:

- Welding process and consumables
- Base-metal composition
- Joint thickness

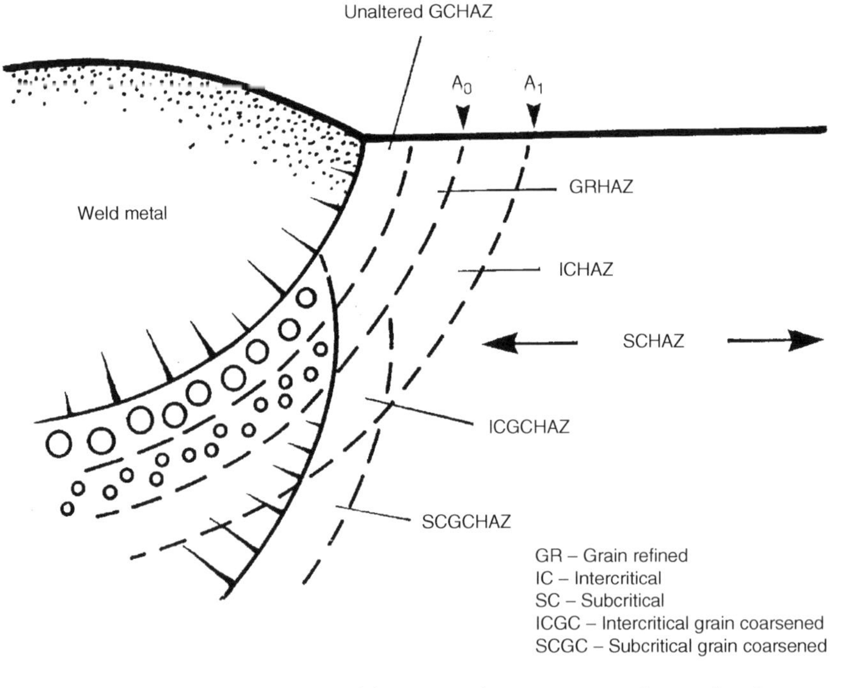

Fig. 30 Schematic diagram of the variety of microstructures that can be obtained in the heat-affected zones (HAZs) of multipass welds. GC, grain coarsened; GR, grain refined; IC, intercritical; SC, subcritical; ICGC, intercritical grain coarsened; SCGC, subcritical grain coarsened. Source: Ref 16

- Preheat and interpass temperature
- Heat input
- Welding position
- Joint configuration
- Restraint
- Postweld heat treatment
- Time between welding and testing

Embrittlement Mechanisms. Localized embrittlement in welded regions can occur by a number of mechanisms, and these can increase the risk of brittle fracture initiation. The three most common mechanisms are

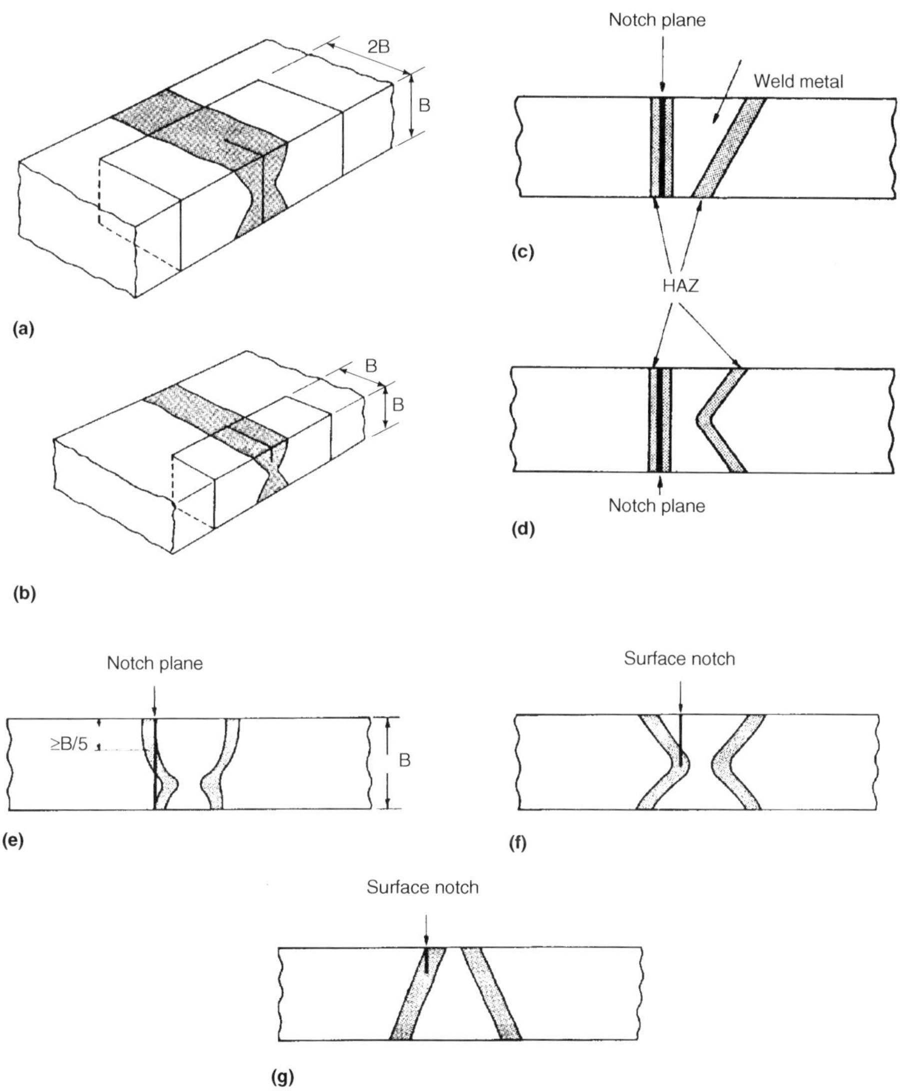

Fig. 31 Examples of notch locations for toughness testing of weld metals (a and b) and heat-affected zone (HAZ) (c through g). Source: Ref 16

hydrogen embrittlement, strain aging, and temper embrittlement. Each of these mechanisms can reduce the toughness of the material, either soon after welding or during service. There are many catastrophic failures associated with these embrittlement mechanisms, and an effective fracture control plan must include their potential effect.

Hydrogen Embrittlement. Hydrogen in steels and welded joints can be introduced in several ways. In welding, hydrogen from the arc atmosphere can dissolve into the liquid weld pool, and if the weld cools rapidly, a significant quantity may be retained in the weldment at low temperatures. Alternatively, in the petrochemical industry, hydrogen can be absorbed by the containment vessel during service when the medium in the vessel contains diffusable hydrogen. Sufficient hydrogen levels will cause cracking in the weld metal and the HAZ, but lower levels of hydrogen can cause other problems, including the formation of "fish eyes" and embrittlement of the material. Hydrogen embrittlement in weld metal often manifests as a loss of ductility and fracture toughness. The susceptibility of a material to hydrogen embrittlement is dependent on the chemistry and microstructure, and therefore, the different regions of the weldment will embrittle differently due to the presence of hydrogen.

Strain Aging. The thermal and strain cycles that accompany welding can give rise to "dynamic" strain aging embrittlement. This is caused by the diffusion of carbides and free nitrogen into dislocations in the steel matrix. These dislocations, generated through the straining mechanism, become pinned (their movement is inhibited), thus increasing the yield strength (effectively the stress required to move the dislocations). Strain aging can also result in an increase in tensile strength, usually accompanied by reduction in toughness. Strain aging in welds is often thought to occur in the root regions, which are highly restrained and receive complex thermal cycling.

Temper embrittlement is the loss in toughness of alloy steels on exposure to temperatures in the range 325 to 575 °C (620 to 1065 °F). It occurs in coarse austenite grain microstructures (e.g., HAZs and weld metal) in which no ferrite is formed on the prior-austenite grain boundaries. It is caused by migration of elements such as manganese, silicon, antimony, arsenic, tin, and phosphorus to high-angle austenite grain boundaries, which become embrittled. The risk of temper embrittlement can be reduced by promoting fine-grained austenite microstructure in both the weld and HAZ regions.

Toughness Requirements for Avoidance of Brittle Fracture. Analysis of early Charpy V-notch data of ship steels indicated that initiation of brittle fracture occurred only in those steels showing less than 13.5 J (10 ft · lbf) Charpy energy, and cracks propagated only in those steels showing less than 27 J (20 ft · lbf). This gave rise to the 20 J (15 ft · lbf) criterion, which, although not universally applicable, has nevertheless been widely used. Today, most codes specify fracture toughness requirements that consider the influences of parameters such as tensile properties, heat treatment condition, constraint, and material grade. All these factors influence the material ductility and the ductile-to-brittle transition temperature. Explicit specifications of toughness requirements for weld metal and HAZ are rare, although some industry codes and standards assume that the toughness specification for base metal also applies to weld metal and HAZ. Material toughness requirements are generally part of comprehensive fracture control plans that include quality control, materials selection and fabrication, and inspection. Account is taken of the in-service environment, degradation of properties during service (e.g., aging), applied loading condition, temperature cycles, and the consequences of catastrophic failure.

ACKNOWLEDGMENTS

Portions of the material for this chapter came from "Fatigue of Mechanically Fastened Joints" by H. Reemsnyder and "Fatigue and Fracture Control of Weldments" by T. Jutla, both in *Fatigue and Fracture*, Vol 19, *ASM Handbook*, ASM International, 1996.

REFERENCES

1. *Structural Steel Design,* Ronald Press, 1974, p 519–551
2. H. Reemsnyder, Fatigue of Mechanically Fastened Joints, *Fatigue and Fracture,* Vol 19, *ASM Handbook,* ASM International, 1996
3. C.W. Lewitt, E. Chesson, Jr., and W.H. Munse, "The Effect of Rivet Bearing on the Fatigue Strength of Riveted Joints," A Progress Report by University of Illinois, Eng. Exp. Station for Illinois Div. of Highways and Dept. of Commerce—Bureau of Public Roads, Urbana, Illinois, Jan 1959
4. R.C.A. Thurston, The Fatigue Strength of Threaded Connections, *Trans. ASME,* Vol 73, Nov 1951, p 1085–1092
5. P.G. Forrest, *Fatigue of Metals,* Pergamon Press, New York, 1962, p 262–280
6. P. Haagensen et al., Size Effects in Machine Components and Welded Joints, *OMAE '88,* 1988, p 386
7. W.J. Jensen, Failures in Mechanical Fasteners, *Failure Analysis and Prevention,* Vol 11, *Metals Handbook,* 9th ed., ASM International, 1986
8. Nomogram for Torque on Bolts, *Des. News,* Feb 1973, p 76–77
9. R.S. Shoberg, Mechanical Testing of Threaded Fasteners and Bolted Joints, *Mechanical Testing and Evaluation,* Vol 8, *ASM Handbook,* ASM International, 2000
10. F.C. Campbell, Machining and Assembly, *Structural Composite Materials,* ASM International, 2010
11. R.B. Heywood, *Designing Against Fatigue in Metals,* Reinhold, 1962, p 168–285

12. F.C. Campbell, Fatigue, *Elements of Metallurgy and Engineering Alloys,* ASM International, 2008
13. F.C. Campbell, *Manufacturing Technology for Aerospace Structural Materials,* Elsevier Scientific, 2006
14. F.C. Campbell, *Elements of Metallurgy and Engineering Alloys*, ASM International, 2008, p. 262
15. "Specification for Structural Joints Using ASTM A-325 or A-490 Bolts," Research Council on Riveted and Bolted Structural Joints, May 8, 1974 (New York)
16. T. Jutla, Fatigue and Fracture Control of Weldments, *Fatigue and Fracture,* Vol 19, *ASM Handbook,* ASM International, 1996
17. I. Huther, L. Promot, H.P. Lieurade, J.J. Janosch, D. Colchen, and S. Debiez, "Weld Quality and the Cyclic Fatigue Strength of Steel Welded Joints," IIW-DOC-XIII-1563-94

SELECTED REFERENCES

- R.T. Barrett, *Fastener Design Manual,* NASA Reference Publication 1228, National Aeronautics and Space Administration, 1990
- F.C. Campbell, *Joining—Understanding the Basics,* ASM International, 2011
- M.J. Cieslak, Cracking Phenomena Associated with Welding, *Welding, Brazing, and Soldering,* Vol 6, *ASM Handbook,* ASM International, 1993
- Classification of Weld Discontinuities, *Welding, Brazing, and Soldering,* Vol 6, *ASM Handbook,* ASM International, 1993
- R.O. Parmley, *Standard Handbook of Fastening and Joining,* 3rd ed., McGraw-Hill, 1996
- M.J. Robin, Mechanical Fasteners, *Mechanical Engineers' Handbook,* 2nd ed., John Wiley and Sons, Inc., 1998

CHAPTER 8

Fracture Control and Damage Tolerance Analysis

FRACTURE CONTROL is the concerted effort to ensure safe operations without catastrophic failure by fracture. Very seldom does a fracture occur due to an unforeseen overload on an undamaged structure. Fractures are usually the end result of crack growth from a small defect or flaw. Due to repeated or sustained "normal" service loads, a crack may develop from a flaw or stress concentration and slowly grow in size. Cracks and defects impair the strength of the component. Thus, during the continuing development of the crack, the structural strength decreases until it becomes so low that the service loads cannot be carried any more, and fracture occurs.

If fracture is to be prevented, the strength should not drop below a certain safe value. This means that cracks must be prevented from growing to a size at which the strength would drop below an acceptable limit. To determine which size of crack is admissible, one must be able to calculate how the structural strength is affected by cracks (as a function of their size); to determine the safe operational life, one must be able to calculate the time in which a crack grows to the permissible size. Damage tolerance analysis is used to obtain this information.

Damage tolerance is the property of a structure to sustain defects or cracks safely, until such time that action is (or can be) taken to eliminate the cracks by repair or by replacing the cracked structure or component. In the design stage, one still has the option to select a more crack-resistant material or improve the structural design to ensure that cracks will not become dangerous during the projected economic service life. Alternatively, periodic inspections may be scheduled, so that cracks can be repaired or components replaced when cracks are detected. Either the time to retirement (replacement) or the inspection interval and type of inspection must be determined from the crack growth time calculated in the damage tolerance analysis.

Inspections can be performed by means of any nondestructive inspection technique, but destructive techniques such as proof testing are essentially inspections. If a burst occurs during hydrostatic testing of a pipeline, for example, then there was apparently a crack of sufficient size to cause the burst. The proof test detects defects above a certain size under controlled circumstances (e.g., with water pressure) to prevent catastrophic failure during operation when the line is filled with oil or gas.

Fracture control is a systematic process to prevent fracture during operation. The extent of the fracture control measures depends on the criticality of the component, the economic consequences of the structure being out of service, and the damage that would be caused by a fracture failure. Fracture control of a hammer may be as simple as selecting a material with sufficient fracture resistance. Fracture control of an airplane includes damage tolerance analysis, tests, and subsequent inspection and repair/replacement plans. Inspections, repairs, and replacements must be scheduled rationally using the information from the damage tolerance analysis.

The mathematical tool employed in damage tolerance analysis is fracture mechanics; it provides the concepts and equations used to determine how cracks grow and affect the strength of a structure. Inaccuracies are due to inaccurate inputs much more often than to inadequacy of the concepts. Although further improvements of fracture mechanics concepts may well be desirable from a fundamental point of view, it is unlikely that damage tolerance analysis can be much improved, because its accuracy is determined mostly by assumptions and (predicted) loads and stresses. Therefore, the accuracy of all assumptions and input data must be clearly understood and identified, and the results of damage tolerance analysis must be used judiciously by engineers. In this context, damage tolerance analysis can give useful answers to questions that hitherto could not be answered at all. The answers may only be preliminary, but a reasonable answer is better than none.

Principles of Fracture Control

Establishment of a fracture control plan requires knowledge of the structural strength as it is affected by cracks, and knowledge of the time involved for cracks to grow to the permissible size. Thus, damage tolerance analysis has two objectives, namely, to determine:

- The effect of cracks on strength (margin against fracture)
- The crack growth as a function of time

The effect of crack size on strength is shown in Fig. 1, where the strength is expressed in terms of the load (P) the structure can carry before fracture occurs (fracture load). Supposing that a new structure has no significant defects ($a = 0$), then the strength of the structure is the design strength (P_u).

In every design a safety factor is used. This factor may be applied in different ways. Usually, the safety factor is applied to the load. For example, if the maximum anticipated service load is P_s, the structure is actually designed to sustain $j \cdot P_s = P_u$, where j is the safety factor. The designer sizes the structure so that the stress is equal to, or slightly less than, the tensile strength when the load is P_u. Alternatively, the safety factor can be applied to the allowable stress: if the actual material strength (tensile strength) is F_{tu}, the structure is sized in such a way that the stress at the highest service load (P_s) is less than or equal to F_{tu}/j. Because load and stress are usually proportional, the structural part is actually capable of carrying $j \cdot P_s = P_u$. The value of j is between 3 (many civil engineering structures) and 1.5 (aircraft).

It is emphasized that P_s is the highest service load. If the service load varies, the load may well be much less than P_s most of the time. For example, the loads on cranes, bridges, off-shore structures, ships, and airplanes are usually much less than P_s. Only in exceptional circumstances does the load reach P_s. At other times, the load is only a fraction of P_s, so that the margin against failure is much larger than j, except in extreme situations.

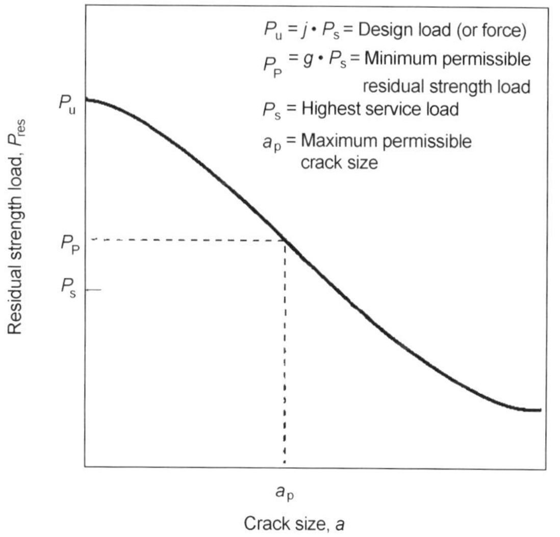

Fig. 1 Residual-strength diagram in terms of load (or force). j, design safety factor; g, safety factor based on residual strength. Source: Ref 1

The new structure has a strength P_u. If the load should reach P_u, the structure fails. The probability of this occurring is not zero, but experience has shown that it is acceptably low if the prescribed safety factor is adhered to. If cracks are present, the strength is less than P_u. This remaining strength under the presence of cracks is generally referred to as the residual strength (P_{res}). Therefore, the diagram in Fig. 1 is called the residual-strength diagram. With a residual strength $P_{res} < P_u$, the safety factor has decreased to $j = P_{res}/P_s$, which is less than $j = P_u/P_s$. In concert, the probability of fracture failure has become higher.

With a crack of size a, the residual strength is P_{res}. Should a load $P = P_{res}$ occur, then fracture takes place. If a load P_{res} does not occur and service loading continues at lower loads, the crack may continue to grow by fatigue. If continued crack growth occurs, then the residual strength decreases further. If nothing is done, fracture will eventually occur under normal service loads. This is what must be prevented.

The previous discussion implies that the limit should be set somewhat above P_s. For example, one may require that the residual strength never be less than $P_p = g \cdot P_s$, where g is the remaining safety factor, and P_p is the minimum permissible residual strength. The design engineer (or user) does not usually decide what should be the initial safety factor j. The factor is usually prescribed by rules and regulations issued by engineering societies (e.g., American Society of Mechanical Engineers, or ASME) or government authorities (e.g., Federal Aviation Administration, Department of Defense). These rules or requirements should also prescribe g. This has not been done for all types of structures yet, and the ASME code, for example, approaches the problem somewhat differently.

Provided that the shape of the residual-strength diagram is known and P_p has been prescribed, the maximum permissible crack size follows from the diagram. For damage tolerance analysis to determine the largest permissible crack, the first objective must be the calculation of the residual-strength diagram of Fig. 1. If a_p can be calculated directly from P_p, it may not be necessary to calculate the entire residual-strength diagram but only the point (a_p, P_p). However, this is seldom possible and rarely time-saving. In general, calculation of the entire diagram is preferable. The residual-strength diagram will be different for different components of a structure and for different crack locations. In addition, permissible crack sizes will be different as well.

Knowing that the crack may not exceed a_p is of little help, unless it is known when the crack may reach a_p. The second objective of the damage tolerance analysis is then the calculation of the crack growth curve, shown diagrammatically in Fig. 2. Under the action of normal service loading, the cracks grow by fatigue at an ever faster rate, leading to the convex curve shown in Fig. 2.

Starting at some crack size a_0, the crack grows to a_p. Provided that one can calculate the curve in Fig. 2, one obtains the time (H) of safe operation

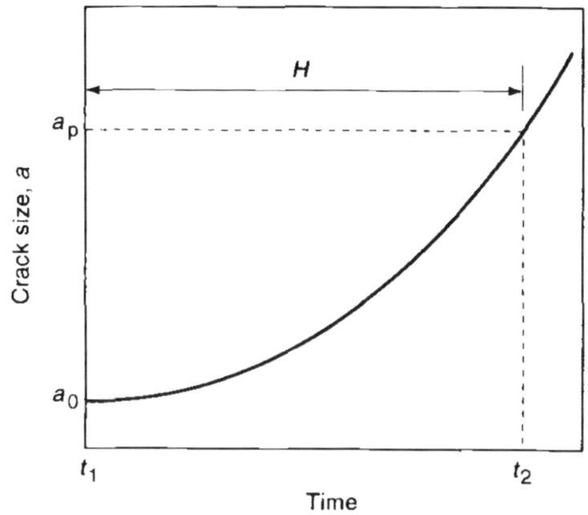

Fig. 2 General crack growth curve showing time interval (H) to reach maximum permissible crack size (a_p). Source: Ref 1

(until a_p is reached). For example, if a_0 is an (assumed or real) initial defect, then the component or structure must be repaired or replaced after a time H. Alternatively, a_0 may be the limit of crack detection by inspection. This crack a_0 will grow to a_p within a time of H. Because crack growth is not allowed beyond a_p, the crack must be detected and repaired or otherwise eliminated before the time H has expired. Therefore, the time between inspections must be less than H; it is often taken as $H/2$. In any case, the time of safe operation by whatever means of fracture control follows from H.

Concepts of Damage Tolerance Analysis

Before any fracture control can be exercised, the residual-strength diagram of Fig. 1 and the crack growth curve of Fig. 2 must be calculated. Although crack growth may occur by other mechanisms than fatigue (e.g., stress corrosion or creep), the discussion is limited to fatigue crack growth.

The first step in damage tolerance analysis is the calculation of a_p, or rather, of the residual-strength diagram. Usually, the residual-strength diagram is expressed in terms of stress rather than load, as in Fig. 3(a). Different fracture criteria may apply (K, J, or others), depending on the material used for the structure. For the purpose of this discussion, an analysis based on K is sufficient.

Using K, the residual strength (σ_c) follows from:

$$\sigma_c = \frac{K_c}{Y\sqrt{\pi a}} \qquad \text{(Eq 1)}$$

where K_c is the toughness of the material, Y is the geometry factor defined by the details of the structure, and a is the crack size. Given the toughness

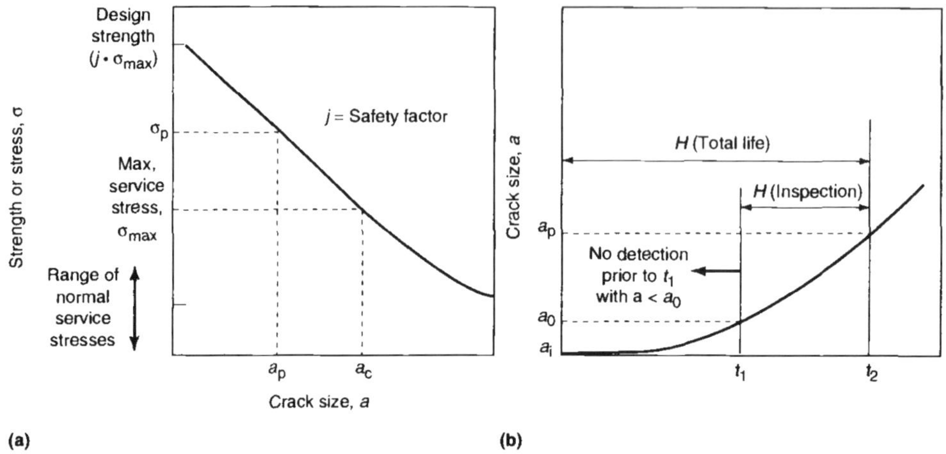

Fig. 3 Key parameters for fracture control. (a) Residual strength in terms of stress. (b) Crack growth and time period when inspection can be performed. Source: Ref 1

(K_c), a residual-strength diagram can be calculated easily by taking different values of a and calculating σ_c. This calculation is adequate if σ_c is sufficiently below the yield stresses. For high stresses, as a approaches zero, the criterion of plastic collapse must be introduced. Once the diagram is established, the permissible crack size (a_p) can be read from the resulting diagram (plot) for the required σ_p (Fig. 3a).

The second step is to calculate the crack growth curve. The rate of crack growth is a function (f) of ΔK and R (the stress ratio) such that:

$$da/dn = f(\Delta K, R) \quad \text{(Eq 2)}$$

The problem is to obtain the crack growth curve by integration of Eq 2, fatigue crack growth rates, as follows:

$$n = \int_{a_0}^{a_p} da / f(\Delta K, R) \quad \text{(Eq 3)}$$

where $\Delta K = Y \Delta\sigma\sqrt{\pi a}$. In the general case, solution of Eq 3 can be accomplished only by numerical integration performed by a computer. For the purpose of the present discussion, it is assumed that the resulting crack growth curve (Fig. 3b) can be obtained.

When the two curves are known, decisions on how to exercise fracture control can be made. The residual-strength analysis provides the permissible crack size a_p, and the crack growth analysis provides the value of H, the time available to exercise fracture control.

Fracture Control Measures

When the residual-strength diagram has been calculated (Fig. 3a), the maximum permissible crack size (a_p) follows from the minimum permissible residual strength. The other information from analysis is the crack propagation curve. It shows how a crack develops by fatigue as a function of time. The maximum permissible crack (a_p), following from the residual-strength analysis of Fig. 3(a), can be plotted on the calculated crack growth curve, as in Fig. 3(b).

There are several ways in which this information can be used to exercise fracture control. In all cases, the time period (H) is the essential information needed. The following options are available for the implementation of fracture control:

- Periodic inspection
- Fail-safe features
- Durability design or mandated retirement (safe-life approach)
- Periodic (destructive) inspection by proof testing; repair after failure in proof test (if feasible)

Damage tolerance requirements sometimes prescribe the fracture control procedure. For example, commercial aircraft requirements prescribe nondestructive periodic inspection.

Periodic Inspection. Safety is ensured when cracks are eliminated before reaching a_p. Therefore, cracks must be discovered before that point by means of periodic inspection. A sample of some of the crack-detection methods that are commonly used by industry are summarized in Table 1. Whatever the inspection method, there is a certain size of crack (a_0) that is not likely to be detected. This implies that discovery and repair must occur in between a_0 and a_p, as shown in Fig. 3(b). For any inspection done before or at time t_1, the crack will be missed; should the next inspection be at t_2, the crack would already be too long (having reached a_p). Hence, the inspection interval (I) must be shorter than H (inspection) = $t_2 - t_1$. It is often taken as $I = H/2$, but a more rational procedure to determine inspection intervals can be employed. Naturally, once a crack is discovered, it must be repaired at the operator's earliest convenience. Because a_p is a permissible and not a critical crack, and because detection commonly occurs at sizes less than a_p, immediate repair may not be necessary, but any complacency will defy all analysis and inspection efforts.

Regardless of how long or short the inspection interval, safety is maintained with some reservation. Whether inspections must be performed every day (i.e., $I = 1$ and $H = 2$ days) or every year (i.e., $I = 1$ and $H = 2$ years), there would always be two inspections between a_0 and a_p. Although a daily inspection may be cumbersome, the achieved safety is not really different in the cases of daily or yearly inspections. If a crack is missed in daily inspections, a potential fracture will occur sooner, but if a crack is missed in yearly inspections, fracture will occur nonetheless before the year is over. If short inspection intervals are undesirable, one has the option of selecting a more refined inspection procedure capable of detecting a smaller a_0. Then H, and hence the inspection intervals, will be longer.

Table 1 Inspection techniques used for cracks

Technique	Principles	Applications and notes
Direct		
Visual	Naked eye, assisted by magnifying glass, low-power microscope, lamps, mirrors	Only at places easily accessible Detection of small cracks requires much experience.
Penetrant	Colored liquid penetrant is brushed on material and allowed to penetrate into cracks. Penetrant is washed off, and a developer is applied. Remnants of penetrant in crack are extracted by developer and give colored line.	Only at places easily accessible Sensitivity on the same order as visual inspection
Magnetic particles	Part to be inspected is covered with a layer of a fluorescent liquid containing iron powder. Part is placed in a strong magnetic field and is observed under ultraviolet light. At cracks, the magnetic field lines are disturbed.	Only applicable to magnetic materials Parts must be dismounted and inspected in a special cabin. Notches and other irregularities give indications. Sensitive method
X-ray	X-rays emitted by portable x-ray tube pass through structure and are caught on film. Cracks, absorbing less x-rays than surrounding material, are delineated by black line on film.	Has great versatility and sensitivity Interpretation problems arise if cracks occur in fillets or at edge of reinforcement. Small surface flaws in thick plates are difficult to detect.
Indirect		
Ultrasonic	Probe (piezoelectric crystal) transmits high-frequency wave into material. The wave is reflected at the ends and also at a crack. The input pulse and the reflections are displayed on an oscilloscope. Distance between first pulse and reflection indicates position of crack. Interpretation: Reflections of crack disappear upon change of direction of wave.	Universal method because a variety of probes and input pulses can be selected. Information about the size and nature of the defect (which may not be a crack) is difficult to obtain.
Eddy current	Coil induces eddy current in the metal. In turn, this induces a current in the coil. Under the presence of a crack, the induction changes; the current in the coil is a measure for the surface condition.	Cheap method (no expensive equipment) and easy to apply. Coils can be made small enough to fit into holes. Sensitive method when applied by skilled person. Provides little or no information about nature and size of defect.
Acoustic emission	Measurement of the intensity of stress waves emitted inside the material as a result of plastic deformation at crack tip and as a result of crack growth	Inspection must be done while structure is under load. Continuous surveillance is possible. Expensive equipment is required. Interpretation of signals is difficult.

Source: Ref 2

It does not matter either at which time the crack initiates (Fig. 4). Inspections scheduled at an $H/2$ interval will always give two opportunities for detection, regardless of when crack growth begins, provided that inspections are scheduled at $H/2$ intervals starting from hour zero, even if initially the chance of a crack is small. Similarly, if the interval is chosen as $H/3$, there will always be three inspections between a_0 and a_p.

Fail safety, in the context of fracture control and damage tolerance, is essentially a variation of fracture control by periodic inspection; cracks or failed members must be detected and repaired. The only difference is that the structure is designed to tolerate more damage than that which is observed. Fail safety can be achieved by means of crack arresters or multiple load paths.

For example, crack arresters are designed specifically to prevent an advancing crack from reaching a critical crack size. Alternatively, a pipeline

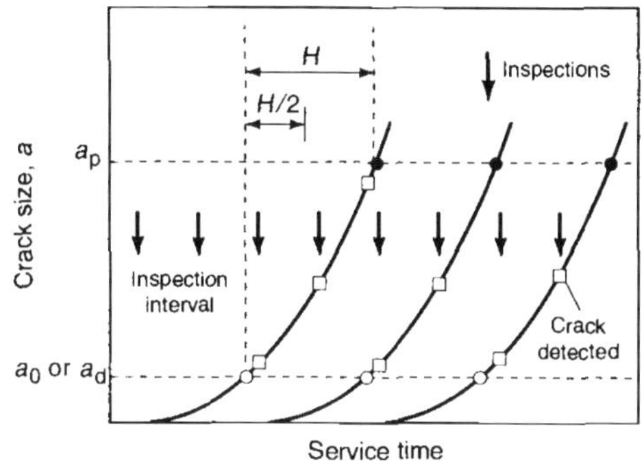

Fig. 4 Two possibilities for crack detection with *H*/2, regardless of when crack starts. Source: Ref 1

or pressure vessel can be designed so that cracks will cause leaks rather than breaks. Because a leak is presumably obvious, no special inspections are necessary other than frequent checks for leaks. The same can be accomplished by providing multiple load paths (e.g., parallel redundant members). When properly designed, the structure can still sustain σ_p when one member fails. Inspection for cracks would not be required, but regular checks for failed members would be. Of course, member failure must be obvious, because a second member will soon develop cracks when it must carry additional load.

Durability (or Safe-Life) Fracture Control. A safe-life approach to fracture control is applied to structures with hard-to-detect cracks (e.g., inaccessible locations) or when initial crack size limits are below inspectability thresholds. If no inspections can or will be done, a small initial crack (a_i) could be assumed to exist initially in the new structure. The time (H) for the crack to grow from a_i to a_p is then the available safe life. In that case, the structure or component must be retired or replaced after, for example, $H/2$ hours. This is called the durability approach.

A durability analysis (based on *S-N* data) and/or tests are used to establish a design life. To account for scatter or uncertainties in the operating conditions, retirement or replacement is then mandated at fraction (e.g., ⅕ to ⅓) of that design life. Crack growth and inspection intervals are not involved. The drawbacks of this approach are:

- The critical dependence on the durability analysis and safety factor (which may be inadequate)
- The high costs of retiring components without an obvious cause

Nonetheless, some components (such as landing gear) are designed to safe-life criteria, because cracks are noninspectable (the materials have low toughness) and proof tests are not practical.

Another problem with the durability approach is how the size of a_i is assumed. The assumption of an initial crack size may be difficult or unclear for high-quality components that may be essentially defect free. In welded structures, the assumption of an initial flaw is more realistic. Welds often contain defects such as porosity or lack of fusion. In particular, the latter is a sharp defect equivalent to a crack of approximately equal size.

Destructive Inspection by Proof Testing. If the toughness is very low, the maximum permissible flaw (a_p) may be smaller than the detectable crack (a_d) (Fig. 5). This can also be the case when the structure is so large that inspections for cracks are impractical (e.g., a 1000 mile pipeline), so that the detectable crack size is effectively infinite. In such situations, proof testing may be another fracture-control option.

As an example, consider a component that is subjected to a proof stress (σ_{proof}). Fracture will occur if a crack (a_{proof}) is present, as shown in Fig. 5(b). Conversely, if no fracture occurs, the maximum possible crack is a_{proof}, so that a safe operational period (H) for growth from a_{proof} to a_p is ensured. If the proof test is repeated every H hours, a period of at least H hours of safe operation is available after each successful proof test. Should failure occur during the proof test (i.e., a crack of size a_{proof} is detected), then a repair or replacement is made. The life can be extended forever, provided that proof tests are always conducted at the proper interval H.

Pipelines and pressure vessels are eminently suitable for the proof-test approach. A line or vessel normally filled with gas or dangerous chemicals

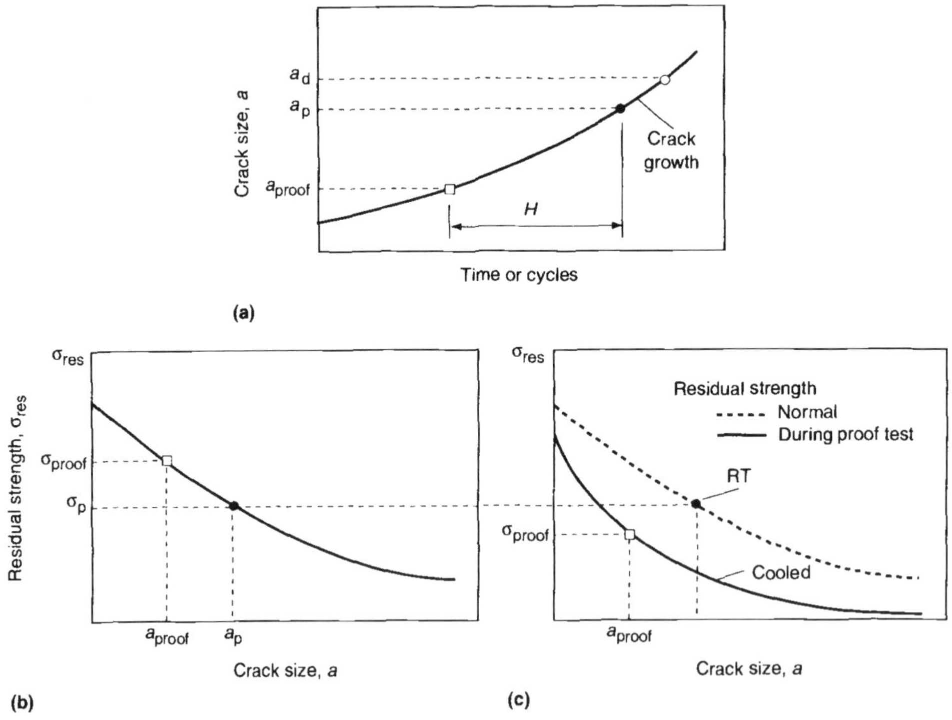

Fig. 5 Destructive inspection with proof test. (a) Crack growth. (b) Detection of cracks equal to or greater than proof size. (c) Lower proof load with cooling. RT, room temperature. Source: Ref 1

can be proof tested (hydrotested) with water. A failure during the proof test would happen under controlled circumstances, causing a water leak only. During proof testing, the large solid propellant rocket motor case shown in Fig. 6 failed. While this was an expensive failure due to the total loss of the hardware, no one was injured or killed. In many cases, hydrotests are already performed anyway. Selecting the proof-stress level and interval on the basis of fracture mechanics analysis, to calculate H, would give these a rational foundation.

Proof tests on structures other than pressurized containers are often hard to perform. However, if the component can be easily loaded or easily removed, the option is available, and it was used to test the wing hinges of the F-111 aircraft. Cooling the structure or component during the proof test causes a drop in toughness, which permits the use of lower proof stress to detect the same a_{proof}, as shown in Fig. 5(c). After the test and warmup, the original toughness and residual strength are restored.

The Probability of Missing the Crack. All information needed to determine H and the inspection interval can be calculated, except for the detectable crack size. The latter must be obtained from inspection experience, which may be difficult to quantify. The length of the inspection interval or the time period available for inspection (H) also is very sensitive

Fig. 6 Catastrophic brittle fracture of a 660 cm (260 in.) diameter solid propellant rocket motor case made of 18% Ni, grade 250, maraging steel. The case fractured at a repaired weld imperfection during a hydrostatic pressure test. Fracture occurred at approximately 57% of the intended proof stress. All welds of the case had been carefully inspected by x-ray and ultrasonic inspection. Arrows indicate origin of fracture. Source: Ref 3

to the choice of detectable crack size, because the slope of the crack growth curve is small for small cracks (Fig. 3b). Consequently, an assumption with regard to the detectable crack size may have more weight in determining the inspection interval than the painstaking and costly damage tolerance analysis. This is unsatisfactory. A more rational procedure for establishing inspection intervals is desirable.

Detection of cracks larger than the detectable size is not a certainty. It is affected by many factors: the skill of the inspector; the specificity of the assignment (e.g., one specific location, as opposed to a whole wing or bridge); the accessibility and viewing angles; exposure (e.g., part of a crack may be hidden behind other structural elements); and possible corrosion products inside the crack. In general, discrimination of flaws depends on detection response levels or crack size and the level of the application noise, as illustrated in Fig. 7. Analysis of signal and signal plus noise is common in electronic devices, optics, and other discrimination processes. However, it is important to recognize that the dominant noise source in a nondestructive evaluation (NDE) process is not electronic

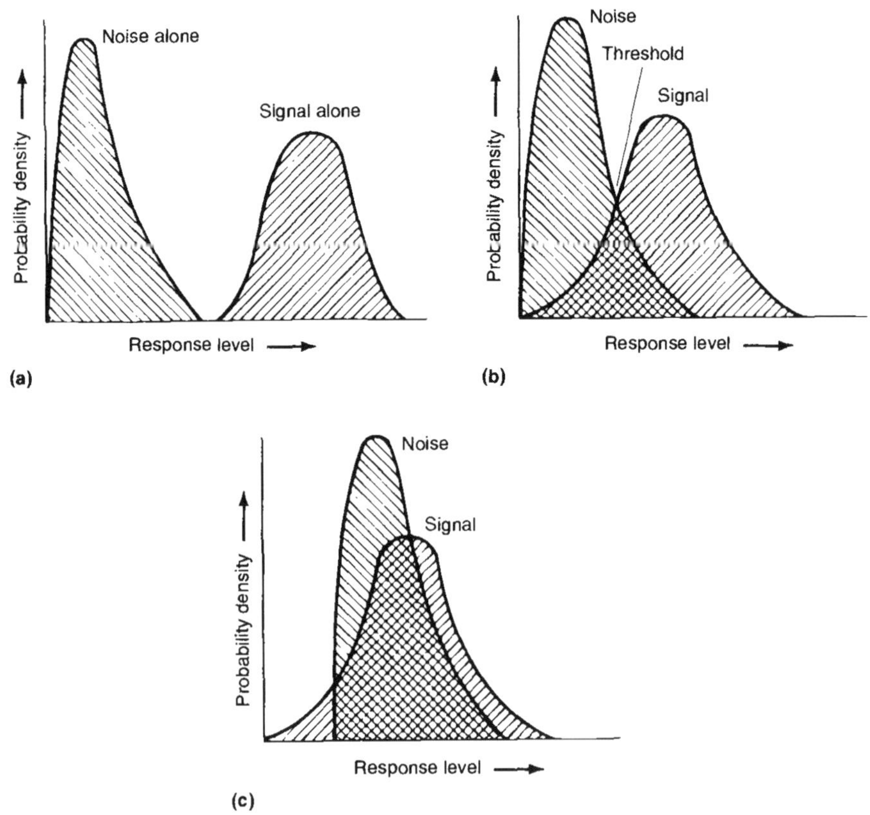

Fig. 7 Signal/noise density distribution for (a) large, (b) medium, and (c) small flaw size. Source: Ref 1

noise that may be reduced by filtering, multiple sampling, and averaging techniques but is instead the noise due to nonrelevant signals generated in applying the NDE procedure to a specific hardware element.

The outcome from an NDE measurement/decision process must be understood as a problem in conditional probability. When an NDE assessment is performed for the purpose of crack detection, the outcome is not a simple accept/reject (binary) process, as is frequently envisioned. It is actually the product of conditional acceptance due to the interdependence of the measurement and decision responses. In general, the four possible outcomes from an inspection process are:

- *True positive:* A crack exists and is detected.
- *False positive:* No crack exists, but one is identified.
- *False negative:* A crack exists but is not detected.
- *True negative:* No crack exists, and none is detected.

Typically, the probability of crack detection depends on crack size in the manner shown in Fig. 8. There is a certain crack size (a_0) below which detection is physically impossible. For example, for visual inspection, this would be determined by the resolution of the eye, or for ultrasonic inspection, by the wavelength. In reality, a_0 is larger than these physical limits. The probability of detection is never equal to 1, even for large cracks; any crack may be missed. It follows that the probability curve must have the general form shown in Fig. 8.

A crack is subject to inspection several times before it reaches the permissible size. At successive inspections, the crack will be longer and the probability of detection higher, but there is still a chance that it will go undetected. Consider 100 cracks growing at equal rate (same population), all in the same stage of growth (same size). Let the probability of detection at a certain inspection be $p = 0.2$. The probability that a crack will be missed is then $q = 1 - p = 0.8$. That means that 80 cracks will go undetected. At the next inspection, the cracks are longer; let the probability of detection then be $p = 0.6$, so that $q = 0.4$. Thus, of the remaining 80 cracks,

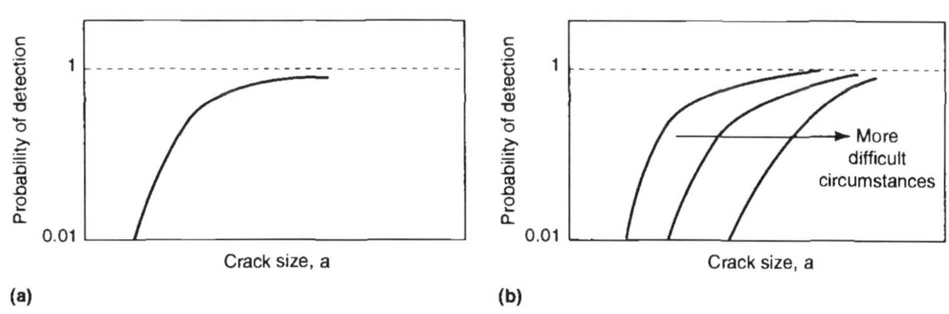

Fig. 8 Probability of crack detection in one inspection. (a) Basic curve. (b) Effect of accessibility, and specificity, or other difficulty factor. Source: Ref 1

$0.4 \times 80 = 32$ cracks will go undetected. Apparently the cumulative probability that a crack will be missed in successive inspection is $Q = q_1 * q_2 * \ldots * q_n$. In the previous example, $Q = 0.8 \times 0.4 = 0.32$. Of the 100 cracks, 32 cracks remain undetected after two inspections. The cumulative probability of detection is $P = 1 - Q$. In the previous example, $P = 0.68$. Of the 100 cracks, 68 were detected after two inspections, but 32 were missed.

Figure 9 shows what happens if inspection intervals are determined as $I = H/2$, where H is the time required for crack growth from a_d to a_p. The detectable crack size (a_d) might be selected as a crack with a certain probability of detection. For example, the detectable crack size could be defined as a_{50}, a crack with 50% probability of detection. Such a criterion certainly has appeal, because it seems consistent, yet it still leads to inconsistencies.

For the case of Fig. 9, either method I or method II could be prescribed. The detectable crack sizes a_{50} lead to different inspection intervals, $H_I/2$ and $H_{II}/2$. Inspections would take place as indicated by the arrows in Fig. 9(b). The two methods would be equivalent but certainly do not provide equal probability of detection, which is unsatisfactory. The probability-of-detection curves are different for the two procedures in Fig. 9. The cracks would be inspected for the first time when they still have a size smaller than a_{50}. At this first inspection, the probability of detection is not zero unless $a < a_0$. There is a distinct probability (p_1) that the crack is already detected during that first inspection; the probability that it is missed is $q_1 = 1 - p_1$. At the next inspection the crack is larger, and the probability of detection is q_2, and so forth. By the time the crack reaches a_p, it has been inspected n times. The cumulative probability of the crack having been

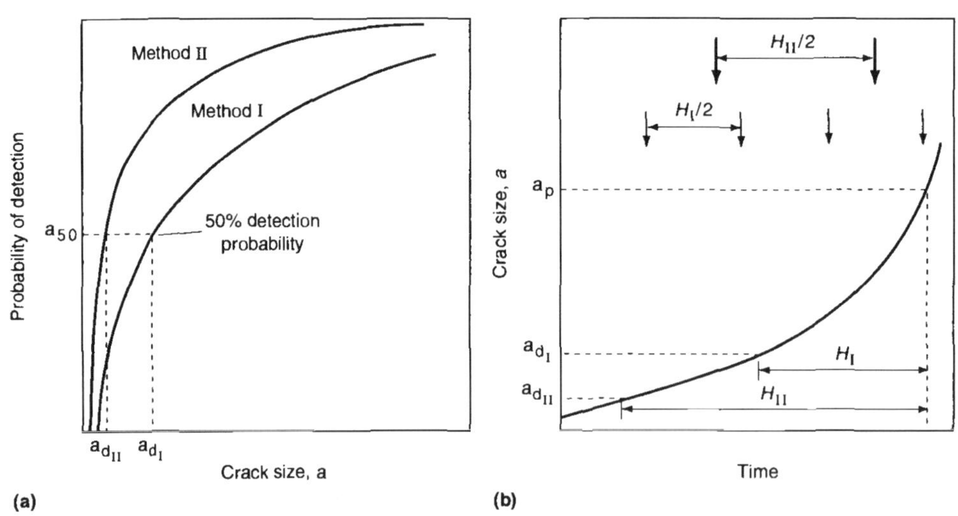

Fig. 9 Inspection intervals based on $H/2$. (a) Probability of detection for two inspection methods. (b) Inspection times on crack growth curve. Source: Ref 1

detected at any one of these inspections can be calculated as in the aforementioned example.

The length of the inspection interval can be established so as to provide a consistent safety level (cumulative probability of detection), independent of the shape of the crack growth curve, the accessibility, and the specificity of the inspection. The target cumulative probability of detection could be set, for example, at 95 or 98% and be specified in damage tolerance requirements. Given the calculated crack growth curve, the permissible crack size, and the probability of detection for the relevant specificity and accessibility, the cumulative probability of detection can be calculated for different lengths of the inspection interval. When the results are plotted, the interval for the desirable probability of detection can be obtained from the curve, as shown in Fig. 10. The interval will be different for different inspection methods, crack growth curves, accessibility, and specificity, but the cumulative probability of detection is always the same (equal safety). The problems with $H/2$ are then eliminated automatically.

A computer can perform the calculation for different interval lengths, provided that the crack growth curve calculated in the damage tolerance analysis and the applicable inspection parameters to the probability-of-detection curve are provided as input. Typical computer results are shown in Fig. 10(a) and (b) for the two crack propagation curves in Fig. 11. The criterion $I = H/2$ would assign the same inspection interval to both cracks. This would lead to different cumulative probabilities of detection for case 1 and case 2. To ensure the same probability in both cases, the intervals must be shorter in case 2. Although refinements can be made, the aforementioned provides a rational procedure to establish inspection intervals for which the probability of detection is independent of the inspection technique, the crack growth curve, and the assignment. This procedure is finding acceptance in the aircraft industry.

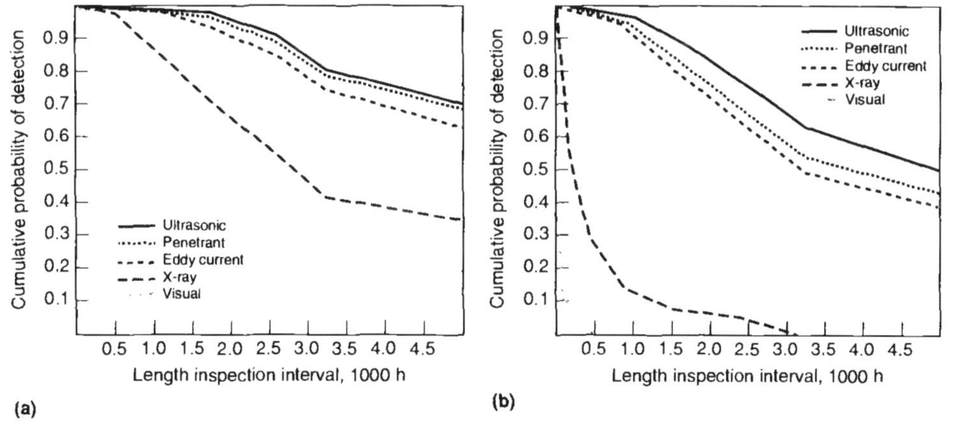

Fig. 10 Cumulative probability of crack detection as a function of the length of the inspection interval. (a) Case 1 of Fig. 11. (b) Case 2 of Fig. 11. Source: Ref 1

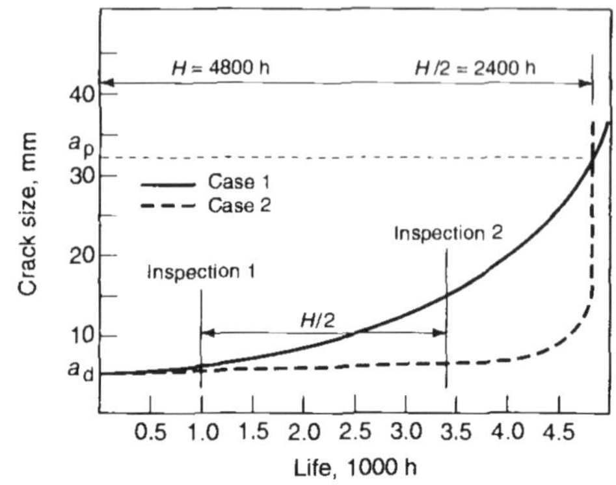

Fig. 11 Two hypothetical crack growth curves used in Fig. 10. Source: Ref 1

Fracture Control Plans

The optimum fracture control plan depends on the consequences of a fracture. If the number of fractures experienced is considered to be an acceptable level with a certain fracture control plan at acceptable costs, the plan is close to optimum. Before implementation of a fracture control program, the objectives must be identified. If a structure can sustain assumed damage under an assumed loading condition, it is not necessarily safe, despite all analysis. Before defining the permissible residual strength or permissible crack size, the desired level of safety should be established, even if only qualitatively. It will appear that every component and structure calls for different fracture control requirements.

Fracture control measures must be in accordance with the acceptable risk. On the basis of rules and regulations issued by governments or engineering societies, the designer or manufacturer prescribes the details of the fracture control plan, and the operator implements this plan through maintenance, inspection, repair, or replacement. The plan must be suitable for a particular structure and for the potential operators. Professional operators of pressure vessels, airplanes, and the like can implement more complex fracture control measures than the general public operating automobiles. Here the concern is with those cases where materials selection alone does not provide adequate safeguards against fractures. Conceptual control plans for a variety of circumstances are given in Table 2.

Detectable Cracks. Table 2 shows the ingredients of fracture control plans for structures in which cracks are detectable by inspection. If initial defects will not grow during service, plan I is applicable. If defects, whether initial or developing later, may grow under service loading, a crack eventually will become critical, unless it is readily discovered and repaired. Inspections should be scheduled in accordance with plan II.

Cracks Not Detectable by Inspection. Cracks may not be detectable, either because their permissible size is so small that it defies inspection,

Table 2 Fracture control plans

Plan	Condition	Fracture control methods
Plans for anticipated cracks detectable by inspection		
Plan I	For initial defect not expected to grow by fatigue	Calculate permissible size of defect. If stress corrosion can occur, calculate which size of defect can be sustained indefinitely given the K_{Iscc} of the material. Inspect once using a technique that can reliably detect defects of the calculated sizes. Eliminate all detected defects larger than those of the calculated sizes.
Plan II	For all defects (initial or initiating later) that will grow during service—these will reach critical size	If possible, show by analysis (or tests) that the structure can sustain without failure such large defects that the damage will be obvious (e.g., readily apparent leak or failed component; fail safety). Repair when damage is discovered. If the above cannot be shown, calculate permissible crack size. Establish crack size that can be detected reliably. Calculate time for crack growth. Implement periodic inspection based on crack growth calculation. Start inspection immediately because time of crack initiation is not known. Repair or replace when crack is detected.
Plan for anticipated cracks not detectable by inspection		
Plan III	For parts where a is so small that it defies inspection	Make best estimate of possible initial defects. Calculate permissible crack size a_p. Calculate crack growth life from initial defect size to a_p. Replace/retire after calculated life expires (using adequate factor).
Plan for structures for which inspection is not feasible, but proof testing is possible		
Plan IV	For components or structures that can be proof tested and where failure during proof testing is not a catastrophe (e.g., leak before break)	Determine feasible proof test pressure or load. Calculate maximum crack size a_{proof} that could be present after proof test. Calculate maximum permissible crack a_p. Calculate crack growth time, H, from a_{proof} to a_p. Repeat proof test before H has expired (using adequate factor).
Plans for cracks discovered in service		
Plan V	For detected cracks for which no analysis is done	Repair or replace unconditionally.
Plan VI	For detected cracks for which analysis is done (if immediate replacement is impractical)	Show by analysis that a larger defect can be sustained. Check growth daily; drill stop hole if permissible (as an interim measure). Prepare for repair or replacement at earliest convenience. Determine exact size and shape of crack. Find materials data; if possible, cut test specimens from structure. Obtain reliable load and stress information. Calculate a_p. Calculate time, H, for growth to a_p. Prepare for repair or replacement before H (with adequate factor) expires. If crack grows faster than calculated, update prognosis and speed up replacement or repair actions. If possible, reduce operational loads. Repair or replace as soon as possible.
Plan VII	For structures identical to those in which a crack was previously detected	Use parts of cracked or failed structure to obtain material properties. Implement Plan II, III, or VI.

Source: Ref 1

the location is not accessible, or the structure is so large that inspections are not feasible. Plan III is applicable in such cases. Plan IV involves destructive inspection by proof testing to show that no cracks larger than a_{proof} are present. If larger cracks are present, a failure will occur during the proof test, but this failure must not be catastrophic in its consequences.

After Crack Detection. Although fracture control calls for immediate repair or replacement when a crack is discovered, this is not always convenient. Large savings may be realized if remedial action can be scheduled for the next major overhaul or shutdown, or if operations can continue until a new part or component has been manufactured and received. Whether this is possible depends upon the fracture control plan in force. A well-conceived plan II already contains information on crack growth and residual strength. Using this information as an initial safeguard, operation can be continued, but the analysis should be updated and plan VI should be put into action. A crack may be discovered accidentally in a structure not subject to a fracture control plan. When no analysis is to be done, plan V is the only possible course, but it is recommended that analysis be done and that plan VI be followed. Recurrence of the incident can be prevented by using plan VII.

Damage Tolerance Requirements

Requirements for Commercial Airplanes. The U.S. Federal Aviation Requirements (FAR.25b), enforced in a similar way in other countries, stipulate that the residual strength with damage present must not fall below the limit load (P_L), so that $P_p = P_L$. In general, the limit load is the load anticipated to occur once in the aircraft life. Given P_p, the residual-strength diagram provides the maximum permissible crack size (a_p).

The aforementioned is the complete requirement; little more is necessary. To satisfy the requirement, the manufacturer is obliged to design in such a manner that cracks can be detected before they reach a_p and to prescribe to the operator where and how often to inspect. Similarly, the operator is obliged to follow the manufacturer inspection instructions. Fracture control by these rules must be exercised by inspection. The excuse that some cracks are noninspectable is not justifiable, except if the components are designed by the safe-life criteria, as in the case of landing gear. Every crack will become detectable if large enough. Thus, the requirement forces structural designs that are tolerant of damage large enough for detection, which promotes fail-safe design with multiple load paths and crack-arrest features.

Although there is a problem in the definition of detectable cracks, an (arbitrary) specification of detectable size would not improve the requirement, because detectability depends on the type of structure, its location, and its accessibility. The best way to determine inspection intervals is based on the cumulative probability of detection. A useful improvement of the requirements would specify the desirable cumulative probability of inspection.

The U.S. Air Force requirements, adopted by some other air forces as well but usually in a modified form, make a distinction between non-fail-safe and fail-safe structures (crack arrest or multiple load paths). More stringent requirements apply to non-fail-safe structures. The military requirements (MIL-A-87221, MIL-A-83444, and MIL-STD-1530) go into great detail with regard to the assumptions to be made in the analysis for initial crack size and crack development. Otherwise, they are very similar to those for commercial aircraft.

ASME and Other Requirements. Other damage tolerance requirements exist for ships and offshore structures and for nuclear pressure vessels. Requirements for ships and offshore structures are issued by shipping bureaus, such as Lloyds of London, Veritas (Norway), and American Bureau of Shipping. Similar requirements exist for military ships. Those for ships are essentially preventive requirements; no analysis is necessary. Ships of a certain size and over must be equipped with so-called arrest strakes, which are located at the gunwale, at the bilge, and sometimes at middecks. They are longitudinal strakes of a higher-toughness material than the normal hull plating.

The damage tolerance requirements for nuclear pressure vessels are contained in the ASME boiler and pressure vessel code, section XI and its appendix A. The requirements essentially provide acceptance limits for cracks detected in service. A variety of possible crack configurations and locations are identified. The requirements then provide the crack sizes for each case that may be left unattended. Should a detected crack exceed the prescribed limits, one has two options: repair weld or perform analysis.

If the option to perform analysis is selected, the following damage tolerance requirements apply: $K <$ arrest toughness/$\sqrt{10}$ for upset conditions (a condition that needs attention but does not impair the operational performance), and $K <$ toughness/$\sqrt{2}$ for emergency and faulty conditions, where K is the stress intensity at these conditions. Strangely enough, these requirements are expressed in terms of the stress intensity and toughness. However, with the fracture condition of $K =$ toughness (in the upset condition, at the stress σ_{cu}), fracture would occur if $Y_p \sigma_p \sqrt{\pi a_p} =$ toughness. If the actual stress intensity must be smaller by a factor of $\sqrt{10}$, it follows that $\sigma_{cu}/\sigma_p \approx \sqrt{10} = 3.16$, assuming $Y_p \approx Y_{cu}$. This means that a safety factor of 3.16 must remain with regard to upset conditions. Thus, the requirement can be stated in terms of the minimum permissible residual strength (σ_p). It follows that $\sigma_p = \sigma_{cu}/3.16$, in accordance with the previous discussions. This case is displayed in Fig. 12(a) using the same nomenclature as before.

At the same time, the stress intensity must be less than the toughness divided by $\sqrt{2}$ for the emergency conditions (stress σ_{cc}). Using the same arguments as before, the minimum permissible residual strength (σ_p) must provide a safety factor of $\sqrt{2} = 1.41$ with regard to upset conditions. This is shown in Fig.12(c).

Different toughnesses are used in the two cases: the arrest toughness and the regular toughness. Because the former is less than the latter, there are effectively two residual-strength curves in play, as shown in Fig. 12.

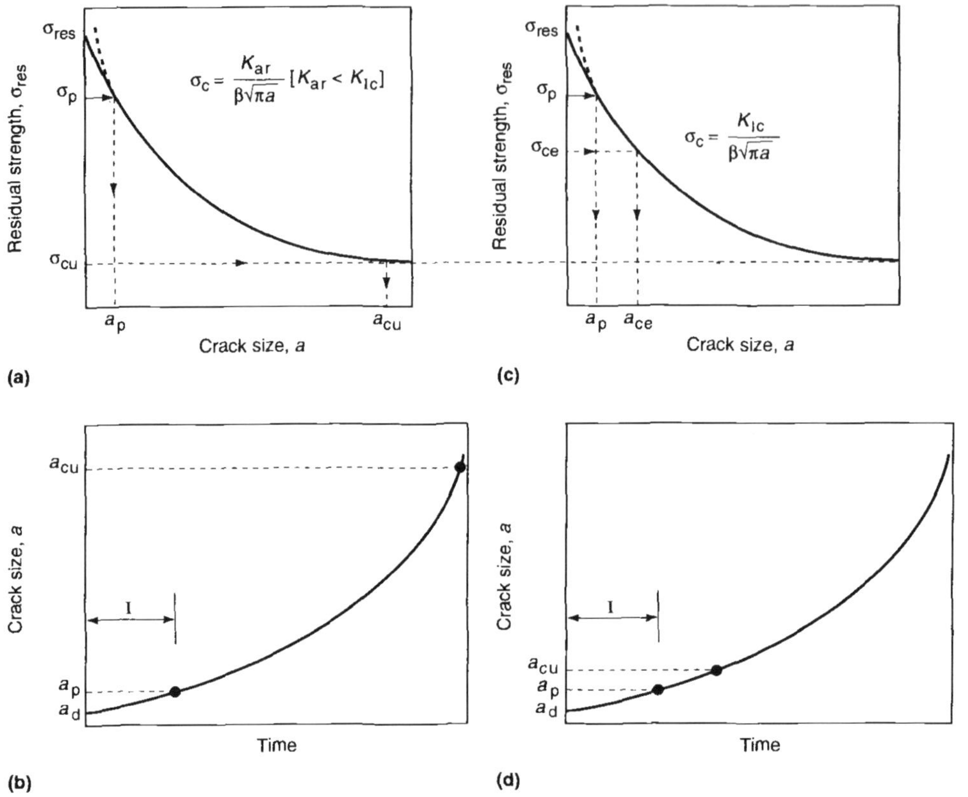

Fig. 12 American Society of Mechanical Engineers (ASME) requirements for (a) permissible crack size (a_p) for upset condition (where K_{ar} is arrest toughness), (b) crack growth for upset condition (I = inspection interval), (c) permissible crack size for emergency condition, and (d) crack growth for emergency condition. (Note: Permissible crack size a_p is actually denoted as a_f in the ASME code.) Source: Ref 1

This does not change the principle of the analysis. In either case, the permissible crack size (a_p) follows from the residual-strength diagram as shown. Obviously, it is impossible to satisfy both requirements exactly. One is always more severe than the other. If it can be foreseen which of the two is the most severe, the requirement can be simplified.

The requirement presents an alternative. Instead of the aforementioned, one may satisfy the following requirements:

- $a_p \leq a_{cu}/10$ upset and operating conditions
- $a_p \leq a_{ce}/2$ emergency conditions

where a_{cu} is the critical crack causing fracture at upset conditions, and a_{ce} is the critical crack in emergency conditions. (In the ASME code, a_p is denoted as a_f.) Thus:

$$a_{cu} = 10a_p \quad \text{(Eq 4)}$$

and

$$K = Y_{cu}\sigma_{cu}\sqrt{\pi(10a_p)} = Y_p\sigma_p\sqrt{\pi a_p} \quad \text{(Eq 5)}$$

so that

$$Y_p\sigma_p / Y_{cu}\sigma_{cu} = \sqrt{10} = 3.16 \quad \text{(Eq 6)}$$

This requirement will be identical to the one stated previously if $Y_p = Y_{cu}$. Because a_{cu} is a longer crack than a_p, in general $Y_{cu} > Y_p$, so that the requirement leads to a safety factor somewhat smaller than 3.16. The same arguments hold for a_p and a_{ce}. Both sets of requirements apparently attempt to set the same conditions, and only one is necessary.

Once a crack is detected and analysis is preferred instead of immediate repair, crack growth must be analyzed as well. Fatigue crack growth must be calculated starting at a_d, which is the crack actually present and discovered, and continuing over the period until the next inspection (shutdown). Over this period, the crack may not grow beyond a_p, as determined by the criteria discussed previously. This condition is shown in Fig. 12(b) and (d). The requirements fully prescribe the analysis procedure as well as the toughness and rate data. Use of other analysis and data is subject to approval by authorities.

The requirement is specifically for a case where a crack is detected in service. One may then prove by analysis that this crack is not dangerous during further operation until the next shutdown. In the ASME requirements, the damage tolerance analysis is used to decide whether a structure with a known crack can be left in service without repair. Besides, analysis is not used to determine the inspection interval. The approach for aircraft has an important difference in that cracks must be repaired, and the analysis is used to ensure detection and repair of a potential crack, not to determine whether it is safe to fly with a known crack. This statement is not meant as criticism but rather as a clarification of the difference in approach.

Fracture Mechanics and Fatigue Design

Fracture mechanics does not provide revolutionary new insights for fatigue design. It merely confirms what has been known for many years. Probably the most important rule in designing for high fatigue performance is that stress concentrations should be kept at a minimum. Fracture mechanics merely emphasizes the significance of this. Without a stress concentration, the stress intensity for a small crack ($Y \approx 1$) is $K = \sigma\sqrt{\pi a}$. For a small crack emanating from a notch or fillet, the geometry factor approximately equals the stress-concentration factor: $Y \approx k_t$, putting the stress intensity at $K = k_t\sigma\sqrt{\pi a}$.

Because fatigue crack growth depends roughly on the third power of K, small cracks at stress concentrations grow faster by approximately k_t^3. Thus, a crack emanating from a simple hole grows about nine times faster than one of equal size starting at a smooth surface. Most service cracks start at stress concentrations, and the implications are self-evident.

The sole objective of damage tolerance analysis is to establish fracture control measures so that cracks can be eliminated before they become dangerous, by either repair or replacement of the component. There is no excuse for a fracture that results from known cracks, regardless of what analysis predicts.

If crack discovery demands repair, a new problem arises. Not only must the repair be adequate to restore strength, it must also be analyzed for damage tolerance again. A repair cannot be treated too casually. A simple cover plate usually does not suffice; such a repair may rather aggravate the situation and cause new cracks in due time. Repairs must be designed to cause gradual transfer of loads to reduce the stress concentration. Fracture control measures must be reinstated for the repair.

The time available for fracture control is the time of safe operation (H), which is governed by the residual strength and the crack growth curve. If H is short, frequent inspections must be scheduled, or the component must be replaced soon. As long as all fracture control decisions are based on H, safety will be maintained. However, long inspection intervals or replacement times are desirable from an economic point of view. The question then is which measures can be taken to improve the situation when H is too small to be economically acceptable. Methods to increase the inspection interval (Fig. 13) are summarized as follows.

Option A—Use of a Material with Better Properties (Fig. 13b). A higher toughness will provide a somewhat larger a_p, but generally speaking, it does not have a great influence on H; most of the life is in the early phase of crack growth. Increasing toughness (a_p) affects only the steeper part of the curve, which in general has only a small effect.

Option B—Selection of a Better Inspection Procedure (Fig. 13c). Improving the inspection technique (e.g., by selecting a more sophisticated inspection procedure) reduces detectable crack sizes. This usually has a significant effect on H because of the small slope of the initial part of the crack growth curve. The penalty will be a more difficult inspection, but the inspection interval will be longer, so fewer inspections will be needed.

Option C—Redesign or Lower Stress (Fig. 13d). The crack growth curve is governed by the stress intensity. Reducing the stress by 15%, for example, will reduce K by 15%. Because crack growth rates are roughly proportional to K^3, a 15% reduction in stress will increase H by a factor $1.15^3 = 1.5$. Such a stress reduction seldom requires a general "beefing up" of the structure; cracks occur where the local stresses are high, and the

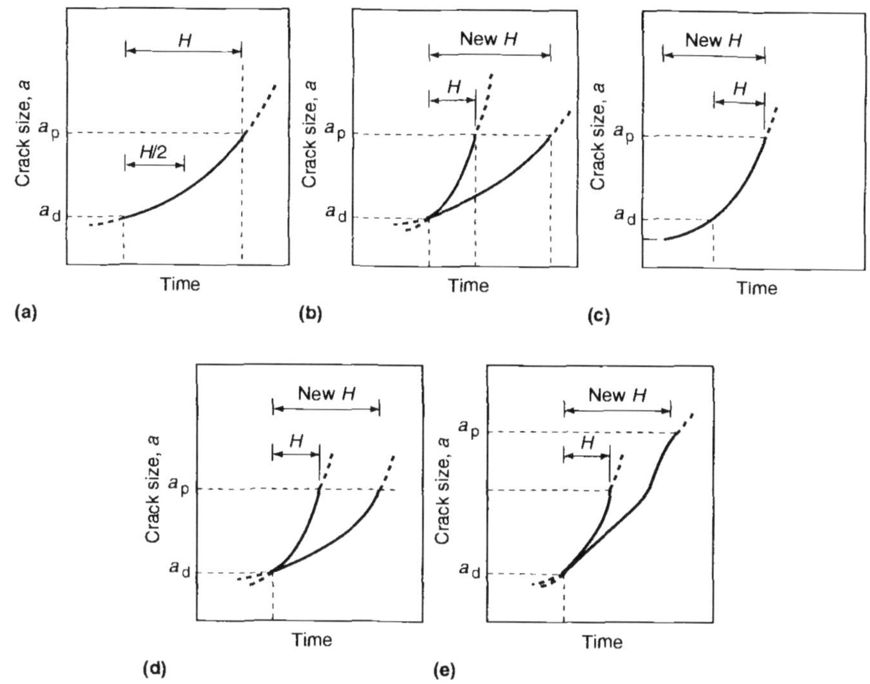

Fig. 13 Ways to increase inspection interval. (a) Base case. (b) Use of better material. (c) Use of more sensitive inspection method. (d) (Detail) design with lower stress. (e) Redundancy (fail safe) or crack arresters. Source: Ref 1

stress reductions are needed only locally. Reduction of stress concentrations, larger fillet radii, and less eccentricity will add hardly any material, cost, or weight.

Redesign may also affect K; a reduction in Y is just as effective as a reduction in σ. In the previous example of the reduced stress concentration, the effect is actually in Y (k_t) instead of in σ (the nominal stress does not change). The redesign may be in the production procedure, so that cracks occur in the cross-grain direction instead of along exposed grain boundaries.

Option D—Providing Redundance and Crack Arresters (Fig. 13e). Building the structure out of more than one element provides multiple load paths. In a well-designed multiple-load-path structure, only inspections for a failed member may be necessary, provided that the fasteners or welds can transfer the load of the failed member by shear.

All of these options can be exercised during design. Therefore, it is crucial that damage tolerance analysis commence in the early design phase when modifications are still possible. Once the design is finalized, the options for improvement are drastically reduced. For finalized designs and existing structures, option B in Fig. 13(c) is often the only recourse, although doublers or arresters can sometimes be added later.

ACKNOWLEDGMENT

The material in this chapter is from "Concepts of Fracture Control and Damage Tolerance Analysis" by D. Broek in *Fatigue and Fracture*, Vol 19, *ASM Handbook*, ASM International, 1996.

REFERENCES

1. D. Broek, Concepts of Fracture Control and Damage Tolerance Analysis, *Fatigue and Fracture,* Vol 19, *ASM Handbook,* ASM International, 1996
2. D. Broek, *Elementary Engineering Fracture Mechanics,* 4th ed., Martinus Nijhoff, 1986
3. D.J. Wulpi, Brittle Fracture, *Understanding How Components Fail,* 2nd ed., ASM International, 1999

CHAPTER 9

Fatigue and Fracture of Ceramics and Polymers

IN PREVIOUS CHAPTERS, the discussion has focused on metals—their mechanical behavior and the analytical methods applicable to them. The structural mechanics and fracture mechanics tools used for metals are believed to be equally applicable to nonmetals, as long as the substance is homogeneous and isotropic. This chapter briefly discusses the essential features associated with ceramics and polymers. Special attention is paid to the characteristics of these materials as compared to metals.

Ceramics

Ceramics are inorganic nonmetallic materials that consist of metallic and nonmetallic elements bonded together with either ionic and/or covalent bonds. Although ceramics can be crystalline or noncrystalline, the important engineering ceramics are all crystalline. Due to the absence of conduction electrons, ceramics are usually good electrical and thermal insulators. In addition, due to the stability of their strong bonds, they normally have high melting temperatures and excellent chemical stability in many hostile environments. However, ceramics are inherently hard and brittle materials that, when loaded in tension, have almost no tolerance for flaws. As a material class, few ceramics have tensile strengths above 172 MPa (25 ksi), while the compressive strengths may be 5 to 10 times higher than the tensile strengths.

Under an applied tensile load at room temperature, ceramics almost always fracture before any plastic deformation can occur. Stress concentrations leading to brittle failure can be minute surface or interior cracks (microcracks), or internal pores, which are virtually impossible to eliminate or control. Plane-strain fracture toughness (K_{Ic}) values for ceramic materials are much lower than for metals, typically below 10 MPa$\sqrt{m}$ (9 ksi$\sqrt{in.}$), while for metals they can exceed 110 MPa$\sqrt{m}$ (100 ksi$\sqrt{in.}$). There is also considerable scatter in the fracture strength for ceramics, which can be explained by the dependence of fracture strength on the probability of the existence of a flaw that is capable of initiating a crack. Therefore, size or volume also influences fracture strength; the larger the size, the greater the probability for a flaw and the lower the fracture strength.

In metals, plastic flow takes place mainly by slip. In metals, due to the nondirectional nature of the metallic bond, dislocations move under relatively low stresses. However, ceramics form either ionic or covalent bonds, both of which restrict dislocation motion and slip. One reason for the hardness and brittleness of ceramics is the difficulty of slip or dislocation motion. While ceramics are inherently strong, because they cannot slip or plastically deform to accommodate even small cracks or imperfections, their inherent strength is never realized in practice. They facture in a premature brittle manner long before their inherent strength is approached.

The nature of the ionic and covalent bonds is shown in Fig. 1. In the ionic bond, the electrons are shared by an electropositive ion (cation) and an electronegative ion (anion). The electropositive ion gives up its valence electrons, and the electronegative ion captures them to produce ions having full electron orbitals or suborbitals. As a consequence, there are no free electrons available to conduct electricity. In ionically bonded ceramics, there are very few slip systems along which dislocations may move. This is a consequence of the electrically charged nature of the ions. For slip in some directions, ions of like charge must be brought into close proximity to each other, and because of electrostatic repulsion, this mode of slip is very restricted. This is not a problem in metals, because all atoms

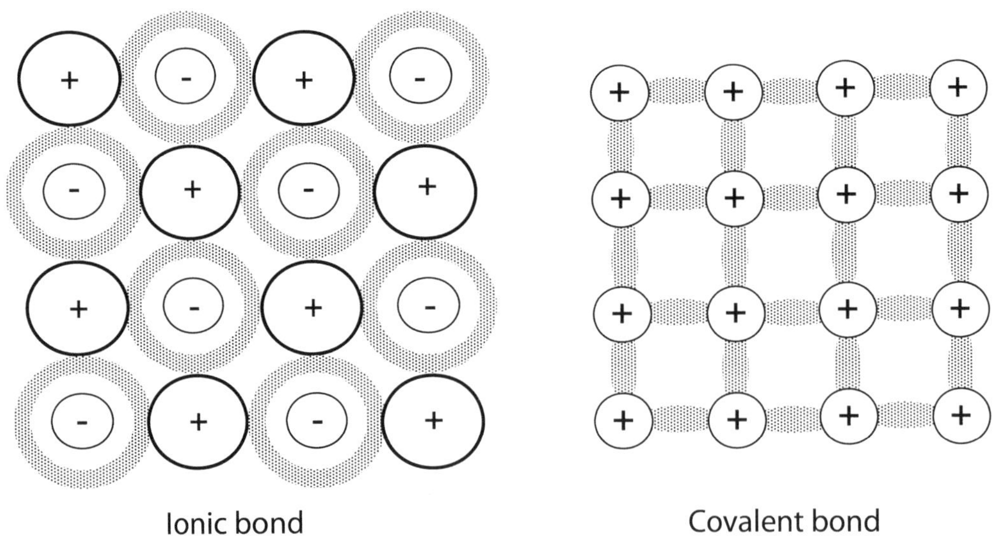

Fig. 1 Ionic and covalent bonding. Source: Ref 1

are electrically neutral. In covalently bonded ceramics, the bonding between atoms is specific and directional, involving the exchange of electron charge between pairs of atoms. Thus, when covalent crystals are stressed to a sufficient extent, they exhibit brittle fracture due to a separation of electron pair bonds, without subsequent reformation. It should be noted that ceramics are rarely either all ionically or covalently bonded; they usually consist of a mix of the two types of bonds. For example, silicon nitride (Si_3N_4) consists of approximately 70% covalent bonds and 30% ionic bonds.

Advanced ceramics are ceramic materials that exhibit superior mechanical properties, corrosion/oxidation resistance, or electrical, optical, and/or magnetic properties. This group includes many monolithic ceramics as well as particulate-, whisker-, and fiber-reinforced glass-, glass-ceramic-, and ceramic-matrix composites. The classification of advanced ceramics according to their function is shown in Table 1. The general term *structural ceramics* refers to a large family of ceramic materials used in an extensive range of applications, including both monolithic ceramics and ceramic-matrix composites. Chemically, structural ceramics include oxides, nitrides, borides, and carbides.

One of the growing uses for advanced ceramics is in the area of structural or load-bearing applications. These applications require materials that have high strengths at room temperature and/or retain high strength at elevated temperatures, resist deformation (slow crack growth or creep), are somewhat damage tolerant, and resist corrosion and oxidation. Ceramics appropriate to these applications offer a significant weight savings over currently used metals. Applications include heat exchangers, automotive engine components such as turbocharger rotors and roller cam followers, power generation components, cutting tools, biomedical implants,

Table 1 Classification of advanced ceramics

Main function	Properties required	Applications (examples)
Thermal	High-temperature and thermal shock resistance, thermal conductivity (high or low, respectively)	High-temperature components, burner nozzles, heat exchanger heating elements, noniron metallurgy insulating parts, thermal barrier coatings
Mechanical	Long-term, high-temperature resistance, fatigue, thermal shock, wear resistance	Wear parts; sealings; bearings; cutting tools; engine, motor, and gas turbine parts; thermal barrier coatings
Chemical, biological	Corrosion resistance, biocompatibility	Corrosion protection, catalyst carriers, environmental protection, sensors, implants (joints, teeth, etc.)
Electrical, magnetic	Electrical conductivity (high or low, respectively), semiconducting, piezo-thermoelectricity, dielectrical properties	Heating elements, insulators, magnets, sensors, integrated circuit packages, substrates, solid electrolytes, piezoelectrics, dielectrics, superconductors
Optical	Low absorption coefficient	Lamps, windows, fiber optics, infrared optics
Nuclear	Irradiation resistance, high absorption coefficient, high-temperature resistance, corrosion resistance	Fuel and breeding elements, absorbers, shields, waste conditioning

Source: Ref 2

and processing equipment used for fabricating a variety of polymer, metal, and ceramic parts. Major obstacles to their more widespread use are their reliability and cost.

Ceramics share many commonalities with metals. Both are crystalline, homogeneous, and isotropic. Ceramics are used in a variety of engineering applications that utilize their hardness, wear resistance, refractoriness, and high compression strength. They are seldom used in tensile-loaded structures, because they are brittle (with minimal ductility) and very sensitive to stress raisers. However, ceramics may be toughened with a more ductile second phase to create a composite or aggregate. This nil ductility and high sensitivity to stress raisers make mechanical testing of ceramics using conventional test methods very difficult.

Toughening Ceramics

Due to the inherently brittle nature of ceramics, much effort has been devoted to finding ways of improving their toughness. The two most noteworthy methods are transformation toughening and ceramic-matrix composites.

Transformation-Toughened Ceramics. One remarkable method of microstructural control to yield improved properties is transformation (martensitic) toughening, first observed in zirconia (ZrO_2) ceramics. Zirconia exhibits phase transformations from cubic to triclinic to monoclinic as its temperature is lowered. If the zirconia is stabilized with materials such as yttria, calcia, or magnesia, the triclinic phase can be preserved in a metastable state at ambient (or other) temperature, depending on the amount of additive, heat treatment, cooling rate, grain size, and grain size distribution. As shown in Fig. 2, the tetragonal zirconia can transform to

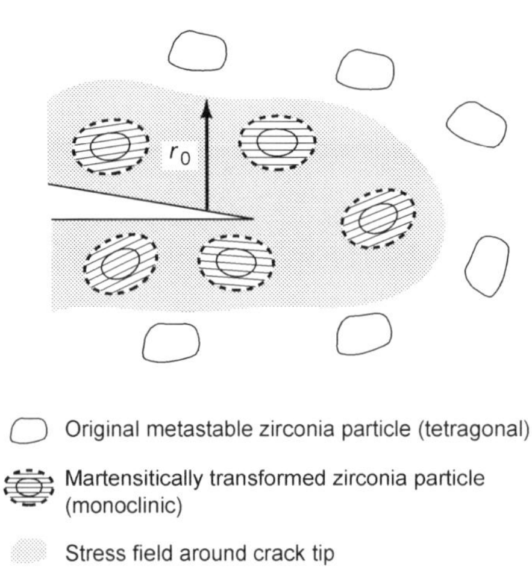

Fig. 2 Stress-induced transformation of metastable ZrO_2 particles in the elastic stress field of a crack. Source: Ref 3

monoclinic under the influence of a crack tip stress. The transformation may cause the crack tip to deflect due to the shear stress (approximately 1 to 7%) or to be impeded by the local compressive stress (>3%). It was found that these structural ceramics exhibit *R*-curve behavior in which the K_{Ic} increases as a crack grows in slow crack growth.

Ceramic-Matrix Composites. While reinforcements such as fibers, whiskers, or particles are used to strengthen polymer- and metal-matrix composites, reinforcements in ceramic-matrix composites are used primarily to increase toughness. The toughness increases afforded by ceramic-matrix composites are due to energy-dissipating mechanisms, such as fiber-to-matrix debonding, crack deflection, fiber bridging, and fiber pull-out. A notional stress-strain curve for a monolithic ceramic and a ceramic-matrix composite is shown in Fig. 3. Because the area under the stress-strain curve is often considered as an indication of toughness, the large increase in toughness for the ceramic-matrix composite is evident. The mechanisms of debonding and fiber pull-out are shown in Fig. 4. For these mechanisms to be effective, there must be a relatively weak bond at the fiber-to-matrix interface. If there is a strong bond, the crack will propagate straight through the fibers, resulting in little or no energy absorption. Therefore, proper control of the interface is critical. Coatings are often applied to protect the fibers during processing and to provide a weak fiber-to-matrix bond.

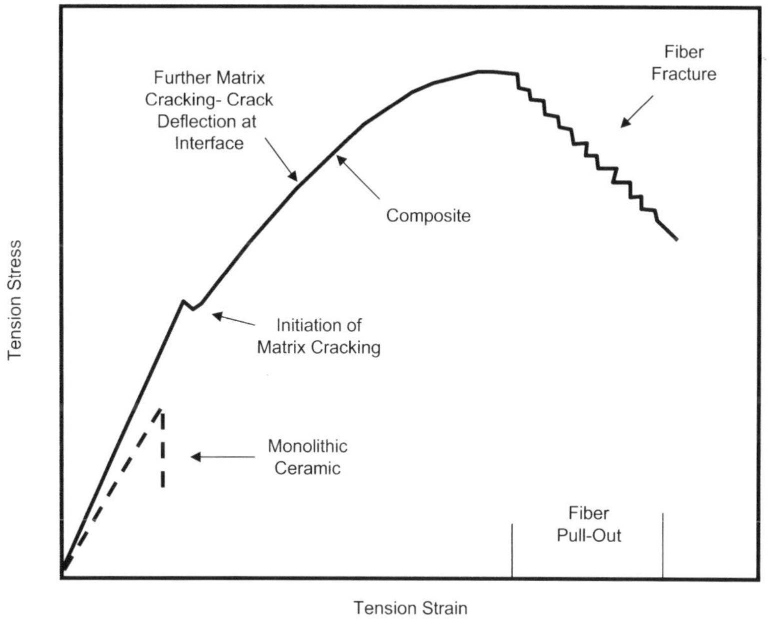

Fig. 3 Stress-strain curves for monolithic ceramics and ceramic-matrix composites. Source: Ref 4

Ceramic Fracture

The brittle nature of ceramics causes great difficulty in detecting flaws or precracking fracture toughness test specimens. Flaws in ceramics are not easily detectable by ordinary nondestructive evaluation methods. Fractographic studies conducted in the 1970s showed that critical flaw sizes in ceramics range from ~1 to ~500 μm (typically 10 to 50 μm). Today's new, toughened ceramics are somewhat improved, but their critical flaw sizes are still very small. These earlier fractographic studies revealed the characteristics of flaw extension in ceramics. The fracture surface of a ceramic material (Fig. 5) normally consists of four distinct regions: the smooth mirror region that separates the original crack (the flaw origin) and the mist region, which is a transitional area preceding the rough/hackle region. A quantitative relationship exists between the stress at frac-

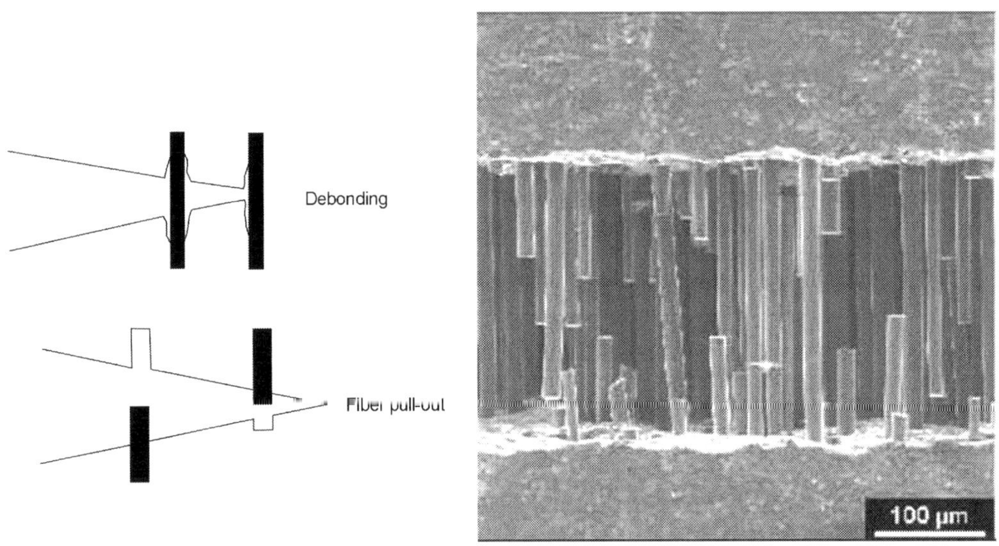

Fig. 4 Crack dissipation mechanisms. Source: Ref 4

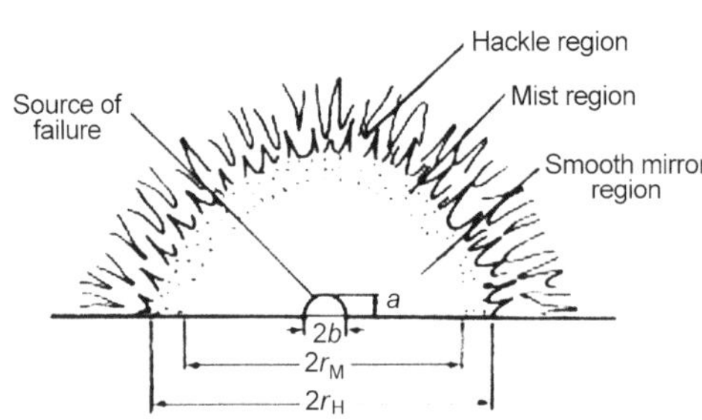

Fig. 5 Schematic of fracture surface features observed on many ceramics. The dimensions *a* and 2*b* denote the minor and major axes of the flaw dimensions, r_M denotes the beginning of the mist region, and r_H denotes the beginning of the hackle region. Source: Ref 5

ture and the distance from the flaw to each of the boundaries. Through the model that follows, one can estimate the critical flaw size and/or the fracture toughness of the material.

Consider a very small flaw originating on a ceramic surface as shown in Fig. 5, which schematically depicts a fracture surface typically found in fractographs of ceramics and glasses.

The mirror region is flat, smooth, and brittle in nature. The mist region is somewhat less smooth. The rough hackle region is easily recognized by the outward divergent lines running along the crack propagation direction. This region is associated with a large amount of strain energy absorption and is thus somewhat ductile in nature. The smooth mirror region is unique. For a wide variety of ceramics, the ratio of the mirror size to the flaw size is a constant with a value of 13. Because the mirror radius (r_M) is measurable, the flaw dimensions a and b can be estimated. The initial and critical flaw sizes are often the same in ceramics; therefore, the critical stress-intensity factor can be calculated by modeling the crack as an elliptical surface flaw. The stress-intensity expression in this case would be:

$$K = \sqrt{1.2\pi\left(\frac{\sqrt{ab}}{r_M}\right)} \cdot \frac{A_M}{\Phi} \qquad \text{(Eq 1)}$$

and

$$A_M = S \cdot r_M^{1/2} \qquad \text{(Eq 2)}$$

where Φ is a crack shape parameter. K becomes K_{Ic} when $a = a_{cr}$ and $b = b_{cr}$. Equation 1 can be used for calculating K_{Ic} or to estimate the critical flaw size when a separately determined K_{Ic} value is available. Due to the brittleness of ceramics, ASTM E 399 test methods are not applicable to fracture and crack growth testing of them. Instead, a number of specially prepared standards have been developed. Typical K_{Ic} values for some common ceramics are available. They are listed along with other mechanical properties in Table 2.

If the resistance of the material to fracture is denoted as R, the conditions for unstable (brittle catastrophic) fracture can be written as:

$$K_I = R \text{ and } \frac{dK}{dc} \geq \frac{dR}{dc} \qquad \text{(Eq 3)}$$

Stable (noncatastrophic) fracture is represented as:

$$K_I \leq R \text{ and/or } \frac{dK}{dc} < \frac{dR}{dc} \qquad \text{(Eq 4)}$$

where dK/dc and dR/dc are the derivatives of the stress-intensity factor K and the fracture resistance of the material R, with respect to the crack length c, respectively. Equations 3 and 4 can be illustrated as a fracture resistance or R-curve, as shown in Fig. 6. Note that the R term can be further described in parts such that:

$$R = R_0 + R(c) \qquad \text{(Eq 5)}$$

where R_0 can be considered the intrinsic fracture resistance, and $R(c)$ is the R-curve component. Note that at some finite length of crack extension, R becomes constant, and $R(c)$ is no longer a function of c. This steady-state

Table 2 Typical properties of selected engineering ceramics

Material	Crystal structure	Theoretical density, g/cm^3	Poisson's ratio	Transverse rupture strength, MPa (ksi)	Fracture toughness, MPa$\sqrt{m}$ (ksi$\sqrt{in.}$)	Young's modulus, GPa (10^6 psi)
Glass-ceramics	Variable	2.4–5.9	0.24	70–350 (10–51)	2.4 (2.2)	83–138 (12–20)
Pyrex glass	Amorphous	2.52	0.2	69 (10)	0.75 (0.7)	70 (10)
TiO_2	Rutile tetragonal	4.25	0.28	69–103 (10–15)	2.5 (2.3)	283 (41)
	Anatase tetragonal	3.84	...	...	...	...
	Brookite orthorhombic	4.17	...	...	...	...
Al_2O_3	Hexagonal	3.97	0.26	76–1034 (40–150)	2.7–4.2 (2.5–3.8)	380 (55)
Cr_2O_3	Hexagonal	5.21	...	>262 (>238)	3.9 (3.5)	>103 (>15)
Mullite	Orthorhombic	2.8	0.25	185 (27)	2.2 (2.0)	145 (21)
Partially stabilized ZrO_2	Cubic, monoclinic, tetragonal	5.70–5.75	0.23	600–700 (87–102)	8–9 (7.3–8.2)(a)	205 (30)
					6–6.5 (5.5–5.9)(b)	
					5 (4.6)(c)	
Fully stabilized ZrO_2	Cubic	5.56–6.1	0.23–0.32	245 (36)	2.8 (2.5)	97–207 (14–30)
Plasma-sprayed ZrO_2	Cubic, monoclinic, tetragonal	5.6–5.7	0.25	6–80 (0.9–12)	1.3–3.2 (1.2–2.9)	48 (7)(d)
CeO_2	Cubic	7.28	0.27–0.31	...	...	172 (25)
TiB_2	Hexagonal	4.5–4.54	0.09–0.13	700–1000 (102–145)	6–8 (5.5–7.3)	514–574 (75–83)
TiC	Cubic	4.92	0.19	241–276 (35-40)	...	430 (62)
TaC	Cubic	14.4–14.5	0.24	97–290 (14–42)	...	285 (41)
Cr_3C_2	Orthorhombic	6.70	...	49 (7.1)	...	373 (54)
Cemented carbides	Variable	5.8–15.2	0.2–0.29	758–3275 (110–475)	5–18 (4.6–16.4)	96–654 (57–95)
SiC	α, hexagonal	3.21	0.19	96–520 (14–75)(e)	4.8 (4.4)(e)	207–483 (30–70)
				250 (36)(f)	2.6–5.0 (2.4–4.6)(f)	
				230–825 (33–120)(g)	4.8–6.1 (4.4–5.6)(g)	
				398–743 (58–108)(h)	4.1–5.0 (3.7–4.6)(h)	
	β, cubic	3.21	...	...	...	...
SiC (CVD)	β, cubic	3.21	0.16	1034–1380 (150–200)(j)	5–7 (4.6–6.4)	415–441 (60–64)
				2060–2400 (300–350)(k)		
Si3N4	α, hexagonal	3.18	0.24	414–650 (60–94)(m)	5.3 (4.8)(m)	304 (44)
				700–1000 (100–145)(n)	4.1–6.0 (3.7–5.5)(n)	
				250–345 (36–50)(p)	3.6 (3.3)(p)	
	β, hexagonal	3.19	...	...	...	...
TiN	Cubic	5.43–5.44	...	...	...	251 (36)

Note: CVD, chemical vapor deposition. (a) At 20 °C (70 °F). (b) At 450 °C (840 °F). (c) At 800 °C, 1470 °F). (d) 21 GPa (3 × 10^6 psi) at 1373 K (1100 °C, or 2010 °F). (e) Sintered at 27 °C (80 °F). (f) Sintered at 1000 °C (1830 °F). (g) Hot pressed at 27 °C (80 °F). (h) Hot pressed at 1000 °C (1830 °F). (j) At 300 K (27 °C, or 80 °F). (k) At 1473 K (1200 °C, or 2190 °F). (m) Sintered. (n) Hot pressed. (p) Reaction bonded. Source: Ref 6

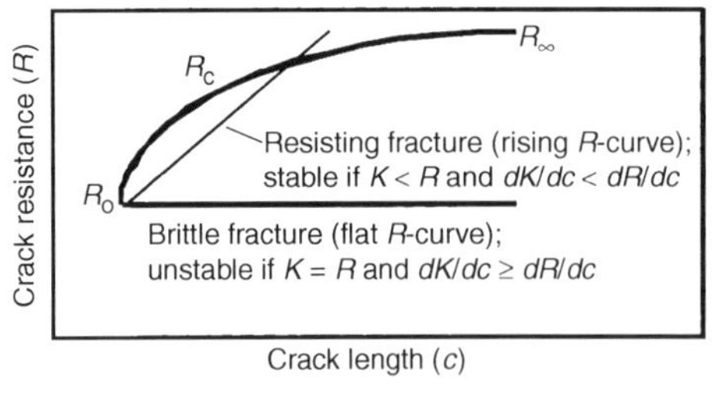

Fig. 6 Schematic representation of R-curve behavior. Source: Ref 7

value of fracture resistance [$R_\infty \neq f(c)$] corresponds to fully developed toughening mechanisms. The *R*-curve behavior in brittle materials typically develops because of microstructural effects, such as those shown in Fig. 7. The *R*-curve effects often confuse and frustrate attempts to experimentally measure R_0 (or K_{Ic}), because the effects of crack growth history add to experimental scatter if *R* (or K_{Ic}) is measured and reported outside the context of the crack extension.

Strength distributions are therefore related to the distributions of these flaws. Intrinsic flaws (e.g., those due to processing) are those that can be treated as microcracks (short cracks) and are on the order of the microstructure. Extrinsic or induced flaws (e.g., those due to service) are those that can be treated as macrocracks (long cracks) and are on the order of component dimensions. Note that sometimes extrinsic flaws may be of the same dimensional order as intrinsic flaws.

Ceramics can exhibit either flat or rising *R*-curve behavior, depending on processing derived microstructure. Figure 8 shows *R*-curves for three different silicon nitride (Si_3N_4) materials, A, B, and C. Material A is a hot-pressed commercial Si_3N_4 (SN-84H by NGK Technical Ceramics) with low fracture toughness that results in flat *R*-curve behavior. Materials B and C, which have relatively higher fracture toughness, are monolithic Si_3N_4 (AS700 by Allied Signal Inc.) prepared by gas pressure sintering green billets, which were formed by cold isostatic pressing. These two Si_3N_4 materials exhibit rising *R*-curve behavior. An *R*-curve from double-cantilever beam tests on a SiC whisker-reinforced alumina-matrix composite is shown in Fig. 9 and also exhibits a rising *R*-curve behavior.

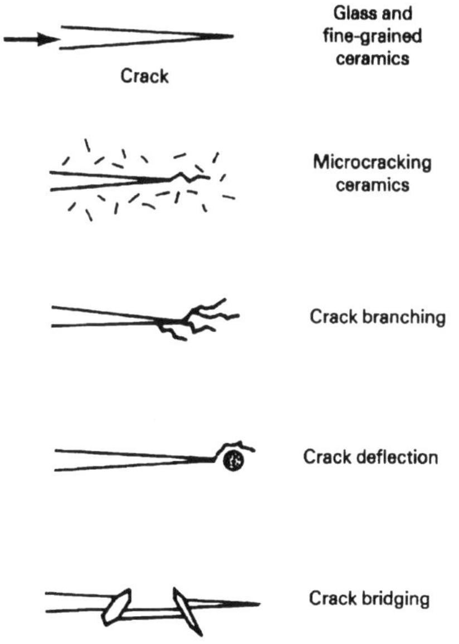

Fig. 7 Microstructural features responsible for fracture resistance as a function of crack length (*R*-curve effects). Source: Ref 7

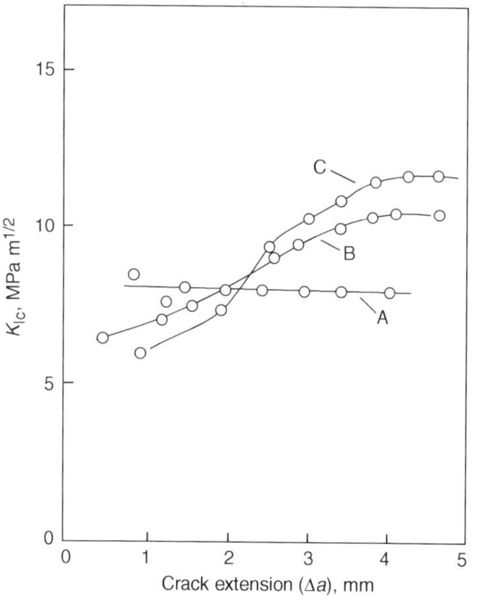

Fig. 8 *R*-curves of three Si_3N_4 ceramic materials measured by the short rod chevron notch technique. Adapted from Ref 8

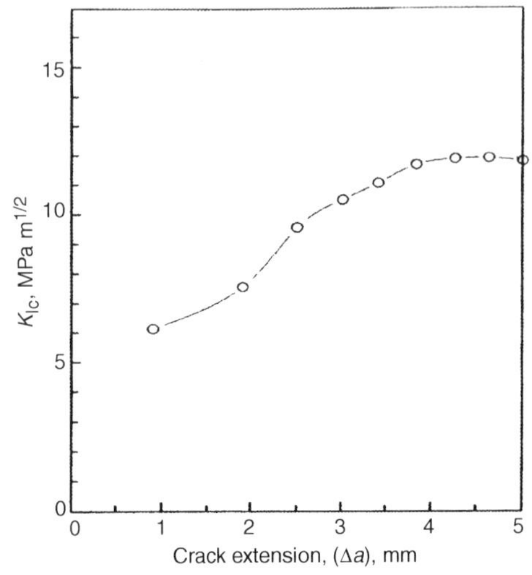

Fig. 9 Rising *R*-curve of a SiC whisker-reinforced alumina-ceramic-matrix composite, measured by the double-cantilever beam technique. Adapted from Ref 8

Fatigue and Subcritical Crack Growth

The *S-N* fatigue data for ceramics are available but are very rare. Stress hysteresis is nonexistent due to lack of plasticity. The term *cyclic fatigue* in ceramics basically refers to fatigue crack propagation *da/dn*. The general method of fatigue crack growth testing of ceramics can be very similar to that given in ASTM Standard E 647 for measuring fatigue crack growth rate in metallics, except that precracking is extremely difficult.

Typical *da/dn* data for ceramics are shown in Fig. 10 and 11. It is interesting to note that these data exhibit very much the same effect and behavior as metals, whether tested in room temperature air or in water. That is, the magnesia partially stabilized zirconia (Mg-PSZ) data in Fig. 10 shows that its fatigue crack growth rates are faster in moist air and much faster in water as compared to in inert nitrogen, resembling the typical corrosion fatigue behavior in metals. Temperature, moisture, water, and water vapor always cause stress-corrosion cracking (SCC)-type crack growth and failure in ceramics. Impurities such as silicate glass in the grain boundaries and interfaces, compounded with nil crack tip plasticity, are responsible for SCC failure in ceramics.

Cyclic slip is the basis for fatigue in metals; however, the strong bonds in highly brittle solids restrict the dislocation movement (slip) for fatigue initiation. Thus, there has been a perception that ceramics are free of true cyclic fatigue effects, and that the cyclic crack growth rates in ceramics are simply manifestations of environmentally assisted crack growth under sustained tensile loading. That means *da/dn* and *da/dt* are interchangeable, after converting the time-based *da/dt* to the cycle-based *da/dn*, and vice versa. Many curve-fitting equations were developed to fit the *da/dt* data

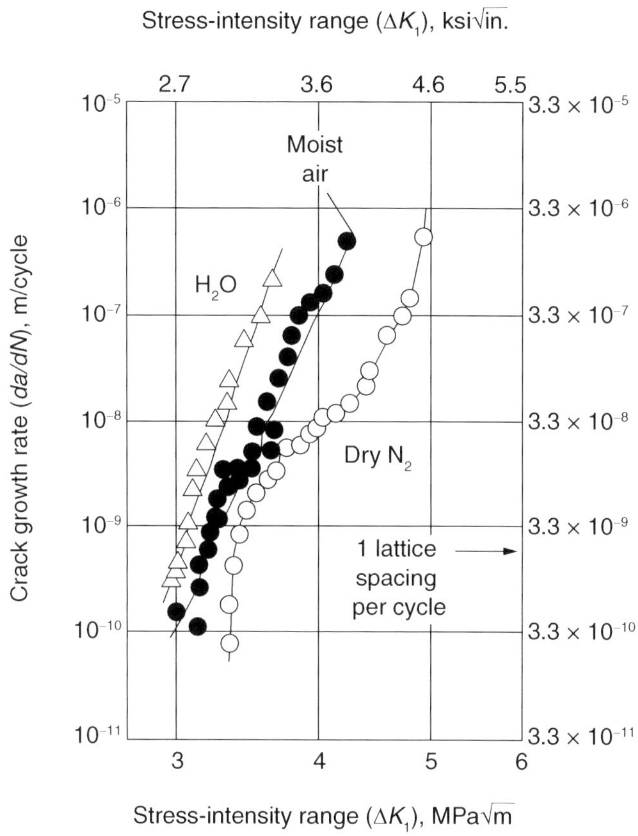

Fig. 10 Cyclic fatigue crack growth rates in low-toughness magnesia partially stabilized zirconia (Mg-PSZ), showing acceleration due to moisture or water. Source: Ref 9

and then convert it to *da/dn* data. However, conclusive evidence of true cyclic crack growth behavior in glasses and ceramics has surfaced in recent studies, as shown in Fig. 10 and 11. Experimental test results for Al_2O_3-33SiC have shown that for the same material, and at the same *K* level, cyclically loaded tests produce much slower crack growth rates compared to sustained load testing. However, it has also been shown that crack growth behavior in midtoughness MgO-PSZ under 55% relative humidity air or in distilled water is just the opposite: *da/dn* is faster than *da/dt*. These discrepancies can be attributed, in part, to ignoring the stress ratio and frequency effects in cyclic crack growth testing. Nevertheless, the practice of using converted crack growth rates will result in unreliable fatigue-life estimates for ceramics.

Polymers

An engineering plastic can be defined as a synthetic polymer capable of being formed into load-bearing shapes and possessing properties that en-

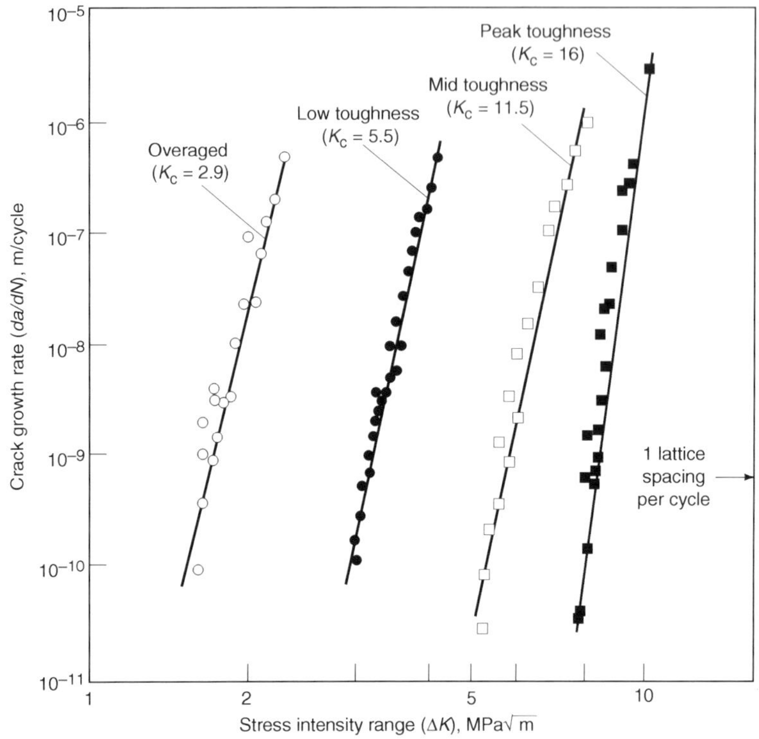

Fig. 11 Cyclic fatigue crack growth rates for MgO-PSZ subeutectoid aged to a range of fracture toughness levels. Data obtained from specimens in room temperature air, 50 Hz, $R = 0.1$. Source: Ref 9

able it to be used in the same manner as traditional materials, and that has high-performance properties that permit it to be used in the same manner as metals and ceramics. This definition includes thermoplastic and thermoset resins. It also includes continuous fiber-reinforced resins (composite materials), as well as the short fiber- and/or particulate-reinforced plastics commonly used for parts and components.

A plastic is a polymeric material composed of molecules made up of many (poly-) repeats of some simpler unit, the mer. What all polymers have in common is that they are chemically constructed of repeats of a basic mer unit, which is chemically bonded to others of its kind to form one-, two-, or three-dimensional molecules. The mer units of polymers are bonded to one another with strong covalent bonds. Most polymers contain mainly carbon in their backbone structures because of the unique ability of carbon to form extensive, stable covalent bonds. While covalent bonds are stronger than metallic bonds, they are highly directional. In thermoplastics, the strong covalent bonds occur within the polymer molecules, but not between them. Only weaker, secondary bonds occur between polymer molecules. Thus, unless the polymer is the type that is bonded together in three dimensions (as in a thermoset), it may be relatively easy to disrupt its structure with moderate heat even though it is difficult to rupture the primary covalent bonds within the polymer molecule itself.

General Characteristics of Plastics

Engineering plastics offer some unique product benefits. These are usually physical properties, or combinations of physical properties, that allow vastly improved product performance, such as:

- *Electrical insulation.* Most combinations of plastic resin and filler are natural electrical insulators.
- *Thermal insulation.* All common resin-filler combinations are excellent thermal insulators, compared to metals and most other nonmetals.
- *Chemical resistance.* Some plastics offer extreme resistance to chemical reagents and solvents.
- *Magnetic inertness.* Most plastics will not respond to an electromagnetic field.
- *Extremely lightweight.* Some resin-reinforcement combinations offer extremely favorable strength-to-weight and stiffness-to-weight ratios.
- *Transparency.* Several plastics offer transparency and a high degree of optical clarity.
- *Toughness.* Many plastics offer toughness, most commonly regarded as freedom from the tendency to fracture.
- *Colorability.* All plastics are colorable to some extent. Many are capable of a nearly unlimited color spectrum while maintaining high visual quality.

Plastics have the following limitations:

- Strength and stiffness is low relative to metals and ceramics.
- Service temperatures are usually limited to only a few hundred degrees Fahrenheit because of the softening of thermoplastic plastics or degradation of thermosetting plastics.
- Some plastics degrade when subjected to sunlight, other forms of radiation, and some chemicals.
- Thermoplastics exhibit viscoelastic properties, which means they creep under relatively low stress levels and can be a distinct limitation in load-bearing applications.

An engineer should immediately disqualify engineering plastics if the application requires maximum efficiency of heat transfer, electrical conductivity, or nonflammable properties. Although some plastics can be formulated to retard flames, their organic nature makes them inherently flammable. Applications under constant stress for which a close tolerance must be held need to be scrutinized in terms of the effects of creep at the application temperature. The engineer also must be cognizant of the poor ultraviolet (UV) resistance of most, though not all, engineering plastics and the resulting requirement for a UV absorber.

Thermosets and Thermoplastics

Polymeric plastics are classified as either thermosets or thermoplastics (Fig. 12). Thermosets are low-molecular-weight, low-viscosity monomers (≈2000 centipoise) that are converted during curing into three-dimensional cross-linked structures that are infusible and insoluble. Cross linking results from chemical reactions that are driven by heat generated by either chemical reactions themselves (i.e., exothermic heat of reaction) or by externally supplied heat. Prior to cure, the resin is a relatively low-molecular-weight semisolid that melts and flows during the initial part of the cure process. As the molecular weight builds during cure, reactions accelerate and the available volume within the molecular arrangement decreases, resulting in less mobility of the molecules and an increase in viscosity. The viscosity increases until the resin gels (solidifies), and then strong covalent bond cross links form during cure. After the resin gels and forms a rubbery solid, it cannot be remelted. Further heating causes additional cross linking until the resin is fully cured. This progression through cure is shown in Fig. 13. Because cure is a thermally-driven event requiring chemical reactions, thermosets are characterized as having rather long processing times. Due to the high cross-link densities obtained for high-performance thermoset systems, they are inherently brittle unless steps are taken to enhance toughness.

Thermoset polymers cure by either addition or condensation reactions. A comparison of these two cure mechanisms is shown in Fig. 14. In the addition reaction shown for an epoxy reacting with an amine curing agent, the epoxy ring opens and reacts with the amine curing agent to form a cross link. The amine shown in this example is what is known as an ali-

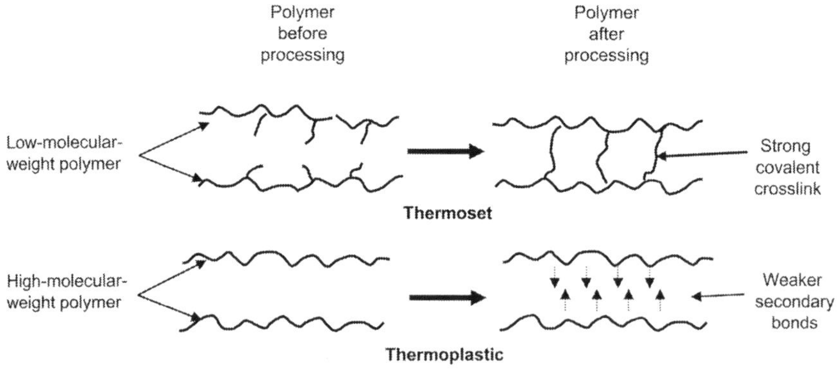

Fig. 12 Comparison of thermoset and thermoplastic polymer structures. Source: Adapted from Ref 10

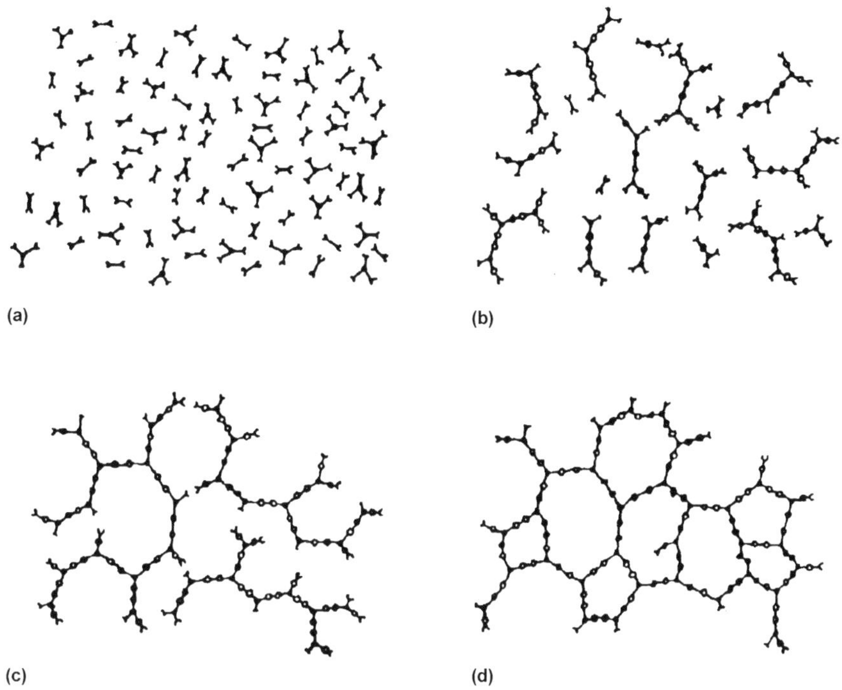

Fig. 13 Stages of cure for thermoset resin. (a) Polymer and curing agent prior to reaction (b) Curing initiated with size of molecules increasing (c) Gelation with full network formed (d) Full cured and crosslinked. Source: Ref 11

phatic amine and would produce a cross-linked structure with only moderate temperature capability. Higher-temperature capabilities can be produced by curing with what are known as aromatic amines that contain the large and bulky benzene ring, which helps to restrict chain movement when the network is heated.

In the lower portion of Fig. 14, two phenol molecules are shown to react with a formaldehyde molecule to form a phenolic linkage in what is known as a condensation reaction. The significant feature of this reaction is the evolution of a water molecule each time the reaction occurs. When thermosets, such as phenolics and polyimides that cure by condensation reactions, are used for molded parts, they often contain high porosity levels, due to the water or alcohol vapors created by the condensation reactions. Addition curing thermosets, such as epoxies, polyesters, vinyl esters, and bismaleimides, are generally much easier to process with low void levels.

An important consideration in selecting any plastic (thermoset or thermoplastic) is the glass transition temperature. The glass transition temperature (T_g) is a good indicator of the temperature capability of the matrix. The glass transition temperature (T_g) of a polymeric material is the

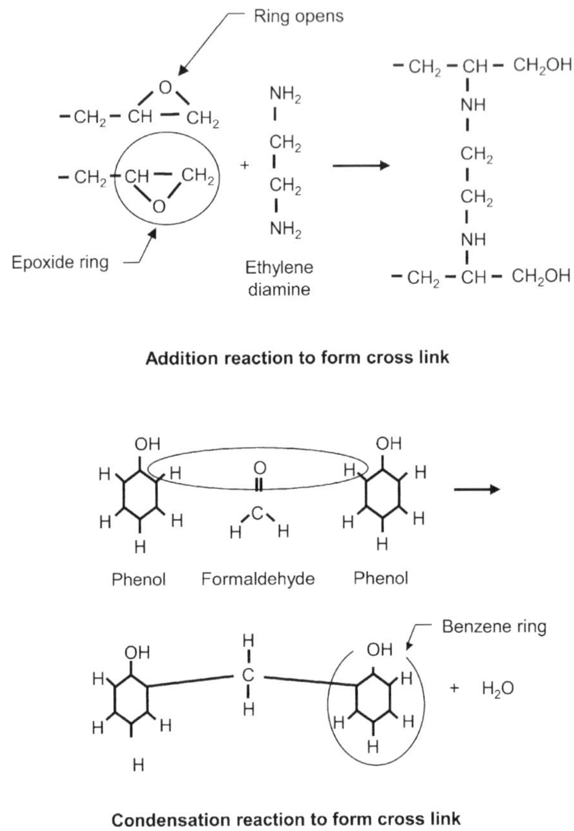

Fig. 14 Thermoset cure reactions. Source: Ref 12

temperature at which it changes from a rigid glassy solid into a softer, semiflexible material. At this point, the polymer structure is still intact, but the cross links are no longer locked in position.

In contrast to thermosets, thermoplastics are high-molecular-weight resins that are fully reacted prior to processing. They melt and flow during processing but do not form cross-linking reactions. Their main chains are held together by relatively weak secondary bonds. When heated to a sufficiently high temperature, these secondary bonds break down, and the thermoplastic eventually reaches the fluid state. On cooling, the secondary bonds reform, and the thermoplastic becomes a solid again. Being high-molecular-weight resins, the viscosities of thermoplastics during processing are orders of magnitude higher than that of thermosets (e.g., 10^4 to 10^7 poise for thermoplastics versus 10 poise for thermosets). Because thermoplastics do not cross link during processing, they can be reprocessed. For example, they can be thermoformed into structural shapes by simply reheating to the processing temperature. Thermosets, due to their highly cross-linked structures, cannot be reprocessed and will thermally degrade and eventually char if heated to high enough temperatures.

Many thermoplastics are polymerized by what is called chain polymerization, as shown in Fig. 15 for the simple thermoplastic polyethylene. Chain polymerization consists of three steps: initiation, propagation, and termination. During initiation, an active polymer capable of propagation is formed by the reaction between an initiator species and a monomer unit. In the figure, R represents the active initiator that contains an unpaired electron (·). Propagation involves the linear growth of the molecule as monomer units become attached to each other in succession, to produce a long chain molecule. Chain growth, or propagation, is very rapid, with the period required to grow a molecule consisting of 1000 repeat units on the order of 100 to 1000 s. Propagation can terminate in one of two ways. In

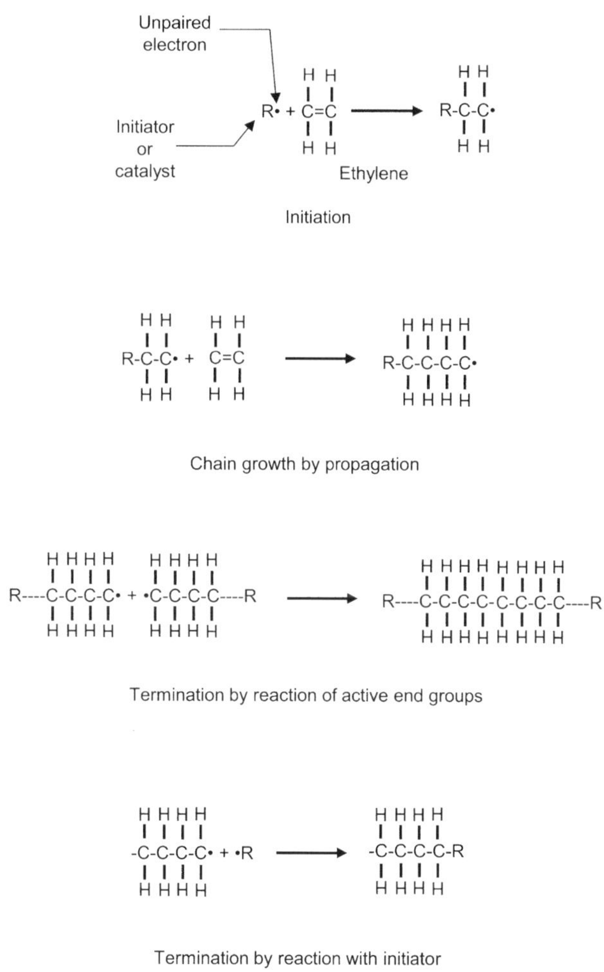

Fig. 15 Chain polymerization of polyethylene. Source: Ref 12

the first, the active ends of two propagating chains react together to form a nonreactive molecule. The second method of termination occurs when an active chain end reacts with an initiator. Polyethylene can normally have anywhere from 3500 to 25,000 of these repeat units.

Thermoplastics are further classified as being either amorphous or semicrystalline. The differences between an amorphous and a semicrystalline thermoplastic are illustrated in Fig. 16. An amorphous thermoplastic contains a massive random array of entangled molecular chains. The chains themselves are held together by strong covalent bonds, while the bonds between the chains are much weaker secondary bonds. When the material is heated to its processing temperature, it is these weak secondary bonds that break down and allow the chains to move and slide past one another. Amorphous thermoplastics exhibit good elongation, toughness, and impact resistance. As the chains become longer, the molecular weight increases, resulting in higher viscosities, higher melting points, and greater chain entanglement, all leading to higher mechanical properties.

Semicrystalline thermoplastics contain areas of tightly folded chains (crystallites) that are connected together with amorphous regions. Amorphous thermoplastics exhibit a gradual softening on heating, while semicrystalline thermoplastics exhibit a sharp melting point when the crystalline regions start to dissolve. As the polymer approaches its melting point, the crystalline lattice breaks down and the molecules are free to rotate and translate, while noncrystalline amorphous thermoplastics exhibit a more gradual transition from a solid to a liquid. In general, the melting point (T_m) increases with increasing chain length, greater attractive forces between the chains, greater chain stiffness, and increasing crystallinity, while the glass transition temperature (T_g) increases with lower free volume, greater attractive forces between the molecules, decreasing chain

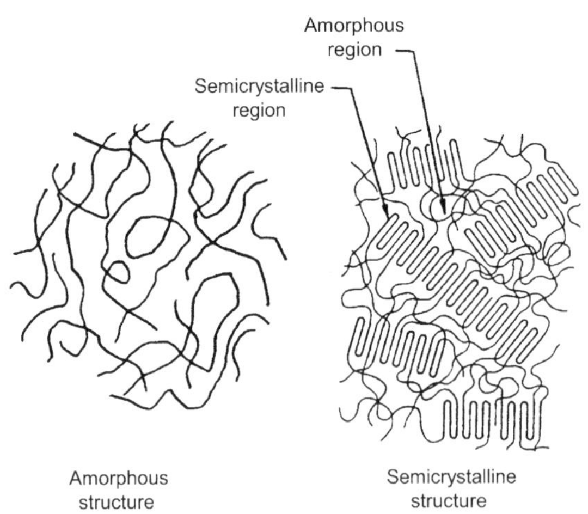

Fig. 16 Comparison of amorphous and semicrystalline thermoplastic structures. Source: Ref 12

mobility, increasing chain stiffness, increasing chain length, and, for thermosets, increasing cross-link density. Crystallinity provides the following attributes to a thermoplastic resin:

- Crystalline regions are held together by amorphous regions. The maximum crystallinity obtainable is approximately 90%, whereas metallic structures are usually 100% crystalline and exhibit much more ordered structures.
- Crystallinity increases density. The density increase helps to explain the improved solvent resistance of semicrystalline thermoplastics, because it becomes more difficult for the solvent molecules to penetrate the tightly packed crystallites.
- Crystallinity increases strength, stiffness, creep resistance, and temperature resistance but usually decreases toughness. The tightly packed crystalline structure behaves somewhat like cross linking in thermosets by decreasing and restricting chain mobility.
- Semicrystalline polymers are either opaque or translucent, while transparent polymers are always amorphous.
- Crystallinity can be increased by mechanical stretching.

It should be noted that all thermoset resins are amorphous but are cross linked to provide strength, stiffness, and temperature stability. As a general class of polymers, thermoplastics are much more widely used than thermosets, accounting for approximately 80% of the polymers produced. A number of industrially important plastics are listed in Table 3, along

Table 3 Abbreviations, chemical names, and structures of select plastics

Material	Chemical name	Structure
ABS	Acrylonitrile-butadiene-styrene	Amorphous
ABS-PC	Acrylonitrile-butadiene-styrene-polycarbonate	Amorphous
DAP	Diallyl phthalate	Amorphous
POM	Polyoxymethylene (acetal)	Semicrystalline
PMMA	Polymethyl methacrylate (acrylic)	Amorphous
PAR	Polyarylate	Amorphous
LCP	Liquid crystal polymer	Semicrystalline
MF	Melamine formaldehyde	Amorphous
Nylon 6	Polyamide	Semicrystalline
Nylon 6/6	Polyamide	Semicrystalline
Nylon 12	Polyamide	Semicrystalline
PAE	Polyaryl ether	Amorphous
PBT	Polybutylene terephthalate	Semicrystalline
PC	Polycarbonate	Amorphous
PBT-PC	Polybutylene terephthalate-polycarbonate	Amorphous
PEEK	Polyetheretherketone	Semicrystalline
PEI	Polyether-imide	Amorphous
PESV	Polyether sulfone	Amorphous
PET	Polyethylene terephthalate	Semicrystalline
PF	Phenol formaldehyde (phenolic)	Amorphous
PPO (mod)	Polyphenylene oxide	Amorphous
PPS	Polyphenylene sulfide	Semicrystalline
PSU	Polysulfone	Amorphous
SMA	Styrene-maleic anhydride	Amorphous
UP	Unsaturated polyester	Amorphous

Source: Ref 12

with their chemical names, and whether they are amorphous or semicrystalline. Typical mechanical properties are shown in Table 4.

Viscoelastic Behavior

Many polymers exhibit viscoelastic behavior. When viscoelastic polymers are stressed, there is an immediate elastic response. This is followed by viscous flow, which decreases with time until a steady state is reached (Fig. 17). If the material is then unloaded, the elastic strain is recovered, followed by time-dependent (delayed) recovery. Some permanent strain, denoted as permanent recovery in Fig. 17, remains.

Table 4 Mechanical properties of selected plastics

Material	Tensile strength		Tensile modulus		Flexural strength		Impact strength		Hardness, Rockwell	Flame rating, UL 94
	MPa	ksi	GPa	10^6 psi	MPa	ksi	J/m	ft · lbf/in.		
ABS	41	6.0	2.3	0.33	724	10.5	347	6.5	R103	HB
ABS-PC	59	8.5	2.6	0.38	89.6	13.0	560	10.5	R117	HB
DAP	48	7.0	10.3	1.50	117	17.0	37	0.7	E80	HB
POM	69.0	10.0	3.2	0.47	98.6	14.3	133	2.5	R120	HB
PMMA	72.4	10.5	3.0	0.43	110	16.0	21	0.4	M68	HB
PAR	68	9.9	2.1	0.30	82.7	12.0	288	5.4	R122	HB
LCP	110	16.0	11.0	1.60	124	18.0	101	1.9	R80	V0
MF	52	7.5	9.65	1.40	93.1	13.5	16	0.3	M120	V0
Nylon 6	81.4	11.8	2.76	0.40	113	16.4	59	1.1	R119	V2
Nylon 6/6	82.7	12.0	2.83	0.41	110	16.0	53	1.0	R121	V2
Nylon 12	81.4	11.8	2.3	0.34	113	16.4	64	1.2	R122	V2
PAE	121	17.6	8.96	1.30	138	20.0	64	1.2	M85	V0
PBT	52	7.5	2.3	0.34	82.7	12.0	53	1.0	R117	HB
PC	69.0	10.0	2.3	0.34	96.5	14.0	694	13.0	R118	V2
PBT-PC	55	8.0	2.2	0.32	86.2	12.5	800	15.0	R115	HB
PEEK	93.8	13.6	3.5	0.51	110	16.0	59	1.1	...	V0
PEI	105	15.2	3.0	0.43	152	22.0	53	1.0	M109	V0
PESV	84.1	12.2	2.6	0.38	129	18.7	75	1.4	M88	V0
PET	159	23.0	8.96	1.30	245	35.5	101	1.9	R120	HB
PF	41	6.0	5.9	0.85	62	9.0	21	0.4	M105	HB
PPO (mod)	54	7.8	2.5	0.36	88.3	12.8	267	5.0	R115	V0
PPS	138	20.0	11.7	1.70	179	26.0	69	1.3	R123	V0
PSU	73.8	10.7	2.5	0.36	106	15.4	64	1.2	M69	HB
SMA	31	4.5	1.9	0.27	55	8.0	133	2.5	R95	HB
UP	41	6.0	5.5	0.80	82.7	12.0	32	0.6	M88	HB

Refer to Table 3 for chemical name. Source: Ref 13

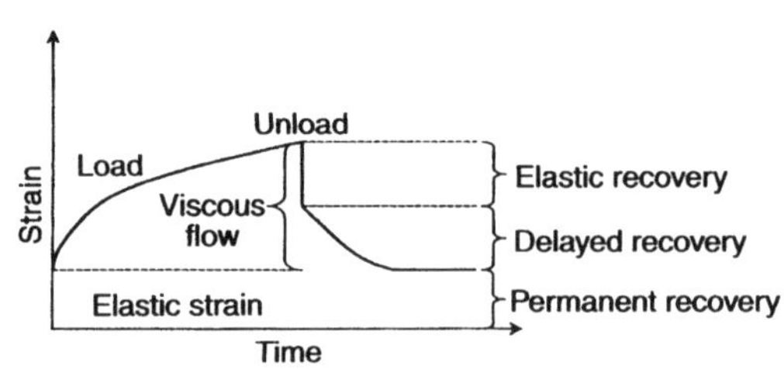

Fig. 17 Schematic representation of viscoelastic behavior of a polymer. Loading produces an immediate elastic strain followed by viscous flow. Unloading produces an immediate elastic recovery followed by additional recovery over a period of time. Source: Ref 14

Because viscoelastic behavior is time dependent, data based on short-term tests may misrepresent the tested polymer in a design application that involves long-term loading. The magnitude of the time dependence of polymers is very temperature dependent. At temperatures well below the glass transition temperature, plastics exhibit a high modulus and are only weakly viscoelastic. At these temperatures, test data based on a time-independent analysis will probably be adequate.

Making use of the time-temperature superposition principle, one may derive long-term design data from a short-term test. This principle states that the mechanical response at long times at some particular temperature is equivalent to the mechanical response at short times but at some higher temperature. By determining shift factors, it is possible to determine which temperature to use in obtaining long-term data from short-term tests.

Crazing

The fracture process in thermoplastics is often dominated by crazing, as shown in the example in Fig. 18. In other words, crack initiation in thermoplastics is preceded by formation of crazes. Whether or not thermosets will craze is inconclusive. It is customarily assumed that thermosets do not craze; that is, their failure mode is all by shear yielding. Geometrically, a craze is a planar defect consisting of bundles of stretched molecules. Crazing begins with microviod formation under the action of the hydrostatic tension component of the stress tensor. A craze is easily visible, especially when viewed at the correct angle with the aid of a directed light source.

Crazing is a mode of plastic deformation rather than mechanical cracking. However, deformation during crazing is constrained laterally; that is, it resists Poisson contraction. Thus, its formation is accompanied by an increase in specimen volume. The upper and lower surfaces of what appears to be a crack in a polymer is not a crack but a craze. These surfaces are interconnected by systems of discrete fibrils in the craze interior, as shown schematically in Fig. 19. Crack surfaces cannot support a load, but craze surfaces can. The density of the material in the craze zone ranges from 40 to 60% of that of the matrix material. For a given polymer and temperature, there is a stress level for craze initiation. A test procedure for determining the stress-to-craze value is specified in ASTM Standard F 484.

After crazing has initiated, the voids increase in size and elongate along the direction of the maximum principal stress. Craze growth occurs by extension of the craze tip into uncrazed material, drawing in new material from the craze flanks. At the same time, the craze thickens by lengthening and disentanglement of the fibrils (Fig. 19, 20). As the craze faces separate while the fibrils increase in length, the fibrils fail at a critical strain, and a true crack is thus formed. Figure 21 schematically illustrates this process.

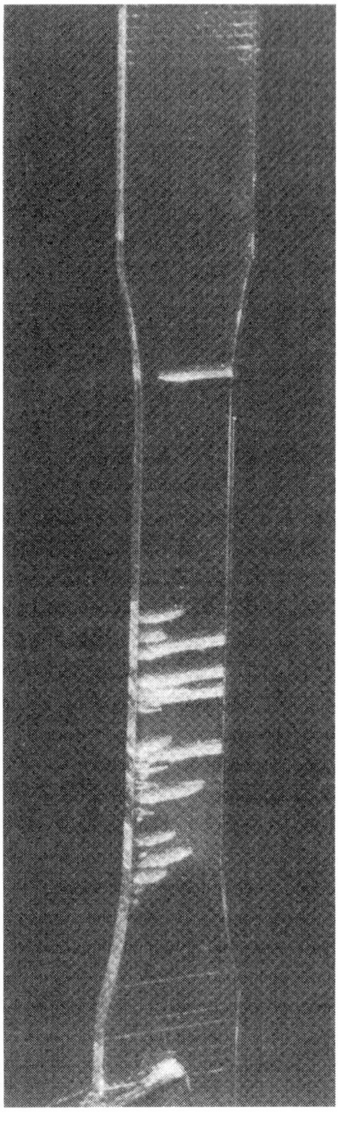

Fig. 18 Craze formation in a polycarbonate polymer in tension under alcohol. Source: Ref 15

Polymer Static Strength

Engineering plastics are not as strong as metals. However, due to their lower density, structural plastics can compete with metals in some applications. The mechanical properties for several common engineering materials are shown for comparison in Table 5. Whether a plastic will behave as brittle or ductile depends on the test temperature or application temperature. A general rule of thumb is that a polymer will be brittle at tem-

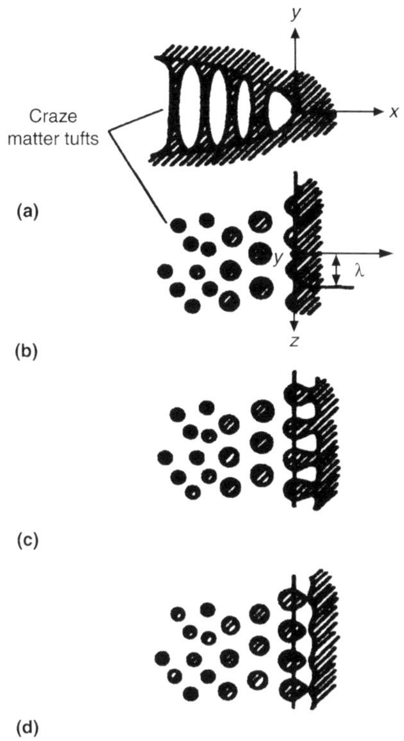

Fig. 19 Schematic of craze formation. (a) Outline of a craze tip and the upper and lower surfaces (side view). (b) Cross section in the craze plane (top view) across craze matter tufts (fibrils). (c) and (d) Advance of the craze front by a completed period of interface convolution. Source: Ref 16

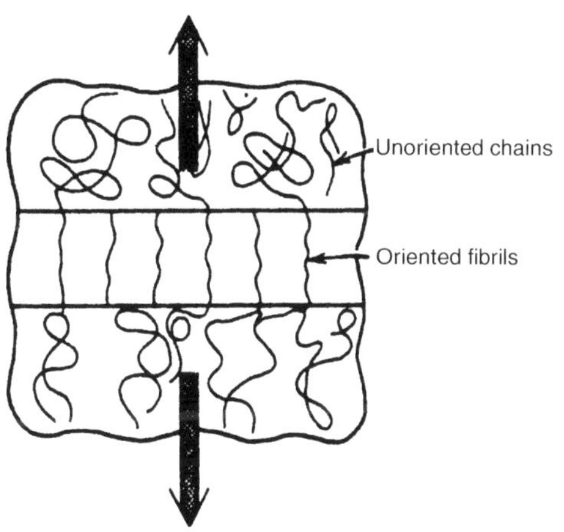

Fig. 20 Schematic showing oriented/lengthened fibrils in between the upper and lower surfaces of a craze. Source: Ref 14

peratures well below its glass transition temperature (T_g); otherwise, it will be ductile. This generalization is based on the fact that the elastic modulus of a plastic abruptly decreases at temperatures near its glass transition temperature.

Figure 22 shows how the modulus of an amorphous polymer changes as temperature increases or decreases through this critical region. At temperatures below T_g, this polymer behaves like a glassy material, with a relatively high modulus. As shown, the same polymer has lower modulus values at temperatures above T_g, and these values continue to decrease as

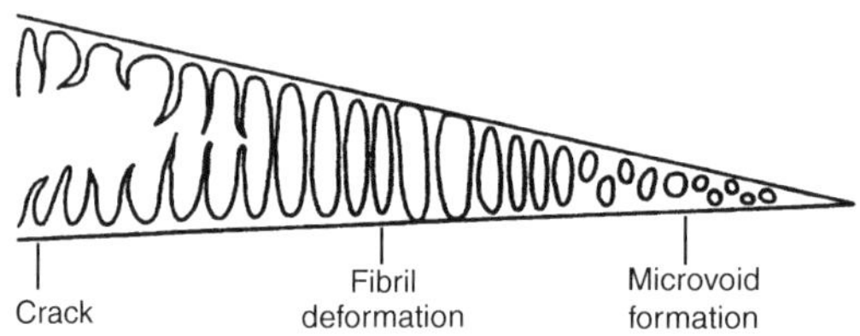

Fig. 21 Schematic crack formation from a craze. Source: Ref 17

Table 5 Range of mechanical properties for common engineering materials

	Elastic modulus		Tensile strength		Maximum strength/density		
Material	GPa	10^6 psi	MPa	ksi	$(km/s)^2$	$(kft/s)^2$	Elongation at break, %
Ductile steel	200	30	350–800	50–120	0.1	1	0.2–0.5
Cast aluminum alloys	65–72	9–10	130–300	19–45	0.1	1	0.01–0.14
Polymers	0.1–21	0.02–3.0	5–190	0.7–28	0.05	0.5	0–0.08
Glasses	40–140	6–20	10–140	1.5–21	0.05	0.5	0
Copper alloys	100–117	15–18	300–1400	45–200	0.17	1.8	0.02–0.65
Moldable glass-filled polymers	11–17	1.6–2.5	55–440	8–64	0.2	2	0.003–0.015
Graphite-epoxy	200	30	1000	150	0.65	1.3	0–0.02

Source: Ref 18, 19

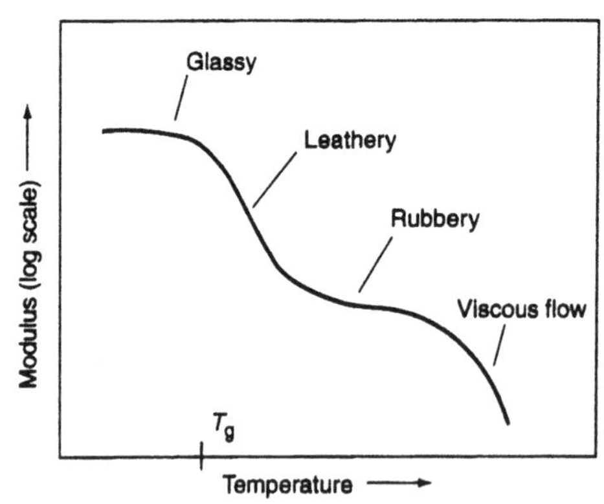

Fig. 22 Schematic modulus versus temperature for a typical amorphous polymer. Source: Ref 14

temperature increases. Its state also changes with increasing temperature—from glassy to leathery, then to rubbery, and finally to a highly viscous liquid. This temperature dependence is generally the same for all types of polymers. The exception is semicrystalline polymers, which will remain in the rubbery state over an extended temperature range. Their transition temperature from rubbery to viscous is higher than for the amorphous polymers. As a result of their strong covalent bonds, cross-linked polymers will not become viscous.

As always, there are exceptions to the rule. The glass transition temperatures for polystyrene (PS) and polymethyl methacrylate (PMMA) are 102 and 107 °C (215 and 225 °F), respectively. According to the glass transition temperature rule, these polymers should be brittle at room temperature—and indeed they are. In contrast, polycarbonate (PC) is ductile at room temperature, even though its glass transition temperature is 150 °C (300 °F). That means that the rule of glass transition temperature does not always work. One possible explanation is that crazing may contribute to brittle behavior. Comparing the minimum crazing stress with the material yield stress of the PC, Petrie (Ref 20) believed that the lower of the two would determine the fracture behavior of the material. Minimum crazing stress is the stress level required for crazing to start. That is, if the minimum craze stress is lower than the tensile yield stress, the polymer will fail in a brittle manner. In other words, if the applied stress reaches the minimum craze stress first, the polymer will be brittle. If the applied stress reaches the yield stress first, as the PC does, the polymer will be ductile. Petrie further states that the minimum crazing stress for PMMA must be lower than the tensile yield stress at ambient temperature, and therefore, brittle behavior prevails.

Many polymers fail in a brittle manner at ambient and low temperatures. Examples of commercial plastics that normally fracture in a brittle manner are PS and PMMA and thermosetting resins such as epoxies and polyesters. On the other hand, polycarbonate (PC), polyethylene (PE), polypropylene (PP), polyamide (PA), polyvinyl chloride (PVC), and polyethylene terephthalate (PET) exhibit ductile fracture behavior. Just like metals, decreasing strain rate and/or increasing temperature will cause a normally brittle material to become ductile (Fig. 23). Thus, it is the combination of temperature, strain rate, and, of course, the material itself. Figure 24(a) shows a group of engineering stress-strain curves (for a plastic) at a constant strain rate but at different temperatures. At low temperatures (e.g., represented by curve 1), the material is completely brittle with an approximately linear relationship between stress and strain. At higher temperatures (e.g., represented by curve 2), a yield point is observed, and the load falls before failure. Sometimes, necking is also observed. However, the breaking strain is still quite low, typically 10 to 20%. At still higher temperatures, approaching the glass transition temperature, the neck grows in a stable manner in some thermoplastics, with material being fed into the neck region from the thicker adjacent regions. This type of deformation in polymers is called drawing or cold drawing. The strains in this case are generally very large and can reach 1000% (curve 3). Finally, at temperatures above the glass transition temperature, the polymer (either thermoset or thermoplastic) undergoes a quasi-rubberlike behavior; the deformations are homogeneous, and the final strain is very large (curve 4).

Actual test data for the PMMA showing the stress-strain behavior at different temperatures are given in Fig. 24(b). Comparing Fig. 24(a) with (b), it is evident that the tests conducted at 4 to 30 °C (39 to 86 °F) exhibit type 1 behavior. The test conducted at 50 °C (122 °F) exhibits type 2 behavior. The test conducted at 60 °C (140 °F) could have behaved like type 2 and probably not quite like type 3, because the test temperature is still much below the material glass transition temperature, which is 107 °C (225 °F).

Comparing the plastic deformation behavior in plastics with metals, both materials neck at temperatures at which both are ductile. The differ-

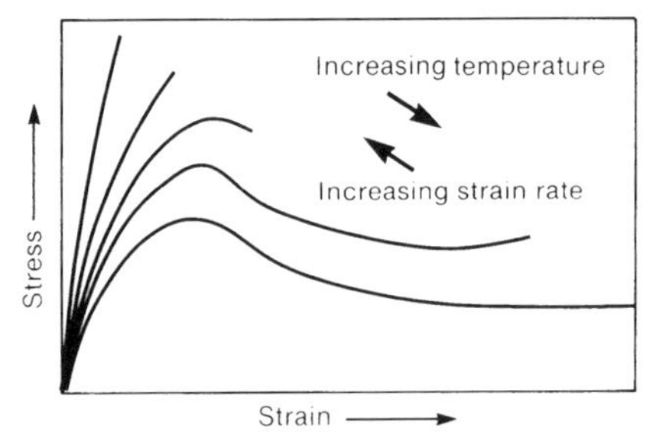

Fig. 23 Schematic tensile stress-strain curves for a plastic, showing the effect of strain rate and temperature. Source: Ref 14

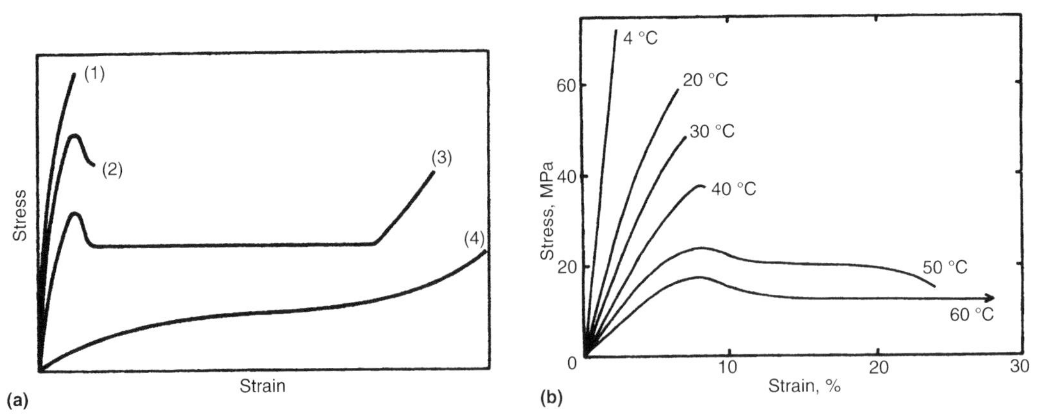

Fig. 24 Tensile engineering stress-strain behavior as a function of temperature. (a) Schematic stress-strain curves for a typical polymer tested at four temperatures while the strain rate is kept constant. 1, low temperature, brittle behavior; 2, intermediate temperature, somewhat ductile; 3, higher temperature (approaching glass transition), necking and cold drawing; 4, above glass transition temperature, homogeneous deformation (quasi-rubberlike behavior). (b) Variation of the stress-strain behavior of polymethyl methacrylate with test temperature. Source: Ref 21

ence is that plastics can sustain much greater strains before fracture. Plastic deformations in metals are evidenced by slip, which takes place along crystallographic planes. However, there is no crystallographic plane in plastics. Even for crystalline polymers, which are inevitably semicrystalline, the plastic deformation mechanism is not clearly identified with some kind of slip system. Therefore, the shear bands formed in polymers are generally more diffused or delocalized than those in metals. The potential for extensive elongation after necking is a consequence of the long chain nature of polymers.

Chains in the necked area can uncoil and align. If the necking can maintain long enough, the chains will become completely aligned. At this point, any added elongation will begin to stretch the bonds in the polymer chain. This stretching will eventually lead to failure.

Polymer Fatigue

Traditional fatigue analysis methods such as those used for metals can be applied to polymers. However, polymers have a significant complication in terms of hysteresis heating and subsequent softening. The temperature of a polymer body will rise during fatigue cycling. Therefore, polymers are highly susceptible to frequency effects. The higher the loading frequency, the lower the fatigue endurance limit. For a given stress level, fatigue life (number of cycles to failure) decreases with increase in cyclic frequency, due to greater heat buildup at higher frequencies. The size of a structural piece also contributes to the extent of thermal heating. For example, for two smooth round bars of the same length, the one with a larger diameter will undergo a greater temperature increase because its outer surface-area-to-volume ratio is smaller. That is, the larger piece heats up more but has a relatively smaller outer surface to release the heat. For the same reason, a flat specimen that is thicker tends to generate more heat, which leads to a greater temperature rise under a given set of test conditions.

Higher stress levels also cause more heat buildup in plastics. The effect of stress level at a given load frequency is shown in Fig. 25. A conventional *S-N* curve can be plotted for this set of test data by connecting the fracture points, marked F, and the fatigue threshold will be the stress level marked U. In fatigue testing of plastics, both mechanical factors (applied stress, *R*-ratio, notch geometries) and thermal factors contribute to the outcome of a test.

Thermohysteresis is why most (if not all) polymers exhibit cyclic strain softening in their cyclic stress-strain curves (Fig. 26). Metals exhibit either cyclic strain hardening or softening. However, all the polymers shown in Fig. 26 exhibit cyclic strain softening regardless of their chemical structure (i.e., semicrystalline, amorphous, or two phase). Note that the yield points that were developed in monotonic tension were eliminated during cyclic stabilization. Moreover, polymer fatigue behavior is generally sen-

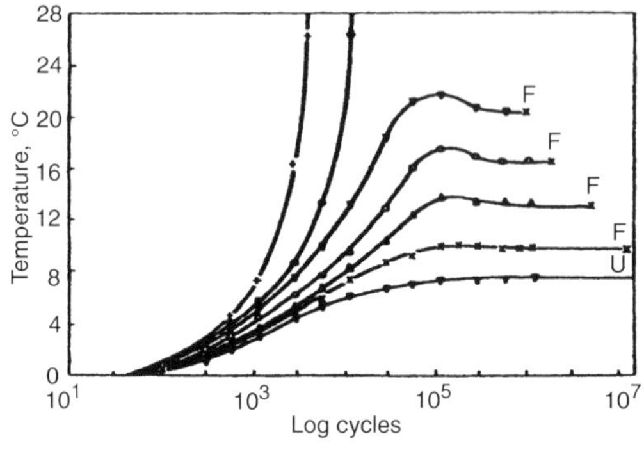

Fig. 25 Temperature increase resulting from uniaxial cycling at 5 Hz with load control, sine wave, and zero mean stress in polyacetal. U, unbroken; F, fracture. The applied stress levels are, counterclockwise starting from U: 15, 16, 17.4, 19.7, 21.6, 22.4, and 27.8 MPa (2.2, 2.3, 2.5, 2.9, 3.1, 3.2, and 4 ksi). Source: Ref 22

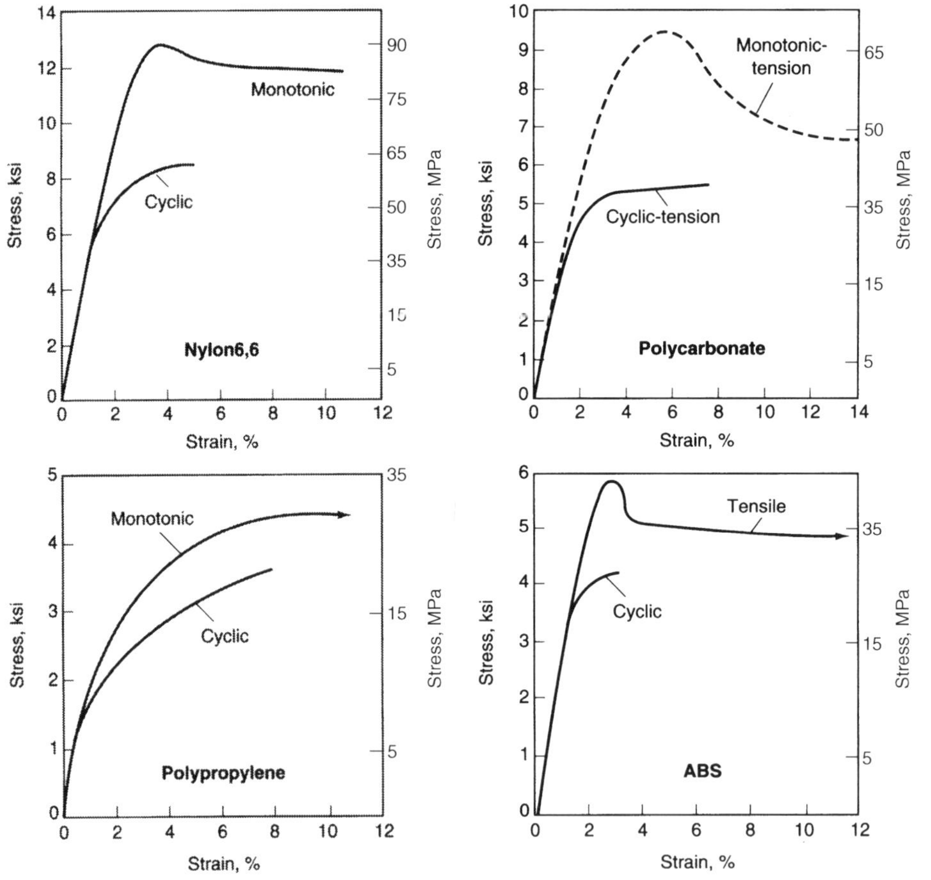

Fig. 26 Comparison of cyclic and monotonic stress-strain curves for several polymers at 25 °C (77 °F). ABS, acrylonitrile-butadiene-styrene. Source: Ref 23

sitive to environment (including test temperature), molecular weight, molecular density, and aging. Additionally, for polymer crystals, fatigue resistance increases as crystallinity increases (Fig. 27). *S-N* curves that do not account for these effects should not be used exclusively without looking at test conditions. In the case of metals, some exhibit an endurance limit below which no fatigue failure will occur. On the basis of experimental observation, we know that the lattice structures of metals are the source for their endurance limits, or the lack thereof. Like metals, polymers may or may not exhibit a traditional fatigue endurance limit (Fig. 28); however, the reason is not known. For those data shown in Fig. 28,

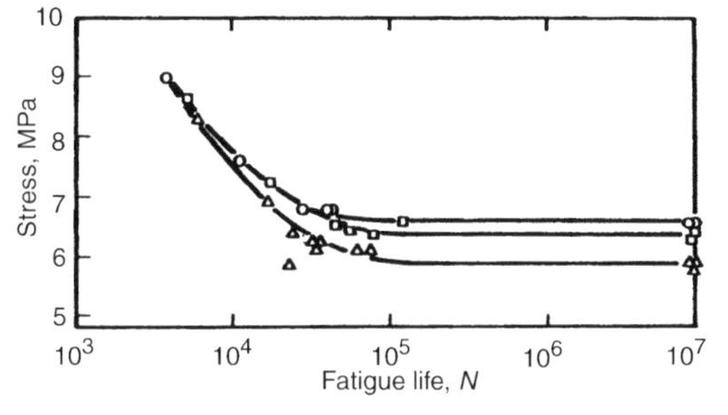

Fig. 27 Fatigue life in polytetrafluoroethylene with increasing crystallinity. Δ, low crystallinity, air quenched; □, medium crystallinity, 33.3 °C/h cooling; O, high crystallinity, 5.6 °C/h cooling. Test frequency, 30 Hz. *R*-ratio not identified. Source: Ref 24

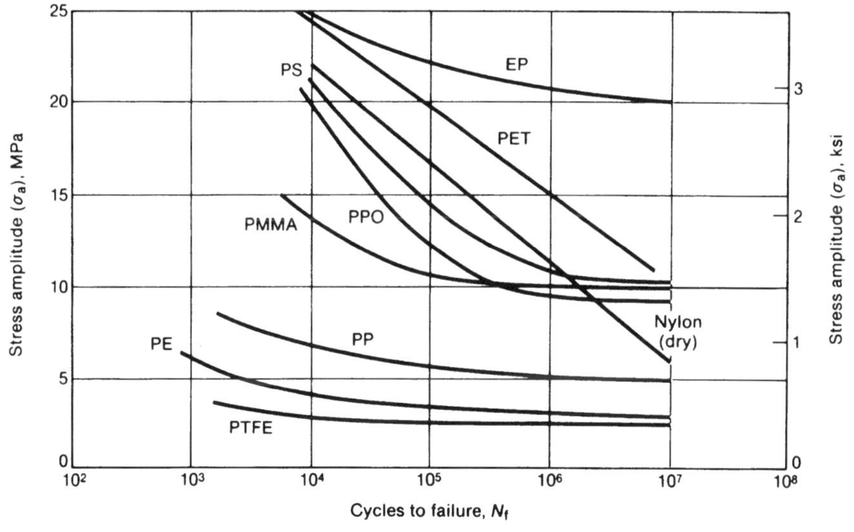

Fig. 28 Typical *S-N* curves for several commodity plastics at 30 Hz. Note: *R*-ratio is missing in this figure. PS, polystyrene; EP, epoxy; PET, polyethylene terephthalate; PMMA, polymethyl methacrylate; PPO, polyphenylene oxide; PE, polyethylene; PP, polypropylene; PTFE, polytetrafluoroethylene. Source: Ref 25

only nylon and polyethylene terephthalate (PET) do not exhibit an endurance limit.

Fatigue Crack Propagation

Polymers display *da*/*dn*-versus-ΔK behavior similar to that in metals. They also display the same frequency effect as in metals; that is, the higher the loading frequency, the lower the fatigue crack growth rate (Fig. 29). This behavior is directly opposite the *S-N* fatigue behavior discussed earlier. One possible reason is that temperature buildup in this case is concentrated in a local area at the crack tip. The material surrounding the crack tip and the free surfaces of the crack may help to dissipate heat from the crack tip. This rationale is in some way supported by the data in Fig. 25, which shows that temperature in each specimen leveled off at the final stage of the test. It could mean that heating was discontinued after crack initiation, or at least when the crack became longer. So, for a fatigue crack growth rate test where a fatigue crack has been inserted into the specimen right from the beginning, temperature buildup during the test may be minimal. Therefore, although some heating does occur at the crack tip, an excessive temperature rise is prevented; that is, fatigue crack propagation is

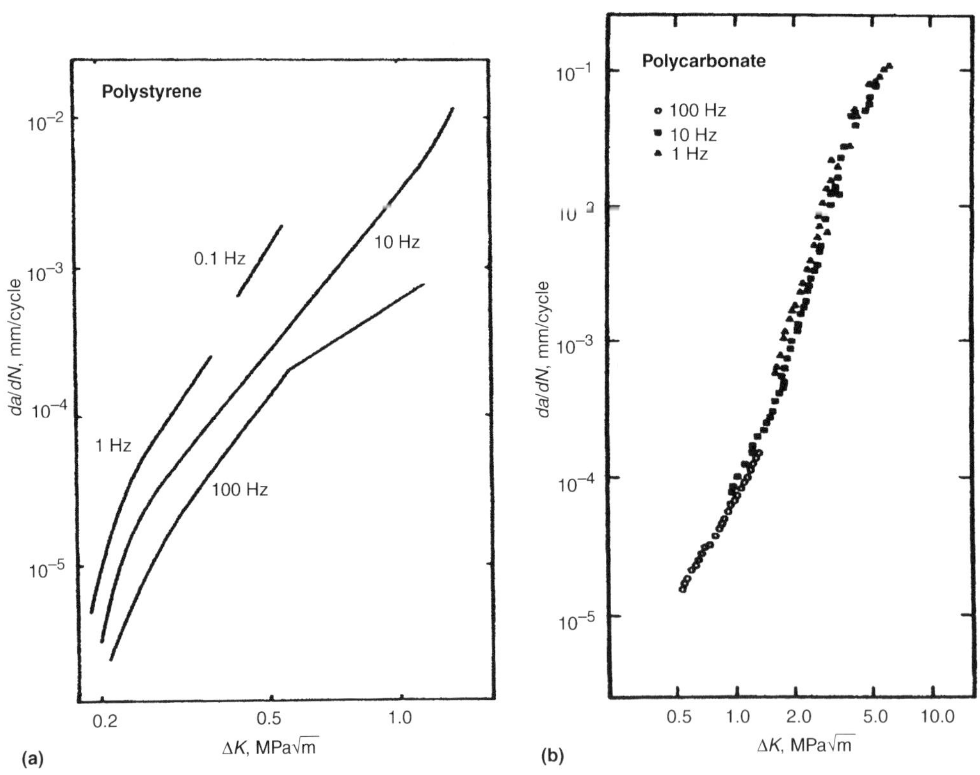

Fig. 29 Effect of cyclic frequency on fatigue crack growth rates in (a) polystyrene and (b) polycarbonate. Source: Ref 26

primarily a mechanical and time-dependent process. Observe in Fig. 29 that PC is much less sensitive to mechanical load frequencies than PS. An explanation has been offered by Hertzberg and Manson (Ref 26), who say that polymers that are prone to crazing are more susceptible to cyclic frequency effect, with PMMA and PS being examples.

The effect of mean stress on fatigue crack growth rate is another matter. Some polymers exhibit a trend similar to that in metals; that is, the higher the R ratio, the higher the fatigue crack growth rate. Some other polymers exhibit just the opposite behavior. In Ref 27, Pruitt examined a vast amount of test data from a number of sources. The results, which identify and classify the polymers examined into two categories, are listed in Table 6.

The appearance of fatigue striations in polymers is similar to that in metals, with one minor exception. Electron fractographs for PMMA and other polymers show that each striation contains a fine linear structure oriented normal to the striation line itself (Fig. 30). Such lines may reflect evidence of excessive material tearing during the striation formation process. Overall, there appears to be no change in morphology from one side of the striation to the other. In theory, as in metals, a striation has a one-to-one correlation with an applied stress cycle. That is, for a given stress

Table 6 Effect of increasing mean stress on polymer fatigue crack propagation

Increasing da/dN with increasing mean stress	Decreasing da/dN with increasing mean stress
High-density polyethylene	Low-density polyethylene
Nylon	Polyvinyl chloride
High-molecular-weight PMMA	Low-molecular-weight PMMA
Polystyrene	Rubber-toughened PMMA
Epoxy	High-impact polystyrene
Polyethylene copolymer	Acrylonitrile-butadiene-styrene
	Polycarbonate
	Toughened polycarbonate copolyester

PMMA, polymethyl methacrylate. Source: Ref 27

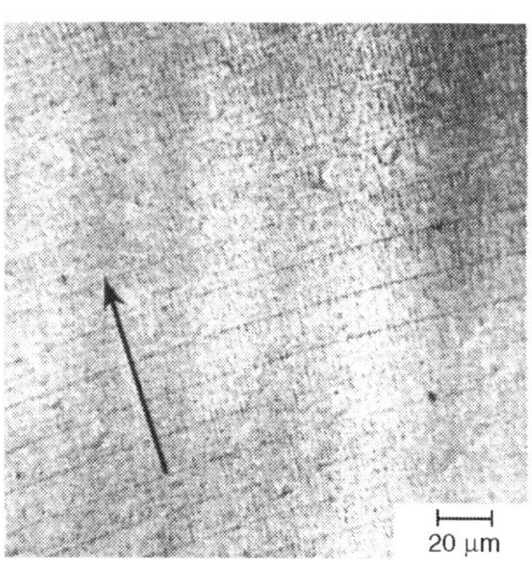

Fig. 30 Fatigue striations in polymethyl methacrylate. Arrow indicates crack growth direction. Source: Ref 28

cycle, the microscopic striation spacing and the macroscopically measured *da*/*dn* should have the same magnitude.

Figure 31 shows comparisons for striations in three types of polymers, thus substantiating that striations can be found on the fracture surfaces of polymers. However, under certain test conditions, fatigue cracks in some polymers progress in a discontinuous manner. The striation-like parallel markings in Fig. 32 are called discontinuous growth bands. Each of these bands is associated with a large number of loading cycles—as many as several hundred. Some polymers exhibit only discontinuous growth bands on the fracture surface (e.g., polyvinyl chloride); some have both (e.g., PC and PS). Examples of discontinuous bands for six different polymers are shown in Fig. 33.

For fatigue crack propagation in polymers, the discontinuous growth band is analogous to the crack tip plastic zone in metals. The mechanism of the discontinuous growth band is described by a model proposed by Hertzberg and Manson (Ref 26). It simply states that the craze at the tip will grow with time during fatigue cycling. Meanwhile, the fatigue crack front does not move. As mentioned earlier, when some critical condition is satisfied (e.g., the craze grows to a critical size with many broken fibrils),

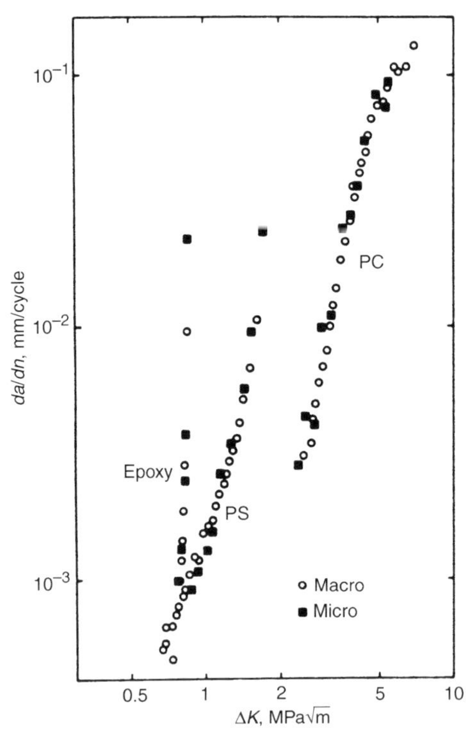

Fig. 31 Comparison of macroscopic fatigue crack growth rates and striation spacing measurements in epoxy, polycarbonate (PC), and polystyrene (PS). □, micro (measured from electron fractograph); O, macro (physical measurement). Source: Ref 26

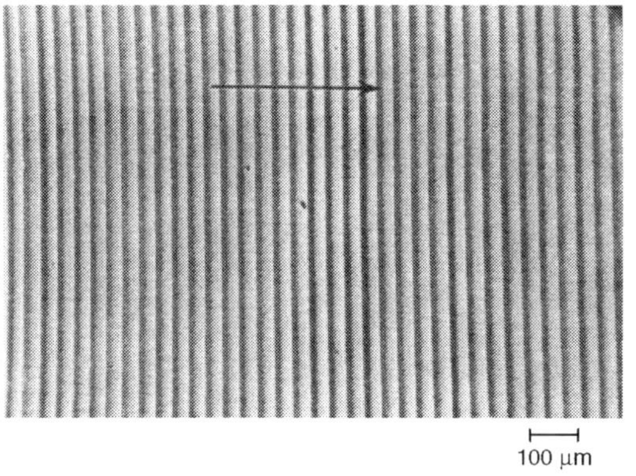

Fig. 32 Parallel discontinuous growth bands in polystyrene. Arrow indicates crack growth direction. Source: Ref 26

the craze becomes a crack and joins with the existing crack. Put another way, the existing crack will jump through the entire broken craze and arrest at the tip of the craze. This can be considered a new crack length. The period that the old crack sat waiting for the craze to grow accounts for the accumulated load cycles.

The series of photographs in Fig. 34(a) demonstrates this discontinuous cracking process, with the arrows in each photograph pointing to the crack tip and craze tip locations. The first four photographs show that while the craze tip advances during fatigue cycling, the crack tip stays at its original position. The last photograph shows that the crack tip moved into the old craze tip location and a new craze tip appeared. A model of the sequence involving continuous craze growth and discontinuous crack growth is shown in Fig. 34(b).

Impact Strength and Fracture Toughness

Fracture toughness of polymers is generally low. The K_{Ic} value for a standard epoxy is as low as 0.454 MPa$\sqrt{m}$ (0.413 ksi$\sqrt{in.}$) for the Hexcel 3501-6 resin. The K_{Ic} values for thermoplastics are much higher. ICI 450G (polyetheretherketone resin) has a K_{Ic} value greater than 6.6 MPa$\sqrt{m}$ (6 ksi$\sqrt{in.}$). Polycarbonate falls in between; its values as functions of moisture and loading rate are shown in Fig. 35. More K_{Ic} values for common engineering plastics at room temperature are given in Table 7. As mentioned previously, ductility of plastics generally increases with temperature. Therefore, fracture toughness and impact strength also increase with temperature. On the basis of impact strength, some common plastics are rated for their brittleness versus temperature (Table 8).

Ductile-to-brittle transition behavior in thermoplastics is very much the same as in metals. Some S-shaped curves that resemble those seen in met-

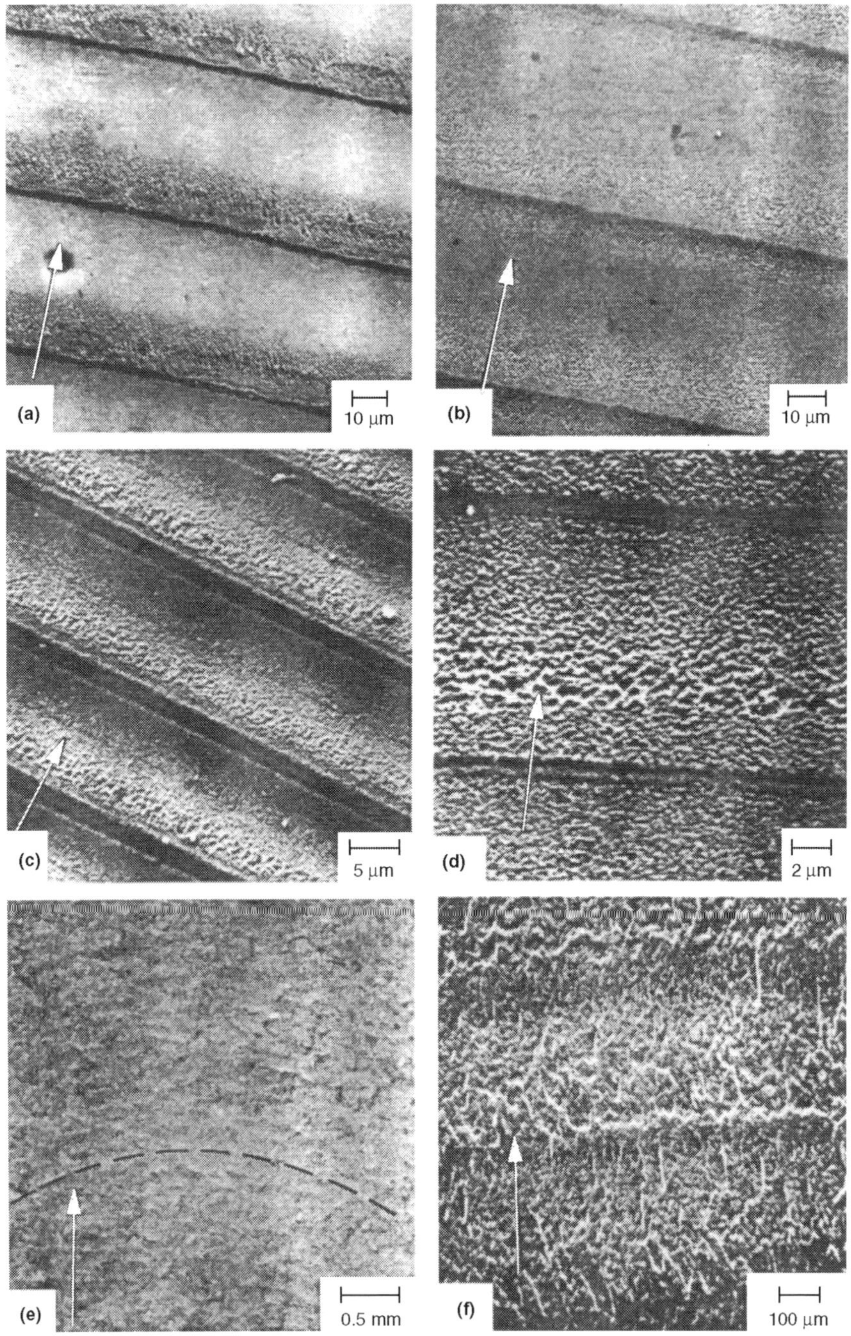

Fig. 33 Scanning electron microscope fractographs of discontinuous growth bands in (a) polyvinyl chloride, (b) polystyrene, (c) polysulfone, (d) polycarbonate, (e) polyamide, and (f) acrylonitrile-butadiene-styrene. Arrows indicate crack growth direction. Source: Ref 28

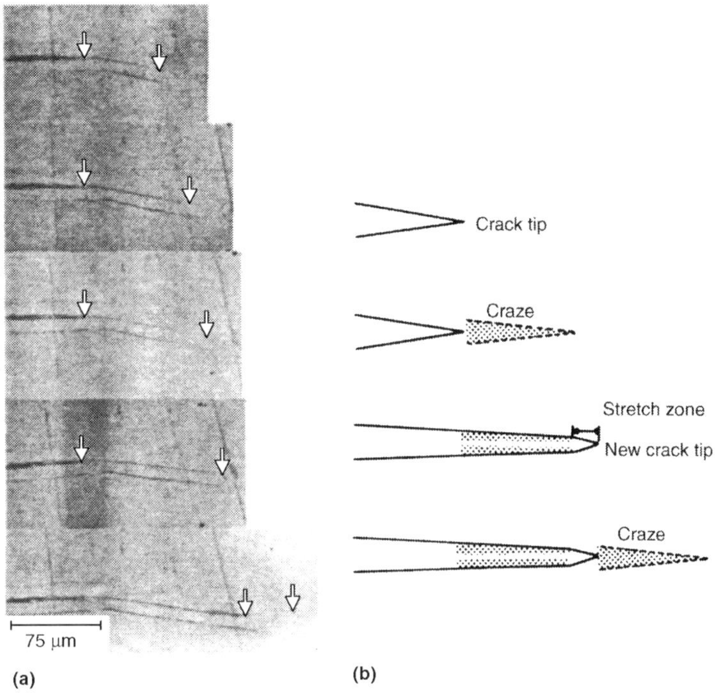

Fig. 34 Discontinuous crack growth process. (a) Composite micrograph of polyvinyl chloride showing positions of crack tip (arrows on the left) and craze tip (arrows on the right) at given cyclic intervals. (b) Model of discontinuous cracking process. Source: Ref 26

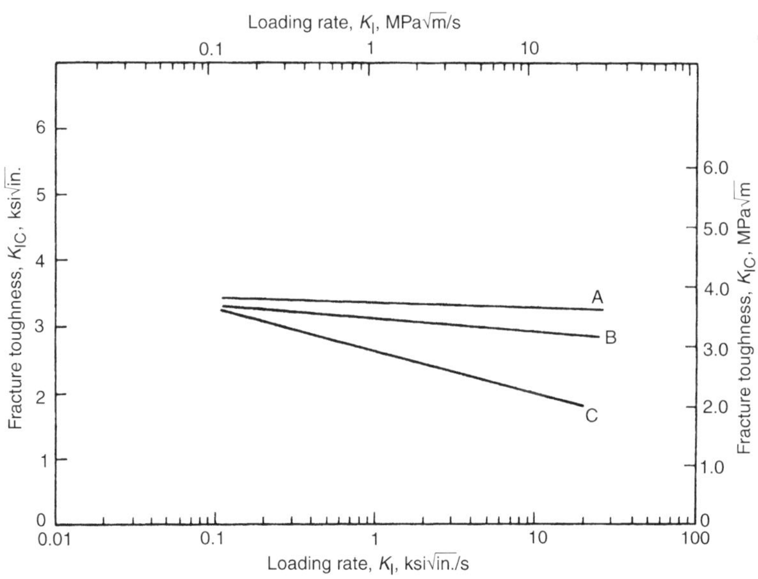

Fig. 35 Fracture toughness values of 13 mm (0.5 in.) thick polycarbonate specimens as functions of environment and loading rate. The lines A, B, and C denote ambient, dry, and wet, respectively. Source: Ref 29

als are shown in Fig. 36 and 37. The K_Q values for a rubber-toughened thermoplastic as a function test temperature are shown in Fig. 36. The K_Q values were developed using compact specimens. The exact material was not identified; nonetheless, a ductile-to-brittle transition behavior is evident. The data in Fig. 37, developed from Izod impact tests, show that impact strength for PS increases at 100 °C (212 °F), which is the material glass transition temperature. According to the glass transition temperature rule discussed earlier, PS is brittle at temperatures below its glass transition temperature and ductile above it. The second material in Fig. 37,

Table 7 Typical K_{Ic} and Izod impact strength values for engineering plastics

	K_{Ic}		Izod impact strength	
Material	MPa$\sqrt{m}$	ksi$\sqrt{in.}$	J/m	ft · lbf/in.
PMMA	0.7–1.6	0.6–1.5	...	...
PS	0.7–1.1	0.6–1.0	13.3–21.3	0.25–0.4
High-impact PS	1.0–2.0	0.9–1.8	...	...
PE	1.0–6.0	0.9–5.5	...	...
ABS	2.0	1.8	213.6	4.0
PC	2.2	2.0	747.6	14
PVC	2.0–4.0	1.8–3.6	...	...
PI	...	...	0.9	0.02
PP	3.0–4.5	2.7–4.1	...	...
PET	5.0	4.5	...	...
PSU	...	...	69.4	1.3
PEI	...	...	53.4	1.0
PEEK	...	...	96.1	1.8
PAI	...	...	133.5	2.5
PTFE	...	...	107–125	2.0–2.4
Nylon 6/6	...	...	53.3–160	1.0–3.0
Nylon 6	...	...	53.3–160	1.0–3.0
Nylon 6/12	...	...	53.3–74.6	1.0–1.4
Nylon 1	...	...	96	1.8
Epoxy	0.6	0.5	...	...
Polyester	0.6	0.5	...	...

Source: Ref 30, 31

Table 8 Fracture behavior of plastics as a function of temperature

	Temperature, °C (°F)							
Plastics	–20 (–4)	–10 (14)	0 (32)	10 (50)	20 (68)	30 (85)	40 (105)	50 (120)
Polystyrene	A	A	A	A	A	A	A	A
Polymethyl methacrylate	A	A	A	A	A	A	A	A
Glass-filled nylon (dry)	A	A	A	A	A	A	A	B
Polypropylene	A	A	A	A	B	B	B	B
Polyethylene terephthalate	B	B	B	B	B	B	B	B
Acetal	B	B	B	B	B	B	B	B
Nylon (dry)	B	B	B	B	B	B	B	B
Polysulfone	B	B	B	B	B	B	B	B
High-density polyethylene	B	B	B	B	B	B	B	B
Rigid polyvinyl chloride	B	B	B	B	B	B	C	C
Polyphenylene oxide	B	B	B	B	B	B	C	C
Acrylonitrile-butadiene-styrene	B	B	B	B	B	B	C	C
Polycarbonate	B	B	B	B	C	C	C	C
Nylon (wet)	B	B	B	C	C	C	C	C
Polytetrafluoroethylene	B	C	C	C	C	C	C	C
Low-density polyethylene	C	C	C	C	C	C	C	C

A. brittle even when unnotched; B, brittle, in the presence of a notch; C, tough. Source: Ref 32

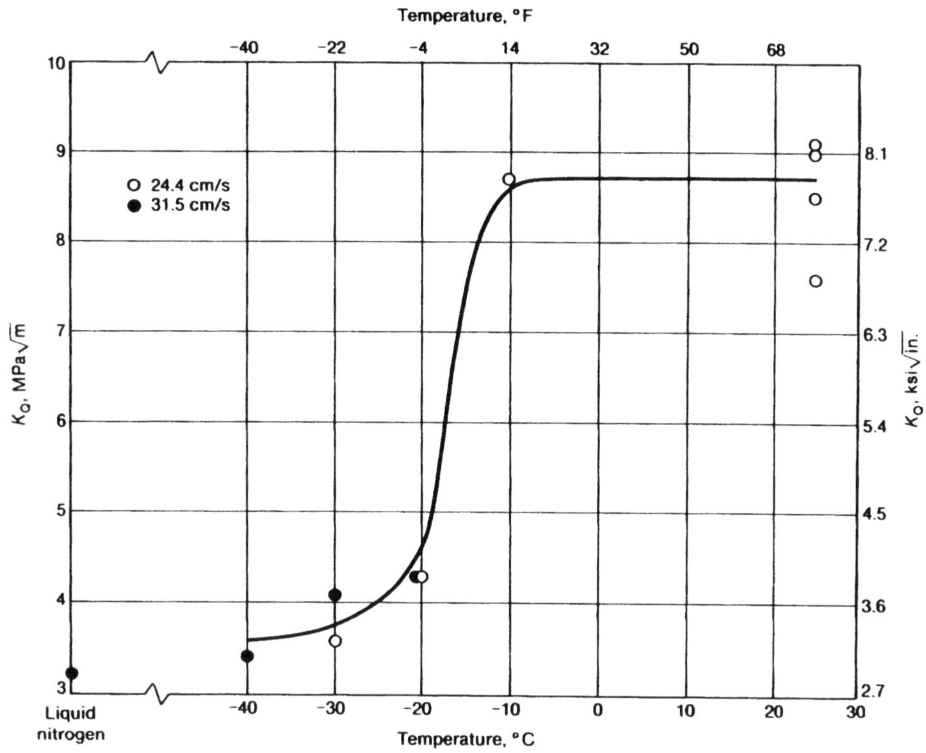

Fig. 36 Fracture toughness (K_Q) as a function of temperature for a rubber-toughened thermoplastic material. Source: Ref 32

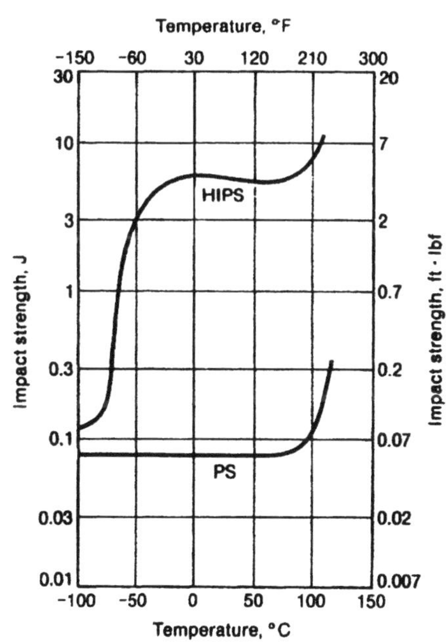

Fig. 37 Izod impact strength as a function of temperature for polystyrene (PS) and high-impact polystyrene (HIPS) thermoplastics. Source: Ref 33

high-impact polystyrene (HIPS), has two transition temperatures. HIPS consists of a second phase of individual rubber particles embedded in a PS-matrix phase. At temperatures below the T_g of rubber, the impact strength is very low. At temperatures above the T_g of rubber, the rubber phase helps to increase the impact strength by absorbing some energy during fracture. An additional increase is observed at 100 °C (212 °F), which is the T_g for the base material. Reference 33 does not further elaborate on this phenomenon. We can only speculate that at temperatures below 100 °C (between 70 to 100 °C, or 95 to 212 °F), the material is not completely brittle.

Puncture testing is another often-used indicator of impact resistance for plastics. The manner of fracture in PC is compared in Fig. 38, with Fig. 38(a) exhibiting a ductile failure and Fig. 38(b) a more brittle failure.

Stress Rupture

Stress-rupture behavior in plastics is significantly different from that in metals. A stress-versus-time curve for plastics (at a given temperature) is shown schematically in Fig. 39. In region I, the plastic undergoes continuous plastic deformation under a constant load. During that time, the specimen cross-sectional area gradually reduces. Final fracture of a specimen occurs when the specimen can no longer support the applied load and thus is ductile. At some lower stress levels and after a long hold time at load, the specimen surface shows signs of damage and perhaps cracking. When that happens, it takes little time to break the piece (region II), or the part breaks almost immediately (region III). The transitions that take place are indicated by the kinks in the graph. It is almost stress-level independent in region III. The switching is quite abrupt. In the last two regions, the fracture behavior is brittle. The surface damage may be very mild, or the crack

(a) (b)

Fig. 38 Comparison of failed polycarbonate disks from puncture tests. (a) Ductile failure at room temperature. (b) Brittle failure at −90 °C (−130 °F). Source: Ref 32 depth may be very shallow, but that is all it takes to change the fracture mode. Quoting the work of Gedde and Ifwarson, Miller (Ref 34) has stated that region II may not exist in some materials. The test data of Gedde and Ifwarson (on cross-linked PE) showed that PE of normal cross-linked density exhibits fracture behavior in all three regions. In contrast, region II is missing in PE of low cross-linked density. In the sample test data (for an unidentified PE) shown in Fig. 40, the tests apparently stopped somewhere

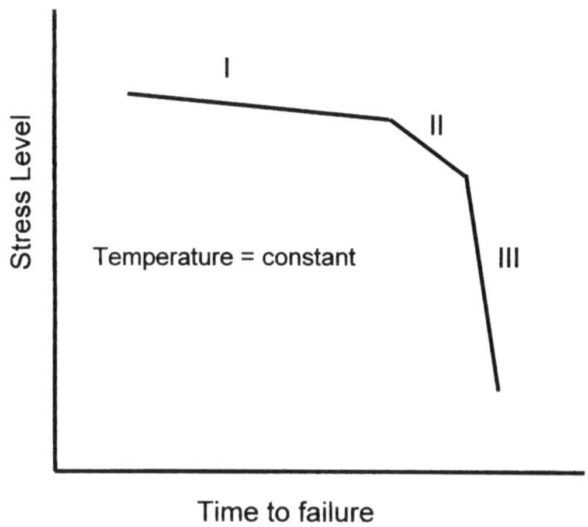

Fig. 39 Schematic stress-rupture behavior of plastics. Source: Ref 14

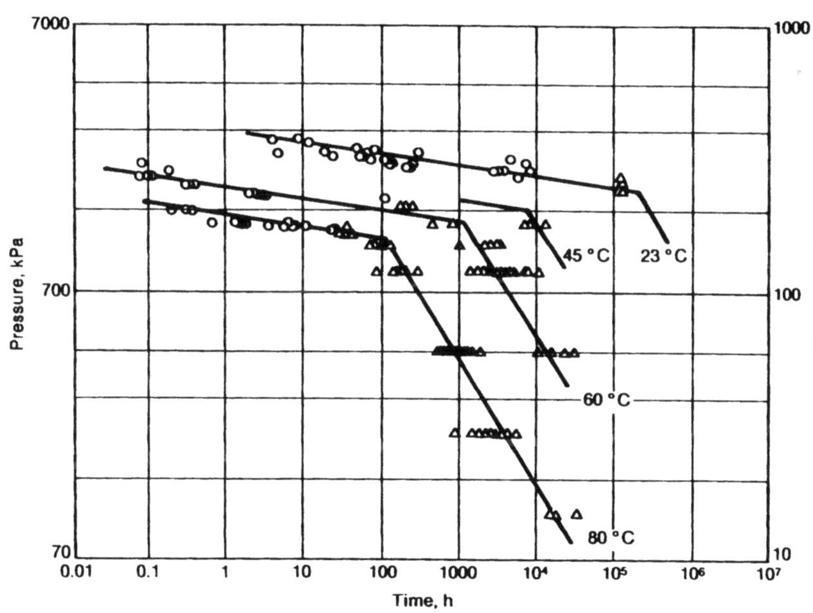

Fig. 40 Effect of internal pressure on time to failure of polyethylene gas pipe at various temperatures, showing ductile (region I) and brittle (region II) behaviors. See Fig. 39 for definition of regions. Source: Ref 35

in region II. This figure also shows that the PE plastic underwent creep and stress rupture at room temperature.

It is interesting to note that the ductile-to-brittle transition behavior is not limited to stress-rupture testing of plastics. For the same reason—that is, exposure to high temperature (or corrosive medium, or radiation) for a long time to accumulate surface damage—the plastic will become brittle in a short-term tension test. Whether or not the fracture strength of a damaged specimen degrades, the reduction in elongation is significant.

Fractography

A typical fracture surface of a PC specimen after impact is shown in Fig. 41. For polymers, the mirror region is the result of nucleation and growth of crazes. The mist region, commonly found in glass, is also observed in some glassy polymers, such as PMMA, PS, and PC. This region is typically flat, smooth, and featureless, except for a slight change in surface texture resembling a fine mist. In polymers, the mist regions are not necessarily confined to the vicinity of the fracture origin. The hackle lines are indications of the final stage of the fracture process. The divergent nature of these lines is useful in locating the crack origin. The Wallner lines, which are absent from the fracture surfaces of ceramics and glasses, are not fatigue striations, nor are they true crack front markings. They form when reflected stress waves intersect a propagating crack. The curved crack front appearance of these markings is another useful feature that helps to locate the fracture origin, which is on the concave side of these markings.

Environmental Performance of Plastics

The mechanical properties of polymeric materials are often segregated into short-term and long-term properties. The category of short-term prop-

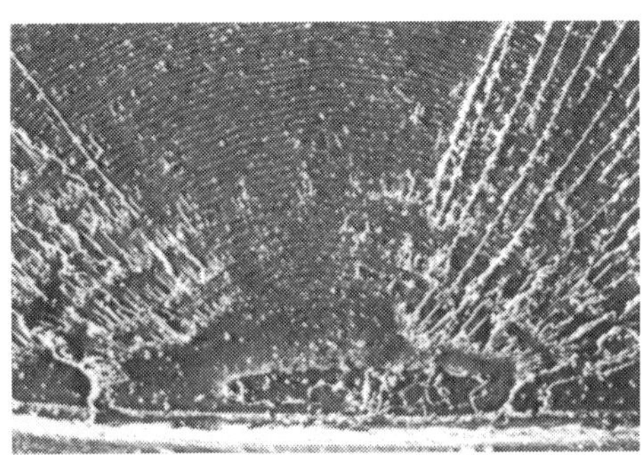

Fig. 41 Fracture surface of a polycarbonate specimen after Izod impact showing the mirror, mist, and hackle regions, along with Wallner lines that spread over the mist and hackle regions. Original magnification: 41×. Source: Ref 17

erties includes such things as tensile and impact strengths. Long-term properties include creep, stress relaxation, and creep (stress) rupture. Both categories of properties are affected by exposure to external chemical environments.

With any polymeric material, chemical exposure may have one or more different effects. Some chemicals act as plasticizers, changing the polymer from one that is hard, stiff, and brittle to one which is softer, more flexible, and tougher. Often these chemicals can dissolve the polymer if they are present in large enough quantity and if the polymer is not cross linked. Other chemicals can induce environmental stress cracking (ESC), an effect in which brittle fracture of a polymer will occur at a level of stress well below that required to cause failure in the absence of the ESC reagent. Finally, there are some chemicals that cause actual degradation of the polymer, breaking the macromolecular chains, reducing molecular weight, and diminishing polymer properties as a result.

Plasticization, Solvation, and Swelling

Certain interactions between liquid chemicals and polymers can be understood through the use of solubility parameters. The Hildebrand solubility parameter, H, is the square root of the cohesive energy density, the latter being the energy required to vaporize 1 mol of a liquid. When linear or branched thermoplastic polymers are exposed to large enough quantities of solvents having solubility parameters within approximately ± 2 H of that of the polymer, dissolution of the polymer will occur. In smaller quantities, these solvents will be adsorbed by the polymer. With polymer-solvent combinations having solubility parameter differences outside this range, some adsorption of the solvent by the polymer may still occur. When large differences between solvent and polymer solubility parameters exist, the solvent will have no apparent effect. Amorphous polymers absorb chemicals more readily than semicrystalline polymers, and the rate varies inversely with the degree of crystallinity. Cross-linked polymers will not dissolve but will swell significantly when exposed to chemicals having similar solubility parameter values.

The impact of these interactions on the mechanical properties and failure of an affected polymer are many. One effect can be plasticization of the polymer by the adsorbed chemical. Plasticization of a polymer can result in the polymer being transformed from a rigid, glassy material to a soft, flexible material. The plasticization effect reduces both tensile strength and stiffness of the affected plastic and also accelerates the creep rate of the material if it is under stress. If an application of a particular plastic requires a certain minimum level of strength or stiffness, unintentional plasticization by exposure to a chemical that the plastic can adsorb could accelerate failure by reducing those properties.

Plasticizer loss from an intentionally plasticized polymer may also have an adverse effect on polymer performance. Plasticizer migration from polyvinyl chloride (PVC) is a well-known phenomenon that results not only in embrittlement and/or a loss of flexibility of the PVC part but also loss of the plasticizer chemical(s) into the environment. Plasticizer migration from flexible PVC products also results in some small amount of shrinkage of the products. This may be problematic if close dimensional tolerances are necessary.

These effects will occur in both amorphous and semicrystalline plastics, but they may not be as visibly evident in semicrystalline ones. Semicrystalline plastics often do not appear to be as affected by interactions with solvents, because diffusion of solvent into the crystalline regions is much more limited. However, adsorption of solvents into the amorphous regions of a semicrystalline polymer will create the discussed effects within those regions of the polymer morphology; that is, this can result in changes in polymer mechanical properties. For example, nylon plastics will absorb moisture from the air. An extremely dry nylon may be rather brittle, while that same nylon exposed to 50% relative humidity for several days can be quite tough. However, the short-term tensile strength and modulus of the hydrated nylon will be somewhat reduced, as will the long-term (creep rupture) strength of the material. Creep deformation of the hydrated nylon will proceed more rapidly than that in the dry material at the same stress level.

Environmental Stress Cracking

Environmental stress cracking of plastics is defined as the failure in surface-initiated brittle fracture of a specimen or part under polyaxial stress in contact with a medium in the absence of which fracture does not occur under the same conditions of stress. Virtually all plastics are stress cracked by some chemical environments. The biggest problem with this is that each plastic has its own set of stress-cracking reagents, and those chemicals that stress crack one type of plastic will have no effect on others. Thus, the potential stress-cracking effect of a specific chemical on a specific plastic must be known from prior work or determined by experiment to know whether or not a problem exists.

Several physical characteristics are typical of ESC failures:

- Failure is always nonductile, even in plastics that would normally exhibit a ductile yielding failure mechanism.
- The brittle fracture is surface initiated.
- The surface on which cracking initiated was in contact with a chemical reagent.
- The plastic was mechanically stressed in some way; both internal (residual) stresses or externally applied stresses qualify.

Chemicals that induce ESC usually have no other apparent effect on the plastic in question, that is, no swelling (or dissolution in large quantities of the chemical) and no physical or chemical changes in the polymer that may be detected by analytical methods. In the absence of a mechanical stress, the ESC chemical has no discernible effect. Conversely, the magnitude of stress that will cause ESC will not cause fracture if imposed in the absence of the stress-cracking reagent. Thus, conventional chemical resistance tests run on plastics, in which unstressed tensile bars are soaked in a chemical and withdrawn periodically for testing, give no indication of the possibility of ESC for any polymer-reagent system. It is only in the presence of both mechanical stress and chemical environment that ESC occurs.

Failures from ESC may occur early or late in the life of a product. In some cases, ESC will occur as soon as a part is loaded, if the reagent is already present on the surface of a previously unstressed part. Stress-cracking reagents also reduce the creep-rupture properties of plastics by shortening the time for brittle fracture to occur over that which exists in the absence of the reagent. Sometimes ESC reagents will create brittle fracture at a low stress level in a polymer such as polyethylene that normally fails in a highly ductile manner. High-density polyethylene exhibits ductile failure (elongations to break of several hundred percent) at stresses near to its yield stress. At lower stresses and longer failure times, a different molecular mechanism controls failure, and brittle fracture occurs at elongations of less than 5%. Presence of an ESC reagent on the surface of a plastic can dramatically shorten the time for failure to occur at a given stress level and change the failure mechanism from highly ductile to macroscopically brittle.

Polymer Degradation by Chemical Reaction

Another effect that chemical environments can have on plastics is to actually degrade the polymer, that is, to break down the polymer chains into lower-molecular-weight compounds that no longer have the original strength or toughness properties. Certain polymer types are more susceptible than others to specific degradation mechanisms, but all polymers can be degraded by at least one mechanism. The most common degradation mechanisms are discussed in the following paragraphs.

Hydrolysis. Polymers created by stepwise reactions, for example, polyesters and nylons, form water as a reaction product along with the polymer. Under certain circumstances of exposure to aqueous environments, the polymerization reaction can essentially be reversed and the polymer broken down. Normally, these hydrolytic degradation reactions occur at extremely slow rates, and certainly nylon and polyester fabrics can be repeatedly washed in water without adverse effects. However, at conditions of either low (<4) or high (>10) pH, the rate of hydrolysis may become perceptible and result in molecular weight reduction and mechanical property losses.

If there is moisture in the material, these polymer types can also degrade during processing (i.e., extrusion or injection molding). Even at neutral pHs, the elevated temperatures used for polymer molding or extrusion (175 to 250 °C, or 350 to 500 °F, or more) will cause hydrolytic degradation if there is moisture in the resin. Because of this, polymer resin manufacturers advise drying of the material just prior to processing to reduce the moisture content to a low enough level that hydrolysis will not occur while the resin is heated in the manufacturing equipment. It is often the case with plastics that are susceptible to hydrolytic degradation that a reduced polymer molecular weight is found during failure analysis. The challenge for the analyst then becomes deciding whether the degradation occurred during fabrication of the part or on exposure to an aqueous service environment.

Thermal Degradation. High-molecular-weight polymers will also break down on exposure to elevated temperatures. Sufficient thermal energy can be input to a polymeric material to break the covalent chemical bonds that hold polymer molecules together. This bond breakage (chain scission), if it occurs in the polymer backbone, will reduce molecular weight. Chain scission of side chain branches may also alter the polymer structure sufficiently to change appearance or mechanical properties enough to create a premature failure.

As with hydrolysis, thermal degradation can occur both in processing and in an end-use environment. In molding or extrusion operations, the molten plastic is exposed not only to elevated temperatures but also to mechanical shearing. The combination of the two may reduce molecular weight to the extent that performance properties will suffer. Thermally degraded plastics also tend to discolor, and sometimes a plastic product is deemed a failure because it no longer has the desired cosmetic appearance due to thermal degradation. These same changes may be observed in end-use environments, albeit at much slower rates due to the lower temperatures at which plastics are normally used.

Thermal degradation of polymers is a chain reaction that begins when an atom (usually a hydrogen atom) is abstracted from the polymer chain, leaving behind an unpaired electron from the broken covalent bond at an atom on the chain. The free radical thus formed may react in several different ways, one of which results in chain scission and molecular weight reduction. Fortunately for plastics usage, there are chemical additives compounded into polymers that will react with the unpaired electron, interrupt the chain reaction, and postpone or at least greatly retard thermal decomposition. If thermal degradation is believed to be a contributing factor to failure, the type and amount of these additives should be checked to be certain that failure was not due to degradation in an unprotected polymer material.

Oxidation. Many polymers, especially the olefins and others with long olefinic segments in the polymer chain backbone, will oxidize when exposed to oxygen-containing environments. As with thermal degradation, oxidation usually commences by formation of a free radical on the polymer chain. An oxygen atom from the environment will then react with the unpaired electron to form a hydroperoxy radical. This will then degrade by one of several reactions, some of which result in chain scission and property loss.

As with thermal degradation, chemical additive stabilizers and antioxidants can be added to the polymer that will break the chain reaction in a variety of ways, preserving polymer properties at least until the additives have been consumed. If oxidative degradation is a possible contributing factor to a premature failure, it becomes necessary to determine what allowed it to occur. It may be that the stabilizers were not present originally in the proper types or amounts. If the polymer resin did contain antioxidants, then it must be determined what caused them to become ineffective. In some cases, stabilizers can "bloom" to the surface of a plastic part and be removed by ablation, dissolution, or evaporation into the environment. In other cases, the additives may simply have been consumed doing the job for which they were intended, and premature oxidation occurred because the service environment was at a higher temperature than the design engineer anticipated. All these can lead to an oxidized polymer with reduced mechanical properties, unacceptable appearance, or other deficiencies.

Photodegradation. With many polymeric materials, ultraviolet (UV) radiation can be the source of energy that will abstract an atom from the polymer backbone and start the degradation process. It is well known that prolonged outdoor exposure of plastics will initially cause color changes that may be undesirable. Oxidation initiated by UV radiation will result in eventual loss of properties as well. Once again, there are chemical additives that will retard these processes, but eventually they will be consumed and degradation will proceed. The plastics design engineer must be certain that the UV radiation stabilizers are present in the proper types and amounts to yield a product that will operate for its intended life without undergoing an inordinate amount of degradation from exposure. There are both accelerated indoor and outdoor test methods that are used to assess the level of stability to UV exposure of a plastic material.

Surface Embrittlement

An adverse effect of polymer degradation on plastic part performance does not require changes in the bulk of the material in that part. In many cases, it is only necessary to cause degradation in a thin surface layer of the part in order for performance to be compromised. This phenomenon has been observed in both the short- and long-term properties of many polymeric materials.

In fact, polymer degradation is often limited, at least initially, to the surfaces of exposed plastic products. Oxidative degradation initiated by either purely thermal means or by UV radiation occurs initially at the surfaces, because that is where oxygen concentration is the greatest. For oxidation to occur deeper in a specimen, it must diffuse in. Because oxygen reacts very rapidly with free radical species, oxidation below the surface of a polymer part is diffusion limited and occurs very slowly compared to surface oxidation. Hydrolytic degradation also occurs first and most rapidly at surfaces, because that is where the concentration of water is the greatest. Since often all that is needed for premature failure to occur is to generate a sufficient level of degradation at the surface, even if the bulk material within the plastic part is unaffected, the fact that degradation is initially limited to the surfaces creates problems for product performance.

The impact of surface degradation on short-term properties has been demonstrated by many authors. Numerous studies have shown that considerable reductions in tensile strength, impact strength, and toughness have been observed for oxidation degradation extending only a short distance into a specimen. This effect in polyethylene is shown in Fig. 42. It has been demonstrated that the creep-rupture behavior of a polyethylene pipe resin could be compromised by a certain level of oxidative degradation occurring only in the first 50 μm below the surface of a 2.5 mm (0.1 in.) thick specimen. Figure 43 illustrates how surface degradation of a plain-strain tension specimen alters the ductile-to-brittle transition in polyethylene creep rupture.

ACKNOWLEDGMENTS

Major portions of this chapter came from *Mechanics and Mechanisms of Fracture: An Introduction* by A.F. Liu, ASM International, 2005, and "Effect of Environment on the Performance of Plastics" by D.E. Duvall,

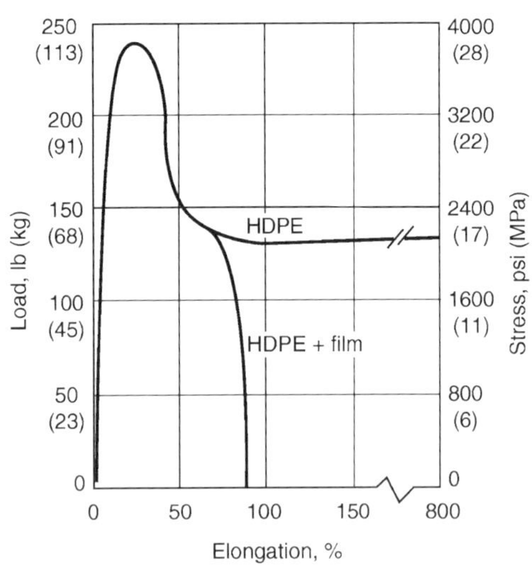

Fig. 42 Effect of thin, brittle film on stress-strain behavior of high-density polyethylene (HDPE). Source: Ref 36

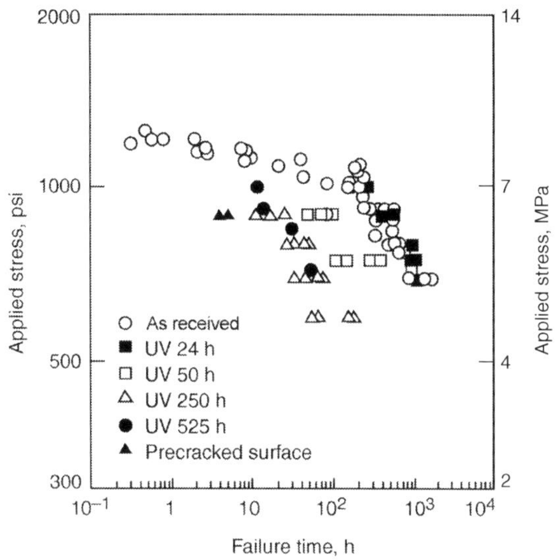

Fig. 43 Effect of surface embrittlement from varied ultraviolet (UV) exposure times on creep-rupture behavior of polyethylene at 80 °C (175 °F). Source: Ref 37

Failure Analysis and Prevention, Vol 11, *ASM Handbook*, ASM International, 2002.

REFERENCES

1. F.C. Campbell, Metallic Structure, *Elements of Metallurgy and Engineering Alloys,* ASM International, 2008
2. I. Birkby, Ed., *Ceramic Technology International,* Sterling Publications Ltd., 1992, p 58
3. V.A. Greenhut, Effects of Composition, Processing, and Structure on Properties of Ceramics and Glasses, *Materials Selection and Design,* Vol 20, *ASM Handbook,* ASM International, 1997
4. F.C. Campbell, Ceramic Matrix Composites, *Structural Composite Materials,* ASM International, 2010
5. S.W. Freiman and S.M.Wiederhorn, Fracture Mechanics Applied to Ceramics, *Fracture Mechanics,* University Press of Virginia, 1978, p 299–316
6. R.L. Lehman, Overview of Ceramic Design and Process Engineering, *Ceramics and Glasses,* Vol 4, *Engineered Materials Handbook,* ASM International, 1991, p 30
7. M. Jenkins and J. Salem, Fracture Resistance Testing of Brittle Solids, *Mechanical Testing,* Vol 8, *ASM Handbook,* ASM International, 2000
8. D.J. Lee, Simple Method to Measure the Crack Resistance of Ceramic Materials, *J. Mater. Sci.*, Vol 30 (No. 8), 1995, p 4617–4622

9. R.H. Dauskardt, D.B. Marshall, and R.O.Ritchie, *J. Am. Ceram. Soc.,* Vol 73 (No. 4), 1990, p 893–903
10. F.C. Campbell, Matrix Resin Systems, *Structural Composite Materials,* ASM International, 2010
11. R.B. Prime, Chapter 5, *Thermal Characterization of Polymeric Materials,* E.A. Turi, Ed., Academic Press, 1981
12. F.C. Campbell, Engineering Plastics, *Lightweight Materials—Understanding the Basics,* ASM International, in publication
13. Mechanical Testing of Polymers, *Engineered Materials Handbook Desk Edition,* ASM International, 1995
14. A.F. Liu, *Mechanics and Mechanisms of Fracture: An Introduction,* ASM International, 2005
15. I.M. Ward, *Mechanical Properties of Solid Polymers,* 2nd ed., John Wiley, New York, 1983
16. A.S. Argon, J.G. Hannoosh, and M.M. Salama, Initiation and Growth of Crazes in Glassy Polymers, *Fracture 1977,* Vol 1, ICF4 (Waterloo), 1977, p 445–470
17. Fracture of Plastics, *Failure Analysis and Prevention,* Vol 11, *ASM Handbook,* ASM International, 2002, p 650–661
18. Mechanical Testing of Polymers and Ceramics, *Mechanical Testing and Evaluation,* Vol 8, *ASM Handbook,* ASM International, 2000, p 26–48
19. H.A. Kuhn, Overview of Mechanical Properties and Testing for Design, *Mechanical Testing and Evaluation,* Vol 8, *ASM Handbook,* ASM International, 2000, p 49–69
20. S.P. Petrie, Crazing and Fracture, *Engineering Plastics,* Vol 2, *Engineered Materials Handbook,* ASM International, 1985, p 734–740
21. E.H. Andrews, *Fracture in Polymers,* Oliver & Boyd, 1968
22. R.J. Crawford and P.P. Benham, *Polymer,* Vol 16, 1975, p 908
23. P. Beardmore and S. Rabinowitz, *Treatise on Material Science and Technology,* Vol 6, 1975, p 267
24. M.N. Riddle, G.P. Koo, and J.L. O'Toole, *Polym. Eng. Sci.,* Vol 6, 1966, p 363
25. M.N. Riddle, *Plast. Eng.,* Vol 30, 1974, p 71
26. R.W. Hertzberg and J.A. Manson, *Fatigue of Engineering Plastics,* Academic Press, 1980
27. L. Pruitt, Fatigue Testing and Behavior of Plastics, *Mechanical Testing and Evaluation,* Vol 8, *ASM Handbook,* ASM International, 2000, p 758–767
28. M.D. Skibo, R.W. Hertzberg, J.A. Manson, and S. Kim, *J. Mater. Sci.,* Vol 12, 1977, p 531
29. S.A. Sutton, J. Tirosh, R.W. Thomas, P.W. Mast, and I. Wolock, The Effect of Loading Rate, Temperature and Moisture on the Fracture Toughness of Polycarbonate, *Proceedings of the 27th National SAMPE Symposium* (San Diego), 1982, p 1003–1021

30. J.M. Margolis, Thermoplastic Resins, *Engineering Plastics,* Vol 2, *Engineered Materials Handbook,* ASM International, 1988, p 618–625
31. S.P. Petrie, Crazing and Fracture, *Engineering Plastics,* Vol 2, *Engineered Materials Handbook,* ASM International, 1988, p 734–740
32. R. Nimmer, Impact Loading, *Engineering Plastics,* Vol 2, *Engineered Materials Handbook,* ASM International, 1988, p 679–700
33. L.R. Pinckney, Phase-Separated Glasses and Glass-Ceramics, *Engineering Plastics,* Vol 2, *Engineered Materials Handbook,* ASM International, 1988, p 433–438
34. E. Miller, *Introduction to Plastics and Composites,* Marcel Dekker, 1996
35. C.G. Bragaw, in *Eighth Plastic Fuel Pipe Symposium,* American Gas Association, 1983, p 40
36. L.J. Broutman, "Surface Embrittlement of Polyethylene," GRI-81-0030, Final Report to the Gas Research Institute, Chicago, Nov 1981
37. S.-W. Choi, "Surface Embrittlement of Polyethylene," Ph.D. dissertation, Illinois Institute of Technology, Chicago, 1992

CHAPTER 10

Fatigue and Fracture of Continuous-Fiber Polymer-Matrix Composites

A COMPOSITE MATERIAL can be defined as a combination of two or more materials that results in better properties than when the individual components are used alone. As opposed to metallic alloys, each material retains its separate chemical, physical, and mechanical properties. The two constituents are the reinforcement and the matrix. The main advantages of composite materials are their high strength and stiffness, combined with a low density, when compared with bulk materials, allowing for a weight reduction in the finished part.

The reinforcing phase provides the strength and stiffness. In most cases, the reinforcement is harder, stronger, and stiffer than the matrix. The reinforcement is usually a fiber or a particulate. Particulate composites have dimensions that are approximately equal in all directions. They may be spherical, platelets, or any other regular or irregular geometry. Particulate composites tend to be much weaker and less stiff than continuous-fiber composites but are usually much less expensive. Particulate-reinforced composites usually contain less reinforcement (up to 40 to 50 vol%) due to processing difficulties and brittleness.

A fiber has a length that is much greater than its diameter. The length-to-diameter (l/d) is known as the aspect ratio and can vary greatly. Continuous fibers have long aspect ratios, while discontinuous fibers have shorter aspect ratios. Continuous-fiber composites normally have a preferred orientation, while discontinuous fibers generally have a random orientation. Examples of continuous reinforcements (Fig. 1) include unidirectional, woven cloth, and helical winding, while the discontinuous reinforcements shown are chopped fibers and random mat. Continuous-fiber composites are often made into laminates by stacking single sheets of continuous fibers in different orientations to obtain the desired strength and stiffness properties, with fiber volumes as high as 60 to 70%. Fibers produce high-strength composites because of their small diameter. They contain much fewer defects—normally surface defects—than if the material were produced in bulk. As a general rule, the smaller the diameter of the fiber, the higher its strength, but often the cost increases as the diameter becomes smaller. In addition, smaller-diameter high-strength fibers have greater flexibility and are more amenable to fabrication processes such as weaving or forming over radii. Typical fibers include glass, aramid, and carbon, which may be continuous or discontinuous.

The continuous phase is the matrix, either a polymer, metal, or ceramic. Polymers have low strength and stiffness, metals have intermediate strength and stiffness but high ductility, while ceramics have high strength and stiffness but are brittle. The matrix, or continuous phase, performs several critical functions. It maintains the fibers in the proper orientation and spacing. It protects them from abrasion and the environment. In polymer- and metal-matrix composites that form a strong bond between the fiber and the matrix, the matrix transmits loads from the matrix to the fibers through shear loading at the interface. In ceramic-matrix composites, the

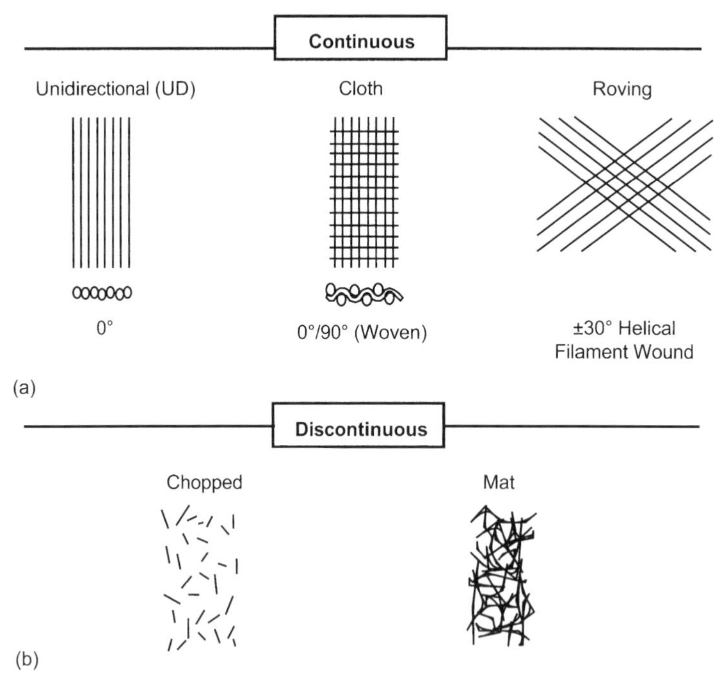

Fig. 1 Typical reinforcement options. Source: Ref 1

objective is usually to increase the toughness rather than the strength and stiffness, and a low-interfacial-strength bond is therefore desirable.

The type and quantity of the reinforcement determines the final properties. As shown in Fig. 2, the highest strength and modulus are obtained with continuous-fiber composites. There is a practical limit of approximately 70 vol% reinforcement that can be added to form a composite. At higher percentages, there is too little matrix to effectively support the fibers. While the theoretical strength of discontinuous-fiber composites can approach those of continuous-fiber composites if their aspect ratios are great enough and they are aligned, it is difficult in practice to maintain good alignment with discontinuous fibers. Discontinuous-fiber composites are normally somewhat random in alignment, which dramatically reduces their strength and modulus. However, discontinuous-fiber composites are generally much less costly than continuous-fiber composites. Therefore, continuous-fiber composites are used where higher strength and stiffness are required, but at a higher cost, and discontinuous composites are used where cost is the main driver and strength and stiffness are less important.

Both the type of reinforcement and the matrix affect processing. The two types of polymer matrices are thermosets and thermoplastics. A thermoset starts with a low-viscosity resin that reacts and cures during processing, forming an intractable solid, while a thermoplastic is a high-viscosity resin that is processed by heating it to above its melting temperature. Because a thermoset resin sets up and cures during processing, it cannot be reprocessed by reheating. On the other hand, thermoplastics can be reheated above their melting temperature for additional processing. For

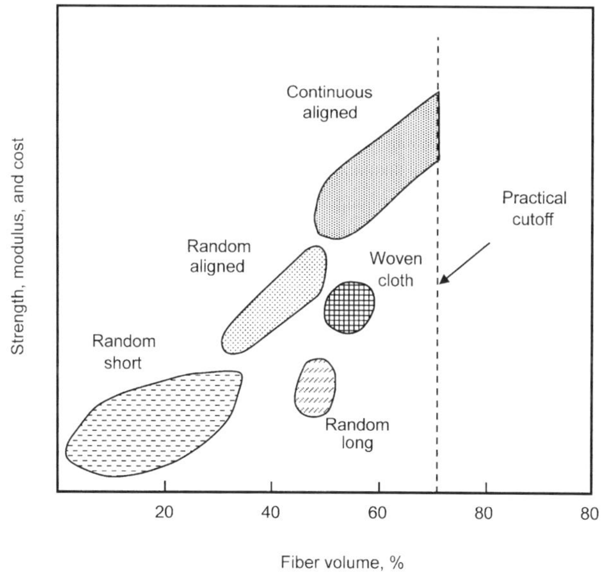

Fig. 2 Influence of reinforcement type and quantity on composite performance. Source: Ref 1

both classes of resins, there are some processes that are more amenable to discontinuous fibers and some that are more amenable to continuous fibers. In general, because metal- and ceramic-matrix composites require very high temperatures and sometimes high pressures for processing, they are normally much more expensive than polymer-matrix composites. However, they have much better thermal stability, a requirement where the application is exposed to high temperatures.

This chapter deals with continuous-fiber polymer-matrix composites, with an emphasis on continuous carbon-fiber composites, which offer the highest structural efficiencies.

Laminates

When dealing with a single ply or a lay-up in which all of the layers or plies are stacked in the same orientation, the lay-up is defined as a lamina, while when the plies are stacked at various angles, the lay-up is called a laminate. Continuous-fiber composites are normally laminated materials (Fig. 3) in which the individual layers, plies, or laminae are oriented in directions that will enhance the strength in the primary load direction. Unidirectional (0°) laminae are extremely strong and stiff in the 0° direction; however, they are also very weak in the 90° direction because the load must be carried by the much weaker polymeric matrix. While a high-strength fiber can have a tensile strength of 3450 MPa (500 ksi) or more, a typical polymeric matrix normally has a tensile strength of only 35 to 70 MPa (5 to 10 ksi), as illustrated in Fig. 4. The longitudinal tension and compression loads are carried by the fibers, while the matrix distributes the loads between the fibers in tension and stabilizes and prevents the fi-

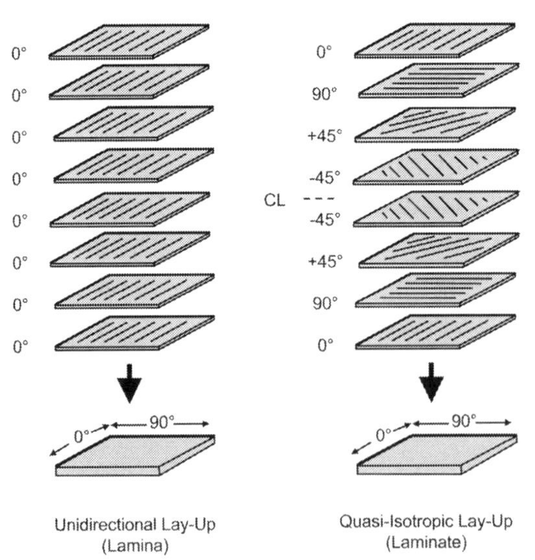

Fig. 3 Lamina and laminate lay-ups. Source: Ref 1

bers from buckling in compression. The matrix is also the primary load carrier for interlaminar shear (i.e., shear between the layers) and transverse (90°) tension.

Because the fiber orientation directly impacts the mechanical properties, it would seem logical to orient as many of the layers as possible in the main load-carrying direction. While this approach may work for some structures, it is usually necessary to balance the load-carrying capability in a number of different directions, such as the 0°, +45°, –45°, and 90° directions. A micrograph of a cross-plied continuous carbon-fiber/epoxy laminate is shown in Fig. 5. A balanced laminate with equal numbers of plies

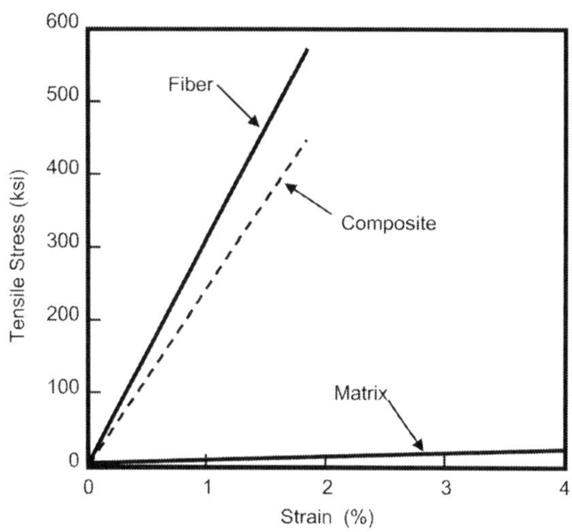

Fig. 4 Tensile properties of fiber, matrix, and composite. Source: Ref 1

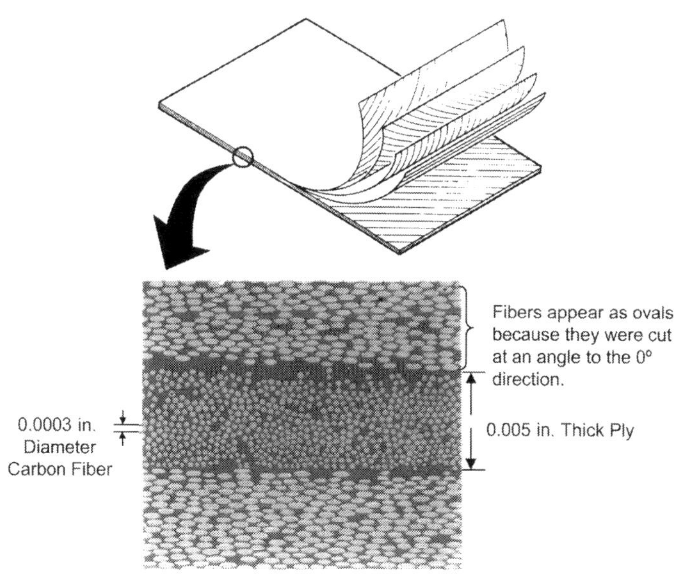

Fig. 5 Laminate construction. Source: Ref 1

in the 0°, +45°, –45°, and 90° directions is called a quasi-isotropic laminate, because it will carry equal loads in all four directions.

Composites versus Metallics

The physical characteristics of composites and metals are significantly different. A comparison of some of the properties of composites and metals are given in Table 1. Because composites are highly anistropic, the in-plane strength and stiffness are usually high and directionally variable, depending on the orientation of the reinforcing fibers. Properties that do not benefit from this reinforcement, at least for polymer-matrix composites, are comparatively low in strength and stiffness. An example of these is the through-the-thickness tensile strength, where the relatively weak matrix is loaded rather than the high-strength fibers. The low through-the-thickness strength of a typical composite laminate compared to aluminum is shown in Fig. 6.

Metals normally have reasonable ductility. When they reach a certain load level, they continue to elongate or compress considerably without

Table 1 Composites vs. metals comparison

Condition		Comparative behavior relative to metals
Load-strain relationship		More linear strain to failure
Notch sensitivity	Static	Greater sensitivity
	Fatigue	Less sensitivity
Transverse properties		Much weaker
Mechanical property variability		Higher
Fatigue strength		Higher
Sensitivity to hydrothermal environment		Greater
Sensitivity to corrosion		Much less
Damage growth mechanism		In-plane delaminations instead of through-thickness cracks

Source: Ref 2

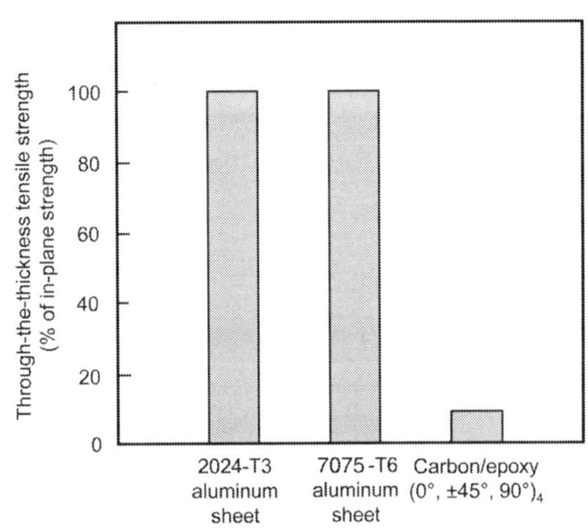

Fig. 6 Through-the-thickness tensile strength comparison. Source: Ref 3

picking up more load and without failure. This ductile yielding has two important benefits. It provides for local load relief by distributing excess load to adjacent material or structure. Consequently, ductile metals have a great capacity to provide relief from stress concentrations when statically loaded. Second, the ductility of metals provides great energy-absorbing capability, as is indicated by the area under a stress-strain curve. As a result, when impacted, a metal structure will typically deform but not actually fracture. In contrast, composites are relatively brittle. A comparison of typical tensile stress-strain curves for the two materials is shown in Fig. 7. The brittleness of the composite is reflected in its poor ability to tolerate stress concentrations, as shown in Fig 8. The characteristically brittle

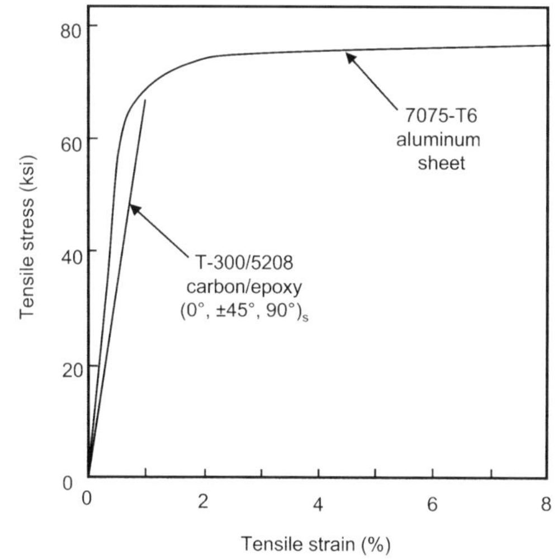

Fig. 7 Stress-strain comparison. Source: Ref 3

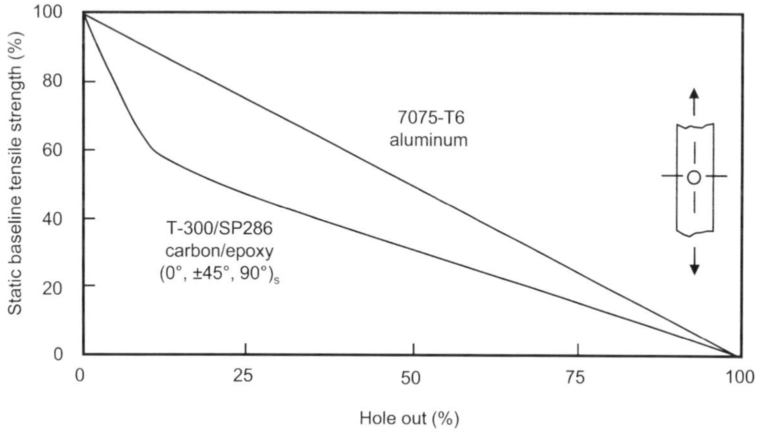

Fig. 8 Stress-concentration comparison. Source: Ref 3

composite material has poor capability to resist impact damage without extensive internal matrix fracture.

The response of damaged composites to cyclic loading is also significantly different from that of metals. In contrast to the poor composite static strength when it has damage or defects, the ability of composites to withstand cyclic loading is far superior to that of metals. The comparison of the normalized notched specimen fatigue response of a common aircraft metal, 7075-T6 aluminum, and a carbon/epoxy laminate is shown in Fig. 9. The fatigue strength of the composite is much higher relative to its static or residual strength. The static or residual-strength requirement for structures is typically much higher than the fatigue requirement. Therefore, because the fatigue threshold of composites is a high percentage of their static or damaged residual strength, they are usually not fatigue critical. In metal structures, fatigue is typically a critical design consideration.

Advantages and Disadvantages of Composite Materials

The advantages of composites are many, including lighter weight, the ability to tailor the lay-up for optimum strength and stiffness, improved fatigue life, corrosion resistance, and, with good design practice, reduced assembly costs due to fewer detail parts and fasteners.

The specific strengths (strength/density) and specific moduli (modulus/density) of high-strength fibers, especially carbon, are higher than other comparable aerospace metallic alloys (Fig. 10). This translates into greater weight savings, resulting in improved performance, greater payloads, longer range, and fuel savings.

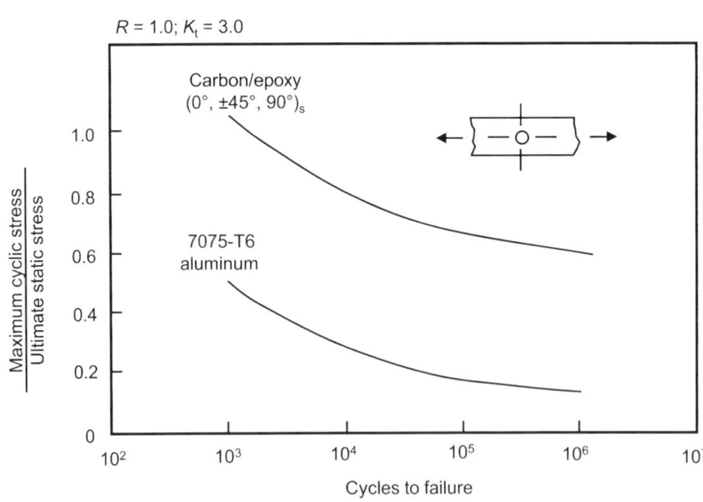

Fig. 9 Comparative notched fatigue strength. Source: Ref 3

The chief engineer of aircraft structures for the U.S. Navy once told the editor that he liked composites because "they don't rot (corrode) and they don't get tired (fatigue)." Corrosion of aluminum alloys is a major cost and constant maintenance problem for both commercial and military aircraft. The corrosion resistance of composites can result in major savings in supportability costs. While carbon-fiber composites will cause galvanic corrosion of aluminum if the fibers are placed in direct contact with the metal surface, bonding a glass fabric electrical insulation layer on all interfaces that contact aluminum eliminates this problem. The fatigue resistance of composites compared to high-strength metals is shown in Fig. 11. As long as reasonable strain levels are used during design, fatigue of carbon-fiber composites should not be a problem.

Assembly costs can account for as much as 50% of the cost of an airframe. Composites offer the opportunity to significantly reduce the amount of assembly labor and fasteners. Detail parts can be combined into a single cured assembly either during initial cure or by secondary adhesive bonding.

Disadvantages of composites include high raw material costs and usually high fabrication and assembly costs; composites are adversely affected by both temperature and moisture; composites are weak in the out-of-plane direction, where the matrix carries the primary load and should not be used where load paths are complex (e.g., lugs and fittings); composites are susceptible to impact damage, and delaminations or ply separations can occur; and composites are more difficult to repair than metallic structure.

The major cost driver in fabrication for a conventional hand layed-up composite part is the cost of laying-up or collating the plies. This cost gen-

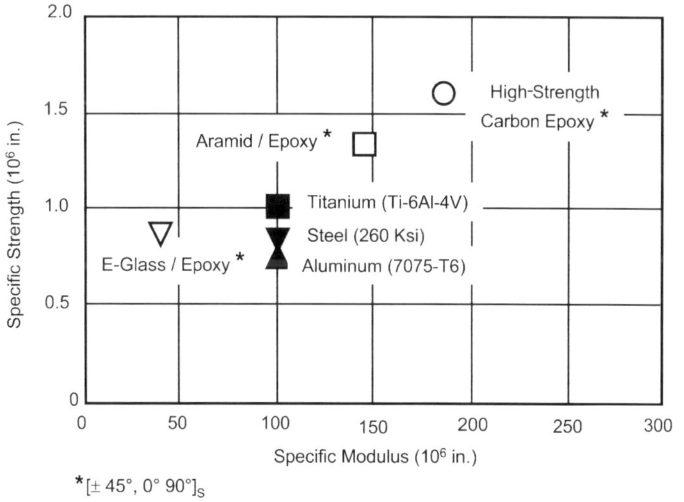

Fig. 10 Specific property comparison. Source: Ref 1

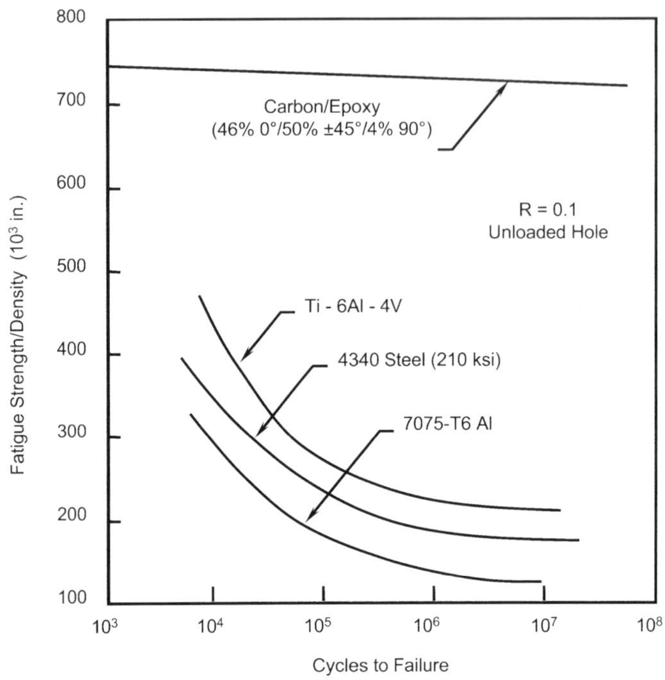

Fig. 11 Fatigue properties of aerospace materials. Source: Ref 1

erally consists of 40 to 60% of the fabrication cost, depending on part complexity. Assembly cost is another major cost driver, accounting for approximately 50% of the total part cost. As previously stated, one of the potential advantages of composites is to cure or bond a number of detail parts together to reduce assembly costs and the number of required fasteners.

Temperature has an effect on composite mechanical properties. Typically, as the temperature increases, the matrix-dominated mechanical properties decrease. Fiber-dominated properties are somewhat affected by cold temperatures, but the effects are not as severe as the effects of elevated temperature on the matrix-dominated properties. As shown in Fig. 12, the design parameters for carbon/epoxy are cold-dry tension and hot-wet compression. The amount of absorbed moisture is dependent on the matrix material and the relative humidity. Elevated temperatures speed the rate of moisture absorption. Absorbed moisture reduces the matrix-dominated mechanical properties. Absorbed moisture also causes the matrix to swell. This swelling relieves locked-in thermal strains from elevated-temperature curing. These strains can be large, and large panels fixed at their edges can buckle due to the swelling strains. During freeze-thaw cycles, the absorbed moisture expands during freezing and can crack the matrix. During thermal spikes, absorbed moisture can turn to steam. When the internal steam pressure exceeds the flatwise tensile (through-the-thickness) strength of the composite, the laminate will delaminate.

Composites are susceptible to delaminations (ply separations) during fabrication, assembly, and in service. During fabrication, foreign materials, such as prepreg backing paper, can be inadvertently left in the lay-up. During assembly, improper part handling or incorrectly installed fasteners can cause delaminations. When in service, low-velocity impact damage from dropped tools or fork lifts running into aircraft can cause damage. The damage may appear as only a small indentation on the surface but can propagate through the laminates, forming a complex network of delaminations and matrix cracks, as depicted in Fig. 13. Depending on the size of the delamination, it can reduce the static and fatigue strength and the compression buckling strength. If it is large enough, it can grow under fatigue loading.

Typically, damage tolerance is a resin-dominated property. The selection of a toughened resin can significantly improve the resistance to impact damage. In addition, S-2 glass and aramid fibers are extremely tough and damage tolerant. During the design phase, it is important to recognize

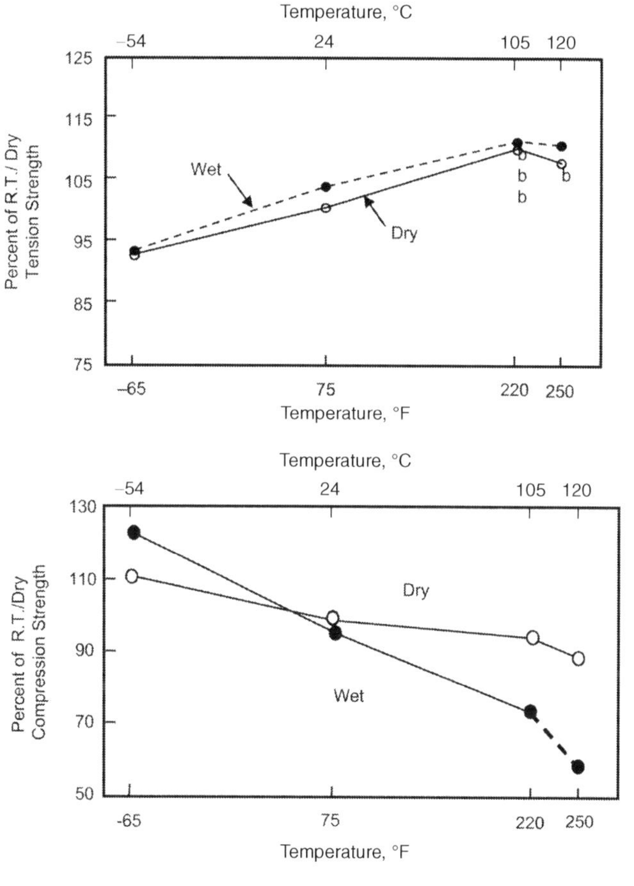

Fig. 12 Effects of temperature and moisture on strength of carbon/epoxy, 0.6% moisture, 42% 0°/50% ±45°/8% 90° plies. RT, room temperature. Source: Ref 1

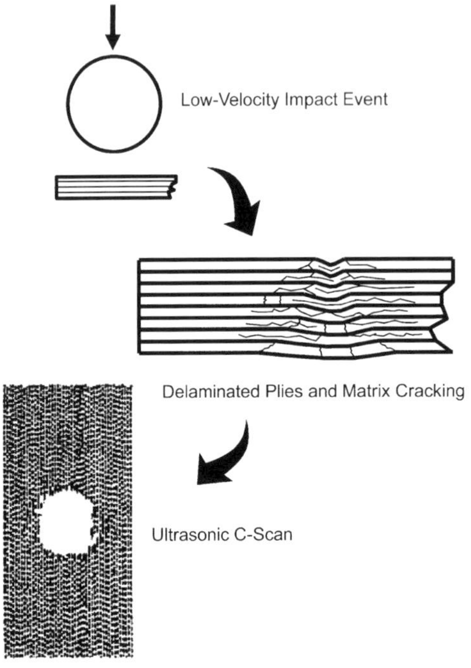

Fig. 13 Delaminations and matrix cracking due to impact damage. Source: Ref 1

the potential for delaminations and use conservative enough design strains so that damaged structure can be repaired.

Static Strength

When examining data sheets for carbon-fiber composites, the unidirectional properties (0° and 90°) are usually the only ones listed. While useful for comparison purposes, most real composite structures have laminate orientations that are quasi-isotropic or near quasi-isotropic. Because the matrix-dependent properties of unidirectional laminates are very low, it is necessary to balance the load-carrying plies fairly evenly in the 0°, ±45°, and 90° directions. In addition, as shown in the Fig. 14 envelope, a fair balance of plies in the 0°, ±45°, and 90° directions is required to carry the bearing loads from mechanical fasteners.

A comparison of the unidirectional and quasi-isotropic properties for a high-strength and intermediate-modulus carbon/epoxy composite are given in Table 2. In addition to the lower properties that result from cross-plying in the quasi-isotropic laminates as compared to the 0° strengths for the unidirectional laminates, note that while the tension properties of the unidirectional laminates have increased significantly with the development of intermediate-modulus carbon fibers, there has been almost no improvement in the compression strength.

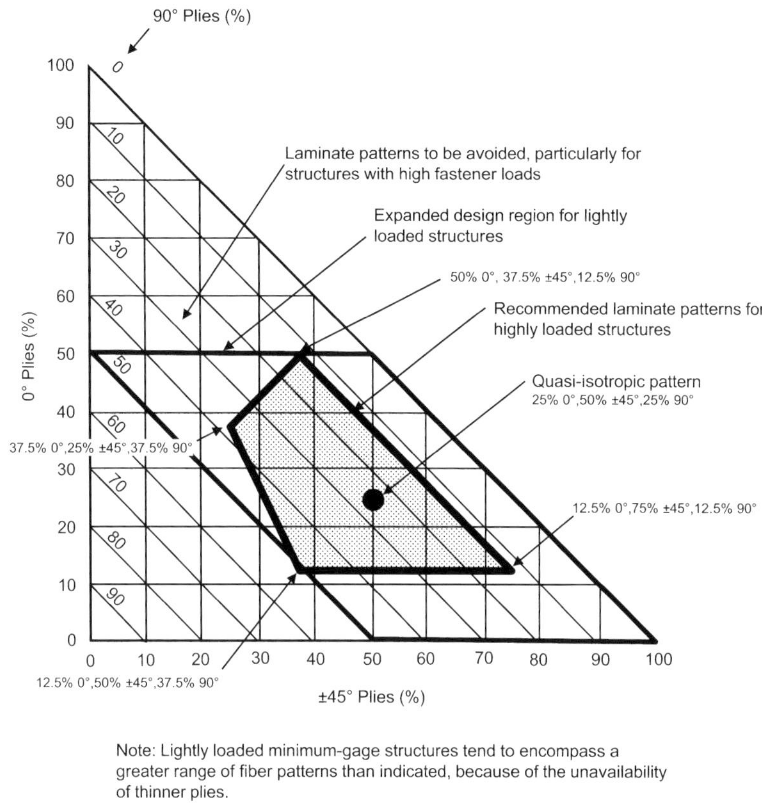

Fig. 14 Recommended design envelope for mechanically fastened composites. Source: Adapted from Ref 4

Table 2 Comparison of unidirectional and quasi-isotropic carbon/epoxy laminates

Property	High-strength carbon/epoxy(a)	Intermediate-modulus carbon/epoxy(a)
0° tensile strength, ksi	290	350
0° tensile modulus, 10^6 psi	19	25
90° tensile strength, ksi	11.5	11.5
90° tensile modulus, 10^6 psi	1.3	1.3
0° compression strength, ksi	190	230
0° compression modulus, 10^6 psi	17	22
Axial tensile strength(b), ksi	85	200
Axial tensile modulus(b), 10^6 psi	8	9
Axial compression strength(b), ksi	85	85
Axial compression modulus(b), 10^6 psi	8	9

(a) 60 fiber vol%. (b) Axial properties are for quasi-isotropic laminates. Source: Ref 5

Tension Failure. When a laminate is loaded in tension to failure, not all plies fail at the same time. The progressive ply failures in a $[0°/45°/90°/-45°]_s$ laminate are shown in Fig. 15. Plies will fail successively in the increasing order of strength in the loading direction. Because the transverse strength of unidirectional plies is lower than the longitudinal strength, the transverse plies will fail first. Failure of the 90° plies is generally followed by other off-angle plies (e.g., ±45°) and finally the much stronger 0° plies. The occurrence of first-ply failure, although a relatively conservative design criterion, is often used in the aerospace industry to signify laminate failure.

Compression Failure. For a longitudinal (0°) specimen loaded in axial compression, microbuckling of the fibers can occur either by what is known as the extensional mode or by the shear mode, as illustrated in Fig. 16. In the extensional mode, the fibers buckle in opposite directions in adjacent fibers. This mode derives its name from the fact that the major deformation of the matrix is an extension of the matrix material in a direc-

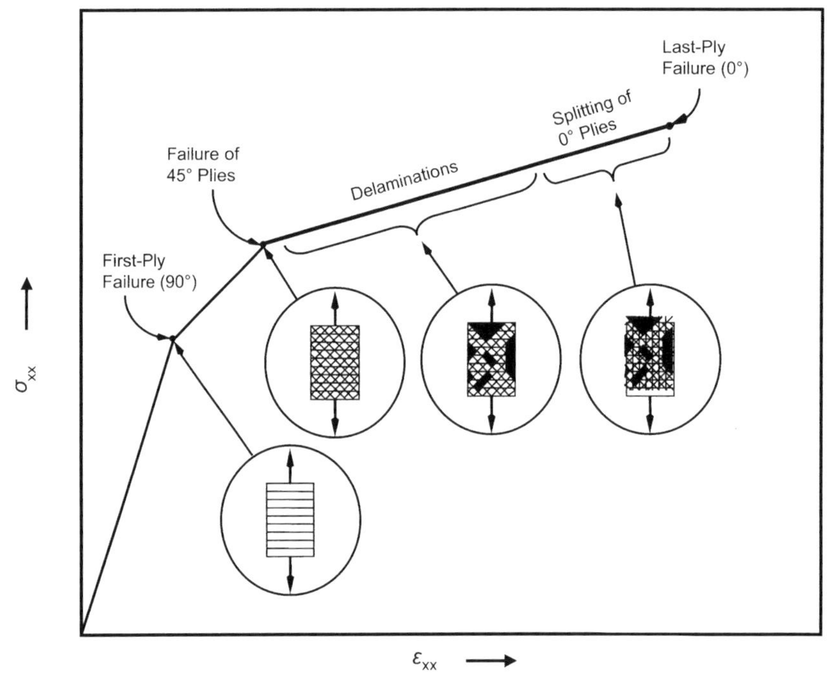

Fig. 15 Progressive ply failure for a $[0°/45°/90°/-45°]_s$ laminate. Source: Ref 6

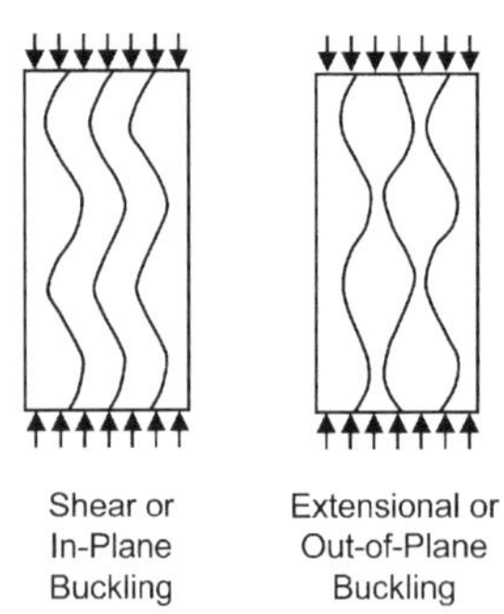

Fig. 16 Compression buckling modes for unidirectional composites. Source: Ref 7

tion perpendicular to the fibers. In the shear mode, the fibers buckle in the same wavelength and in phase with one another, so that the matrix material between the adjacent fibers deforms primarily by shear. The shear mode occurs in composites with a higher volume fraction of fibers and is therefore the predominant failure mode in high-strength carbon-fiber composites. Tests have shown that the modulus of elasticity of the matrix is also important, with stiffer matrices providing better support to the fibers and thus higher compression strengths.

Fatigue Behavior

The fatigue failure mechanisms in metals and composites are vastly different. Fatigue failure in metals normally results from a single crack that slowly propagates through the section until it becomes so long that the remaining section can no longer support the load, and failure occurs. Fatigue of composites is considerably different. As opposed to one discrete crack, many different forms of damage form at different locations and eventually link up to cause failure. There are five major damage mechanisms: matrix cracking, fiber breaking, crack coupling, delamination initiation, and delamination growth.

The damage state versus percent of life in a quasi-isotropic subjected to tension-compression fatigue loading during the three phases of damage progression is illustrated in Fig. 17.

Phase I: Matrix Cracking. During the first phase, matrix cracks develop due to tension loads in the off-axis plies, initially in the 90° plies and then in the 45° plies. Matrix cracking in a $[0°/90°/\pm45°]_s$ laminate is

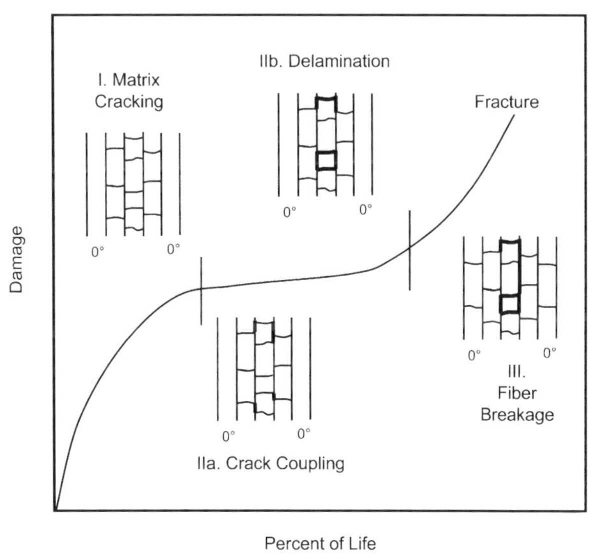

Fig. 17 Progressive fatigue damage. Source: Ref 8

shown in Fig. 18. More and more matrix cracks develop until they reach a steady-state or saturation level that is a function of ply thickness and material properties. The damage during this phase is relatively small and occurs during the first 10 to 25% of life. Although the strength loss is negligible and the stiffness loss is less than 10%, matrix cracks are the precursors for future damage events. In addition, if the matrix cracks are open to the surface (and they almost always are), they can provide paths for moisture ingression, which can accelerate the failure process.

Phase II: Fiber Fracture, Crack Coupling, and Delamination Initiation. As the matrix cracks grow in the off-axis plies, they eventually intersect the main load-bearing 0° plies. The 0° plies blunt the crack growth, but this results in stress concentrations that initiate fiber fracture in the adjacent plies and induce longitudinal cracks running parallel to the 0° fibers and perpendicular to the matrix cracks. This combination of longitudinal and matrix cracking is called crack coupling. Crack coupling is illustrated in Fig. 19 in a cross-ply laminate that has been subjected to a fatigue loading. In this figure, transverse matrix cracks and longitudinal cracks are visible. During the latter stages of phase II, as a result of extensive crack coupling, large interlaminar stresses develop at the intersection of the matrix cracks and the longitudinal cracks, leading to the initiation of delaminations. Phase II damage occurs at a slower rate than phase I damage and accounts for the next 70 to 80% of the life.

As the load increases, stress concentrations develop at the intersection of the matrix cracks and the 0° fibers where these cracks are blunted.

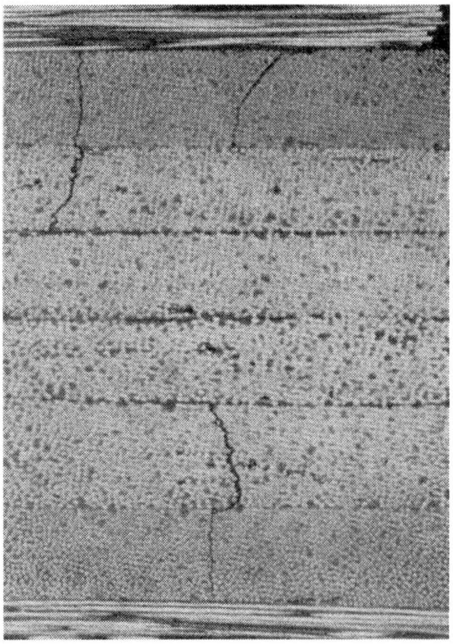

Fig. 18 Micrograph showing crack pattern in matrix of $[0°/90°/\pm45°]_s$ carbon/epoxy laminate. Cracks in 90° plies have linked to those in ±45° plies but have not propagated to the 0° plies. Source: Ref 9

These stress concentrations cause fibers to fracture in the adjacent plies. Large tensile stresses located at the tip of the matrix cracks induce longitudinal cracks in the direction of the 0° fibers and perpendicular to the matrix cracks. This damage mechanism is called crack coupling and includes the combination of matrix cracks and longitudinal cracks. In the

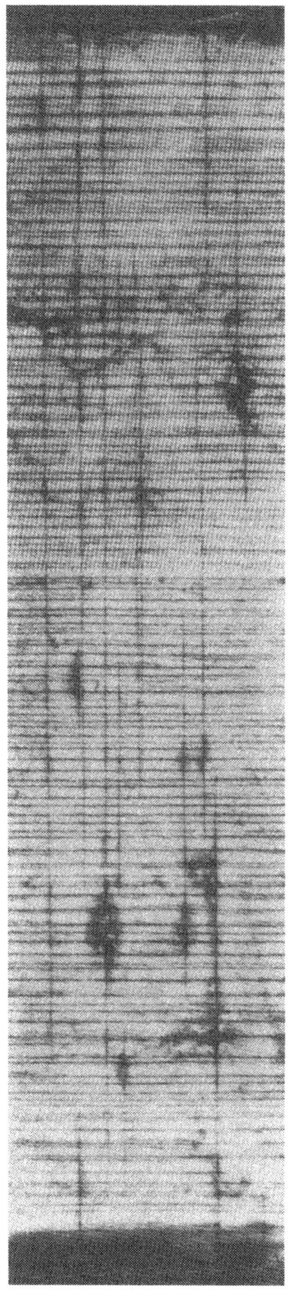

Fig. 19 Radiograph of carbon/epoxy laminate using image-enhancing penetrant to reveal transverse and longitudinal cracking and local delaminations. Source: Ref 9

later stages of phase II, delamination initiation becomes the dominant damage mechanism.

Phase III: Delamination Growth and Fracture. The final phase is dominated by delamination growth and the eventual failure of the 0° plies. The delaminations initiated during phase II start growing between the ply interfaces, eventually isolating the 0° plies. When the delaminations become large enough, the laminate section is essentially divided into a number of sublaminates, resulting in a large reduction in stiffness. The presence of compression loads during phase III is especially detrimental, because the compression loads cause the sublaminates to start buckling. Eventually, the strength and stiffness become so degraded that gross section failure occurs.

The direction of load relative to the fiber orientation of each individual lamina is also important. This effect on carbon/epoxy under tension-tension loading is shown in Fig. 20. The slope of the *S-N* curve for the 0° ply laminate is relatively flat because of the fatigue insensitivity of the carbon fibers. Consequently, the slope of the *S-N* curve tends to become more negative as the content of the 0° ply decreases, because the angle plies are subjected to matrix cracking failures.

A different trend is observed when fatigue tests are conducted at negative *R*-ratios. Comparing the *S-N* curves in Fig. 21 and 22, the slopes of the multidirectional ply laminates for the negative *R*-ratios are less negative compared to those for positive *R*-ratios. However, a greater reduction in the fatigue strength is seen in the cases of negative *R*-ratios. Another phenomenon worth noting is that the fatigue strength of multidirectional laminates is closely related to their tensile strengths. When the ratio of laminate fatigue strength to unidirectional fatigue strength is plotted against the ratio of laminate static strength to unidirectional static strength, a nearly one-to-one correlation is obtained (Fig. 22).

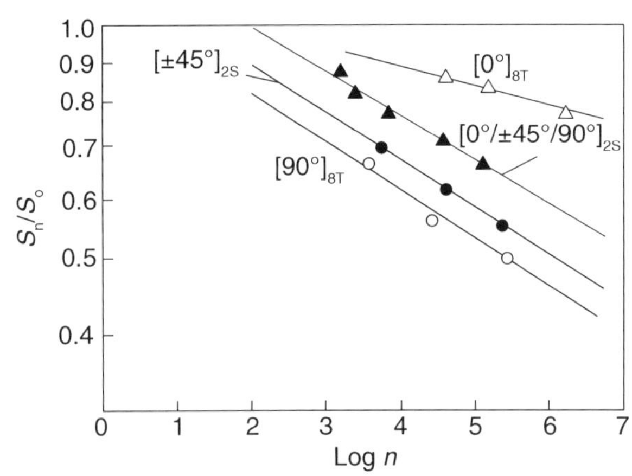

Fig. 20 *S-N* plot showing the effect of fiber orientation on fatigue performance of AS4/3502 carbon/epoxy laminates. The ordinate represents the fatigue strength ratio, which is the ratio of fatigue stress to static strength. Source: Ref 10

Unlike static strength, tension fatigue is not a threat to notched composite laminates. A test program revealed that tension fatigue strength is usually very high, only slightly lower than static strength. For $R = 0.05$ and maximum stress equal to 90% of the composite ultimate tensile strength, the tests ran for 10^6 cycles without specimen failure. However, typical *S-N* fatigue behavior shows up in those tests having $R = -1$ (complete reverse) or $R = 10$ (compression-compression). A set of fatigue data generated at $R = -1$ is shown in Fig. 23. Thus, the compressive stress component of a fatigue spectrum is the dominating parameter in determining the fatigue

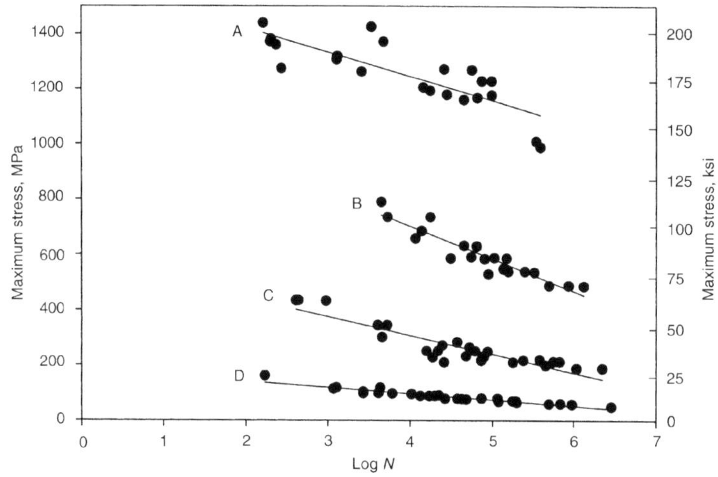

Fig. 21 *S-N* curves for T300/934 carbon/epoxy laminates, $R = -1$. A, unidirectional $[0]_{16}$; B, cross-ply $[0/90]_{4S}$; C, quasi-isotropic $[0/45/90]_{2S}$; D, angle-ply $[45]_{4S}$. Source: Ref 11

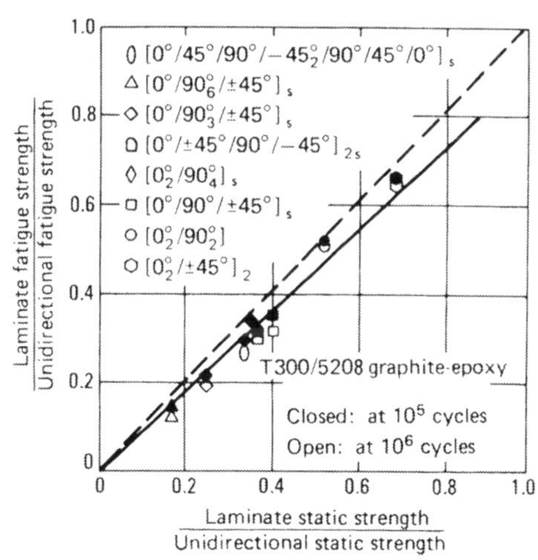

Fig. 22 Fatigue and static strength data for the T300/5208 carbon/epoxy showing a one-to-one correlation between the ratio of laminate fatigue strength to unidirectional fatigue strength and the ratio of laminate static strength to unidirectional static strength. Source: Ref 12

life of composites with holes. Figure 23 also shows that there will be no further degradation in static strength after prolonged fatigue cycling. Again, this observation agrees with the set of test data that showed that the residual strength of notched, fatigued laminates exceeded the static strength of those not fatigued.

A primary concern in the fatigue of composites is delamination. However, *S-N* data representing in-plane axial loading provide little information in characterizing interlaminar failure. Thus, it is necessary to develop *S-N* relationships between interlaminar stresses and interlaminar failure. Experience has shown that three- or four-point short-beam shear testing can be used to correlate cyclic interlaminar shear stress with delamination. The axial tension-tension and short-beam shear *S-N* curves for unidirectional carbon/epoxy laminates are shown in Fig. 24.

When cyclic tests are conducted at ambient temperatures, experience has shown that, within reasonable limits, the results are relatively insensitive to cyclic rate. For example, test results on panels tested at 5 Hz have correlated well with results from panels tested at 25 cycles per minute (0.5 Hz). The latter simulates a loading rate that may be applied to spectrum cycling of a large full-scale test component.

Delaminations and Impact Resistance

Delaminations can occur both during manufacturing operations or later when the part is placed in service. Manufacturing-induced delaminations can occur during part fabrication or assembly. Examples of fabrication delaminations are foreign objects left in the lay-up (e.g., release paper), those caused by insufficient pressure during cure due to vacuum bag rup-

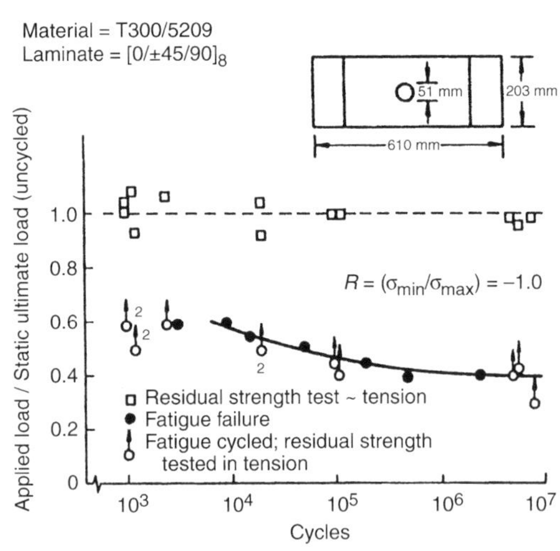

Fig. 23 *S-N* curve for T300/5209 carbon/epoxy laminates and residual strength of specimens after fatigue cycling. Source: Ref 13

ture, or poorly fitting tool details. During assembly, improper hole drilling can cause surface splintering of the plies, and even more serious are delaminations caused by unshimmed gaps during fastener installation. The clamping force that is applied when the fastener is torqued can cause matrix cracking and delaminations at gap locations. When the part is placed in service, delaminations can occur from runway debris, hail storms, bird impacts, dropped tools, or other unintended maintenance impacts.

When a delamination occurs, the main concern is in compression loading. Because a delamination splits the single laminate into one or more sublaminates, the section at the delamination is more susceptible to buckling during compression loading. Another key issue is whether or not the delamination will grow and become bigger under fatigue loading, with compression loading again being the prime concern.

Low-velocity impacts are a prime concern because they not only create delaminations but also result in a network of matrix cracking and some crushing of fibers, depending on the energy level of the impact. The delaminations normally occur at the ply interfaces, as shown in the impact-damaged carbon/epoxy laminate in Fig. 25. Extensive diagonal shear cracks occur in the matrix between the transverse-oriented fibers, that is, the fibers oriented normal to the plane. Extensive delamination has also taken place in the matrix between the plies. In most instances of low-velocity impact, the damage is confined to the matrix, and little fiber damage occurs. Therefore, the in-plane tension strength of the laminate may not be seriously degraded. However, even with impact levels that leave little indication of damage on the surface, the matrix damage may be significant, and therefore, its ability to stabilize the fibers in compression

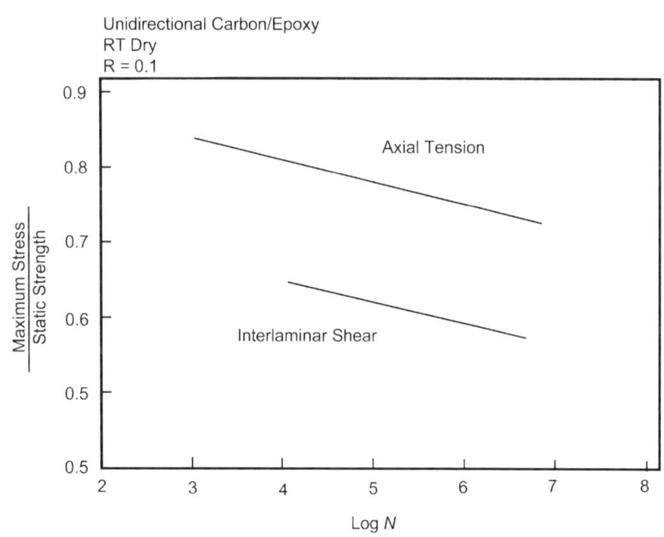

Fig. 24 *S-N* curves for axial tension and interlaminar shear. RT, room temperature. Source: Ref 14

may be seriously degraded. Because of this, tolerance to impact damage is often the critical design consideration, and compression is the critical loading mode.

Take another look at Fig. 25. Note that the damage produces very little evidence of damage on the surface impacted. This could be difficult to detect visually. The impact spreads out like a pine tree at the point of impact, with most of the damage being internal; higher-energy impacts produce broken fibers on the exit side. Thus, one of the main concerns is that the structure could sustain a low-velocity impact that no one is aware of that could then grow under either static or fatigue loading, eventually leading to failure. This concern has led to the concept of barely visible impact damage (BVID). BVID is a somewhat arbitrary term because the capability to detect an impact is dependent on lighting conditions, the type of surface finish on the part, the observer's distance from the impact, and differing capabilities of the human eye. Nevertheless, a lot of work has been conducted to correlate impacts with differing energy levels and the capability to visually detect them.

A comparison of the compressive static and fatigue degradation for delaminations and impact damage of carbon/epoxy laminates is shown in Fig. 26 and 27. The degradation due to a single-layer delamination is less than that of an impact, because the damage is confined to a single layer. In addition, delaminations that occur during manufacturing are likely to be detected during routine nondestructive testing. Although impacts can occur during manufacturing, they are more likely to occur in service where, unless it is visible to an inspectable surface, it can remain undetected for a long time or even the life of the structure. The concern with impact damage is apparent in Fig. 26, where a laminate can lose 60 to 65% of its undamaged static compression strength with an impact event that is essentially nonvisible. The loss in cyclic load-carrying capability is also greatest when an impact occurs. Thus, while fatigue and corrosion are the Achilles heels of metals, delaminations and impact resistance of carbon-fiber composites are their weak points.

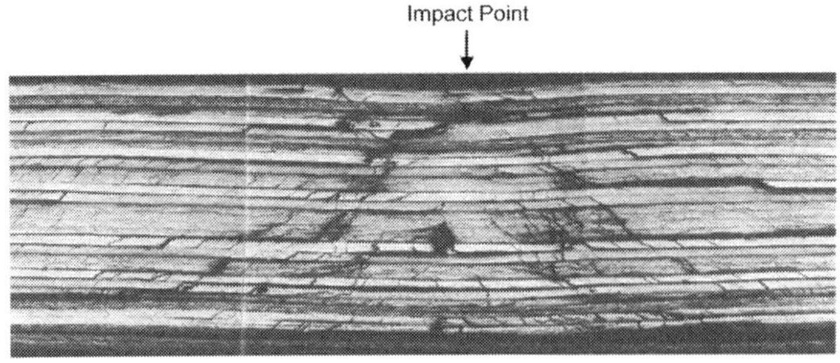

Fig. 25 Impact damage in carbon/epoxy laminate. Source: Ref 7

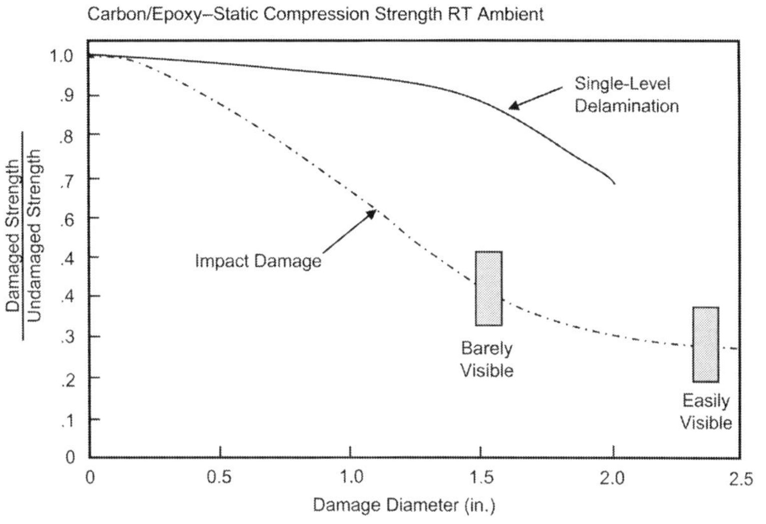

Fig. 26 Static strength loss due to impact damage. RT, room temperature. Source: Ref 7

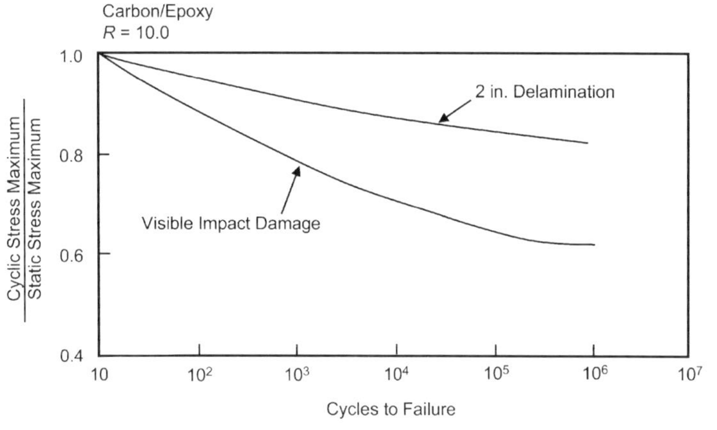

Fig. 27 Relative severity of defects on compression fatigue strength. Source: Ref 7

The type of fiber, the matrix, and the material product form can all affect the impact performance of composite laminates. The impact resistance of glass, in particular S-glass, and aramid composites is quite good (Fig. 28). S-glass/epoxy composites are approximately 7 times more impact resistant than high-strength carbon/epoxy and approximately 35 times more resistant than high-modulus carbon/epoxy. Glass/epoxy and Kevlar/epoxy composites are even more impact resistant than 4330 steel and 7075-T6 aluminum, a common aircraft grade of aluminum. Both glass/epoxy and aramid-fiber composites are used where a high resistance to impact damage is required, such as gas pressure vessels subject to rough handling. The ballistic performance of S-glass and aramid-fiber composites is similar, as measured by the parameter V50. The V50 parameter is the velocity at which there is a 50% probability that a projectile will penetrate the target. This type of evaluation shows that both S-glass and aramid-fiber composites provide a similar level of protection, and both are superior to E-glass and aluminum.

The early epoxy systems used during the 1970s through the mid-1980s, now referred to as first-generation epoxies, were highly cross-linked systems that were fairly brittle. A comparison of the compression strength after impact (CAI) for these systems and some of the newer toughened epoxies and thermoplastics is shown in Fig. 29. Several points are noteworthy here. First, even for the first-generation epoxies, the impact resistance for woven cloth is somewhat better than for tape material. Second, the newer thermoplastic-toughened epoxies have much higher CAI values, some of which approach the CAI performance of thermoplastics. Third, the CAI performance of the 93 °C (200 °F) system is higher than the 121 °C (250 °F) system. In other words, greater toughness can be obtained for lower-operating-temperature systems. Fourth, the impact performance is somewhat improved with the newer intermediate-modulus carbon fibers, although other design factors will determine whether a high-strength or intermediate-modulus fiber is selected.

A third method of improving damage tolerance is product form selection. Through-the-thickness reinforcement has been shown to improve the delamination resistance and impact performance of composites. Various methods have been developed, including stitching, three-dimensional weaving, and three-dimensional braiding, usually in conjunction with

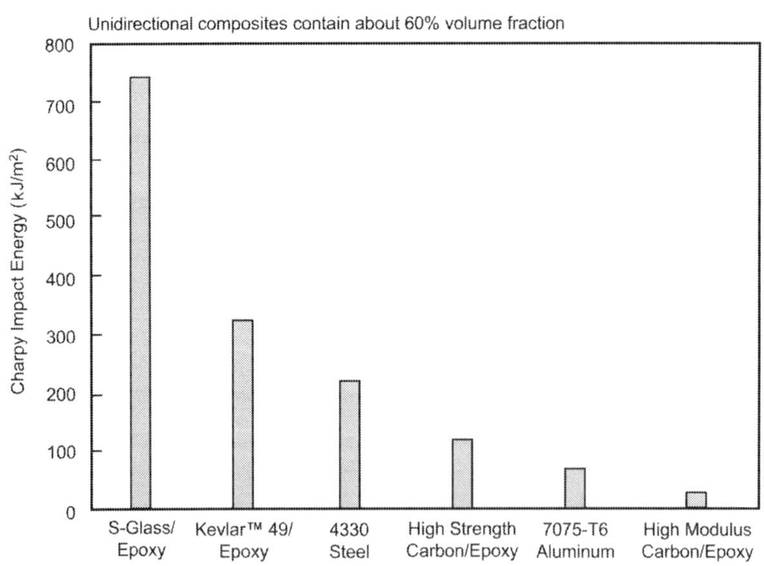

Fig. 28 Impact properties of various engineering materials. Source: Ref 15

liquid molding techniques. An example of the improvement due to through-the-thickness stitching is shown in Fig. 30. However, through-the-thickness reinforcement usually results in a reduction in the in-plane properties.

Damage Tolerance Considerations

Damage tolerance is the ability of a structure to sustain a certain level of damage, or to have a certain size of preexisting defect, and still be able to successfully perform for the duration of its life. The structure must have adequate residual strength and stiffness to continue service safely until the damage can be detected during a scheduled maintenance inspection or, if the damage is never detected, for the remainder of the structural life. Therefore, safety is the primary goal of damage tolerance. Both static load and durability-related damage tolerance must be determined experimentally because there are few, if any, accurate analytical methods. The normal design approach to ensuring that these requirements can be met is to

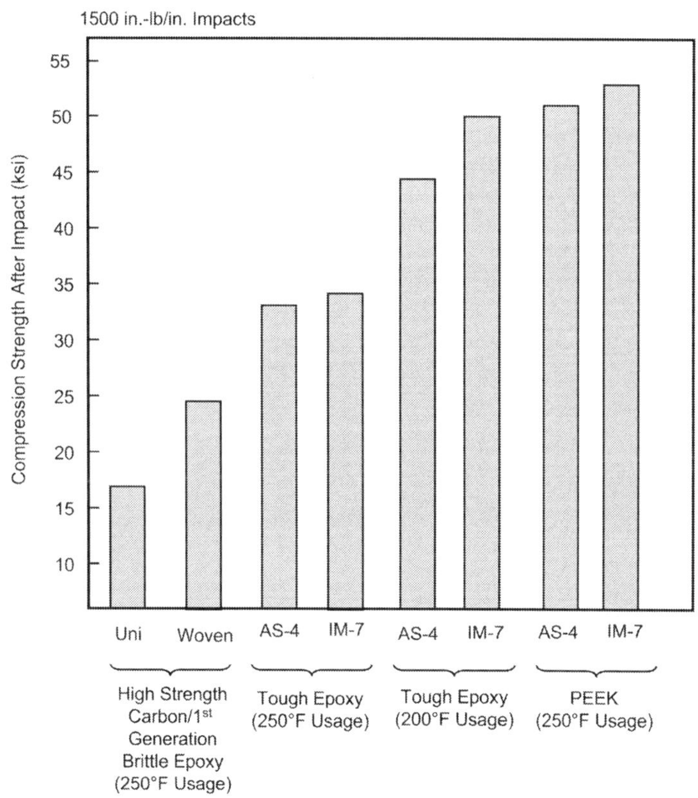

Fig. 29 Relative compression strengths after impact of carbon-reinforced composite systems. PEEK, polyetheretherketone (thermoplastic). Source: Ref 7

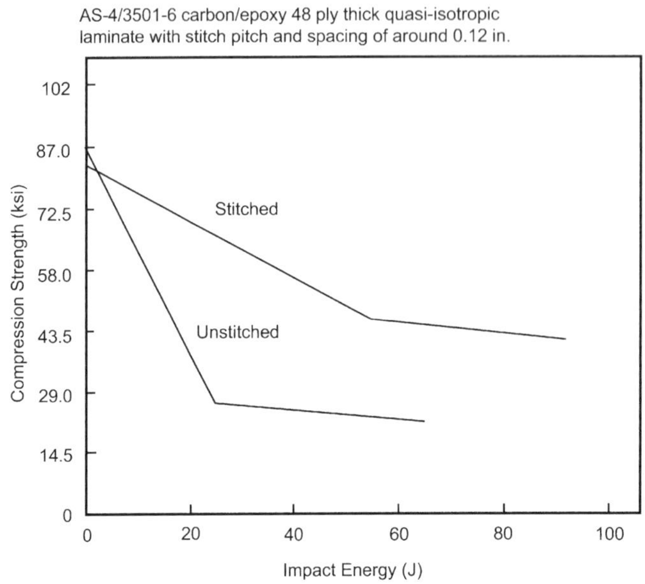

Fig. 30 Use of stitching to improve damage tolerance. Source: Ref 16

keep normal operating stresses and strains in the structure low enough that if damage does occur, the structure will still have sufficient residual strength to resist failure.

There are basically two types of damage: defects that occur during the manufacturing process, and damage that the part suffers when placed in service. While large defects that occur during the manufacturing process should be detected during normal nondestructive testing, some "escapes" beyond acceptable limits may not be detected. Consequently, their occurrence must be assumed during the design process and subsequent static and fatigue testing performed to verify structural integrity.

In-service damage can originate from a variety of sources and includes runway debris hitting control surfaces, hailstones from storms, tools dropped by maintenance mechanics, and collisions with ground handling equipment such as forklifts. The major damage tolerance concern for composites is delaminations due to low-velocity impact damage (LVID). When the impact contains sufficient energy, it can cause internal delaminations and extensive matrix cracking that, depending on the energy level and laminate construction, may or may not be visible on the surface. Because damage areas can be sizeable without being clearly visible, the part must be designed to perform its function when it contains nonvisible damage. LVID tests are usually performed with 12.7 mm (0.5 in.) diameter impactors, which tend to limit fiber breakage and maximize delamination damage. A full range of laminate thickness and stacking sequences should be tested at various impact energies to correctly establish the relationship between panel attributes, impact energy, dent depth, and visibility.

Most criteria involve the concept of defining undesirable damage as that which is clearly visible by a trained observer at a set distance, for example 1.5 m (5 ft). This allows composite surfaces to be visually inspected during routine walk-around inspections. Although visibility is somewhat subjective, based on the viewing conditions and surface finish, often a dent depth criterion can be established and matched to a visibility requirement. An upper limit on impact energy is usually established based on an analysis of threats found in the operating environment (Fig. 31). In the aircraft industry, once a level of nondetectable damage is established, it is verified by fatigue testing that it will not endanger the normal operation of the aircraft structure for two lifetimes. After the fatigue tests are completed, the structure is statically loaded to failure.

Other properties of composites affect their damage tolerance characteristics. One factor is ply orientation. Impacted laminates with stiff ply orientations, that is, a higher percentage of plies oriented parallel to the loading direction, typically fail at lower strains than those with softer orientations (Fig. 32). However, structural weight is more significantly related to stress capability than to strain. Thus, a laminate with a higher modulus may be able to operate to a higher stress even though its strain capability may be less. Each specific case must be carefully examined. The potential advantage of using a soft laminate, especially in damage-prone areas or in areas where there are stress concentrations, should be considered. For example, a soft laminate may be used effectively for an outer skin, whereas the interior stiffeners, which are exposed to less damage, may advantageously use a stiffer and hence more weight-efficient but less damage-tolerant lay-up.

Single-layer delaminations are also important; however, unless they are very large (more than 50 mm, or 2 in., in diameter), the problem is mostly

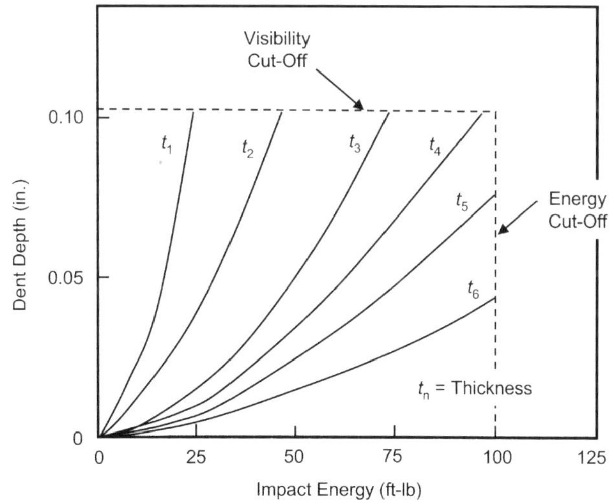

Fig. 31 Impact energy cut-off diagram. Source: Ref 17

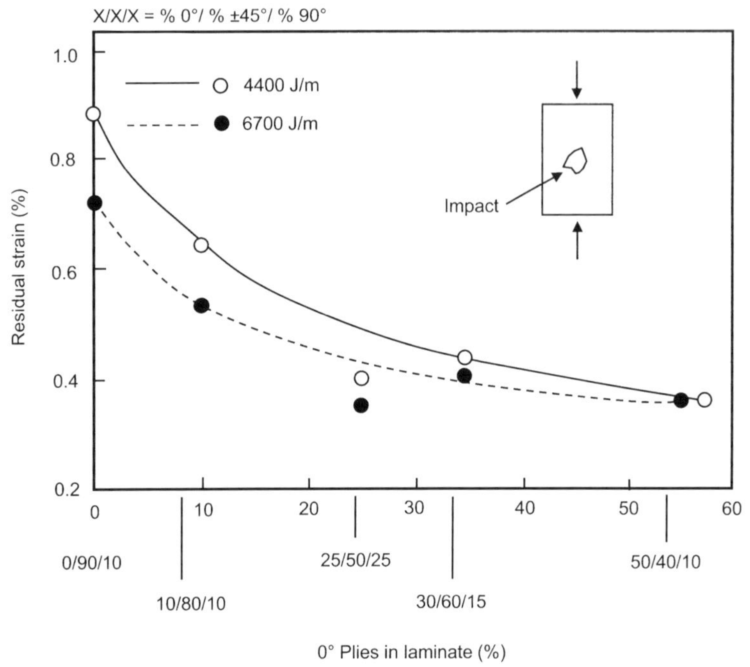

Fig. 32 Effect of laminate orientation on residual compression strain after impact. Source: Ref 3

with thin laminates. Effects of manufacturing defects such as porosity and flawed fastener holes that are slightly in excess of the maximum allowable are usually less severe. They are generally accounted for by the use of design-allowable properties that have been obtained by testing specimens with stress concentrations, for example, open- and filled-hole tension and compression tests. These specimens normally have a 6.4 mm (0.25 in.) diameter hole located at the center of the specimen, which is left open (no fastener) or filled (fastener installed). Because open holes are more critical than filled holes in compression loading, open-hole specimens are typically used for compression tests. For tension loading, filled holes may be more critical, especially for laminates with ply orientations with a predominant number of plies oriented in the load direction. Therefore, the design allowables generated with these tests can be used to account for a nominal design stress concentration caused by an installed or missing fastener, at least to a 6.4 mm (0.25 in.) diameter.

Effects of Defects

Besides delaminations and impact-induced damage, other types of defects can degrade the performance of composite parts. Again, the major effect is with the compression properties. Three types of manufacturing-related defects are discussed in this section: voids and porosity, fiber distortion, and fastener hole defects.

Voids

Voids and porosity are one of the most detrimental defects that can occur during manufacturing. The two main reasons for porosity in continuous-fiber polymer-matrix composites that are autoclave processed are air that becomes trapped between the plies during lay-up that cannot be successfully evacuated later during the initial part of the cure cycle, and dissolved water or other entrapped volatiles that come out of solution as the cure temperature increases and are then locked in place when the matrix solidifies or gels. Porosity can form either at the ply interfaces (interply porosity) or within the individual plies (intraply porosity), as shown in Fig. 33. It can form in isolated areas, or it can be pervasive throughout the part. Although the terms *voids* and *porosity* are used fairly interchangeably, porosity is usually smaller and more extensive, while voids are usually larger and fairly discrete.

Porosity adversely affects almost all of the matrix-dependent mechanical properties. The curves in Fig. 34 and 35 illustrate the reduction in interlaminar shear and compression strength due to porosity. Note that as little as 2% porosity causes a significant reduction in properties. A properly processed autoclave-cured laminate should contain 1% or less of porosity. Fiber-dominated longitudinal tensile strength is minimally affected, but the matrix-dependent properties, such as flexural strength, interlaminar shear strength, and compression strength, degrade with increasing percentages of porosity. Porosity adversely affects both the static and fatigue strength.

One of the problems in studying the effects of porosity is accurately determining the porosity level and its extent. One method that has been

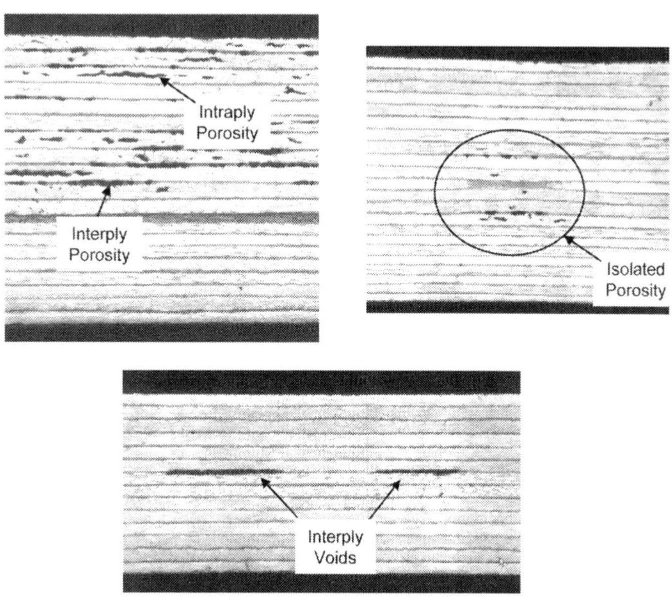

Fig. 33 Various forms of voids and porosity. Source: Ref 7

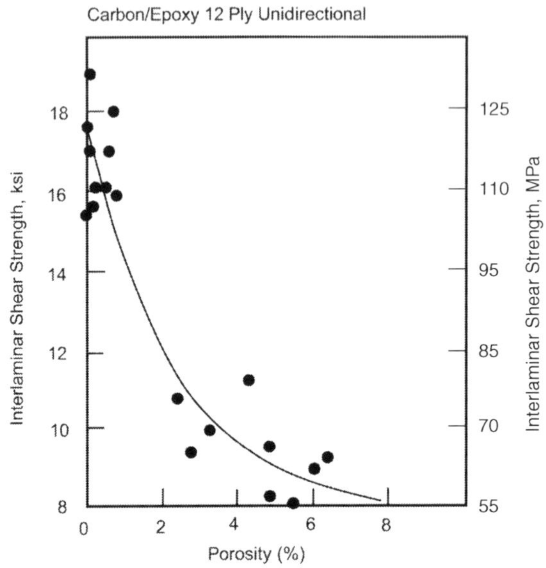

Fig. 34 Effect of porosity on interlaminar shear strength of carbon/epoxy. Source: Ref 18

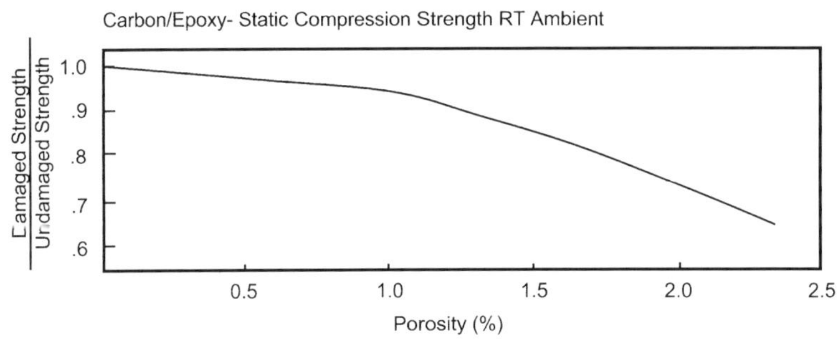

Fig. 35 Strength loss due to porosity. RT, room temperature. Source: Ref 7

employed is to use acid digestion to dissolve the matrix and then use the constituent weights and densities to calculate the porosity. However, this method is subject to an assumed fiber and resin density. Due to inaccuracies in these assumptions, negative porosity levels are even sometimes reported. In addition, while this method gives a gross porosity level, it tells nothing about how the porosity is distributed. Another method is to conduct materiallographic examination, such as the micrographs shown in Fig. 33. The problem with this method is that only a small area is examined. A third method is to use ultrasonic inspection. In the ultrasonic attenuation method, the more porous the laminate, the greater the amount of sound that is attenuated (loss) due to the presence of porosity. If one knows what the sound attenuation is for a porosity-free laminate, then it is assumed that attenuation above that level is due to porosity, and the greater the attenuation, the greater the amount of porosity. For example, if a sound laminate attenuates to a level of 20 db, and a similar laminate with porosity attenuates at 36 db, it can be assumed this extra attenuation (Δ16 db) is due to porosity, and it is assumed that a similar laminate that attenuates at 56 db is worse than the one at 36 db. Of course, to confirm that these assumptions are correct, extensive cross sections are taken from areas with different attenuation levels and examined for porosity. In what is called an "effect of defects" test program, matrix-dependent mechanical properties are measured on both sound laminates and those containing various degrees of porosity. When this work is completed, it provides a way of determining the severity of porosity in production parts. A curve showing the reduction in interlaminar shear strength of carbon/epoxy with increasing Δdb levels is shown in Fig. 36. In the example in the figure, if a 3.8 mm (0.15 in.) thick laminate had an attenuation of 40 db over the baseline for this thickness of laminate, the interlaminar shear strength would be reduced to approximately 80% of that of a nonporous laminate.

Fiber Distortion

While fibers used in high-performance composites are strong and stiff in tension, they also have very small diameters that make them rather feeble in compression when they are not supported by a fully cured matrix. Thus, during curing when the matrix is a fairly low-viscosity fluid, the fibers can be subjected to movement and rearrangement due to the hydraulic pressures caused by the flowing resin. In addition, mismatches in tooling details can result in fiber wrinkling and distortion, also known as fiber marcelling. A number of examples of fiber distortion caused by tool mark-

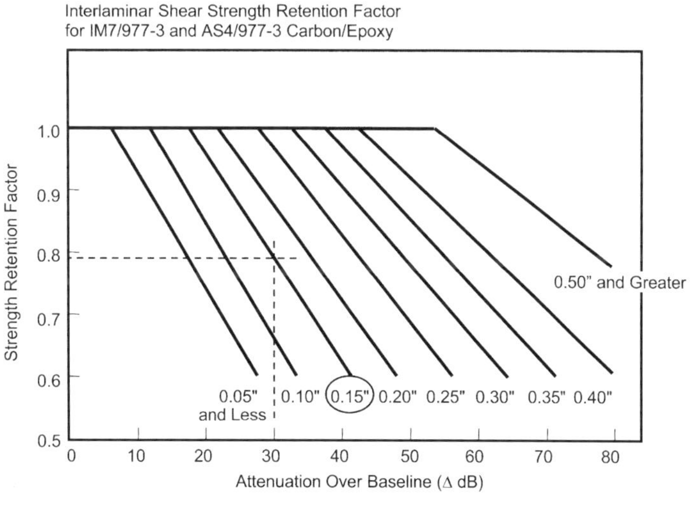

Fig. 36 Ultrasonic attenuation for porosity characterization. Source: Ref 7

off are shown in Fig. 37. Mark-off that is caused by a vacuum bag wrinkle will usually have some resin buildup on the surface. These resin lumps can make the mark-off appear deeper than it really is. Therefore, the resin buildups can be removed by lightly sanding to obtain a true mark-off depth.

While wrinkled fibers will reduce almost all of the mechanical properties to some extent, it is the compression strength that is the most adversely affected, as shown for the carbon/epoxy laminate in Fig. 38. Because the wrinkled carbon fibers are already prebuckled, it should not be surprising that the compression strength is adversely affected. Note that in the example shown in the figure, the compression strength was reduced by almost 80% for a 40% indentation. It is obvious that when this type of defect is detected, it must be fully characterized and corrective action immediately taken. Unfortunately, ply wrinkling can be difficult to detect. During ultrasonic inspection, because the wrinkled plies scatter sound in a manner similar to porosity, wrinkled plies will often be mistaken for porosity.

Fastener Hole Defects

Fastener hole preparation and fastener installation in composites is more error prone than for metals. Composites are subject to delaminations on both the entrance and exit sides of holes, can be damaged by excessive heat generated during machining or drilling, and fasteners that are not installed in properly prepared holes can further induce delaminations. While interference-fit fasteners are often used in metallic structures to improve fatigue life, normal interference-fit fasteners installed in composite structures have often resulted in matrix crushing and delaminations. The effects of a number of these defects on the tension and compression strength of carbon/epoxy are shown in Table 3. Two of the most severe defects are

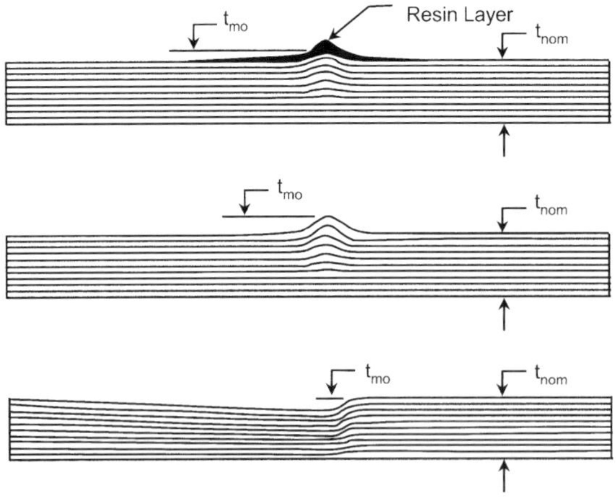

Fig. 37 Examples of tool mark-off causing ply distortion. Source: Ref 7

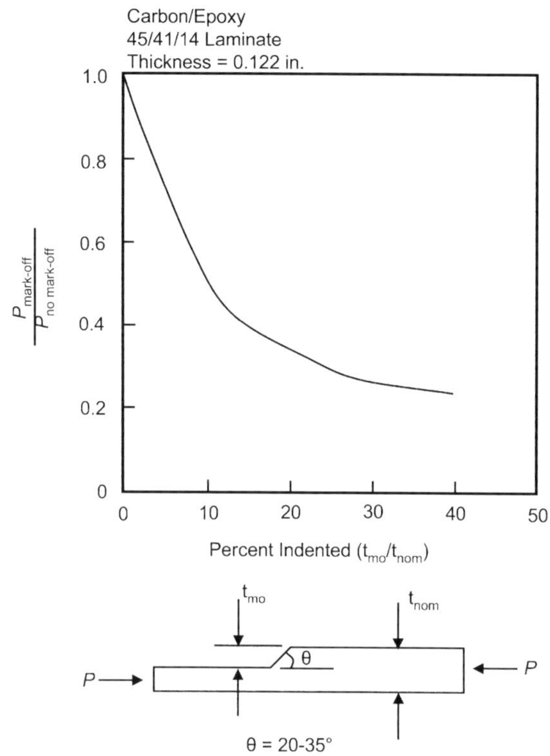

Fig. 38 Effect of tool mark-off on compression strength of carbon/epoxy. Source: Ref 7

Table 3 Strength degradation due to fastener hole defects

	RTD tension	Compression RT(a)	Compression 250 °F(a)
Out-of-round holes			
50/40/10 laminate	(b)	—	—
30/60/10 laminate	−4.8	—	—
Broken fibers exit side of hole			
Severe	−7.3	−8.4	−9.2
Moderate	−1.4	−3.2	−4.2
Porosity around hole			
Severe	(b)	−10.3	−30.8
Severe with freeze-thaw	—	−11.6	—
Moderate	—	−7.1	−13.3
Moderate with freeze-thaw	—	−8.4	—
Improper fastener seating depth			
80% thickness	−16.4	—	—
100% thickness	−34.3	—	—
Tilted countersinks			
Away from bearing surface	(b)	—	−16.7
Toward bearing surface	−21.4	—	−16.7
Interference-fit tolerance (in.)			
50/40/10 at 0.003	(b)	—	+9.1(c)
at 0.008	(b)	—	+9.1(c)
30/60/10 at 0.003	(b)	—	(b)(c)
at 0.008	(b)	—	(b)(c)
Fastener removal and reinstallation			
100 cycles	(b)	—	−8.3

Notes: RTD, room-temperature dry. Dash indicates no test. (a) Moisture content = 0.86%. (b) Less than 2% change. (c) Tensile loading. Source: Ref 7 improper fastener seating depth and tilted countersinks. These are detrimental because they tend to cause delaminations in the plies adjacent to the holes when placed under load. Severe porosity around the hole is especially detrimental in elevated-temperature compression loading.

Building-Block Approach

The building-block approach to testing and certification is used extensively in the aerospace industry. It starts with large numbers of simple coupon tests and, at each succeeding level, builds on the level below with fewer but more complicated tests. The pyramid in Fig. 39 illustrates the building-block concept. At the lowest levels, the basic materials properties are determined using large quantities of small test specimens from multibatches (usually five or six) of material. At each succeeding level, progressively more complicated specimens and structures are built and tested, and the failure modes and loads are predicted by analysis based on the lower-level data. When more data are obtained, the structural analysis models are

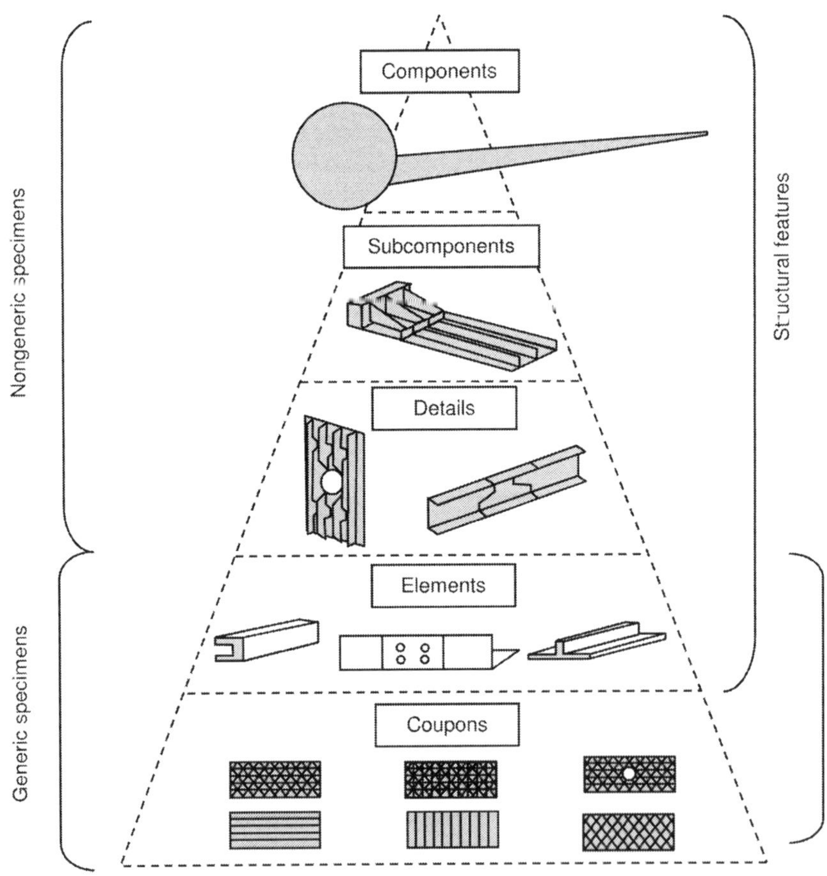

Fig. 39 Aircraft industry building-block approach. Source: Ref 19

refined as needed to agree with the test results. As the test structures become more complex, the number of replicate specimens of the same type and environment and the number of environments are reduced. The culmination of the building-block pyramid is often a single confirming test of a full-scale component or a full-scale structural assembly.

Large full-scale tests are not normally performed at the worst-case design environment; that is, it would be extremely difficult to environmentally expose a large structure, such as an entire airframe, and then conduct the testing at elevated temperature. Data from the lower levels of the building-block test program are used to establish environmental compensation values that are applied to the loads of higher-level room-temperature tests. Similarly, lower-level fatigue tests are used to determine truncation approaches for fatigue spectra and compensation for fatigue scatter at the full-scale level.

The building-block approach consists of the following steps:

1. Order materials to a specification that includes controls on the fibers, resins, prepreg, chemical, and physical properties. Normally, five or six separate batches are required using fibers from different lots and resins from different batches of raw ingredients.
2. Conduct coupon testing at the lamina and laminate levels on each batch of prepreg. Calculate preliminary design allowables using statistical analysis. Lamina tests are conducted to establish allowables that can be used along with classical lamination theory to predict laminate properties. Conduct laminate testing to confirm the predictions from classical lamination theory. The laminate orientations tested should be the major ones to be used in the final structure.
3. Based on analysis of the structure, select critical areas and design features for subsequent verification tests. For each critical design feature, determine the load and environment combination that is expected to produce a given failure mode with the lowest margin. Special attention should be given to matrix-sensitive failure modes, such as compression, shear, bondlines, and potential "hot spots" caused by out-of-plane loads.
4. Design and test a series of test specimens, each of which simulates a single selected failure mode and load and environment condition. Compare the test results to previous analytical predictions and adjust analysis models or design properties as necessary.
5. Conduct increasingly more complicated tests that evaluate more complicated loading situations with the possibility of failure from several different potential failure modes. Compare to analytical predictions and adjust analysis models as necessary.
6. Conduct, as required, full-scale component or assembly static and fatigue testing for final validation of internal loads and structural integrity. Compare the results to analysis. In the aircraft industry, this means

conducting a full-scale static and a full-scale fatigue test of two separate airframes.

The building-block approach is more than just a test program. During all levels of increasing complexity, manufacturing has the opportunity to develop and refine tooling and processing approaches that will later be used during production. Because this is an expensive and lengthy process, small-scale testing should be conducted on candidate materials prior to full-scale qualification. While this is a very expensive approach involving thousands of tests, it has been used successfully to design and build sophisticated systems, such as aircraft, where safety is paramount. Such an extensive test program should not be required for many applications where unexpected product failure would not result in injury or death.

ACKNOWLEDGMENTS

Portions of the material for this chapter came from *Structural Composite Materials* by F.C. Campbell, ASM International, 2010. Portions of the section on fatigue came from *Mechanics and Mechanisms of Fracture: An Introduction* by A.F. Liu, ASM International, 2005.

REFERENCES

1. F.C. Campbell, Introduction to Composite Materials, *Structural Composite Materials,* ASM International, 2010
2. M.C.Y. Niu, *Composite Airframe Structures,* 2nd ed., Hong Kong Conmilit Press Limited, 2000
3. R.E. Horton and J.E. McCarty, Damage Tolerance of Composites, *Composites,* Vol 1, *Engineered Materials Handbook,* ASM International, 1987
4. L.J. Hart-Smith, Bolted Joint Analysis for Fibrous Composite Structures—Current Empirical Methods and Future Scientific Prospects, *Joining and Repair of Composite Structures,* STP 1455, ASTM International, 2004
5. F.C. Campbell, Polymer Matrix Composites, *Lightweight Materials—Understanding the Basics,* ASM International, 2012
6. F.C. Campbell, Structural Analysis, *Structural Composite Materials,* ASM International, 2010
7. F.C. Campbell, Composite Properties, *Structural Composite Materials,* ASM International, 2010
8. R.L. Reifsnider, *Composite Materials Series: Fatigue of Composite Materials,* Vol 4, Elsevier Science Publishing Company, New York, 1991
9. J.R. Schaff, Fatigue and Life Prediction, *Composites,* Vol 21, *ASM Handbook,* ASM International, 2001

10. *Composites,* Vol 1, *Engineered Materials Handbook,* ASM International, 1987, p 43
11. A. Rotem and H.G. Nelson, Residual Strength of Composite Laminates Subjected to Tension-Compressive Fatigue Loading, *J. Compos. Technol. Res.,* Vol 12, 1990, p 76–84
12. J.M. Whitney, "Fatigue Characterization of Composite Materials," Report AFWALTR-79–4111, Air Force Materials Laboratory, 1979
13. R.W. Walter, R.W. Johnson, R.R. June, and J.E. McCarthy, *Designing for Integrity in Long-Life Composite Aircraft Structures,* STP 636, ASTM, 1977, p 228–247
14. J.R. Schaff and A. Dobyns, "Fatigue Analysis of Helicopter Tail Rotor Spar," AIAA Report 98-1738, American Institute of Aeronautics and Astronautics—ASM Symposium (Long Beach), 1998
15. S.K. Mazumdar, *Composites Manufacturing: Materials, Product, and Process Engineering,* CRC Press, 2002
16. C.C. Poe, H.B. Dexter, and I.S. Raju, Review of NASA Textile Composites Research, *J. Aircr.,* Vol 36, 1999, p 876–884
17. "General Specification for Aircraft Structures," MIL-A-872221, U.S. Air Force, Feb 1985
18. M.J. Yokota, In-Process Controlled Curing of Resin Matrix Composites, *SAMPE J.,* Vol 11, 1978
19. R.E. Fields, Overview of Testing and Certification, *Composites,* Vol 21, *ASM Handbook,* ASM International, 2001

SELECTED REFERENCES

- J. Bottle, F. Burzesi, and L. Fiorini, Design Guidelines, *Composites,* Vol 21, *ASM Handbook,* ASM International, 2001
- MIL-HNBK-17-1F, *Guidelines for Characterization of Structural Materials,* Vol 1, *Polymer Matrix Composites,* U.S. Department of Defense, 2001
- MIL-HNBK-17-2F, *Polymer Matrix Composites Materials Properties,* Vol 2, *Polymer Matrix Composites,* U.S. Department of Defense, 2001
- MIL-HNBK-17-3F, *Materials Usage, Design, and Analysis,* Vol 3, *Polymer Matrix Composites,* U.S. Department of Defense, 2001
- M.R. Woodward and R. Stover, Damage Tolerance, *Composites,* Vol 21, *ASM Handbook,* ASM International, 2001

CHAPTER 11

High-Temperature Failures

FOR METALS AT LOW TEMPERATURES, yield strength is usually the limiting design factor. However, at high temperatures, permanent deformation can occur over a period of time at stresses well below the yield strength. This time-dependent deformation is known as creep and occurs at temperatures greater than approximately 0.3 to 0.5 of the absolute melting point. In creep, thermal activation enables plastic deformation at stresses below those needed to deform the lattice without thermal activation. When a metal is placed under a constant load, it stretches elastically but also gradually extends plastically. Thus, a metal subjected to a constant tensile load at elevated temperature will creep and undergo a time-dependent increase in length.

Because the mobility of atoms increases with increasing temperature, diffusion-controlled mechanisms become active. Dislocation mobility increases, slip becomes easier, new slip systems become available, and dislocation climb is aided by both increases in temperature and by the presence of a greater number of vacancies. Deformation at grain boundaries also becomes a possibility. The metallurgical stability of the alloy can become an issue. For example, precipitation-hardened alloys may undergo overaging, with a resulting strength loss. Oxidation and intergranular attack can occur.

Creep occurs in any metal or alloy at a temperature where atoms become sufficiently mobile to allow the time-dependent rearrangement of structure. Because the elevated-temperature strength of metals is closely related to their melting points, it is normal practice to specify the homologous temperature, which is the ratio of the exposure temperature to the melting point (T/T_m), on an absolute scale. Creep behavior of a polycrystalline metal or alloy often is considered to begin at approximately ⅓ to ½ of its melting point (~0.3 to 0.5 T_m). However, creep becomes important when the mechanical strength of a metal becomes limited by creep rather than by its yield strength. Because this transition in engineering design is not directly related to melting temperature, the temperature at which the mechanical strength of a metal becomes creep limited must be determined individually for each metal or alloy.

Approximate temperatures at which creep behavior begins for several metals and alloys are listed in Table 1. Low-melting-point metals, such as lead and tin, may deform by creep at or a little above room temperature. In contrast, refractory and nickel-base superalloys require temperatures exceeding 1000 °C (1800 °F) to activate the onset of creep deformation. Creep is critical in a number of high-temperature applications, such as power and chemical plants and turbine components. Typical materials and application temperatures of some creep-resistant alloys are listed in Table 2. Often, the main concern is dimensional stability, although prolonged creep will eventually lead to rupture.

The Creep Curve

Two types of elevated-temperature time-dependent tests are normally conducted to evaluate metals for long-term elevated-temperature service. In a creep test, the time-dependent strain, or extension, is measured under long-term elevated temperature. A shorter test, the stress-rupture test, measures the time it takes to fail a metal at a given stress at elevated temperature.

Table 1 Approximate temperatures for onset of creep

Metal or alloy	°F	T as ratio of T_m, K
Aluminum alloys	300–400	0.48–0.54 T_m
Titanium alloys	600	0.3 T_m
Low-alloy steels	700	0.36 T_m
Austenitic iron-base heat-resisting alloys	1000	0.49 T_m
Nickel- and cobalt-base heat-resisting alloys	1200	0.56 T_m
Refractory metals and alloys	1800–2800	0.4–0.45 T_m

Source: Ref 1

Table 2 Typical elevated temperatures in engineering applications

Application	Typical materials	Typical temperatures, °C	Homologous temperatures, T/T_m
Rotors and piping for steam turbines	Cr-Mo-V steels	550–700	0.45–0.50
Pressure vessels and piping in nuclear reactors	316 stainless steel	375–475	0.35–0.40
Reactor skirts in nuclear reactors	316 stainless steel	575–675	0.45–0.55
Gas turbine blades	Nickel-base superalloys	500–650	0.45–0.60
Burner cans for gas turbine engines	Oxide dispersion-strengthened nickel-base alloys	1075–1125	0.55–0.65

Source: Ref 1

Engineering creep is measured by applying a constant load on a tensile specimen at a constant temperature and measuring the strain, or extension, of the specimen as a function of time. A schematic of a test setup for a creep test is shown in Fig. 1. Creep tests may be run for as short as several months to as long as up to 10 years. Because a creep test is usually con-

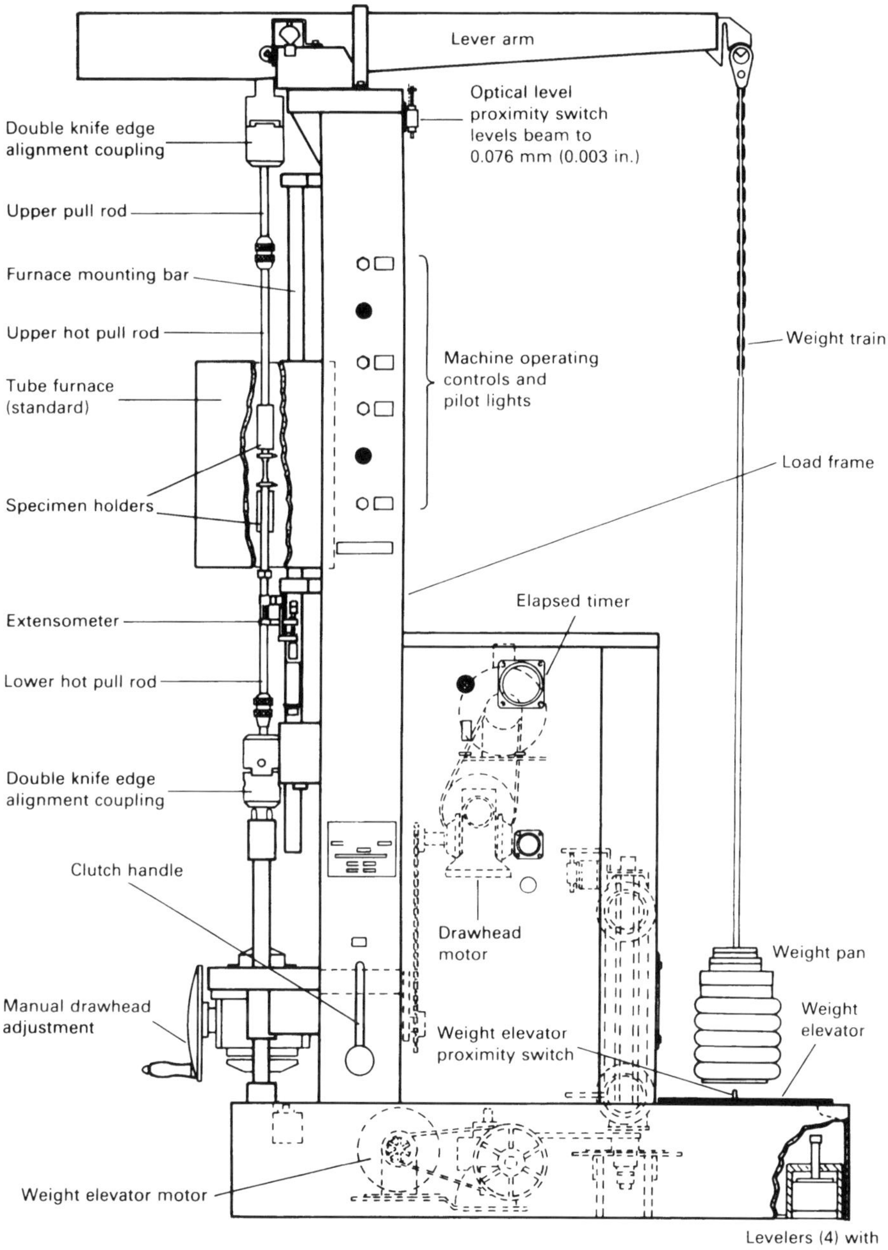

Fig. 1 Schematic of a test stand used for creep and stress-rupture testing. Source: Ref 2

ducted at a constant load rather than a constant stress, as the specimen elongates and the cross-sectional area decreases, the applied stress increases with time. Constant-load creep curves typically, but not always, consist of three distinct stages, as shown in Fig. 2.

Primary Creep. During primary creep, the specimen undergoes an initial elongation (ε_o), and then the creep rate ($\dot{\varepsilon} = d\varepsilon/dt$) rapidly decreases with time. Primary creep, also known as transient creep, represents a stage of adjustment in the metal during which rapid, thermally activated plastic strain occurs. The competing processes of strain hardening and recovery eventually lead to a somewhat stable dislocation configuration. Primary creep occurs in the first few moments after initial strain and decreases in rate as crystallographic imperfections within the metal undergo realignment. This realignment leads to secondary creep.

Secondary Creep. Following primary creep is the region of secondary creep, where the creep rate is nominally constant at a minimum rate, generally known as the minimum creep rate, as shown in Fig. 2. Secondary creep, also known as steady-state creep, occurs when there is a balance between the competing processes of strain hardening and recovery. Secondary creep often occupies the major portion of the duration of the creep test, and the strain rate in this region for many creep-resistant materials is sufficiently constant to be considered as a steady-state creep rate. For these materials, the minimum creep rate is a steady-state value that can be empirically related to rupture life and is widely used in engineering analyses.

During creep, significant microstructural changes occur on all levels. On the atomic scale, dislocations are created and forced to move through the material. This leads to work hardening as the dislocation density increases and the dislocations encounter barriers to their motion. At low temperatures, an ever-diminishing creep rate results. However, if the temperature is sufficiently high, dislocations can rearrange and annihilate

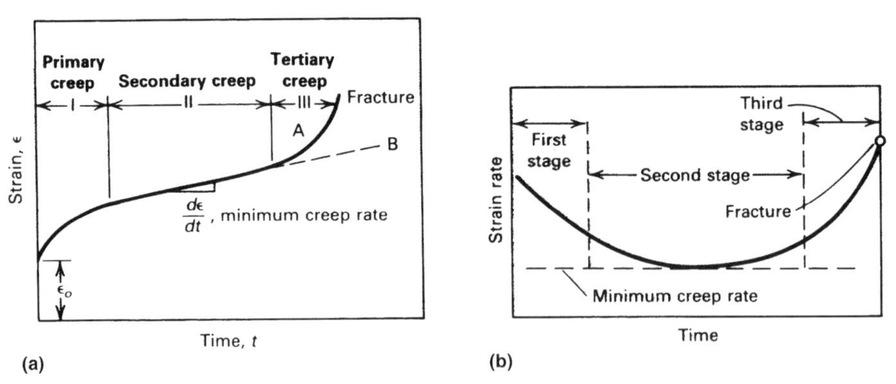

Fig. 2 Stages of creep deformation. (a) Strain curve for the three stages of creep under constant-load testing (curve A) and constant-stress testing (curve B). (b) Relationship of strain rate, or creep rate, and time during a constant-load creep test. The minimum creep rate is attained during second-stage creep. Source: Ref 1

themselves through recovery. During creep deformation, the material also is progressively degraded or damaged as the amount of creep strain increases over time.

Power Law Model of Steady-State Creep Rates. In the intermediate-temperature regime ($0.4\ T_m < T < 0.6\ T_m$), the creep rate varies nonlinearly with stress, as either a power function or an exponential function of stress. At stresses and temperatures of interest to the engineer, the following behavior is generally obeyed:

$$\dot{\varepsilon} = A\sigma^n \quad \text{(Eq 1)}$$

where A and n are stress-independent constants. In addition, because one contribution to creep is a thermally activated diffusion process, its temperature sensitivity would be expected to obey an Arrhenius-type expression, with a characteristic activation energy (Q) for the rate-controlling mechanism. Therefore, Eq 1 can be rewritten as:

$$\dot{\varepsilon} = A\sigma^n e^{-\left(Q/RT\right)} \quad \text{(Eq 2)}$$

where Q is the activation energy for creep, A is a pre-exponential constant, n is a constant usually between 3 and 10, R is the universal gas constant, and T is the absolute temperature. A correlation of creep and diffusion data for pure metals shows that the activation energy for creep (Q) is equal to the activation energy for self-diffusion (Fig. 3).

Equation 2 is a power law relation that provides a basis for modeling creep rates as a function of temperature and stress. However, A, n, and Q are parameters that depend on the particular material condition, stress level, and temperature. For example, the parameter A includes microstructural factors that influence the modes of creep deformation. Therefore, the parameter A must be determined from data in order to calculate absolute

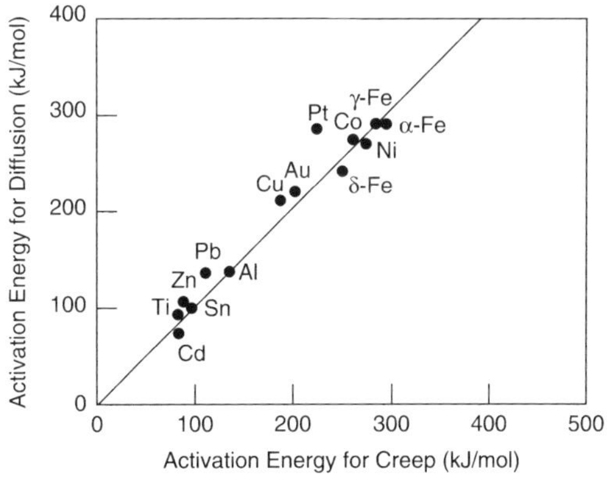

Fig. 3 Correlation between diffusion and creep activation energies. Source: Ref 3

values of creep rates. At high stress levels, where creep rates begin to increase more rapidly with stress, this relationship no longer holds and is commonly referred to as power law breakdown.

Creep deformation results also indicate that the values for n and Q in Eq 2 are both variable with respect to stress and temperature. An example of the change in the value of n is shown in Fig. 4 for a normalized and tempered 1.25Cr-0.50Mo steel. A distinct break in the curve is evident, with $n = 4$ at low stresses and $n = 10$ at higher stresses. The breaks in the curves occurred at stresses at which the fracture mode changed from intergranular to transgranular at high stresses. Values of n ranging from $n = 1$ at low stresses to $n = 14$ at high stresses have been reported. Although many investigators report a distinct break in the curve, others view the value of n as continuously changing with stress and temperature.

To account for these changes, it has been suggested that the effective stress changes with test conditions due to changes in an internal back stress (σ_0) and that the stress term in Eq 2 should be modified to $(\sigma - \sigma_0)^n$. The internal back stress represents a resisting force and is postulated to arise from a variety of microstructural factors such as dislocation configurations, precipitate dispersion, and solid-solution effects. While discussions continue regarding the natures of n and Q and the reasons for their variations, industrial practice has continued to ignore these controversies and to use a simple power law (Eq 2) with discretely chosen values of n and Q. Because variations in n and Q are generally interrelated and self-

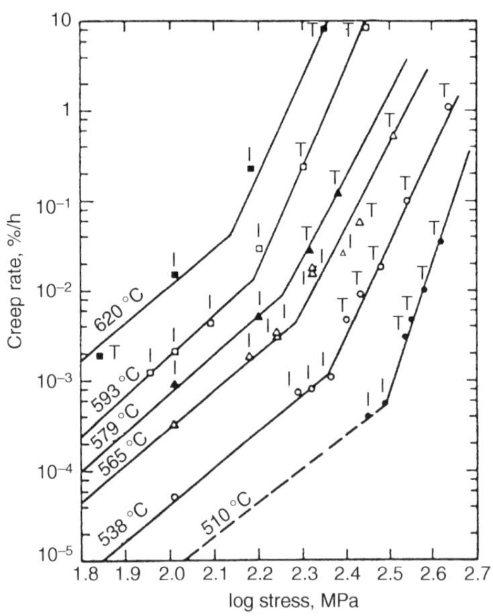

Fig. 4 Variation of minimum creep rate with stress for a normalized and tempered 1.25Cr-0.50Mo steel. T and I denote transgranular and intergranular failure, respectively. Source: Ref 4

compensating, no major discrepancies in the end results have been noted in the low-stress region. At higher stress levels, the power law model of creep strain rates begins to break down.

The minimum creep rate during steady-state creep is usually the most important design parameter obtained during a creep test. Two criteria are common: the stress to produce a creep rate of 1% in 10,000 h, or the stress to produce a creep rate of 1% in 100,000 h. The first criterion is often used for jet engine components and the second for steam turbines.

Power Law Breakdown. Steady-state creep rate at high temperatures can be described well by a power law (Eq 2), where the stress exponent (n) ranges from 4 to 5 for face-centered cubic metals (Fig. 5) and approximately 3 for body-centered cubic metals (Fig. 6), and where the activation energy (Q) for creep is equal to that for self-diffusion. The data presented in Fig. 5 and 6 indicate that this holds for a wide variety of pure metals for normalized stress values below approximately $\sigma_s/G = 10^{-3}$.

At normalized stresses above $(\sigma_s/G) = 10^{-3}$, creep rates begin to increase more strongly with stress, and the power law model breaks down. In this high-stress region, an exponential or sink function appears to be more applicable. For example, an exponential relationship, although not generally used, has been proposed to explain the behavior at very high stresses:

$$\dot{\varepsilon} = A\exp(C_7\sigma) \qquad \text{(Eq 3)}$$

where A and C_7 are stress-independent constants.

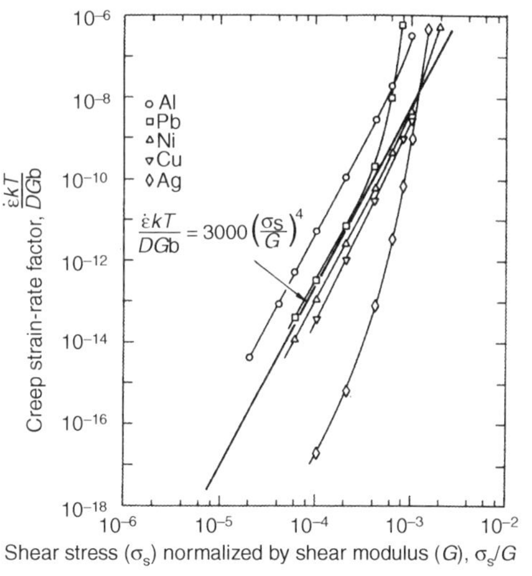

Fig. 5 Creep data for several face-centered cubic metals plotted as a function of normalized shear stress (σ_s/G) compared with a power law stress exponent of $n = 4$. b, Burgers vector. Source: Ref 5

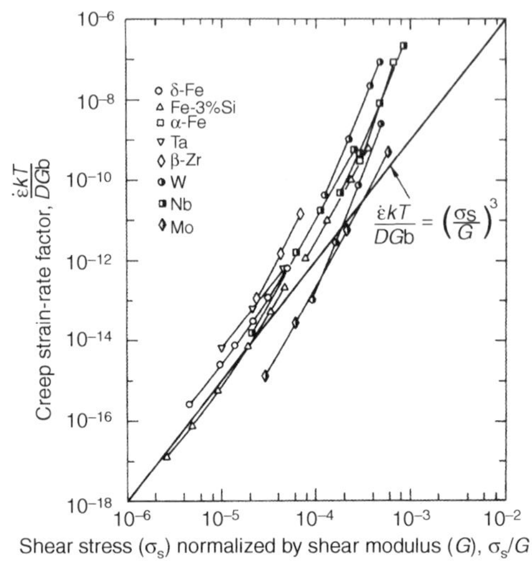

Fig. 6 Creep data for several body-centered cubic metals plotted as a function of normalized shear stress (σ_s/G) compared with a power law stress exponent of $n = 3$. Source: Ref 5

Tertiary creep is a region of drastically increasing strain rate with rapid extension to fracture. Tertiary creep is dominated by a number of weakening metallurgical instabilities, such as localized necking, corrosion, intercrystalline fracture, microvoid formation, precipitation of brittle second-phase particles, and dissolution of second phases that originally contributed to strengthening of the alloy. In addition, recrystallization of the strain-hardened grains can destroy the balance between the material hardening and softening processes. As opposed to constant-load tests, constant-stress tests do not often show tertiary behavior.

During service or during creep testing, tertiary creep may be accelerated by a reduction in cross-sectional area resulting from cracking or necking. Environmental effects, such as oxidation, that reduce the cross section may also initiate tertiary creep or increase the tertiary creep rate. In many commercial creep-resistant alloys, tertiary creep is apparently caused by inherent deformation processes and occurs at creep strains of 0.5% or less. In designing components for service at elevated temperatures, the steady-state creep rate is usually the significant design parameter. However, the duration of tertiary creep is also important, because it constitutes a safety factor that may allow detection of a failing component before catastrophic fracture.

Under certain conditions, some metals also may not exhibit all three stages of plastic extension, as depicted in Fig. 7. For example, at high stresses or temperatures, the absence of primary creep is not uncommon, with secondary creep or, in extreme cases, tertiary creep following immediately upon loading. At the other extreme, notably in cast alloys, no tertiary creep can be observed, and fracture may occur with only minimum

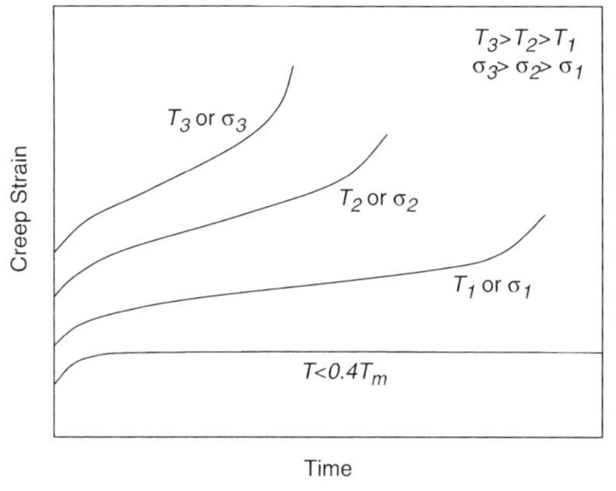

Fig. 7 Effects of temperature and stress on creep curve. Source: Ref 3

extension. In general, the higher the temperature and/or the greater the stress, the greater will be the creep rate.

It is important to note that the region of initial elastic strain under load (ε_o) is not usually identified as a stage of creep deformation, and it is common practice to ignore this contribution to total strain when plotting creep curves. Therefore, creep curves generally show only the time-dependent plastic strain that follows this initial elastic strain. Although this procedure is acceptable for research, the initial strain may amount to a substantial fraction of the total strain and should not be omitted from design analysis. In addition, while creep tests are normally conducted at constant temperatures and loads, components in service hardly ever operate under constant conditions. Start-stop cycles, reduced power operation, thermal gradients, and other factors result in variations in stresses and temperatures.

Nonclassical Creep Behavior. Although the classical pattern of creep deformation can be made to fit many materials and test conditions, the relative duration of the three periods differs widely with materials and conditions. For example, in many superalloys and other materials in which a strengthening precipitate continues to age at creep temperatures, brief primary creep often shows transition to a long, upward sweep of creep rate, with only a point of inflection for the secondary period.

Aging of normalized and tempered 0.5Cr-0.5Mo-0.25V steel during creep under 80 MPa (11.6 ksi) stress at 565 °C (1050 °F) has been reported to cause the creep curve to effectively exhibit only a continuously increasing creep rate to fracture. For twice the amount of stress, the creep curve in this case followed the classical trends. In other alloys, such as titanium alloys, with limited elongation before fracture, the tertiary stage may be brief and may show little increase in creep rate before rupture occurs. A more obvious departure from classical behavior develops during the early portion of many tests, when precise creep measurements are taken. When 34 ferritic steels were studied for as long as 100,000 h at temperatures ranging from 450 to 600 °C (842 to 1112 °F), step-form irregularities were observed, with an extended period of secondary creep preceded by a lower creep rate of shorter duration during primary creep.

An entirely different source of variation from classical creep may occur from environmental reactions at high temperature. For example, tests longer than 50 h with 80Ni-20Cr alloys at 816 and 982 °C (1500 and 1800 °F) showed a deceleration of creep after the normal tertiary stage was reached, resulting in a second period of steady-state creep and later, another period of last-stage creep. This behavior is due to oxide strengthening, which prolonged rupture life and caused a slope decrease in curves of log stress versus log rupture life. Oxides and nitride formed on the surfaces of the intercrystalline cracks that occur extensively during tertiary creep. The interconnection of the bulk of these cracks added substantially to strengthening against creep deformation in the late stages of the tests. This effect also has been observed in 99.8% Ni tested at 816 °C (1500 °F) under 20.7 MPa (3000 psi) stress.

Stress-Rupture Testing

In the stress-rupture test, the specimen is normally loaded at higher stresses than in the creep test, and the test is taken to failure. In the creep test, the total strain is often less than 0.5%, while in the stress-rupture test the strain can approach 50%. An advantage of the stress-rupture test is time; a stress-rupture test can often be terminated at approximately 1000 h. As shown in Fig. 8, the stresses at different temperatures are plotted against the time to failure. Changes in the slope of the curve often indicate that some type of metallurgical instability has taken place, such as a phase change, internal oxidation, or the change from transgranular to intergranular failure.

Stress-rupture ductility is an important consideration when stress concentrations and localized defects, such as notches, are a design factor. Most engineering alloys can lose ductility during high-temperature service, because the diffusion of impurities to the grain boundaries becomes

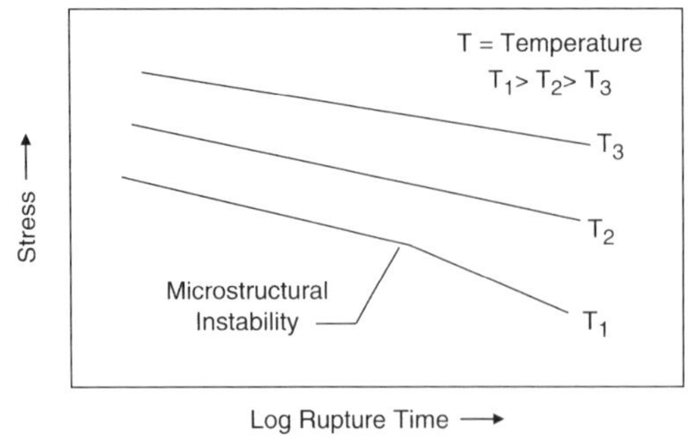

Fig. 8 Stress-rupture curves. Source: Ref 3

more pronounced. In creep curves, there are two meaningful measures of elongation: true elongation, which is defined as the elongation at the end of the second stage of creep, and total elongation, which is the elongation at fracture. The common practice is to plot both elongations versus rupture life. As an example, a plot of both total and true elongation at two different testing temperatures for the high-temperature alloy S-590 is shown in Fig. 9. At both temperatures, total elongation shows appreciable data scatter. Because true elongation is not influenced by the localized creep mechanisms (e.g., necking) that can occur during tertiary creep, it provides a more accurate representation of ductility in metals at elevated temperatures. The large differences between total and true elongation are a function of crack volume and distribution during tertiary creep. Stress-rupture ductility is an important factor in alloy selection, because premature failures have resulted from lack of ductility during tertiary creep. Because component designs are frequently based on 1% creep, a low stress-rupture ductility may prevent using the alloy to its full strength potential. As shown by the schematic creep curves in Fig. 10, a higher rupture ductility for the same load and temperature conditions provides a higher safety margin.

Because the creep and stress-rupture tests are similar, there are a number of empirical relationships that relate steady-state creep rate ($\dot{\varepsilon}_s$) to rupture life (t_R). Monkman and Grant developed one that implies an inverse relationship between the two:

$$\log t_R + m \log \dot{\varepsilon}_s = B \quad \text{(Eq 4)}$$

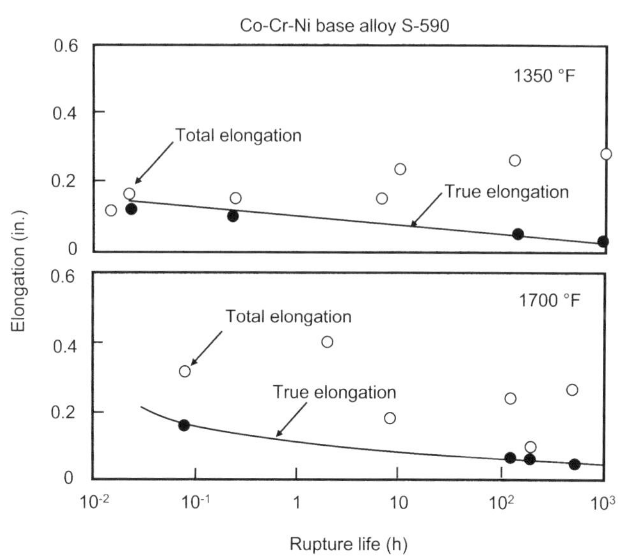

Fig. 9 Comparison of true and total elongation. Source: Ref 1

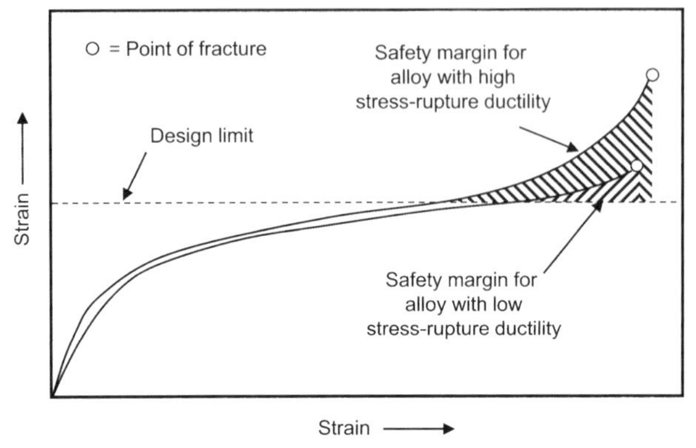

Fig. 10 Stress-rupture curves with high and low rupture ductility. Source: Ref 6

where t_R is the rupture life, $\dot{\varepsilon}_s$ is the steady-state creep rate, and m and B are constants.

For a number of aluminum-, copper-, titanium-, iron-, and nickel-base alloys, Monkman and Grant found that $0.77 < m < 0.93$ and $0.48 < B < 1.3$. As long as the validity of the relationship can be established along with the values of the constants, the rupture life can be estimated from the steady-state creep rate.

Creep Deformation Mechanisms

The major classes of creep mechanisms are those that are governed by dislocation motion and those that are diffusion controlled. The dominating mechanism is determined by both the stress and temperature; however, several mechanisms may be active at the same time. In general, the ones governed by dislocation motion are more prevalent at lower temperatures and higher stresses, while those controlled by diffusion occur at higher temperatures and lower stresses.

Dislocation creep is the result of dislocations, in combination with vacancies and thermal activation, that climb over obstacles that would normally impede their motion at lower temperatures. The relevant dislocation mechanisms are dislocation glide and climb. As the temperature is increased, slip systems that were not available at room temperature become active, promoting dislocation glide. When dislocations encounter an obstacle, they are blocked and tend to pile up against the obstacle (Fig. 11). At low stress levels, the applied stress is insufficient to enable the dislocations to bow around, or cut through, the obstacle. However, at elevated temperature, a dislocation may climb by diffusion to a parallel slip plane. Having climbed, the dislocation proceeds along the new slip plane until it encounters another resistant obstacle, whereupon it climbs (or descends) to another parallel plane and the process repeats. Because dislocation motion depends on both dislocation glide and climb, the term *climb-glide creep* is used to describe this form of creep. Climb-glide creep depends more strongly on stress than does diffusion creep. At temperatures in excess of 0.5 T_m, dislocations can escape from the obstacles by climbing out of the slip plane and thus can continue to glide. The rate-determining step is the climb process, but the strain is produced during the glide to the next obstacle. In creep-resistant alloys, precipitates or dispersions are added to prevent dislocation glide.

Diffusion creep is often the dominating mechanism at high temperatures and low stresses. Two types of diffusion creep occur: Nabarro-Herring creep and Coble creep. Under the driving force of an applied stress, atoms diffuse from the sides of the grains to the tops and bottoms in the manner shown in Fig. 12. The grain becomes longer as the applied stress does work, and the process will be faster at high temperatures because there are more vacancies. It should be noted that atomic diffusion in one direction is the same as vacancy diffusion in the opposite direction.

Nabarro-Herring creep results from diffusion of vacancies in the grains from grain boundaries that are undergoing tensile stress to areas that are in compression. At the same time, atoms flow in the opposite direction, resulting in grain elongation in the direction of axial load. It is intrinsically a very slow process but becomes operable at stresses too low for dislocation motion to be activated and at relatively high temperatures where the diffusion is fast enough to produce a measurable creep rate.

Coble creep is similar, except that it occurs at lower temperatures, and all of the flow, vacancies and atoms, occurs within the grain boundaries

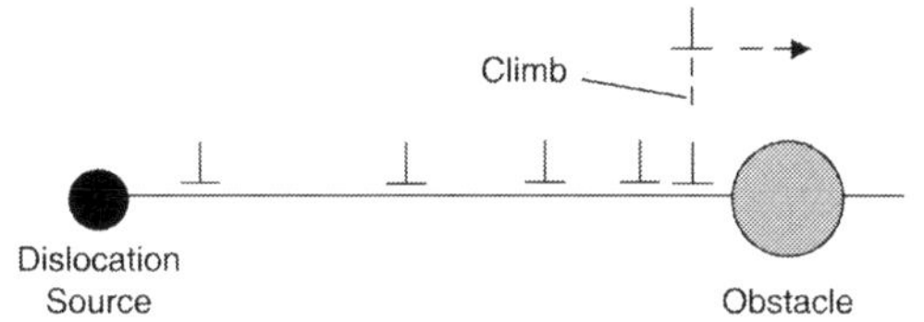

Fig. 11 Thermally activated dislocation climb. Source: Ref 3

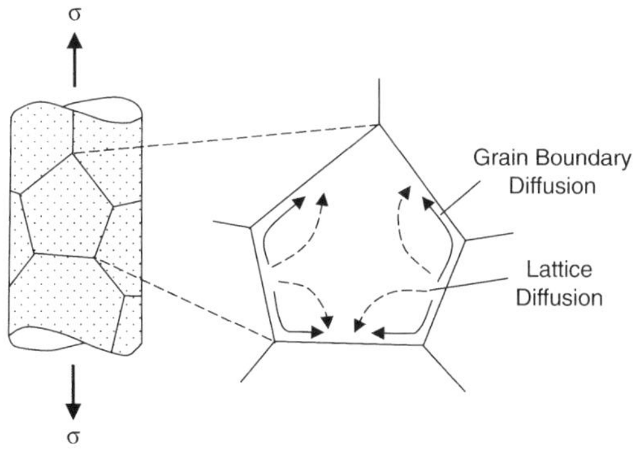

Fig. 12 Diffusion creep mechanisms. Source: Ref 3

themselves. Because diffusion is very sensitive to temperature, at lower temperatures the main diffusion path is along the grain boundaries, since the activation energy for grain-boundary diffusion is considerably less than that for bulk diffusion.

Grain-boundary sliding is often observed in the final stages of creep, just prior to failure. As the grains change shape, relative movement of the grain centers is necessary to maintain continuity at the grain boundaries. The grains actually start rotating and elongating in the direction of load. This often leads to intergranular failure modes. Grain-boundary precipitates are often used to inhibit grain-boundary sliding.

Grain size plays a role in creep of metals. The finer the grain size, the more rapid the mass transport causing permanent deformation. Thus, under conditions where creep is solely due to diffusion, the creep resistance is improved by increasing the grain size. Note that this is different from low-temperature behavior, where a fine grain size is almost always beneficial to strength and ductility.

The dominant creep mechanism at a specific stress-temperature combination is found by calculating the creep rate for the several mechanisms and determining the maximum rate. Although the net creep rate is the sum of the several rates, the maximum one is usually much greater than the others. Deformation mechanism maps having axes of stress and temperature can be used to display the results of such calculations. Regions in the diagram essentially show stress-temperature combinations in which a given creep mechanism dominates. Ashby and his coworkers have developed deformation mechanism maps that are useful in determining the dominant creep mechanism at different combinations of stress and temperature. Two deformation mechanism maps for the turbine blade alloy MAR-M-200 are shown in Fig. 13. In these maps, the shear stress (τ) normalized by the shear modulus (G) is plotted against the homologous temperature (T/T_m). In these maps, it is seen that diffusion creep mechanisms operate at low stresses and high temperatures, while at intermediate stresses, dislocation creep or power law creep is operative. By comparing Fig. 13(a) and (b), it can also be seen that increasing the grain size from 0.1 to 10 mm (3.9 mils to 0.39 in.) expands the power law creep regime and appreciably decreases the creep rate. These two maps for the same material with different grain sizes illustrate the wide variation in creep rates at a given combination of stress and temperature and help to explain why large grain sizes are favorable for many creep-limited applications.

Elevated-Temperature Fracture

Creep deformation can produce sufficiently large changes in the dimensions of a component to either render it useless for further service or cause fracture. When excessive creep deformation causes the material to reach or exceed some design limit on strain, the term *creep failure* is used. For example, a creep failure of a cobalt-base alloy turbine vane is shown in

Fig. 14. The bowing is the result of a reduction in creep strength at the higher temperatures from overheating.

Creep deformation can also result in the complete separation or fracture of a material and the breach of a boundary or structural support. Fracture can occur from either localized creep damage or more widespread bulk

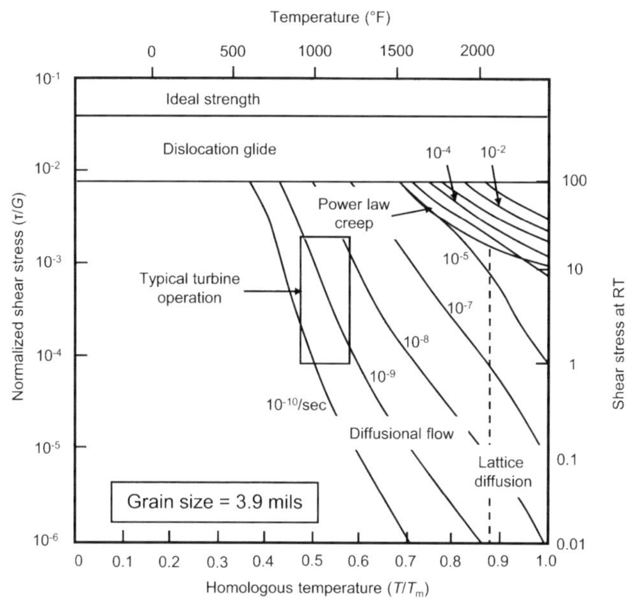

(a)

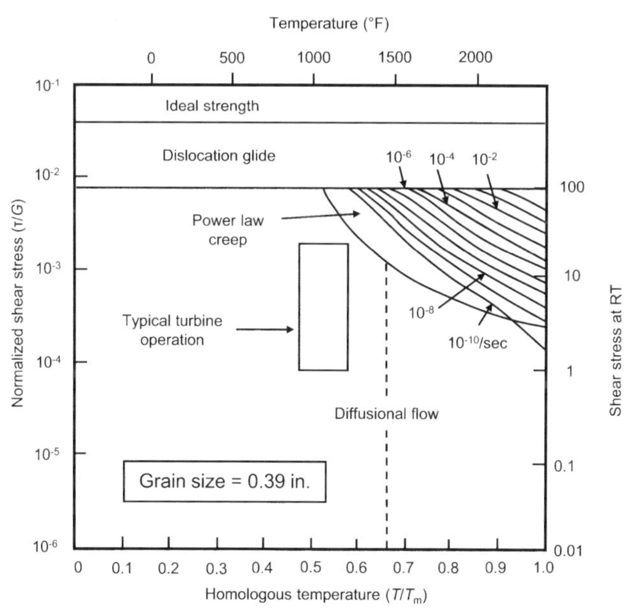

(b)

Fig. 13 Deformation mechanism maps for MAR-M-200 superalloy. RT, room temperature. Source: Ref 2

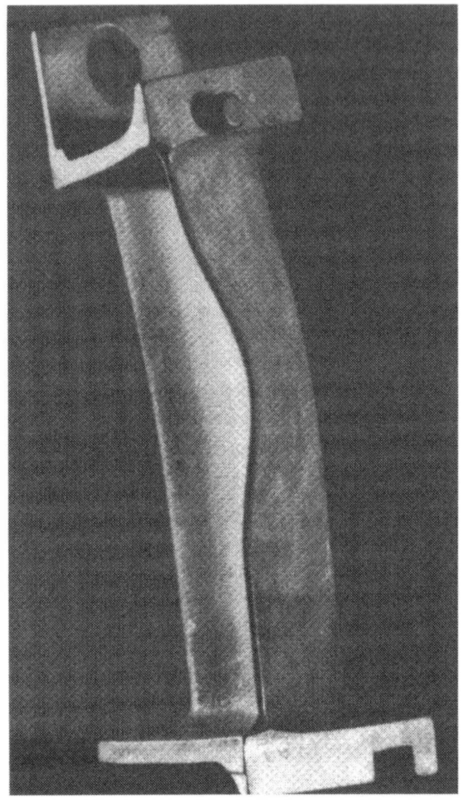

Fig. 14 Creep damage (bowing) of a cobalt-base alloy turbine vane from overheating. Source: Ref 1

damage caused by the accumulation of creep strains over time. Structural components that are vulnerable to bulk creep damage typically are subjected to uniform loading and uniform temperature distribution during service. The life of such a component is related to the creep-rupture properties, and the type of failure is referred to as stress rupture or creep rupture. Stress or creep rupture is apt to occur when damage is widespread with uniform stress and temperature exposure, as in the situation of thin-section components, such as steam pipes or boiler tubes.

On the other hand, creep damage can also be localized, particularly for thick-section components that are subjected to gradients in stress (strain) and temperature. An example of a creep-related crack is shown in Fig. 15. Cracks can develop at a critical location and propagate to failure before the end of the predicted creep-rupture life. Creep cracking can also originate at a stress concentration or at pre-existing defects in the component. In these cases, most of the life of the component is spent in crack growth. This involves assessment of fracture resistance rather than a strength assessment based on bulk creep rates and time to stress rupture. Therefore, creep-life assessment should involve evaluation of both creep strength

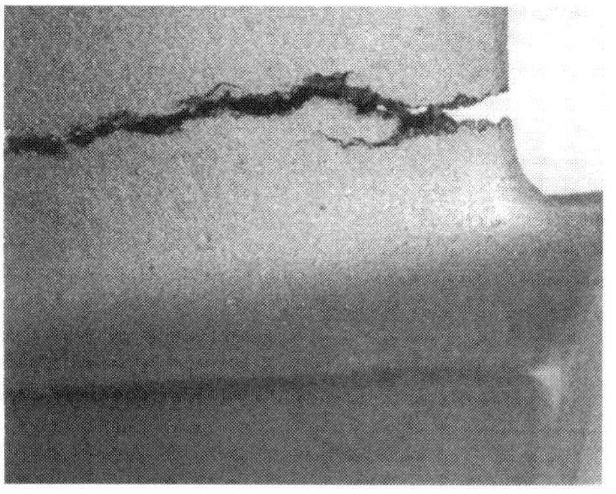

Fig. 15 Creep crack in a turbine vane. Source: Ref 1

(i.e., creep rate, stress rupture) and resistance to fracture under creep conditions.

Stress-rupture fractures are typically characterized by a multiplicity of creep voids adjacent to the main fracture. The voids are generally easy to identify by optical examination of microsections. An example of a heater tube that failed by stress rupture and microvoids that formed near the fracture surface is shown in Fig. 16.

Depending on the alloy, temperature, and strain rate, stress-rupture fracture can be macroscopically either ductile or brittle. A macroscale brittle fracture is usually intergranular and occurs with little elongation or necking. In general, lower creep rates, longer rupture times, or higher temperatures promote intergranular fractures. Ductile fractures are transgranular and are typically accompanied by more pronounced elongation and necking. Transgranular creep ruptures, which generally result from high applied stresses (high strain rates), fail by a void-forming process similar to that of microvoid coalescence in dimple rupture. Fractographs of some stress-rupture fractures exhibit both transgranular and intergranular fracture paths. In such instances, it is usually found that the transgranular fractures were initiated by prior intergranular fissures that decreased the cross-sectional area and raised the stress level.

The three primary modes of elevated fracture are rupture, transgranular fracture, and intergranular fracture, as illustrated in Fig. 17. Rupture, which is characterized by a reduction of area approaching 100%, occurs at high stresses and high temperatures. This type of failure can occur during hot working at high strain rates. Dynamic recovery and recrystallization are usually associated with rupture failures. Because it is not really a creep-dominated failure mode, it will not be considered further in this section. Transgranular fractures can occur during creep when the stress levels and strain levels are fairly high. Voids nucleate, usually around inclusions, and then grow and coalesce until fracture occurs. This type of fracture is very similar to ductile fracture modes experienced at room temperature,

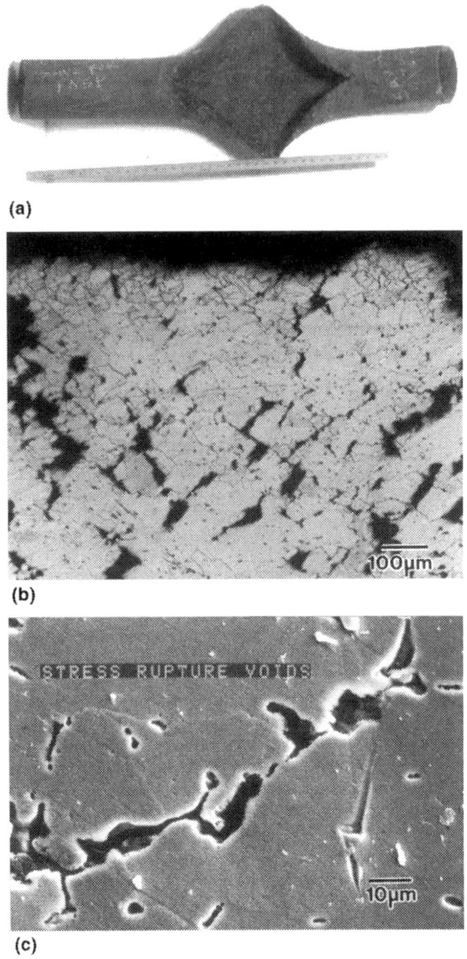

Fig. 16 Stress rupture of heater tube. (a) Heater tube that failed due to stress rupture. (b) and (c) Stress-rupture voids near the fracture. Source: Ref 6

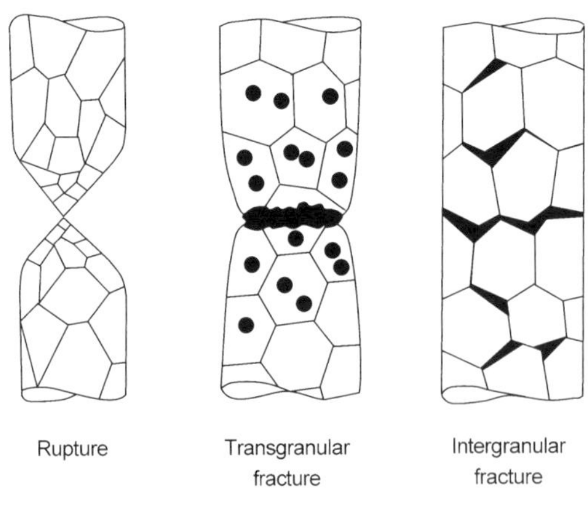

Fig. 17 High-temperature failure modes. Source: Ref 7

except that at elevated temperature, void formation and growth is aided by diffusion mechanisms. Intergranular fracture is a creep fracture mode that occurs at lower stresses and longer times. Again, voids nucleate, grow, and coalesce, but their formation and growth is primarily restricted to the grain boundaries themselves. There is very little macroscopic plastic flow, and the failure appears brittle in nature.

As the temperature is increased, metals typically undergo a transition from transgranular fracture to intergranular fracture. In other words, at low temperatures, the grain boundaries are usually stronger than the grains themselves, while at high temperatures, the grains are stronger than the grain boundaries. An example of an intergranular fracture is shown in Fig. 18 for Inconel 751 that was stress-rupture tested at 730 °C (1350 °F) and 380 MPa (55 ksi) for 125 h. At the equicohesive temperature (ECT), the strength of the grains and the grain boundaries are the same. The ECT is not a fixed temperature. In addition to the obvious effects of stress and temperature, the strain rate is important. Decreasing the strain rate lowers the ECT and increases the tendency for intergranular failure. Because the total grain-boundary area decreases with increasing grain size, a material with a larger grain size will have a higher strength above the ECT but a lower strength below the ECT.

Intergranular fracture is the more common fracture path for most stress-rupture failures, depending on stress and temperature. However, sometimes intergranular cracking may not be readily discernible on the surface of a brittle stress-rupture fracture due to a buildup of surface oxides. Inter-

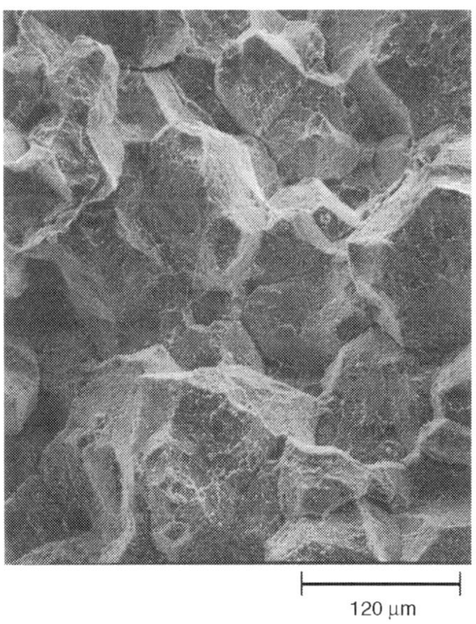

Fig. 18 Intergranular failure in nickel-base alloy. Inconel 751, stress rupture at 1350 °F, 55 ksi, 125 h. Source: Ref 8

granular creep fracture depends on the nucleation, growth, and subsequent linking of voids on grain boundaries to form two types of cavities: wedge-type (w-type) cavities or isolated, rounded-type (r-type) cavities (Fig. 19). Cavitation damage, or the nucleation and growth of voids, is the most common microstructural change during creep and stress-rupture testing. Stress-rupture fractures are characterized by multiple creep voids adjacent to the main fracture. Wedge-shaped (w-type) cracks initiate along the grain boundaries that aligned in shear, while round or elliptical cavities (r-type) form along the boundaries that are aligned in tension. Under low stress conditions, intergranular fractures occur by void formation at the grain boundaries. These cavities form along grain edges rather than at grain corners. Because they appear to be round or spherical on metallographic cross sections, these voids are sometimes referred to as r-type cavities.

Wedge or w-type are usually associated with cracking at grain-boundary triple points and are also referred to as triple-point or grain-corner cracks. Wedge-type cracks form at triple points due to grain-boundary sliding and may be promoted by decohesion at interfaces between grain-boundary precipitates and the matrix. High stresses and lower temperatures promote wedge crack formation. Wedge cracking at triple points produces a rough fracture surface with identifiable grain-boundary precipitates.

Other significant microstructural features also are associated with stress-rupture fracture. When intergranular cracking occurs, not only do the crack paths follow grain boundaries at and beneath the fracture surfaces, but also the grains appear equiaxed even after considerable plastic deformation and total elongation, which implies dynamic recrystallization. In contrast, transgranular fracture frequently exhibits severely elongated grains in the vicinity of the fracture, with no recrystallization.

Metallurgical Instabilities

Stress, time, temperature, and environment can change the metallurgical structure during service and hence change the strength. The micro-

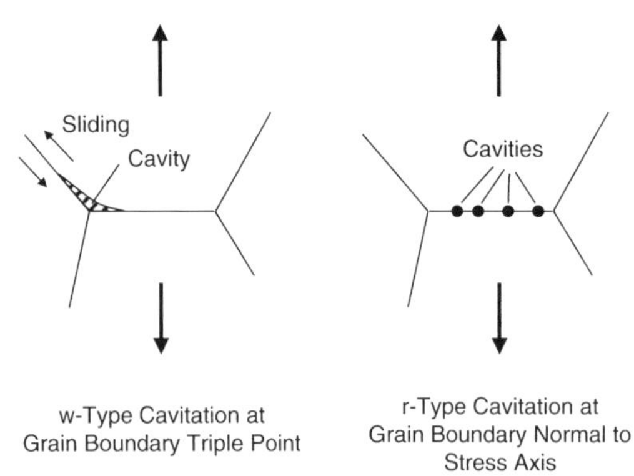

Fig. 19 Creep cavitation mechanisms. Source: Ref 9

structure of a creep or stress-rupture specimen after elevated-temperature exposure rarely resembles the initial microstructure, and prolonged exposure under creep conditions can result in microstructural changes, such as the precipitation of new phases, dissolution or growth of desired phases, grain growth, and so on.

A sharp change of slope in a rupture-life curve can result from metallurgical instabilities, such as transgranular-to-intergranular fracture transition, recrystallization, aging or overaging (phase precipitation or decomposition of carbides, borides, or nitrides), intermetallic phase precipitation, delayed transformation to equilibrium phases, order-disorder transition, general oxidation, intergranular corrosion, stress-corrosion cracking, slag-enhanced corrosion, and contamination by trace elements. Other metallurgical changes, such as spheroidization and graphitization, and corrosion effects can also occur during long-term exposure at elevated temperature.

Transgranular-Intergranular Fracture Transition. The principal metallurgical factor in stress-rupture behavior is the transition from transgranular fracture to intergranular fracture at the ECT. The ECT varies with exposure time and stress. For each combination of stress and rupture life, there is a temperature above which all stress-rupture fractures are intergranular. For shorter rupture lives and lower stresses, this transgranular-to-intergranular transition occurs at higher temperatures. Longer rupture lives and higher stresses occur at lower temperatures. This effect is illustrated by points A and B in Fig. 20. Points N, O, and Y represent other types of metallurgical instabilities. Under certain conditions, both transgranular and intergranular fractures are found. Consequently, an analysis of rupture-life data or component failure is not complete without a thorough metallographic examination to establish the initial failure mechanism.

Aging. Age-hardening alloys, which include certain stainless steels, heat-resisting alloys, and some aluminum alloys, are characteristically unstable when they are in a state of transition to the stable (equilibrium) condition. Overaging in alloys can occur when they are exposed to ele-

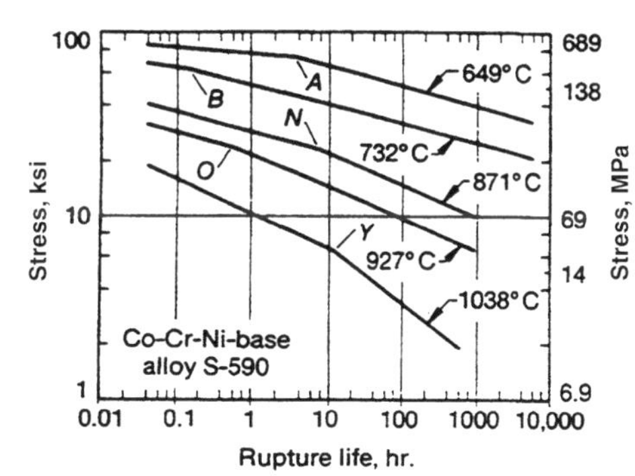

Fig. 20 Logarithmic plot of stress-rupture stress versus rupture life for Co-Cr-Ni-base alloy S-590. The significance of inflection points A, B, N, O, and Y is explained in the text. Source: Ref 6

vated operating temperatures and without being loaded. Under creep conditions, temperature-induced and stress-induced atomic migration usually causes aging to continue, resulting in reduced strength. The extent and nature of this change depends on several factors, including the condition of the alloy prior to creep and the temperature, stress, and time of exposure.

Some of the more common high-temperature structural alloys that harden as a result of decomposition of highly supersaturated solid solutions include the Nimonic alloys (Ni-Cr-Al-Ti), austenitic steels that do not contain strong carbide formers, and secondary hardening ferritic steels. These alloys are widely used for their creep resistance, but they are not immune to reduced rupture life due to overaging.

Intermetallic Phase Precipitation. Topologically close-packed (tcp) phases, such as sigma, mu, and Laves phases, form at elevated temperatures in austenitic high-temperature alloys. The morphology (shape) of the precipitates determines the effect they have on creep strength. For example, needlelike precipates reduce the toughness and creep strength. Not all of the effects of such phases on rupture life are well known, but some are well understood, such as the behavior of certain nickel-base alloys. The effect of sigma phase (a hard, brittle intermediate phase) on alloy U-700 creep strength at 815 °C (1500 °F) is shown in Fig. 21. Here, a pronounced break is shown in the rupture curve starting at approximately 1000 h. Sigma phase was identified in this alloy system and was clearly associated with the failure. However, it was found that sigma did not have a similar effect in certain other nickel-base alloys. Therefore, it must be concluded that sigma phase does not seem to have a universally deleterious effect on stress-rupture behavior. The amount, location, and shape of sigma-phase precipitation determine whether sigma strengthens or weakens an alloy, or whether it has no effect. Sigma and other intermetallics can severely reduce ductility and toughness on subsequent cooling to room temperature.

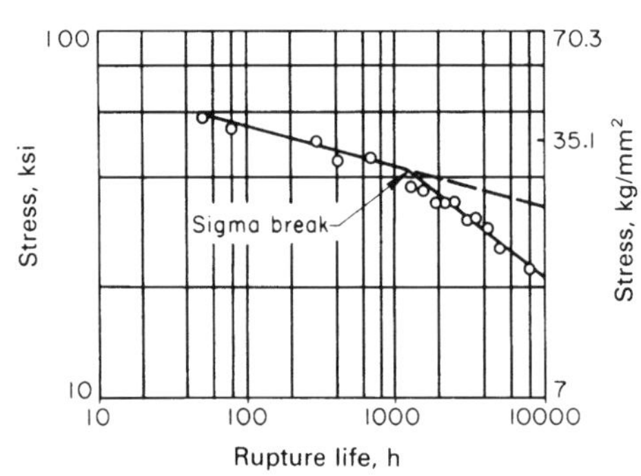

Fig. 21 Logarithmic plot of stress-rupture stress versus rupture life for nickel-base alloy U-700 at 815 °C (1500 °F). The increasing slope of the curve to the right of the sigma break is caused by sigma-phase formation. Source: Ref 1

The inconsistency of the effect of sigma-phase formation on creep and stress-rupture properties may arise from the simultaneous presence of other phases, such as carbides. The shape, distribution, and chemistry of carbide particles can influence crack initiation and propagation, and hence the resultant stress-rupture ductility and rupture life, in a pronounced manner. It is improbable that sigma phase affects ductility to a significant degree at low strain rates. Consequently, sigma need not always result in deterioration of creep and stress-rupture properties, unless it is present in relatively large amounts. Because the presence of sigma does not automatically result in decreased rupture life, careful metallographic work must be performed on failures to ensure discrimination between sigma-promoted failures and other types of failures in which sigma or other tcp phases are merely present.

Carbide Reactions. Several types of carbides are found in steels as well as in heat-resisting alloys. Although temperature and stress affect carbides both within grains and in the grain boundaries, the effects of grain-boundary carbides usually are a more significant factor in altering creep behavior. The presence of carbides is considered necessary for optimal creep and stress-rupture resistance in polycrystalline metals, although subsequent alteration in their shape or breakdown and transition to other carbide forms may be a source of property degradation and resulting failure.

In acicular form, grain-boundary carbides do not appear to act as brittle notch formers that may directly affect rupture life at elevated temperatures, but they may reduce impact strength. Indirectly, compositional changes in the vicinity of carbides can alter rupture strength. In general, acicular M_6C carbides are not believed to affect the properties of nickel-base high-temperature alloys greatly, unless the alloying elements involved in the carbide reaction alter the matrix composition noticeably. However, continuous carbide films formed at grain boundaries can decrease rupture life. An electron micrograph of grain-boundary carbide films in a Waspaloy forging is shown in Fig. 22. These films substantially reduced the stress-rupture life of the alloy.

Carbides, for example, MC, form in steel when carbon combines with reactive elements, such as titanium, tantalum, hafnium, and niobium. During heat treatment and service, the MC carbides can decompose and generate other carbides, such as $M_{23}C_6$, which tends to form along the grain boundary. Specific transitions are known for common Fe-Cr-Mo alloys, such as the sequence of carbide formation in Fig. 23 for 2.25Cr-1Mo steel.

Long-term exposure to elevated temperature can affect either short-term or long-term properties. For example, the initial microstructure of creep-resistant chromium-molybdenum steels consists of bainite- and ferrite-containing Fe_3C carbides, ε-carbides, and fine M_2C carbides. Although a number of different carbides may be present, the principal carbide phase responsible for strengthening is a fine dispersion of M_2C

carbides, where M is essentially molybdenum. With increasing elevated-temperature aging in service, a series of transformations of the carbide phases takes place that eventually transforms M_2C into M_6C and $M_{23}C$, where the M in the latter two metal carbides is mostly chromium. Such an evolution of the carbide structure results in coarsening of the carbides,

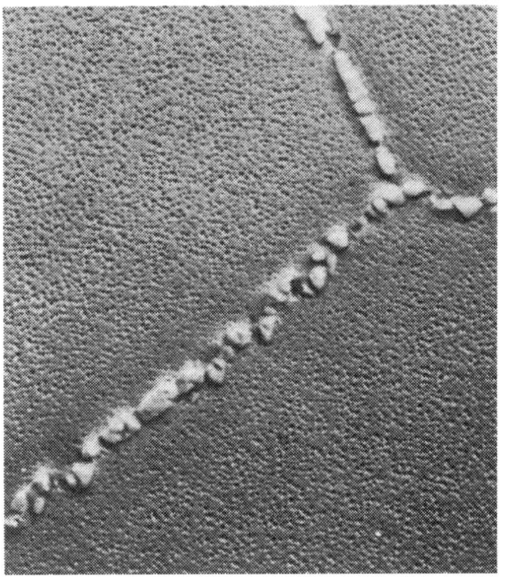

Fig. 22 Grain-boundary carbide films in a Waspaloy forging. The films substantially reduced stress-rupture life. Transmission electron micrograph. Original magnification: 4000×. Source: Ref 1

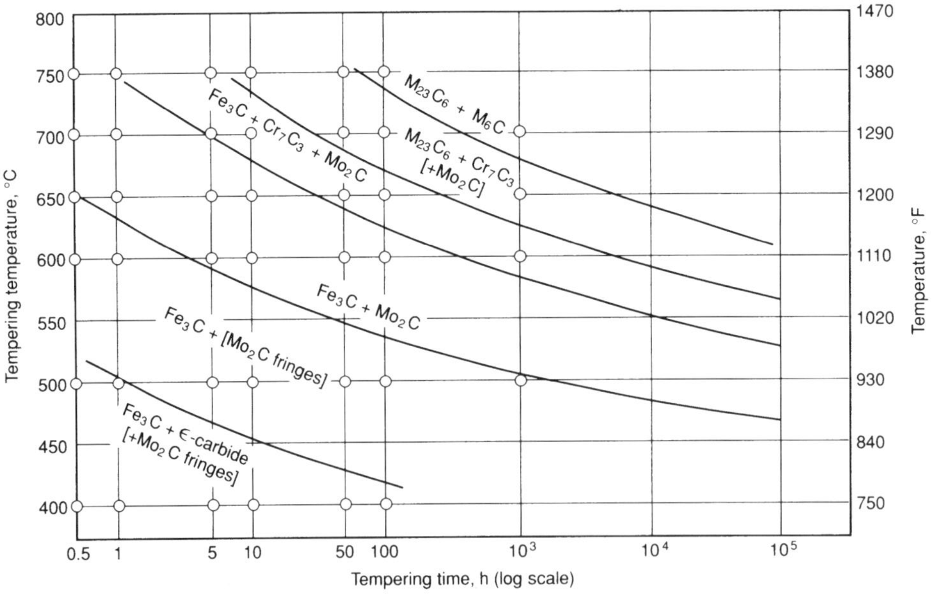

Fig. 23 Isothermal diagram showing the sequence of carbide formation on tempering of normalized 2.25Cr-1Mo steel. Source: Ref 10

changes in the matrix composition, and an overall decrease in creep strength. The effect of exposure on the stress-rupture strength of two chromium-molybdenum steels is shown in Fig. 24.

Interaction of Precipitation Processes. The inconsistent behavior of phase precipitation in alloys at elevated temperature is very often due to the fact that carbide and tcp phases do not always react independently but may interact with each other. Because of these reactions, which can become extremely complex, a thorough microstructural examination by the failure analyst must be made (usually with the aid of an electron microscope) before accurate conclusions can be drawn.

Creep-Rupture Embrittlement. The accumulation of creep damage, in conjunction with metallurgical changes, can cause embrittlement and a reduction in rupture ductility. In low-alloy steels, it has similarities with temper embrittlement but is irreversible. In low-alloy steels, the temperature range of creep embrittlement is 425 to 590 °C (800 to 1100 °F) and appears closely related to formation of fine intergranular precipitates during creep. Impurities such as phosphorus, sulfur, copper, arsenic, antimony, and tin have been shown to reduce rupture ductility.

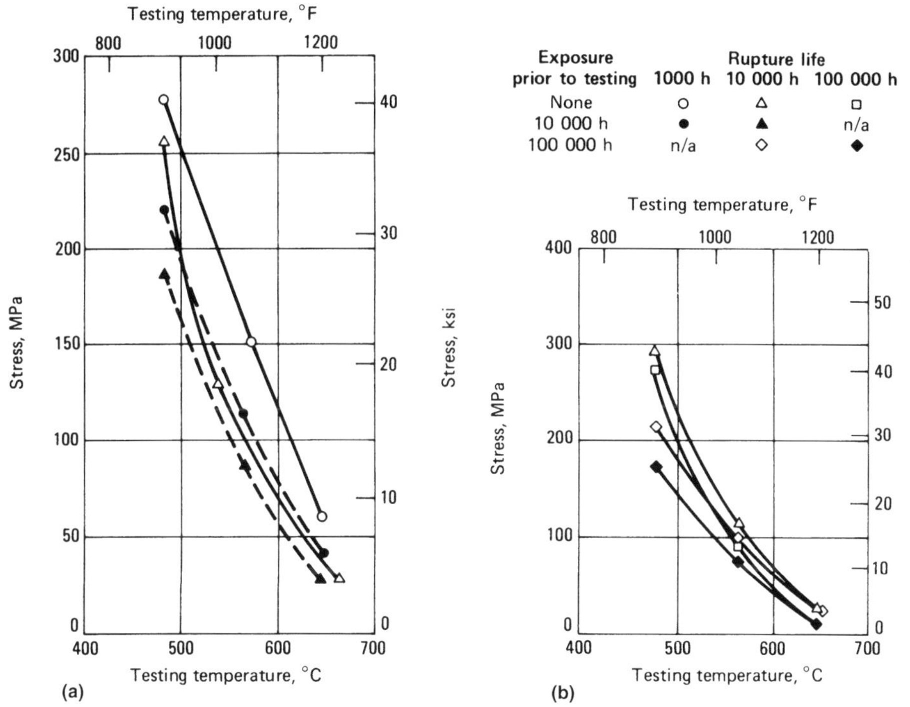

Fig. 24 Effect of elevated-temperature exposure on stress-rupture behavior of (a) normalized and tempered 2Cr-1Mo steel and (b) annealed 9Cr-1Mo steel. Exposure prior to stress-rupture testing was at the indicated test temperatures (without stress) and was 10,000 h long for the 2Cr-1Mo steel and 100,000 h long for the 9Cr-1Mo steel. n/a, data not available at indicated exposure and rupture life. Source: Ref 1

Environmental Effects

It has long been known that test environment can affect creep-rupture behavior. Until recently, however, the work has been largely empirical, with creep tests being conducted in various atmospheres and differences noted in creep rates and rupture lives. Often, the effects on ductility were not reported, and there were very few studies of crack propagation.

A renewed interest developed in the 1970s as a result of observations of a dominant role played by the environment in high-temperature fatigue crack growth of superalloys. There were subsequent studies of sustained load crack propagation that also showed very strong effects. In some cases at high stresses, the test environment was so severe, as in the case of sulfur, that profound changes were seen in smooth-bar rupture life. Such an example is shown in Fig. 25, in which the time to rupture in common superalloys was reduced by several orders of magnitude in tests in a sulfate/chloride mixture at 705 °C (1300 °F). These results were explained in terms of grain-boundary penetration of sulfur, which leads to rapid crack propagation. The coated specimen was far less susceptible, and the addition of a grain-boundary modifier, in this case boron, in Udimet 720 gave an enormous improvement relative to Udimet 710.

Embrittling Effects of Oxygen. At about the same time that the ideas on environmental attack at an intergranular crack tip were being developed, it was also shown that short-term prior exposure in air at high tem-

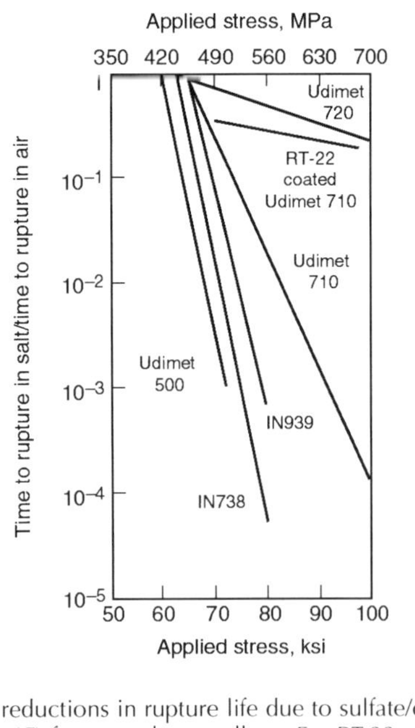

Fig. 25 Relative reductions in rupture life due to sulfate/chloride salt at 705 °C (1300 °F) for several superalloys. For RT-22-coated Udimet 710, rupture time in salt for coated alloy divided by time in air for uncoated alloy. Source: Ref 11

perature (greater than approximately 900 °C, or 1650 °F) could lead to profound embrittlement at intermediate temperatures (700 to 800 °C, or 1290 to 1470 °F). This was shown to be caused by intergranular diffusion of oxygen that penetrated on the order of millimeters in a few hours at 1000 °C (1830 °F). The embrittlement was monitored using measurements of tensile ductility at intermediate temperatures in iron-, nickel-, and cobalt-base alloys. An example of the results for the Fe-Ni-Co alloy (IN903A) is shown in Fig. 26, which also confirms the extent of damage penetration from tests in which the specimen diameter of 2.54 mm (0.1 in.) was reduced by half. Postexposure tests on cast alloys showed that this embrittlement could also lead to a reduction in rupture life of several orders of magnitude. An example for alloy IN738 is shown in Fig. 27.

Using model alloys based on nickel, it was shown that oxygen in the elemental form in high-purity nickel did not embrittle; that is, a chemical reaction was necessary. Three embrittling reactions were confirmed: a reaction with carbon to form carbon dioxide gas bubbles; a reaction with

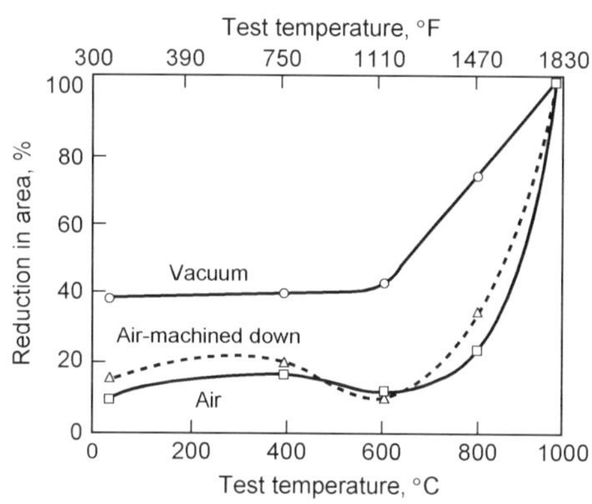

Fig. 26 Tensile ductility of IN903A after air and vacuum exposures at 1000 °C (1830 °F) for 100 h as a function of test temperature in vacuum tests. Embrittlement remained after reducing to half the initial diameter. Source: Ref 12

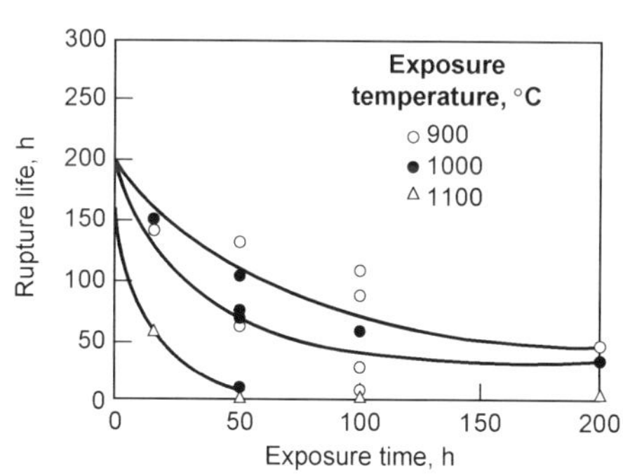

Fig. 27 Effect of exposure in air at various temperatures on stress-rupture life of IN738 at 800 °C (1470 °F) and 400 MPa (58 ksi). Source: Ref 13

sulfides on grain boundaries to release sulfur, which embrittles in the elemental form; and a reaction with oxide formers to form fine oxides that act to pin grain boundaries. These phenomena are believed to be the same processes that serve to embrittle the region ahead of a crack tip. Thus, oxygen attack may occur dynamically to account for the accelerated advance of a crack in air tests compared with inert environment tests, and it may occur during higher-temperature exposure with or without an applied stress to set up an embrittlement situation. Thermal fatigue in combustion turbines is a particularly challenging situation for oxygen attack, because maximum strains develop at intermediate temperatures in the cycle, but holding may be at the maximum temperature.

Combined Effects of Oxygen and Carbon. Of special interest is the reaction between diffusing oxygen and carbon. In nickel, it was found that if this reaction were prevented, creep cavitation could not develop during creep tests. Prevention was achieved either by removing the carbon (decarburizing) or by applying an environmental protective diffusion or overlay coating. Air tests at 800 °C (1470 °F) and at various stresses showed an enormous increase in rupture life (Fig. 28) if the gas bubble formation did not occur. Because nickel has been used as an archetypical metal for the study of creep cavitation, the confirmation that the cavities are nucleated as gas bubbles now solves the problem of nucleation. The observation that even in superalloys gas bubbles are frequently nucleated at carbides, which may serve as cavity nuclei, and that cavitation during creep is often concentrated near the specimen surface points to the likelihood

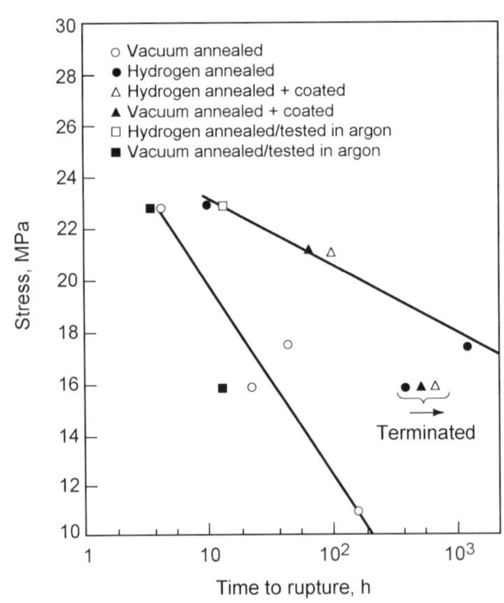

Fig. 28 Effect of environmental interaction on rupture life of Ni270 at 800 °C (1470 °F). Longer lives are obtained by preventing cavitation nucleation from carbon dioxide gas formation. This is achieved by decarburizing to eliminate carbon or by coating to prevent oxygen penetration. Source: Ref 14

that oxygen attack may be invariably associated with creep cavitation. The gas pressures developed in the bubbles appear to be quite adequate for nucleation.

Effect of Other Gaseous Elements. Hydrogen, chlorine, and sulfur can also embrittle as a result of penetration. Sulfur is particularly aggressive in that it diffuses more rapidly and embrittles more severely than does oxygen. It is frequently found in coal gasification and oil refining processes, as well as industrial gas turbines operating on impure fuel.

Creep-Life Prediction

A frequent problem encountered in alloy design is that it may be desirable to use a new alloy in a high-temperature structure and the alloy is so new that there is not sufficient data to substantiate its long-term creep resistance. Therefore, it is necessary to extrapolate higher-temperature data to much longer times at lower temperatures. The creep test is accelerated by raising the temperature, and the time difference is accounted for by using the activation energy, assuming that it is the same at all temperatures. Data are required at a range of temperatures and stresses. To extrapolate the data to the required times, a considerable amount of effort has been expended in developing empirical prediction methods. One of the most widely used is the Larson-Miller parameter, P:

$$P = T(\ln t + C) = \frac{Q}{R} \quad \text{(Eq 5)}$$

where T is the absolute temperature, t is the time to rupture in hours, Q is the activation energy for creep, R is the universal gas constant, and C is the Larson-Miller constant, which typically has a value in the range of 30 of 65. A value of 46 is often used for metals when a specific value is not determined.

If P is evaluated for pairs of t and T for a number of different stress levels, a single master curve can be plotted for the material. An actual set of stress-rupture data is shown for the nickel-base alloy Inconel 718 as log stress versus log time to rupture in Fig. 29(a). The data are then replotted as constant time curves on coordinates of stress versus temperature in Fig. 29(b). Dashed horizontal lines have been added at stress levels of 550, 620, 760, and 830 MPa (80, 90, 110, and 120 ksi). Values for T at the intercepts of these dashed lines and the constant-time curves are then determined and plotted in Fig. 29(c) on coordinates of log t versus $10^4/T_A$. By extending the data in Fig. 29(c), a set of converging isostress lines meeting on the ordinate at a value of log $t = -25$ can be obtained. The Larson-Miller equation for this set of data is:

$$P = T_A(\log t_r + 25) \quad \text{(Eq 6)}$$

where absolute temperature, T_A, is in units of the Kelvin or Rankine temperature scales. The final master Larson-Miller curve is shown in Fig. 29(d).

One of the dangers of any of the extrapolation methods is an unanticipated structural instability in the material that may go undetected. Because the instabilities usually occur at higher temperatures, several tests should be conducted at substantially higher temperatures than the actual part will see in service. It should also be pointed out that the Larson-Miller parameter is only one of many methods for extrapolating stress-rupture data.

Design against Creep

Design against creep can take several approaches. First, it makes sense to use a material with a high melting temperature in high-temperature applications, because to a good approximation, diffusion coefficients on

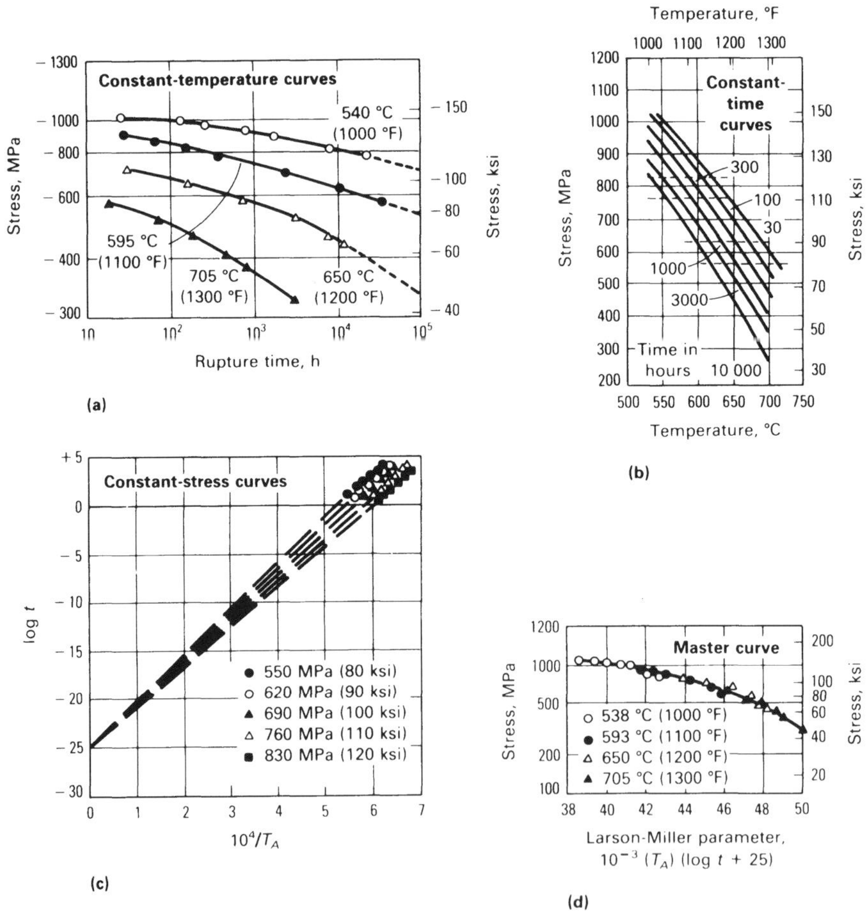

Fig. 29 Larson-Miller curve generation for Inconel 718. Source: Ref 15

which the creep rate depend scale with the homologous temperature (T/T_m). Face-centered cubic (fcc) metals generally have superior creep resistance to body-centered cubic (bcc) metals at equivalent homologous temperatures, because the slightly more open bcc structure results in greater diffusivities.

Design against creep may also involve first determining the dominant creep mechanism. If diffusion creep dominates, then increasing the grain size is a beneficial structural alteration. Further, for diffusion creep, improved creep resistance can sometimes be obtained by placing inert particles, such as carbides, on grain boundaries. This helps to pin the grain boundaries, enhancing creep resistance. Creep-resistant alloys include carbon steels, chromium-molybdenum steels, chromium-molybdenum-vanadium steels, stainless steels, nickel and cobalt alloys, and superalloys. A comparison of the stress-rupture properties for a number of different metallic alloy families is shown in Fig. 30.

Depending on the operating stress, carbon steels can be used at operating temperatures between 400 and 480 °C (750 and 900 °F). Above approximately 480 °C (900 °F), the pearlite in carbon steels begins to transform to graphite, which leads to embrittlement. Chromium-molybdenum steels contain 0.5 to 9.0% Cr and either 0.5 or 1.0% Mo. The chromium content increases oxidation resistance, while molybdenum increases elevated-temperature strength and forms carbides that help prevent graphitization. Chromium-molybdenum-vanadium steels contain up to 0.5% C, 5% Cr, 1% Mo, and 0.9% V. The vanadium addition increases the softening point of the steel. Both the chromium-molybdenum and the chromium-

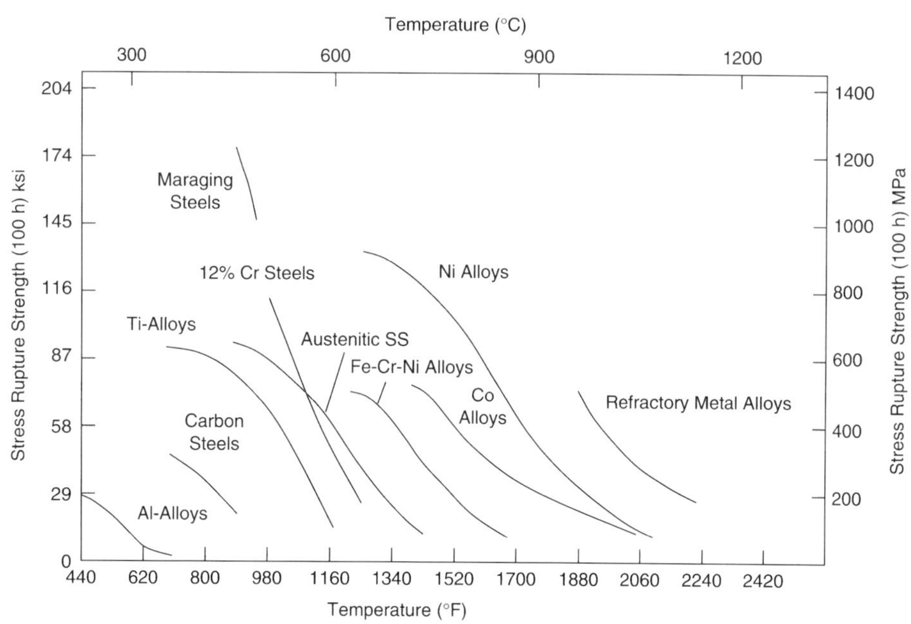

Fig. 30 Stress-rupture comparison for several classes of alloys. Source: Ref 16

molybdenum-vanadium steels extend the operating range up to 540 °C (1000 °F).

Stainless steels are also used for elevated-temperature applications. Martensitic stainless steels are used at temperatures up to 540 to 650 °C (1000 to 1200 °F) but must be tempered at approximately 56 °C (100 °F) higher than the operating temperature to prevent softening in service. Ferritic stainless steels, due to their high chromium contents, have the best scaling resistance but are limited to approximately 370 °C (700 °F) due to the precipitation of embrittling phases at higher temperatures. Austenitic stainless steels have the best creep resistance of the stainless steels and are used at temperatures up to 870 °C (1600 °F). However, a major problem with the austenitic grades is their high coefficients of thermal expansion, which must be compensated for during design.

Nickel-base alloys are frequently used for high-temperature service in corrosive environments. These alloys contain alloying elements such as chromium, molybdenum, and iron. The addition of chromium and molybdenum to nickel produces alloys with resistance to both oxidizing and reducing corrosives. Resistance to oxidizing environments is provided by the chromium, while molybdenum gives resistance to reducing environments. These alloys are the most corrosion resistant of the nickel alloys. The Ni-Cr-Mo alloys are used in the chemical processing, pollution control, and waste treatment industries to utilize their excellent combination of heat and corrosion resistance.

Superalloys are heat-resistant alloys of nickel, iron-nickel, and cobalt that frequently operate at temperatures exceeding 540 °C (1000 °F). However, some superalloys are capable of being used in load-bearing applications in excess of 85% of their incipient melting temperatures. They exhibit the best combination of high strength, good fatigue and creep resistance, good corrosion resistance, and the ability to operate at elevated temperatures for extended periods of time (i.e., metallurgical stability). Their combination of elevated-temperature strength and resistance to surface degradation is unmatched by other metallic materials.

Nickel-base superalloys have the best creep resistance of the superalloys and are used as turbine blades in engines operating at temperatures as high as 1290 °C (2350 °F) or $T/T_m = 0.9$. The microstructure consists of fine Ni_3Al particles dispersed in an fcc nickel-rich solid solution. These alloys are remarkably resistant to particle coarsening. Solid-solution hardening is also used to strengthen the matrix, and when used in polycrystalline form, carbides are dispersed on the grain boundaries. Increases in engine operating temperatures over the last 50 years have been gradual but significant, from approximately 400 °C (750 °F) to above 1290 °C (2350 °F). This increase in operating temperature translates into improved engine efficiency. Improved high-temperature material properties are largely responsible for the increase in the engine operating temperature. Some improved material performance has come about from minor alterations in alloy chemistry, and some has resulted from processing changes. For example, the major stress axis in a turbine blade is parallel to the blade axis. In a polycrystalline blade, this stress is normal to some grain boundaries, and this causes voids, precursors to fracture, to initiate on the boundaries. Directional solidification (i.e., solidifying the material sequentially from the blade bottom to its top) results in columnar grains having boundaries aligned along the blade axis. This significantly reduces the cavitation problem. Further improvements came with the casting of single-crystal blades. Increasing the superalloy grain size from 0.1 to 10 mm (0.004 to 0.4 in.), on the order of the thickness of a turbine blade, reduces the creep rate by approximately 6 orders of magnitude when Coble creep is the dominant creep mechanism, as it often is under typical blade operating conditions.

High-Temperature Fatigue

The fatigue life normally increases with decreasing temperatures. Although steels become more notch sensitive at low temperatures, there is no experimental evidence that indicates that any sudden change in fatigue properties occurs at temperatures below the ductile-to-brittle transition temperature.

The *S-N* curves for a number of metals tested in fully reversed bending at different temperatures (Fig. 31) show that the fatigue strength decreases with increasing temperature. An exception is mild steel, which exhibits a maximum in fatigue strength between 250 and 350 °C (480 and 660 °F) due to strain aging. In addition, the fatigue crack growth rate increases at

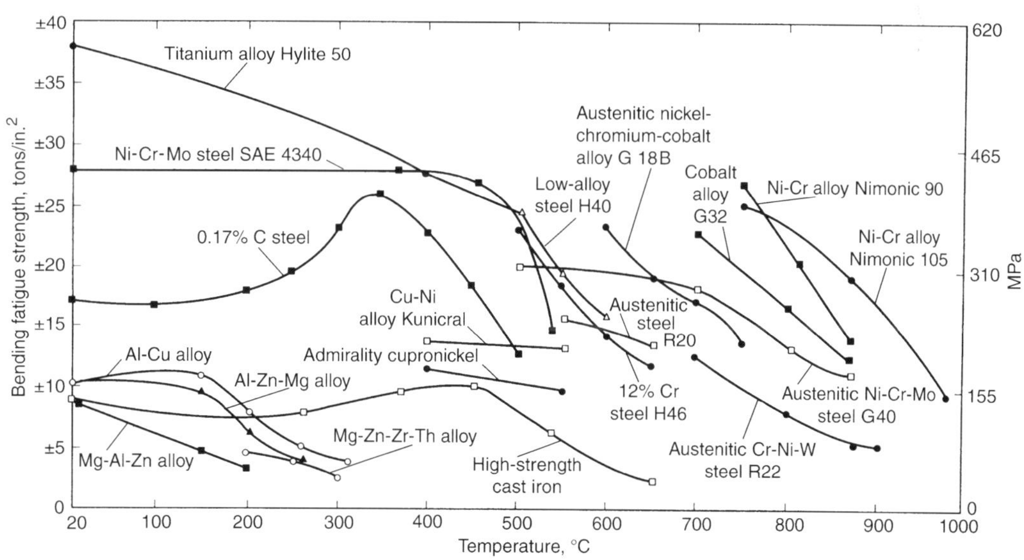

Fig. 31 Temperature influence on fully reversed fatigue strengths of metals. Source: Ref 17

elevated temperatures, as shown for the 2.25Cr-1Mo steel in Fig. 32. As the temperature exceeds approximately ½ T_m, creep can become the dominant cause of failure. Ferrous alloys, which usually exhibit an endurance limit at room temperature, will no longer have an endurance limit when the temperature exceeds approximately 425 °C (800 °F). In general, the higher the creep strength of an alloy, the higher will be its high-temperature fatigue strength. However, there are conflicting requirements, with fine grain sizes being beneficial to fatigue properties, while coarse grain sizes are preferred in certain creep regimes. In addition, fatigue-improvement mechanisms, such as shot peening, which are effective at room temperature, may be annealed out at elevated temperatures.

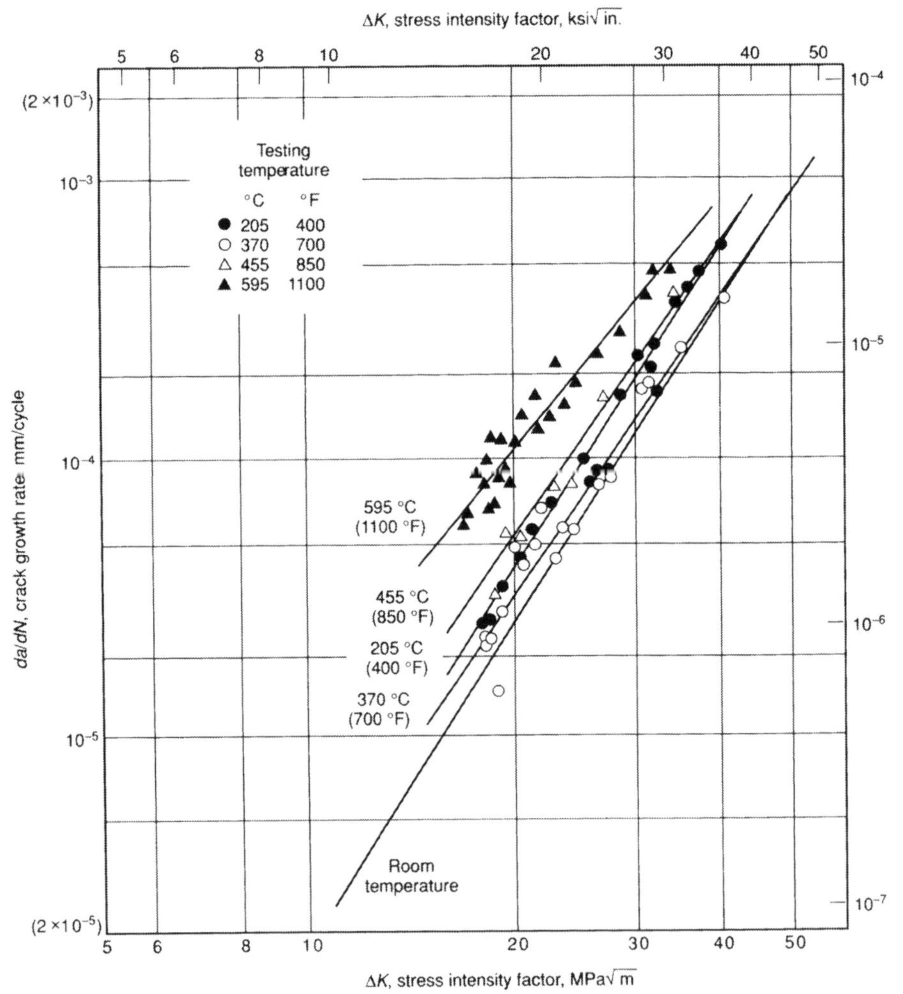

Fig. 32 Effect of temperature on fatigue crack growth rate for 2.25Cr-1Mo steel tested in air. $R = 0.05$; cyclic frequency of 400/min. Source: Ref 18

Creep-Fatigue Interaction

In elevated-temperature structures, it is quite common to have both creep and fatigue conditions operating at the same time. A possible creep-fatigue interaction is shown in the elevated-temperature constant-strain fatigue curves of Fig. 33, where a dwell at both the high and low stress levels produces creep that reduces the stress. While either creep or fatigue is complicated enough when they are operating separately, combining creep and fatigue produces a truly complex situation.

There are number of approaches to addressing this combined problem, but one of the simplest is the damage accumulation approach. This approach is essentially an extension of the Palmgren-Miner rule for predicting cumulative fatigue damage. In other words, the independent damage from fatigue and creep is summed and compared to the limit of damage the material can withstand:

$$\sum_{\mathrm{j=l}}^{\mathrm{j=k}}\left(\frac{n}{N_{\mathrm{d}}}\right)_{\mathrm{j}}+\sum_{\mathrm{l=l}}^{\mathrm{l=m}}\left(\frac{t}{t_{\mathrm{d}}}\right)_{\mathrm{l}}\leq D \qquad \text{(Eq 7)}$$

where n is the number of fatigue cycles applied at loading condition j, N_d is the number of design-allowable fatigue cycles at loading condition j, t is the time under the applied creep load condition l, t_d is the allowable time at load condition l, and D is the total allowable creep-fatigue damage, which is usually taken as 1.0 but sometimes less than 1.0.

A plot of the combined effects of creep and fatigue is shown in Fig. 34. Here, there are three types of interaction shown. When the contributions of creep and fatigue are the same, Eq 7 is obeyed, D is equal to 1.0, and a

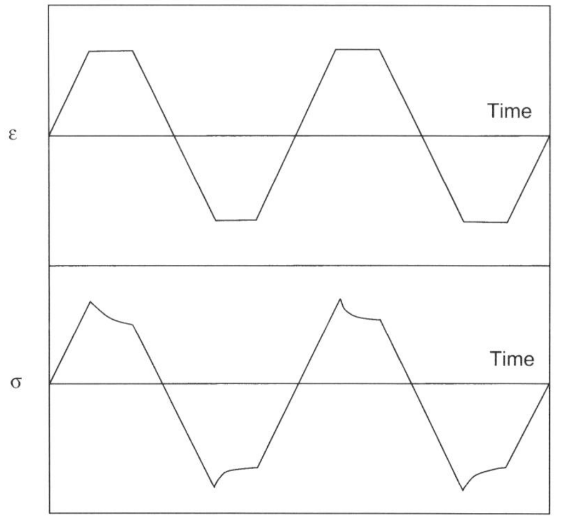

Fig. 33 Elevated-temperature fatigue curves with high- and low-stress dwells. Source: Ref 3

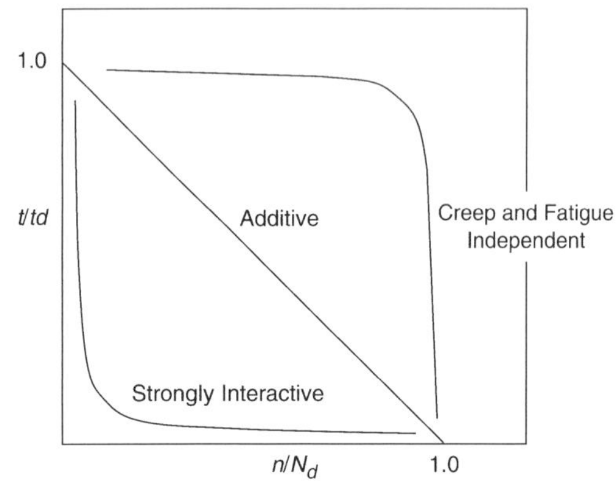

Fig. 34 Example of fatigue-creep interaction. Source: Ref 3

straight-line interaction is obtained. In essence, this results in a halving of life in a case where creep and fatigue would be expected to result in failure at the same number of cycles. For the case where there is no interaction, creep and fatigue have no effect on each other, and their respective lifetimes are unchanged by damage from the other mechanism. Finally, if there is a strong interaction, the lifetimes for both creep and fatigue are much shorter as a result of the interaction of the two. In this case, if life is being predicted on the basis of the creep and fatigue data separately, there will be a serious problem.

Thermal Fatigue

Structural alloys are commonly subjected to a variety of thermal and thermomechanical loads. If the stresses in a component develop under thermal cycling without external loading, the term *thermal fatigue* or *thermal stress fatigue* is used. This process can be caused by steep temperature gradients in a component or across a section and can occur in a perfectly homogeneous isotropic material. For example, when the surface is heated, it is constrained by the cooler material beneath the surface, and the surface undergoes compressive stresses. On cooling, the deformation is in the reverse direction, and tensile stresses could develop. Under heat/cool cycles, the surface will undergo thermal fatigue damage. Examples of thermal fatigue are encountered in railroad wheels subjected to brake-shoe action, which generates temperature gradients and consequently internal stresses.

On the other hand, thermal fatigue can develop even under conditions of uniform specimen temperature, instead caused by internal constraints such as different grain orientations at the microlevel or anisotropy of the thermal expansion coefficient of certain crystals (noncubic). Internal strains and stresses can be of sufficiently high magnitude to cause growth, distortion, and surface irregularities in the material. Consequently, thermal cycling results in damage and deterioration of the microstructure. This behavior has been observed in pure metals such as uranium-, tin-, and cadmium-base alloys and in duplex steels with ferritic/martensitic microstructures.

The tendency for thermal fatigue has been related to the parameter:

$$\sigma_f k / E\alpha \qquad \text{(Eq 8)}$$

where σ_f is the mean fatigue strength, and k is the thermal conductivity. A high value of this parameter indicates good resistance to thermal fatigue.

Thermomechanical Fatigue

The term *thermomechanical fatigue* describes fatigue under simultaneous changes in temperature and mechanical strain. Mechanical strain is defined by subtracting the thermal strain from the net strain, which should be uniform in a specimen. The mechanical strain arises from external constraints or externally applied loading. For example, if a specimen is held between two rigid walls and subjected to thermal cycling and is not permitted to expand, it undergoes external compressive mechanical strain. Examples of thermomechanical fatigue can be found in pressure vessels and piping; in the electric power industry, where structures experience pressure loadings and thermal transients with temperature gradients in the thickness direction; and in the aeronautical industry, where turbine blades and turbine disks undergo temperature gradients superimposed on stresses due to rotation. In the railroad application, when external loading due to rail/wheel contact is considered, then the material undergoes the more general case of thermomechanical fatigue. The temperature rise on the surfaces of cylinders and pistons in automotive engines combined with applied cylinder pressures also represents thermomechanical fatigue. Based on the mechanical strain range, the results of thermal fatigue and thermomechanical fatigue tests should correlate well.

A distinction must be drawn between isothermal high-temperature fatigue as cyclic straining under constant nominal temperature conditions versus thermomechanical fatigue. As such, isothermal fatigue can be considered a special case of thermomechanical fatigue. However, the deformation and fatigue damage under thermomechanical fatigue cannot be predicted based on isothermal fatigue data.

Sometimes, the term *low-cycle thermal fatigue* or *low-cycle thermomechanical fatigue* is used. Low-cycle fatigue can be identified two ways: high-strain cycling where the inelastic strain range in the cycle exceeds the elastic strain range, and where the inelastic strains are of sufficient magnitude that they are spread uniformly over the microstructure. Fatigue damage at high temperatures develops as a result of this inelastic deformation where the strains are nonrecoverable. In low-cycle cases, the material suffers from damage in a finite (short) number of cycles. Thermomechanical fatigue is often a low-cycle fatigue issue. For example, in railroad wheels, only severe braking applications, occurring infrequently over thousands of miles, contribute to damage, with fewer than 10,000 cycles taking place during a wheel lifetime. Similarly, the largest thermal gradients and transients in jet engines develop during startup and shutdown. The total number of takeoffs and landings for an aircraft is fewer than 30,000 cycles over the lifetime of an aircraft. In the laboratory, investigations often are conducted under low-cycle conditions to complete the experiments in a reasonable period of time.

The inability to predict thermomechanical fatigue damage from isothermal fatigue data continues to challenge engineers and researchers. Thermomechanical fatigue encompasses several mechanisms in addition to "pure" fatigue damage, including high-temperature creep and oxidation, which directly contribute to damage. These mechanisms differ, depending on the strain-temperature history. They are different from those predicted by creep tests (with no reversals) and by stress-free (or constant-stress) oxidation tests. Microstructural degradation can occur under thermomechanical fatigue in the form of overaging, such as coarsening of precipitates or lamellae; strain aging in the case of solute-hardened systems; precipitation of second-phase particles; and phase transformation within the temperature limits of the cycle. Also, variations in mechanical properties or thermal expansion coefficients between the matrix and strengthening particles present in many alloys result in local stresses and cracking. These mechanisms influence the deformation characteristics of the material, which inevitably couples with the damage processes.

A distinction must be drawn between thermomechanical fatigue and thermal shock. Thermal shock involves a very rapid and sudden application of temperature due to surface heating or internal heat generation, and the resulting stresses are often different from those produced under slow heating and cooling (i.e., quasi-static) conditions. Physical properties, such as specific heat and conductivity, which do not appear in low-strain-rate cases, appear explicitly in the thermal shock case. The rate of strain influences the material response and should be considered in damage due to thermal shock or in selection of materials for better thermal shock resistance.

Finally, if the body is subjected to thermal cycling conditions with superimposed net section loads, the component will undergo thermal ratcheting, which is the gradual accumulation of inelastic strains with cycles. Failure due to thermal ratcheting involves both fatigue and ductile rupture mechanisms. Thermal ratcheting sometimes occurs unintentionally in thermomechanical tests when a region of the specimen is hotter than the surrounding regions, resulting in a bulge in the hot region.

Mechanical Strain and Thermal Strain. Free (unrestrained) thermal expansion and contraction produce no stresses. When the thermal expansion of a body is restrained during uniform heating, thermal stresses develop. Consider the case where a bar is held between two rigid walls and subjected to thermal cycling. The length of the bar cannot change during heating and cooling. Let T_0 be the reference temperature at which the bar was placed under total constraint. The compatibility equation for this bar is given as:

$$\varepsilon_{net} = \varepsilon_{th} + \varepsilon_{mech} = \alpha(T - T_0) + \varepsilon_{mech} \quad \text{(Eq 9)}$$

In this case, the net strain is zero, and all of the thermal strain is converted to mechanical strain. The thermal strain is defined as the product of coefficient of thermal expansion and the temperature range $T - T_0$, where T is the current temperature. Then:

$$\varepsilon_{mech} = -\alpha(T - T_0) \quad \text{(Eq 10)}$$

Sometimes the total constraint case is identified as $\dot{\varepsilon}_{th}/\dot{\varepsilon}_{mech} = -1$. When this ratio is larger than −1, some free expansion and contraction occur, and the term *partial constraint* is used. If the $\dot{\varepsilon}_{th}/\dot{\varepsilon}_{mech}$ ratio is less than −1, the condition is known as overconstraint. Therefore, the constraint influences the mechanical strain for a given thermal strain. Mechanical strain comprises elastic strain and inelastic strain (once the yield stress is reached) and is the key parameter in thermomechanical fatigue studies. The stress/mechanical-strain behavior shown in Fig. 35 is highly idealized; the material exhibits no hardening after yielding, the tension and compression

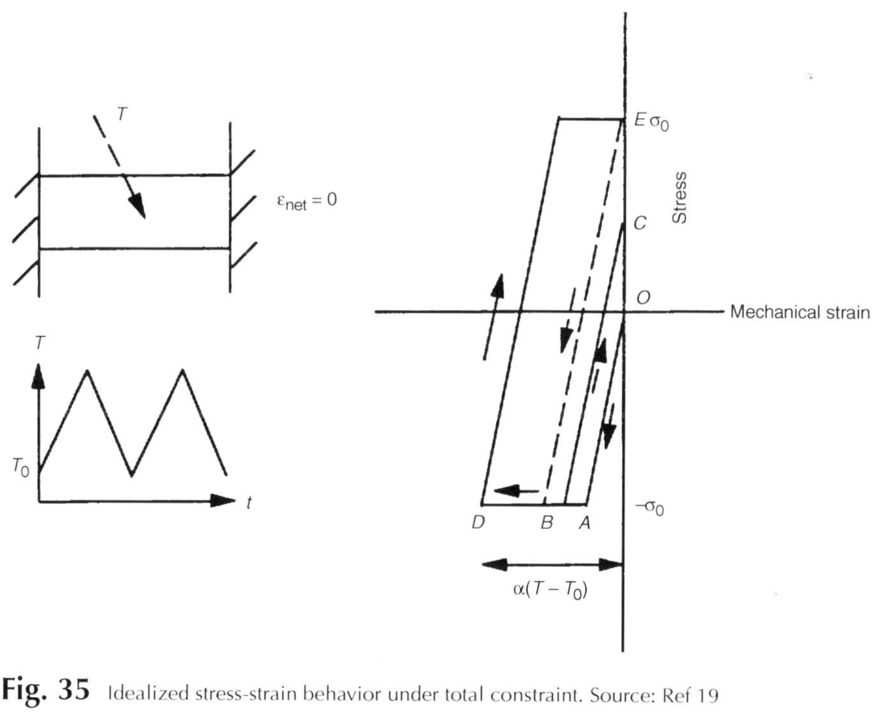

Fig. 35 Idealized stress-strain behavior under total constraint. Source: Ref 19

strength are the same, and elastic modulus is independent of temperature. On heating, the bar is elastic and follows the stress-strain curve along *OA*. At *A*, the bar yields in compression, and on further increases in temperature, the mechanical strain on the bar increases along *AB*. The bar accumulates inelastic strain along *AB*. If the bar is cooled from *B*, it will deform in the reverse (i.e., tensile) direction. When the initial temperature is reached, the bar will return to zero mechanical strain, but a residual tensile stress will exist in the bar at point *C*. If the bar is again heated to the maximum temperature, the material will cycle between the stress points *B* and *C*. The bar is operating within the "shakedown" regime. It is unlikely that the bar will fail under these conditions, because there is no plastic flow after the first reversal.

Next, consider the case when the thermal strain in the first heating portion of the cycle exceeded twice the elastic strain, and a mechanical strain corresponding to point *D* is reached. On cooling back to the initial temperature, T_0, the bar will yield in tension, and inelastic flow will occur until point *E* is reached. On reheating, the bar will deform in the reverse direction (dashed line) until it reaches point *D* in compression. A hysteresis loop develops as a result of this thermal cycle. Under alternate heat/cool cycles, forward and reverse yielding will occur during every cycle, resulting in failure in a finite number of cycles. The constrained bar model is conceptually easy to visualize, but in real structures the condition can be different from total constraint.

In-Phase versus Out-of-Phase Thermomechanical Fatigue. The mechanical strain/temperature waveform is classified according to the phase relation between mechanical strain and temperature. In-phase thermomechanical fatigue means that peak strain coincides with maximum temperature, while out-of-phase thermomechanical fatigue means that peak strain coincides with minimum temperature. These two cases are shown in Fig. 36(a), along with the isothermal fatigue case. Generic hysteresis loops corresponding to the out-of-phase thermomechanical fatigue and in-phase thermomechanical fatigue cases are shown in Fig. 36(b) and (c), respectively. For a thermomechanical fatigue cycle, the hysteresis loops are unbalanced in tension versus compression. In the out-of-phase thermomechanical fatigue case, considerably more inelastic strains develop in compression relative to tension. The opposite behavior occurs in the in-phase thermomechanical fatigue case. Some thermomechanical fatigue experiments have been conducted under $R\varepsilon = 1$ (i.e., completely reversed) conditions. Other thermomechanical fatigue experiments have been conducted under $R\varepsilon = -$infinity (maximum mechanical strain is zero; Fig. 35) and $R\varepsilon = 0$ conditions (minimum mechanical strain is zero).

The inelastic strain (ε_{in}) is defined by subtracting the elastic strain from the mechanical strain:

$$\varepsilon_{in} = \varepsilon_{mech} - \frac{\sigma}{E\{T(t)\}} \quad \text{(Eq 11)}$$

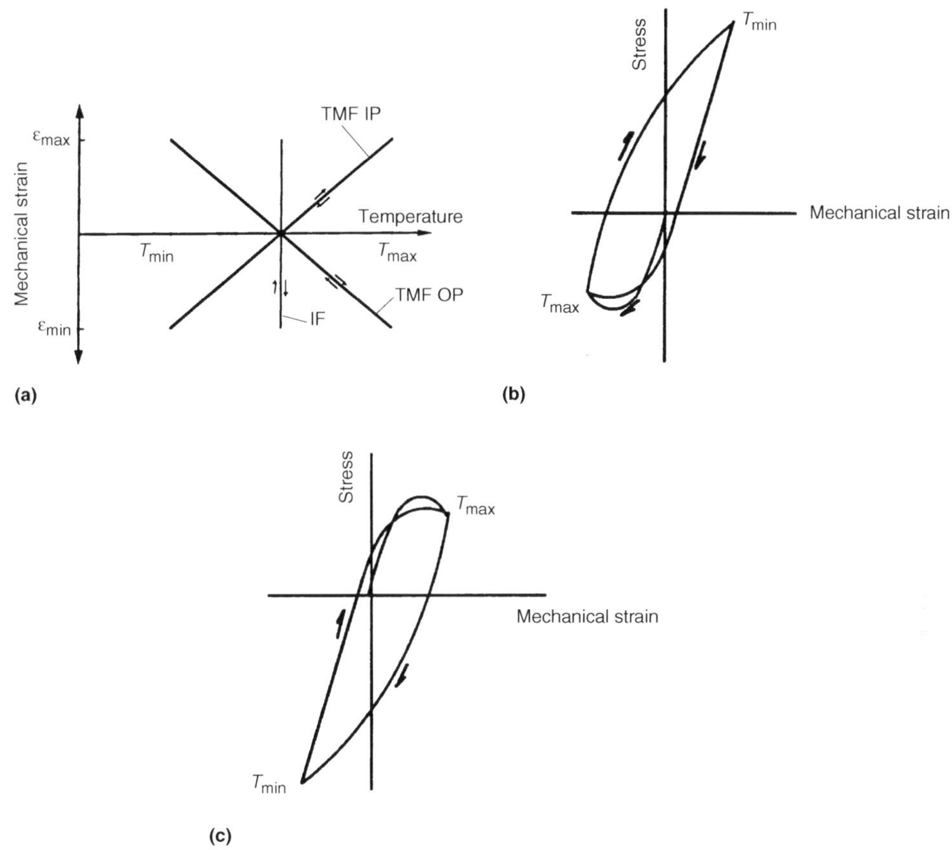

Fig. 36 (a) Mechanical strain/temperature variation in out-of-phase thermomechanical (TMF OP), in-phase thermomechanical (TMF IP), and isothermal fatigue (IF). (b) Out-of-phase thermomechanical stress-strain response. (c) In-phase thermomechanical stress-strain response. Source: Ref 19

For computational purposes, pairs of stress and temperature data points are needed. The variation in elastic modulus, $E\{T(t)\}$, as a function of temperature should be determined from isothermal experiments. A stress-inelastic strain hysteresis loop can then be constructed using Eq 11. If there are hold periods during the thermomechanical fatigue cycle, the equation will still be valid. The mechanical strain range $\Delta\dot{\varepsilon}_{mech}$ is shown in Fig. 37. The stress range in a thermomechanical fatigue cycle is also shown for the out-of-phase case. The loop for the in-phase case is similar, but reversed. Note that at the minimum strain (point B), the stress is not necessarily a minimum. Inelastic deformation with softening due to a decrease in strength with increasing temperature is observed during AB. At B, the maximum temperature is reached. On cooling, the behavior is elastic, followed by plastic deformation at the low-temperature end.

For engineering purposes, the inelastic strain range of a thermomechanical cycle can be determined to a first approximation by subtracting the elastic strains computed at the maximum and minimum strain levels. This gives:

$$\Delta\varepsilon_{in} \approx \Delta\varepsilon_{mech} - \left[\frac{|\sigma_B|}{E_B} + \frac{|\sigma_C|}{E_C}\right] \quad \text{(Eq 12)}$$

where E_C is the elastic modulus at the maximum strain, and E_B is the elastic modulus corresponding to the minimum strain. Equation 12 slightly underestimates the inelastic strain range compared to the more exact equation. Note that the inelastic strain range includes the plastic strain, creep strain, and other strain components (e.g., transformation strain). Separation of plastic and creep strains in a thermomechanical fatigue cycle is not straightforward. If needed, it can be done experimentally by a stress hold at selected points of the hysteresis loop or by constitutive models that include plasticity and creep.

Just as in isothermal fatigue conditions, the thermomechanical fatigue response of engineering materials involves cyclic hardening, cyclic softening, or cyclically stable behavior, depending on the microstructure, the maximum temperature level, and the phasing of strain and temperature. However, the behavior can be somewhat complex because of strain-temperature interaction. A material can harden, soften, or be cyclically stable at T_{max} of the cycle. Likewise, at T_{min}, the material can cyclically soften, harden, or be stable. These two possibilities are shown in Fig. 38. In Fig. 38(a), the material softens at T_{max} and remains cyclically stable at T_{min}. The material can cyclically soften at high temperature due to thermal recovery, causing coarsening of the microstructure, and in this case, the hysteresis loops appear to climb in the tensile direction. Therefore, the tensile mean stress increases with increasing number of cycles. The microstructural coarsening could subsequently affect the strength at T_{min}, with the maximum stress in the cycle dropping with increasing number of cycles. Thereafter, the climbing of the hysteresis loops stops, and the range of stress in the cycle decreases. In the second example (Fig. 38b), stable be-

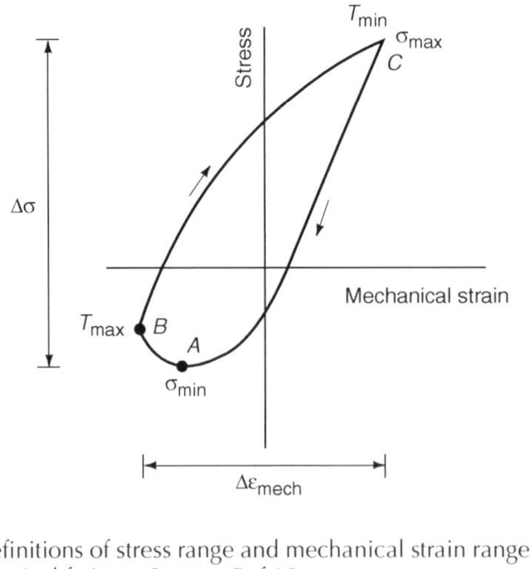

Fig. 37 Definitions of stress range and mechanical strain range in thermomechanical fatigue. Source: Ref 19

havior is observed at T_{max}, but the strength at T_{min} increases because of dynamic or static strain aging effects. In this case, the hysteresis loops also climb in the tensile direction, and, at the same time, the overall stress range increases.

The hysteresis response and life on 1010 steel subjected to thermal cycling in the range of 95 to 540 °C (200 to 1000 °F) is shown in Fig. 39. The mechanical strain range versus life is plotted for out-of-phase thermomechanical fatigue, in-phase thermomechanical fatigue, and isothermal fatigue cases. Cyclic hardening occurs when the maximum temperature is below 425 °C (800 °F), possibly due to strain-aging effects. The thermomechanical fatigue lives are significantly shorter than isothermal fatigue lives, even when the isothermal fatigue results from the maximum tem-

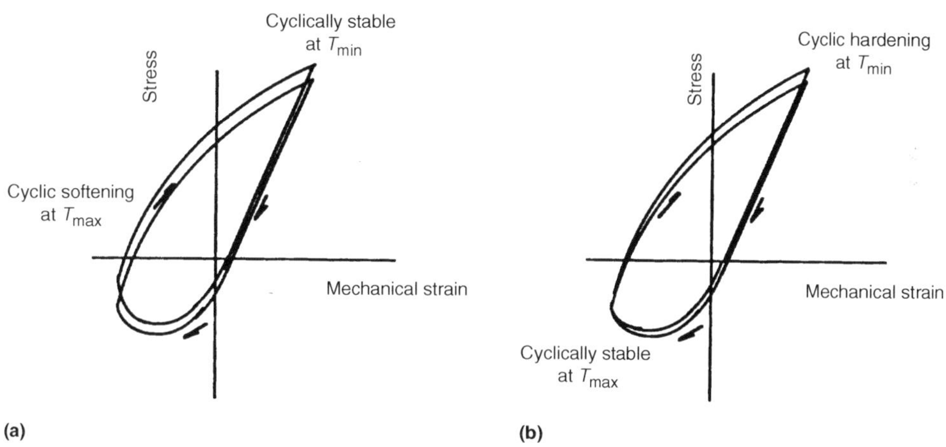

Fig. 38 Stress-strain response under (a) cyclic softening or (b) cyclic hardening conditions. Source: Ref 19

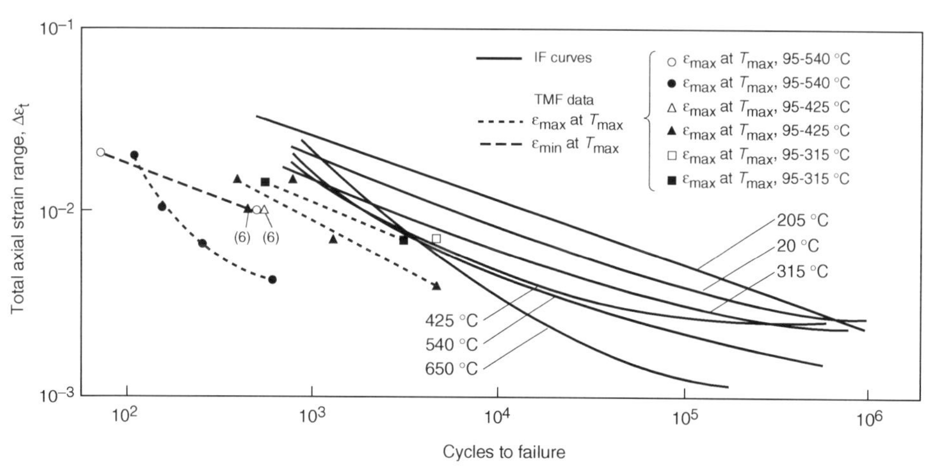

Fig. 39 Isothermal (IF) and thermomechanical fatigue (TMF) data of 1010 carbon steel. Note: (6) indicates a 6 min hold time at maximum temperature. Source: Ref 20

perature are considered. There appears to be crossover in lives between out-of-phase thermomechanical fatigue and in-phase thermomechanical fatigue slightly below a strain range of 0.02. All of the thermomechanical fatigue data shown fall below the isothermal curves.

During long-term thermomechanical fatigue cycling, metallic alloys are subject to many of the same metallurgical instabilities and environmental degradation mechanisms that occur during creep exposure.

ACKNOWLEDGMENTS

Portions of this chapter came from "Creep and Creep-Rupture Testing," *Mechanical Testing and Evaluation*, Vol 8, *ASM Handbook*, ASM International, 2000; "Creep and Stress-Rupture Failures," *Failure Analysis and Prevention*, Vol 11, *ASM Handbook*, ASM International, 2002; "Design for High-Temperature Applications" by D. A. Woodford, *Materials Selection and Design*, Vol 20, *ASM Handbook*, ASM International, 1997; "Thermal and Thermomechanical Fatigue of Structural Alloys" by H. Sehitoglu, *Fatigue and Fracture*, Vol 19, *ASM Handbook*, ASM International, 1996; and "Creep" by F.C. Campbell, *Elements of Metallurgy and Engineering Alloys*, ASM International, 2008.

REFERENCES

1. Creep and Stress-Rupture Failures, *Failure Analysis and Prevention,* Vol 11, *ASM Handbook,* ASM International, 2002
2. Creep and Creep-Rupture Testing, *Mechanical Testing and Evaluation,* Vol 8, *ASM Handbook,* ASM International, 2000
3. F.C. Campbell, Creep, *Elements of Metallurgy and Engineering Alloys,* ASM International, 2008
4. R. Viswanathan, The Effect of Stress and Temperature on the Creep and Rupture Behavior of a 1.25 pct. Chromium-0.5 pct. Molybdenum Steel, *Metall. Trans. A,* Vol 8, 1977, p 877–883
5. O.D. Sherby, *Acta Metall.,* Vol 10, 1962, p 135–147
6. D.J. Benac, Elevated-Temperature Failures, Lesson 7, *Principles of Failure Analysis,* Materials Engineering Institute, ASM International
7. T.H. Courtney, *Mechanical Behavior of Materials,* 2nd ed., McGraw-Hill Book Co., 2000
8. S. Lampman, Intergranular Fracture, *Failure Analysis and Prevention,* Vol 11, *ASM Handbook,* ASM International, 2002
9. M.A. Meyers and K.K. Chawla, *Mechanical Metallurgy—Principles and Applications,* Prentice-Hall Inc., 1984
10. R.G. Baker and J. Nutting, *J. Iron Steel Inst.,* Vol 192, 1959, p 257–268
11. G.A. Whitlow, C.G. Beck, R. Viswanathan, and E.A. Crombie, The Effects of a Liquid Sulfate/Chloride Environment on Superalloys, *Metall. Trans. A,* Vol 15, 1984, p 23–28

12. D.A. Woodford and R.H. Bricknell, Environmental Embrittlement of High Temperature Alloys by Oxygen, *Embrittlement of Engineering Alloys,* C.L. Briant and S.K. Banerji, Ed., Academic Press, 1983, p 157
13. D.A. Woodford, Environmental Damage of a Cast Nickel Base Superalloy, *Metall. Trans. A,* Vol 12, 1981, p 299–308
14. R.H. Bricknell and D.A. Woodford, Cavitation in Nickel during Oxidation and Creep, *Int. Conf. on Creep and Fracture of Engineering Materials and Structures,* B. Wilshire and R.W. Evans, Ed., Pineridge Press, Inst. of Metals, 1991, p 249–262
15. H.R. Voorhees and M. Prager, Assessment and Use of Creep-Rupture Properties, *Mechanical Testing and Evaluation,* Vol 8, *ASM Handbook,* ASM International, 2000
16. A. Weronski and T. Hejwowski, *Thermal Fatigue of Metals,* Marcel Dekker, Inc., 1991, p 83
17. P.G. Forrest, *Fatigue of Metals,* Pergamon Press, 1962
18. C.R. Brinkman, W.R. Corwin, M.K. Booker, T.L. Hebble, and R.L. Klueh, "Time Dependent Mechanical Properties of 2.25-1Mo Steel for Use in Steam Generator Design," ORNL-5125, Oak Ridge National Laboratory, 1976
19. H. Sehitoglu, Thermal and Thermomechanical Fatigue of Structural Alloys, *Fatigue and Fracture,* Vol 19, *ASM Handbook,* ASM International, 1996
20. C.E. Jaske, Thermal-Mechanical, Low-Cycle Fatigue of AISI 1010 Steel, *Thermal Fatigue of Materials and Components,* STP 612, D.A. Spera and D.F. Mowbray, Ed., ASTM, 1976, p 170–198

CHAPTER 12

Wear Failures

WEAR IS USUALLY DEFINED as the undesired removal of material from contacting surfaces by mechanical action. Although it is usually not as serious a service problem as fracture, wear is an enormously expensive problem. Wear is usually a foreseeable type of deterioration. We expect various rubbing surfaces in any machine to eventually "wear out." In many cases, this type of deterioration can be minimized by proper lubrication, filtering, materials engineering, and proper design, among other factors. In many respects, wear is similar to corrosion. Both have many types and subtypes, of which usually at least two are progressing simultaneously. Both are somewhat foreseeable unless the environment changes. Both are extremely difficult to test and evaluate in accelerated laboratory or service tests, with rankings of materials subject to change depending on seemingly minor changes in the test conditions. Finally, both are of enormous economic importance. When studying any failure where wear is known or suspected, it is necessary to have a good understanding of the history and operation of the part or mechanism involved. In many cases, it is not possible to conduct a good investigation by simply examining the worn part itself. Because wear involves the interaction of other parts and/or materials, these must also be studied. Because wear is predominantly a surface phenomenon, anything that affects the surface is likely to affect the wear behavior.

Because the definition of wear is quite broad and the various categories or types of wear are not precisely defined, authors on the subject may organize the different types in different ways. One way to organize the rather overwhelming subject of wear is by the following categories:

- Abrasive wear
- Erosive wear
- Grinding wear
- Gouging wear

- Adhesive wear
- Fretting wear

and where fatigue plays a major role:

- Contact-stress fatigue
- Subsurface-origin fatigue
- Surface-origin fatigue
- Subcase-origin fatigue
- Cavitation fatigue

Two of the aforementioned categories—fretting wear and cavitation fatigue—are usually included in the subject of corrosion. Because both involve some chemical changes, there is some justification for this grouping. However, because both are primarily the result of undesired removal of material from contacting surfaces by mechanical action—which is the definition of wear—they are included with the other wear phenomena.

Abrasive Wear

The general category of abrasive wear can be characterized by a single key word: *cutting*. Abrasive wear occurs when hard particles suspended in a fluid or projections from one surface roll or slide under pressure against another surface, as shown in Fig. 1, thereby cutting the other surface. Indeed, machining would fall into the category of abrasive wear, except that it is usually not undesirable, which is a condition of the wear definition. A machine tool, even a hand file, sliding under pressure across a softer metal cuts the metal, usually causing microscopic distortion of the surface structure and forming a distorted chip or fragment of the metal removed. Another very important characteristic of abrasive wear is the heat that is generated by friction between the two materials.

Abrasion is typically categorized according to types of contact, as well as contact environment. Types of contact include two-body and three-body wear. Two-body wear occurs when an abrasive slides along a surface, and three-body wear occurs when an abrasive is caught between two surfaces. Contact environments (Fig. 2) are classified as either open (free) or closed (constrained). For a given load and path length of wear, the wear rate is approximately the same for both open and closed systems. However, measurements of the loss in closed systems will often appear higher than the loss in open systems, probably because most closed systems experience higher loads.

Several mechanisms have been proposed to explain how material is removed from a surface during abrasion. These mechanisms include fracture, fatigue, and melting. Because of the complexity of abrasion, no one mechanism completely accounts for all the loss. Some of the processes that are possible when a single abrasive tip traverses a surface are shown in Fig. 3, including plowing, wedge formation, cutting, microfatigue, and microcracking.

Plowing is the process of displacing material from a groove to the sides. This occurs under light loads and does not result in any real material loss. Damage occurs to the near surface of the material in the form of a buildup of dislocations through cold work. If later scratches occur on this cold-

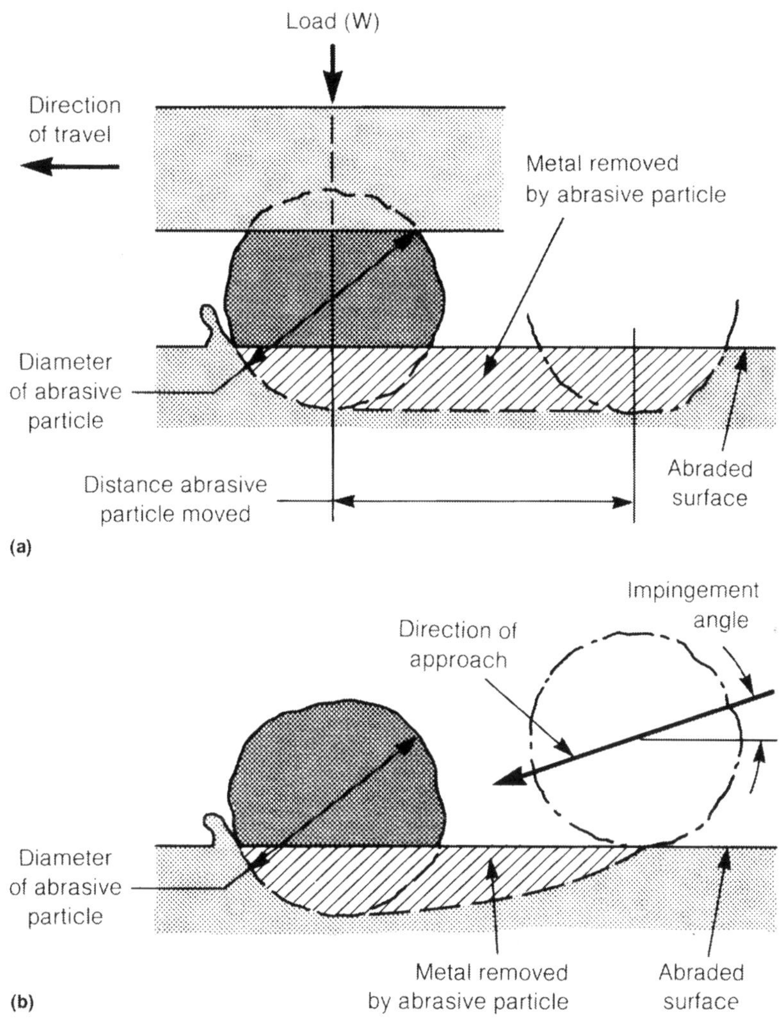

Fig. 1 Idealized representations of the two types of force applications on abrasive wear particles. (a) Represents the cutting or plowing action of a contained particle under pressure. That is, the particle is not free but is under pressure from other particles or a solid object. This is characteristic of grinding and gouging abrasion, in which the hard particles are forced to scratch or cut the metal surface. (b) Represents the cutting or plowing action of a loose particle flowing across the metal surface after impinging upon the surface. This is characteristic of erosive wear, in which free particles strike the surface at an angle, then slide across the surface. Source: Ref 1

worked surface, then the additional work could result in loss through microfatigue.

When the ratio of shear strength of the contact interface relative to the shear strength of the bulk rises to a high enough level (from 0.5 to 1.0), it has been found that a wedge can develop on the front of an abrasive tip. In this case, the total amount of material displaced from the groove is greater than the material displaced to the sides. This wedge formation is still a fairly mild form of abrasive wear.

The most severe form of wear for ductile material is cutting. During the cutting process, the abrasive tip removes a chip, much like a machine tool. This results in removed material but very little displaced material relative to the size of the groove. For a sharp abrasive particle, a critical angle exists, for which there is a transition from plowing to cutting. This angle depends on the material being abraded. Examples of critical angles range from 45° for copper to 85° for aluminum.

For ductile materials, the mechanisms of plowing, wedge formation, and cutting have been observed (Fig. 4). It was found that the degree of penetration was critical to the transition from plowing and wedge formation to cutting. When the degree of penetration, defined as depth of penetration divided by the contact area, exceeded approximately 0.2, cutting was the predominant mode of wear.

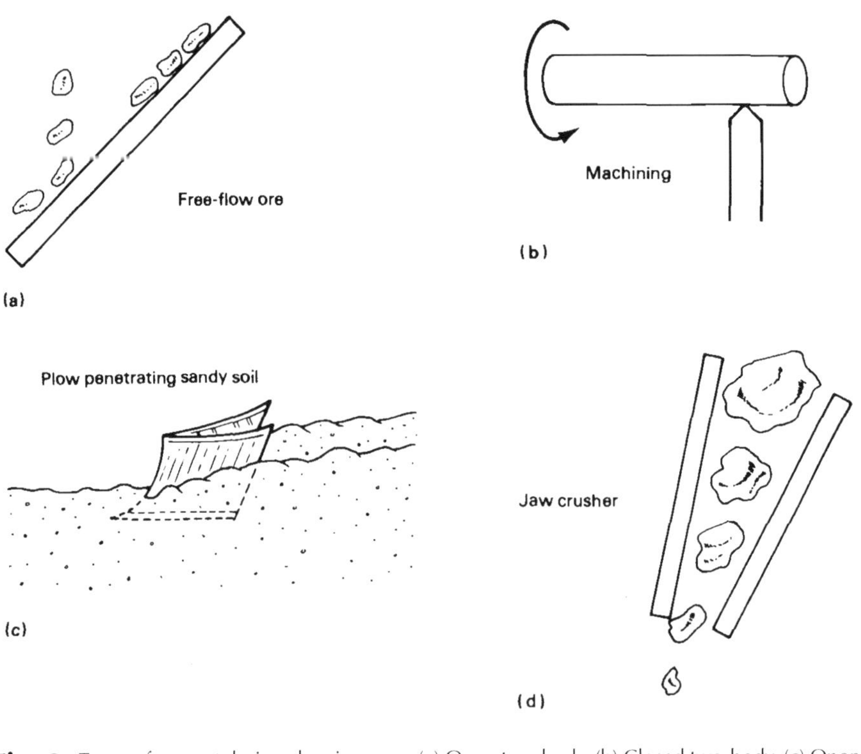

Fig. 2 Types of contact during abrasive wear. (a) Open two-body. (b) Closed two-body. (c) Open three-body. (d) Closed three-body. Source: Ref 2

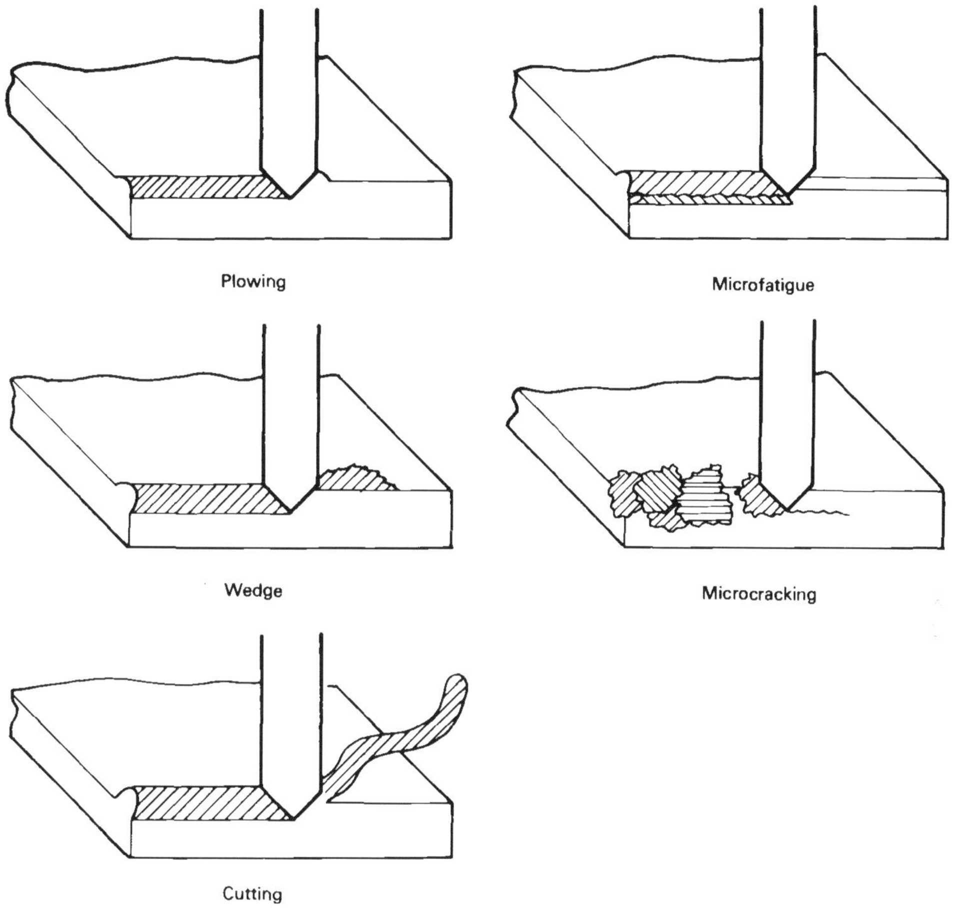

Fig. 3 Five processes of abrasive wear. Source: Ref 2

When an abrasive grain abrades while cutting a surface, the maximum volume of wear that can occur is described by:

$$W = Ad \qquad \text{(Eq 1)}$$

where W is the volume of material removed, A is the cross-sectional area of the groove, and d is the distance slid. The cross-sectional area of the groove (A) is dependent on the abrasive grain shape and the depth of penetration (p):

$$A = k_1 p \qquad \text{(Eq 2)}$$

where k_1 is a constant dependent on the shape. In turn, the depth of penetration (p) is again dependent on the shape of the grain, the load (L), and the hardness (H), of the material:

$$p = k_2 \frac{L}{H} \quad \text{(Eq 3)}$$

Many factors affect k_1: the possibility of plowing rather than cutting, the abrasive grain may roll and avoid wear, the abrasive grain may break down and not be effective during the latter part of its contact path, and others. Equations 1, 2, and 3 can be combined, forming:

$$W = k_3 \frac{Ld}{H} \quad \text{(Eq 4)}$$

This is commonly known as Archard's equation, which was derived for adhesive wear but has proven very useful in abrasive wear, as well.

Commonly, materials are described as having good or bad wear resistance (R), which is simply defined as the reciprocal of wear volume:

$$R = \frac{1}{W} \quad \text{(Eq 5)}$$

Brittle materials have an additional mode of abrasive wear, namely, microfracture. This occurs when forces applied by the abrasive grain exceed

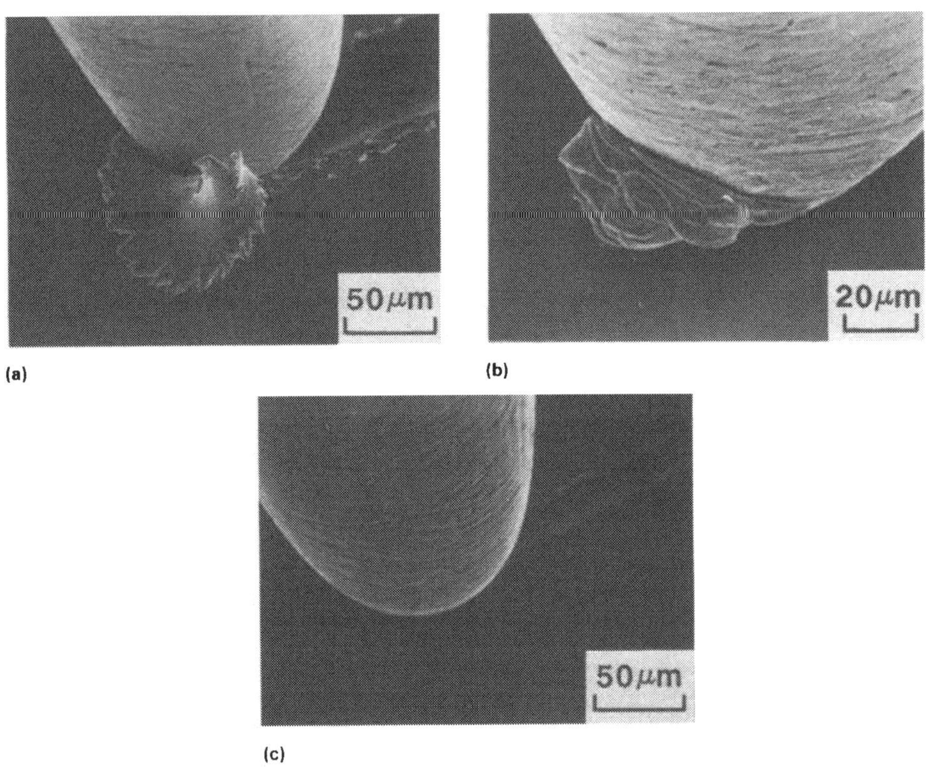

Fig. 4 Examples of three processes of abrasive wear, observed using a scanning electron microscope. (a) Cutting. (b) Wedge formation. (c) Plowing. Source: Ref 3

the fracture toughness of the material. This is often the predominant mode of severe wear for ceramic materials, and it is active in brittle materials such as white cast irons.

In general, abrasive wear may sometimes be reduced or dealt with by any of several methods, which may or may not be practical in individual circumstances:

- *Increase surface hardness.* This is a rather obvious solution to abrasive wear problems; however, it may not always be the answer to a specific problem. In cutting tools, such as various types of knives, blades, and the like, increasing the hardness may indeed make the cutting tool more resistant to dulling of the sharp edge. However, increasing the hardness also increases the chance of brittle fracture of the cutting tool itself. Brittle fracture would be a much more serious problem than the dulling from abrasive wear, because the dull tool can always be resharpened and reused, while a broken tool may cause injury to persons or to machines after fracture.
- *Remove foreign particles.* If hard, abrasive foreign particles are causing abrasive wear, it again seems obvious that if the particles are trapped and removed, the wear cannot take place. This is exactly the reason that filters for air, water, and oil are used in various types of machinery. An automotive engine is a typical example in which air, oil, and fuel filters are used to prevent entry of external foreign particles and to trap and collect internal foreign particles before they can damage the engine. In heavy-duty engines, filters for the cooling system also are used to minimize abrasive wear of various parts, particularly the impeller that circulates the coolant. In other cases, the foreign particles cannot be removed; this is particularly true with erosive wear, where many high-speed particles slide and roll across a metal surface.
- *Replace worn part.* One of the most practical ways by which we live with wear is simply to design parts and assemblies that are subject to abrasive wear in such a way that they can be replaced when they are worn out. This is one of the simplest and most common ways of dealing with the problem. However, replacement may not be practical in a given situation because of inaccessibility, excessive labor cost or downtime, unavailability of replacement parts in an emergency, or other problems.

The general solutions mentioned may or may not be effective, depending on the circumstances of the particular wear problem.

Erosive Wear

Erosive wear (or erosion) occurs when particles in a fluid or other carrier slide and roll at relatively high velocity against a surface. Each particle contacting the surface cuts a tiny particle from the surface. Individually, each particle removed is insignificant, but a large number of particles removed over a long period of time can produce staggering degrees of erosion. The classic example is the Grand Canyon of the Colorado River. Whenever the dirt particles carried by the river current come in contact with the relatively soft rock of the riverbed, small amounts of the rock are removed. Over millions of years, this giant chasm has been cut through the rock by the erosive action of dirt particles in the river, which flows rapidly in some places but is more placid in others.

Erosive wear can be expected in metal parts and assemblies where the aforementioned conditions are present. Common problem areas are found in pumps and impellers, fans, steam lines and nozzles, the inside of sharp bends in tubes and pipes, sand and shot blasting equipment, and similar areas where there is considerable relative motion between the metal and the particles.

Erosive wear can be recognized by any or all of the following conditions, depending on the parts involved.

General Removal of Soft Surface Coatings or Material. This is a common form of wear for fan and propeller blades. In automotive applications, for example, the paint on the rear, or concave, side of the blade is usually removed by the scouring or cutting action of dust and dirt particles in the air. The concave side of a rotating fan blade has a positive pressure, while the convex side has negative pressure, and the positive pressure forces the particles against the surface, thus leading to erosive wear.

Grooving or Channeling of the Material. This type of erosive wear is common in assemblies involving fluids (liquids or gases), where the design of the parts is such that the fluid flows faster or is changed in direction at certain locations. Examples are in pumps or impellers in which vanes push the particle-laden fluid into various passages. The inside of tubes or pipes is often damaged at curves because the inertia of the particles and the fluid forces them against the outer side of the curve. Obviously, sudden, sharp curves or bends cause more erosion problems than gentle curves. In textile machinery, even high-velocity thread or yarn can cause erosion; a sudden change in direction of yarn caused the grooving and erosive wear in the eyelet in Fig. 5. Grooving and channeling are also quite common in various types of nozzles where high-speed or high-pressure fluids scour and cut their way through the metal. Liquid droplets can lead to erosive wear, as is frequently seen on the leading edges of high-speed aircraft.

Rounding of Corners. Erosive wear can change the shape of impellers, turbine blades, and vanes in such a way as to cause substantially impaired operating efficiency. An example of this type of damage is shown in Fig. 6 with before-and-after views. If service had been continued, the vanes would have been completely eroded away, leading to zero efficiency in pumping the liquid.

Fig. 5 Abrasive wear of a yarn eyelet made of hardened and tempered 1095 steel. Grooving was caused by a sharp change in direction of the yarn as it came out of the hole. Source: Ref 4

For suggested solutions to erosive wear, see the general solutions in the section "Abrasive Wear" in this chapter.

Erosion-Corrosion

Erosion-corrosion is the acceleration of corrosive attack due to the simultaneous action of corrosion and erosion. The erosion-corrosion of mild steel in flowing water is illustrated in Fig. 7. The attack is more severe than if just corrosion or erosion alone were acting. The erosive action removes metal from the surface as dissolved ions, as particles of solid corrosion products, or as elemental metal. Erosion-corrosion can primarily be erosive attack or primarily chemical attack, or somewhere in between. Both gases and liquids can cause attack. It is encountered when particles in a liquid or gas impinge on a metal surface, causing the removal of protective surface films, such as protective oxide films or adherent corrosion products, thus exposing new reactive surfaces that are anodic to noneroded neighboring areas on the surface. This results in rapid localized corrosion of the exposed areas in the form of smooth-bottomed shallow recesses. As the temperature rises, the protective film may become more soluble and/or less resistant to abrasion. Hence, the same flow rates, but at higher temperatures, can cause an increase in the corrosion rate. Hot gases may oxi-

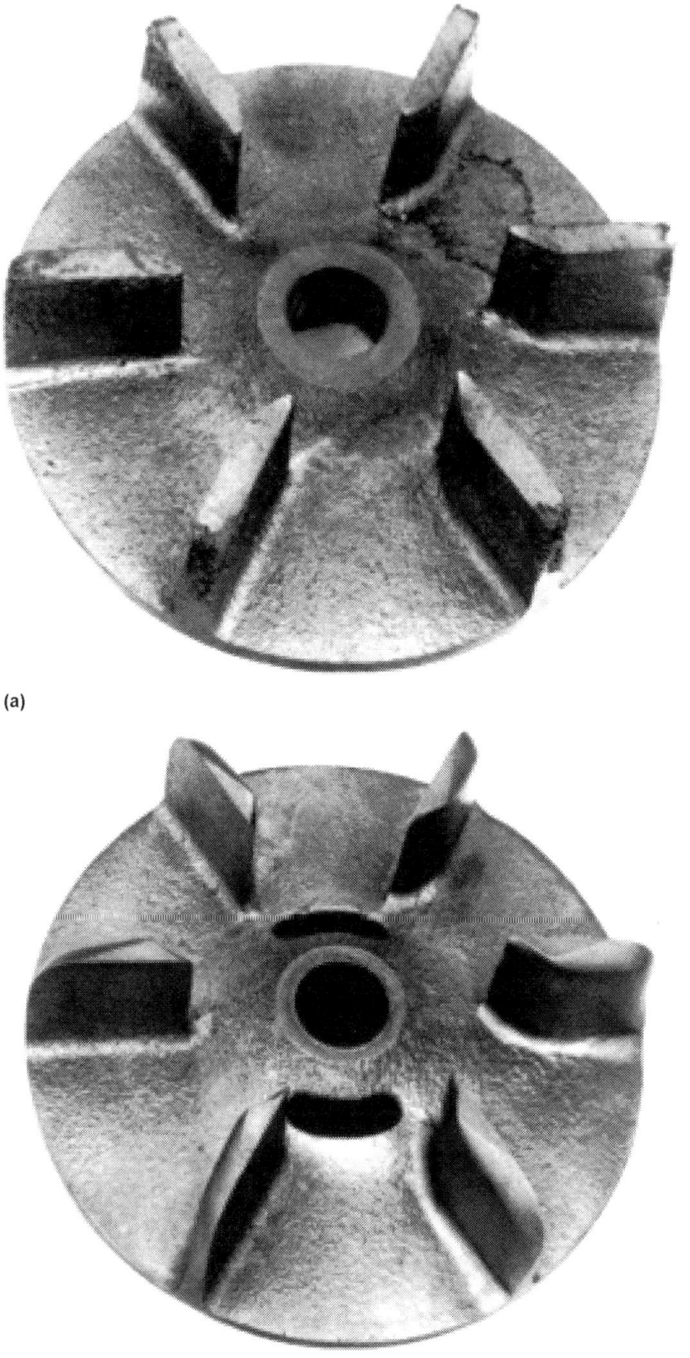

(a)

(b)

Fig. 6 Erosive wear of a gray cast iron water pump impeller. The sharp corners of the (a) new impeller have been (b) completely rounded off by the abrasive wear of sand in the cooling system. Source: Ref 4

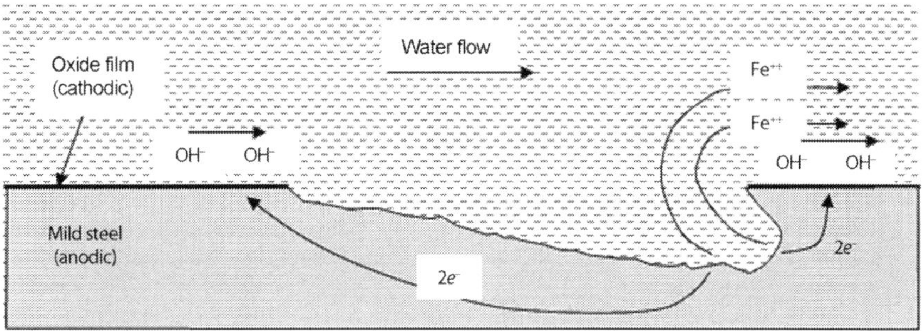

Fig. 7 Erosion-corrosion of mild steel. Source: Ref 5

dize a metal and then at high velocities blow the protective film off of the metal. Slurries, present in solids as liquid suspensions, are particularly aggressive media.

Nearly all turbulent corrosive media can cause erosion-corrosion. The attack can exhibit a directional pattern related to the path taken by the corrosive as it moves over the surface of the metal. Erosion-corrosion is characterized in appearance by grooves, waves, rounded holes, and/or horseshoe-shaped grooves. Affected areas are usually free of deposits and corrosion products, although corrosion products can sometimes be found if erosion-corrosion occurs intermittently and/or the liquid flow rate is relatively low.

Most metals are susceptible to erosion-corrosion under specific conditions. Metals that develop thick protective coatings of a corrosion product are often susceptible because thick coatings frequently exhibit poorer adhesion than thin coatings. Thin-film oxide coatings, such as those that form in stainless steel and titanium alloys, are relatively immune to erosion-corrosion. On the other hand, soft metals, such as copper and lead, are quite susceptible to attack.

Impingement corrosion is a severe form of erosion-corrosion. It frequently occurs in turns or elbows of tubes and pipes and on the surfaces of impellers or turbines. It occurs as deep, clean, horseshoe-shaped pits with the deep, or undercut, end pointing in the flow direction. Impingement corrosion attack can also occur as the result of partial blockage of a tube. A stone, a piece of wood, or some other object can cause the main flow to deflect against the wall of the tube. The impinging stream can rapidly perforate tube walls. Water that carries sand, silt, or mud will have an additional severely erosive effect on tubes. The energy transfer is a function of the rate of change of the momentum of the flowing medium. As energy transfer takes place over a smaller and smaller area per unit time, the energy or power density of the process becomes damaging to the substrate. Steam erosion is another form of impingement corrosion, occurring when high-velocity wet steam impacts a metal surface. The resulting attack usually produces a roughened surface showing a large number of small cones with the points facing in the flow direction.

Erosion-corrosion can be prevented, or reduced, through improved design, such as increasing the diameter and using streamlined bends in pipes, by altering the environment by deaeration or the addition of inhibitors, and by using hard and tough protective coatings.

Grinding Wear

The principal characteristics of grinding wear are that it is caused primarily by particles under high stress that cut, or plow, many small grooves at relatively low speed across a metal surface. This high-stress, low-speed operation is characteristic of tillage tools (e.g., plows, cultivators, and rakes) and other ground-contact parts, such as bulldozer track shoes and the cutting edges of blades. In other industries, there are many other operations that have similar effects on metal parts. These tend to dull cutting edges, changing their shape to make them perform their function less efficiently or not at all and generally causing unsatisfactory service. Thus, grinding wear can be recognized if the type of service that caused it is known and if the wear occurs at high-stress locations, particularly points and edges, causing a general shape change of the part or parts involved. When two hard metal surfaces slide against each other, frequently in the presence of a lubricant, each may tend to smooth the other, particularly if fine abrasive particles are present. When properly controlled, this process may be very useful as a lapping, or polishing, process in which there is only one metal surface together with the fine abrasive material that is used for the polishing operation, such as in preparation of metallographic specimens for microscopic examination.

When considering means to prevent grinding wear, an increase of hardness is always an obvious thought. However, this may lead to more serious problems in certain types of parts. The judicious use of certain hardening methods and of surface coatings may lead to some improvement in managing this type of wear.

Hardfacing by welding, metal spraying, or other means of deposition is frequently used to improve the resistance to grinding wear. Usually the deposits contain large quantities of alloy carbides, such as those of tungsten, titanium, chromium, molybdenum, vanadium, and others. In certain applications, oxides, borides, or nitrides may be more satisfactory for the conditions of the service, and the environment is extremely important in selecting the best alternative. In many cases, diffusion treatments, such as carburizing, carbonitriding, nitriding, chromizing, or boronizing, are sufficient for the purpose. The economics and practical aspects of the situation are usually overriding factors in the choice of a method of improving wear resistance.

Evaluating the best type of coating or diffusion treatment for a given application can be fraught with frustration, because simulated laboratory tests, usually accelerated to reduce testing time, often give misleading results. Actual service testing is usually the best way to evaluate, or rank, wear resistance of various materials or processes. However, even this can be misleading, because the combination that is best in one type of abrasive environment may perform poorly in another type of environment. The same abrasive material used under different environmental conditions may produce contradictory results because of differences in packing and thermal characteristics. Sand, for example, tends to cool a cutting edge better when wet than when dry. Therefore, a cutting edge with greater hot hardness may be more efficient, depending on the local frictional temperatures reached at the cutting edge.

Slight, controlled grinding wear may sometimes be used to advantage to maintain a self-sharpening behavior in certain cutting tools; hardened surfaces can be used with soft surfaces to keep the tool sharp. For example, plowshares (the cutting edges of plows) can be made self-sharpening if the front high-stress surface is soft and the rear low-stress side is hardfaced with an appropriate material. During service, the soft high-stress surface is worn, the hard surface remains relatively undamaged, and the plowshare stays sharp, as shown in Fig. 8. Electric carving knives, which have two blades sliding back and forth against each other, are sometimes hardsurfaced on the outer low-stress sides. As the blades slide back and forth against each other while cutting, slight metal wear occurs on the soft high-stress inner surfaces, keeping the blades sharp. In the mining indus-

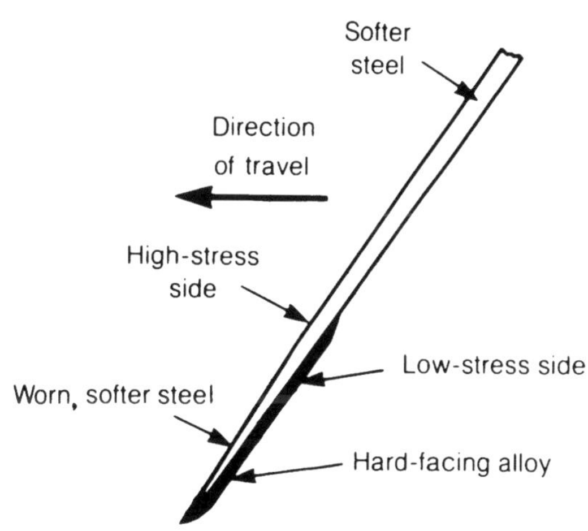

Fig. 8 Diagram of self-sharpening plowshare. As the plowshare cuts through the soil from the right to left, the relatively soft steel on the forward high-stress side is gradually worn away, but the hardfacing applied to the rear low-stress side is continually exposed at the sharp tip. Source: Ref 4

try, digging tools are sometimes hardfaced on only one side to maintain the same self-sharpening action, as shown in Fig. 9.

Gouging Wear

This type of wear is caused by extremely high-stress battering or impact that tends to cut or gouge large wear fragments from the surface of the metal. This is encountered in certain applications in the fields of earthmoving, mining, quarrying, oil well drilling, steelmaking, cement and clay product manufacture, railroading, dredging and lumbering, and undoubtedly other industries. When hard abrasive products are crushed, battered, or pounded under extremely high stress, rapid deterioration of the contact surfaces can be expected unless specific steps are taken to prevent this problem.

In certain cases, it may be more economical to use replaceable parts, such as the teeth for backhoe buckets. Figure 10 shows gouging wear on a tooth made of moderately hard medium-carbon alloy steel on the flat, originally sharp point. The original condition is shown in Fig. 10(a). The soft top of the tooth (Fig. 10b), made of 1010 steel, wore considerably more during operation in rocky, frozen soil than did the flat opposite side of 8640 medium-hard steel (Fig. 10c). In other cases, parts are not easily replaceable and must be made from a more resistant material.

A type of steel invented over one hundred years ago by Sir Robert Hadfield in England has been used successfully in many applications, such as railway crossing frogs and switches, rock crushers, grinding mills, dredge buckets, power shovel buckets and teeth, and pumps for handling gravel and rocks. This specialty steel, usually called Hadfield's steel, is an aus-

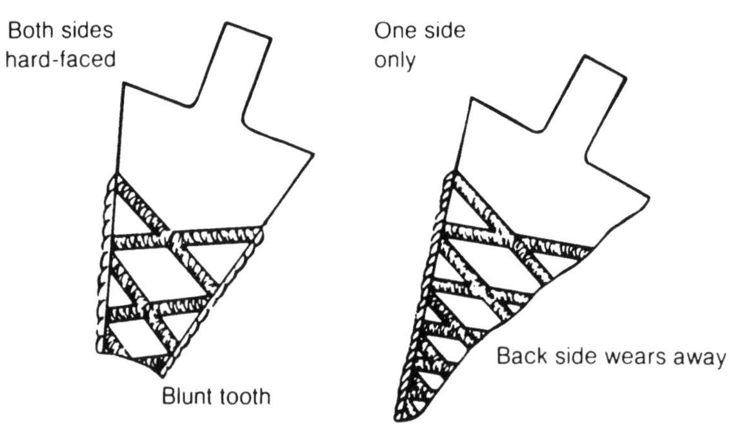

Fig. 9 Self-sharpening of a digging tooth from ground-contact equipment by controlled wear through selective hardfacing. The pattern of hardfacing can be varied to suit the condition, but note that the blunt tooth is hardened on both sides, while the self-sharpening tooth is hardened on only one side. Source: Ref 6

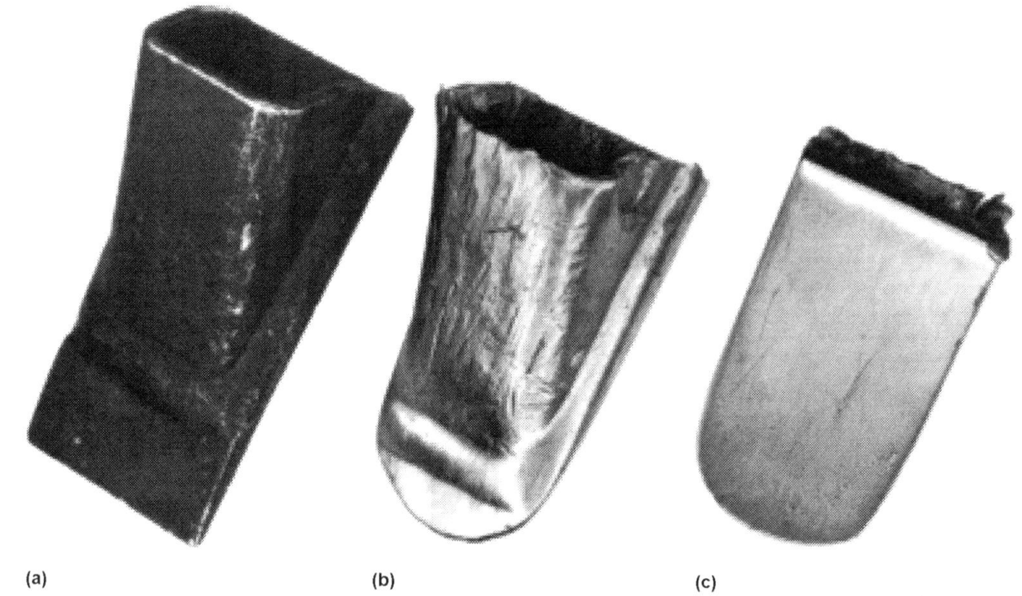

Fig. 10 Tooth for a backhoe bucket. (a) Original condition. (b) The soft top of the tooth, made of 1010 steel, wore considerably more during operation in rocky, frozen soil than did (c), the flat opposite side of 8640 medium-hard steel. The tooth is a replaceable part that is pinched over a stub to hold it in position. Source: Ref 4

tenitic manganese steel normally used either in the form of castings or welded to a steel base. It is machined only with difficulty, because the machining operation work hardens the austenitic steel, as does the necessary battering during high-stress service. In some cases, it is possible to preharden this material by submitting it to heavy battering or hammering before putting it into service. This type of steel is not intended for resistance to erosive wear or most kinds of grinding wear.

The remedies for gouging abrasive wear, as with other types, usually come down to a combination of economics, availability, accessibility, and design. Frequently, there are several ways to improve a product, but only one is chosen because it provides the optimum properties at the lowest cost.

Adhesive Wear

Like abrasive wear, adhesive wear can also be characterized by a single word. In the case of adhesive wear, the word is *welding* or, more precisely, *microwelding*. The actual micromechanism is well described by the term *adhesive wear*. Other terms are sometimes used—for example, *scoring*, *scuffing*, *galling*, and *seizing*—but these do not accurately characterize the mechanism of failure and should be avoided.

An exaggerated view of two surfaces that are sliding with respect to each other is shown in Fig. 11. The surfaces may or may not be separated by a lubricant. When a peak, or asperity (from a Latin word meaning rough), from one surface comes in contact with a peak from the other surface, there may be instantaneous microwelding due to the frictional heat, as shown in Fig. 11(a). Continued relative sliding between the two surfaces fractures one side of the welded junction, as shown in Fig. 11(b), making the asperity on one side higher and the asperity on the other side lower than the original height. The higher peak is now available to contact another on the opposite side, as shown in Fig. 11(c). The tip may either be fractured by the new contact or rewelded to the opposite side, and the cycle repeated. In either case, adhesive wear frequently starts out on a small scale but rapidly escalates as the two sides alternately weld and tear metal from each other's surfaces. Also, the debris may be carried by the lubricant, if one is present, to other parts of the mechanism. Extreme wear can result, as shown in Fig. 12 and 13, and complete failure of the mechanism may result. In severe adhesive wear, the debris is composed of free metallic particles. In mild cases, the much finer particles can react with the environment to form debris that is largely free oxide particles.

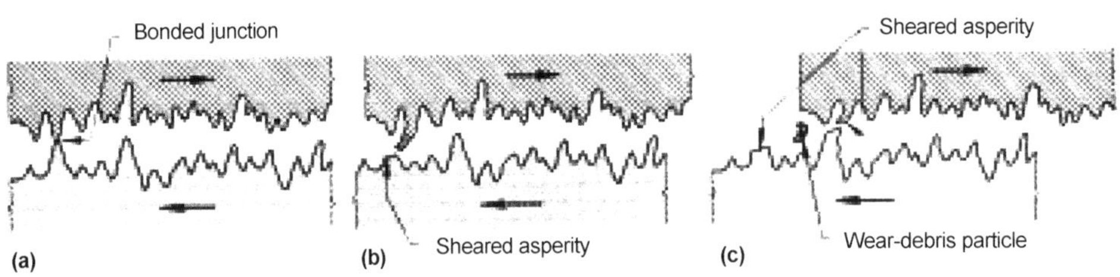

Fig. 11 Schematic illustration of one process by which a particle of wear debris is detached during adhesive wear. As the surfaces slide across each other, (a) a bonded junction is (b) torn from one peak, or asperity, then is (c) sheared off by impact with a larger, adjacent peak to form a particle of wear debris. Metal also may be transferred from one surface to another by the microwelding process. Arrows indicate direction of sliding. Source: Ref 1

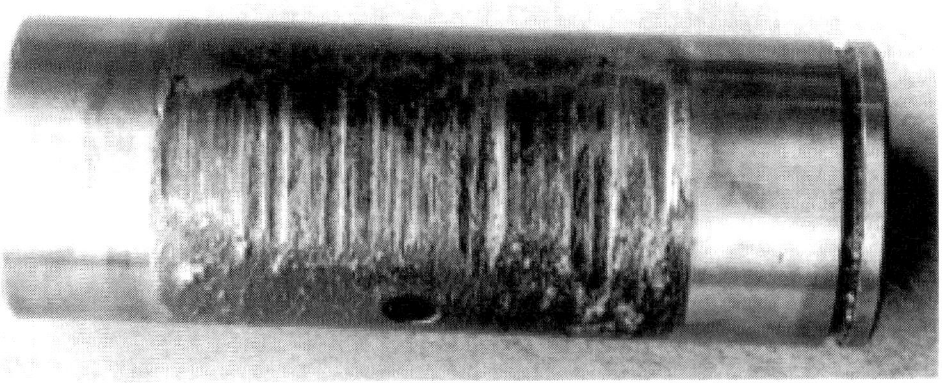

Fig. 12 Severe adhesive wear on a stationary shaft upon which a planetary gear rotated in the presence of an inadequate lubricant. Because the radial force was on only one side of the shaft, the adhesive wear was only on one side. However, the entire bore of the gear was damaged by adhesive wear. Both parts were carburized and hardened steel. Source: Ref 4

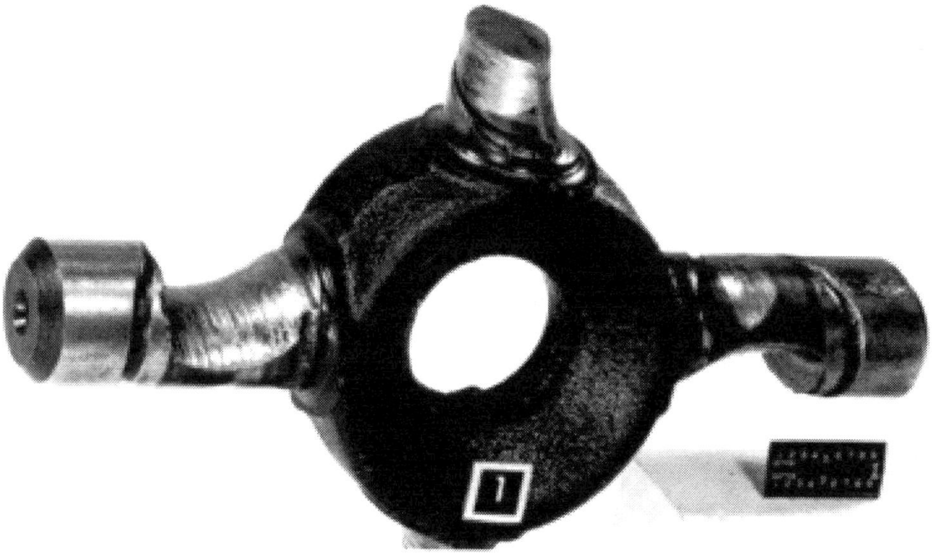

Fig. 13 Destructive adhesive wear of a differential cross following fracture of one severely worn trunnion. This vehicle was operated primarily in forward speeds; consequently, only the forward drive side was damaged severely by the rotation of differential pinions on the trunnions. The lubricant was inadequate for this application. The differential cross, or spider, and the differential pinions were made of carburized and hardened steel. Source: Ref 4

The interface between two sliding surfaces is an extremely complex system, consisting of two metal surfaces (each with its own metallurgical, mechanical, chemical, and topographical characteristics) and usually a lubricant, which also is an extremely complex blend of physical and chemical characteristics that change with the temperature. In other words, there are both good and poor combinations of metals, and also good and poor lubricants in a given application. The ideal situation is for the lubricant to achieve complete separation between every part of the two metal surfaces. Unfortunately, this does not always occur, and there may be problems with adhesive wear.

The heat generated by friction is, locally, high enough to cause microwelding. This means that the temperature is also high enough to cause unintended local heat treatment of the surface metal. Adhesive wear is quite similar to a grinding burn in that both can cause tempering of the subsurface regions and actual rehardening of steel microstructures, resulting in untempered martensite, which is extremely susceptible to cracking because of its brittleness. These cracks can lead to either brittle fracture or fatigue fracture, depending on the part and the application.

A practical way of checking for the presence of adhesive wear in hardened steels is to use the Tarasov etching technique in which two etching solutions give a high-contrast nondestructive means of checking for adhesive wear in hardened steels. Other etchants, such as the conventional nital (a weak solution of nitric acid in alcohol) may also be used, but the contrast and sensitivity are usually not as good as with Tarasov etching. Metallographic examination on a cross section of the surface will also show evidence of adhesive wear when viewed at moderate to high magnification, depending on the amount of thermal damage.

Prevention of adhesive wear usually can be achieved by use of any, or all, of the following methods:

- *Keep the lubricant cool.* Because adhesive wear is caused by locally high temperatures, the bulk temperature of the lubricant must be kept relatively cool. This is the reason for the use of transmission oil coolers in racing cars. Obviously, the lower the bulk oil temperature, the lower the interfacial temperature should be.
- *Use contacting metals that are insoluble in each other.* Because adhesive wear is a microwelding process, it follows that if two metals are not weldable to each other, there can be no adhesive wear. This is exactly the principle that is used in developing alloys for sliding bearings; for example, do not use steel against steel in sliding bearings (at least, not intentionally). However, there are many applications where steel is used against steel in gears, cams, and the like, and special precautions must be taken against adhesive wear. The antiwelding principle is frequently used in very high-speed gears by plating one of the gears with silver or gold. Because these metals are not soluble in steel, thin plates are usually sufficient to prevent adhesive wear with reasonable lubrication. For obvious economic reasons, use of this type of plating is limited, but it may be extremely useful in preventing downtime in plant equipment.
- *Use smooth surfaces, because if there are no projections to penetrate the lubricant film, there is a reduced probability of adhesive wear.* If two smooth surfaces are separated by a thin oil film, they will essentially glide across each other without contact. However, if one of the surfaces has projections that can rupture the lubricant film, adhesive wear is more likely. However, in some cases, roughness or waviness may be desirable so that the depressions can act as reservoirs to retain lubricant.
- *Contaminate surfaces to keep them chemically "dirty."* Chemical films frequently are used to prevent the like metal-to-metal contact that leads to adhesive wear. Phosphate coatings help separate metal surfaces, particularly during the early phases of operation. During the wearing-in or breaking-in period, the projections (or asperities) are removed from mating surfaces, because the phosphate coating crystals help to retain the lubricant. Eventually, however, the phosphate crystals may be gradually worn, or polished, away and the parts should enter a long period of service without problems. Special oils and other lubricants have been developed that form monomolecular surface films on steel surfaces. These are the extreme-pressure lubricants that are used in appli-

cations where there are high sliding velocities, such as in hypoid gear sets in automotive axles. These lubricants form extremely thin compounds on the surfaces to prevent metal-to-metal contact.

Fretting Wear

Fretting wear is quite similar to adhesive wear in that microwelding occurs on mating surfaces. The difference is that adhesive wear is related to interfaces that are sliding across each other, while fretting wear is related to interfaces that are essentially stationary with respect to each other. However, when minute elastic deflections or slight motion actually occurs, the cyclic motion of extremely small amplitude is enough to cause microwelding on both surfaces, as is shown in Fig. 14.

Fretting wear is also known as fretting corrosion, false brinelling, friction oxidation, chafing fatigue, and wear oxidation. Fretting frequently occurs in stationary joints that are "fixed" from shrinking or pressing by interference fits or by bolts, pins, rivets, or other mechanisms, and also at the various contact points in antifriction, or rolling element, bearings. This

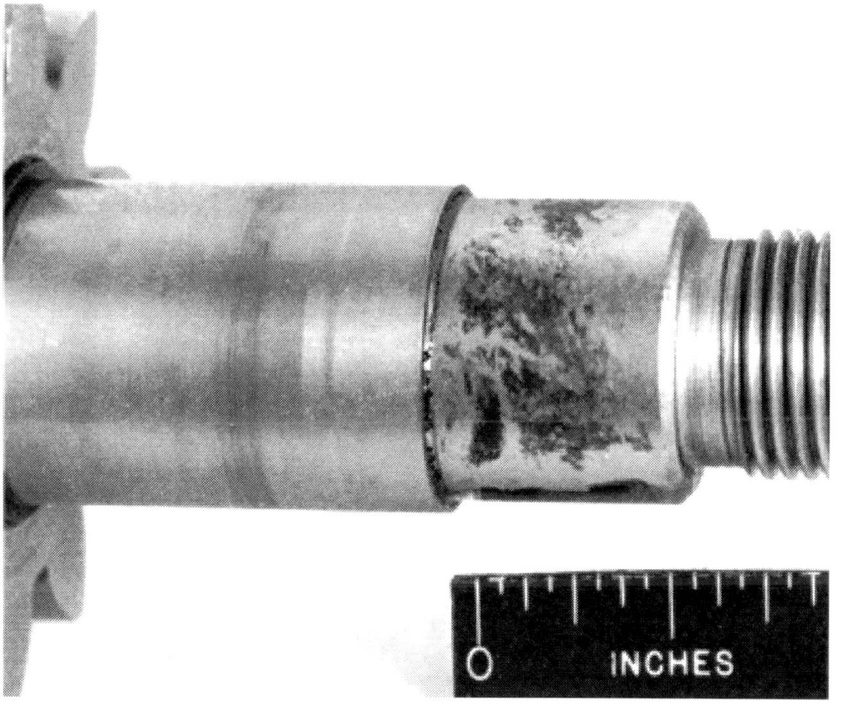

Fig. 14 Fretting wear on a steel shaft at the interface with the hub intended to be a press fit. The same fretting also appeared on the bore of the hub. This is typical of damage in a joint that is nominally stationary but in reality has slight movement between the hub and the shaft. Source: Ref 4

means that nonrotating antifriction bearings that are subject to vibration over a period of time may have fretting wear wherever a ball or roller contacts a raceway under load. If the bearings subsequently rotate in normal service, they may be noisy because of the wear patterns and small indentations that are present in the raceways and the corresponding flat spots on the rolling elements. Fretting is also a serious problem on parts such as shafts, where it can initiate fatigue cracking on the contacting surfaces. In fact, many fatigue fractures of shafts are caused directly by fretting.

Because fretting wear is essentially a stationary phenomenon, the debris that is formed is retained at or near the location where it was formed. The debris usually consists of oxides of the contacting metals; with ferrous metals, it is brown, reddish, or black, depending on the type of iron oxide formed. For this reason, with ferrous metals the debris is sometimes called "cocoa" or "red mud" when it is mixed with oil or grease. Aluminum alloys form a black powder when fretting wear is present.

Prevention of fretting wear is not easy. However, its damage can sometimes be minimized with one or more of the following measures:

- *Eliminate or reduce the vibration.* This can sometimes be accomplished with the aid of vibration-damping pads or by stiffening certain members to increase the natural frequency of vibration. Occasionally, however, neither of these measures is effective, and movement must be increased greatly to improve lubrication.
- *Eliminate or reduce the slip at the interface.* This can sometimes be accomplished by trying to "lock" rough mating surfaces together by increasing the pressure between them. However, if the slip is not completely eliminated, the fretting wear may increase because of the increased contact stress between the mating surfaces.
- *Use an elastomeric material in the joint.* Complete redesign of the joint to include an elastomeric bushing or sleeve may be necessary. Vibration and minute movement still may be present, but the elastomeric material absorbs the motion and prevents metal-to-metal contact.
- *Lubricate the joint.* Because the joint is essentially stationary, liquid lubricant cannot flow through the interface as it can where there is continual sliding motion. Certain greases, solid-film lubricants (such as molybdenum disulfide), and oils are intended to reduce or delay fretting.
- *Prevent fracture.* Fracture resulting from fretting wear may be prevented by inducing compressive residual stresses or by certain heat treatments, although these probably will not eliminate the fretting. One of the most effective means of preventing fatigue fracture is the use of mechanical prestressing by shot peening, surface rolling, or any other method of prestressing. Because fatigue cracks cannot propagate easily through a compressive residual-stress barrier, these methods can be used to prevent fracture unless the part is used at a temperature

high enough to stress relieve the material. In another process, the elements diffused into the metal by proprietary salt bath or gaseous nitriding methods form compounds resistant to adhesive and fretting wear. Epsilon iron nitride is one of the most effective surface compounds in preventing fretting wear and possible fatigue fracture.

Contact-Stress Fatigue

Fatigue, in conjunction with wear, can produce cavities, or pits, in either of the two surfaces that contact each other primarily by rolling and/or sliding action, or in the case of cavitation pitting fatigue, in a metal surface in contact with a liquid. The cavities themselves are serious because they frequently act as stress concentrations that can cause fracture of the major part. This is particularly true with respect to gear teeth. Also, the metal removed from the cavities usually is very hard and brittle. Thus, it is readily crushed and fragmented into much smaller particles, which can cause abrasive wear as well as other damage when carried by the lubricant to other parts of the mechanism. Both rolling and sliding types of bearings, gears, and parts, such as pumps, impellers, and propellers, are subject to this type of damage.

As discussed in this chapter, fatigue is the mechanism that results from cyclic slip under repetitive load applications for many thousands or millions of load cycles. The only difference is that, instead of causing gross fractures of parts, only fragments of the surface are removed, at least initially. These fragments, when lost, result in pits or cavities in the surfaces. This type of pitting wear, or contact-stress fatigue, is frequently the limiting factor in load-carrying ability.

The pits that occur on the surfaces of contacting parts as a result of contact-stress fatigue seem to have three different types of behavior. Some start as microscopic cavities and may stay microscopic throughout the life of the part. They cause only a dull, frosted appearance on an otherwise bright surface. The second type starts the same way with microscopic cavities. However, under continued service of rolling and sliding under load, these pits gradually become larger. The third type of pits is large to start with and rapidly becomes even larger. The latter two types are completely destructive of the surfaces of hardened steel gears, rolling element bearings, roller cams, and other parts or assemblies where there is a combination of rolling and sliding motion.

The parts subject to this type of failure generally have two convex, or counterformal, surfaces in contact under load, such as gear teeth and various types of antifriction bearings. However, the same type of failure can occur where a convex shape fits within a concave shape, such as a shaft within a sliding bearing or balls in a ball bearing race. These situations are shown in Fig. 15. The reason that pitting fatigue occurs on one or both mating surfaces under compressive load is that the contact areas are concen-

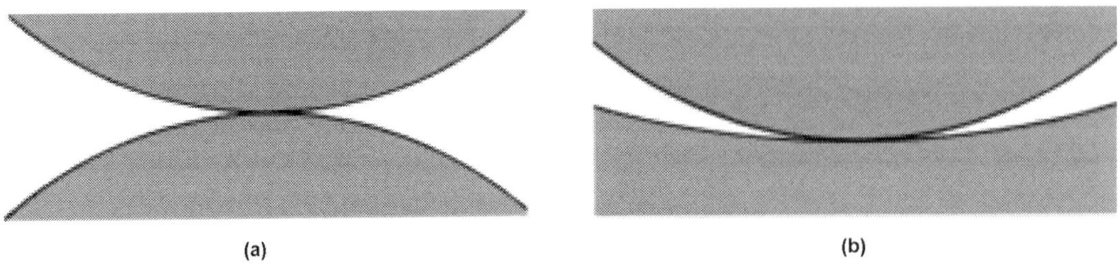

Fig. 15 (a) Sketch of counterformal, or convex, surfaces in contact; examples are gear teeth and roller or needle bearings rolling on a shaft, an inner raceway, or a flat surface. (b) Sketch of conformal surfaces, where a convex surface is in contact with a concave surface; examples are ball bearings in contact with an inner or outer raceway, roller or needle bearings in contact with an outer raceway, and a shaft in contact with a sliding bearing or on a flat surface. Source: Ref 7

trated into either a line or point contact, depending on the geometry. Actually, there will be either a broadened line or an elliptical contact area. Because the instantaneous contact areas may be quite small under heavy loads, the compressive and shear stresses that are formed can be extremely high.

Convex surfaces under heavy pressure deform elastically to form a bulge at the ends of contact. Also, surfaces in contact have the maximum shear stress a short distance below the surfaces when the members are either stationary or rolling with respect to each other. Because fatigue fractures are caused by shear stresses, this location is of primary concern in rolling element components.

When dealing with contact-stress fatigue, it is necessary to understand the difference between pure rolling and rolling plus sliding contact. Pure rolling of metals under compressive load is quite elusive and is difficult, if not impossible, to achieve because of the elasticity of the metals, since all metals are elastic and deform elastically under load. In fact, the harder and stronger metals, such as those typically used in gears and rolling contact bearings, can deform elastically to a greater degree than relatively soft, lower-strength metals. Elasticity under heavy compressive loads results in sliding, or shearing, forces at the interface between the contacting metals. This sliding under load generates heat that must be dissipated, usually by a lubricant. Elasticity under load also causes internal friction within the metals, generating additional heat. In fact, heat dissipation is one of the primary functions of a lubricant, in addition to reduction of friction at the interface.

Because of the many variations in rolling and sliding contact, as well as in metallurgical and geometrical variables, there are several types of fatigue failures that can occur in rolling/sliding elements. In a real sense, there is a competition between the different modes of failure to determine which mode will dominate and cause failure of the element, the assembly, and sometimes the entire mechanism, depending on the degree of failure. In fact, it is a race to failure, because all modes are progressing simultaneously, with usually only one mode causing total failure.

All of the different types of fatigue lead to the same final result: complete destruction of the original surface. However, the factors that cause failure are very different, and there are various ways of trying to prevent this type of failure. The fractures that occur cause pieces of metal to separate from the surface, leaving cavities, or pits, in the surface. Contact-stress fatigue is covered as follows:

- Subsurface-origin fatigue
- Surface-origin fatigue
- Subcase-origin fatigue (spalling fatigue)
- Cavitation fatigue

Subsurface-Origin Fatigue

Pitting in hardened steel as a result of subsurface-origin fatigue occurs during essentially pure rolling motion of one element across or around another. This is most common in antifriction, or rolling element, bearings such as ball and roller bearings, needle bearings, and roller cams but is also found on gear teeth. As pointed out previously, pure rolling is really a misnomer, because there is always some degree of sliding, due to elastic deflection, of the parts under load. Also, because the maximum shear stress is located at a relatively short distance below the surface, this is the normal location for fatigue fractures to originate. Stress concentrations slightly below the surface are the usual origins. Because most geometrical stress concentrations are at the surface of metal parts, the most common geometrical stress concentrations within the metal are various types of inclusions inherent in the steel. There are many types of inclusions, but the most serious are the hard, brittle inclusions that are often angular in shape. Because inclusions are distributed at random within the steel, only those within the high-shear stress region are likely to cause subsurface-origin pitting fatigue. Figure 16(a) shows two rolling elements under pressure, each of which has the same contact (compressive) stress at the surface. However, the depth of the maximum shear stress can differ, depending on the relative geometry of the parts involved. Any damaging inclusions within the high-shear stress region can cause fatigue cracks to originate within the metal parallel to the surface. However, continued rolling across the damaged area will eventually cause cracks to reach the surface. The same procedure may be occurring at several locations within the steel, and eventually, a volume of metal will be surrounded by cracks. When this happens, the particle of metal is lost from the surface, and an irregular-shaped cavity, or pit, is left in the surface. At first, the pit has sides perpendicular to the surface, but continued operation under the pressure of the mating roller rapidly causes the sides to fracture and break down each time the mating member rolls across the pit. Thus, the pit does not stay in the original steep-sided shape very long, unless the rolling action is stopped.

An enlarged cross section through a pit in formation is shown in the metallograph of Fig. 16(b). The horizontal crack is approximately 0.2 mm (0.007 in.) below the surface, and the vertical crack goes up to the surface, but a volume of metal had not yet been surrounded by cracks. It is very difficult to locate the inclusions that cause this behavior. The chances of finding them in a random cut through the metal are very slight, and they may fall out during cutting or preparation. A pit, in which the 25 mm (1 in.) diameter test roller was stopped immediately after the pit was formed, is shown in Fig. 17.

Because hard, brittle inclusions or other internal stress concentrations are often the cause of this type of pitting, one may wonder why the antifriction bearing industry does not use cleaner steel to minimize subsurface pitting and permit higher service loads. The answer is that they do use cleaner steels. This problem has been recognized for many years, and the bearing industry is always alert to use high-quality vacuum-melted steels to try to eliminate the problem. However, each time that an advance is made in eliminating inclusions and providing better-quality steel, the load and speed ratings are increased by each company that uses the improved metal. Customers are told that they can use a higher load on the same size bearing, use a smaller bearing to do a certain job, use higher speeds with the same size bearing, or all of the aforementioned. This means that if the users do these things, the problem will always be with us, because it is impossible to eliminate all of the offending particles and other imperfections from the steel.

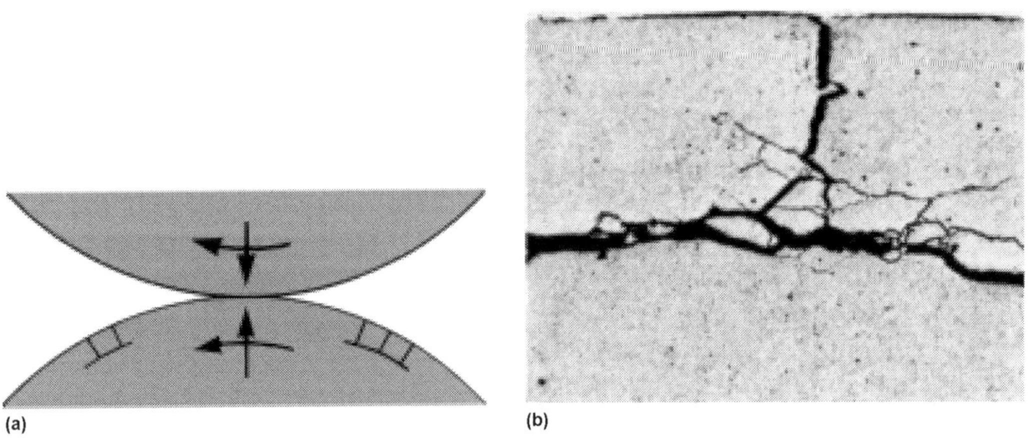

(a) (b)

Fig. 16 Subsurface-origin pitting fatigue. (a) Sketch showing usual origin slightly below the surface where the shear stress is high. The fatigue cracks, which usually originate at stress concentrations such as hard, brittle inclusions, propagate parallel and perpendicular to the surface. When a small volume of metal is surrounded by cracks, it falls from the surface, leaving a cavity or pit with steep-sided walls and a flat bottom, until continued operation breaks down the sides of the walls. (b) Micrograph showing fatigue cracks parallel to the surface, and perpendicular cracks going to the surface of the specimen. The total depth of this crack system was approximately 0.2 mm (0.007 in.), and the metal was case-hardened steel with a surface hardness near 60 HRC. The specimen was a 25 mm (1 in.) diameter roller, tested at a calculated compressive stress of 2900 MPa (425,000 psi) in pure rolling. Source: Ref 7

Fig. 17 Subsurface-origin pit in a carburized and hardened alloy steel test roller caused by fatigue in the manner shown in Fig. 16. When this specimen was tested in essentially pure rolling, a steep-sided, irregularly shaped pit was formed, and the test was stopped. The extremely high force needed to cause subsurface pitting is shown by the plastic deformation at the sides of the wear track formed by the mating roller, which was 125 mm (5 in.) in diameter. Source: Ref 7

Positive identification of subsurface-origin fatigue can only be made by metallographic examination. The microstructure at the surface is not distorted, and there are frequently many cracks a short distance below the surface. It is also necessary to point out that a rolling element bearing does not have an unlimited life—a fundamental fact that does not appear to be generally appreciated. Even if a bearing is run under the recommended conditions of load, speed, and lubrication and is protected against adverse external influences that otherwise would tend to reduce its life, failure will ultimately result by some process such as fatigue, wear, or corrosion.

Surface-Origin Fatigue

Surface-origin pitting of hardened steel surfaces is also destructive to the surface, and, in the case of gear teeth, may also cause bending fatigue fracture of the gear teeth. When sliding is added to the rolling action, an entirely different and complicating set of circumstances arises. The maximum shear stress is no longer located below the surface of the steel but is brought up to the surface because of the influence of sliding friction and the associated traction forces.

To understand the state of stress at the surface of a rolling/sliding interface, it is necessary to analyze separately the relative motions of rolling and sliding. Rolling is shown schematically in Fig. 18, in which a small roller rotates counterclockwise against a larger roller rotating clockwise. There will be pure rolling (ignoring elastic deformation) if the surface velocities (not revolutions per minute) of the two rollers are identical. Points A, B, and C on each roller come into contact successively, as shown in Fig.18(a). However, the same motion relative to the small roller can be achieved if it is held stationary and the large roller rolls around it with a clockwise planetary motion, as in Fig. 18(b). Again, points A, B, and C successively come into contact. Thus, the rolling direction, the direction of the contact point, is to the right, or clockwise, on the small roller in each example.

It must also be noted that the direction of rolling is opposite to the direction of rotation. That this is always true may be shown by an analogy: Consider the "contact point" of the sun as it "rolls" around the Earth. (The "contact point" is that part of the Earth directly closest to the sun, i.e., high noon.) The Earth actually rotates from west to east, but the contact point "rolls" around the Earth from east to west, in the opposite direction.

The previous discussion assumes that there is only pure rolling between the two elements. The situation becomes more complicated if the rollers do not have the same surface velocity, that is, if there is also sliding at the interface. Assume that the larger roller is rotating with a higher surface velocity than is the smaller roller, as shown in Fig. 19(a). The difference in the sliding tends to drag the surface of the small roller to the left, or counterclockwise, while the surface of the larger roller is dragged to the right, also counterclockwise on that roller. But note that the direction of sliding is opposite to the direction of rolling on the small roller, while the rolling and sliding directions are the same on the larger, faster roller.

The term *negative sliding* is used when rolling and sliding are in opposite directions, as on the small roller in this example, and positive sliding occurs when the rolling and sliding directions are the same, as on the larger roller. Because the surface of the small negative-sliding roller is, in effect, rolling in one direction and simultaneously being dragged in the opposite direction, the frictional, thermal, and shear stresses tend to be higher in this member than in the larger positive-sliding roller. In addition,

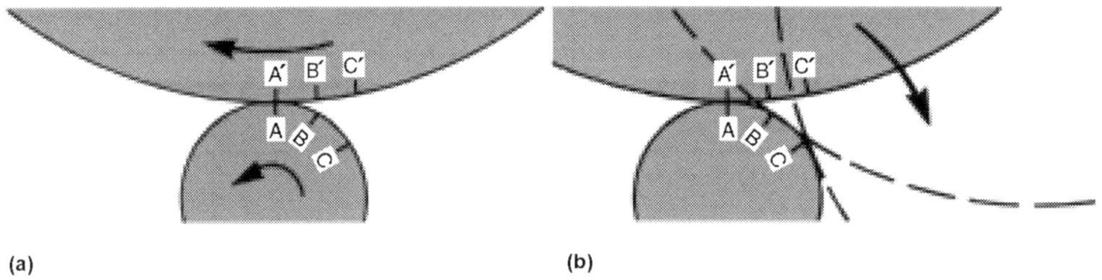

Fig. 18 Schematic of pure rolling (ignoring elastic deformation). (a) When two rollers of different sizes but the same surface velocity (not revolutions per minute) contact in pure rolling, point A will contact point A′, then point B will contact B′, point C will contact point C′; other points around the periphery will contact in succession. However, note that the direction of rotation is toward the left and that the direction of rolling (or of the point of contact) is to the right, in the opposite direction to that of rotation. (b) This concept may be simplified by stopping one roller, such as the lower roller, as shown. If the mating roller rolls around the lower roller in a planetary motion, as shown, again point A, B, and C contact successively the corresponding points on the other roller. The effect on each roller is the same as before, because the relative motion is the same. Source: Ref 7

the smaller member has more load on each surface point than does the larger member. Therefore, in this example, the small roller is far more likely to undergo destructive surface-origin pitting than the larger roller.

In devices that undergo combined rolling and sliding, knowledge of the relative velocities and directions of rolling and sliding is necessary for definition of the wear mechanism. The direction of rolling is defined as the direction in which the point of contact moves; the direction of rolling is always opposite to the direction of rotation of a rolling element. On a given surface, a condition of positive sliding exists if the direction of sliding is the same as the direction of rolling. Negative sliding occurs on the mating surface, where the directions of rolling and sliding are opposite to each other. Most surface fatigue failures originate in regions of negative sliding, because the shear stresses there are usually more severe than in regions of positive sliding. Negative sliding occurs on the dedenda of the gear teeth (i.e., the regions below the pitch line), on the cam follower riding on a cam and, in other devices, on the part that has the lower surface velocity in a rolling-sliding system.

The reasons for the severe stress conditions in negative sliding become obvious with study of Fig. 19(b). Note that the lower, slower roller is being dragged to the left, making a bulge on the exit side of the contact. At the same time, the point of contact of the upper roller is moving to the right, trying to make a bulge on the entrance side of contact at the right.

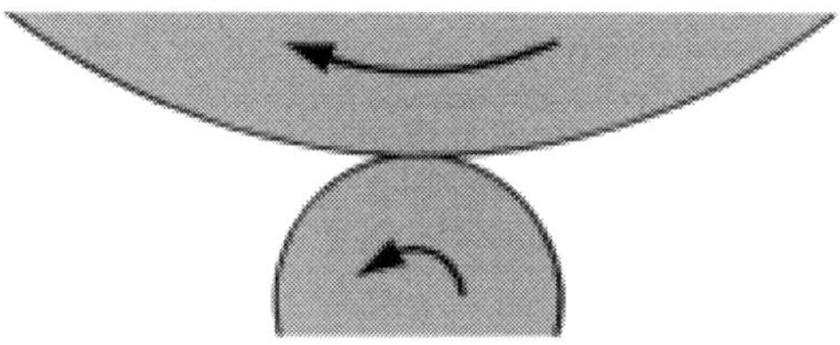

(a)

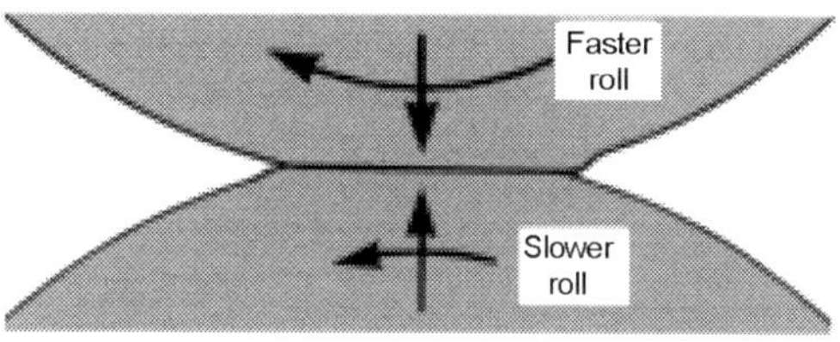

(b)

Characteristic	Lower roller	Upper roller
Surface velocity	Slower	Faster
Rotation	Counterclockwise	Clockwise
Rolling	Clockwise	Counterclockwise
Sliding	Counterclockwise	Counterclockwise
Positive/negative sliding	Negative	Positive

Fig. 19 Schematic of rolling/sliding contact. (a) The situation shown in Fig. 18 changes drastically if the rollers are externally driven and forced to rotate with different surface velocities. In this figure, the upper roller is driven at a higher surface velocity than the lower roller, which introduces sliding into the interface. Now there is rolling-sliding contact, which greatly complicates the situation compared with the pure rolling shown in Fig. 18. (b) A closer look at the interface of (a) after heavy compression forces are applied to the two rollers shows that the faster upper roller tends to drag the surface of the slower lower roller to the left, increasing the bulge on that side and reducing the bulge on the opposite side of the slower roller. However, the surface of the faster upper roller is itself held back by the action of the slower lower roller, increasing the bulge on the right side of the upper roller and reducing the bulge on the opposite side of the upper roller. The elastic deformation of each roller progresses like a wave around them, alternately applying shear stresses to and removing them from the surface of each. However, in the slower roller, the rolling and sliding are in opposite directions, creating the condition of negative sliding, which is most likely to cause surface pitting fatigue. Source: Ref 7

The complex stress conditions that result from this negative sliding are the reasons why the maximum compressive stress that can be carried is less in a rolling/sliding situation than in a pure rolling situation. In other words, the addition of sliding reduces the load-carrying capacity. At approximately 65% rolling and 35% sliding, hardened steels commonly pit at 20 to 30 million stress cycles when the calculated contact stress is approximately 2400 MPa (350,000 psi). When rolling alone is involved, the calculated contact stress must be 4100 to 4500 MPa (600,000 to 650,000 psi) to cause pitting in the same number of stress cycles.

Figure 20(a) shows a metallographic section through a nearly complete surface pit in a case-hardened steel roller test specimen. In this view, the direction of rotation and sliding is to the left, while the direction of rolling is to the right, the condition of negative sliding. Note that the surface has several small, diagonal surface-origin fatigue cracks slanting toward the lower right, because the surface was dragged to the left. The major crack system has progressed inward to a depth of approximately 0.25 mm (0.010 in.) and has started to come back toward the surface.

Meanwhile, the view from outside the roller shows that the surface, which usually has a frosted appearance to the unaided eye, actually has many tiny V-shaped cracks when examined in a metallurgical microscope, as in Fig. 20(b). The V's point in the direction of rotation and are actually the outer view of the small diagonal cracks shown on the cross section in Fig. 20(a). These are potential origins of continuously growing fatigue cracks that may form larger-size pits. Each time the mating roller rolls across from left to right, the lubricant is forced down into the crack and can act as a hydraulic wedge to help push the fatigue crack deeper, in addition to the cantilever beam effect that exists. As the crack expands in the V-shape on the surface and gradually becomes deeper, a V-shaped volume of metal falls out, leaving an arrowhead-shaped pit pointing in the direction of rotation, as shown in Fig. 21. With continued battering and high-stress contact from the mating surface, however, this V-shape rapidly breaks down. Thus, these shapes are not often seen in actual operating parts that have run for some time after the original pit is formed.

Gear teeth are the principal hardened steel parts subject to surface origin pitting. Because they are essentially carefully shaped cantilever beams, gear teeth have complex rolling/sliding motions, as shown in Fig. 22 and summarized in Table 1. It is significant that the dedenda of all gear teeth are in negative sliding, and this is where surface-origin pitting is most likely to occur, other conditions being equal. Pure rolling occurs only on the pitch line of spur, bevel, and helical gears but not at all on worms, spiral bevel, and hypoid gears and pinions. In fact, the direction of sliding undergoes a reversal at the pitch line, as shown in the table.

The driving gear is usually the smaller of the pair of gears and receives many more load applications per tooth, depending on the ratio between the gears. For this reason, the dedendum of the smaller gear is where pit-

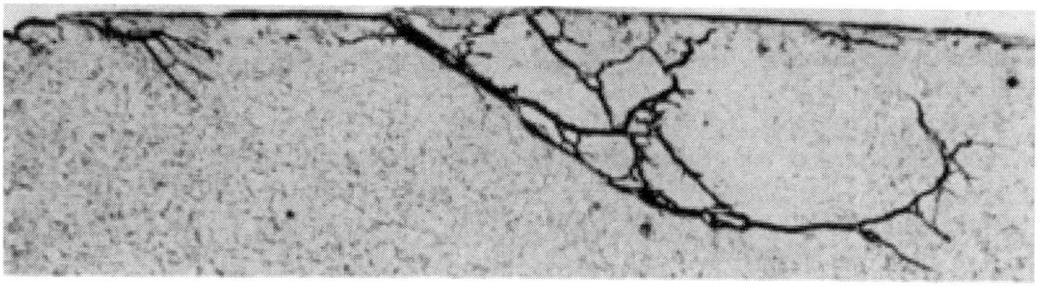

(a)

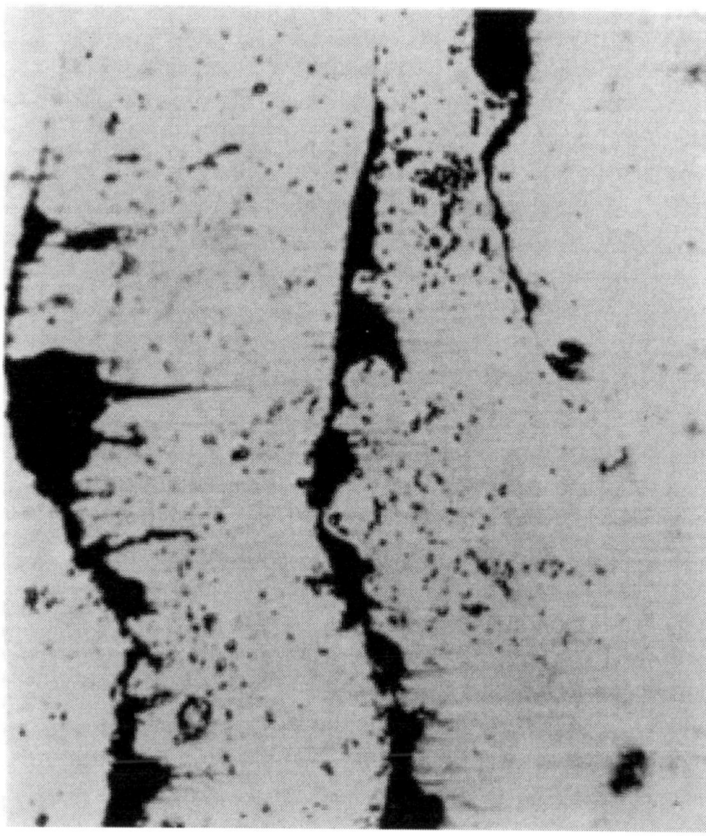

(b)

Fig. 20 (a) Cross section through a nearly complete surface pit (max depth approximately 0.25 mm, or 0.010 in.) in a roller test specimen. The negative-sliding member (lower and slower roller in Fig. 19b) forms small diagonal cracks from the surface inward at an angle. Some of these cracks may proceed deeper into the hardened steel, as seen in this section. The orientation is the same as in Fig. 19(b), in which the sliding is to the left and the rolling is to the right, producing negative sliding. (b) Highly magnified view (1000×) of the surface of a negative-sliding roller. The many small, V-shaped cracks are the outer view of diagonal surface cracks like those seen in (a). As a crack progresses deeper into the steel, the V widens on the surface. Eventually a V-shaped volume of metal becomes totally surrounded by crack, falls out, and leaves an arrowhead-shaped pit, which rapidly changes shape under continued loading. Sliding direction is to the left, rolling to the right. Source: Ref 7 ting fatigue will originate if the parts have the same metallurgical properties. For this reason, the smaller gear often is made slightly harder than the larger gear to compensate for the difference in the number of load applications.

Because sliding is always present in the operation of gears, the lubricant is of extreme importance for the survival of any set of heavily loaded

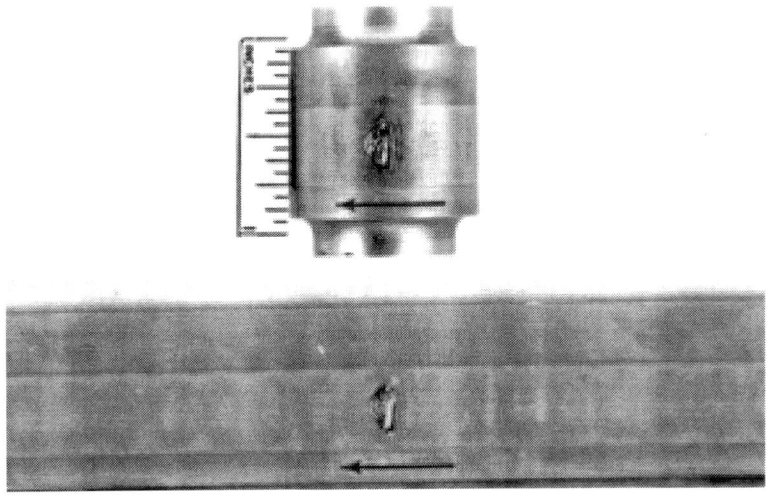

Fig. 21 Two views of a roller, showing typical arrowhead-shaped surface pit. The arrows indicate the direction of rotation; the direction of rolling of the mating roller is to the right, while the direction of sliding is to the left. The upper view is a normal photograph of the pitted area, while the lower view shows the entire surface "unwrapped" by a special photographic method. This specimen was tested at a calculated compressive stress of approximately 3100 MPa (450,000 psi) and ran for 14,229,000 revolutions prior to pitting failure. This was a carburized and hardened alloy steel, with a surface hardness near 60 HRC. Source: Ref 7

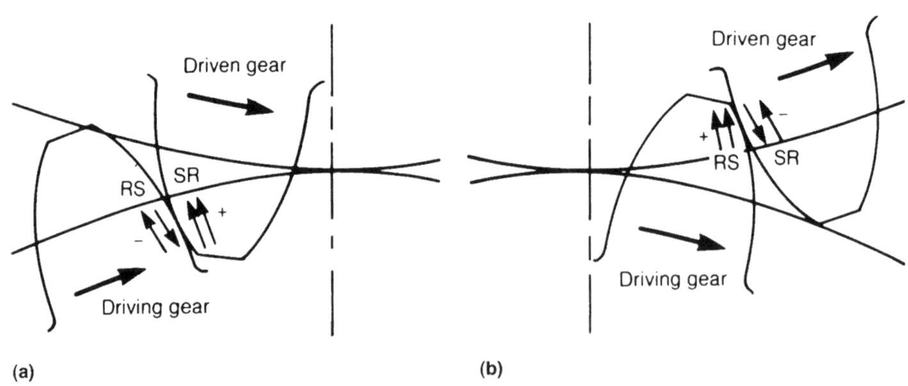

Fig. 22 Schematic of rolling-sliding action inherent in gear teeth. As gear teeth contact, rolling (R) and sliding (S) stresses are formed by the relative movement. Pure rolling occurs only at the pitch line, and on each gear the direction of sliding reverses at the pitch line. Analysis of the relative motions of gears reveals that on both the driving and driven gears there is negative sliding below the pitch line and positive sliding above the pitch line. Because negative sliding is more likely to cause surface-origin pitting, damage is most likely to occur below the pitch line on the smaller gear, which is usually the driving gear. (a) Beginning of contact. (b) End of contact. Source: Ref 7

Table 1 Movements in mating gears

Gear	Rolling direction(a)	Sliding direction(a) On dedendum (below pitch line)(b)	On addendum (above pitch line)(c)	With resect to pitch line
Driving	Up	Down	Up	Away
Driven	Down	Up	Down	Toward

(a) Up, toward top; down, toward root. (b) Negative sliding: rolling and sliding are in opposite directions. (c) Positive sliding: rolling and sliding are in the same direction. Source: Ref 7

gears. In addition to adhesive wear, the reduction of surface friction is of critical importance in the effort to resist pitting fatigue. Because of the difficulty in increasing the life and/or load-carrying capacity of gear systems, the development of pits usually represents the limit of gear service. Extensive research with roller test specimens has indicated that the optimum conditions for the long-term resistance to pitting fatigue on gears are provided with a surface hardness near 60 HRC, smooth surface finishes, and at least 10 to 20% retained austenite on the surface of case-hardened steel. The function of the retained austenite is to permit some degree of plastic deformation by increasing the contact area in order to reduce the actual compressive stress. However, bending fatigue strength is reduced by excessive retained austenite. The damage that surface-origin pitting fatigue can cause to operating gear teeth is shown in Fig. 23.

Another type of surface-origin pitting fatigue occurs on conformal surfaces, such as a shaft rotating within a sliding bearing. Fatigue failures of sliding bearings is usually the result of long, hard service under severe repetitive, compressive forces, such as occur on the upper half of engine connecting rod bearings (which transmit the explosive forces to the crankshaft) or the lower half of the main crankshaft bearings (which resist bending of the crankshaft due to the explosive forces). The locations of highest stress are approximately 35° on each side of top center of the upper halves and approximately 35° on each side of the bottom center of the lower halves. From these origins, the fatigue pits usually spread out to wider locations on the bearings until the bearings are completely destroyed.

In addition to the explosive forces imposed on the bearings, there are also centrifugal and inertial forces that, together, cause the total load spectrum on the individual bearings. Fatigue of soft, nonferrous bearing metals, such as tin-base or lead-base Babbitts, copper-lead alloys, bronzes, aluminum alloys, certain powder metal alloys, and trimetal bearings (consisting of three layers of metals), normally originates at the surface because that is where the shearing stresses are highest due to the sliding action on the surface. The lubricant system and geometrical characteristics of the bearing and the mating surface are critical so that the proper lubricant film thickness is maintained. Obviously, if an oil film can prevent the surfaces from contact, neither fatigue failure nor other problems associated with sliding contact can occur.

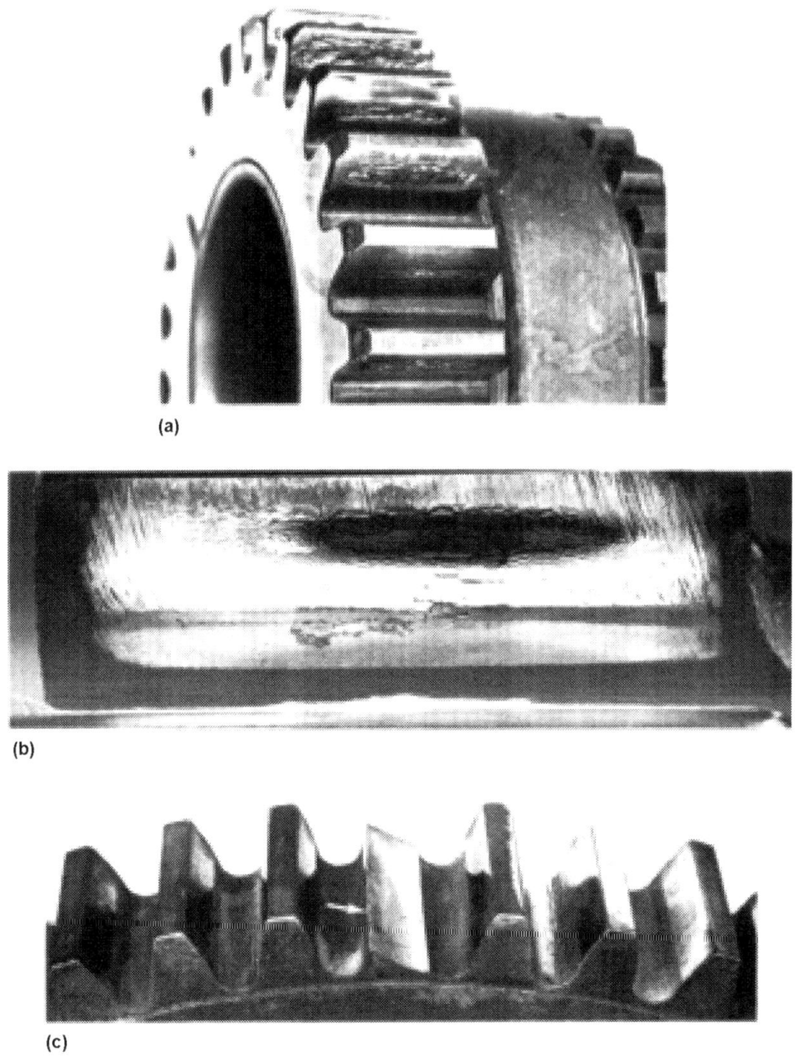

Fig. 23 Surface-origin pitting fatigue. (a) Typical surface deterioration due to pitting fatigue on gear teeth; in a standard gear system, the pitch line is near the center of the height of the teeth. Pitting fatigue usually starts slightly below the pitch line, then rapidly spreads to adjacent areas, causing complete surface deterioration. This heavily loaded gear was made of case-hardened alloy steel, with a surface hardness near 60 HRC. (b) Heavily loaded final drive pinion with surface pitting near the pitch line, which was low on this gear tooth. The rippling shown is plastic deformation caused by the sliding action under heavy contact pressure and is a warning that the metal is close to failure. The parallel diagonal marks are tool marks resulting from the shaving operation and are not involved with the service in any way. (c) Fracture of gear tooth at pitch line. Another consequence of pitting fatigue on the active profile of gear teeth is the fact that the pit itself is a stress concentration that can cause abnormal bending fatigue fracture of gear teeth, which are really carefully shaped cantilever beams. This gear tooth pitted just below the pitch line, then fractured as a result of bending fatigue. The normal location for bending fatigue fractures of gear teeth is at the root fillet, not midway up the tooth, as in this example. This gear was made from a medium-carbon steel, induction hardened to approximately 55 HRC. Source: Ref 7

Subcase-Origin Fatigue

Subcase-origin fatigue damage to case-hardened rolling/sliding surfaces, such as gear teeth and certain roller mechanisms, can completely destroy the contacting surfaces. In this type of failure, very large pieces are suddenly lost from the surface, and extensive damage can result. However, this type of failure is relatively easy to prevent, once it is identified. Subcase-origin fatigue is also known as spalling fatigue or case crushing. However, the term *subcase fatigue* is more descriptive of the mechanism involved and is preferred. *Case crushing* implies static fracture, which may be accurate in instances of severe overloading, but this is not a fatigue mechanism.

As shown in Fig. 24, subcase fatigue is somewhat similar to subsurface fatigue. However, the difference is in the scale of magnitude. Subsurface fatigue resulting in pits originates a few thousandths of an inch from the surface, whereas subcase fatigue may originate at much deeper regions, usually slightly below the case depth, which may be 1 mm (0.040 in.) or more from the surface, depending on the heat treatment of the particular part. At any rate, fatigue can originate deep within the part as a result of contact-stress fatigue, causing fatigue cracks parallel and perpendicular to the surface. These cracks tend to be very long, as shown on the hypoid rear axle pinions in Fig. 24. Surface cracking is the first obvious indication of the subsurface fatigue, although hidden cracks could be detected by ultrasonic instrumentation. Continuing service rapidly leads to the severe destruction shown, in which the long-undermined pieces have come out as large fragments. These leave large, long, gouged-out cavities that completely destroy the surfaces involved.

The correction of this type of problem is relatively simple, because this type of failure indicates that the shear strength is inadequate below the case. The strength can be raised and failures prevented by increasing the case depth and/or the core hardness (and strength) by using a steel with higher carbon or alloy content. However, any metallurgical changes must be made carefully, because if the case depth or core hardness is increased too much, the teeth could be through hardened, which could lead to brittle fracture, depending on the shape of the teeth and other regions of the part. Also, the residual-stress pattern would be changed drastically, potentially causing additional problems. The characteristics of contact-stress fatigue on mating metal surfaces are summarized in Table 2.

Cavitation Fatigue

Because the general title of this section is "Contact-Stress Fatigue," it seems appropriate to include a type of pitting fatigue that is caused by vibration and movement in various liquids, of which water is the most common. Because many liquids are corrosive in some ways to most metals, the problem of environmental reaction becomes entwined with the problem of contact-stress fatigue.

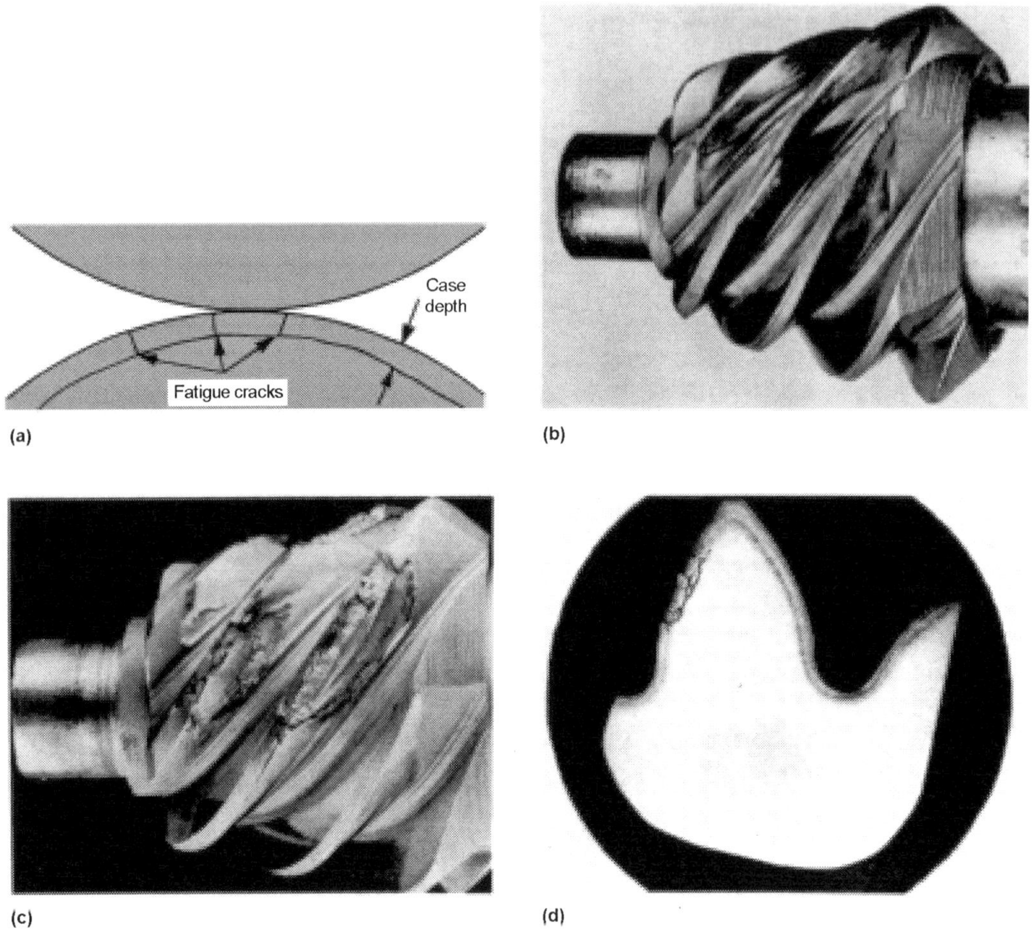

Fig. 24 Subcase-origin fatigue. (a) As the name implies, subcase fatigue cracks originate deep within the steel at the region below the case, where the core metal is comparatively soft in relation to the case itself. The fatigue cracks spread laterally, parallel to the surface, then join and cause cracks that come out to the surface. By the time surface cracks are observed, however, the fatigue is well developed within the gear teeth. Large chunks can come from the surface at one time; deterioration is not gradual as in surface-origin fatigue. (b) Hypoid pinion with many long cracks on the surface and tips of the teeth. Although the teeth in this photograph appear to be relatively intact, they are actually undermined with fatigue cracks. The cracks themselves are several inches long. (c) Hypoid pinion similar to that shown in (b). Long, finger-length pieces of metal have come out of the surface, leaving a longitudinally ridged and gouged appearance. The ridges are the locations where two adjacent fatigue cracks joined and sent a crack through the case to the surface. This and the gear in (b) were made of low-carbon alloy steel, case carburized to a surface hardness near 60 HRC. (d) Cross section through a gear tooth showing subcase-origin fatigue cracks similar to those sketched in (a). The original fatigue cracks are those parallel to the surface below the dark etched case. When these cracks join, the cracks to the surface then complete removal of the large fragments of metal. Core hardness was low on this gear, contributing to this type of failure. Source: Ref 7

Cavitation pitting fatigue can be a serious problem in marine propellers of all sizes, diesel engine cylinder liners, pump impellers, hydraulic pumps and equipment, turbines, torque converters, and miscellaneous other parts that contact or vibrate in various liquids.

The pits can range in size from very small to very large, from pinhead size to golf ball size, or even larger in some cases. The pits can completely penetrate the thickness of the metal, which may be several inches thick. Obviously, this can result in catastrophic damage to the structure, in addition to destroying the functional efficiency of the parts involved. Typical examples of cavitation pitting fatigue are shown in Fig. 25.

Cavitation pitting is characterized by the fact that it appears to be pitting fatigue, sometimes aggravated by corrosion, and occurs in low-pressure regions at a rapidly vibrating liquid-metal interface. Cavitation pitting fatigue can be visualized by studying Fig. 26 and imagining that the events shown are taking place in ultraslow motion. Actually, they occur in microseconds, depending on the vibration frequency of the parts involved and the vapor pressure of the liquid.

Assume that the metal vibrates from side to side, with the liquid to the right of the metal wall. In Fig. 26(a), the metal wall is moving to the right, against the inertia of the liquid. In Fig. 26(b) the metal has reached the end of its travel to the right, but the liquid is still being pushed to the right. As the metal moves back to the left in Fig. 26(c), the liquid is still moving to the right because of its inertia. Small cavities, or negative-pressure bubbles, form in the liquid at the interface with the metal as the two materials momentarily move away from each other. At Fig. 26(d), the metal reaches the end of its travel to the left, pulling the cavities along with it. In Fig. 26(e), the metal is again moving to the right but collides with the liquid moving to the left. The cavities collapse violently, or implode, because of the inertia of both the metal and the liquid as they move toward each other.

The collapsing cavities implode on the metal with compressive stresses estimated at several hundred atmospheres, equivalent to many thousands of pounds per square inch. Because the geometry of the vibrating system and the properties of the liquid are relatively constant, the cavities form in clusters at certain preferred locations. With constant repetition of the pounding, the fatigue mechanism progresses until pits form in the metal surface at these locations.

Corrosion may enter the picture if surface films are formed on the clean, unprotected, virgin metal in the pit when the system is at rest. Then, when the motion resumes, any surface film is rapidly destroyed by the very high

Table 2 Characteristics of contact-stress fatigue

Characteristic	Surface pitting	Subsurface pitting	Subcase fatigue
Location of origin	Surface, often at "micropits"	Short distance below surface, usually at a nonmetallic inclusion	Near case-core boundary in case-hardened parts
Initial size	Small	Small	Large
Initial area:depth ratio	Small	Small	Large
Initial shape	Arrowhead, then irregular	Irregular	Gouged and ridged
Crack angle with respect to surface	Acute	Roughly parallel at bottom, perpendicular at sides	Roughly parallel at bottom, perpendicular at sides
Apparent occurrence	Gradual	Sudden	Sudden

Source: Ref 7

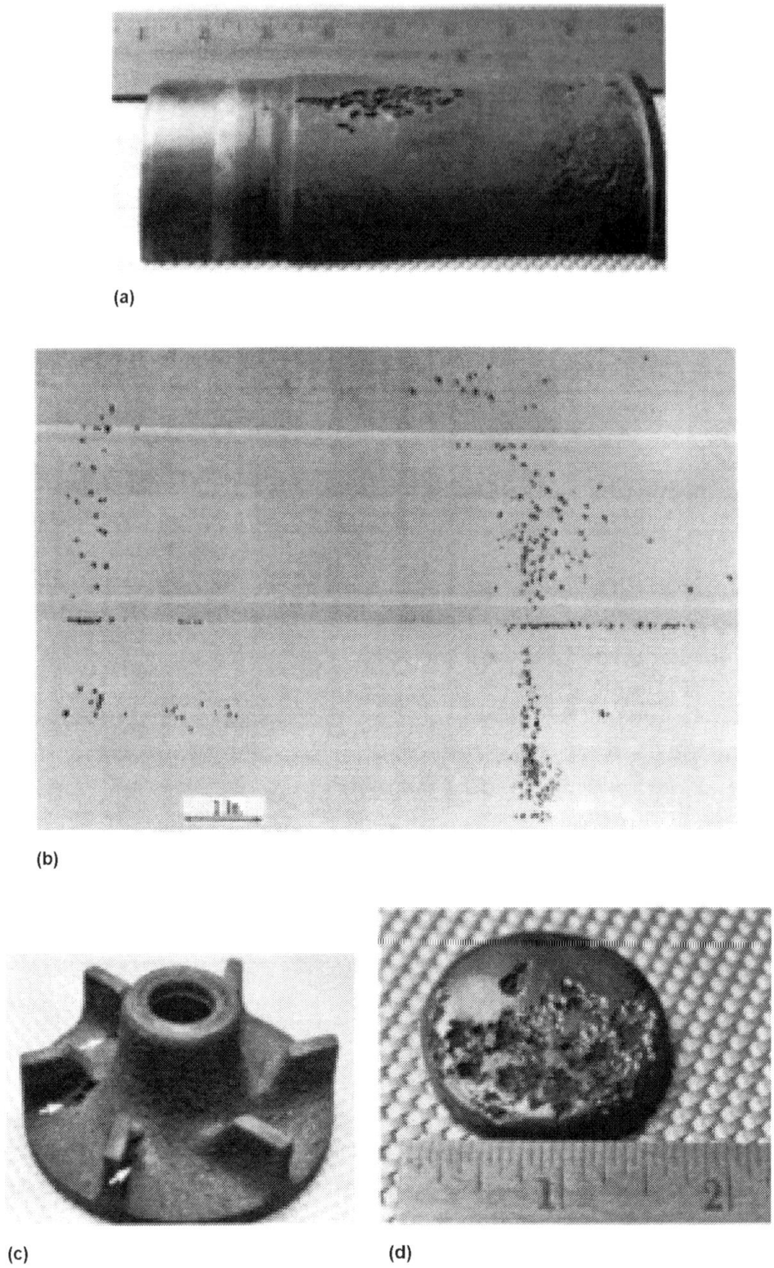

Fig. 25 Cavitation pitting fatigue. (a) Cavitation pitting on a gray cast iron diesel engine cylinder sleeve. The pitted area is several inches long, and the pits nearly penetrated the thickness of the sleeve. Note the clustered appearance of the pits at preferred locations. (b) Cavitation pitting on another gray cast iron diesel engine cylinder sleeve unwrapped by a special photographic process. Again, note the clustered locations, with the most severe pitting on the thrust side of the sleeve, against which the piston slides on the power stroke of the combustion cycle. The lighter pitting at left is on the opposite, or antithrust, side of the sleeve. (c) Cavitation pitting at preferred locations on the vanes of a gray cast iron water pump impeller. This impeller rotated in a clockwise direction; the arrows show some of the pits that were formed in the metal on the suction side of the vanes. (d) Cavitation pitting that perforated this steel freeze plug from a gasoline engine, causing leakage of coolant that could have damaged the engine. Vibration of the wall of the engine block at this location caused this type of damage on the coolant side. Source: Ref 7

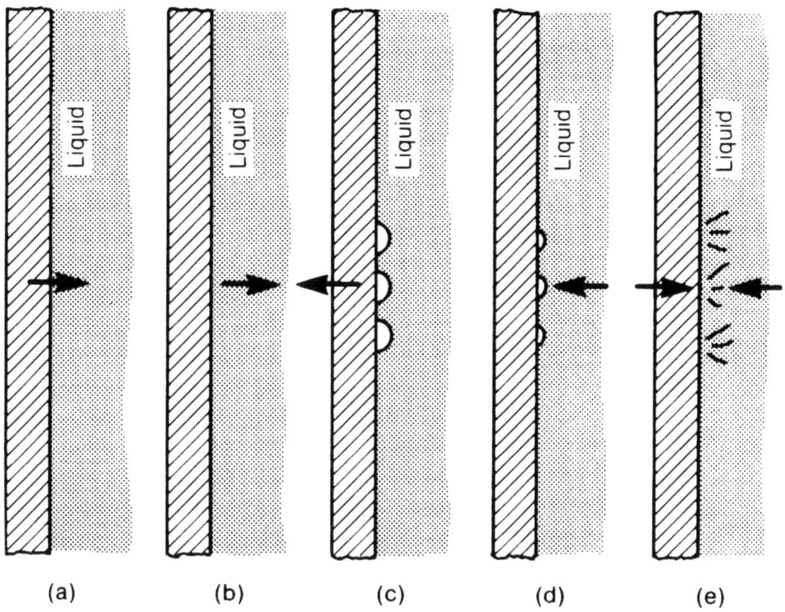

Fig. 26 Mechanism of cavitation pitting fatigue. Serial sketches show a metal wall vibrating to right and left against a liquid, which in all cases is to the right of the wall. The events shown can occur in a very short time, on the order of microseconds. (a) The metal moves to the right against the stationary liquid, which resists movement because of its inertia. (b) The metal reaches the end of its travel, but the inertia of the liquid causes it to continue to move to the right. (c) The metal starts moving toward the left, away from the liquid, which cannot catch up to the wall because of its inertia. Consequently, cavities or voids are formed in the liquid on the surface of the metal. These are essentially negative-pressure bubbles. (d) The metal reaches the end of its travel to the left, and the liquid tries to catch up to the metal and collapse the cavities. (e) The metal starts moving to the right as the cavities collapse violently, or implode, against the metal. As the vibration of the metal continues, the formation and collapse of these cavities in the liquid can cause pitting fatigue on the surface of the metal. If the pounding continues at the same locations on the metal, pits will eventually lead to complete perforation of the metal wall. Source: Ref 7

compressive forces encountered, and corrosion then becomes a factor in the pitting process. However, even glass and ceramics can have cavitation pitting in inert liquids. Apparently, corrosion is not necessary for cavitation pitting fatigue.

It is not easy to prevent cavitation fatigue in a liquid-metal system. The reason is that cavitation pitting fatigue is a function of metal properties, design of the part and vibrational characteristics of the entire mechanism, and the pressure and flow characteristics of the liquid. Some possible ways to try to reduce the cavitation fatigue problem are as follows, in no particular order:

- *Increase the stiffness of the part.* This should reduce its amplitude of vibration and increase the natural vibration frequency of the part. It may be possible to increase wall thickness or to add stiffening ribs to change its vibration characteristics.

- *Increase the smoothness of the surface.* Because the cavities tend to cluster in certain low-pressure areas, the effect may be dispersed if there are no surface peaks and valleys.
- *Increase the hardness and strength of the metal.* Because cavitation pitting is essentially a fatigue phenomenon, the metal properties are important, although an increase in hardness and strength may only delay the problem, not prevent it. Gray cast iron is a frequent victim of cavitation pitting; because of its softness, it has poor resistance to pitting but is often used in parts subject to this type of failure.
- *Streamline the fluid flow.* It may be possible to eliminate corners on trailing edges or in low-pressure regions that could cause cavitation pitting. This has obvious relevance to the design of marine propellers, which may pit in the low-pressure regions, usually near the trailing edge on the forward side. In a closed system such as a diesel engine, for example, it may be possible to increase the coolant flow around cylinder sleeves to disperse the cavities so that they collapse on different locations of the surface.
- *Increase the pressure on the liquid or use inhibitors to increase the vapor pressure on the liquid.* If the cavities (or bubbles) cannot form in the liquid, they cannot cause damage to the metal.
- *Live with the problem.* If none of the aforementioned possible solutions is practical for a given application, it may be necessary to simply live with the problem and replace pitted parts during regularly scheduled maintenance. This is the way that we live with other inevitable fatigue and wear problems such as occur in bearings, tires, shoes, and the like.

ACKNOWLEDGMENTS

The material in this chapter is largely from *Understanding How Components Fail*, 2nd ed., by D.J. Wulpi, ASM International, 1999. A portion of the section on abrasive wear came from "Abrasive Wear" by J.H. Tylczak, *Friction, Lubrication, and Wear Technology*, Vol 18, *ASM Handbook*, ASM International, 1992.

REFERENCES

1. *Failure Analysis and Prevention,* Vol 11, *ASM Handbook,* ASM International, 1986, p 145–162, 497–498
2. J.H. Tylczak, Abrasive Wear, *Friction, Lubrication, and Wear Technology,* Vol 18, *ASM Handbook,* ASM International, 1992
3. K. Kato, Wear Mode Transitions, *Scr. Metall.,* Vol 24, 1990, p 815–820
4. D.J. Wulpi, Wear Failures—Abrasive and Adhesive, *Understanding How Components Fail,* 2nd ed., ASM International, 1999

5. R.A. Higgins, *Engineering Metallurgy—Applied Physical Metallurgy,* 6th ed., Arnold, 1993
6. C.R. Morin, K.F. Packer, and J.E. Slater, Failure Analysis Associated with Mining and Heavy Mechanical Equipment, *Metallography in Failure Analysis,* J.L. McCall and P.M. French, Ed., Plenum Press, 1978, p 191–205
7. D.J. Wulpi, Wear Failures—Fatigue, *Understanding How Components Fail,* 2nd ed., ASM International, 1999

SELECTED REFERENCES

- R.G. Bayer, *Mechanical Wear Prediction and Prevention,* Marcel Dekker, 1994
- F.P. Bowden and D. Tabor, *Friction and Wear,* Methuen and Co., 1967
- *Erosion, Wear, and Interfaces with Corrosion,* STP 567, ASTM, 1974
- M.B. Peterson and W.O. Winer, Ed., *Wear Control Handbook,* American Society of Mechanical Engineers, 1980
- E. Rabinowicz, *Friction and Wear of Materials,* 2nd ed., John Wiley & Sons, 1995
- D.A. Rigney, Ed., *Fundamentals of Friction and Wear of Materials,* American Society for Metals, 1981
- D.A. Rigney and WA. Glaeser, Ed., *Source Book on Wear Control Technology,* American Society for Metals, 1978
- J.A. Schey, *Tribology in Metalworking: Friction, Lubrication, and Wear,* American Society for Metals, 1981
- *Wear and Fracture Prevention,* American Society for Metals, 1981

CHAPTER **13**

Environmentally-Induced Failures

CORROSION is the gradual degradation of a material due to the environment. In fact, the word *corrode* is derived from the Latin *corrodere*, which means to gnaw to pieces. Metallic corrosion is a chemical or electrochemical process in which surface atoms of a solid metal react with a substance in contact with the exposed surface. The corroding medium is usually a liquid substance, but gases and even solids can also act as corroding media. Corrosion can manifest itself in numerous ways that may or may not be obvious. For example, while the rusting of a steel surface is fairly obvious, the intergranular corrosion of stainless steel is less obvious but just as damaging.

Although the principles of corrosion are fairly well understood, corrosion continues to cost billions of dollars a year in the United States alone. The cost in the United States exceeds $300 billion per year. Fortunately or unfortunately, depending on how one looks at it, approximately 35% of the total cost could be avoided by proper prevention methods.

This chapter covers some basic principles of electrochemical corrosion and then some of the various types of corrosion, with an emphasis on stress-corrosion cracking and corrosion fatigue. This is followed by a short section on corrosion control. The last section deals with high-temperature oxidation, which usually occurs in the absence of moisture.

Basics of Electrochemical Corrosion

Electrochemical corrosion is a process resulting in part or all of the metal being transformed from the metallic to the ionic state. Electrochemical corrosion in metals is caused by a flow of electricity from one metal to another metal or from one part of a metal surface to another part of the same surface. For a current to flow, a complete electrical circuit is required. In a corroding system (Fig. 1), this circuit is made up of four components:

- The anode is the electrode of an electrolytic cell at which oxidation is the principal reaction. Electrons flow away from the anode in the external circuit. It is the anode at which corrosion occurs and metal ions enter solution.
- The electrolyte is an electrical-conducting solution that contains ions, which are atomic particles or radicals possessing an electrical charge. The electrolyte is a conductive liquid through which the current is carried by positively charged ions (cations) to the cathode. Negatively charged ions (anions) are simultaneously attracted to the anode. Charged ions are present in solutions of acids, alkalis, and salts. Water, especially salt water, is an excellent electrolyte. In pure water, there are positively charged hydrogen ions (H^+) and negatively charged hydroxyl ions (OH^-) in equal amounts. The metal undergoing electrochemical corrosion need not be immersed in a liquid but may be in contact with moist soil or may just have moist areas on the metal surface.
- The cathode is the electrode of an electrolytic cell at which reduction is the principal reaction. Electrons flow toward the cathode in the external circuit. The cathode does not corrode.
- A metallic path is an external circuit to complete the connection between the anode and cathode.

If any one of these four conditions is absent, corrosion cannot occur. In essence, the components are arranged so that there is a closed electrical circuit.

It is important to note that separate anode and cathode metals are not required for corrosion. As shown in Fig. 2, distinct anode and cathode areas can be caused by inhomogeneities within a single piece of metal.

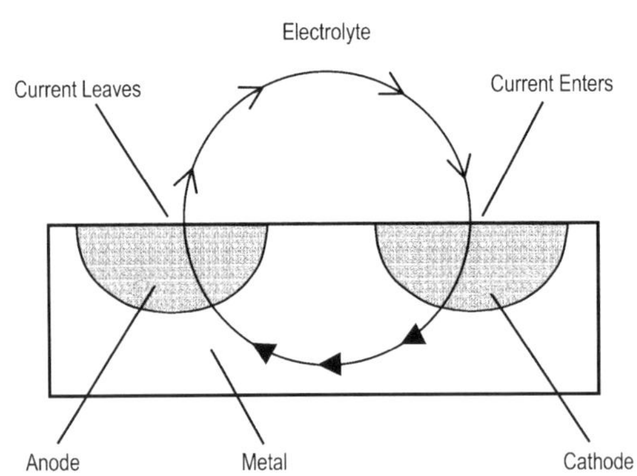

Fig. 1 A corroding system. Source: Ref 1

Second phases in alloys are regions with an electrode potential different from that of the bulk metal and can therefore cause localized areas of corrosion. Other inhomogeneities include grain boundaries, segregation of impurities, cold-worked areas, and nonuniformly heat treated areas. Discrete anodes and cathodes can also be caused by variations in the electrolyte, such as temperature differences or concentration gradients in the solution, or by variations of ions in the solution or of dissolved gases such as oxygen.

A simple corrosion cell is shown in Fig. 3. The cell contains an anode, a cathode, and an electrolyte. If a voltmeter is connected to the circuit, a potential difference between the anode and cathode can be measured, indicating that a direct current is flowing between the two. Anodic reactions are always oxidation reactions and therefore tend to destroy the anode

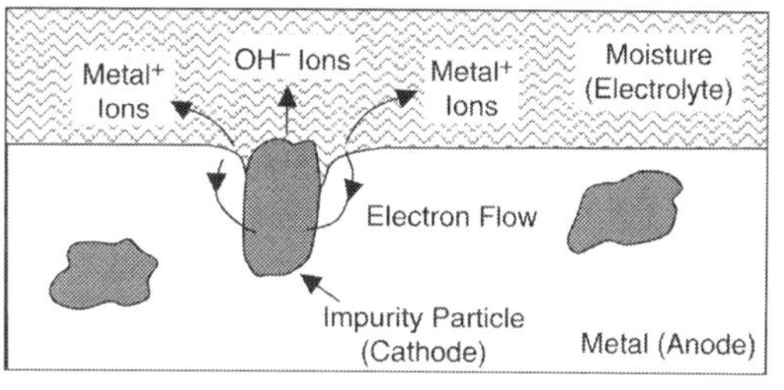

Fig. 2 Electrolytic corrosion of steel. Source: Ref 2

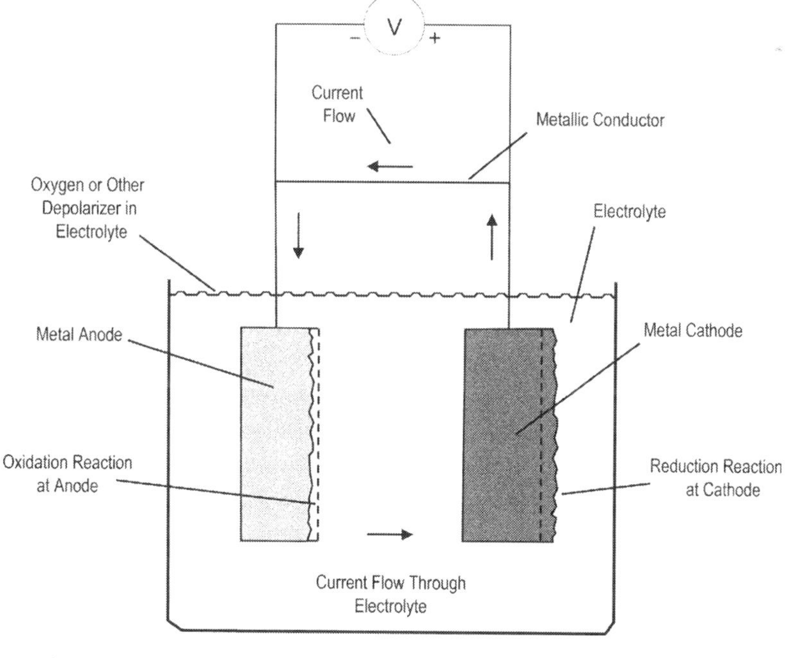

Fig. 3 Basic electrochemical cell. Source: Ref 3

metal by causing it to dissolve as an ion or to revert to a combined state such as an oxide. Therefore, in a corrosion cell, corrosion always takes place at the anode which is the metal (M_a) that undergoes an oxidation reaction and gives up electrons to the circuit:

$$M_a \rightarrow M^{n+} + n^{e-} \quad \text{(Eq 1)}$$

where M^{n+} is the number of positive metal ions, and n^{e-} is the number of negative electrons.

A reduction reaction, which is the reverse of the anode reaction, occurs at the cathode:

$$M^{n+} + n^{e-} \rightarrow M_c \quad \text{(Eq 2)}$$

Cathodic reactions are always reduction reactions and usually do not affect the cathode metal. During metallic corrosion, the rate of oxidation equals the rate of reduction.

As an example, iron immersed in water corrodes according to the mechanism shown in Fig. 4. The iron contains discrete areas that are anodic to the rest of the metal surface. At these areas, iron is oxidized according to the equation:

$$Fe \rightarrow Fe^{++} + 2e^{-} \quad \text{(Eq 3)}$$

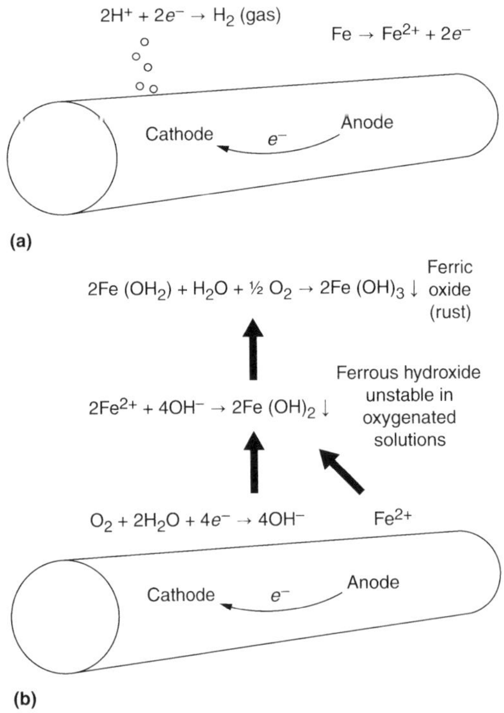

Fig. 4 Corrosion of iron in (a) unaerated and (b) aerated water. Source: Ref 3

If the water is pure and contains no dissolved oxygen, the cathodic reaction is the reduction of ionic hydrogen:

$$2H^{+} + 2e^{-} \rightarrow H_2 \qquad \text{(Eq 4)}$$

To maintain overall electrical neutrality, these reactions must proceed in balance. Therefore, two hydrogen ions must be reduced for every iron atom that corrodes. In pure water, only one water molecule in approximately 10 million dissociates to produce hydrogen and hydroxide ions, so that the supply of hydrogen ions is quite limited. For this reason, the cathodic reaction is quite slow and the corrosion rate is very low. Cases in which the rate of attack is limited by the speed of the cathodic reaction are known as cathodically controlled reactions.

If, as is usually the case, dissolved oxygen is present in the water, another cathodic reaction can occur, the reduction of oxygen:

$$\tfrac{1}{2}O_2 + H_2O + 2e^{-} \rightarrow 2OH^{-} \qquad \text{(Eq 5)}$$

This reaction will support a more rapid rate of attack because it depends only on the diffusion of oxygen to cathodic areas of the iron. When iron corrodes in water, the hydroxyl ions from the cathodic reaction react with the ferrous ions from the anodic reaction to form ferrous hydroxide [$Fe(OH)_2$]. The ferrous hydroxide further reacts with oxygen to form ferric hydroxide [$Fe(OH)_3$], which is familiar as reddish-brown rust. Unfortunately, this corrosion product is porous and not very adherent and therefore does not prevent further corrosion.

In practice, a metal often exhibits a high initial rate of corrosion. However, the rate can often diminish with time, an effect known as polarization. Polarization can result from reactions at either the anode or the cathode. In some corrosion reactions (e.g., iron in aerated water) that produce anodic polarization, the corrosion rate diminishes due to an accumulation of insoluble corrosion products that become somewhat protective of the iron anode. Conversely, cathodic polarization can result from reactions at the cathode. For example, when iron is immersed in nonaerated neutral water, the absence of dissolved oxygen allows the development of an adsorbed film of hydrogen that quickly covers the surface:

$$Fe + 2H^{+} \rightarrow Fe^{++} + 2H \qquad \text{(Eq 6)}$$

Because oxygen is not present, it is difficult for the hydrogen atoms to combine to form hydrogen gas and escape from the surface. When oxygen is introduced and the hydrogen atoms can escape the surface of the iron, depolarization occurs and corrosion resumes. Thus, in this case, the corrosion rate of iron is not controlled by the primary reaction with hydrogen ions but by the depolarization reaction involving oxygen.

Another characteristic of oxygen and oxidizing media is their ability to make certain metals and alloys passive by forming complex oxide films on the surface. A metal is less reactive and corrosion prone in the passive condition than in the normal or active condition. As an example, notice in the galvanic series (Table 1) that stainless steels are listed in both the passive and active conditions, with the passive conditions being more cathodic than the active conditions. Chromium in stainless steels oxidizes and forms a thin, tightly adherent layer of oxide (Cr_2O_3) on the surface that normally prevents corrosion in salt water. However, if a pit develops

Table 1 Galvanic series in seawater at 25 °C (77 °F)

Anodic end (most easily corroded)
Magnesium
Magnesium alloys
Zinc
Galvanized steel or galvanized wrought iron
Aluminum alloys 5052, 3004, 3003,1100, 6053 (in this order)
Cadmium
Aluminum alloys 2117, 2017, 2024 (in this order)
Low-carbon steel
Wrought iron
Cast iron
Ni-Resist (high-nickel cast iron)
Type 410 stainless steel (active)
50-50 lead-tin solder
Type 304 stainless steel (active)
Type 316 stainless steel (active)
Lead
Tin
Copper alloy 280 (Muntz metal, 60%)
Copper alloy 675 (manganese bronze A)
Copper alloys 464, 465, 466, 467 (naval brass)
Nickel 200 (active)
Inconel alloy 600 (active)
Hastelloy B
Chlorimet 2
Copper alloy 270 (yellow brass, 65%)
Copper alloys 443, 444, 445 (admiralty brass)
Copper alloys 608, 614 (aluminum bronze)
Copper alloy 230 (red brass, 85%)
Copper 110 (electrolytic tough pitch copper)
Copper alloys 651, 655 (silicon bronze)
Copper alloy 715 (copper nickel, 30%)
Copper alloy 923, cast (leaded tin bronze G)
Copper alloy 922, cast (leaded tin bronze M)
Nickel 200 (passive)
Inconel alloy 600 (passive)
Monel alloy 400
Type 410 stainless steel (passive)
Type 304 stainless steel (passive)
Type 316 stainless steel (passive)
Incoloy alloy 825
Inconel alloy 625
Hastelloy C
Chlorimet 3
Silver
Titanium
Graphite
Gold
Platinum
Cathodic end (least easily corroded)

Source: Ref 4

on the surface and destroys the oxide layer, it forms a local anodic area and corrosion is accelerated.

One of the most important factors influencing corrosion is the difference in electrical potential of two different metals when they are coupled together and immersed in an electrolyte. The electrode potential, measured in volts, provides an electrical measure of a metal to give up electrons. Electrode potentials are measured with two half-cells, one for the corroding metal and the other for a standard hydrogen half-cell consisting of gaseous hydrogen (H^+) at unit concentration with a specially prepared platinum electrode. The electromotive series is shown in Table 2. However, the electromotive series is determined under ideal laboratory conditions that may not reflect reality. Therefore, the galvanic series for seawater (Table 1) is often more useful. Again, the more anodic metal will corrode. For example, when iron and copper are coupled together in an electrolyte, iron will corrode because it is more anodic than copper. However, if iron is coupled to zinc, the iron is now protected because it is cathodic compared to zinc. The further away from each other two metals are on the galvanic series, the greater will be the tendency for corrosion. On the other hand, a metal coupled with another close to it on the series will usually corrode more slowly than when coupled with one further away from it.

Forms of Corrosion

There are many different types of corrosion, and more than one corrosion mechanism may be operating at the same time. Some types of corro-

Table 2 Electromotive series

	Electrode potential	Standard electrode potential (E^0), V
↑	$Au^{+++} + 3e^- \rightarrow Au$	+1.420
	$O_2 + 4H^+ + 4e^- \rightarrow H_2O$	+1.229
	$Pt^{++} + 2e^- \rightarrow Pt$	~+1.2
	$Ag^+ + e^- \rightarrow Ag$	+0.800
	$Fe^{+++} + 3e^- \rightarrow Fe^{++}$	+0.771
Increasingly cathodic	$O_2 + H_2O + 4e^- \rightarrow 4(OH^-)$	+0.401
(inert)	$Cu^{++} + 2e^- \rightarrow Cu$	+0.340
	$2H^+ + 2e^- \rightarrow H_2$	0.000
	$Pb^{++} + 2e^- \rightarrow Pb$	−0.126
	$Sn^{++} + 2e^- \rightarrow Sn$	−0.136
	$Ni^{++} + 2e^- \rightarrow Ni$	−0.250
	$Co^{++} + 2e^- \rightarrow Co$	−0.277
	$Cd^{++} + 2e^- \rightarrow Cd$	−0.403
Increasingly anodic	$Fe^{++} + 2e^- \rightarrow Fe$	−0.440
(active)	$Cr^{+++} + 3e^- \rightarrow Cr$	−0.744
	$Zn^{++} + 2e^- \rightarrow Zn$	−0.763
	$Al^{+++} + 3e^- \rightarrow Al$	−1.662
↓	$Mg^{++} + 2e^- \rightarrow Mg$	−2.363
	$Na^+ + e^- \rightarrow Na$	−2.714
	$K^+ + e^- \rightarrow K$	−2.924

Source: Ref 3

sion depend only on the environment, while others need mechanical or microbiological assistance. Some are unique to certain metals and alloys, while others attack many if not most metals and alloys. In this section, some of the more common types of corrosion are discussed. A number of these are illustrated in Fig. 5.

Uniform Corrosion

Uniform corrosion is the fairly uniform attack of the entire metal surface, resulting in the gradual thinning of the metal. It is by far the most common form of attack, accounting for the greatest corrosion loss of metal. Because the attack is relatively linear with time, the life of equipment can be predicted with reasonable accuracy. Uniform corrosion is uniform because it results from the formation and dissolution of multiple anodic and cathodic areas that move around the surface with time. Uniform corrosion is often caused by exposure to the atmosphere but can be aggravated by industrial pollution, brackish and salt waters, and soils and chemicals.

Uniform corrosion rates are measured as the average metal-thickness loss with time, in mils per year. A convenient rating for metals subject to uniform attack based on corrosion rates is as follows:

- *Excellent.* Rate of less than 2 mils/yr. Metals suitable for making critical parts
- *Satisfactory.* Rate of 2 to 20 mils/yr. Metals generally suitable for noncritical parts, where a higher rate of attack can be tolerated
- *Acceptable.* Rate of 20 to 50 mils/yr. A rate tolerable for massive equipment with a generous corrosion allowance

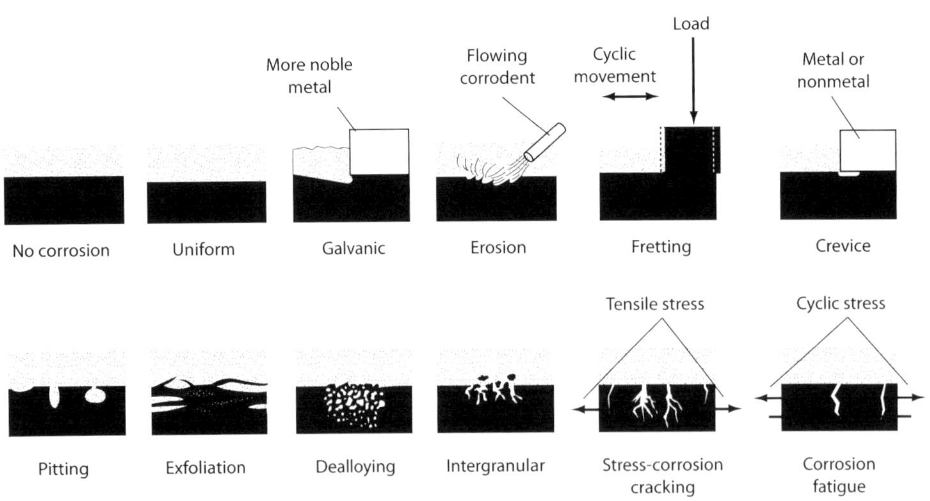

Fig. 5 Various forms of corrosion. Source: Ref 3

- *Unsatisfactory.* Rates of over 50 mils/yr. Metals are usually not acceptable in the environment.

Some metals form protective passive films on their surface when exposed to air that inhibits further attack. For example, stainless steel forms a protective oxide of Cr_2O_3, while aluminum forms a protective film of alumina (Al_2O_3). Unfortunately, although steel also produces a surface film, the film is porous and does not adhere very well, so the corrosion continues.

Controls for uniform corrosion are the relatively simple measures that follow:

- Use a more suitable material, such as a more noble metal or stainless steel. However, other considerations, such as mechanical or physical properties of the metal, as well as economics and availability, are usually the controlling factors in choice of material.
- Use coatings of various types to protect the metal. These may be:
  a. Paint or another type of coating to prevent contact of the environment with the metal surface. Painting is a time-honored method of corrosion protection and is remarkably effective as long as the paint film is intact. However, if corrosion originates on the opposite, unprotected side of thin, painted sheet steel, repainting the painted side will have no long-term benefit.
  b. Various oxide coatings are frequently used for improved corrosion resistance. An example is anodized aluminum, which is essentially aluminum with a relatively thick aluminum oxide film formed on the surface. A thin oxide film naturally forms on aluminum when exposed to air but is easily removed and destroyed, although it will reform. Various types of oxide coatings are frequently applied to steel for decorative and corrosion-resistance purposes. However, like certain other coatings, oxides are ineffective if damaged and are subject to pitting by certain environments, particularly in chlorides and other halides.
  c. Plating with a more active sacrificial metal is usually quite effective for relatively long periods. Plating with a less active (more cathodic) metal can be very effective, provided that the base metal is completely and uniformly coated without pinholes, cracks, scratches, abrasions, and the like. For example, plating steel with lead or tin is very effective in corrosion resistance, unless the coating has some locations that are open to the base metal. If an electrolyte can reach the steel through a small opening, the steel will corrode rapidly because the steel now is the anode and also because of the small anodic area exposed to the electrolyte.
  d. Cladding of one flat metal with a different alloy is another method of protection that is very effective in certain cases. Cladding may

consist of two or three metals flat rolled together, as in coins that consist of two nickel alloy outer layers on each side of a copper alloy, in a "sandwich" construction. Two aluminum alloys of different properties are frequently rolled together; a layer with better corrosion resistance may be joined to another layer that may have better mechanical properties.

- Another way to cope with the relatively predictable nature of uniform corrosion is to let the part corrode. This may be the most economical way of handling the problem. Railroad rails, for example, are not protected from corrosion by painting, plating, or the like but may become covered with an oily or greasy coating that tends to resist rusting on the sides and bottom. However, the upper contact surface remains completely unprotected because of the heavy contact stress of the steel railway car wheels on the steel rails.

Galvanic Corrosion

When two different metals are placed in electrical contact in the presence of an electrolyte, a potential or voltage difference is established. This potential difference causes a current to flow, and the less noble, or more anodic, metal corrodes while the more noble, or cathodic metal, is unaffected. The rate of attack depends on the relative voltage difference between the two metals, the relative areas of each metal that are exposed, and the particular corrosive environment.

The relative tendency for galvanic corrosion is given by the galvanic series shown in Table 1. Metals close to each other in the series generally do not have a tendency to react. The further the two metals are separated on the series, the greater is the tendency for the more anodic metal to corrode. It should be noted that it is possible for some metals to reverse their positions in certain environments; however, their normal positions are maintained in natural water and normal atmospheres.

The ratio of the anodic to cathodic areas is important in galvanic corrosion. Galvanic corrosion is accelerated when the anodic area is small in relation to the cathodic area; that is, the surface area ratio (Sa/Sc) is small, where Sa and Sc are the anodic and cathodic surface areas, respectively. This results in a high current density and causes severe corrosion of the anode. For example, a large area of stainless steel in contact with a small surface area of carbon steel is undesirable. The potential difference will tend to corrode the carbon steel, and the very large area of stainless steel will make that corrosion occur more quickly. The reverse condition is preferred. That is, a small area of stainless steel (or more noble metal) may be coupled with a much larger area of carbon steel (anodic) with a much slower rate of attack.

Although the galvanic series indicates the potential for corrosion, actual corrosion is difficult to predict. Electrolytes may be poor conductors, long distances may increase the resistance to the point that corrosion does not occur, or the reaction rate may be very sluggish. Corrosion products can also form a partially insulating layer over the anode. A cathode having a layer of adsorbed gas bubbles resulting from a corrosion reaction can become polarized and reduce the corrosion rate.

The passivity of stainless steels is a result of either the presence of a corrosion-resistant oxide film (Cr_2O_3) or an oxygen-caused polarizing effect, durable only as long as there is sufficient oxygen to maintain the effect. In most natural environments, stainless steels will remain in a passive state and thus tend to be cathodic to ordinary iron and steel. A change to an active state usually occurs only where chloride concentrations are high, as in seawater, or in reducing solutions. Oxygen starvation also produces a change in the active state, leading to accelerated corrosion. This occurs where the oxygen supply is limited, as in crevices and beneath contamination on partially biologically fouled surfaces.

Galvanic corrosion may be prevented, or minimized, by any or all of the following measures:

- Prevent electrical current flow by physical separation or by insulating the dissimilar metals from each other by using nonconductive, nonabsorbent materials such as plastics, waxy coatings, certain heavy greases, paint, and the like.
- Try to eliminate the electrolyte; if it is not present, there can be no galvanic corrosion. This is the reason for little or no galvanic corrosion in dry, desert atmospheres. If the humidity is below approximately 30 to 35%, atmospheric corrosion of carbon steel usually cannot occur.
- If different metals must be used together, choose those that are close together in the galvanic series. For example, contact between aluminum and steel in an electrolyte may cause gradual pitting and deterioration of the aluminum. However, if brass or copper replaces the steel, deterioration of the aluminum will be much more rapid in the same environment.
- Use a large anode metal and a small cathode metal to take advantage of the area effect. For example, steel rivets through aluminum sheet or plate may be satisfactory, but aluminum rivets through steel will corrode rapidly in an electrolyte. Similarly, copper rivets in steel plate may be satisfactory, but steel rivets through copper plate will corrode rapidly in an electrolyte.
- In a closed system, it may be possible to use corrosion inhibitors in the electrolyte. This is the principle used in automotive engine antifreezes, because the cooling system of an automotive engine may contain such dissimilar metals as gray cast iron, aluminum, copper, brass, tin-lead solder, and steel. The antifreezes periodically should be drained and replaced, because the inhibitors tend to become less effective with time.

- The principle of galvanic corrosion can be used to protect a structural metal by contact with a sacrificial (expendable), more anodic metal. The most common example is the use of zinc to protect iron or steel. The steel may be dipped into molten zinc (galvanized), electroplated with zinc, or coated with a zinc-rich primer or zinc-rich polymeric coating. In any case, the zinc is gradually corroded, or sacrificed, in order to protect the steel. Other active metals, such as aluminum and cadmium, also tend to protect iron or steel from corrosion but are used less frequently for various reasons. Coatings of aluminum and aluminum-zinc alloys on steel, however, are widely used for resistance to both high temperatures and corrosion, as are required in automotive exhaust systems, for example. The same cathodic protection principle is used with magnesium anodes buried in the ground to protect steel pipelines from corrosion. The anodes must be replaced periodically, because they are sacrificed to protect the steel. Magnesium anodes are also used within glass-lined water heaters to help retard corrosion of the steel shell if the glass lining is cracked. Replaceable zinc anodes are attached to the hulls of steel ships for protection of the steel in the same way.

Pitting

Pitting is a form of highly localized attack characterized by the formation of small pits on the surface. Several pits in an austenitic stainless steel thin-walled bellows are shown in Fig. 6. Pitting is considered to be more dangerous than uniform corrosion damage because it is more difficult to detect, predict, and design against. Pitting occurs when discrete areas of a material undergo rapid attack, while most of the adjacent surface remains virtually unaffected. Although the total metal loss may be small, the part may be rendered useless due to perforation. In addition to the localized loss of thickness, corrosion pits can also act as stress raisers, leading to fatigue or stress-corrosion cracking.

Pitting occurs when the anodic or corroding area is small in relation to the cathodic or protected area. Pitting can occur in protected metals when

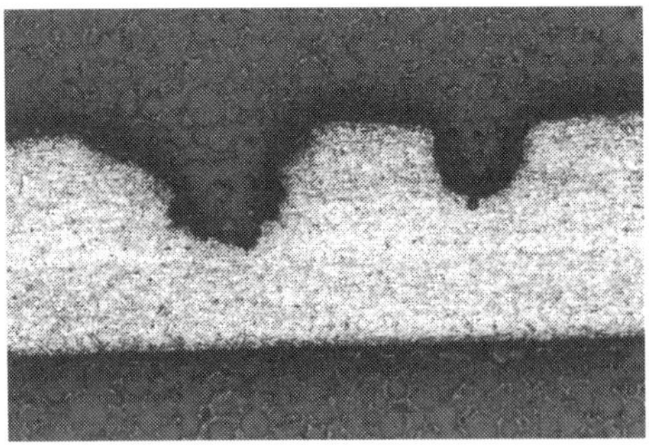

Fig. 6 Corrosion pits in thin-walled austenitic stainless steel sheet approximately 0.5 mm (0.02 in.). Courtesy of M.D. Chaudhari. Source: Ref 5

there are small breaks in the continuity of the metal coating. Pitting can also occur on bare, clean metal surfaces as a result of irregularities in their physical or chemical structure. Pitting can result in perforation of steel parts, rendering them unserviceable, even though less than 5% of the total metal has been lost through rusting. The rate of penetration into the metal by pitting can be 10 to 100 times that caused by general corrosion. Pitting can cause structural failure from localized weakening, while considerable sound metal still remains.

Pitting usually requires a rather long initiation period before the attack becomes visible. However, once a pit has started, the attack continues at an accelerating rate. Pits tend to grow in a manner that undermines or undercuts the surface. Typically, a very small hole is seen on the surface. Poking at the hole with a sharp instrument may reveal a much larger hole under what had looked like solid metal. Pitting can cause visible pits, or they may be covered with a semipermeable membrane of corrosion products. Pitting corrosion can assume different shapes. Pits can be either hemispherical or cup-shaped. In some cases they are flat-walled, revealing the crystal structure of the metal, or they may have a completely irregular shape.

Pitting normally occurs in a stagnant environment. Concentration cells can accelerate pitting. Concentration cells are areas on the metal surface where the oxygen or conductive salt concentrations in water differ. As a pit becomes deeper, an oxygen concentration cell is started by depletion of oxygen in the pit. The rate of penetration of such pits is accelerated proportionately as the bottom of the pit becomes more anodic. Pitting attack increases with temperature. Variations in soil conditions can also trigger pitting.

The depth of pitting can be expressed by the pitting factor (p/d), as defined in Fig. 7. A value of 1 would represent uniform corrosion. The maximum depth of penetration (p) can be measured by several methods, including metallographic examination, machining, use of a micrometer, or a microscope. The average penetration depth (d) is calculated from the weight lost by the sample. The maximum penetration depth is extremely significant if the metal is part of a barrier or tank or is part of a pressurized system. For a mechanical component, the density of pits (number per unit surface area) and size may be a more critical characteristic than the maximum depth. The loss of effective cross section can decrease the strength

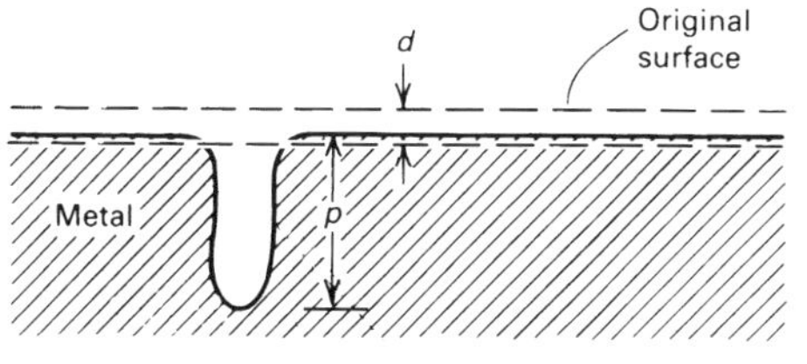

Fig. 7 Pitting factor p/d. Source: Ref 5

of the component, and pits can become sites of stress concentrations, leading to either static overload or fatigue failures.

Pitting occurs in most commonly used metals and alloys. While iron buried in the soil corrodes with the formation of shallow pits, carbon steels in contact with hydrochloric acid or certain stainless steels immersed in seawater develop deep pits. Mill scale is cathodic to steel and is a common cause of pitting. The potential difference between steel and mill scale often amounts to 0.2 to 0.3 V, enough to cause serious attack. When the anodic area is relatively small, the metal loss is concentrated and may be very serious. As the size of the anodic area decreases, the degree of penetration increases. A pit can form wherever there is a break in the mill scale. Aluminum tends to pit in waters containing chloride ions, particularly in stagnant water.

Despite their good resistance to general corrosion, stainless steels are susceptible to pitting. High-alloy stainless steels containing chromium, nickel, and molybdenum are more resistant to pitting but are not immune under all service conditions. The pitting resistance equivalent number (PREN), or pitting index, can be used to quantify and compare the resistance of stainless steel alloys to pitting:

$$\text{PREN} = \%\text{Cr} + 3.3(\%\text{Mo}) + 16(\%\text{N}) \qquad \text{(Eq 7)}$$

The higher the number, the more resistant the alloy is to pitting and crevice corrosion. A number of different PREN equations have been developed for specific alloy and weld metal groups. Equations include the effects of nickel, manganese, nitrogen, tungsten, and carbon.

Typical approaches to alleviating or minimizing pitting corrosion include the following: using defect-free barrier coatings; reducing the aggressiveness of the environment, for example, chloride ion concentrations, temperature, acidity, and oxidizing agents; using more-resistant metals, for example, using molybdenum-containing (4 to 6% Mo) stainless steels or Mo-W-Ni-base alloys; using corrosion-resistant alloy linings; modifying the design of the system, for example, avoiding crevices and the formation of deposits; circulating/stirring to eliminate stagnant solutions; and by ensuring proper drainage. When contact between dissimilar metals is unavoidable and the surface is painted, it is advisable to paint both metals. If only one surface is painted, it should be the cathode. If only the anode is coated, any weak points such as pinholes in the coating can result in intense pitting.

Crevice Corrosion

Crevice corrosion is a form of localized attack that occurs at narrow openings, spaces, or gaps between metal-to-metal or nonmetal-to-metal components (Fig. 8). Attack results from a concentration cell formed between the electrolyte within the crevice, which is oxygen starved, and the

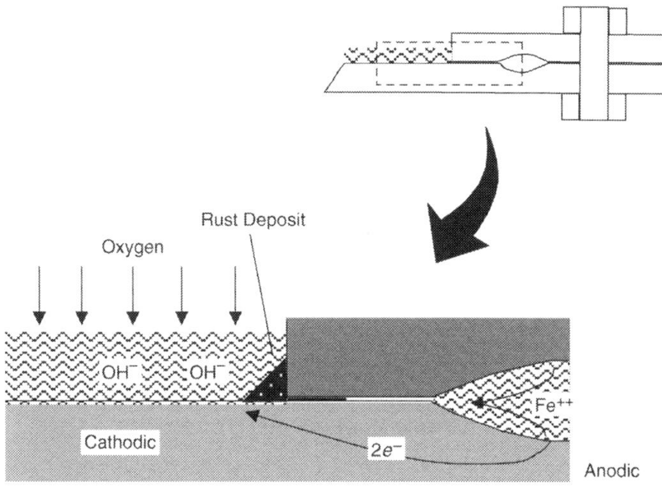

Fig. 8 Crevice corrosion of steel plates. Source: Ref 2

electrolyte outside the crevice, where oxygen is more plentiful. The material within the oxygen-starved crevice becomes the anode, while the exterior material becomes the cathode. The attack usually occurs in small volumes of stagnant solution under gasket surfaces, lap joints, marine fouling, solid deposits, in the crevices under bolt heads, and the mating surfaces of male and female threads. Crevice corrosion can progress very rapidly, on the order of tens to hundreds of times faster than the normal rate of general corrosion in the same given solution. Susceptibility to crevice corrosion increases rapidly with increases in temperature.

Crevices are produced by both design and accident. Examples of crevices created by design include gaskets, flanges, washers, bolt holes, threaded joints, riveted seams, lap joints, or anywhere close-fitting surfaces are present. Unintentional crevices include cracks, seams, and metallurgical defects. Although crevice corrosion affects both active and passive metals, the attack is often more severe for passive alloys, particularly those in the stainless steel group. Breakdown of the passive film within a restricted area leads to rapid metal loss and penetration in that area. Any layer of solid material on the surface of a metal that can exclude oxygen from the surface or allows the accumulation of metal ions beneath the deposit, because of restricted diffusion, is a candidate site for crevice corrosion.

Riveted and bolted joints are prime sites for crevice corrosion; therefore, they require careful attention in design and assembly to minimize crevices, as well as provisions to ensure uniform aeration and moderate, but not excessive, flow rates at the joints. Not only does the geometry of the joint affect crevice corrosion, but corners and cracks can collect debris, which increases the potential for corrosion. Replacement with welded joints can eliminate crevice corrosion, provided special care is taken in welding and subsequent finishing of the welds to provide smooth, defect-free joints.

Once the nature of crevice corrosion is understood, the prevention methods become rather obvious, even though they may not be easy to put into practice. Consider the following possible solutions:

- Avoid bolted or riveted joints unless the metals are coated, ideally both before and after joining. However, if the alternative is some form of welding, the heat of welding may destroy the coated surfaces. In either case, it may be necessary to spray or dip the completed assembly with paint or a waxy or greasy corrosion-resistant material that has the ability to flow and cover any uncoated areas. Also, a sacrificial anodic metal may be used to protect the structural metal.
- Close or seal existing crevices, because if moisture or another electrolyte cannot reach the crevice, there can be no crevice corrosion. However, in many cases, it is impossible to keep moisture from entering the crevices. Then, it is necessary to provide drain holes or tubes to continuously or periodically drain the moisture.
- Inspect and remove deposits frequently. If foreign particles such as dirt, rust, or sand are not present, they cannot cause crevice corrosion. Another possible solution is to use filters, traps, or settling tanks to remove particles from the system. However, all of these require periodic maintenance to remove the debris that accumulates.
- Use solid, nonabsorbent gaskets or seals, such as those made from solid rubber or plastic. These tend to seal a joint and keep the electrolyte out. Obviously, the surfaces must be smooth to promote sealing, and the clamping force must be adequate for the application.

As can be seen from the preceding possible solutions, there is no single way in which crevice corrosion can be eliminated. Prevention differs with the individual case and the conditions.

Dealloying Corrosion

Dealloying, also known as selective leaching or parting corrosion, is a destructive process in which the more-anodic alloying element is selectively removed from the alloy, leaving behind a porous, spongy mass. Specific categories of dealloying are often known by the name of the dissolved element. The preferential leaching of zinc from brass is known as dezincification, while the loss of iron from gray iron is called graphitic corrosion. During dealloying, typically one of two processes occurs: alloy dissolution and replating of the more-cathodic element, or selective leaching of the more-anodic alloying element.

Dezincification occurs in copper-zinc brasses containing more than 15% Zn. Dezincification leaves behind a porous and weak layer of copper and copper oxide. High potentials and low pH values favor selective removal of zinc. At negative potentials and acidic conditions, copper and zinc dissolve. At zero to slightly positive potentials and acidic conditions, a region exists in which copper is expected to redeposit. Both dezincification mechanisms can occur independently or in conjunction, depending on the given environment conditions.

Two types of damage can be characterized: one type of dezincification is uniform, and the second is plug type. Uniform or layer-type dezincification results in a relatively uniform zone of dezincified material, with the underlying material remaining unaffected. Brasses with high zinc content in an acidic environment are prone to uniform dezincification. Plug-type dezincification, shown in Fig. 9, results in localized penetrations of dezincified areas that progress through the wall thickness of the material. The overall dimensions of the material do not change. The dezincified areas are weakened or, in some cases, perforated. Plug-type corrosion is most likely to occur in basic or neutral environments and at elevated temperatures.

Prevention of dezincification can be achieved most readily by proper alloy selection. Alloys containing greater than 85% Cu are considered resistant to dezincification. Tin is added to act as an inhibitor. In addition, inhibited copper-zinc alloys containing 0.020 to 0.6% As, Sb, or P are also considered resistant to dezincification. This is likely due to the formation of a redeposited film of protective elements. Other possible remedies for dezincification include the use of cathodic protection, liners, or coatings.

Graphitic corrosion is a form of dealloying that occurs in cast irons. This corrosion mechanism is usually found in gray cast irons and is associated with the presence of graphite flakes. The graphite flakes are cathodic to the iron matrix. Exposure to an electrolyte results in selective leaching of the iron matrix, leaving behind a porous mass of graphite flakes. Graphitic corrosion is generally a long-term mechanism, resulting

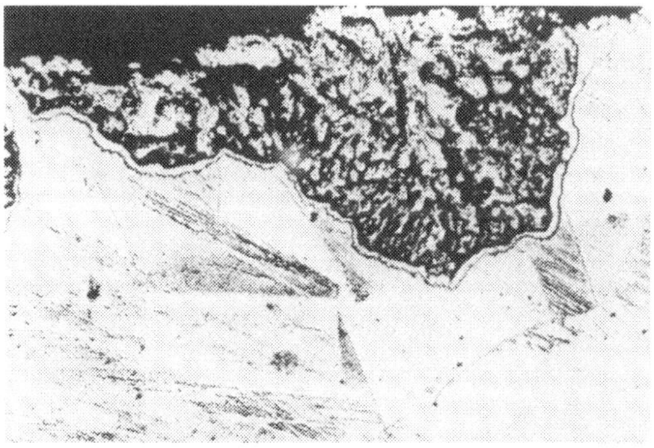

Fig. 9 Plug-type dezincification in an α-brass (70Cu-30Zn) exposed for 79 days in 1 N NaCl at room temperature. Note porous structure within the plug. Dark line surrounding the plug is an etching artifact. Total width shown is 0.56 mm (22 mils). Source: Ref 3

from exposures of 50 years or more. Pipelines made of cast iron, especially those buried in clay-based soils and soils containing sulfates, are susceptible. In cases where graphitic corrosion has caused extensive wall loss, a reduction in component strength will occur. Thus, it is not unusual for cracking to accompany graphitic corrosion. In some cases, graphitic corrosion has been observed on the fracture surfaces, indicating the long-term existence of the crack. Graphitic corrosion is reduced by alloy substitution, for example, by the use of a ductile or alloyed iron rather than gray iron, by raising the water pH to neutral or slightly alkaline levels, by the addition of inhibitors, and by avoiding stagnant water conditions.

Other forms of dealloying include dealuminification of aluminum bronze and nickel-aluminum bronze alloys; denickelification involving the removal of nickel from copper-nickel alloys; destanification or the removal of tin in cast tin bronzes; desiliconification of silicon bronzes; and dealloying in copper-gold and silver-gold alloys.

Intergranular Corrosion

Intergranular corrosion is the selective attack of grain boundaries, or the immediate adjacent regions, usually with only slight or negligible attack of the grains themselves. In essence, the grain-boundary area becomes anodic to that of the grain interiors. These differences can occur during manufacturing or in-service exposure. When the grain boundaries become anodic, the metal is said to be sensitized and is susceptible to intergranular attack in a corrosive environment. The classic example of intergranular corrosion occurs when stainless steels become sensitized by the diffusion of chromium and trace amounts of carbon to the grain boundaries during elevated temperatures, resulting in the precipitation of chromium carbides (Fig. 10). Some aluminum alloys also exhibit a similar behavior, in particular, those containing copper that precipitate $CuAl_2$, resulting in precipitate-free zones.

When the attack is severe, entire grains can be dislodged due to the complete deterioration of the grain boundaries. When an alloy is undergo-

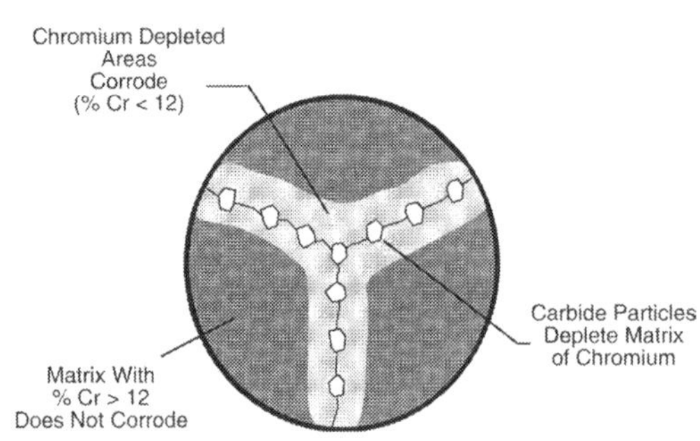

Fig. 10 Intergranular corrosion of stainless steel. Source: Ref 6

ing intergranular corrosion, its rate of weight loss can accelerate with time. As the grain-boundary area dissolves, the unaffected grains are undermined and fall out, thus increasing the weight loss. Intergranular corrosion can occur in a number of alloy systems under specific circumstances. The susceptibility depends on the corrosive environment and on alloy composition, fabrication, and heat treatment parameters. Susceptibility of a component to intergranular corrosion can be corrected by proper heat treatment to distribute alloying elements more uniformly, by modification of the alloy, or by the use of a completely different alloy.

Although stainless steels provide resistance against general corrosion and pitting, the 300 and 400 series of stainless steels can be susceptible to intergranular corrosion by sensitization. Susceptible stainless steels are those that have normal carbon contents (generally >0.04%) and do not contain titanium and niobium carbide stabilizing elements. Sensitization is caused by the precipitation of chromium carbides at grain boundaries during exposure to temperatures from 455 to 870 °C (850 to 1600 °F), with the maximum effect occurring near 675 °C (1250 °F). As shown in Fig. 11, the resulting depletion in chromium adjacent to the chromium-rich carbides provides a selective path for intergranular corrosion. Precipitation can occur from the heat of welding, from slow cooling after annealing, or from prolonged exposure to intermediate temperatures in service. For exposures at very long times or at the high end of the temperature range, diffusion of chromium back into the depleted zone can restore the corrosion resistance. An effective means of combating intergranular corrosion in stainless steels is to restrict the carbon content of the alloy. In the stainless L-grades, limiting the carbon content to a maximum of 0.03% is often sufficient. High chromium and molybdenum additions also reduce the chance of intergranular attack. However, even better performance can be obtained from the stabilized types, which contain sufficient titanium and niobium that combine preferentially with carbon to form titanium and niobium carbides.

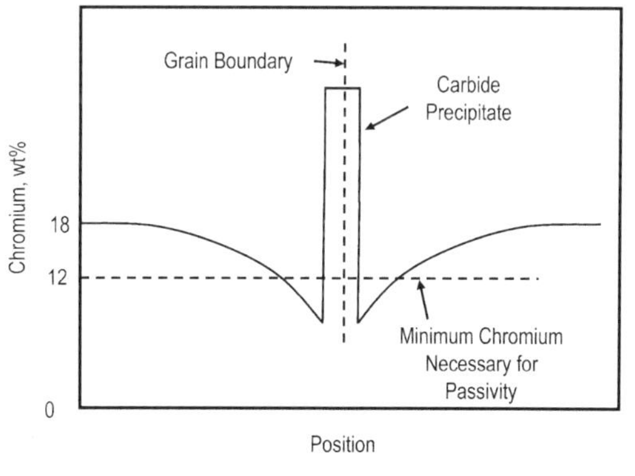

Fig. 11 Precipitation of chromium carbide at grain boundaries. Source: Ref 3

The typical appearance of intergranular corrosion of stainless steels is shown in Fig. 12 for sensitized type 304 stainless steel attacked by a water solution containing a low concentration of fluorides at 82 °C (180 °F). Intergranular corrosion of this type is more or less randomly oriented and does not have highly localized propagation, as does intergranular stress corrosion in which cracking progresses in a direction normal to applied or residual stresses. For austenitic stainless steels, the susceptibility to intergranular corrosion is mitigated by solution heat treating at 1065 to 1120 °C (1950 to 2050 °F), followed by water quenching. In this treatment, the chromium carbides are redissolved in solid solution and then retained in solid solution as a result of the quench. However, a solution heat treatment may be difficult on many welded assemblies and is generally impracticable on large equipment or when making repairs.

Under severe conditions, such as multipass welding, even the stabilized alloys will sensitize. They are also susceptible to a highly localized form of intergranular corrosion known as knifeline attack, which occurs in the base metal immediately adjacent to the weld fusion line. In some cases, these alloys are given stabilizing heat treatments after solution heat treatment for maximum resistance in the as-welded condition. For example, type 321 stainless steel is given a stabilizing anneal at 900 °C (1650 °F) for 2 h before fabrication to avoid knifeline attack. However, type 321 may still be susceptible because titanium has a tendency to form an oxide during welding; therefore, its role as a carbide stabilizer may be diminished. For this reason, type 321 is always welded with a niobium-stabilized weld filler metal, such as type 347 stainless.

Sensitization and intergranular corrosion can also occur in ferritic stainless steels. A wider range of corrosive environments can produce inter-

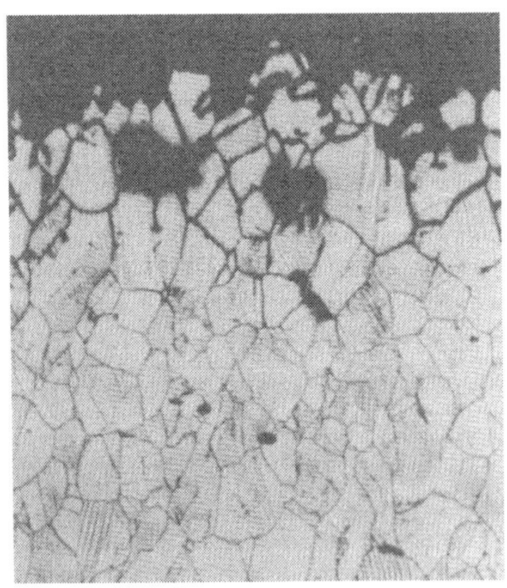

Fig. 12 Intergranular corrosion in type 304 stainless steel. Original magnification: 100×. Source: Ref 3

granular attack in ferritic grades than for austenitic grades. The thermal processes causing intergranular corrosion in ferritic stainless steels are also different from those in austenitic stainless steels. In the case of welds, the attacked region is usually larger for ferritic grades than for austenitic grades, because temperatures above 925 °C (1700 °F) are involved in causing sensitization. However, ferritic grades with less than 15% Cr are not susceptible. New grades of ferritic stainless steels are also alternatives to the more common 300- and 400-series stainless steels.

In the 2*xxx* and 7*xxx* alloys, the presence of copper can be a problem. These alloys can precipitate $CuAl_2$ at the grain boundaries, which is more anodic than the grains themselves. During exposure to chloride solutions, particularly salt water, galvanic couples form between the precipitate-rich grain boundaries and the grain interiors. One approach to mitigating this type of intergranular attack is to employ special heat treatments that control the distribution of the precipitate particles.

Although many types of intergranular corrosion are not associated with a potential difference between the grain-boundary region and the adjacent grains, intergranular corrosion of aluminum alloys can occur due to potential differences. In 2*xxx*-series alloys, a narrow band on either side of the boundary is depleted of copper; in 5*xxx*-series alloys, it is the anodic constituent Mg_2Al_3 that forms a continuous path along a grain boundary; in copper-free 7*xxx*-series alloys, it is generally considered to be the anodic zinc- and magnesium-bearing constituents at the grain boundary; and in the copper-bearing 7*xxx*-series alloys, it appears to be the copper-depleted bands along the grain boundaries. The 6*xxx*-series alloys generally resist this type of corrosion, although slight intergranular attack has been observed in aggressive environments.

Exfoliation is a form of intergranular corrosion that primarily affects aluminum alloys in industrial or marine environments. Corrosion proceeds laterally from initiation sites on the surface and generally proceeds intergranularly along planes parallel to the surface, as shown in the Fig. 13 example. The corrosion products that form in the grain boundaries force metal away from the underlying base material, resulting in a layered or flakelike appearance.

Resistance to exfoliation corrosion is attained through proper alloy and temper selection. The most susceptible alloys are the high-strength heat treatable 2*xxx* and 7*xxx* alloys. Exfoliation corrosion in these alloys is usually confined to relatively thin sections of highly worked products.

Stress-Corrosion Cracking

The phenomenon of stress-corrosion cracking (SCC) is a great problem in many industries and types of parts because it can result in brittle fracture of a normally ductile metal. Stress-corrosion cracking is a failure process that occurs because of the simultaneous presence of a tensile stress, a specific environment, and a susceptible material. Removal of or changes in any one of these three factors will often eliminate or reduce the susceptibility to SCC. Stress-corrosion cracking occurs by subcritical crack growth involving crack initiation at selected sites, crack propagation, and finally overload fracture of the remaining section. Failure by SCC is frequently encountered in seemingly mild chemical environments at tensile

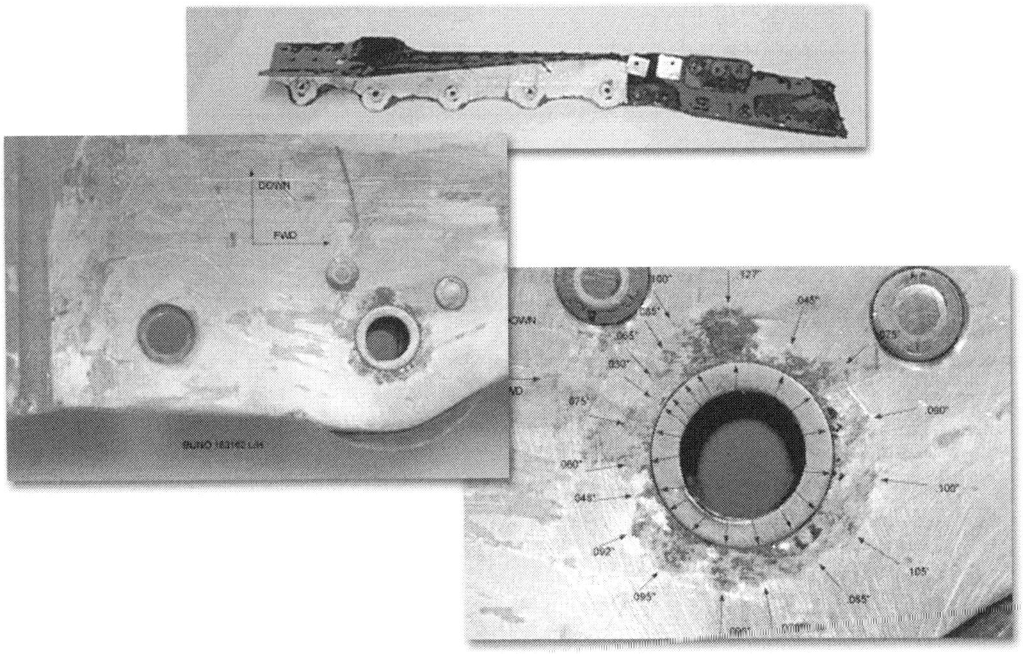

(a)

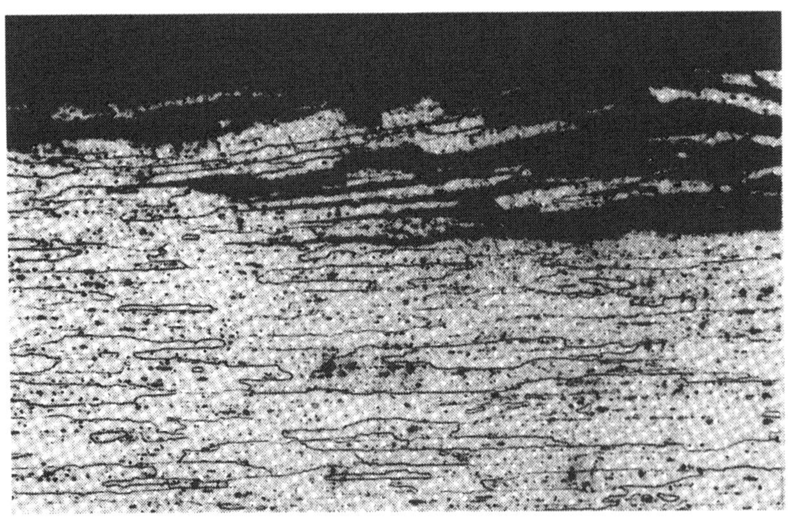

(b)

Fig. 13 Exfoliation corrosion. (a) Around a fastener hole in 7049-T73 aluminum alloy longeron. (Radial lines indicate measurements taken to assess damage.) Source: Ref 7. (b) In aluminum plate, 7178-T651, exposed to maritime environment. Source: Ref 3 stresses well below the yield strength. Failures often take the form of fine cracks that penetrate deeply into the metal, with little or no evidence of corrosion. Therefore, during casual inspection, no macroscopic evidence of impending failure is seen. Stress-corrosion cracking continues to be a cause of significant service failures. It is very likely that for every alloy there is an environment that will cause SCC, but most of the ones of industrial significance are known and avoidable. Materials selection is the first line of defense. Lowering of the applied stresses and elimination of residual stresses can also go a long way toward eliminating problems. Sometimes, minor changes or additions to the environment can help.

The classic example of SCC is the so-called season cracking of brass cartridge cases, as shown in Fig. 14. The term arose during the Indian campaigns of the British army during the 1800s, when serious problems resulted from the spontaneous cracking of the thin-wall necks of cartridge cases, which were stored in ammunition dumps near the horse corrals during the monsoon seasons. High temperature and humidity, plus traces of ammonia in the air, caused SCC in the severely deformed thin sections, that is, at locations that were subjected to high tensile hoop stresses when the bullets were inserted. It is now known that most zinc-containing copper alloys, such as the 70% Cu, 30% Zn alloy called cartridge brass, are susceptible to SCC when the surface is tensile stressed and is in the presence of certain chemicals, such as moist ammonia, mercurous nitrate, and amines.

Stress-corrosion cracking is a progressive type of fracture, somewhat similar to fatigue. The crack, or cracks, grow gradually over a period of time until a critical size is reached; the stress concentration can then cause a sudden brittle fracture of the remaining metal. In other instances, as in a cartridge case, the crack will grow away from the highly stressed origin, then stop when it is no longer highly stressed in tension.

Stress-corrosion cracks ordinarily undergo extensive branching and proceed in a general direction perpendicular to the stresses contributing to their initiation and propagation, as illustrated in Fig. 15 for intergranular

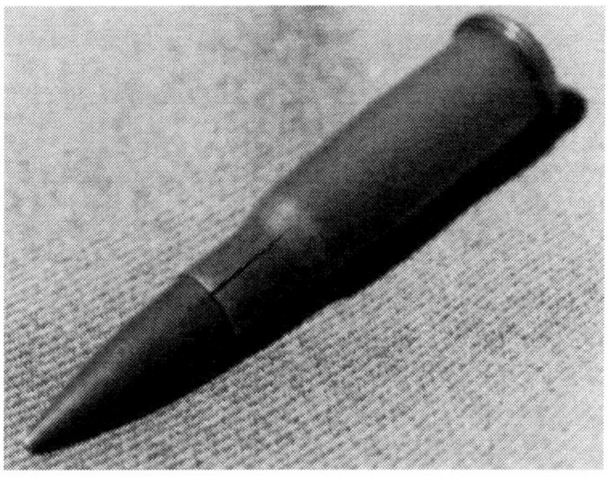

Fig. 14 Stress-corrosion crack in thin neck of a cartridge case. Source: Ref 8

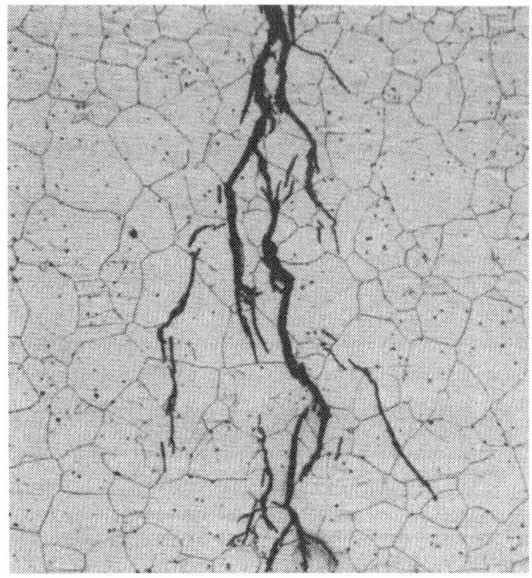

Fig. 15 Chloride-induced stress-corrosion cracking of type 316 stainless steel pipe. Source: Ref 9

SCC of 316L stainless steel. The surfaces of some stress-corrosion cracks resemble those of brittle mechanical fractures, although they are actually the result of local corrosion in combination with a tensile stress. In some metals, cracking propagates intergranularly and in others transgranularly. In certain metals, such as high-nickel alloys, iron-chromium alloys, and brasses, either type of cracking can occur, depending on the specific metal-environment combination.

A number of mechanisms have been proposed to account for the subcritical crack-propagation phase of SCC. The following is a brief review of a few of these mechanisms:

- When fresh metal is exposed by plastic deformation at the crack tip, there is a competition between dissolution, which would blunt the crack, and passivation, which would prevent the environment from reaching the metal. Between these two extremes there exists a range where the fresh metal at the crack tip is attacked, but the crack walls are passivated.
- The crack tip repassivates, but the passive layer eventually cracks and the fresh surface is attacked by anodic dissolution, which then repassivates again. This process gives discontinuous growth that could produce the striations observed in some systems.
- The crack progresses by repassivation and cracking of the passive layer, with no attack of the base metal. This is called the tarnish rupture model.
- There is preferential dissolution of one microstructural constituent, such as a grain boundary, that is anodic with respect to the main body of the metal.

- Adsorption of damaging ions weakens the atomic bonds at the crack tip, allowing them to be pulled apart at a lower-than-normal applied stress. This is known as the stress-sorption theory.
- Adsorption of damaging ions promotes enhanced, highly localized plasticity that promotes slip at the crack tip, leading to crack growth. This theory is called adsorption-enhanced plasticity.
- Hydrogen is generated at local cathodes, enters the metal, diffuses to the crack tip region, and causes propagation by hydrogen stress cracking. This mechanism is no doubt involved in the environmental embrittlement of high-hardness, high-strength carbon, alloy, and stainless steels.

While each of these mechanisms explains some of the observed facts and may be active in one or more alloy/corrodent systems, no single theory has been generally accepted as the universal mechanism completely explaining stress-corrosion crack propagation in all alloy systems and environments.

A simplified mechanism for SCC is shown in Fig. 16. Stress causes rupture of the oxide film at the crack tip, which exposes fresh metal that corrodes and forms another thin oxide film. The oxide ruptures again, allowing more corrosion, and the crack slowly propagates through the alloy until the crack reaches a critical length and failure occurs. Because the cathodic reaction during corrosion can often produce hydrogen, hydrogen can contribute to SCC, often making it difficult to distinguish between SCC and hydrogen embrittlement.

A characteristic of SCC is the existence of a minimum tensile stress for failure, or threshold stress, for smooth components, and a threshold stress intensity for crack propagation for precracked components. The threshold stress is that stress below which the probability for cracking is extremely low, and it depends on the temperature, composition, metallurgical structure of the alloy, and the environment. In some tests, cracking has oc-

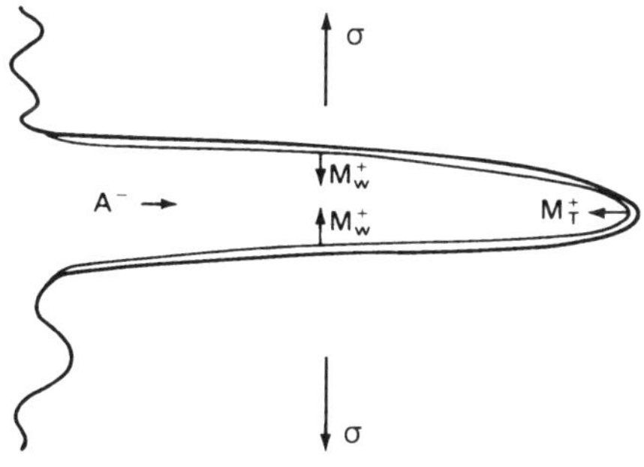

Fig. 16 Schematic of stress-corrosion crack showing important transport and corrosion reactions. A– represents negatively charged anions migrating to the crack tip; M_w^+ represents metal ions entering the crack solution from the crack walls; and M_T^+ indicates metal ions entering the crack solution from the crack tip. Source: Ref 10

curred at an applied stress as low as approximately 10% of the yield strength, and for other metal-environment combinations, the threshold stress is as high as 70% of yield strength. The effect of alloy composition on threshold stress is shown in Fig. 17, which illustrates the relationship between applied stress and average time to fracture in boiling 42% magnesium chloride solution for two 18-8 stainless steels (AISI types 304 and 304L) and two more highly alloyed stainless steels (AISI types 310 and 314). As indicated by the nearly level portions of the curves, the threshold stress in this environment is approximately 240 MPa (35 ksi) for high-alloy stainless steels and only 83 MPa (12 ksi) for the 18-8-type stainless steels.

Because SCC is the result of the combination of a static tensile stress and a particular environment, let us examine some of the ways that tensile stresses may be generated. It is usually felt that tensile residual stresses (including assembly stresses) are more often the cause of SCC than are applied tensile stresses. Residual stresses are frequently the result of welding. Tensile stresses are also generated in other ways, such as in shrink fits, bending or torsion during assembly, crimping, and the like. The only requirement for SCC is that there be a tensile stress on the surface of a metal in a critical environment. The stress need not exceed the yield strength of the metal, but the higher the stress, the less critical the environment, and vice versa.

In certain metals, particularly many austenitic stainless steels, the heat of welding causes sensitization, or depletion of chromium due to forma-

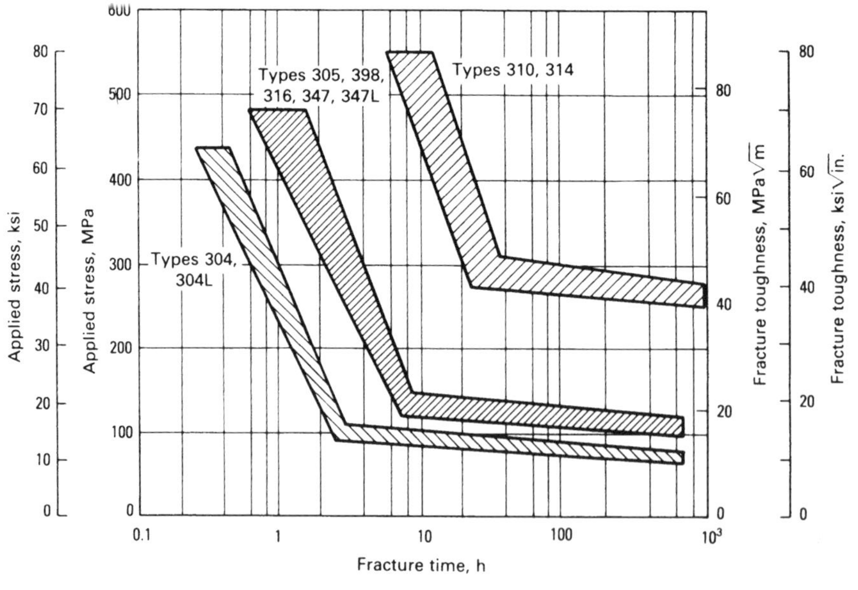

Fig. 17 Relative stress-corrosion cracking behavior of austenitic stainless steels in boiling magnesium chloride. Source: Ref 11

tion of complex chromium carbides in the grain boundaries. Because chromium is the major element that makes stainless steels corrosion resistant, SCC can occur alongside the carbides in the grain boundaries, where there is little or no chromium. To solve this problem, it is necessary to do one of two things: use a stainless steel of very low carbon content (0.03% or less) so that there is little or no carbon to deplete the chromium from the grain boundaries, or use a stainless steel containing an element that forms carbides even more readily than does chromium. Those commonly used are titanium or niobium plus tantalum, which are the bases for the type 321 and type 347 variations, respectively, of the standard type 304 stainless steel.

If the right combination of stress and environment are present, almost every metal can be prone to SCC. However, only specific combinations of alloys and environments result in SCC. Susceptibility of a given metal to SCC in a specific environment depends on its condition, that is, its overall and local chemical composition and its metallurgical structure, as determined by thermal processing and cold working. Some aspects of metallurgical conditions that are significant include phase distribution, grain size and shape, grain-boundary precipitation, grain-boundary segregation, cold work, and inclusion type and distribution. Stress-corrosion cracking frequently occurs in commercial alloys such as low-carbon steels, high-strength steels, austenitic stainless steels and other austenitic alloys (especially in the sensitized condition), high-strength aluminum alloys, and brasses and certain other copper alloys. A partial list of alloy and environmental combinations exhibiting SCC is given in Table 3.

Stress-corrosion cracking can be described within a fracture mechanics framework. In this approach, it is assumed that a crack already exists and

Table 3 Some environments that can cause stress-corrosion cracking of metals and alloys under certain conditions

Material	Environment	Material	Environment
Aluminum alloys	NaCl-H_2O_2 solutions	Carbon and alloy steels	NaOH solutions
	NaCl solutions		NaOH-Na_2SiO_2 solutions
	Seawater		Calcium, ammonium, and sodium nitride solutions
	Air, water vapor		
Copper alloys	Ammonia vapors and solutions		Mixed acids (H_2SO_4-HNO_3)
	Amines		HCN solutions
	Water, water vapor		Acidic H_2S solutions
Gold alloys	$FeCl_3$ solutions		Moist H_2S gas
	Acetic acid-salt solutions		Seawater
Inconel	Caustic soda solutions		Molten Na-Pb alloys
Lead	Lead acetate solutions	Stainless steels	Acid chloride solutions such as $MgCl_2$ and $BaCl_2$
Magnesium alloys	NaCl-K_2CrO_4 solutions		
	Rural and coastal atmospheres		NaCl-H_2O_2 solutions
	Distilled water		Seawater
Monel	Fused caustic soda		H_2S
	Hydrofluoric acid		NaOH-H_2S solutions
	Hydrofluorosilicic acid		Condensing stream from chloride waters
Nickel	Fused caustic soda		
		Titanium	Red fuming nitric acid

Source: Ref 12

it is the extension of the crack that governs the life of the component. Under such conditions, it would be expected that the rate of crack growth would correlate with the magnitude of the applied stress-intensity parameter. In fact, it has been demonstrated that SCC rates of cracked specimens correlate with the stress-intensity parameter (K) and that any attempt to correlate the crack growth rate with stress alone leads to logical inconsistencies.

The crack propagation behavior of SCC systems is often depicted by the crack growth rate (da/dt) as a function of the applied stress intensity (K), as depicted in the Fig. 18 curve. Three stages are characteristic of these curves:

- *Stage 1:* Crack growth rate increases rapidly with stress intensity above the threshold K_{ISCC}.
- *Stage 2:* A plateau where velocity is independent of stress intensity
- *Stage 3:* Velocity again increases rapidly with increasing K as K_{Ic} is approached.

The threshold is defined in Fig. 18 by the minimum detectable crack growth rate. The threshold stress intensity is generally associated with the development of a plastic zone at the crack tip. Stage I crack growth shows a rapid increase in crack growth rate, while in stage II, the crack growth rate is independent of the stress intensity.

The existence of stage 2, the regime where the crack propagation velocity is independent of the driving force, is of particular interest. The plateau

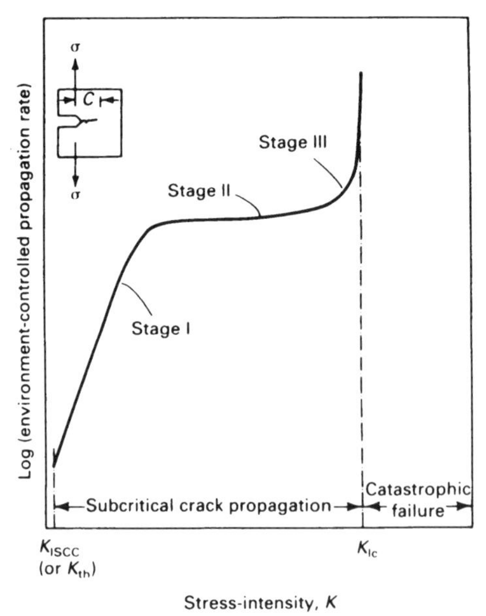

Fig. 18 Schematic diagram of stress-corrosion crack velocity as a function of stress-intensity factor (K). Source: Ref 13

velocity is a function of the particular alloy-environment system and may vary by orders of magnitude with the alloy composition and heat treatment, pH, and electrochemical potential. Independence of the driving force indicates that the crack growth rate is controlled by the rate of a chemical reaction, the rate of supply of the embrittling anion or cation to the crack tip, or some other nonmechanical factor.

Threshold stresses and stress-intensity factors, the presence of a stress-independent crack growth regime, and the dependence of cracking to strain rate are important features in determining the susceptibility of alloys to SCC. The threshold stress is typically the stress value obtained from constant-load testing below which SCC does not occur and can serve as a simple measure for susceptibility of a material to SCC in a certain environment.

In general, there is a threshold for SCC, denoted K_{ISCC}, below which crack growth is not observed. Above this level, increases in stress intensity produce increases in the real-time crack growth rate (da/dt). Depending on the material and the mechanism of crack extension, further increases in the stress intensity do not produce significant increases in the crack growth rate. This can be understood in terms of a transport-controlled step in which the maximum potential for acceleration of the crack growth rate by the crack tip stress field has been reached. At this point, the rate is governed by how rapidly the corrosive media can be transported to the crack tip or the limiting value of diffusion of a damaging species into the bulk material. The crack growth rates are more sensitive to parameters such as temperature, pressure, and pH levels. As K approaches K_{Ic}, tensile fracture mechanisms begin to appear, the process becomes mechanically dominated, and the crack growth rates again increase with K (stage III in Fig. 18). Final failure occurs at K_{Ic}, at which point the crack is moving at a substantial fraction of the velocity of sound and is essentially free of the environment. Actual experimental data are shown in Fig. 19.

The life of a cracked component in an aggressive environment can be calculated by an integration technique. To compute the life, the crack growth rate, as a function of K, must be known, as well as other factors such as operating stresses, initial crack length, fracture toughness of the material, and the geometry of the part in which the crack is embedded. These concepts are illustrated in the following example.

Example: Estimation of Life in Corrosive Environment. The crack growth rate of a high-strength steel in H_2O at 80 °C (175 °F) is given by:

$$\frac{da}{dt} = 9.32 \times 10^{-5}\, K - 2.08 \times 10^{-3}$$

Here K is in MPa$\sqrt{\text{m}}$, a is in meters, and t is in days. Calculate the life of the part assuming that edge cracks of length 2.02×10^{-3} m are initially present. Assume that K_{Ic} is 55 MPa$\sqrt{\text{m}}$ and that the part is subjected to a constant stress of 345 MPa.

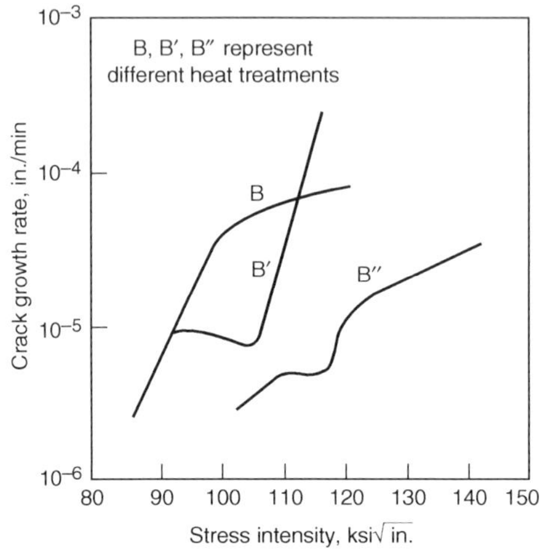

Fig. 19 Experimental crack growth rate as a function of stress intensity for a high-strength steel in humid air. Source: Ref 14

Solution. The initial stress intensity from a handbook is:

$$K_i = 1.12 \times 345 \times \left(\pi \times 2.02 \times 10^{-3}\right)^{1/2} = 30.8 \text{ MPa}\sqrt{\text{m}}$$

The threshold stress intensity may be calculated assuming:

$$\frac{da}{dt} = 0 \text{ at } K = K_{ISCC}$$

This gives a value of approximately 22.3 MPa$\sqrt{\text{m}}$ for K_{ISCC}. Because $K_i > K_{ISCC}$, the crack will grow by a SCC process.

The rate of growth is given by:

$$\frac{da}{dt} = 9.32 \times 10^{-5} \times 1.12 \times 345\sqrt{\pi a} - 2.08 \times 10^{-3} = 0.0640a^{1/2} = 2.08 \times 10^{-3}$$

The preceding equation can be integrated:

$$t_f = \int_{t=0}^{t=t_f} dt = \int_{a=a_0}^{a_f} \frac{da}{0.064a^{1/2} - 2.08 \times 10^{-3}}$$

The crack length at fracture may be computed by noting that at this point $K = K_{Ic}$:

$$a_f = \frac{1}{\pi}\left[\frac{K_{Ic}}{1.12\sigma}\right]^2 = \frac{1}{\pi}\left[\frac{55}{386.40}\right]^2 = 6.45 \times 10^{-3}$$

Using tables of integrals, $t_f \approx 2.5$ days.

Preventing SCC should be simple: remove either the tensile stress or the corrosive environment, because both are required for the problem to occur. In the real world, this is easier said than done. One way that has been found to be very effective is to form compressive residual stresses on the surface of the part by mechanical methods, such as shot peening, surface rolling, and the like. This will increase resistance to cracking and ultimate fracture. If a residual tensile stress is high, it may be possible to stress relieve the part or assembly by a thermal treatment. In fact, warming to subcritical temperatures, depending on the metal involved, may be all that is necessary.

Identification of SCC is not always easy, because it may be confused with another type of fracture. For example, Fig. 20 shows a stress-corrosion crack in a high-strength steel part that has a fracture pattern that could be mistaken for fatigue fracture, because it appears to have fatigue beachmarks. However, because the part had not been cyclically stressed, it could not be fatigue. The beachmarklike pattern observed is the result of differences in the rate of penetration of corrosion on the surface as the crack advanced. As frequently happens, the crack progressed relatively slowly until it reached the critical size, then fractured suddenly in a brittle manner. The texture of the fracture surface is sometimes helpful in differentiating a stress-corrosion fracture from a fatigue fracture, because the surface texture in a fatigue fracture usually is quite smooth in the origin region, gradually becoming rougher toward the final rupture. This change is usually not seen in a stress-corrosion fracture.

Stress-corrosion cracking is also frequently confused with hydrogen embrittlement cracking. It is difficult and sometimes impossible to distin-

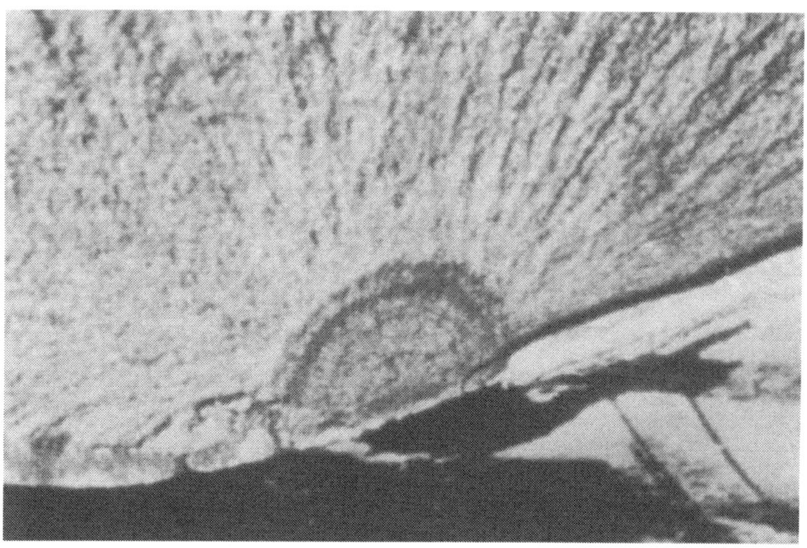

Fig. 20 Stress-corrosion crack in a high-strength steel part. Fracture surface appears to have the characteristic beachmark pattern of a fatigue fracture. However, this was a stress-corrosion fracture in which the pattern was caused by differences in the rate of corrosion penetration. Final fracture was brittle. Original magnification: 4×. Source: Ref 15

guish with certainty between hydrogen-induced and SCC failures that have occurred in service by exposure to hydrogen gas, hydrogen sulfide, and water and dilute aqueous solutions. Several basic characteristics to be observed in investigating failures of these types are history of the metal or part, crack origin, crack pattern, evidence of little or no corrosion on the fracture surfaces, and microscopic features.

The complexity of SCC and of the many factors, such as alloy, heat treatment, microstructure, stress system, part geometry, time, environmental conditions, and temperature, make it obvious that suspected instances of environment-related fractures must be carefully studied and analyzed, taking into consideration all available information before deciding on the dominant failure mode.

Hydrogen Damage

Hydrogen damage, like stress corrosion, is another insidious type of attack that can strike without warning. In addition, there are a number of different mechanisms associated with hydrogen damage. The problem with hydrogen is that it is an extremely small molecule that can quite easily penetrate into the metallic lattice. As opposed to stress corrosion, hydrogen damage can occur without an applied or residual tensile stress, although the presence of tensile stresses can often make a bad situation worse. Different mechanisms operate in different metals. Specific types of hydrogen damage include:

- *Hydrogen embrittlement.* This most often occurs in high-strength steels, primarily quenched and tempered and precipitation-hardened steels with tensile strengths greater than approximately 1035 MPa (150 ksi).
- *Hydrogen-induced blistering.* Also commonly referred to as hydrogen-induced cracking, this type of damage occurs in unhardened lower-strength steels, typically with tensile strengths less than approximately 550 MPa (80 ksi).
- *Cracking from precipitation of internal hydrogen.* During cooling from the melt, hydrogen diffuses and precipitates in voids and discontinuities. Examples include shatter cracks, flakes, and fisheyes found in steel forgings, weldments, and castings.
- *Hydrogen attack.* This is a high-pressure, high-temperature form of hydrogen damage commonly experienced in steels used in petrochemical plant equipment that often handles hydrogen and hydrogen-hydrocarbon streams at pressures as high as 21 MPa (3 ksi) and temperatures up to 540 °C (1000 °F).
- *Hydride formation.* This type of damage occurs when excess hydrogen is picked up during melting or welding of titanium, tantalum, zirco-

nium, uranium, and thorium. Brittle needlelike hydride particles cause a significant loss in strength and large losses in ductility and toughness.

Hydrogen embrittlement has been a problem particularly in heat treated high-strength steels. In general, the higher the strength level of the steel, the greater the susceptibility to hydrogen embrittlement. Hydrogen embrittlement occurs primarily in body-centered cubic and hexagonal close-packed metals, while face-centered cubic metals are generally not susceptible.

Hydrogen embrittlement results in sudden failures at stress levels below the yield strength. It is normally a delayed failure, in which an appreciable amount of time passes between the time hydrogen is introduced into the metal and the failure occurs. Hydrogen embrittlement is a complex process, and different mechanisms may operate in different metals under different environments and operating stresses. However, hydrogen is a small molecule that can dissociate into monatomic hydrogen that readily diffuses into the crystalline structure. Very small amounts of hydrogen can cause damage; for example, as little as 0.0001% hydrogen can cause cracking in steel. Typical sources of hydrogen include melting operations, heat treatments, welding, pickling, and plating. In addition, the cathodic reaction during in-service corrosion can also produce hydrogen.

Characteristics of hydrogen embrittlement include a strain-rate sensitivity, a temperature dependence, and delayed fracture. As opposed to many forms of brittle fracture, hydrogen embrittlement is enhanced by slow strain rates. In addition, it does not occur at low or high temperatures but occurs at intermediate temperature ranges. For steels, the most susceptible temperature is near room temperature.

A comparison between hydrogen-free notched tensile specimens and ones charged with hydrogen in a static tensile test is shown in Fig. 21. Note that there is a time delay before failure occurs, hence the term *static fatigue*. Also, below a certain stress level, failure does not occur. The higher the hydrogen content, the lower the stress level that can be endured

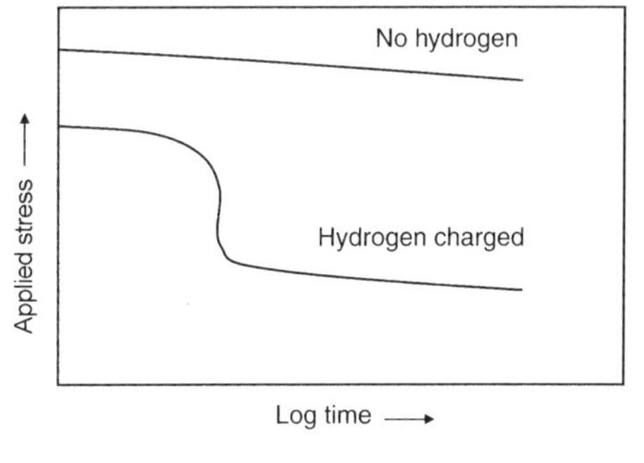

Fig. 21 Hydrogen effect on static tensile strength. Source: Ref 3

before failure. There is also a large reduction in ductility associated with embrittlement (Fig. 22).

There is no single fracture mode associated with hydrogen embrittlement. Fracture can be transgranular or intergranular (Fig. 23) and can exhibit characteristics of both brittle and ductile failure modes. If a steel part is not under stress when it contains hydrogen, then hydrogen can usually be safely removed without damage to the part by baking the part at ele-

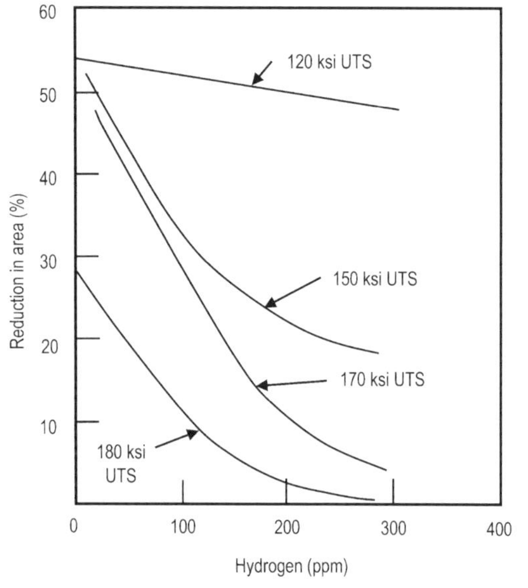

Fig. 22 Effect of hydrogen on ductility of steels. UTS, ultimate tensile strength. Source: Ref 3

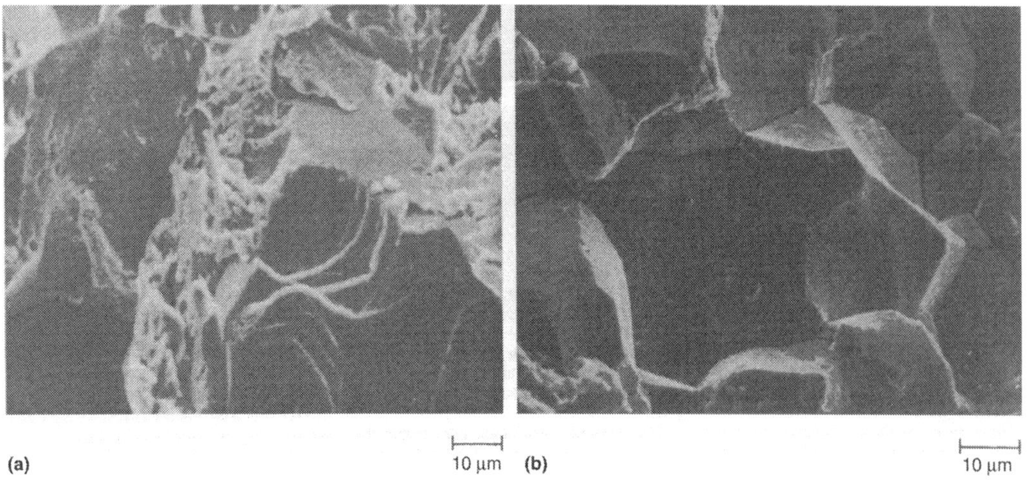

Fig. 23 Transgranular and intergranular hydrogen embrittlement fractures. (a) Transgranular cleavage fracture in hydrogen embrittled annealed type 301 austenitic stainless steel; (b) Intergranular decohesion fracture in hydrogen embrittled 4130 steel heat treated to 1275 MPa (185 ksi). Source: Ref 16

vated temperature. The use of a vacuum during baking is even more effective. High-strength steels are usually baked at 185 to 195 °C (365 to 385 °F) for at least 8 to 24 h to remove any hydrogen after chromium or cadmium plating operations.

The primary factors controlling hydrogen damage are material, stress, and environment. Hydrogen damage can often be prevented by using a more resistant material, changing the manufacturing processes, modifying the design to lower stresses, or changing the environment. Inhibitors and postprocessing bake-out treatments can also be used. Baking of electroplated high-strength steel parts reduces the possibility of hydrogen embrittlement by removing the hydrogen from the metal.

Corrosion Fatigue

Corrosion fatigue occurs in metals as a result of the combined action of a cyclic stress and a corrosive environment. Corrosion fatigue is dependent on the interactions among loading, environmental, and metallurgical factors. In the absence of fatigue loading, corrosive attack can often cause pitting of a metal surface. These pits can then act as stress concentrations that will initiate fatigue cracking. When corrosion and fatigue occur simultaneously, the chemical attack greatly accelerates fatigue crack growth. Materials, such as steels, that show a definite endurance limit when tested in air do not exhibit a definite endurance limit when tested in a corrosive environment. Frequently, fatiguing will remove protective films of corrosion products, which would normally slow the attack, allowing continued and accelerated attack. Because corrosion is time dependent, slow cycling rates will normally exhibit greater life reductions than accelerated fatigue cycling. Another possible effect of corrosion environment on fatigue is the absence of the stable stage II crack growth that occurs according to the Paris law. A reduction in fatigue life of 50% or more is not unusual. The effect varies widely, depending primarily on the particular metal-environment combination. The environment may affect the probability of fatigue crack initiation, the fatigue crack growth rate, or both. An example of corrosion fatigue failure in a Ti-6Al- 4V alloy is shown in Fig. 24.

Corrosion Fatigue Crack Initiation. The influence of an aggressive environment on fatigue crack initiation of a material is illustrated in Fig. 25, which compares the smooth specimen stress-life (*S-N*) curves obtained from inert and aggressive environments. Because as much as 95% of the structure life is spent on fatigue crack initiation, *S-N* curve comparison provides a good indication of the effect of environment on crack initiation. As shown in Fig. 25, an aggressive environment can promote crack initiation and can shorten the fatigue life of the structure. Corrosion fatigue cracks are always initiated at the surface, unless there are near-surface defects that act as stress-concentration sites and facilitate subsurface crack initiation. Surface features at origins of corrosion fatigue cracks vary with

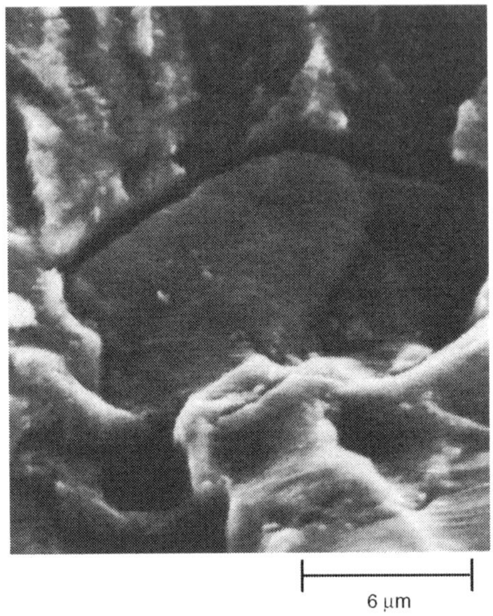

Fig. 24 Corrosion fatigue of a Ti-6Al-4V alloy tested in ambient air. Intergranular cracking and fatigue striations are evident on the fracture surface; the grain appears to have separated from the rest of the microstructure. Source: Ref 17

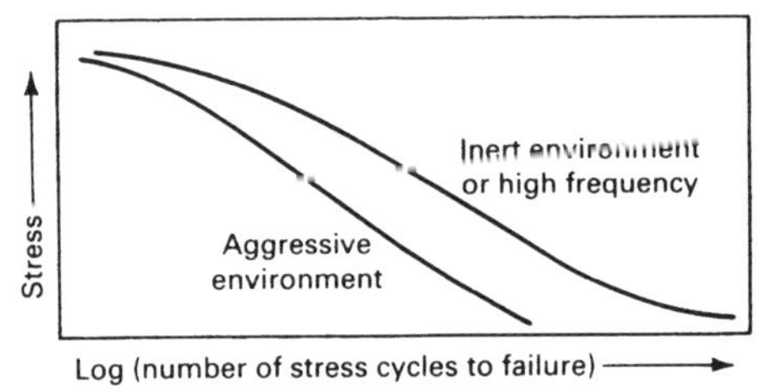

Fig. 25 Comparison of *S-N* curves for a material in an inert environment (top curve) and an aggressive environment (lower curve). Source: Ref 18

the alloy and with specific environmental conditions. In carbon steels, cracks often originate at hemispherical corrosion pits and often contain significant amounts of corrosion products. The cracks are often transgranular and may exhibit a slight amount of branching. However, surface pitting is not a prerequisite for corrosion fatigue cracking of carbon steels nor is the transgranular fracture path. Corrosion fatigue cracks sometimes occur in the absence of pits and follow grain boundaries or prior-austenite grain boundaries.

In aluminum alloys exposed to aqueous chloride solutions, corrosion fatigue cracks originate frequently at sites of pitting or intergranular corrosion. Initial crack propagation is normal to the axis of principal stress. This is contrary to the behavior of fatigue cracks initiated in dry air, where initial growth follows crystallographic planes. Initial corrosion fatigue cracking normal to the axis of principal stress also occurs in aluminum alloys exposed to humid air, but pitting is not a requisite for crack initiation.

Corrosion fatigue cracks in copper and various copper alloys initiate and propagate intergranularly. Corrosive environments have little additional effect on the fatigue life of pure copper over that observed in air, although they change the fatigue crack path from transgranular to intergranular. However, copper-zinc and copper-aluminum alloys exhibit a marked reduction in fatigue resistance, particularly in aqueous chloride solutions.

This type of failure is difficult to distinguish from SCC, except that it may occur in environments that normally do not cause failures under static stress, such as sodium chloride or sodium sulfate solutions.

Environmental effects can usually be identified by the presence of corrosion damage or corrosion products on fracture surfaces or within growing cracks. However, corrosion products may not always be present. For example, corrosion fatigue cracking of high-strength steel exposed to a hydrogen-producing gas, such as water vapor, may be difficult to differentiate from some other forms of hydrogen damage. At sufficiently high frequencies, the fracture surface features produced by corrosion fatigue crack initiation and propagation do not differ significantly from those produced by fatigue in nonaggressive environments.

Corrosion Fatigue Crack Propagation. Although corrosion fatigue phenomena are diverse, several variables are known to repeatedly influence crack growth rate:

- Stress-intensity range
- Loading frequency
- Stress ratio
- Aqueous environment electrode potential
- Environment
- Metallurgical variables

Effects of such variables as temperature, load history and waveform, stress state, and environment composition are unique to specific materials and environments.

Stress-Intensity Range. For embrittling environments, crack growth generally increases with increasing stress intensity (ΔK); however, the precise dependence varies markedly. Materials that are extremely environment sensitive, such as ultrahigh-strength steel in distilled water (Fig. 26), are characterized by high growth rates that depend on ΔK to a reduced power. Time-dependent corrosion fatigue crack growth occurs mainly above the threshold stress intensity for static load cracking and is modeled through linear superposition of SCC and inert environment fatigue rates.

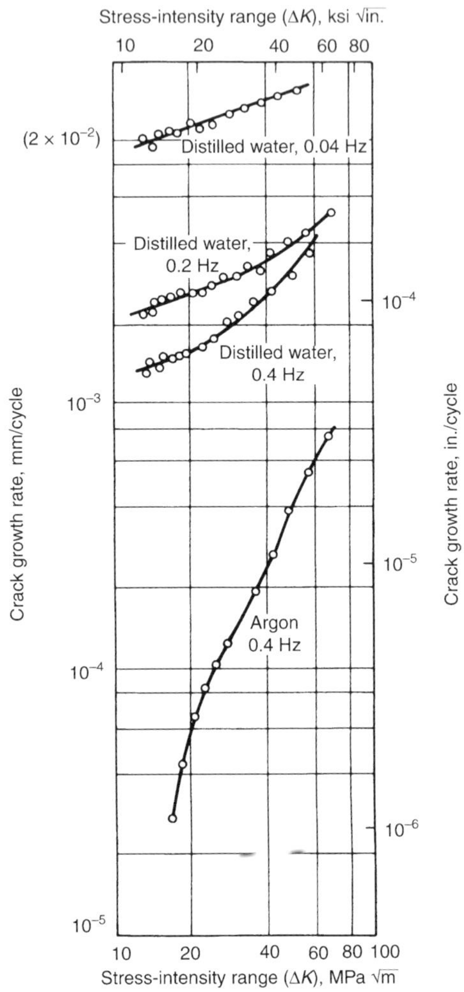

Fig. 26 Effect of stress-intensity range and loading frequency on corrosion fatigue crack growth in ultrahigh-strength 4340 steel exposed to distilled water at 23 °C (73 °F). Source: Ref 18

Loading Frequency. Cyclic load frequency is the most important variable that influences corrosion fatigue for most material, environment, and stress-intensity conditions. The rate of brittle cracking above that produced in vacuum generally decreases with increasing frequency. Frequencies exist above which corrosion fatigue is eliminated. The dominance of frequency is related directly to the time dependence of the mass transport and chemical reaction steps required for brittle cracking. Basically, insufficient time is available for chemical embrittlement at rapid loading rates, and damage is purely mechanical, equivalent to the crack growth in vacuum.

Stress Ratio. Rates of corrosion fatigue crack propagation generally are enhanced by increased stress ratio (R). Stress ratio has only a slight influence on fatigue crack growth rates in a benign environment.

Electrode potential, similar to loading frequency, strongly influences corrosion fatigue crack propagation rates in aqueous environments. Controlled changes in the potential of a specimen can result in either the complete elimination or the dramatic enhancement of brittle fatigue cracking. The precise influence depends on the mechanism of the environmental effect and on the anodic or cathodic magnitude of the applied potential.

Environment. Increasing the chemical activity of the environment, for example, by lowering the pH of a solution, by increasing the concentration of the corrosion species, or by increasing the pressure of a gaseous environment, generally decreases the resistance of a material to corrosion fatigue. Decreasing the chemical activity of the environment improves resistance to corrosion fatigue. For example, in aluminum alloys and high-strength steels, corrosion fatigue behavior is related to the relative humidity or partial pressure of water vapor in the air. Corrosion fatigue crack growth rates for these materials generally increase with increasing water vapor pressure until a saturation condition is reached. The appearance of the fracture surfaces of an aluminum alloy fatigue tested in argon and in air with water vapor present is compared in Fig. 27. Temperature can have a significant effect on corrosion fatigue. The effect is complex and depends on temperature range and the particular material-environment combination in question, among other factors. However, the general tendency is for fatigue crack growth rates to increase with increasing temperature.

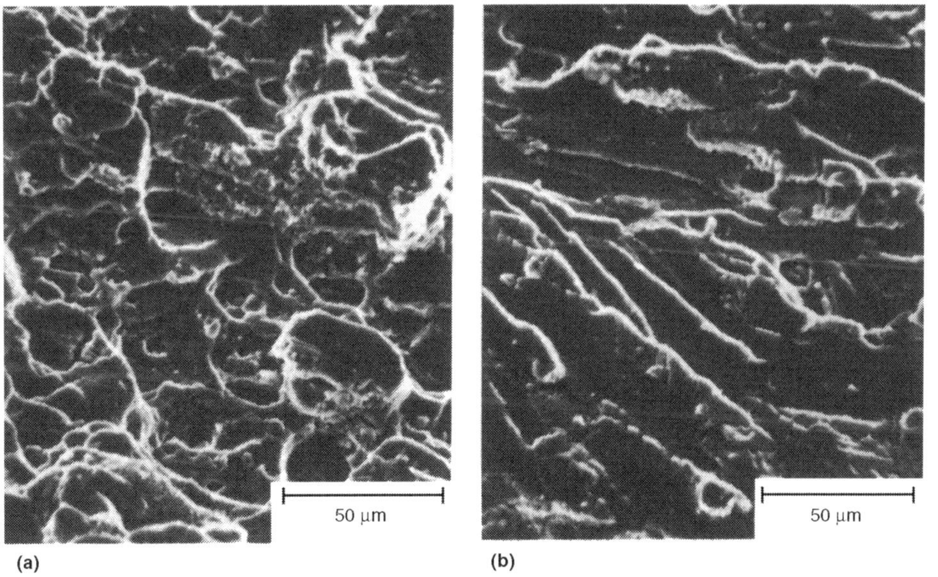

Fig. 27 Effect of water vapor on the fracture surface appearance of aluminum alloy 2219-T851 fatigue tested (a) in 0.101 MPa (1 atm) dry argon and (b) in 27 Pa (0.2 torr) water vapor. Testing conditions were the same except for frequency, which was 20 Hz in (a) and 5 Hz in (b). The magnifications, while too low to resolve fatigue striations clearly, indicate the general change in fracture morphology. Crack propagation was from left to right. Source: Ref 19

Other factors, including the metallurgical condition of the material (such as composition and heat treatment) and the loading mode (such as uniaxial), affect corrosion fatigue crack propagation.

Relationship between Corrosion Fatigue, Hydrogen Embrittlement, and SCC. The relationship between corrosion fatigue and two other environmental cracking mechanisms, SCC and hydrogen embrittlement, is shown in Fig. 28. Many investigations have attempted to link the mechanisms of corrosion fatigue and other environmental cracking processes (especially SCC). There are, however, many unanswered questions about the mechanisms of these phenomena.

Effective measures for minimizing corrosion fatigue include the following:

- Reduce or eliminate corrosion by any of the conventional means, such as painting or plating, if these are practical for the application. It may be possible to reduce the aggressiveness of the environment by adding inhibitors or by changing the concentrations of the solution in a closed system.
- Change the material to one more resistant to the environment, such as a stainless steel, or a different nonferrous alloy system. However, this should be done only as a last resort, because many other potential problems arise, such as economics, availability, general engineering suitability, and manufacturing difficulties.
- Reduce the resultant tensile stress that is causing the fatigue problem. In some cases, it may be possible to reduce the applied stress by decreasing the load applied to the part. In other cases, it may be necessary to redesign the part to increase the section size, but again, this is a major decision.
- Reduce the resultant tensile stress by increasing the compressive residual stresses on the critical surface(s). There are many ways in which this may be accomplished. Some of the simplest ways involve using mechanical prestressing, such as shot peening or surface rolling. Shot

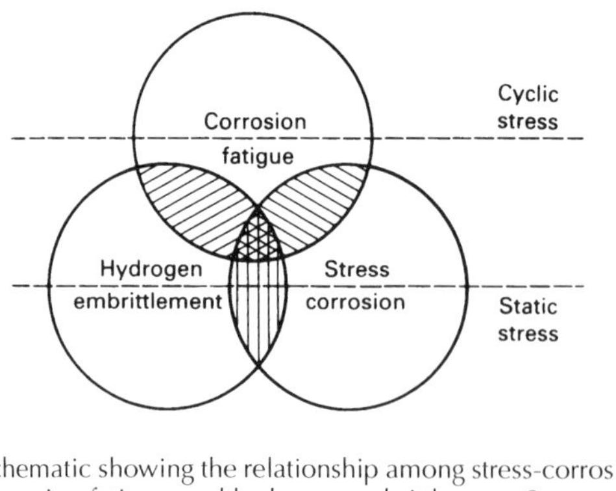

Fig. 28 Schematic showing the relationship among stress-corrosion cracking, corrosion fatigue, and hydrogen embrittlement. Source: Ref 20

peening puts into compression a relatively shallow layer below the surface (usually several thousandths of an inch), which may be inadequate if the corrosion makes pits that penetrate this layer. Deeper penetration may be accomplished by surface rolling or by pressing of grooves in critical areas. However, both of these are limited by geometry. For example, surface rolling is usually done on propeller shafts of ships at the location where the hub for the propeller fits. This is a critical location because the rotating propeller hub must be separated from the stationary rear support or bearing by a sealing system, which often does not work properly. As a result, water, often seawater, corrodes the steel shaft, sometimes resulting in corrosion fatigue that could mean disastrous loss of the propeller at sea.

As with most types of failures, corrosion fatigue problems may be solved by a variety of approaches. The particular solutions chosen depend on the individual situation, the critical nature of the part, current and proposed metals, availability, engineering and manufacturing problems, and economics.

Corrosion Prevention

The first line of defense against corrosion is good design practices. During the previous discussion on the various forms of corrosion, a number of good design practices were mentioned for the different types of corrosion. In addition, by retarding either the anodic or cathodic reactions, the rate of corrosion can be reduced. This can be achieved in several ways: conditioning of the metal, conditioning the environment, and by electrochemical control.

Conditioning the Metal

Conditioning the metal can be subdivided into two main groups: coating the metal or using a more corrosion-resistant alloy. Coating the metal provides a corrosion-resistant coating between the metal and the environment. The coating may consist of another metal; a protective coating derived from the metal itself, such as a protective oxide; or an organic coating, such as resins, plastics, paints, and enamels.

Protective coatings are widely used to control corrosion. Protective coatings include organic coatings such as epoxy, polyesters, polyurethanes, vinyl, or chlorinated rubber. Special primers are used to provide passivation, galvanic protection, corrosion inhibition, or mechanical or electrical barriers to corrosive action. A slightly soluble inhibitor incorporated into the primer coat can have a considerable protective influence. Most paint primers contain a partially soluble inhibitive pigment such as zinc chromate, which reacts with the steel substrate to form iron salts that slow the corrosion of steel. Chromates, phosphates, molybdates, borates, and silicates are also commonly used for this purpose. Some pigments add alkalinity, slowing chemical attack on steel. Alkaline pigments are effective, provided that the environment is not too aggressive. In addition, many newer pigments have been introduced to the paint industry, such as zinc phosphosilicate and zinc flake.

Metallic coatings include tin-plated and zinc-plated (galvanized) steel. A continuous coating of either isolates the steel from the electrolyte. However, when the coating is scratched or otherwise penetrated, exposing the underlying steel, the two coatings behave differently (Fig. 29). Because zinc is anodic to steel, it continues to be effective in protecting the steel. Because the area of the exposed steel cathode is small, the zinc coating corrodes at a very slow rate, and the underlying steel is protected. On the other hand, steel is anodic to tin, so that a small steel anode is created when the tin plate is breached, leading to rapid corrosion of the underlying steel.

Alloying is often used to produce a more corrosion-resistant alloy. Not all metals exhibit passivity, but the ones that do are among the most widely used corrosion-resisting materials. Nickel, chromium, titanium, and zirconium spontaneously react with the oxygen in air to form protective films. In the case of steels, this often means using a stainless steel instead of ordinary plain carbon steel. The chromium in stainless steel forms a thin oxide of Cr_2O_3 that protects the metal. Chromium additions to nickel-base alloys also generally enhance resistance.

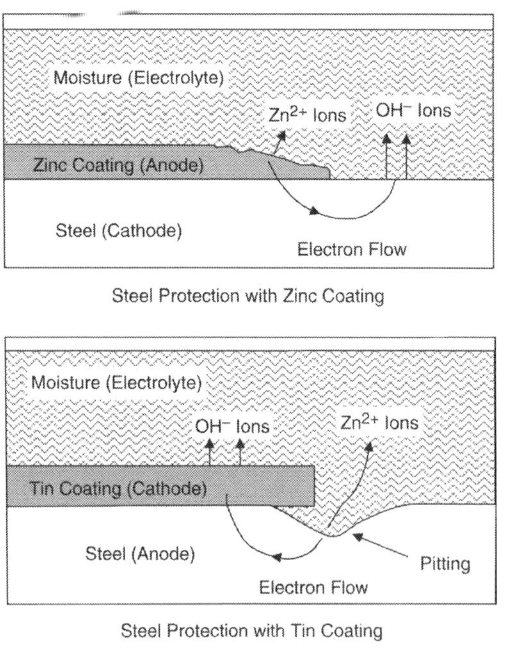

Fig. 29 Comparison of zinc and tin plating on steel. Source: Ref 2

Conditioning the Corrosive Environment

A corrosion inhibitor is a chemical additive that, when added to a corrosive aqueous environment, reduces the corrosion rate. Corrosion inhibitors can consist of anodic inhibitors, cathodic inhibitors, adsorption type, or mixed inhibitors.

Anodic Inhibitors. As their name implies, anodic inhibitors interfere with the anodic reaction:

$$Fe \rightarrow Fe^{++} + 2e^{-} \qquad \text{(Eq 8)}$$

However, if an anodic inhibitor is not present at a concentration level sufficient to block off all the anodic sites, localized attack, such as pitting, can become a serious problem due to the oxidizing nature of the inhibitor, which raises the metal potential and encourages the anodic reaction. Thus, anodic inhibitors are often classified as dangerous inhibitors.

Cathodic Inhibitors. The major cathodic reaction in cooling systems is the reduction of oxygen:

$$\tfrac{1}{2}O_2 + H_2O + 2e^{-} \rightarrow 2OH^{-} \qquad \text{(Eq 9)}$$

Cathodic inhibitors include other additives that suppress the cathodic reaction. They function by reducing the available area for the cathodic reaction. This is often achieved by precipitating an insoluble species onto the cathodic sites. For example, zinc ions are used as cathodic inhibitors because of the precipitation of $Zn(OH)_2$ at cathodic sites as a consequence of the localized high pH. Cathodic inhibitors are classed as "safe" because they do not cause localized corrosion.

Mixed Inhibitors. Because of the danger of pitting when using anodic inhibitors alone, a common practice incorporates a cathodic inhibitor along with an anodic inhibitor. A common formulation would consist of a mix of zinc and chromate ions.

Adsorption-Type Inhibitors. Many organic inhibitors work by an adsorption mechanism. The resultant film of chemisorbed inhibitor is then responsible for protection, either by physically blocking the surface from the corrosion environment or by retarding the electrochemical processes. The main functional groups capable of forming chemisorbed bonds with metal surfaces are aminos ($-NH_2$), carboxyls (-COOH), and phosphonates ($-PO_3H_2$), although other functional groups or atoms can also form coordinate bonds with metal surfaces.

In general, inhibitors added to the environment work best in closed systems that can be monitored. For example, the antifreeze that is added to water in a car radiator contains corrosion inhibitors. It is important to change the antifreeze according to the manufacturer's instructions, because the inhibitors become depleted over time.

Electrochemical Control

Steel pipes buried in moist soils or immersed in water are often protected by cathodic protection. Cathodic protection can be achieved by using a direct current power supply (impressed current) or by obtaining electrons from the anodic dissolution of a metal lower in the galvanic series, such as aluminum, zinc, or magnesium sacrificial anodes. Both of these approaches are shown schematically in Fig. 30.

Protective coatings are normally used in conjunction with cathodic protection. Because the cathodic protection current must protect only the bare or poorly insulated areas of the surface, coatings that are highly insulating, very durable, and free of discontinuities lower the current requirements and system costs. A good coating also enables a single impressed-current installation to protect many miles of piping. Coal tar enamel, epoxy powder coatings, and vinyl resin are examples of coatings that are most suitable for use with cathodic protection

High-Temperature Oxidation and Corrosion

Oxidation usually refers to the reaction of metals with oxygen at high temperatures, usually in the absence of moisture. The nature of the oxide film usually takes one of three forms:

- The oxide is unstable, such as in gold, and oxidation does not occur.
- The oxide is volatile, such as in the case of refractory metals, and oxidation occurs at a constant, relatively high rate.
- More commonly, one or more oxides form a layer, or layers, on the metal surface.

When their thickness is less than approximately 3000Å, surface oxide layers are called films, and when thicker, they are referred to as scales. Thick oxide or scale layers are divided into two categories, protective and

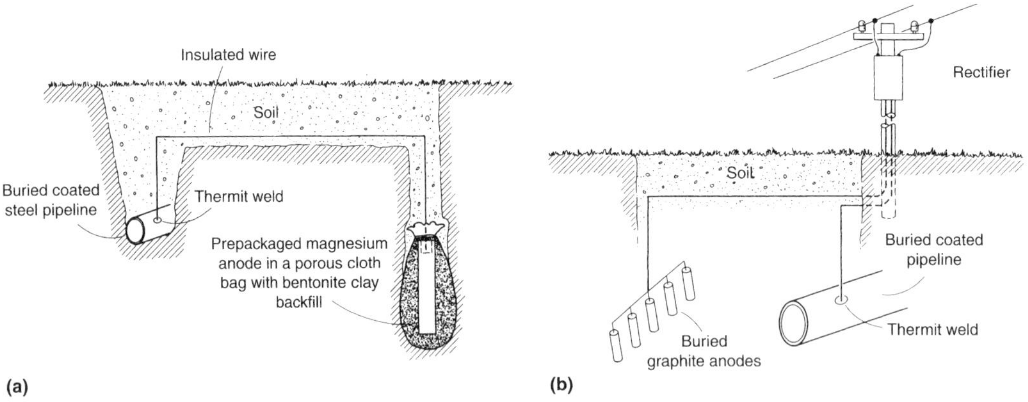

Fig. 30 Cathodic protection of a buried pipeline. (a) Using a buried magnesium anode. (b) Impressed-current cathodic protection of a buried pipeline using graphite anodes. Source: Ref 3

nonprotective, on the basis of the Pilling-Bedworth (P-B) ratio. According to the P-B ratio, the oxide is protective if the volume of the oxide is at least as great as the volume of metal from which it formed. If the volume of oxide is less than this amount, the scale is not continuous and is not effective in blocking oxygen from the surface. Although there are many exceptions to the P-B ratio, it is a useful guide when the oxidation characteristics of a metal are unknown.

Three common types of oxide growth, shown in Fig. 31, are linear, parabolic, and logarithmic. Metals that have nonprotective oxides tend to increase the weight (W) of their scale at a linear rate according to:

$$W = At \qquad \text{(Eq 10)}$$

where A is a constant, and t is time. In linear growth, oxygen usually passes right through pores or fissures in the oxide layer. When a protective oxide forms on the surface, diffusion must occur for additional growth (Fig. 32). The metal ionizes at the surface, and then both the metal ion (M^{++}) and the electrons diffuse through the oxide layer to the oxygen surface. Electrons aid in forming the oxygen ion. The ion reacts near the oxygen surface to form the MO oxide. In this case, the rate of oxidation is slower than that for linear growth and occurs by parabolic growth rate:

$$W^2 = Bt \qquad \text{(Eq 11)}$$

where the constant B depends on the temperature. Some metals that form protective oxides have reaction rates that diminish more rapidly with time than a parabolic relationship predicts. A logarithmic increase in the weight of oxide follows a relationship of the form:

$$W = C \log\left(Dt + E\right) \qquad \text{(Eq 12)}$$

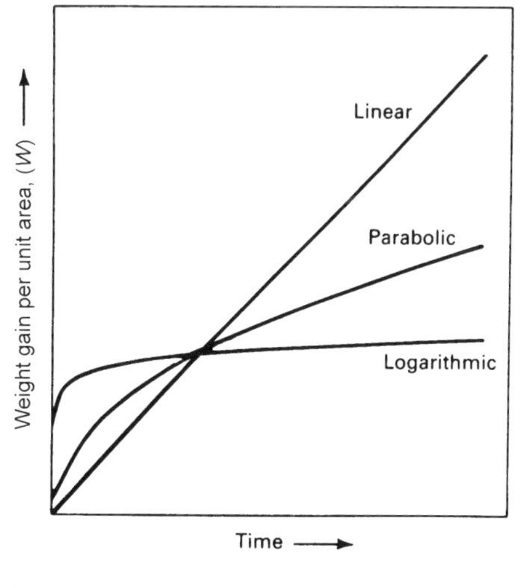

Fig. 31 Oxidation growth rate curves. Ref 3

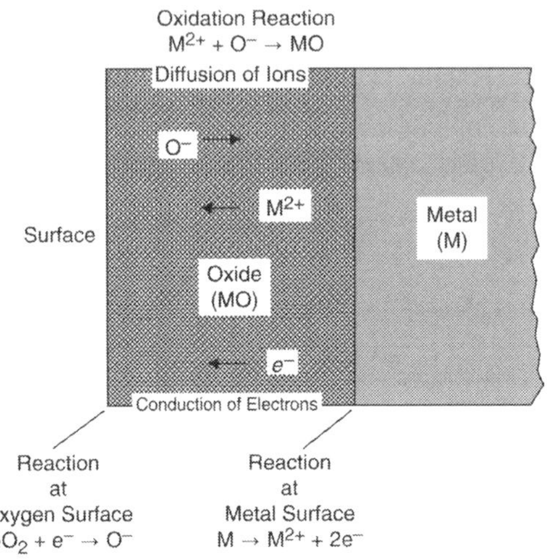

Fig. 32 Oxidation of metal through oxide layer. Source: Ref 21

where the constants *C*, *D*, and *E* depend on temperature.

When an alloy, rather than a pure metal, undergoes oxidation, additional factors must be considered. The separate oxides may form a solid solution, a multiphase scale may appear, or only a single component of the alloy may undergo oxidation, a process called selective oxidation. In the case of selective oxidation, benefit can be derived from oxide films if they are adherent, have poor electrical conductivity, and posses a complex crystal structure that hinders diffusion through them. Under certain conditions, oxidation of an alloying element can occur below the surface of the base metal. This internal oxidation can be a problem with some copper and silver alloys that are given high-temperature treatments in mildly oxidizing environments. Compared with ordinary oxide films, the subscales are difficult to remove with conventional cleaning methods.

Oxidation rarely occurs in only one of these ways described; instead, a combination of two or three types of reactions may occur simultaneously in different parts of the metal. The stress condition and the orientation of the oxide layer may vary with time or thickness, and discontinuous cracking or spalling of the oxide layer can cause sudden changes in the reaction rate. If more than one oxide layer is stable under oxidation conditions, a series of oxide layers may be formed on the metal surface.

The two major environmental effects on superalloys are oxidation and hot corrosion. At temperatures of approximately 870 °C (1600 °F) and lower, oxidation of superalloys is not a major problem; however, at higher temperatures, oxidation can rapidly occur. Because Cr_2O_3 forms as a protective oxide, the level of oxidation resistance at temperatures below 980 °C (1800 °F) is a function of the chromium content. At temperatures above 980 °C (1800 °F), the aluminum content becomes more important as Al_2O_3 becomes the dominant oxide protector. Chromium and aluminum can contribute in an interactive manner to provide oxidation protection.

For example, the higher the chromium content, the less aluminum that may be required. However, the alloy content of many superalloys is insufficient to provide long-term protection, and protective coatings are usually required to provide satisfactory life.

Hot corrosion of superalloys, often referred to as sulfidation, is classified as either type I or II, depending on the temperature. Type I occurs at higher temperatures (900 to 1050 °C, or 1650 to 1920 °F), while type II occurs at lower temperatures (680 to 750 °C, or 1255 to 1380 °F). Both are triggered by the presence of sulfur in fuels combining with salt from the environment. Hot corrosion is an accelerated, often catastrophic, surface attack of parts in the hot gas path. It is believed that the presence of alkali metal salts (i.e., Na_2SO_4) is a prerequisite for hot corrosion.

ACKNOWLEDGMENTS

Portions of this chapter came from "Forms of Corrosion," *Failure Analysis and Prevention*, Vol 11, *ASM Handbook*, ASM International, 2002; "Forms of Mechanically Assisted Degradation" by W. Glaeser and I.G. Wright, *Corrosion: Fundamentals, Testing, and Protection*, Vol 13A, *ASM Handbook*, ASM International, 2003: "Corrosion Failures" by D.J. Wulpi, *Understanding How Components Fail*, 2nd ed., ASM International, 1999; and "Corrosion" by F.C. Campbell, *Elements of Metallurgy and Engineering Alloys*, ASM International, 2008.

REFERENCES

1. "High-Performance Alloys for Resistance to Aqueous Corrosion," Special Metals Corporation
2. R.A. Higgins, *Engineering Metallurgy—Applied Physical Metallurgy,* 6th ed., Arnold, 1993
3. F.C. Campbell, Corrosion, *Elements of Metallurgy and Engineering Alloys,* ASM International, 2008
4. *Corrosion,* Vol 13, *ASM Handbook,* ASM International, 1987, p 4, 83–87, 291–302
5. Forms of Corrosion, *Failure Analysis and Prevention,* Vol 11, *ASM Handbook,* ASM International, 2002
6. D.R. Askeland, *The Science and Engineering of Materials,* 2nd ed., PWS-Kent Publishing Co., 1989
7. K.K. Sankaran, R. Perez, and H. Smith, Military Aircraft Corrosion Fatigue, *Corrosion: Environments and Industries,* Vol 13C, *ASM Handbook,* ASM International, 2006, p 195–204
8. D.J. Wulpi, Corrosion Failures, *Understanding How Components Fail,* 2nd ed., ASM International, 1999
9. Atlas of Fractographs, Austenitic Stainless Steels, *Fractography,* Vol 12, *ASM Handbook,* ASM International, 1987
10. R.H. Jones, Stress-Corrosion Cracking, *Corrosion: Fundamentals,*

Testing, and Protection, Vol 13A, *ASM Handbook,* ASM International, 2003, p 346–366
11. E. Denhard, Effect of Composition and Heat Treatment on the Stress Corrosion Cracking of Austenitic Stainless Steels, *Corrosion,* Vol 16 (No. 7), 1960, p 131–141
12. M. Fontana, Stress Corrosion, Lesson 5, *Corrosion,* Metals Engineering Institute Course, American Society for Metals, 1968
13. G.H. Koch, Stress-Corrosion Cracking and Hydrogen Embrittlement, *Fatigue and Fracture,* Vol 19, *ASM Handbook,* ASM International, 1996
14. S.D. Antolovich and G.R. Chanani, *Eng. Fract. Mech.,* Vol 4, 1972, p 765
15. *Fractography and Atlas of Fractographs,* Vol 9, *Metals Handbook,* 8th ed., American Society for Metals, 1974, p 31, 91
16. V. Kerlins and A. Phillips, Effect of Environment on Failure Modes, *Fractography,* Vol 12, *ASM Handbook,* ASM International, 1987
17. C.J. Heathcock and A. Ball, Cavitation Erosion of Cobalt-Base Stellite Alloys, Cemented Carbides and Surface Treated Low Alloy Steels, *Wear,* Vol 74, 1981–1982, p 11–26
18. W. Glaeser and I.G. Wright, Forms of Mechanically Assisted Degradation, *Corrosion: Fundamentals, Testing, and Protection,* Vol 13A, *ASM Handbook,* ASM International, 2003
19. A.J. McEvily and R.P. Wei, Fracture Mechanics and Corrosion Fatigue, *Corrosion Fatigue: Chemistry, Mechanics, and Microstructure,* O. Deveraux, A.J. McEvily, and R.W. Staehle, Ed., NACE, 1973, p 381–395
20. R.P. Wei and G. Shim, Fracture Mechanics and Corrosion Fatigue, *Corrosion Fatigue: Mechanics, Metallurgy, Electrochemistry and Engineering,* STP 801, T.W. Crooker and B.N. Leis, Ed., ASTM, 1984, p 5–25
21. A.G. Guy and J.J. Hren, *Elements of Physical Metallurgy,* 3rd ed., Addison-Wesley Publishing, 1974

CHAPTER 14

The Failure Analysis Process

IN THE STUDY OF ANY FAILURE, the analyst must consider a broad spectrum of possibilities or reasons for the occurrence. Often a large number of factors, frequently interrelated, must be understood to determine the cause of the original, or primary, failure. The analyst is in the position of Sherlock Holmes attempting to solve a baffling case. Like the great detective, the analyst must carefully examine and evaluate all evidence available, then prepare a hypothesis—or possible chain of events—that could have caused the "crime." The analyst may also be compared to a coroner performing an autopsy on a person who suffered an unnatural death, except that the failure analyst works on parts or assemblies that have had an unnatural or premature demise. If the failure can be duplicated under controlled simulated service conditions in the laboratory, much can be learned about how the failure actually occurred. If this is not possible, there may be factors about the service of the part or assembly that are not well understood.

For a complete evaluation, the sequence of stages in the investigation and analysis of failure is as follows:

1. Collection of background data and selection of samples
2. Preliminary examination of the failed part (visual examination and record keeping)
3. Nondestructive testing
4. Mechanical testing (including hardness and toughness testing)
5. Selection, identification, preservation, and/or cleaning of specimens (and comparison with parts that have not failed)
6. Macroscopic examination and analysis and photographic documentation (fracture surfaces, secondary cracks, and other surface phenomena)

7. Microscopic examination and analysis (electron microscopy may be necessary)
8. Selection and preparation of metallographic sections
9. Examination and analysis of metallographic specimens
10. Determination of failure mechanism
11. Chemical analysis (bulk, local, surface corrosion products, deposits or coatings, and microprobe analysis)
12. Analysis by fracture mechanics
13. Testing under simulated service conditions (special tests)
14. Analysis of all the evidence, formulation of conclusions, and writing the report (including recommendations). Writing a report may not be necessary in many product litigation cases; it is best to follow the advice of the attorney or client with whom the analyst is working.

This list is quite extensive, and many of these steps may not be necessary for all failure analysis investigations. However, it is important not to "jump to a conclusion" too early in the analysis.

In cases that involve personal injury or will most likely involve legal pursuit of compensation from another company, care must be taken in preserving the scene and physical evidence. Accidental or deliberate destruction of evidence can result in diverting the legal liability of a failure to the person or company destroying the evidence, even though they may not have caused the original failure.

Collection of Background Data and Samples

The failure investigation should include gaining an acquaintance with all pertinent details relating to the failure; collecting the available information regarding the design, manufacture, processing, and service histories of the failed component or structure; and reconstructing, insofar as possible, the sequence of events leading to the failure. Collection of background data on the manufacturing and fabricating history of a component should begin with obtaining specifications and drawings and should encompass all the design aspects of the failed part as well as all manufacturing and fabrication details: machining, welding, heat treating, coating, quality-control records, and pertinent purchase specifications.

Collecting Data and Samples

On-Site Investigation. In the investigation of failures, it is also often desirable for the analyst to visit the scene, but for the analysis of some components, it may be impractical or impossible for the failure analyst to visit the failure site. Under these circumstances, data and samples may be collected at the site by field engineers or by other personnel under the direction of the failure analyst. A field failure report sheet or checklist can be used to ensure that all pertinent information regarding the failure is recorded.

There are also situations where it is essential to perform failure analyses on the site. While it is recommended that examination be done in a laboratory, the requirements for on-site testing may involve the use of portable laboratories with metallographic equipment for grinding, mechanical polishing, and etching. Small specimens can be cut from a part on the site for preparation, examination, and photography immediately or upon return to a fully equipped laboratory. Photography is, of course, essential; it should be performed by the analyst or perhaps a professional photographer in the case of a large-scale accident scene. It is also frequently desirable to make acetate tape replicas or room-temperature vulcanized (RTV) rubber replicas of fracture surfaces or of wear patterns of large parts during an on-site failure analysis. Several replicas should be made of the fracture origin region using acetate tape softened in acetone, dried, then carefully stripped from the fracture surface. Upon return to the laboratory, the replicas may be gold coated and examined with a scanning electron microscope. Foreign particles removed from the fracture surface may also be analyzed. The RTV rubber replica can be applied over a rather large area with less chance of missing a critical spot. A combination of acetate tape and RTV rubber replicas can assure the investigator of better coverage of the area in question. Room-temperature vulcanized rubber does not provide the sensitivity of an acetate replica, and a setup time of several hours is required. However, the added area can be very important in an investigation.

Hardness testing with a portable hardness testing instrument may also be performed during on-site failure analysis. Several different types of testers are available and, in general, are either electronic or mechanical in principle. Obviously, small size and light weight are advantages in portable testers.

The major components of the portable laboratory may include:

- A custom-made machine, plus auxiliary materials, for grinding and polishing small, mounted or unmounted metal specimens
- A right-angle head, electric drill motor with attachments and materials for grinding and polishing selected spots on large parts or assemblies. It is also used for driving the grinding and polishing machine described in the previous item.
- A portable microscope, with camera attachment and film for use in photographing metallographic specimens
- Equipment and materials for mounting and etching specimens
- A handheld single-lens reflex 35 mm camera, with macrolenses and film
- A pocket-sized magnifier and a ruler or scale
- A hacksaw and blades for cutting specimens
- Portable hardness tester

- Acetate tape, acetone, and containers
- RTV rubber for replicas

Service History. The availability of a complete service history depends on how detailed and thorough the record keeping was prior to the failure. A complete service record greatly simplifies the assignment of the failure analyst. In collecting service histories, special attention should be given to environmental details such as normal and abnormal loading, accidental overloads, cyclic loads, temperature variations, temperature gradients, and operation in a corrosive environment. However, in most instances, complete service records are not available, forcing the analyst to work from fragmentary service information. When service data are sparse, the analyst must, to the best of his or her ability, deduce the service conditions. Much depends on the analyst's skill and judgment, because a misleading deduction can be more harmful than the absence of information.

Photographic Records. Photographs of the failed component or structure are often critical to an accurate analysis. A detail that appears almost inconsequential in a preliminary investigation may later be found to have serious consequences; thus, a complete, detailed photographic record of the scene and failed component can be essential. Photographs should be of professional quality, but this is not always possible. For the analyst who does his own photography, a single-lens reflex 35 mm or larger camera with a macrolens, extension bellows, and battery flash unit is capable of producing excellent results. It may be desirable to supplement the 35 mm equipment with an instant camera and close-up lenses. When accurate color rendition is required, the subject should be photographed with a color chart, which should be sent to the photographic studio for use as a guide in developing and printing. Some indication of size, such as a scale, coin, hand, and so forth, should be included in the photograph. Samples should be selected judiciously before starting the examination, especially if the investigation is to be lengthy or involved. As with photographs, the analyst is responsible for ensuring that the samples will be suitable for the intended purpose and that they adequately represent the characteristics of the failure. It is advisable to look for additional evidence of damage beyond that which is immediately apparent. For failures involving large structures or key machinery, there is often a financially urgent need to remove the damaged structure or repair the machine for return to production. This is a valid reason to move evidence, but a reasonable attempt must be made to allow other parties, who may become involved in a potential legal case, to inspect the site. All concerned parties then can agree on the critical samples and the best way to remove them. If all parties are not available, care must be taken not to damage or alter critical elements to avoid spoiling evidence. Guidelines governing sample collection are covered in ASTM E 620, E 860, E 1020, and especially E 678. It is also recommended that samples be taken from other parts of the failed equipment, because they may display supportive damage.

It is often necessary to compare failed components with similar components that did not fail to determine whether the failure was brought about by service conditions or was the result of an error in manufacture. For example, if a boiler tube fails and overheating is suspected to be the cause, and if investigation reveals a spheroidized structure in the boiler tube at the failure site, which may be indicative of overheating in service, then comparison with an unexposed tube will determine if the tubes were supplied in the spheroidized condition.

As another example, in the case of a bolt failure, it is desirable to examine the nuts and other associated parts that may have contributed to the failures. Also, in failures involving corrosion, stress corrosion, or corrosion fatigue, a sample of the fluid that has been in contact with the metal, or of any deposits that have formed, will often be required for analysis.

Abnormal Conditions and Wreckage Analysis. In addition to developing a history of the failed part, it is also advisable to determine if any abnormal conditions prevailed, such as an accident occurred in service, that may have initiated the failure, or if any recent repairs or overhauls had been carried out and why. In addition, it is also necessary to inquire whether or not the failure was an isolated example or if others have occurred, either in the component under consideration or in another of a similar design. In the routine examination of a brittle fracture, it is important to know if, at the time of the accident or failure, the prevailing temperature was low and/or if some measure of shock loading was involved. When dealing with failures of crankshafts or other shafts, it is generally desirable to ascertain the conditions of the bearings and whether any misalignment existed, either within the machine or between the driving and driven components.

In an analysis where multiple components and structures are involved, it is essential that the position of each piece be documented before any of the pieces are touched or moved. Such recording usually requires extensive photography, the preparation of suitable sketches, and the taking and tabulation of appropriate measurements of the pieces.

Next, it may be necessary to take an inventory to determine if all of the pieces or fragments are present at the site of the accident. For example, an investigation of an aircraft accident involves the development of a considerable inventory, including listing the number of engines, flaps, landing gear, and the various parts of the fuselage and wings. It is essential to establish whether all the primary parts of the aircraft were aboard at the time that it crashed. Providing an inventory, although painstaking, is often invaluable. An experienced investigator determined the cause of a complex aircraft accident when he observed that a portion of one wing tip was missing from the main impact site. This fragment was subsequently located several miles back on the flight path of the aircraft. The fragment provided evidence of a fatigue failure and was the first component that separated from the aircraft, thus accounting for the crash.

The most common problem encountered in examining wreckage involves the establishment of the sequence of fractures to determine the origin of the initial failure. Usually, the direction of crack growth can be detected from marks on a fracture surface, such as V-shaped chevron marks. The typical sequence of fractures is shown in Fig. 1(b), where "A" and "B" represent fractures that intersect at approximately 90°. Here, the sequence of fractures is clearly discernible from crack branching. Obviously, fracture A must have occurred prior to fracture B, because the presence of fracture A served to arrest cracking at fracture B. This method of sequencing is called the T-junction procedure and is an important technique in wreckage analysis.

Provided the fragments are not permitted to contact each other, it is also helpful to carefully fit together the fragments of broken components that, when assembled and photographed, may indicate the sequence in which fractures occurred. Figure 2 shows a lug that was part of a pin-joint as-

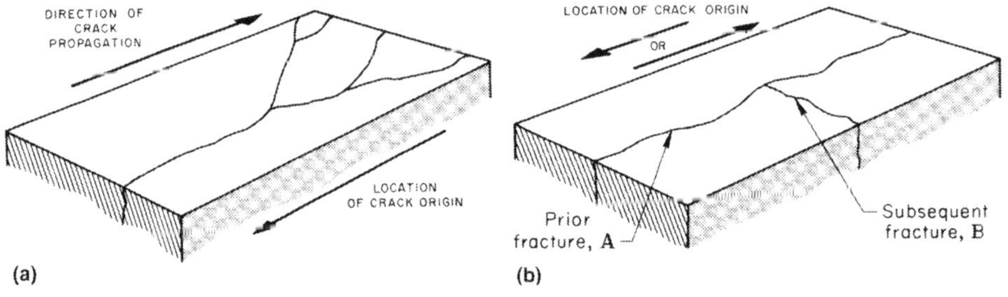

Fig. 1 General features to locate origin from crack paths. (a) Branching and (b) sequencing of cracking by the T-junction procedure, where fracture A precedes and arrests fracture B. Source: Ref 1

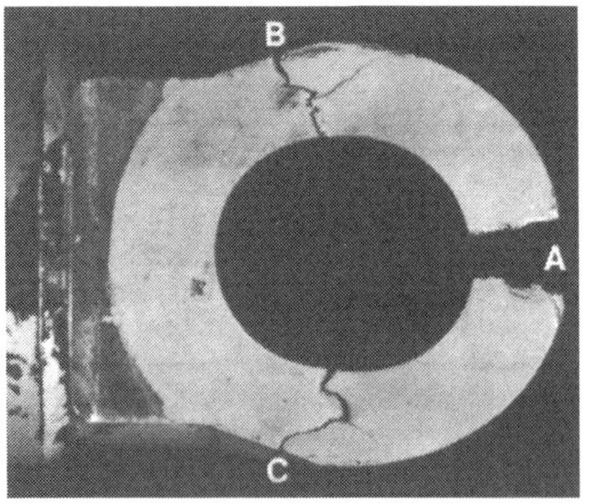

Fig. 2 Fractured lug, part of a pin-joint assembly, showing sequence of fracture. Fracture A preceded fractures B and C. Source: Ref 1

sembly; failure occurred when the pin broke out of the lug. With the broken pieces of the lug fitted together, it is apparent from the deformation that fracture A must have preceded fractures B and C. However, parts will not fit well together because of plastic deformation that occurred before or during the fracture process.

Preliminary Examination

The failed part, including all its fragments, should be subjected to a thorough visual examination before any cleaning is undertaken. Often, soils and debris found on the part provide useful evidence in establishing the cause of failure or in determining a sequence of events leading to the failure. For example, traces of paint or corrosion found on a portion of a fracture surface may provide evidence that the crack was present in the surface for some time before complete fracture occurred. Such evidence should be recorded photographically.

Visual Inspection. The preliminary examination should begin with an unaided visual inspection. The unaided eye has exceptional depth of focus and the ability to examine large areas rapidly and to detect changes of color and texture. Some of these advantages are lost when any optical or electron-optical device is used. Particular attention should be given to the surfaces of fractures and to the paths of cracks. The significance of any indications of abnormal conditions or abuse in service should be observed and assessed, and a general assessment of the basic design and workmanship of the part should also be made. Each important feature, including dimensions, should be recorded, either in writing or by sketches or photographs.

It cannot be emphasized too strongly that the examination should be performed as carefully as possible, because clues to the cause of breakdown are often present but may be missed if the observer is not vigilant. Inspection of the topographic features of the failed component should start with an unaided visual examination and proceed to higher and higher magnification. A magnifying glass followed by a low-power microscope is an invaluable aid in detection of small details of the failed part.

Examination and Photography of the Damaged/Failed Part or Sample. The next step should be a preliminary examination and general photography of the entire part and damaged or failed regions. Where fractures are involved, the entire fractured part, including broken pieces, should be examined and photographed to record their size and condition and to show how the fracture is related to the components. This should be followed by careful examination of the fracture. The examination should begin with the use of direct lighting and proceed at various angles of oblique lighting to delineate and emphasize fracture characteristics. This should also assist in determining which areas of the fracture are of prime interest and which magnifications will be possible (for a given picture size) to bring out fine details. When this evaluation has been completed, it is appropriate to proceed with photography of the fracture, recording what each photograph shows, its magnification, and how it relates to the other photographs.

Nondestructive Examination

Although often used as quality-control tools, several nondestructive tests are useful in failure analysis: magnetic-particle inspection of ferrous metals, liquid penetrant inspection, ultrasonic inspection, and sometimes eddy-current inspection. All these tests are used to detect surface cracks and discontinuities. Radiography is used mainly for internal examination. A photographic record of the results of nondestructive inspection is a necessary part of record keeping in the investigation.

Magnetic-particle inspection uses magnetic fields to locate surface and subsurface discontinuities in ferromagnetic materials. When the material or part to be tested is magnetized, discontinuities that generally lie transverse to the direction of the magnetic field will cause a leakage field to be formed at and above the surface of the part. This leakage field, and therefore the presence of the discontinuity, is detected by means of fine ferromagnetic particles applied over the surface, some of which are gathered and held by the leakage field. The magnetically held collection of particles forms an outline of the discontinuity and indicates its size, shape, and extent. Frequently, a fluorescent material is combined with the particles so that discontinuities can be detected visually under ultraviolet light. This method reveals surface cracks that are not visible to the naked eye.

Liquid penetrant inspection is used to detect surface flaws in materials. It is used mainly, but not exclusively, with nonmagnetic materials, on which magnetic-particle inspection cannot be used. This technique involves the spreading of a liquid penetrant on the sample. Liquid penetrants can seep into small cracks and flaws (as fine as 1 μm) in the surface of the sample by capillary action. The excess liquid is wiped from the surface, and a developer is applied that causes the liquid to be drawn from the cracks or flaws that are open at the surface. The liquid itself is usually a very bright color or contains fluorescent particles that, under ultraviolet light, cause discontinuities in the material to stand out. The main advantages of the liquid penetrant method are its ability to be used on nonmagnetic materials, its low cost, its portability, and the ease with which results can be interpreted. The principal limitations of the liquid penetrant method include:

- Discontinuities must be open to the surface.
- Testpieces must be cleaned before and after testing because the liquid penetrant may corrode the metal.
- Surface films may prevent detection of discontinuities.
- Penetrant may be a source of contamination that masks results in subsequent chemical analysis of fracture surfaces.

- The process is generally not suited to inspection of low-density powder metallurgy parts or other porous materials.

Ultrasonic inspection methods depend on sound waves of very high frequency being transmitted through metal and reflected at any boundary, such as a metal/air boundary at the surface of the metal or a metal/crack boundary at a discontinuity within the part or component. High-frequency sound waves can detect small irregularities, but they are easily absorbed, particularly by coarse-grained materials. The application of ultrasonic testing is limited in failure analysis because accurate interpretations depend on reference standards to isolate the variables. In some instances, ultrasonic testing has proved to be a useful tool in failure analysis, particularly in the investigation of large castings and forgings. Cracks, laminations, shrinkage cavities, bursts, flakes, pores, disbonds, and other discontinuities that produce reflective interfaces can easily be detected. Inclusions and other inhomogeneities can also be detected by causing partial reflection or scattering of the ultrasonic waves or by producing some other detectable effect on the ultrasonic waves. The disadvantages of ultrasonic inspection include:

- Manual operation requires careful attention by experienced technicians.
- Extensive technical knowledge is required for the development of inspection procedures.
- Parts that are rough, irregular in shape, very small or thin, or not homogeneous are difficult to inspect.
- Discontinuities that are present in a shallow layer immediately beneath the surface may not be detectable.
- Couplants are needed to provide effective transfer of ultrasonic wave energy between transducers and parts being inspected.
- Reference standards are needed, both for calibrating the equipment and for characterizing flaws.

Radiography uses x-rays or gamma rays, which are directed through the sample to a photographic film. After the film has been developed, it can be examined by placing it in front of a light source. The intensity of the light passing through the film will be proportional to the density of the sample and the path length of the radiation. Thus, lighter areas on the plate correspond to the denser areas of the sample, whereas darker areas indicate a crack or defect running in the direction of the incident beam. The main advantages of radiography are its ability to detect internal discontinuities and to provide permanent photographic records. However, certain types of flaws are difficult to detect by radiography. Laminar defects, such as cracks, present problems unless they are essentially parallel to the radiation beam. Tight, meandering cracks in thick sections usually cannot be detected even when properly oriented. Minute discontinuities, such as inclusions in wrought material, flakes, microporosity, and microfissures, cannot be detected unless they are sufficiently segregated to yield a detectable gross effect. Laminations normally are not detectable by radiography because of their unfavorable orientation, usually parallel to the surface. Laminations seldom yield differences in absorption that enable laminated areas to be distinguished from lamination-free areas.

Eddy-current inspection can be used on all materials that conduct electricity. If a coil conducting an alternating current is placed around or near the surface of the sample, it will set up eddy currents within the material by electromagnetic induction. These eddy currents affect the impedance in the exciting coil or any other pickup coil that is nearby. Cracks or flaws within the sample will cause distortions in the eddy currents, which in turn cause distortion in the impedance of the coil. The resulting change in impedance can be detected by attaching the appropriate electrical circuits and a meter. Flaws or cracks will show up as some deflection or fluctuation on the meter. The advantages of electromagnetic inspection include:

- Both surface and subsurface defects are detectable.
- No special operator skills are required.
- The process is adaptable to continuous monitoring.
- The process may be substantially automated and is capable of high speeds.
- No probe contact is needed.

Limitations of electromagnetic inspection include:

- Depth of penetration is shallow.
- Materials to be inspected must be electrically conductive.
- Indications are influenced by more than one variable.
- Reference standards are required.

Residual-Stress Analysis. X-ray diffraction is the most common method for direct, nondestructive measurement of residual (internal) stresses in metals. Stresses are determined by measuring the submicroscopic distortion of crystalline lattice structures by tensile or compressive residual stresses. However, it should be pointed out that measurement of residual stresses near fractures or cracks may be erroneous because the residual stresses may have already been relieved by the fracture and cracks. Testing of undamaged similar, or exemplar, parts is frequently used as the only alternative in order to understand the residual-stress system in the failed part prior to failure.

Acoustic emission inspection detects and analyzes minute acoustic emission signals generated by discontinuities in materials under applied stress. Proper analysis of these signals can provide information concerning the location and structural significance of the detected discontinuities. Some of the significant applications of acoustic emission inspection are:

- Continuous surveillance of pressure vessels and nuclear primary pressure boundaries for the detection and location of active flaws
- Detection of incipient fatigue fracture in aircraft structures
- Monitoring of both fusion and resistance weldments during welding and cooling
- Determination of the onset of stress-corrosion cracking and hydrogen damage in susceptible structures
- Use as a study tool for the investigation of fracture mechanisms and of behavior of materials
- Periodic inspection of tanks and aerial device booms made of composite materials

This type of data may be useful background information in a failure analysis, or the technique may be used in evaluation of stress effects. Sources of acoustic emission that generate stress waves in material include local dynamic movements, such as the initiation and propagation of cracks, twinning, slip, sudden reorientation of grain boundaries, and bubble formation during boiling. This energy may originate from stored elastic energy, as in crack propagation, or from stored chemical free energy, as in a phase transformation.

Experimental stress analysis can be done by several methods, all of which may be valuable in determining machine loads and component stresses that can cause failures. Stress coatings can be used effectively for locating small areas of high strains, determining the directions of the principal strains, and measuring the approximate magnitude of tensile and compressive strains. Gages can then be placed at the high-strain areas and in the principal strain directions to measure the strain accurately on gage lengths 0.5 to 150 mm (0.02 to 6 in.). Although there are many mechanical, optical, and electrical devices capable of accurate strain measurements, the bonded electrical resistance strain gage has become the standard tool for general laboratory and field use. Photoelastic coatings have also been used for laboratory stress measurements. For this technique, a birefringent coating of controlled thickness is bonded to the testpiece with a reflective cement. Optical analysis is similar to conventional analysis but requires special equipment. The analysis may be recorded on color film with a single-frame or movie camera.

Fracture Origin

Establishing the origin of a fracture is essential in failure analysis, and the location of the origin determines which measures should be taken to prevent a repetition of the fracture. The fracture surface characteristics that show the direction of crack propagation (and conversely, the direction toward the origin) include features such as chevron marks, crack branching, and river patterns. Features that help identify the crack origin include concentric fibrous marks, radial marks, and beach marks. By a study of these features, crack progress can be traced back to the point of origin, and then it can be ascertained whether the crack was initiated by an inclusion, a porous region, a segregated phase, a corrosion pit, a machined notch, a forging lap, a nick, a mar, or another type of discontinuity, or was simply the result of overloading. However, the time employed in ascertaining all the circumstances of a failure is extremely important. When a broken component is received for examination, the investigator is sometimes inclined to prepare specimens immediately without devising an investigative procedure. To proceed without forethought may destroy important evidence and waste time. Some of the questions that should be raised concerning the nature, history, functions, and properties of the fractured part, and the manner in which it interacts with other parts, include:

- *Loading.* Were the nature, rate, and magnitude of the applied load correctly anticipated in the design of the part? Were repeated or cyclic loadings involved? What was the direction of the principal stress relative to the shape of the part? Were residual stresses present to an undesirable degree?
- *Material.* Was the recommended alloy used? Were its mechanical properties at the level expected? Were surface or internal discontinuities present that could have contributed to failure? Did the microstructure conform to that prescribed?
- *Shape.* Did the part comply with all pertinent dimensional requirements of the specification? Did the part have sufficient section thickness to prevent local overloading? Were fillets formed with sufficiently large radii? Were there adequate clearances between interacting parts? Were any of the contours deformed during service? Was there evidence of mechanical surface damage?
- *Environment.* Was the part exposed to a corrosive environment or to excessively high or low temperatures? Was the surface of the part suitably protected? Were the properties of the part altered by the exposure? Was there interaction (for example, galvanic) between the material of the part and that of adjacent components?

Examination of a fracture begins with visual scrutiny, which establishes:

- Whether there is gross evidence of mechanical abuse
- Whether there are indications of excessive corrosion
- Whether the part is deformed
- Whether there are obvious secondary fractures
- Whether the origin of the crack can be readily identified
- Whether the direction of crack propagation can be easily recognized

Often, it is helpful to have an undamaged part of the same design as the fractured part available during this portion of the examination. The findings of this scrutiny will permit many deductions to be drawn concerning the service conditions existing prior to and at the time of fracture. These findings can then be extended by an examination of the fracture surface at low magnification with a stereomicroscope and then at high magnification by electron microscopy, metallography (occasionally), or some combination of these examination techniques. A survey at low magnification is important for identification of those areas that need further inspection at high magnification. The salient features are recorded in fractographs of appropriate magnification for report purposes and for future reference should subsequent handling or sectioning destroy evidence needed for failure analysis.

Fracture of a part in service is often intimately associated with the type of environment to which the part was exposed. Active chemical environments include water, salt air, salt water, acid solutions, alkaline solutions, molten metal, and even solid metal. Thermal environments that affect metal properties and fracture include exposure to low (cryogenic, for example) and high temperatures.

Selection and Preservation of Fracture Surfaces

The proper selection, preservation, and cleaning of the fracture surfaces is vital to prevent important evidence from being destroyed or obscured. Fracture surfaces may suffer either mechanical or chemical damage. Mechanical damage may arise from several sources, including the striking of the surface of the fracture by other objects. This can occur during actual fracture in service or when removing or transporting a fractured part for analysis.

Protection. Usually, the surface of a fracture can be protected during shipment by a cloth or cotton covering, but this may remove some loosely adhering material that might contain the primary clue to the cause of the fracture. Touching or rubbing the surface of a fracture with the fingers should definitely be avoided. Also, no attempt should be made to fit together the sections of a fractured part by placing them in contact. This generally accomplishes nothing and almost always causes damage to the fracture surface. The use of corrosion-inhibiting paper to package samples should be considered. Chemical (corrosion) damage to a fracture specimen can be prevented in several ways. For instance, because the identification of foreign material present on a fracture surface may be important in the overall determination of the cause of the fracture, many laboratories prefer not to use corrosion-preventive coatings on a fracture specimen. When possible, it is best to dry the fracture specimen, preferably using a jet of dry, compressed air (which will also blow extraneous foreign material from the surface), and then to place it in a dessicator or pack it with a suitable dessicant. However, clean, fresh fracture surfaces should be coated when they cannot be protected from the elements. After failure of large structures, several days may be required to remove critical specimens, and so coating the fracture surfaces would be the prudent decision.

Cleaning of fractured surfaces should be avoided in general but must be done for scanning electron microscopy (SEM) examination and often to reveal macroscopic fractographic features. Cleaning should proceed in stages using the least aggressive procedure first, then proceeding to more aggressive procedures if needed. Washing the fracture surface with water should especially be avoided. However, specimens contaminated with seawater or with fire-extinguishing fluids require thorough washing, usually with water, followed by a rinse with acetone or alcohol before storage in a dessicator or coating with a dessicant. Sometimes, cleaning may also be required for removal of obliterating debris and dirt, or to prepare the fracture surface for SEM examination. Other acceptable cleaning procedures include use of a dry air blast or of a soft-hair artist's brush; treating with inorganic solvents, either by immersion or by jet; treating with mild acid or alkaline solutions (depending on the metal) that will attack deposits but to which the base metal is essentially inert; ultrasonic cleaning; and application and stripping of plastic replicas.

Cleaning with cellulose acetate tape is one of the most widely used methods, particularly when the surface of a fracture has been affected by corrosion. A strip of acetate approximately 0.1 mm (0.005 in.) thick and of suitable size is softened by immersion in acetone and placed on the fracture surface. The initial strip is backed by a piece of unsoftened acetate, and then the replica is pressed hard onto the surface of the fracture using a finger. The drying time will depend on the extent to which the replicating material was softened, and this in turn will be governed by the texture of the surface of the fracture. Drying times of not less than 15 to 30 min are recommended. The dry replica is lifted from the fracture, using a scalpel or tweezers. The replicating procedure must be repeated several times if the fracture is badly contaminated. When a clean and uncontaminated replica is obtained, the process is complete. An advantage of this method is that the debris removed from the fracture is preserved for any subsequent examination that may be necessary for identification by x-ray or electron diffraction techniques. To be complete, the analyst should filter solvents used for cleaning to recapture insoluble particulates.

Sectioning. Because examination tools, including hardness testers and optical and electron microscopes, are limited as to the size of specimen they can accept, it is often necessary to remove from a failed component a fracture-containing portion or section that is of a size convenient to handle and examine. This is a destructive process and may spoil evidence in potential litigation cases. Before cutting or sectioning, the fracture area should be carefully protected. All cutting should be done so that surfaces of fractures and areas adjacent to them are not damaged or altered; this includes keeping the fracture surface dry, whenever possible. For large parts, the common method of removing specimens is by flame cutting. Cutting must be done at a sufficient distance from the fracture site so that the microstructure of the metal underlying the surface of the fracture is not altered by the heat of the flame, and so that none of the molten metal from flame cutting is deposited on the surface of the fracture. Heat from any source can affect metal properties and microstructures during cutting. Therefore, dry abrasive cutoff wheels should never be used near critical surfaces that will be examined microscopically. Therefore, sectioning should be performed with jewelers' saws; precision diamond-edged, thin cutoff wheels; hacksaws; band saws; or soft abrasive cutoff wheels flooded with water-based soluble oil solution to keep metal surfaces cool and corrosion free. Dry cutting with an air-driven abrasive disk may also be used with care to remove small specimens from large parts if kept cool, along with coating the fracture surface for protection.

Secondary Cracks. When the primary fracture has been damaged or corroded to such a degree that most of the information relevant to the cause of the failure is obliterated, it is desirable to open any secondary cracks to expose their fracture surfaces for examination and study. These cracks may provide more information than the primary fracture. In opening cracks for examination, care must be exercised to prevent damage, primarily mechanical, to the surface of the fracture. This can usually be accomplished if opening is done in such a way that the two surfaces of the fracture are moved in opposite directions, normal to the fracture plane. Generally, a saw cut can be made from the back of the fractured part to a point near the tip of the crack, using extreme care to avoid actually reaching the tip of the crack. This saw cut will reduce the amount of solid metal that must be broken. The final breaking of the specimen can be done in several ways:

- By clamping the two sides of the fractured part in a tensile testing machine, if the shape permits, and pulling
- By placing the specimen in a vise and bending one half away from the other by striking it with a hammer in a manner that will avoid damage to the surfaces of the crack
- By gripping the halves of the fracture in pliers or vise grips and bending or pulling them apart

Cooling the part with liquid nitrogen often reduces the force and plastic deformation necessary to fracture the part. Fortunately, there is little confusion during subsequent examination as to which part of the fracture surface was obtained in opening the crack.

It is recommended that both the crack separation and the visible crack length be measured prior to opening. The analyst may have to use dye penetrant or other nondestructive evaluation techniques to actually see the length of a tightly closed crack. Often, the amount of strain that occurred in the specimen can be determined from a measurement of the separation between the adjacent halves of a fracture. This should be done before preparation for opening a secondary crack has begun. The lengths of cracks may also be important for analyses of fatigue fractures or for consideration for the application of fracture mechanics.

Macroscopic Examination of Fracture Surfaces

One very important part of any failure analysis is the macroscopic examination of fracture surfaces. Performed at magnifications from 1 to 50 or 100×, it may be conducted by the unaided eye, a hand lens or magnifier, a low-power stereoscopic microscope, or an SEM. Macroscopic photography of up to 20× magnification requires a high-quality camera and special lenses; alternatively, a large magnifying glass may be used to enlarge a specific area in the photo, such as a crack or other small detail. A metallographic microscope with macro-objectives and lights may be used for somewhat higher magnifications. However, the depth of field becomes extremely limited with light optics. For a much greater depth of field, an SEM may be used for low-magnification photography as well as higher-magnification work. Stereo or three-dimensional photographs may also be made to reveal the topographic features of a fracture or other surface. Frequently, a specimen may be too large or too heavy for the stage of the metallograph or the chamber of an SEM, and cutting or sectioning the specimen may be difficult or not allowed because of legal limitations at the time. In these instances, excellent results can be achieved by examining and, where appropriate, photographing replicas made by the method for cleaning fractures. These replicas can be coated with a thin layer (approximately 20 nm, or 2×10^{-8} m, thick) of vacuum-deposited gold or aluminum to improve their reflectivity, or they may be shadowed at an angle to increase the contrast of fine detail. The replicas may be examined by incident light or transmitted light microscopy. Because they are electrically conductive, the coated replicas may also be examined with an SEM. The amount of information that can be obtained from examination of a fracture surface at low-power magnification is extensive. A careful scan of the exterior surface of the part in the area adjacent to the fracture is required to determine whether specific stress raisers are present of a type that could have initiated the fracture. If any marks possess sharp reentrant angles, they constitute sites of stress concentration, a frequent cause of crack initiation. In this situation, the obvious remedy is more careful handling procedures and better inspection. Tool marks are another source of stress concentration. A fillet that has too small a radius, even though the surface of the fillet may be an excellent example of high-grade machining, is a recognized initiation site for fatigue cracks. Sharp-bottomed tool marks can initiate fatigue fractures even though the general contour of the area has a generous radius.

The shape, size, and cross section of a specimen or structural component can have a large effect on both the macroscopic and the microscopic appearance of the fracture surface, especially when pronounced stress raisers are present. Holes, corners, notches, machining marks, and, most of all, preexisting cracks actively influence fracture appearance. Pronounced stress raisers are more likely to be contained in a large part than in a small part, because large parts have greater volumes and surface areas.

The orientation of the fracture surfaces must be consistent with the proposed mode of failure and the known loads on the failed part. Failure in monotonic tension produces a flat (square) fracture normal (perpendicular) to the maximum tensile stress and frequently a slant (shear) fracture at approximately 45°. This 45° slant fracture is often called a shear lip. Many fractures are flat at the center but surrounded by a "picture frame" of slant fracture. An example of this behavior is to be found in the familiar cup-and-cone fracture of a round tensile test bar. In thin sheets, tube walls, and small-diameter rods, a slant shear fracture may occur because through-the-thickness stresses are minimized; that is, even though there may be a plane-strain condition, there may be minimal triaxial (hydrostatic) tensile stresses.

Macroscopic examination can usually determine the direction of crack growth and hence the origin of failure. With brittle, flat fractures, determination depends largely on the fracture surface showing a radial fanlike pattern of the type shown in Fig. 3. Cracks propagate parallel to shear lips if they are present. Where fracture surfaces show both flat and slant surfaces, this can be the terminal end of a fast-moving brittle fracture where the crack speed has slowed significantly.

Crack extension can relax the stress so that final fracture occurs by slant shear fracture. Conversely, if a fracture has begun at a free surface, the fracture origin area is usually characterized by a total absence of slant fracture or shear lip.

Low-power examinations of fracture surfaces often reveal regions having a texture different from the region of final fracture. Fatigue, stress-corrosion, and hydrogen embrittlement fractures may also show these differences because the final failure is due to overload after the cross section is reduced by one of the crack initiation modes. Figure 4(a) shows the fracture surface of a steel tube and is an excellent example of the type of information that can be obtained by macroscopic examination. The V-shaped chevron marks and fanlike marks clearly indicate that the fracture origin is at the point marked by the arrow. This region, unlike the rest of the fracture, has no shear lip. The flat fracture surface suggests that the stress causing the failure was tension parallel to the length of the tube. The origin of the fracture, as seen at higher magnification in Fig. 4(b), shows

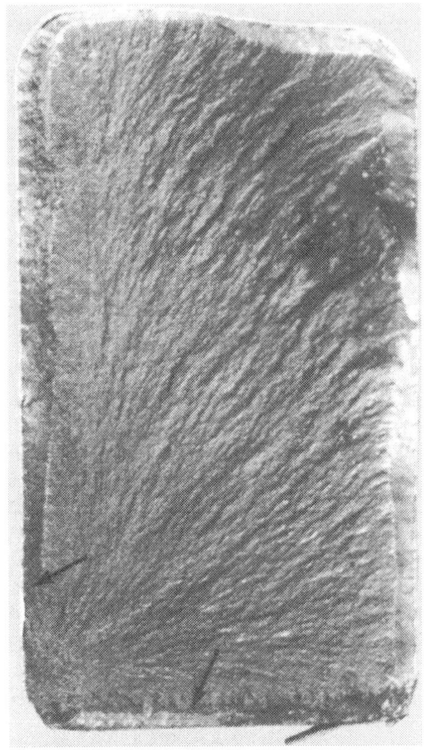

Fig. 3 Surface of a fatigue fracture in a 4330V steel part. Chevron marks point to origin of fatigue in lower left corner. Arrows identify shear rupture along the periphery. Source: Ref 1

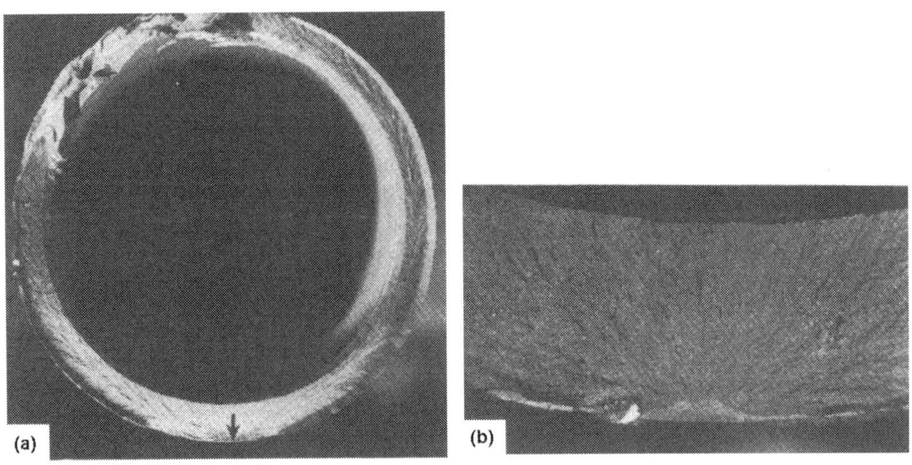

Fig. 4 Fracture of a steel tube. (a) Fracture surface at approximately actual size, showing point of crack initiation (at arrow), chevron and fanlike marks, and development of shear lips. (b) Fracture-origin area at 5×; note that fracture nuclei differ in texture from the main fracture surface. Source: Ref 1

several small fracture origins having a texture different from that of the remainder of the fracture surface.

Microscopic Examination of Fracture Surfaces

Microscopic examination of fracture surfaces is typically done with an SEM. An SEM has an advantage over light microscopy because of the large depth of field and very high magnifications attainable, typically 5000 to 10,000×. In addition, SEMs are often equipped with microanalytical capabilities, for example, energy-dispersive x-ray spectroscopes. Chemical analysis can be helpful in confirming the chemistry of microstructural features that may be confused with fracture features.

The primary limitation of SEM analysis is sample size. An SEM analysis must be conducted in a vacuum, so the sample must be put into a chamber that typically holds a sample less than 20 cm (8 in.) in diameter. Although there are some fracture surface features that are commonly associated with particular failure modes, the novice failure analyst must be very careful in fractographic analyses. Some of the more classic examples of fracture surface topography that indicates a fracture mode are:

- Dimpled rupture typical of overstress failures of ductile metals and alloys
- Cleavage facets, typical of transgranular brittle fracture of body-centered cubic and hexagonal close-packed metals and alloys
- Brittle intergranular fracture typical of temper-embrittled steel, where fracture is due to segregation of an embrittling species to grain boundaries (such as oxygen in iron or nickel), due to intergranular stress-corrosion cracking or due to hydrogen embrittlement
- Stage II striations, typical of some (but not all) fatigue failures

Stress Analysis

It is sometimes quite apparent that an excessively high load or stress level was the direct cause or contributed significantly to the failure. Even so, an accurate stress analysis of the magnitude and type (axial, torsion, bending) of stress is required to substantiate the role of stress. In other failure analyses, the analyst may have strong evidence that the cause of a failure is related to excessively high static stresses (or cyclic stresses in the case of fatigue). In these cases, an analysis of the stress during normal operation (or abnormal operation if identified) must be conducted. Analytical, closed-form calculations based on engineering mechanics are often used by designers to predict stress levels in the early design stages. This method of using known machine design and structural formulas to predict the stress under a given load is also helpful to the failure analyst, especially in cases where this step may not have been used in the original design of the part.

It is not uncommon to find products that have no record of any stress analysis in product design. Even if such calculations were made, the failure analyst may not have access to them. The analyst must answer the questions "Was the component sized properly by the design stress analysis? Did the material have the properties assumed in the design? Did the part fail in a manner consistent with that assumed in design, or did it fail in a way not anticipated in the original design?"

In cases where unusual or abnormal loading is suspected, direct calculation of stresses will fall short and predict incorrect stress levels. In these cases, experimental stress analysis is used for determining machine loads and component stresses. This technique normally involves attachment of strain gages to similar parts in critical areas or typical areas where the failure has occurred. The strain gages are connected to a monitoring device either directly with wires or indirectly by radio signals for monitoring moving or rotating parts. In this way, the actual dynamic stresses can be determined. For products with very complex shapes and high thermal gradients, a finite-element analysis (FEA) may be required to estimate the level of stress that most likely existed in the failed component. These analyses can stand alone or can be used to help select critical locations for strain gage attachment. Finite-element analyses can be time-consuming and expensive, but they are necessary for an accurate assessment of stress levels in areas of complex geometry of some components. This type of analysis is almost essential for determining stresses caused by thermal gradients, such as those found in welding.

Overload failures are often a result of improper design or improper operation. A design analysis is essential in determining which of these is the root cause. Sometimes, improper design is a result of incorrect information passed to the designer. In these cases, a failure is the only indication that the wrong input was used for the design. This is also true for fatigue failures. For proper design of rotating or moving parts, a detailed stress analysis is essential, because it is much more difficult to predict dynamic stresses than static stresses.

Fracture Modes

Because the initial steps in failure analysis of a fracture involve visual and macroscopic observation, the first impressions should be based on obvious visual evidence. The simplest and most important observations relate to deformation: Was the metal obviously deformed? If it was deformed prior to fracture, yielding and fracture have occurred due to one or more gross overloads. It is predominantly a ductile fracture or a very high-stress, low-cycle fatigue fracture, as can be demonstrated by repeated manual bending of a paper clip or wire coat hanger. The deformation is directly related to the type of stress causing fracture: tension (stretched), bending (bent), torsion (twisted), or compression (shortened or buckled), or a combination of these stress types.

The absence of gross deformation of the failed part indicates that the fracture is predominantly brittle. A brittle fracture should not be confused with brittle material. The shape or geometry of the part made from a ductile metal can result in an overload failure with little overall shape change, or a failure mechanism can operate to start and grow a crack, such as a fatigue crack or stress-corrosion crack. When such a crack grows to the point that the remaining cross-sectional area of the part is overloaded by the normal loads, the final overload failure has little macroscale deformation associated with it. Thus, on a macroscale, the failure of a ductile metal can appear brittle. Of course, overload failures of brittle material always appear brittle on a macroscale.

It is usually more difficult to analyze a brittle fracture because there are a large number of possible mechanisms that can cause fracture with little or no obvious deformation. For single overload fractures, these include such factors as stress concentrations, low temperatures, high rates of loading, high metal strength and hardness, stress-corrosion cracking, hydrogen embrittlement, temper embrittlement, large section size, and others. For fatigue fractures, causative factors can include stress concentrations, tensile residual stresses, large stress amplitudes, large numbers of load applications, corrosive environments, high temperatures, low metal strength and hardness, wear, and others.

From this discussion, it should become clear that proper failure analysis is not simple but can become exceedingly complex, requiring considerable thought, examination, questioning, and reference to other sources of information in the literature. However, identifying the failure mode is the key step in a failure analysis, and it is the essential part of determining the root cause.

Ductile Fracture. Overload fractures of many metals and alloys occurs by ductile fracture. Overloading in tension is perhaps the least complex of the overload fractures, although essentially the same processes operate in bending and torsion as well as under the complex states of stress that may have produced a given service failure. The classic example of ductile failure is a tensile test. In this fracture process, considerable elongation, that is, deformation, takes place before the geometric instability, necking, begins. Even after the deformation is localized at the neck, significant deformation occurs at the neck before fracture. After the neck forms, the curvature of the neck creates a region of tensile hydrostatic (or triaxial) stress. This leads to initiation of an internal crack near the center of the necked region. In commercial-grade alloys, discontinuities such as inclusions or second-phase particles are sources of early void formation by separation of the matrix and the particle. Some of these voids coalesce to develop a crack, which is perpendicular to the tensile axis. The crack spreads until the state of stress, ductility of the metal, and flow condition reach a condition that favors a shear displacement. The crack path then shifts to a maximum shear plane, which is at an angle to the tensile axis (close to 45° in cylindrical specimens). Sometimes, this shear lip forms only on one side of the initial flat crack. When this occurs, the resulting fracture surface has a macroscopic appearance known as cup-and-cone. For brittle materials, the majority of the fracture surface is perpendicular to the tensile axis, with little or no fracture surface lying on a plane of shear.

Ductile failures in biaxially loaded sheet and plate structures often consist entirely of a shear lip. Pipe and pressure vessels are examples of biaxially stressed components. Often, failures in these components may first appear brittle with limited ductility; however, a close inspection usually reveals some general thickness reduction but no necking at the fracture surface. High-magnification examination of ductile fracture surfaces usually reveals dimples, which tend to be equiaxed when fractures occur under tensile load. Slant fractures or ductile fracture on planes of high shear stress generate elongated dimples. Ductile fractures (i.e., those with macroscopic deformation) are usually transgranular.

Brittle Fracture. There are two general types of brittle fracture caused by a single overload: transgranular cleavage and intergranular separation. Each has distinct features that make identification relatively simple. Transgranular cleavage can occur in body-centered cubic metals and their alloys (for example, ferritic steels, iron, tungsten, molybdenum, and chromium) and some hexagonal close-packed metals (for example, zinc, magnesium, and beryllium). Face-centered cubic metals and alloys (such as aluminum and austenitic stainless steels) are usually regarded as immune from this fracture mechanism.

Iron and low-carbon steels show a ductile-to-brittle transition with decreasing temperature that arises from a strong dependence of the yield stress on temperature. Brittle fracture of normally ductile metals depends on several physical factors, including specimen shape and size, temperature, and strain rate. Thus, a component or structure that has given satisfactory service may fracture unexpectedly; the catastrophic brittle fracture of ships in heavy seas and the failure of bridges on unusually cold days are examples. Metallurgical changes, especially strain aging, may cause the brittle fracture of such items as crane hooks and chain links after long periods of satisfactory operation.

Cleavage fracture is not difficult to diagnose because the fracture path is, by definition, crystallographic. In polycrystalline specimens, this often produces a pattern of brightly reflecting crystal facets, and such fractures are often described as crystalline (improperly because metals are, by definition, crystalline). The general plane of fracture is approximately normal (perpendicular) to the axis of maximum tensile stress, and a shear lip is often present as a "picture frame" around the fracture. The local absence of a shear lip or slant fracture suggests a possible location for fracture initiation, because shear lips form during the final stages of the fracture process. The fractography of cleavage fracture in low-carbon steels, iron, and other single-phase, body-centered cubic metals and alloys is fairly well established. Polycrystalline specimens contain numerous fan-shaped cleavage plateaus. The most characteristic feature of these plateaus is the presence of a pattern of river marks, which consist of cleavage steps or tear ridges and indicate the local direction of crack growth. The rule is that, if the tributaries of the "river lines" are regarded as flowing into the main stream, then the direction of crack growth is downstream. This is in contrast to macroscopic chevron marks, where the direction of crack growth, using the river analogy, would be upstream.

Other fractographic features that may be observed include the presence of cleavage on conjugate planes, tear ridges, ductile tears joining cleavage planes at different levels, and tongues, which result from fracture in mechanical twins formed ahead of the advancing crack. Cleavage fracture in pearlitic and martensitic steels is less easily interpreted because microstructure tends to modify the fracture surface. In fact, cleavage fracture surfaces of pearlitic steel have characteristics similar to fatigue striations, so one must be careful not to confuse the fracture mode.

Intergranular fracture can usually be recognized, but determining the primary cause of the fracture may be difficult. Fractographic and microscopic examination can readily identify the presence of second-phase particles at grain boundaries. Unfortunately, the segregation of a layer a few atoms thick of some element or compound that produces intergranular fracture often cannot be detected by fractography. Auger analysis and sometimes energy-dispersive x-ray spectrometry are useful for very thin layers. Some causes of intergranular brittle fracture are given as follows, but the list is not exhaustive. However, it does indicate some of the possibilities that must be considered, and either eliminated or confirmed, as contributing to the fracture:

- The presence at a grain boundary of a large area of second-phase particles, such as carbides in Fe-Ni-Cr alloys or MnS particles in an overheated steel
- Segregation of a specific element or compound to a grain boundary, where a layer a few atoms thick is sufficient to cause embrittlement. Embrittlement caused by the presence of oxygen in high-purity iron, oxygen in nickel, or antimony in copper and temper embrittlement of certain steels are examples of intergranular embrittlement where detection of a second phase at grain boundaries is difficult.

The conditions under which a progressively growing crack may follow an intergranular path before final fracture occurs include stress-corrosion cracking, embrittlement by liquid metals, hydrogen embrittlement, and creep- and stress-rupture failures.

Fracture Mechanics Applied to Failure Analysis

The application of fracture mechanics is often pertinent to the investigation of failures, as well as to the formulation of preventive measures. In general, there are two types of conditions that may lead to structural failure:

- Net-section instability, where the overall structural cross section can no longer support the applied load
- The critical flaw size (a_c) is exceeded by some preexisting discontinuity, or when subcritical cracking mechanisms (for example, fatigue, stress-corrosion cracking, creep) reach the critical crack size.

Failures due to net-section instability typically occur when a damage process such as corrosion or wear reduces the thickness of a structural section. This type of failure can be evaluated by traditional stress analysis or FEA, which are effective methods in evaluating the effects of loading and geometric conditions on the distribution of stress and strain in a body or structural system.

However, stress analyses by traditional methods or FEA do not easily account for crack propagation from preexisting cracks or sharp discontinuities in the material. When a preexisting crack or discontinuity is present, the concentration of stresses at the crack tip becomes asymptotic (infinite) when using the conventional theory of elasticity. In this regard, fracture mechanics is a useful tool, because it is a method that quantifies stresses at a crack tip in terms of a stress-intensity parameter (K).

Linear elastic fracture mechanics is a useful tool in failure analysis, because many (and perhaps most) structural failures occur by the combined processes of crack initiation followed by subcritical crack growth mechanism (for example, fatigue, stress corrosion, creep) until a critical crack (a_c) is reached. In this regard, fracture mechanics is an effective tool for evaluating critical flaw size (a_c) that leads to rapid unstable fracture and can help answer questions during a failure analysis, such as:

- Where should one look for the transition from subcritical crack growth to unstable rapid fracture?
- What was the load on the component at the time of failure?
- Was the correct material used, and was manufacturing/processing sound?
- Was the part designed properly?
- Did the environment influence the failure?

Of course, many situations may involve thin sections and/or very ductile materials, where the conditional constraint for linear elastic fracture mechanics may not apply. In this case, the measure of toughness is plane-stress fracture toughness (K_c) and requires the use of elastic-plastic fracture mechanics, because the process of unstable fracture involves some plasticity. Plane-stress fracture toughness (K_c) is higher than plane-strain fracture toughness (K_{Ic}), but when thinner sections and more ductile materials are involved, net-section instability becomes a factor.

Metallographic Examination

Metallographic examination of polished, and of polished and etched, sections by optical microscopy and by electron-optical techniques is a vital part of failure investigation and should be carried out as a routine procedure when possible. Metallographic examination provides the investigator with a good indication of the class of material involved and its structure. If abnormalities are present, these may be associated with undesirable characteristics that caused an early failure. It is sometimes possible to relate them to an unsuitable composition or to the effects of service, such as aging in low-carbon steel that has caused precipitation of iron nitride or gassing in copper. Microstructural examination may also provide information as to the method of manufacture of the part under investigation. It can reveal the heat treatment and possible deficiencies in heat treatment, such as decarburization at the surface. Microstructural inspection can also reveal possible overheating through coarsening of carbides of superalloys and solution and precipitation of manganese sulfide in steels. Other service effects, such as corrosion, oxidation, and severe work hardening of surfaces, also are revealed, and their extent can be investigated. The topographical characteristics of any cracks, particularly their mode of propagation, can be determined, for example, transgranular or intergranular. This provides information that can be helpful in distinguishing between different modes of failure. For example, fatigue cracks always propagate perpendicular to the maximum cyclic tensile stress, while stress-corrosion cracking may propagate along grain boundaries.

Only a few general directions can be given as to the best location from which to take specimens for microscopic examination, because almost every failure has individual features to be taken into account. However, in most examinations, it must be determined whether the structure of a specimen taken adjacent to a fracture surface or a region at which a service defect has developed is representative of the component as a whole. This can only be done by the examination of specimens taken from the failure region and specimens taken from other locations. For instance, in the case of ruptured or bulged boiler tubes in which failure is usually restricted to one portion only, it is desirable to examine specimens taken from both sides of the fracture, from a location opposite the affected zone, and also from an area as remote from the failure as the size of the sample permits in order to determine whether the failure has been due to a material defect or to overheating and, if the latter, whether this was of a general or localized nature. In investigations involving general overheating, sometimes the original condition of the material can only be ascertained from a sample cut from a part of the tube many feet away from the affected zone.

Metallographic specimens should be taken perpendicular to the fracture surface, showing the fracture surface in edge view. In cases where metal cleanliness may be an issue, the specimen orientation must be selected properly to determine inclusion density and morphology. This type of examination must be performed on the unetched metallographic specimen.

In the investigation of fatigue cracks, it may be desirable to take a specimen from the region where the fracture originated to ascertain if the initial development was associated with an abnormality, such as a weld defect, a decarburized surface, a zone rich in inclusions, or, in castings, a zone containing severe porosity. Multiple fatigue crack initiations are very typical of both fretting and corrosion fatigue and may form in areas where there is constant stress across a section. Similarly, with surface marks, where the origin cannot be identified from outward appearances, a microscopic examination will show whether they occurred in rolling or arose from ingot defects, such as scabs, laps, or seams. In brittle fractures, it is useful to examine a specimen cut from where the failure originated, if this can be located with certainty. Failures by brittle fracture may be associated with locally work-hardened surfaces, arc strikes, local untempered martensite, and so forth.

For good edge retention when looking at a fracture surface, it is usually best to plate the surface of a specimen with a metal, such as nickel, prior to mounting and sectioning, so that the fracture edge is supported during grinding and polishing and can be included in the examination. Alternative means include hard metal or nonmetal particles embedded in the mount adjacent to the edges.

Analysis of Metallographic Sections. As with hardness testing and macroscopic examination, the examination of metallographic sections with a microscope is standard practice in most failure analyses, because of the outstanding capability of the microscope to reveal material imperfections caused during processing and of detecting the results of a variety of in-service operating conditions and environments that may have contributed to failure. Inclusions, microstructural segregation, decarburization, carbon pickup, improper heat treatment, untempered white martensite, second phases in nickel-base superalloys, and intergranular corrosion are among the many metallurgical imperfections and undesirable conditions that can be detected and analyzed by microscopic examination of metallographic sections.

Even in the absence of a specific metallurgical imperfection, examination of metallographic sections is invaluable to the investigator in the measurement of microstructural parameters such as case depth, grain size, thickness of plated coatings, and heat-affected zone (HAZ) size, all of which may have a bearing on the cause of failure.

Mechanical Testing

Hardness testing is the simplest of the mechanical tests and is often the most versatile tool available to the failure analyst. Among its many applications, hardness testing can be used to assist in evaluating heat treatment (comparing the hardness of the failed component with that prescribed by specification), to provide an estimate of the tensile strength of steel, to detect work hardening, or to detect softening or hardening caused by overheating, decarburization, or by carbon or nitrogen pickup. Hardness testing is also essentially nondestructive, except when preparation of a special hardness test specimen is required, as in microhardness testing. Portable hardness testers are useful for field examination, but the type of hardness test must be appropriate for the sample. For example, Brinell is preferred over Rockwell for a gray cast iron part. One must ensure the proper load is used for the test specimen thickness.

Other mechanical tests are useful in confirming that the failed component conforms to specification or in evaluating the effects of surface conditions on mechanical properties. Where appropriate, tensile and impact tests should be carried out, provided sufficient material for the fabrication of test specimens is available. After all photography, fractography, and nondestructive testing have been performed, the tensile properties of the failed component(s) may be tested. (If litigation is involved, it may be necessary to obtain permission for this destructive testing.) This involves sectioning the component and machining test specimens. In all phases, from rough sectioning to the application of load and measurement of the final dimensions, it is advisable to photodocument the process. It is especially important to understand that the failed component may have been exposed to environmental conditions not experienced by exemplar components. For example, if the component was involved in a crash, there may have been a fire that exposed the component to temperatures that would have altered the mechanical properties. Likewise, the forces that acted to produce an overload failure in one area of the component may have also plastically deformed the component in other areas. To determine the mechanical properties of the component material, a location for testing must be chosen that has not been exposed to detrimental conditions. Here again, exemplar testing is a valuable tool for the failure analyst and should be employed when data from a failed component are skewed by environmental conditions. Exemplar components are components that match the failed component (i.e., the same part number and hopefully the same lot or batch). The use of exemplars can range from simply being a visual reference in a demonstration to being used for testing specimens to provide mechanical property data about a particular part or batch of parts. The closer the exemplars are to the failed component (geometrically and chronologically), the more reliable the comparison of test data.

If a raw material discrepancy is suspected, it may be that other components of the same batch or lot have the same properties. It may also be useful to have exemplars from other batches for comparison testing. The failure analyst should exercise care in interpreting mechanical test results. If a material has a tensile strength 5 to 10% below the minimum specified, this does not mean that low hardness or strength is the cause of its failure in service. Also, it should be understood that laboratory tests on small specimens may not adequately represent the behavior of a much larger structure or component in service. For instance, it is possible for a brittle fracture of a large structure to occur at or near ambient temperature, while subsequent laboratory tests of Charpy or Izod specimens show a transition temperature well below −18 °C (0 °F). The effects of size in fatigue, stress-corrosion, and hydrogen embrittlement testing are not well understood. However, on the basis of the limited evidence available, it appears that resistance to these failure processes decreases as specimen size increases. Several investigators have found correlation problems of transition temperature-type impact tests with service performance.

Occasionally, the mechanical properties may be acceptable over most of the component but may vary at a bend or other discontinuity. Castings can have significant variations in properties from one location to another depending on the solidification practice for the casting. Thus, the location of the test specimen within the component can also be significant. Mechanical property tests for cast components are frequently performed on coupons separately cast for this purpose. Therefore, results from samples from the casting itself may not be directly comparable. Other factors may affect material properties results. Material may have been tested prior to forming or other deformation. Subsequent coatings or case hardening may have improved or degraded mechanical properties. Variations from one location to another due to local material processing variations may help explain differences in mechanical properties between the bulk material and the failure origin.

Tensile tests, in many failure analysis investigations, do not provide enough useful information, because relatively few failures result from metal that is deficient in tensile strength. Furthermore, samples cut from components that have failed in a brittle manner generally show adequate ductility under the conditions imposed during a tensile test. However, sometimes there is justification for tensile testing of failed components to eliminate poor-quality material as a possible cause of failure. Often, these tensile tests for determining material quality are carried out by manufacturers and suppliers when examining components that have been returned to them for analysis.

The role of directionality in tensile testing of wrought metals should also be considered. Specimens cut transversely to the longitudinal axis of a component (such as a shaft, plate, or sheet) usually give lower tensile and ductility values than those cut along the longitudinal axis. This is due to the marked directionality and the resulting anisotropy produced during rolling or forging. When sectioning tension test specimens from the failed component, special attention should be paid to the orientation of the specimen. Some components may have quality-assurance notes on the drawings that indicate where tensile specimens are to be taken. This is especially true of critical aircraft components. Anisotropic materials have properties that vary with test specimen orientation. Tensile strength and yield strength specifications are usually given in the longitudinal and transverse directions. Typically, and unless otherwise specified, the tensile specimen should be taken with its major axis parallel to the direction of grain flow; however, test specimens are typically taken in two of the three directions: longitudinal and long transverse. The short transverse direction is typically not tested because it is difficult to obtain specimens of sufficient length in that direction. It may be necessary to lightly polish the component surface and use a macroetchant for the material being worked with to determine the grain flow. The specimens are removed from the parent component by first cutting a rough specimen shape, then final machining to the specified form.

Residual stresses in the component may result in warped specimens or pinched cutting tools during rough machining. Care should be taken to document the residual-stress observations, so that comparison to exemplar parts sectioned in the same manner can be made. Hopefully, final machining of the testpiece will correct the warped shape while still allowing the specimen to run parallel to the grain flow. Straightening of a warped tensile specimen during the test will result in a nonlinear indication on the initial portion of the stress-strain curve. This can be corrected to a line by using curve-fitting software or by drawing a line by eye back to the abscissa, as shown in Fig. 5. There may also be nonuniformities in the specimen microstructure that affect the properties of the material, for example, a change in grain size due to cold work or changes in temper due to the HAZ of a weld.

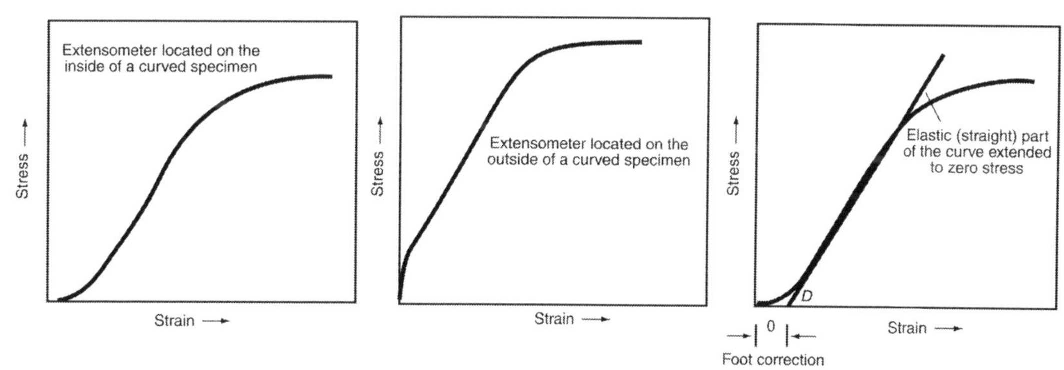

Fig. 5 Examples of stress-strain curves requiring foot correction. *D*, point where the extension of the straight (elastic) part diverges from the stress-strain curve. Source: Ref 2

It may be necessary to make some tests either at slightly elevated or at low temperatures to simulate service conditions. Also, it may be helpful to test specimens after they have been subjected to particular heat treatments simulating those of the failed component in service to determine how this treatment has modified mechanical properties. For example, treating a steel at a temperature in the embrittling range for approximately 1 h prior to impact testing will indicate any tendency to strain age embrittlement. The determination of the ductile-to-brittle transition temperature may be useful in investigating brittle fracture of a low-carbon steel.

Component Proof Testing. Critical components are sometimes proof tested after manufacture. This is especially true of critical components that will be subjected to large loads. Proof testing involves loading a component past the recommended operating limits, and possibly slightly into the yield zone. This helps to ensure there are no manufacturing or materials defects that would cause premature failure. Proof loads are usually determined on a case-by-case basis and may be expressed in several forms, such as proof load to twice the operating load or proof load to 90% of the determined yield strength. Extreme care should be taken not to damage the component in the specification of a proof load. For example, stretching a component may alter a designed preload by relieving residual compressive stresses introduced during manufacture. There are usually documents with details of specific methods, tools, and loads used in proof testing and results of tests on critical components. If performing proof testing on exemplar parts from a given manufacturer, it will lend credibility to the results if the manufacturer's test procedures are followed. If the proof tests cite standard methods, such as ASTM International, International Organization for Standardization (ISO), or other industry-accepted procedure, acquire a copy of the method and follow it closely. If results vary from those reported by the manufacturer, it may be necessary to prove the results are valid.

Conductivity of Aluminum Alloys. During a failure analysis investigation of a heat treated aluminum alloy, conductivity testing may be performed to evaluate proper heat treat condition, to assess areas for heat damage, or to estimate tensile strength. Conductivity is the reciprocal of electrical resistivity and is directly proportional to the mean free path of an electron in the crystal structure of the material. The mean free path is affected by the microstructure of the material, which is affected by heat treat condition.

Solution-annealed structures have soluble constituents that have precipitated out of solution, providing for a material with high conductivity. Aging causes fine precipitation of a second phase and thus decreases the conductivity of the material. Natural aging yields the lowest conductivity and intermediate strength, while artificial aging results in intermediate conductivity and highest hardness/strength. Overaging of the alloy causes the conductivity and hardness values to approach those of the solution-annealed condition.

The manufacturer does not typically specify conductivity values of an aluminum alloy component. However, with proper design and materials selection, the alloy and temper of the component are specified, and corresponding values of conductivity may be found in reference data such as specification SAE-MIL-H-6088. Conductivity measurements must be paired with hardness measurements to determine if the failed component meets alloy and heat treatment requirements. An aluminum alloy and its temper can be determined by measuring hardness and conductivity and verifying against reference data.

Chemical Analysis

In a failure investigation, routine analysis of the material is usually recommended. Often, it is done last because an analysis usually involves destroying a certain amount of material. There are instances where the wrong material was used, under which conditions the material may be the major cause of failure. In many cases, however, the difficulties are caused by factors other than material composition.

In most instances, slight deviations from specified compositions are not likely to be of major importance in failure analysis. However, small deviations in aluminum content can lead to strain aging in steel, and small quantities of impurities can lead to temper embrittlement. In specific investigations, particularly where corrosion and stress corrosion are involved, chemical analysis of any deposit, scale, or corrosion product or a substance with which the affected material has been in contact is required to assist in establishing the primary cause of failure.

Where analysis shows that the content of a particular element is slightly greater than that required in the specifications, it should not be inferred that such deviation is responsible for the failure. Often, it is doubtful whether such a deviation has played even a contributory part in the failure. For example, sulfur and phosphorus in structural steels are limited to 0.04% in many specifications, but rarely can a failure in service be attributed to sulfur content slightly in excess of 0.04%. Within limits, the distribution of the microstructural constituents in a material is of more importance than their exact proportions. An analysis (except a spectrographic analysis restricted to a limited region of the surface) is usually made on drillings representing a considerable volume of material and therefore provides no indication of possible local deviation due to segregation and similar effects.

Also, certain gaseous elements, or interstitials, normally not reported in a chemical analysis, have profound effects on the mechanical properties of metals. In steel, for example, the effects of oxygen, nitrogen, and hydrogen are of major importance. Oxygen and nitrogen may give rise to strain aging and quench aging. Hydrogen can induce brittleness, particularly when absorbed during welding, cathodic cleaning, electroplating, or pickling. Hydrogen is also responsible for the characteristic halos or fisheyes on the fracture surfaces of welds in steels, in which instance the presence of hydrogen often is due to the use of damp electrodes. These halos are indications of local rupture that has taken place under the bursting microstresses induced by the molecular hydrogen, which diffuses through the metal in the atomic state and collects under pressure in pores and other discontinuities. Various effects due to gas absorption are found in other metals and alloys. For example, excessive levels of nitrogen in superalloys can lead to brittle nitride phases that cause failures of highly stressed parts.

Various analytical techniques can be used to determine elemental concentrations and to identify compounds in alloys, bulky deposits, and samples of environmental fluids, lubricants, and suspensions. Semiquantitative emission spectrography, spectrophotometry, and atomic absorption spectroscopy can be used to determine dissolved metals (as in analysis of an alloy), with wet chemical methods used where greater accuracy is needed to determine the concentration of metals. Wet chemical analysis methods are employed for determining the presence and concentration of anions such as Cl^-, NO_3^-, and S^-. These methods are very sensitive. Combustion methods are ordinarily used for determining the concentration of carbon, sulfur, nitrogen, hydrogen, and oxygen.

X-ray diffraction identifies crystalline compounds either on the metal surface or as a mass of particles and can be used to analyze corrosion products and other surface deposits. Minor and trace elements capable of being dissolved can be determined by atomic absorption spectroscopy of the solution. X-ray fluorescence spectrography can be used to analyze both crystalline and amorphous solids, as well as liquids and gases.

Infrared and ultraviolet spectroscopy are used in analyzing organic materials. When the organic materials are present in a complex mixture, such as, for example, solvents, oils, greases, rubber, and plastics, the mixture is first separated into its components by gas chromatography.

Analysis of Surfaces and Deposits. Wavelength-dispersive x-ray spectrometers and energy-dispersive x-ray spectroscopes are frequently used for providing information regarding the chemical composition of surface constituents. They are employed as accessories for SEMs and permit simultaneous viewing and chemical analysis of a surface. The Auger electron spectrometer is useful for detecting the elements in extremely thin surface layers. The Auger electron spectrometer can provide semiquantitative determinations of elements with atomic numbers down to 3 (lithium). The size of the area examined varies greatly with the test conditions; it may be from 1 to 50 μm in diameter.

For chemical analysis of surface areas as small as 1 μm in diameter, the electron microprobe analyzer is widely used. This instrument can determine the concentration of all but the low-atomic-number elements, with a limit of detection below 0.1%. The area examined with the ion microprobe analyzer is a few micrometers in diameter larger than that examined with the electron microprobe analyzer. The ion microprobe analyzer has the advantage of being able to detect nearly all elements (including those of low atomic weights) in concentrations as low as 100 ppm. It is sometimes used to volatilize materials, which are then passed through a mass spectrometer.

The instruments discussed previously are used for direct analysis of surfaces; other techniques can be used for analyzing material that has been removed from the surface. For example, if material is removed in a replica (perhaps chemically extracted), it can be analyzed structurally by x-ray diffraction or electron diffraction. Also, depending on the quantity of material extracted, many of the routine chemical analysis techniques may be applicable.

Spot testing uses chemical tests to identify the metal, the alloying elements present, deposits, corrosion products, and soil. Spot tests can be performed both in the laboratory and in the field; they do not require extensive training in analytical chemistry. The only requirement is that the substance be dissolvable; hydrochloric acid or even aqua regia may be used to dissolve the material. Spot tests for metallic elements such as chromium, nickel, cobalt, iron, and molybdenum are usually done by dissolving a small amount of the alloy in acid and mixing a drop of the resulting solution with a drop of a specific reagent on absorbent paper or a porcelain plate. Spot colorings produced in this way indicate the presence or absence of the metallic radical under test. Samples may be removed from gross surfaces by spotting the specimen with a suitable acid, allowing time for solution, and collecting the acid spot with an eyedropper.

Simulated Service Testing

During the concluding stages of an investigation, it may be necessary to conduct tests that simulate the conditions under which failure is believed to have occurred. Often, simulated service testing may not be practical because elaborate equipment is required, and even where practical, it is possible that not all of the service conditions are fully known or understood. For example, corrosion failures are difficult to reproduce in a laboratory, and some attempts to reproduce them have given misleading results. Serious errors can arise when attempts are made to reduce the time required for a test by artificially increasing the severity of one of the factors, such as the corrosive medium or the operating temperature. Similar problems are encountered in wear testing.

On the other hand, when its limitations are clearly understood, the simulated testing and statistical experimental design analysis of the effects of certain selected variables encountered in service may be helpful in planning corrective action or, at least, may extend service life. The evaluation of the efficacy of special additives to lubricants is an example of the successful application of simulated service testing. The aircraft industry has made successful use of devices such as the wind tunnel to simulate some of the conditions encountered in flight, and naval architects have employed tank tests to evaluate hull modifications, power requirements, steerage, and other variables that may forestall component failure or promote safety at sea.

Taken singly, most of the metallurgical phenomena involved in failures can be satisfactorily reproduced on a laboratory scale, and the information derived from such experiments can be helpful to the investigator, provided the limitations of the tests are fully recognized. However, many company managers prefer to conduct trials to verify improvements before major changes are approved. This is a more conservative approach, but it only takes one improper recommendation that results in an adverse result to justify trials of major changes.

A Final Word

Failure analysis is an extremely complex subject and involves areas of mechanics, physics, metallurgy, chemistry and electrochemistry, manufacturing processes, stress analysis, design analysis, and fracture mechanics, to name a few specialties. Because it is nearly impossible for any one person to be an expert in all these fields, it is extremely important to know when to seek help. In any situation, it is very important not to leap to conclusions, because a misstep can be extremely hazardous for all concerned.

ACKNOWLEDGMENT

The material for this chapter came from "Practices in Failure Analysis" in *Failure Analysis and Prevention*, Vol 11, *ASM Handbook*, ASM International, 2002.

REFERENCES

1. Practices in Failure Analysis, *Failure Analysis and Prevention,* Vol 11, *ASM Handbook,* ASM International, 2002
2. P.M. Mumford, Test Methodology and Data Analysis, *Tensile Testing,* P. Han, Ed., ASM International, 1992, p 55

SELECTED REFERENCES

- D. Aliya, The Failure Analysis Process: An Overview, *Failure Analysis and Prevention,* Vol 11, *ASM Handbook,* ASM International, 2002

- J.J. Scutti and W.J. McBrine, Introduction to Failure Analysis and Prevention, *Failure Analysis and Prevention,* Vol 11, *ASM Handbook,* ASM International, 2002
- D.J. Wulpi, *Understanding How Components Fail,* 2nd ed., ASM International, 1999

APPENDIX A

Defects Leading to Failure

CATEGORIZING DEFECTS leading to failure vary among failure analysis practitioners, quality engineers, other engineers, and managers, as well as legal and insurance professionals. Grouping failure causes into a few fundamental categories is advantageous and informative because it defines which aspect of a product or system requires corrective action and prevention strategies. Systematic analysis of equipment failures reveals physical root causes that fall into one of three fundamental categories:

- Design deficiencies
- Material and manufacturing defects
- Service-life anomalies

An effective graphical representation of the impact of defects on the service life of a component or system is provided in the application-life diagram shown in Fig. 1. The diagram is constructed by plotting the service lives of components having specific characteristics in the design/configuration, as related to the severity of a specific service condition that is anticipated for the application. Typical characteristics include strength, corrosion resistance, heat treatment condition, flaw size, surface finish, bend radius, void content (i.e., in castings), and degree of sensitization. Examples of service conditions include magnitude of stress (either static or cyclic), exposure temperature, and aggressiveness of the environment.

Design Deficiencies

Failures that stem from design deficiencies are unacceptable features of a product or system that are a result of the design process. The design process encompasses the original concept development, the general configu-

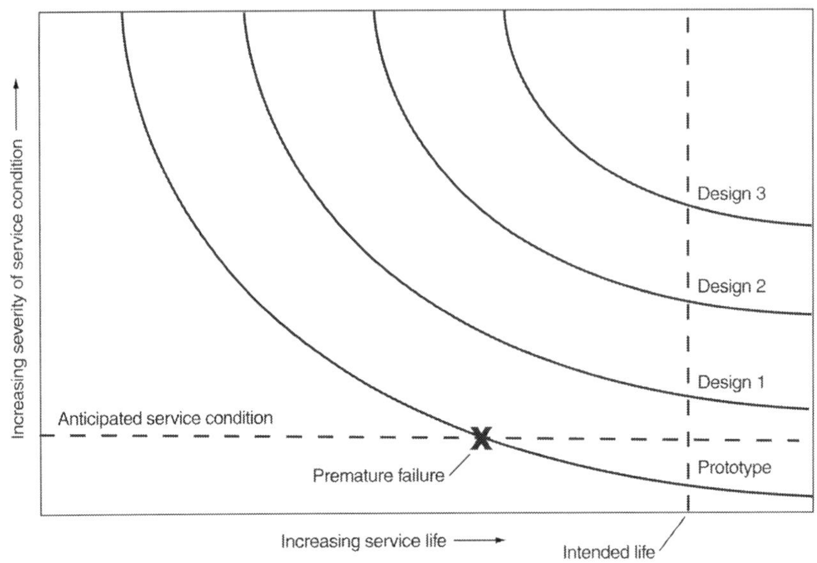

Fig. 1 Application-life diagram comparing the severity of a service condition with the service lives of products having a variable characteristic. Source: Ref 1

ration definition, and the detail design, including selection and specification of materials and manufacturing processes. Design involves identifying and defining a need for the product or system, followed by definition of the performance requirements, anticipated service conditions, the constraints on the design, and the criticality or risks associated with failure.

Some examples of design deficiencies include unintended stress raisers due to excessively sharp notches (e.g., in keyways on shafts) or insufficient radii (e.g., on shafts at bearing journals). Other examples include unanticipated residual stresses associated with heat treated parts designed with complex geometries, or assembly stresses from configurations that contain unwanted interference. Inappropriate surface treatments can result in failures, such as the use of cadmium plating on an A286 superalloy fastener subjected to service temperatures above 315 °C (600 °F), because the melting temperature of cadmium is 320 °C (610 °F). Two metals specified for use in a wear application can sustain galling if the metals are incompatible, such as sliding wear of components made from 300-series stainless steels.

Selection of a material that is incapable of providing adequate mechanical properties for the application (including strength, fatigue resistance, fracture toughness, corrosion resistance, and elevated-temperature resistance) is also a design deficiency. Materials can exhibit anisotropy, or variability in properties within a product, such as between the thick and thin portions of a casting or between the longitudinal and transverse properties in a wrought material. Note that a material can be shown to meet the properties required or specified (e.g., a separately cast tensile bar used to certify a casting, or the longitudinal tensile properties to certify a complex aluminum extrusion), but the specific properties required for the application may rely on the strength, toughness, or stress-corrosion cracking resistance in a direction other than longitudinal.

Design-caused failures include inappropriate geometries defined on the engineering drawing, which may lead to a compromise of component or system capabilities. Examples of inappropriate geometries include improper joint preparation for welding or brazing, such as an insufficient or a missing groove for a groove weld, insufficient fit-up relief in a socket weld, or inadequate joint overlap in a brazed joint. Other geometry-caused failures can result from insufficient section thickness for a failure based on gross yielding, excessive section thickness in the presence of a flaw for a material of limited fracture toughness, or a fabrication configuration with an excessively sharp forming bend, with the resulting high residual stresses causing a reduction in the fatigue life.

For example, an excessively tight cold-formed bend radius, an application-life diagram can be constructed as shown in Fig. 2. The service condition considered is stress, and the characteristic that is varied is the radius of the cold-formed bend. Upon examination of the relationship between the characteristic curves and the intended service life, the components having the large and moderate bend radii are found to meet the intended service life at the severity of stress that is anticipated in the specific application. However, in this illustration, the component with the small bend radius sustained a premature failure at the anticipated stress level in the

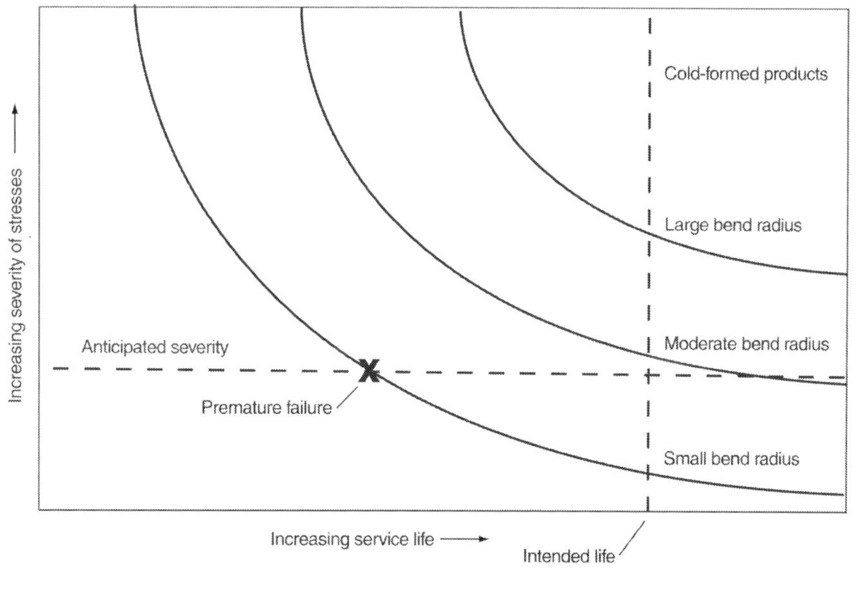

Fig. 2 Application-life diagram for design deficiency of a cold-formed part. Source: Ref 1

application, because the curve intersects the severity-of-stress line prior to reaching the intended service life.

Material and Manufacturing Defects

Unacceptable imperfections or discontinuities in materials are defects, and some types of imperfections may be generally detrimental to the performance or appearance of a product or system. Material defects can be generally described as discontinuities that degrade the performance of a product in some way. Despite measures taken to control, document, measure, analyze, and improve the processes involved in manufacturing the metal product (such as in total quality management and Six Sigma systems), material defects occur. Many defective products are prevented from leaving the mill, foundry, or forge through diligence in adhering to internal procedures and quality-assurance systems. Yet, defective materials are sometimes delivered. Depending on the criticality, periodic field inspections may be required and may reveal defects not previously identified.

Ingot-Related Defects

Many flaws in wrought products can be traced back to the pouring and solidification of the hot metal in molds during production of the ingot. Primary deformation processes, such as hot rolling, tube piercing, extrusion, and open-die forging, are then used for converting the cast ingot structure. Many large open-die forgings are forged directly from ingots, while most closed-die and upset forgings are produced from billets, bar stock, or a preform that has received some previous mechanical working.

During hot working, subsurface and core discontinuities can occur. Subsurface flaws often (but not always) originate from the as-cast ingot due to shrinkage, voids, and porosity that form during solidification. For example, this is shown schematically in Fig. 3 for rolled bar that contains ingot porosity and pipe imperfections. These imperfections can serve as sites for crack initiation during fabrication or in service.

In addition to the primary pipe near the top of the ingot, secondary regions of piping and centerline shrinkage may extend deeper into an ingot (Fig. 4a). Primary piping is generally an economic concern, but if it extends sufficiently deep enough into the ingot body and goes undetected, it can eventually result in a defective product, as shown in the secondary pipe in the as-forged steel bar in Fig. 4(b).

Factors that work against obtaining a perfect homogeneous product include:

- Fast shrinkage as the molten metal cools (roughly 5% in volume for steel)
- Gaseous products that are trapped by the solidifying metal as they try to escape from the liquid and semisolid metal

- Small crevices in the mold walls, which cause the metal to tear during the stripping operation
- Spatter during pouring, which produces globs of metal frozen on the mold walls because of the great difference in temperature of the mold surfaces and the liquid metal

A certain degree of alloy segregation occurs in all wrought products, and hot working can alleviate some of the inhomogeneity. However, if the ingot is badly segregated, hot working just tends to alter the shape of the segregation region into a banded structure. Banding from a carbon-rich centerline condition in a hot-rolled 1041 steel is shown in Fig. 5a, while Fig. 5(b) shows an extreme example of banding in a hot-rolled plain carbon steel (1022) in which alternate layers of ferrite and pearlite have formed along the rolling direction.

The metal-producing industry attempts to control nonmetallic inclusions, but inclusions cannot be removed with complete assurance. In demanding applications, inclusion shape-control methods are used to produce inclusions predominantly with a spherical shape, which is less of a stress concentrator with the surrounding matrix and less damaging than inclusions with more angular shapes. The principle of inclusion shape control is to reduce the surface energy between the inclusion and the metal, normally through chemistry modification of the metal. For example, sulfide inclusions have been shape controlled by the addition of cal-

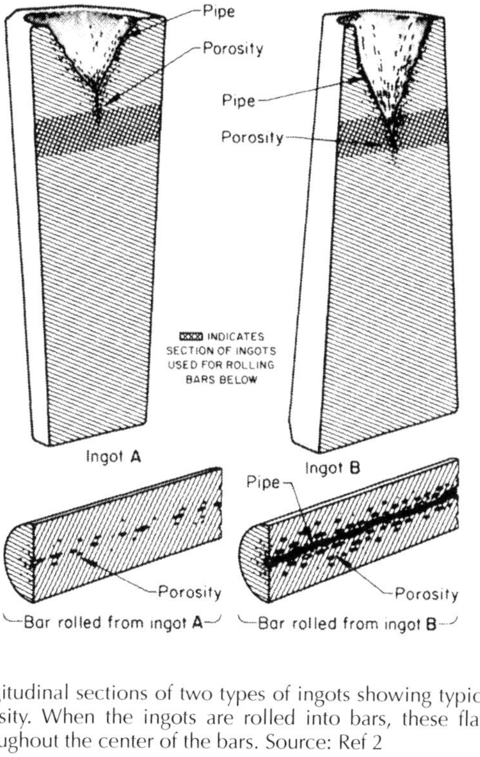

Fig. 3 Longitudinal sections of two types of ingots showing typical pipe and porosity. When the ingots are rolled into bars, these flaws become elongated throughout the center of the bars. Source: Ref 2

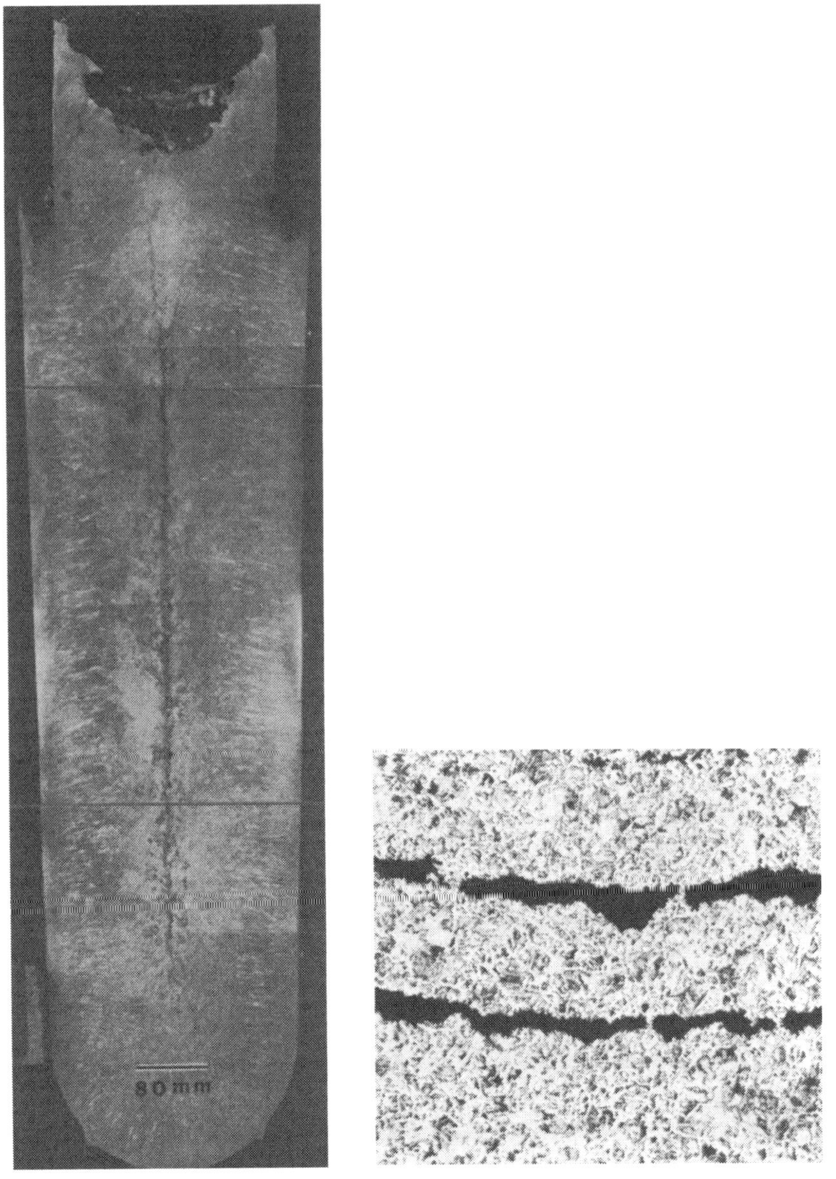

(a) (b)

Fig. 4 (a) Longitudinal section through an ingot showing extensive centerline shrinkage. (b) 1038 steel bar, as-forged. Longitudinal section displays secondary pipe (black areas) that was carried along from the original bar stock into the forged piece. Gray areas are pearlite; white areas, ferrite. 2% nital. Original magnification: 50×. Source: Ref 2

cium or rare earth metal treatments. These inclusions are not plastic or deformable at typical hot working temperatures and retain their globular shape. Therefore, they are less injurious with respect to ductility, toughness, and fatigue life in finished wrought products.

(a)

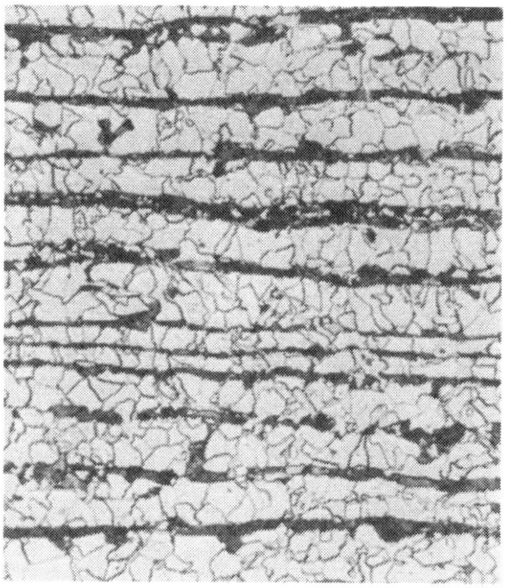

(b)

Fig. 5 Segregation banding in hot rolled steel that originated from the ingot. (a) Longitudinal section through a hot-rolled 1041 steel bar showing a carbon-rich centerline (dark horizontal bands) that resulted from segregation in the ingot. Picral. Original magnification: 3×. Courtesy of J.R. Kilpatrick. (b) Hot-rolled 1022 steel showing severe banding. Bands of pearlite (dark) and ferrite were caused by segregation of carbon and other elements during solidification and later decomposition of austenite. Nital. Original magnification: 250×. Courtesy of J.R. Kilpatrick. Source: Ref 2

With the number and size of inclusions held roughly constant, inclusion shape control reduces the likelihood of crack initiation from an inclusion. This is shown in Fig. 6(a) for a rolled constructional steel with calcium treatment, which provides longer life under cyclic loading due to inclusion shape modification. The advantage of a spherical inclusion shape is also evident in Charpy V-notch impact data for construction steels (Fig. 6b). Scanning electron microscope fractographs also show a pronounced difference in deformation mode with an inclusion shape modification (Fig. 6c). When the shape is spherical (Fig. 6cii), local ductility increases adjacent to the inclusions.

Various terms are used to describe surface and subsurface imperfections in wrought product. The meaning of terms can vary by location and industry, but Fig. 7 is a schematic illustration of some terms used to describe flaws in rolled bar stock. This schematic is not a complete summary of possible imperfections; for example, die scratches of cold-worked product are not included. However, the schematic of Fig. 7 shows typical terms for surface and subsurface flaws that may occur in wrought products. For example, *inclusions* (Fig. 7a) and *seams* (Fig. 7h) are commonly used terms to describe imperfections in wrought form. These imperfections, whether at or below the surface, can adversely affect performance of a part by creating a notch of unknown severity and serve as a crack-initiation site during fabrication or in service. Corrosion and wear damage can also be assisted by discontinuities, especially at the surface. These flaws may occur from the melting practices and solidification of ingot, the primary or secondary working of the material, or the metallurgical characteristic of a particular alloy system.

Forging Imperfections

In many cases, the internal and surface imperfections that occur during forging are the same as, or at least similar to, those that may occur during hot working of the ingots or billets, as briefly introduced in the previous

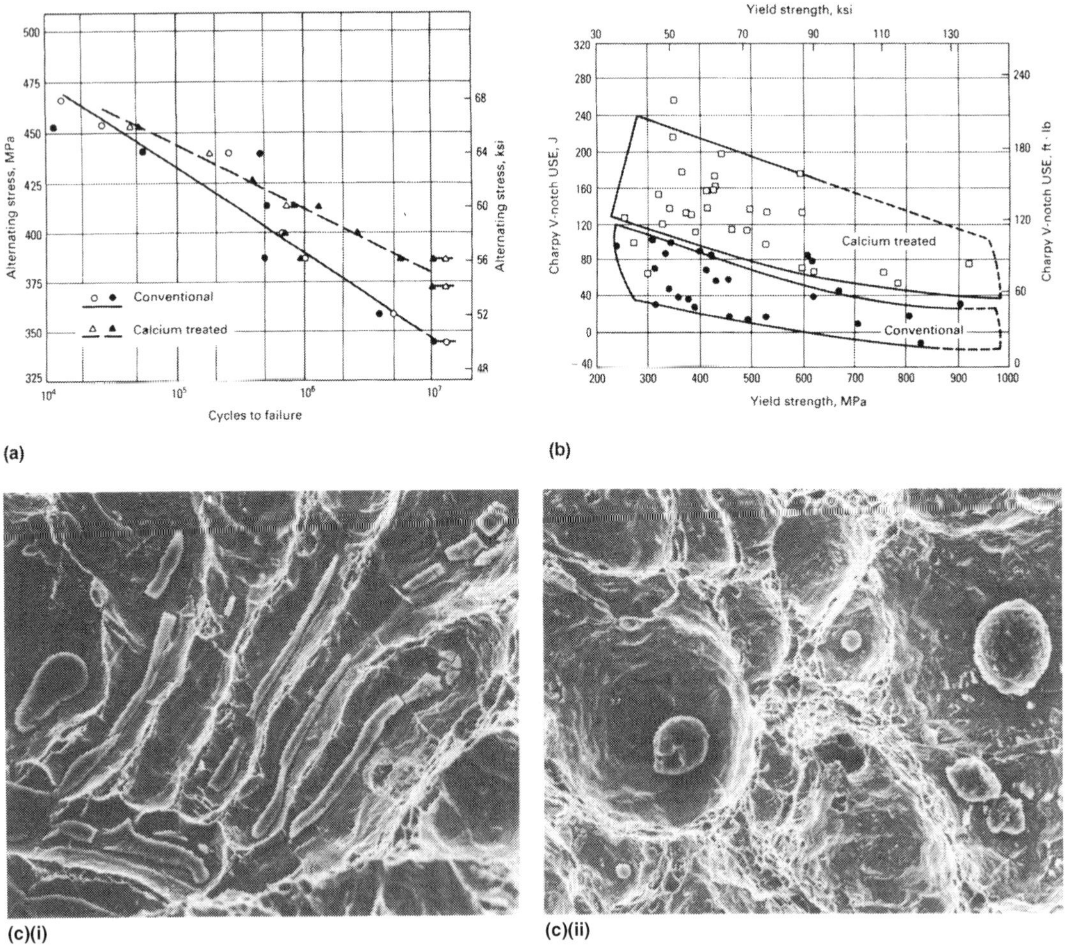

Fig. 6 Property improvement due to inclusion control in steel. (a) Axial fatigue data. Comparison of axial fatigue data for untreated and calcium-treated rolled ASTM A516 steel. 51 mm (2 in.) thick plates tested with alternating stress ratio of 0.1. Source: Ref 3. (b) Charpy V-notch data. Comparison of Charpy V-notch upper-shelf energies (USEs) for several grades and thicknesses of untreated and calcium-treated steel. Source: Ref 4. (c) Scanning electron microscope (SEM) fractographs showing the effect of calcium treatment on the fracture morphology of ASTM A633C steel impact specimens. (i) Untreated steel with type II manganese sulfide inclusions showing evidence of brittle fracture. (ii) Calcium-treated steel with spherical inclusions; the fracture is ductile. Courtesy of A.D. Wilson, Lukens Steel Company. Source: Ref 2.

section. A number of forging imperfections are shown in Fig. 8. A brief review for structural alloys follows:

- *Steel alloys.* The most common internal imperfections found in steel forgings are pipe, segregation, nonmetallic inclusions, and stringers. Internal flaws caused by forgings also include cracks or tears, which may result either from forging with too light a hammer or from continuing forging after the metal has cooled down below a safe forging temperature. Other flaws in steel forgings that can be produced by improper die design or maintenance are internal cracks and splits. If the material is moved abnormally during forging, these flaws may be formed without any evidence on the surface of the forging. Bursts can also occur.
- *Aluminum alloys.* The internal discontinuities that occur in aluminum alloy forgings are ruptures, cracks, inclusions, segregation, and occasionally porosity. Ruptures and cracks are associated with temperature control during preheating or forging or with excessive reduction during a single forging operation. Cracks can also occur in stock that has been excessively reduced in one operation. Inclusions, segregation, and porosity result from forging stock that contains these types of discontinuities.

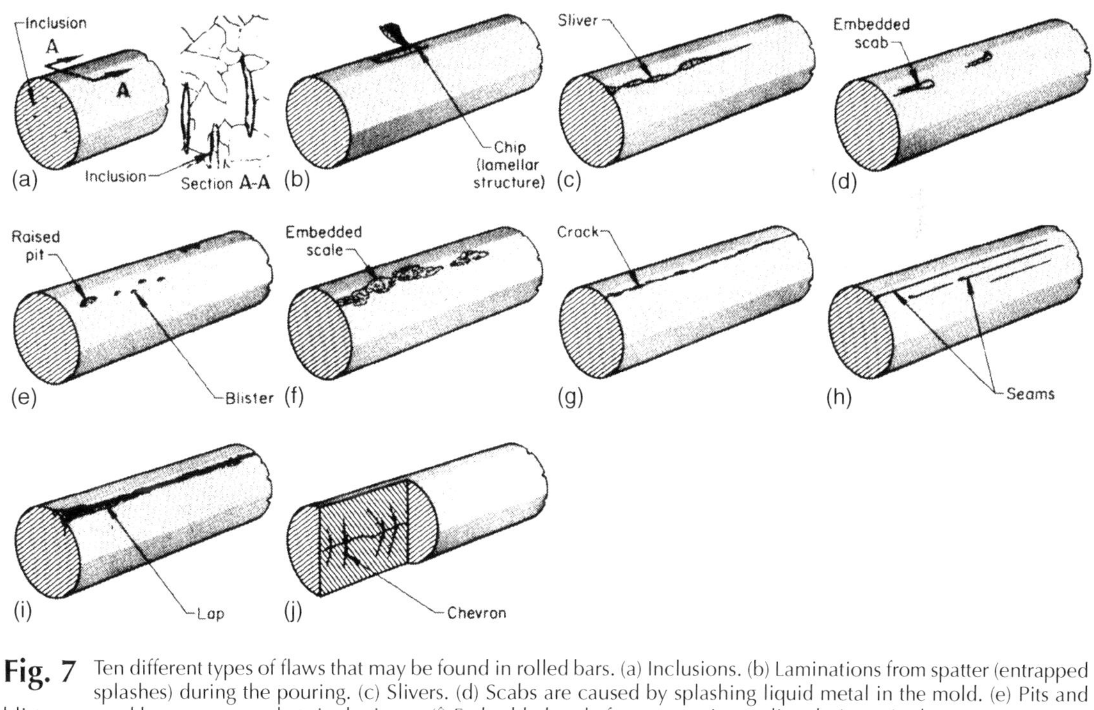

Fig. 7 Ten different types of flaws that may be found in rolled bars. (a) Inclusions. (b) Laminations from spatter (entrapped splashes) during the pouring. (c) Slivers. (d) Scabs are caused by splashing liquid metal in the mold. (e) Pits and blisters caused by gaseous pockets in the ingot. (f) Embedded scale from excessive scaling during prior heating operations. (g) Cracks with little or no oxide present on their edges when the metal cools in the mold, setting up highly stressed areas. (h) Seams that develop from elongated trapped-gas pockets or from cracks during working. (i) Laps when excessive material is squeezed out and turned back into the material. (j) Chevron or internal bursts. Source: Ref 2

- *Titanium alloys.* Discontinuities that are most likely to occur in titanium alloy forgings are usually carried over in the bar or billet. Typical discontinuities in titanium alloy forgings are alpha-stabilized voids, macrostructural defects, unsealed center conditions, clean voids, and forging imperfections. Alpha-stabilized voids are among the most common discontinuities found in forgings of titanium alloys. Three principal macrodefects are commonly found in ingot, forged billet, or other semifinished product forms. These include high-aluminum defects (type II defects), high-interstitial defects (type I defects or low-density interstitial defects), and beta flecks.

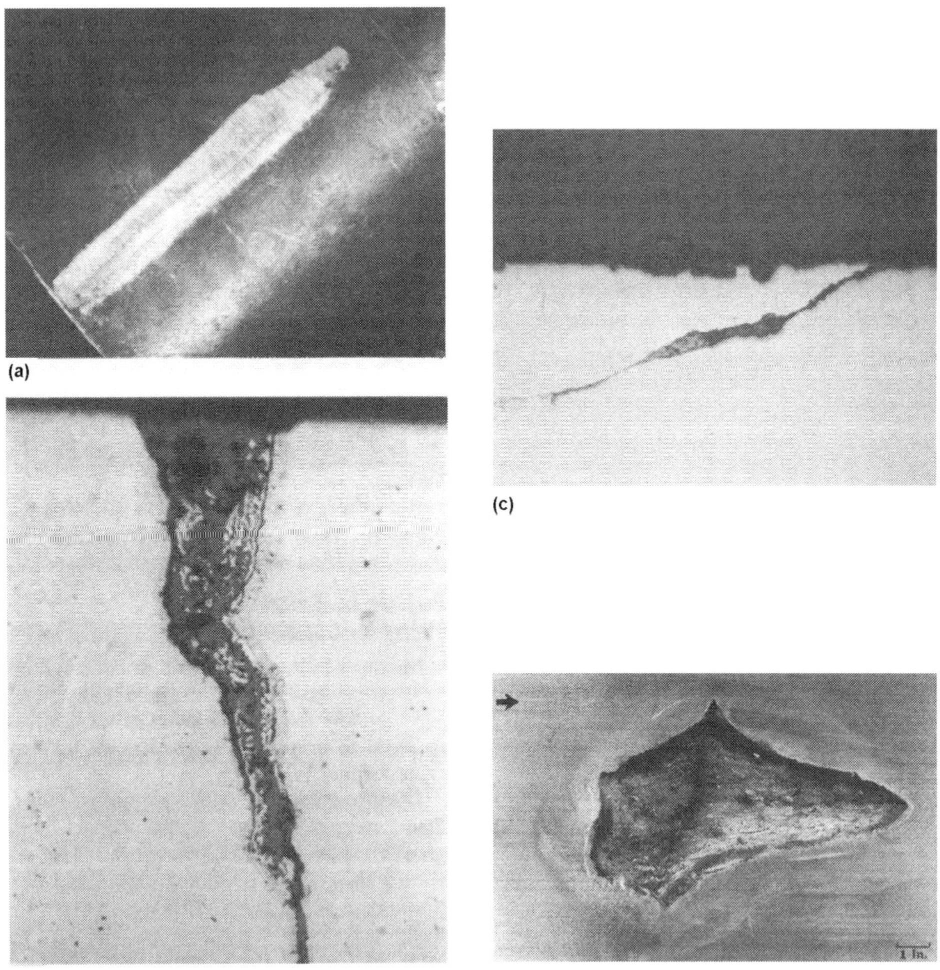

Fig. 8 Typical forging defects. Seam in rolled 4130 steel bar. (a) Closeup of seam. Note the linear characteristics of this flaw. (b) Micrograph of seam in (a) showing cross section of the bar. Seam is normal to the surface and filled with oxide. Original magnification: 30×. (c) Micrograph of a forging lap. Note the included oxide material in the lap. Original magnification: 20×. (d) Cross section of a forged bar showing a forging burst. The burst is located approximately at the centerline of the workpiece. Arrow indicates the direction of working. Source: Ref 2

- *Nickel-base heat-resistant alloys.* Most of the flaws found in forgings of heat-resistant alloys can be categorized as those related to scrap selection, melting, or primary conversion to bar or billet, or those that occur during forging or heat treatment. Nickel-base superalloys are highly susceptible to surface contamination during heating for forging. Fuel oils containing sulfur induce grain-boundary attack, which causes subsequent rupturing during forging. Paint or marking crayons with high levels of similar contaminants cause similar areas of grain-boundary contamination.

A number of surface flaws also can be produced by the forging operation. These flaws are often caused by the movement of metal over or on another surface without actual welding or fusing of the surfaces. The most common surface flaws in steel forgings are seams, laps, and slivers. Laps and seams (Fig. 8a–c) are surface discontinuities that are caused by folding over of metal without fusion. They are usually filled with scale and, on steel components, are enclosed by a layer of decarburized metal. Laps, seams, and other surface defects formed during manufacture of the blanks can also lead to major problems during cold forming of the actual parts. Once any surface defect is exposed to the atmosphere, its surfaces become oxidized, providing an initiation point for failure during future operations.

Other surface flaws include rolled-in scale, ferrite fingers, fins, overfills, and underfills. Surface flaws weaken forgings and can usually be eliminated by correct die design, proper heating, and correct sequencing and positioning of the workpieces in the dies. When such cracks form early in the steel manufacturing process, the surfaces of such breaks can become completely decarburized, forming what is known as a ferrite finger. Even if the original crack about which they formed is then removed by conditioning, the remaining ferrite is extremely low in tensile strength and often splits during the metal movement necessary during cold forming.

Cold shuts often occur in closed-die forgings. They are junctures of two adjoining surfaces caused by incomplete metal fill and incomplete fusion of the surfaces. Shear cracks often occur in steel forgings; they are diagonal cracks occurring on the trimmed edges and are caused by shear stresses. Proper design and condition of trimming dies to remove forging flash are required for the prevention of shear cracks.

The common failure modes occurring in deformation processing are summarized in Table 1. Three temperature regions are common: cold working, warm working, and hot working. The general types of fracture in bulk working are:

- Free-surface fracture
- Die-contact surface fracture
- Internal fracture (see centerline burst in Fig. 8d)

Table 1 Common failure modes in deformation processing

	Metallurgical defects in:	
Temperature regime	Cast grain structure	Wrought (recrystallized) grain structure
Cold working	(a)	Free surface fracture, dead metal zones (shear bands, shear cracks), center bursts, galling
Warm working	(b)	Triple-point cracks/fractures, grain-boundary cavitation/fracture
Hot working	Hot shortness, center bursts, triple-point cracks/fractures, grain-boundary cavitation/fracture, shear bands/fractures	Shear bands/fractures, triple-point cracks/fractures, grain-boundary cavitation/fracture, hot shortness

(a) Cold working of cast structures is typically performed only for very ductile metals (such as dental alloys) and usually involves many stages of working with intermediate-recrystallization anneals. (b) Warm working of cast structures is rare. Source: Ref 2

Sheet Forming Imperfections

The major problems encountered in sheet metal forming are fracturing, buckling and wrinkling, shape distortion, loose metal, and undesirable surface textures. The occurrence of any one or a combination of these conditions can render the sheet metal part unusable.

Fracturing occurs when a sheet metal blank is subjected to stretching or shearing (drawing) forces that exceed the failure limits of the material for a given strain history, strain state, strain rate, and temperature. In stretching, the sheet initially thins uniformly, at least in a local area. Eventually, a point is reached at which deformation concentrates and causes a band of localized thinning, known as a neck, which ultimately fractures. The formation of a neck is generally regarded as failure, because it produces a visible defect and a structural weakness. Most current formability tests are concerned with fracture occurring in stretching operations.

In shearing, fracture can take place without prior thinning. The most common examples of this type of fracture occur in slitting, blanking, and trimming. In these operations, sheets are sheared by knife edges that apply forces normal to the plane of the sheet. Shearing failures are sometimes produced in stamping operations by shearing forces in the plane of the sheet, but they are much less common than stretching failures.

Buckling and Wrinkling. In a typical stamping operation, the punch contacts the blank, stretches it, and starts to pull it through the blankholder ring. The edges of the blank are pulled into regions with progressively smaller perimeters. This produces compressive stresses in the circumferential direction. If these stresses reach a critical level characteristic of the material and its thickness, they cause slight undulations known as buckles. Buckles may develop into more pronounced undulations or waves, known as wrinkles, if the blankholder pressure is not sufficiently high. This effect can also cause wrinkles in other locations, particularly in regions with abrupt changes in section and in regions where the metal is unsupported or contacted on one side only. In extreme cases, folds and double or triple metal may develop. These may, in turn, lead to splitting in another location by preventing metal flow or by locking the metal out. Therefore, increasing the blankholder pressure often corrects a splitting problem.

Shape Distortion. On removal of the external forces, the internal elastic stresses relax. In some locations, they can relax completely, with only a very slight change in the dimensions of the part. However, in areas subjected to bending, through-the-thickness gradients in the elastic stresses occur; that is, the stresses on the outer surfaces are different from those on the inner surfaces. If these stresses are not constrained or locked in by the geometry of the part, relaxation causes a change in the part shape known as shape distortion or springback. Springback can be compensated for in die design for a specific set of material properties but may still be a problem if there are large material property or process variations from blank to blank.

Loose metal occurs in undeformed regions and is undesirable, because it can be easily deflected. A phenomenon usually referred to as oil canning, in which a local area can be either concave or convex, can also be encountered. In stampings with two or more sharp bends of the same sign in roughly the same direction, such as a pair of feature lines, a tendency exists for the metal between them to be loose because of the difficulty involved in pulling metal across a sharp radius. It is sometimes possible to avoid the problem by ensuring that the metal is not contacted by both lines at the same time; thus, some stretching can occur before the second line is contacted. There is a tendency for loose metal to occur toward the center of large, flat, or slightly curved parts. Increasing the restraining forces on the blank edges usually improves this condition.

Undesirable Surface Textures. Heavily deformed sheet metal, particularly if it is coarse grained, often develops a rough surface texture commonly known as "orange peel." This is usually unacceptable in parts that are visible in service. Another source of surface problems occurs in metals that have a pronounced yield point elongation; that is, materials that stretch several percent without an increase in load after yielding. In these metals, deformation at low strain levels is concentrated in irregular bands known as Lüders lines (or bands) or stretcher strains. These defects disappear at moderate and high strain levels. However, almost all parts have some low strain regions. These defects are unsightly and are not concealed by painting. Aged rimmed steels and some aluminum-magnesium alloys develop severe Lüders lines.

Casting Defects

Poor casting design can interfere with the ability of the foundry to use the best techniques to produce reliable castings. The designer also specifies the quality requirements that ensure that the cast component will perform as desired. Overspecification causes needless expense and can be avoided by understanding the effect of discontinuities on casting performance and the effect of casting design on the tendency for discontinuities to form during the casting process. Important types of casting discontinuities include porosity, inclusions, oxide films, second phases, hot tears, metal penetration, and surface defects. A number of typical casting defects are shown in Fig. 9.

Porosity is a common defect in castings and takes many forms. An example of gas porosity is shown in Fig. 9(b). Pores may be connected to the surface, where they can be detected by dye penetrant techniques, or they may be wholly internal, where they require radiographic techniques to detect. Macroporosity refers to pores that are large enough to see with the unaided eye on radiographic inspection, while microporosity refers to pores that are not visible without magnification. Both macroporosity and microporosity are caused by the combined action of metal shrinkage and gas evolution during solidification. It has been shown that nucleation of

Fig. 9 Typical casting defects. (a) Inclusion (arrow) on machined surface of a casting. (b) Typical micrograph of gas porosity. Original magnification: 100×. (c) Micrograph of low-alloy steel shrinkage crack. Original magnification: 7.5×. (d) Optical micrograph of a hot tear in a casting. Original magnification: 200×. Source: Ref 5

pores is difficult in the absence of some sort of substrate, such as a nonmetallic inclusion, a grain refiner, or a second-phase particle. This is why numerous investigations have shown that clean castings, those castings that are free from inclusions, have fewer pores than castings that contain inclusions. Microporosity is found not only in castings but also in heavy-section forgings that have not been worked sufficiently to close it up.

When the shrinkage and the gas combine to form macroporosity, properties are deleteriously affected. Static properties are reduced at least by the portion of the cross-sectional area that is taken up with the pores; because there is no metal in the pores, there is no metal to support the load there, and the section acts as though its area was reduced. Because the pores may also cause a stress concentration in the remaining material, static properties may be reduced by more than the percentage of cross-sectional area that is caused by the macroporosity. Dynamic properties are also affected by porosity. A study of aluminum alloys showed that fatigue properties in some were reduced 11% when specimens having x-ray quality equivalent to ASTM E 155 level 4 were tested, and that they were reduced 17% when specimens having quality of ASTM E 155 level 8 were tested.

Static properties are mostly unaffected by microporosity. Microporosity is found between dendrites and, like macroporosity, is caused by the inability of feed metal to reach the interdendritic areas of the casting where shrinkage is occurring and where gas is being evolved. However, because this type of porosity occurs late in solidification, particularly in long-range-freezing (mushy-freezing) alloys, it is particularly difficult to eliminate. The most effective method is to increase the thermal gradient, often accomplished by increasing the solidification rate, which decreases the length of the mushy zone. This technique may be limited by alloy and mold thermal properties and by casting geometry, that is, the design of the casting.

As long as the micropores are less than 0.2 mm (0.008 in.) in length, there is no effect on dynamic properties; fatigue properties of castings with pores that size or smaller are in the same range as those of castings where no micropores were found. The shape of the micropore is as important as its size, with elongated pores having a greater effect than round pores. Microporosity can be healed by hot isostatic pressing (HIP). In one study comparing HIP and non-HIP samples, no difference was found in fatigue lives of HIP and non-HIP samples. However, the HIP samples showed a lower crack growth rate than non-HIP samples. In another study, HIP improved fatigue crack growth resistance only close to threshold levels. The design of the casting directly affects its tendency to solidify in a progressive manner, thereby affecting both the quality and the price of the cast component. Porosity and casting costs are minimized in casting designs that emphasize progressive solidification toward a gate or riser, tapered walls, and the avoidance of hot spots.

Inclusions are nonmetallic particles that are found in the casting. They may form during solidification as some elements (notably manganese and sulfur in steel) precipitate from solution in the liquid. More frequently, they are formed before solidification begins. The former are sometimes called indigenous inclusions, and the latter are called exogenous inclusions. Inclusions are ceramic phases that have little ductility. A crack may form in the inclusion and propagate from the inclusion into the metal, or a crack may form at the interface between the metal and the inclusion. In addition, because the inclusion and the metal have different coefficients of thermal expansion, thermally induced stresses may appear in the metal surrounding the inclusion during solidification. As a result, the inclusion acts as a stress-concentration point and reduces dynamic properties. As in the case of microporosity, the size of the inclusion and its location determine its effect. Small inclusions that are located well within the center of the cross section of the casting have little effect, whereas larger inclusions and those located near the surface of the casting may be particularly detrimental to properties. Inclusions may also be a problem when machining surfaces (Fig. 9a), causing excessive tool wear and tool breakage.

Exogenous inclusions are mostly oxides or mixtures of oxides and are primarily slag or dross particles, which are the oxides that result when the metal reacts with oxygen in the air during melting. These are removed from the melt before pouring by filtration. Most inclusions found in steel castings arise from the oxidation of metal during the pouring operation. This is known as reoxidation, and it takes place when the turbulent flow of the metal in the gating system causes the metal to break up into small droplets, which then react with the oxygen in the air in the gating system or casting cavity to form oxides. Metal casters use computer analysis of gating systems to indicate when reoxidation can be expected in a gating system and to eliminate them. However, casting designs that require molten metal to "jet" through a section of the casting to fill other sections will recreate these inclusions and should be avoided.

Oxide films are similar to inclusions and have been found to reduce casting properties. These form on the surface of the molten metal as it fills the mold. If this surface film is trapped within the casting instead of being carried into a riser, it is a linear discontinuity and an obvious site for crack initiation. It has been shown that elimination of oxide films, in addition to substantially improving static properties, results in a fivefold improvement of fatigue life in axial tension-tension tests.

Oxide films are of particular concern in nonferrous castings, although they also must be controlled in steel and stainless steel castings. If the film folds over on itself as a result of turbulent flow or "waterfalling" when molten metal falls to a lower level in the casting during mold filling, the effects are particularly damaging. Casting design influences how the metal fills the mold, and features of the design that require the metal to fall from one level to another while the mold is filling should be avoided so that waterfalls are eliminated. Oxide films are avoided by filling the casting from the bottom, in a controlled manner, by pumping the metal into the mold using pneumatic or electromagnetic pumps. If the casting is poorly designed, waterfalling will result, as shown in the example in Fig. 10.

Second phases, which form during solidification, can also nucleate cracks if they have the proper size and morphology. An example is aluminum-silicon alloys, where the silicon eutectic is present as large platelets, which can nucleate cracks and along which cracks propagate. The size of these platelets may be significantly reduced by modifying the alloy with additions of sodium or strontium. However, such additions increase the size of micropores, and for this reason, many foundrymen rely on accelerated solidification of the casting to refine the silicon. As solidification rates increase, the structure is refined in thin sections. Heavy sections are to be avoided if a fine structure is desired.

Hot tears form when casting sections are constrained by the mold from shrinking as they cool near the end of solidification. These discontinuities are fairly large and are most often weld repaired. If not repaired, their effect is not readily predictable. While generally they are detrimental to casting properties, under some circumstances they do not affect them. Hot tears (Fig. 9d) are caused by a combination of factors, including alloy type, metal cleanliness, and mold and core hardness. However, poor casting design is the primary cause. Castings should be designed so that solidifying sections are not subjected to tensile forces caused by shrinkage during solidification, because the solidifying alloy has little strength before it solidifies. An example is given in Fig. 11.

Metal Penetration. Molten metal may penetrate the surface of the mold, forming a rough surface or, in extreme cases, actually becoming intimately mixed with the sand in the mold. In iron castings, this is normally the result of the combination of metallostatic head (the pressure exerted on the molten iron at the bottom of the mold by the weight of the metal on top of it) and the surface tension relationship between the liquid iron and molding materials. In cast iron, it is also frequently the result of

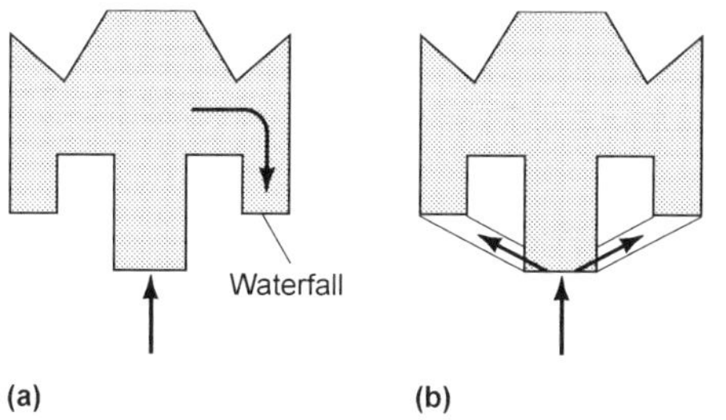

Fig. 10 Redesign of a casting to avoid waterfalling. (a) In this design, waterfalling results when casting is filled from the bottom. (b) Improved design provides a path for the metal to follow as it fills the mold. Source: Ref 6

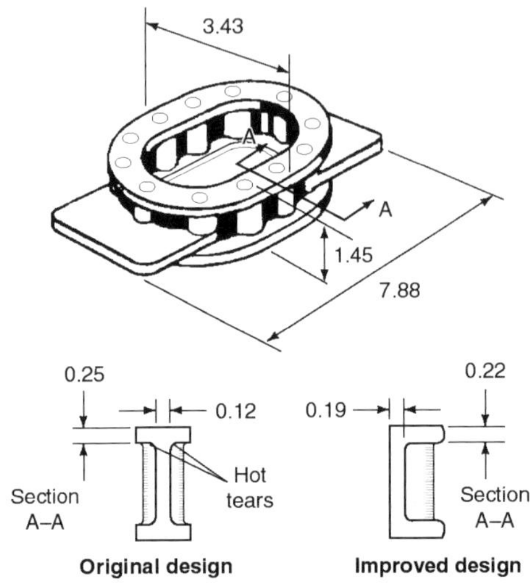

Fig. 11 Redesign of a casting to eliminate hot tears. Mold restraint coupled with nonuniform freezing of the various sections of this aluminum alloy 356 casting resulted in hot tears. Moving the wall and increasing its thickness corrected the problem. Part dimensions in inches. Source: Ref 7

the expansion of graphite at the end of solidification, forcing liquid metal into the mold if the casting is not properly designed with a tapered wall to promote directional solidification and avoid hot spots.

Surface Defects. Surface finish is also an important specification. Surface discontinuities affect fatigue life, and obviously, smoother surfaces are superior to rough surfaces. Designers should be certain that fatigue data used in design calculations have been taken from as-cast surfaces rather than machined surfaces, because most surfaces on castings where stress concentrations may be expected are not machined. Surface finish in castings is controlled by the application of coatings to the mold as well as proper selection of mold materials. Metal mold casting processes generally produce better surfaces than sand casting processes.

Design and Service Considerations. The existence of casting discontinuities does not, in and of itself, indicate that casting performance in service will be affected. Equally important are the size, location, and distribution of these discontinuities. Those discontinuities that are small and located near the center of the casting have little effect, while those located at or near the surface of the casting are usually damaging. Clustered discontinuities and those that occur in a regular array have a greater effect on properties than those that are isolated and randomly distributed. In specifying acceptable levels of discontinuities, such as microporosity and inclusion sizes and distribution, the designer should determine the critical flaw size that will deleteriously affect performance in service. This permits the foundry to design a casting practice that will eliminate such discontinuities at minimum cost.

Heat Treating Defects

Most of the problems in heat treated parts are attributed to faulty heat treatment practices, such as overheating and burning, nonuniform heating and quenching, a deficiency in the steel grade used, part defects, improper grinding, and/or poor part design.

Overheating and Burning of Low-Alloy Steels

When low-alloy steels are preheated to a high temperature (usually >1200 °C, or 2200 °F) prior to hot working (e.g., forging) for long times (overheated), the room-temperature mechanical properties, particularly tensile ductility and impact strength, can deteriorate after the steel is given a final heat treatment (i.e., reaustenitizing, quenching, and tempering). When hardened overheated steels are tested in impact at a temperature that would normally produce a ductile fracture, they fail in a brittle intergranular manner.

When a low-alloy steel is preheated prior to hot working at too high a temperature (normally >1400 °C, or 2550 °F), local melting occurs at the austenite grain boundaries (known as burning) as a result of the segregation of phosphorus, sulfur, and carbon. During cooling, dendritic sulfides initially form within the austenite grain boundaries, which then transform to ferrite. This results in excessively weak boundaries. Subsequent heat treatment provides a very poor impact strength and almost completely intergranular fracture surface on impact failure. Thus, burning occurs at a higher temperature than overheating. An example of severely overheated 1038 steel is shown in Fig. 12. If this occurs during forging, the forging will often break during cooling or subsequent heat treatment.

Residual Stresses

Heat treatment often causes stress- and strain-related problems such as residual stress, quench cracks, and deformation and/or distortion. Residual stresses can be defined as the self-equilibrating internal or locked-in stress remaining within a body with no applied (external) force, external constraint, or temperature gradient.

The major effects of residual stress include dimensional changes and resistance to crack initiation. Dimensional changes occur when the residual stresses (or a portion of it) in a body are eliminated. In terms of crack initiation, residual stresses can be either beneficial or detrimental, depending on whether the stresses are compressive or tensile.

Compressive Residual Stress. Because residual stresses are additive to any applied stresses, residual compressive stresses in the surface layers are generally helpful; they can reduce applied tensile stresses that may produce cracking or failure. Thus, compressive stresses contribute to the improvement of fatigue strength and the resistance to stress-corrosion cracking.

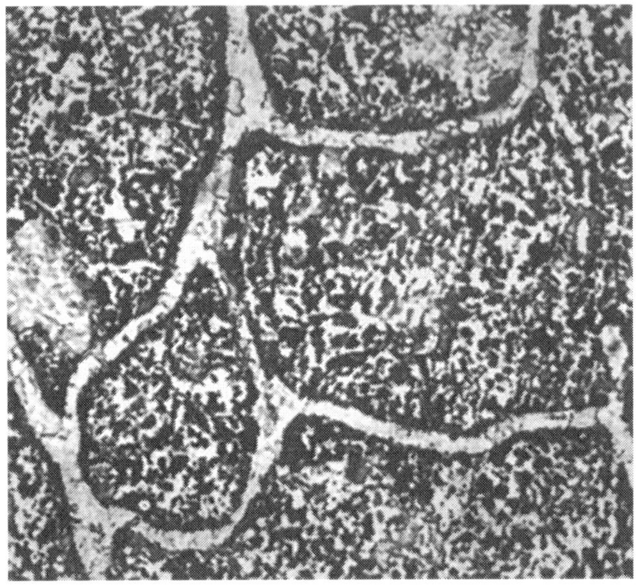

Fig. 12 Severely overheated 1038 steel showing initial stage of burning. Ferrite (white) outlines prior coarse austenite grain boundaries; matrix consists of ferrite (white) and pearlite (black). Source: Ref 8

As shown in Fig. 13, the endurance fatigue limit of the steels increases with increases in surface residual compressive stresses that resulted from heat treatment and/or surface processing. These fatigue improvements are significant in components that contain stress concentrations, such as notches, keyways, oil holes, crankshafts, and half-shafts. Methods to produce residual compressive stresses on the surfaces of parts include shrink fits, interference fits, coined holes, and mechanical surface-working processes such as shot peening and surface roiling.

Residual tensile stresses at the surface of a part are usually undesirable because they can effectively increase the stress levels and may cause quench cracking, grinding cracks, and unpredicted stress-corrosion cracking as well as reduce the fatigue life and strength. In extreme cases, the residual stresses may be so high that they are close to, or even larger than, the strength of the material.

Residual tensile stresses in the interior of a component may also be damaging because of the existence of internal defects. Delayed cracking, in the absence of adverse environments and large applied stresses, is attributed to the action of residual stresses on minute defects in the material. For example, a 17.5 cm (6.9 in.) diameter × 125 cm (49.2 in.) long steel shaft exploded into several pieces while just setting on the floor. Under normal loading, it would have required a tensile stress of greater than 150 MPa (22 ksi) to rupture the shaft.

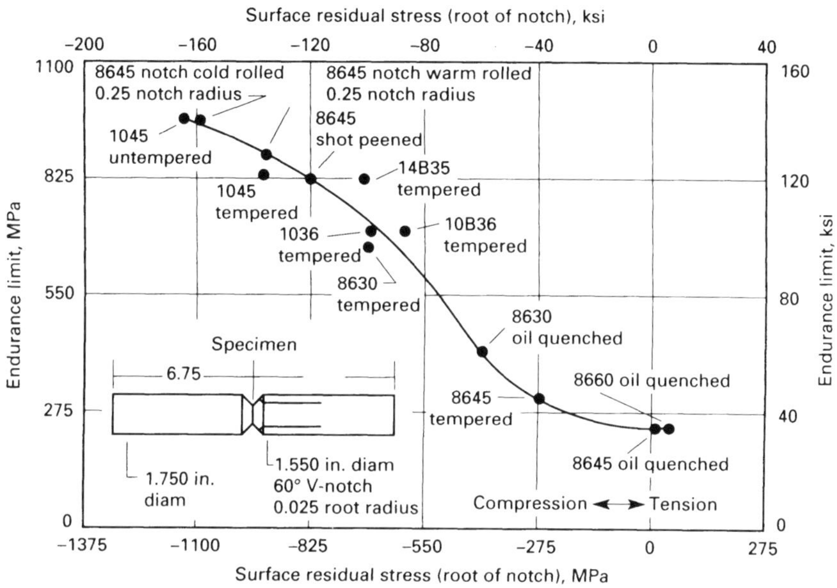

Fig. 13 Effect of surface residual stress on the endurance limit of selected steel. All samples were water quenched except as shown, and all specimen dimensions are given in inches. Source: Ref 9, 10

Development of Residual Stress in Processed Parts

Variations in stresses, temperature, and chemical species within the body during processing cause residual stresses. Various manufacturing processes, such as forming, machining, heat treatment, shot peening, casting, welding, flame cutting, and plating, can cause various residual-stress patterns. A summary of compressive and tensile residual stresses at the surface of parts fabricated by common manufacturing processes is given in Table 2.

In heat treated parts, residual stresses can be classified as those caused by a thermal gradient alone or a thermal gradient in combination with a structural change (e.g., a phase transformation). When a steel part is quenched from the austenitizing temperature to room temperature, a residual-stress pattern is established due to a combination of thermal gradients and any local transformation-induced volume expansion. Thermal contraction develops nonuniform thermal (or quenching) stresses due to different rates of cooling experienced by the part surface and interior. In steels, transformational volume expansion induces transformation stresses due to the transformation of austenite into martensite or other transformation products.

Thermal Contraction. The relation between the thermal stresses (σ_{th}) during cooling and the corresponding temperature gradient is given by:

$$\sigma_{th} = \alpha E \Delta T \tag{Eq 1}$$

Table 2 Summary of compressive and tensile residual stresses at the surface of the parts created by common manufacturing processes

Compression at the surface	Tension at the surface
Surface working: shot peening, surface rolling, lapping, and so on	Rod or wire drawing with deep penetration
Rod or wire drawing with shallow penetration(a)	Rolling with deep penetration
Rolling with shallow penetration(a)	Swaging with deep penetration
Swaging with shallow penetration(a)	Tube sinking of the outer surface
Tube sinking of the inner surface	Plastic bending of the shortened side
Coining around holes	Grinding: normal practice and abusive conditions
Plastic bending of the stretched side	Direct-hardening steel (through-hardened)(b)
Grinding under gentle conditions	Decarburization of steel surface
Hammer peening	Weldment (last portion to reach room temperature)
Quenching without phase transformation	Machining: turning, milling
Direct-hardening steel (not through-hardened)	Built-up surface of shaft
Case-hardening steel	Electrical discharge machining
Induction and flame hardening	Flame cutting
Prestressing	

(a) Shallow penetration refers to ≤1% reduction in area of thickness; deep penetration refers to ≥1%. (b) Depends on the efficiency of the quenching medium. Source: Ref 11

where E is the modulus of elasticity, and α is the thermal coefficient of expansion. It is apparent that thermal stresses are greatest for materials with a high elastic modulus and a high coefficient of thermal expansion. The temperature gradient is also a function of thermal conductivity. Hence, it is quite unlikely to develop high-temperature gradients in materials that are good thermal conductors (e.g., copper and aluminum), but it is much more likely with materials with lower thermal conductivities, such as steel and titanium.

Residual-Stress Pattern due to Thermal Contraction. Even if any solid-state phase changes are ignored, thermal volume changes during quenching cause residual stresses. The development of longitudinal thermal and residual stresses in a 100 mm (4 in.) diameter steel bar on water quenching from the austenitizing temperature (850 °C, or 1560 °F) are shown in Fig. 14. At the start of cooling, the surface temperature (S) falls drastically as compared to the center temperature (C) (Fig. 14a). At time w, the temperature difference between the surface and core is at a maximum of approximately 550 °C (1020 °F), corresponding to a thermal stress of 1200 MPa (174 ksi) due to a linear differential contraction of approximately 0.6%. Under these conditions, tensile stresses are developed at the surface with a maximum value of a (Fig. 14b), corresponding to time w in Fig. 14(a), and the core will contract, producing compressive stresses with a maximum of c. The combined effects of tensile and compressive stresses on the surface and core, respectively, will result in residual stresses as indicated by curve C, where a complete neutralization of stress will occur at some lower temperature u. Therefore, further decreases in temperature produces longitudinal, compressive residual stresses at the surface and the tensile stresses at the core, as shown in Fig. 14(c). A schematic illustration of the distribution of residual stress over the diameter of

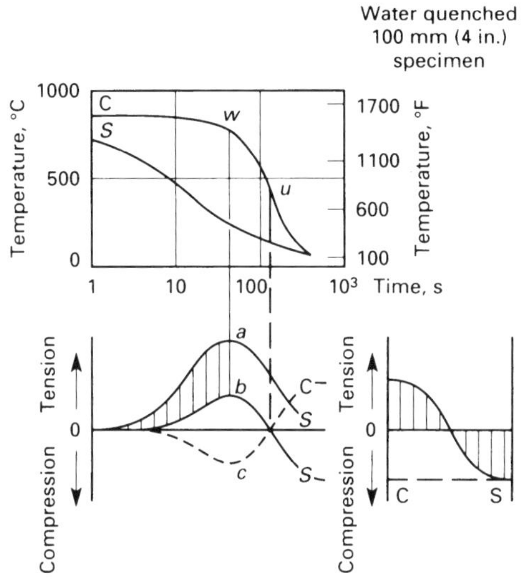

Fig. 14 Development of thermal and residual stresses in the longitudinal direction in a 100 mm (4 in.) diameter steel bar on water quenching from the austenitizing temperature (850 °C, or 1560 °F). Transformation stresses are not taken into consideration. Source: Ref 12

a quenched bar due solely to thermal contraction in the longitudinal, tangential, and radial directions is shown in Fig. 15(a). The maximum residual stress produced during quenching increases as the quenching temperature and quenching power of the coolant are increased.

Residual-Stress Pattern due to Thermal and Transformational Volume Changes. During quench hardening of a steel, hard martensite forms at the surface layers, producing a volume expansion, whereas the remainder of the part is still hot and ductile austenite. Later, the remainder austenite transforms to martensite, but its volumetric expansion is restricted by the hardened surface layer. This restraint causes the central portion to be under compression, with the outer surface under tension. Figure 15(c) illustrates the residual-stress distribution over the diameter of a quenched bar showing volume expansion associated with phase transformation in the longitudinal, tangential, and radial directions. At the same time, during final cooling of the interior, its contraction is hindered by the hardened surface layers. This restraint in contraction produces tensile stresses in the interior and compressive stresses at the outer surface. However, the situation shown in Fig. 15(c) prevails, provided that the net volumetric expansion in the interior, after the surface has hardened, is larger than the remaining thermal contraction. In some instances, these volumetric changes can produce sufficiently large residual stresses that can cause plastic deformation on cooling, leading to warping or distortion of the steel part. While plastic deformation appears to reduce the severity of quenching stresses, if the quench is severe, the quenching stresses are so high that they are not sufficiently relieved by plastic deformation. Consequently, the large residual stress remaining may reach or even exceed the fracture stress of steel. This localized rupture or fracture is called quench cracking.

It should be emphasized that for a given grade of steel, both a large part size and higher quenching speed contribute to a larger value of thermal contraction, as compared to the volumetric expansion of martensite. In contrast, when the parts are thin and the quenching rate is not high, thermal contraction of the part subsequent to the hardening of the surface will be smaller than the volumetric expansion of martensite. Similarly, for a given quenching rate, the temperature gradients decrease with decreasing section thickness, and consequently, the thermal component of residual stress also decreases.

Although shallower-hardening steels exhibit higher surface compressive stresses, deep-hardening steels may develop moderately high surface compressive stresses with severe water quenching. When these deep-hardening steels are through-hardened in a less efficient quenchant, they

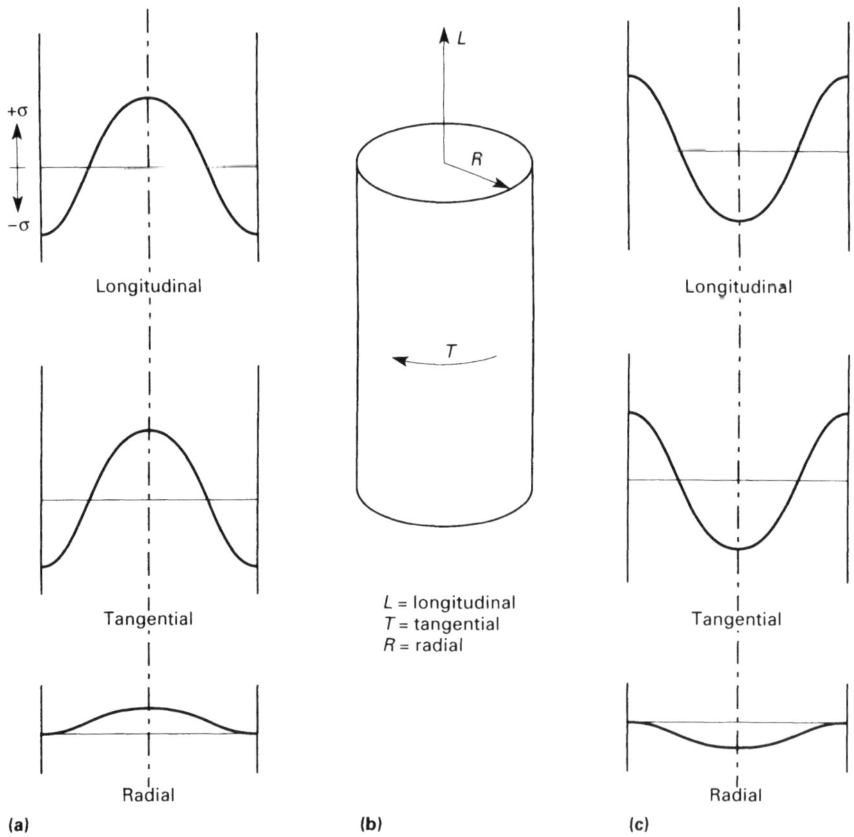

Fig. 15 Schematic illustration of the distribution of residual stress over the diameter of a quenched bar in the longitudinal, tangential, and radial directions due to (a) thermal contraction and (c) both thermal and transformational volume changes. (b) Schematic illustration of orientation of directions. Source: Ref 12

may exhibit surface tensile stresses. In general, thermochemical and thermal surface-hardening treatments, such as carburizing or nitriding, produce beneficial compressive residual stresses at the surface.

Quench Cracking

Anything that produces excessive quenching stresses is the basic cause of cracking. Quench cracking is primarily intergranular, and it may be related to some of the same factors that cause intergranular fracture in overheated and burned steels. Typical quench cracks are shown in Fig. 16. The main factors contributing to cracking during heat treatment are part design, steel grades, part defects, heat treating practice, and tempering practice.

Part Design. Features such as sharp corners, the number, location, and size of holes, deep keyways, splines, and abrupt changes in section thickness within a part (i.e., a badly unbalanced section) enhance crack formation, because while the thin areas are cooling quickly in the quenchant, the adjacent thick areas are cooling very slowly. One solution to this problem is to change the material so that a less drastic quenchant (e.g., oil) can be used.

Steel Grades. In general, the carbon content of steel should not exceed the required level; otherwise, the risk of cracking will increase. For example, a decrease in carbon content from 0.72 to 0.61% has been shown to

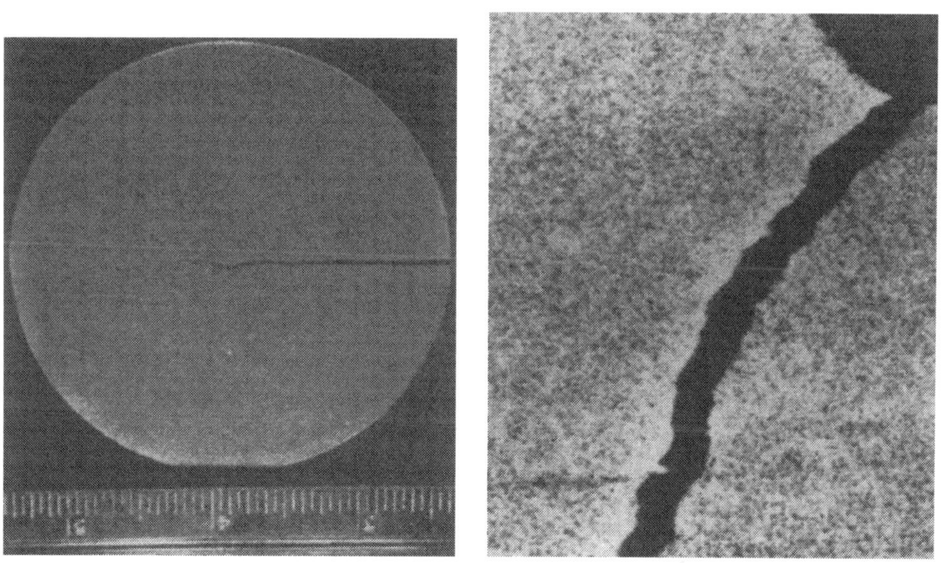

(a) (b)

Fig. 16 Examples of quench cracks. (a) Micrograph of AISI 4340 quenched and tempered steel illustrating a macroetched pure quench crack. (b) Micrograph of AISI 4142 steel as-quenched and tempered. The microstructure is tempered martensite with quench cracking at the fillet radius. Original magnification: 100×; 3% nital etch. Source: Ref 13 slightly increase the crack resistance of rim steel railroad wheels. Because of the segregation of carbon and alloying elements, some steels are more prone than others to quench cracking. Dirty steels, that is, steels with more than 0.05% S, are more susceptible to cracking than low-sulfur grades. The reasons for this are that they are more segregated in alloying elements, the surface of this hot-rolled high-sulfur steel has a greater tendency to form seams, which act as stress raisers during quenching, and they are usually coarse grained for better machinability, which increases brittleness and therefore promotes cracking. If these high-sulfur grades are replaced by calcium-treated steels or cold-finished leaded steels, this problem can be eliminated.

Part Defects. Surface defects or weakness in the material can also cause cracking. For example, deep surface seams or nonmetallic stringers in both hot-rolled and cold-finished bars can contribute to cracking. Other defects are inclusions and stamp marks. For large seam depths, it is advisable to use turned bars or screen them using magnetic-particle inspection. Forging defects in small forgings, such as seams, laps, flash line, or shear cracks, as well as hydrogen flakes and internal ruptures in heavy forgings aggravate cracking.

Heat Treating Practice. Higher austenitizing temperatures increase the tendency toward quench cracking. In addition, steels with a coarser grain size are more prone to cracking than fine-grained steels, because fine-grained steels possess more grain-boundary area that acts as crack barriers, and grain boundaries help to absorb and redistribute residual stresses. A contributor to severe cracking is improper heat treating practice, for example, nonuniform heating and nonuniform cooling. It is a good heat treating practice to anneal alloy steels prior to the hardening treatment, because this produces grain-refined microstructures and relieves stresses.

Water-hardening steels are most susceptible to cracks if they are not handled properly. Soft spots are most likely to occur in the water-hardening steels, especially where the part is grabbed with tongs for quenching. Normally, clean surfaces show adequate hardening and scaled surfaces show insufficient hardening. Most large parts that are water hardened contain some soft spots.

Decarburized Steel. Decarburization is usually caused from insufficient atmospheric protection during heat treatment. This can result from plant failures (e.g., defective furnace or container seals), poor process control (e.g., insufficient atmosphere-monitoring equipment), or the existence of decarburizing agents in the furnace atmosphere (e.g., CO_2, water vapor, and H_2 in endogas).

A partially decarburized surface during hardening also contributes to cracking, because the martensite transformation on the surface is completed well before the formation of martensite in the core. Decarburized surfaces can also reduce hardness, which will lead to premature wear and scuffing. Partial decarburization must be avoided, especially for deep-hardening steels, either by providing some type of protective atmosphere during the heating operation, stock removal by grinding, or by a carbon-restoration process. In addition to a protective atmosphere, salt baths, inert packs, or vacuum furnaces can be used to obtain the desired surface chemistry.

Carburized Alloy Steel. Two types of cracking occur in the carburized and hardened case of carburized alloy steels: microcracking and tip cracking. Microcracking of quenched steels is small cracks that occur across or alongside martensite plates and prior-austenite grain boundaries. They mostly form on quenched steel parts that contain chromium and/or molybdenum as the major alloying elements and where the hardening is done by direct quenching.

Microcracks are more frequently observed in coarse-grained structures, such as large martensite plates. This is presumably because of more impingements of the larger plates of martensite by other large plates. Another cause of microcracking is the increased carbon content of martensite (that is, increased hardenability), which is a function of austenitizing temperature and/or time. This problem can be avoided by selecting a steel with less hardenability (that is, with a lower austenitizing temperature). Another solution is to change the heat treating cycle to carburizing, slow cooling to black temperature, reheating to, for example, 815 or 845 °C (1500 or 1550 °F), and then quenching.

Microcracking in case-hardened surfaces can be aggravated by the presence of hydrogen, which it tends to absorb during carburizing. However, hydrogen-enhanced microcracking can be eliminated by tempering carburized parts at 150 °C (300 °F) immediately after quenching. Tempering has an additional beneficial effect in that it has the ability to heal the microcracks due to the volume changes and associated plastic flow that develop during the first stage of tempering. No adverse effect of the influence of microcracks on mechanical properties has been reported.

Tip cracking is cracking that occurs in the teeth of carburized and quenched gears and runs partly or fully to the ends of the teeth in a direction parallel to the axis of the part. Many heat treaters have solved this problem by decreasing the carbon content and case depth to the minimum acceptable design level or by copper plating the outer diameter of the gear blank prior to hobbing.

Nitrided Steels. Nitrided cases are very brittle. Consequently, cracking may occur in service prior to realizing any improved wear and galling resistance. This can be avoided by a proper design, for example, incorporating all section changes with a minimum radius of 3 mm (0.125 in.).

Tempering Practice. The longer the time the steel is kept at temperatures between room temperature and 100 °C (212 °F) after the complete transformation of martensite in the core, the more likely the occurrence of quench cracking. This is caused by the continuing volumetric expansion caused by the isothermal transformation of retained austenite into martensite at room temperature.

There are two tempering practices that lead to cracking problems: tempering too soon after quenching, that is, before the steel has finished transforming to martensite during quenching, and skin tempering, usually observed in heavy sections (≥50 mm, or 2 in., thick in plates and >75 mm, or 3 in., in diameter in round bars).

It is normal practice to temper immediately after quenching. In this case, some restraint must be exercised, especially for large sections (>75 mm, or 3 in.) in deep-hardening alloy steels. The reason is that the core has not yet completed its transformation to martensite and is still expanding, whereas the surface and/or projections, such as flanges, begin to temper and try to shrink. This simultaneous volume change produces radial cracks. This problem can become severe if rapid heating (e.g., induction, flame, lead, or molten salt bath) is used for tempering. Therefore, very large and very intricate steel parts should be removed from the quenching medium, and tempering should be started when they are warm enough to comfortably hold with bare hands (~50 °C, or 120 °F).

Skin tempering occurs in heavy-section parts when the final hardness is >360 HB. This is due to insufficient tempering time and is usually determined when the surface hardness falls by 5 or more HRC points from the core hardness. Cracking often occurs several hours after the component has cooled from the tempering temperature and often runs through the entire cross section. This problem can be eliminated by retempering for 3 h at the original tempering temperature, which will produce a change in hardness of ~2 HRC points.

Distortion in Heat Treatment

Distortion can be defined as an irreversible and usually unpredictable dimensional change in the component during heat treatment and from temperature variations and loading in service. The term *dimensional change* is used to denote changes in both size and shape. Therefore, *heat treatment distortion* is a term often used by engineers to describe an uncontrolled movement that has occurred in a component as a result of heat treatment. Although it is recognized as one of the most difficult and troublesome problems confronting the heat treatment industries on a daily basis, it is only in the simplest thermal heat treatment methods that the mechanisms of distortion are understood. Changes in size and shape of steel parts can be either reversible or irreversible.

Reversible changes, which are produced by applying stress in the elastic range or by temperature variations, do not induce stresses that are above the elastic limit nor cause changes in the metallurgical structure. In this situation, the initial dimensional values can be restored to their stress state. Irreversible changes in size and shape of steel parts are those that are caused by stresses that exceed the elastic limit or by changes in the metallurgical structure (e.g., phase changes). These dimensional changes sometimes can be corrected by mechanical processing to remove extra and unwanted material or to redistribute residual stresses, or by heat treatment by annealing or tempering.

When heat treated parts suffer from distortion beyond the permissible limits, it may lead to scrapping of the article, rendering it useless for the service for which it was intended, or it may require correction. Allowable distortion limits vary to a large extent, depending on service applications. In cases where very little distortion can be tolerated, specialty steels are used. These steels possess metallurgical characteristics that minimize distortion.

Types of Distortion

Distortion is a general term that involves all irreversible dimensional change produced during heat treatment. This can be classified into two categories: (1) size distortion, which is the net change in specific volume between the parent and transformation products produced by phase transformations without a change in geometrical form, and (2) shape distortion or warpage, which is a change in geometrical form or shape and is revealed by changes in curvature, bending, twisting, and/or nonsymmetrical dimensional changes without any volume change. Usually both types of distortion occur during a heat treatment cycle.

Dimensional Changes Caused by Changes in Metallurgical Structure during Heat Treatment. Various dimensional changes produced by a change in metallurgical structure during the heat treatment cycle of steels are described as follows.

Heating (Austenitizing). When annealed steel is heated from room temperature, thermal expansion occurs continuously up to the temperature at which austenite begins to form on heating (A_{c1}), where the steel contracts as it transforms from body-centered cubic ferrite to face-centered cubic austenite. Further heating expands the newly formed austenite.

Hardening. When austenite is quickly cooled, martensite forms; at intermediate cooling rates, bainite forms; and at slow cooling rates, pearlite precipitates. The volume increase is a maximum when austenite transforms to martensite, intermediate with lower bainite, and least with upper bainite and pearlite. The volume increases associated with the transformation of austenite to martensite in 1 and 1.5% carbon steels are 4.1 and 3.84%, respectively, and the volume increases in the transformation of austenite to pearlite in the same steels are 2.4 and 1.33%, respectively. Volume increases are less in alloy steels. It should be noted that plastic deformation (or strain) occurs during such transformations at stresses that are lower than the yield stress for the phases present. The occurrence of this plastic deformation, called the transformation plasticity effect, influences the development of stresses during the hardening of steel parts. During quenching from the austenite range, the steel contracts until the martensite start (M_s) temperature is reached, then expands during martensitic transformation. Finally, thermal contraction occurs on further cooling to room temperature. As the hardening temperature increases, a greater amount of carbide goes into solution, and consequently, both the grain size and the amount of retained austenite are increased. This also increases the hardenability of steel.

More trouble with distortion comes from the quenching or hardening operation than during heating for hardening, in which the faster the cooling rate (that is, the more severe the quenching), the greater the danger of distortion. When the milder quenchants are used, the distortion is less. Thus, the severity of quenching influences distortion.

The dependence of volume increases, particularly in steels of different dimensions, on grain size (or hardenability) is another important factor. Variations in volume during quenching of a fine-grained shallow-hardening steel in all but small sections is less than a coarse-grained deep-hardening steel of the same composition.

Tempering. There is a certain correlation between the tempering temperature and volume changes. Tempering reduces the volume of martensite but not enough to completely equalize the prior volume increase as a result of martensitic transformation. In low-alloy and plain medium- and high-carbon steels, during the first and third stages of tempering, a decrease in volume occurs that is associated with the decomposition of high-carbon martensite into low-carbon martensite plus ε-carbide during the first stage, and aggregate of low-carbon martensite and ε-carbide into ferrite plus cementite in the third stage. However, in the second stage, an increase in volume takes place (due to the decomposition of retained austenite into bainite) that tends to compensate for the early volume reduction. As the tempering temperature is increased further toward A_{c1}, a more pronounced volume reduction occurs. In some highly alloyed steel compositions, the volume changes during martensite formation are less pronounced because of the large proportion of retained austenite and the resistance to tempering of the alloy-rich martensite. These hardened steels show sharp increases in both hardness and volume between 500 and 600 °C (930 and 1110 °F), owing to the precipitation of very finely dispersed alloy carbides from the retained austenite. This produces a depleted matrix in alloy content, raising the M_s temperature of retained austenite. During cooling from the tempering temperature, further transformation of retained austenite into martensite will occur with an additional increase in volume.

Shape Distortion or Warpage. This is sometimes called straightness or angularity change. It occurs in nonsymmetrical components during heat treatment. From a practical viewpoint, warpage in water- or oil-hardening steels is normally of greater magnitude than is size distortion and is more of a problem because it is usually not predictable. This is caused by the sum effect of more than one of the following factors:

- Rapid heating (or overheating), drastic (or careless) quenching, or nonuniform heating and cooling cause severe shape distortion. Slow heating, as well as preheating, prior to heating to the austenitizing temperature yields the most satisfactory result. Rapid quenching produces thermal and mechanical stresses due to the martensitic transformation. In the case of low- and high-hardenability steels, respectively, this problem becomes severe or very small.
- Residual stresses present in the component before heat treating. These arise from machining, grinding, straightening, welding, casting, spinning, forging, and rolling operations, which will contribute to a shape change.
- Applied stress causing plastic deformation. Sagging and creep of the components can occur during heat treatment as a result of improper support of components or a warped hearth in the hardening furnace. Hence, large, long, and complex parts must be properly supported at critical positions to avoid sagging or, preferably, are hung with the long axis in a vertical orientation.
- Nonuniform agitation/quenching or nonuniform circulation of quenchant around a part results in an assortment of cooling rates that creates shape distortion. Uneven hardening, with the formation of soft spots, increases warpage. Similarly, an increase in case depth, particularly uneven case depths in case-hardening steels, increases warpage on quenching.
- Tight, thin, and highly adherent scale and decarburization, at least in certain areas. Tight scale is usually a problem encountered in forgings hardened from direct-fired gas furnaces having high-pressure burners. Quenching in areas with a tight scale is extremely retarded compared to the areas where the scale comes off. This produces soft spots and, in some cases, severe unpredicted distortion. Some heat treaters coat the components with a scale-loosening chemical prior to placing in the furnace. Similarly, the areas beneath the decarburized surface do not harden as completely as the areas below the nondecarburized surface. The decarburized layer also varies in depth and produces an inconsistent softer region as compared to the region with full carbon. All these factors can cause a condition of unbalanced stresses with resultant distortion.
- Long parts with small cross sections ($>L = 5d$ for water quenching, $>L = 8d$ for oil quenching, and $>L = 10d$ for austempering, where L is the length of the part, and d is its diameter or thickness)
- Thin parts with larger areas ($>A = 50t$, where A is the area of the part, and t is its thickness)
- Unevenness or great variations in section

Weld Discontinuities

Weld discontinuities are interruptions in the desirable physical structure of a weld. A discontinuity constituting a danger to the fitness-for-service of a weld is a defect. By definition, a defect is a condition that must be removed or corrected. Neither construction materials nor engineered structures are free from imperfections, and welds and weld repairs are not exceptions. Weld acceptance standards are used when a discontinuity has been clearly located, identified, sized, its orientation determined, and its structural significance questioned. Critical engineering assessments of weld discontinuities are performed to define acceptable, harmless discontinuities in a structure that will not sacrifice weldment reliability. One of the major reasons for understanding the engineering meaning of weld discontinuities is to decrease the cost of welded structures by avoiding unnecessary repairs of harmless weld discontinuities. The commonly encountered inclusions, as well as cracking, which is the most serious of weld defects, are discussed in this section. The fatigue of welded joints and the methods for improving fatigue life are discussed in Chapter 7, "Metallic Joints—Mechanically Fastened and Welded," in this book.

Gas Porosity

Gas porosity can occur on or just below the surface of a weld. Pores are characterized by a rounded or elongated teardrop shape with or without a sharp point. Pores can be uniformly distributed throughout the weld or isolated in small groups. They can also be concentrated at the root or toe of the weld. Porosity in welds is caused by gas entrapment in the molten metal, by too much moisture on the base or filler metal, or by improper cleaning of the joint during preparation for welding. The type of porosity within a weld is usually designated by the amount and distribution of the pores. Types include uniformly scattered porosity, cluster porosity, linear porosity, elongated porosity, and wormhole porosity. Radiography is the most widely used nondestructive method for detecting subsurface gas porosity in weldments.

Slag Inclusions

Slag inclusions can occur when using welding processes that employ a slag covering for shielding purposes. With other processes, the oxide present on the metal surface before welding may also become entrapped. Slag inclusions can be found near the surface and in the root of a weld (Fig. 17a), between weld beads in multipass welds (Fig. 17b), and at the side of a weld near the root (Fig. 17c).

During welding, slag may spill ahead of the arc and subsequently be covered by the weld pool because of poor joint fitup, incorrect electrode manipulation, or forward arc blow. Slag trapped in this manner is gener-

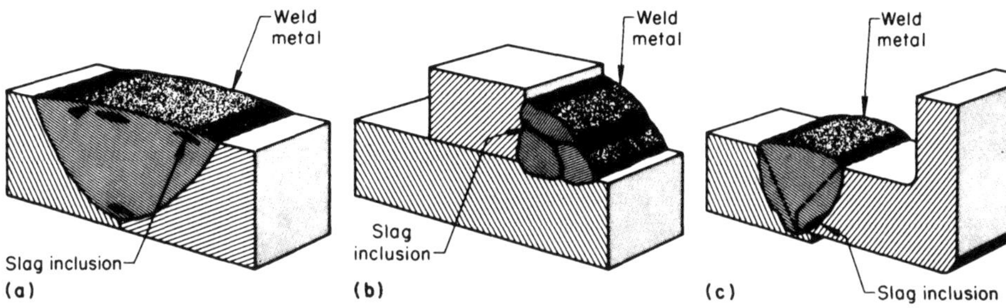

Fig. 17 Sections showing locations of slag inclusions in weld metal. (a) Near the surface and in the root of a single-pass weld. (b) Between weld beads in a multipass weld. (c) At the side of a weld near the root. Source: Ref 14

ally located near the root. Radical motions of the electrode, such as wide weaving, can also cause slag entrapment on the sides or near the top of the weld after the slag spills into a portion of the joint that has not been filled by the molten pool. Incomplete removal of the slag from the previous pass in multipass welding is another common cause of entrapment. In multipass welds, slag can be entrapped any number of places in the weld between passes. Slag inclusions are generally oriented along the direction of welding. Three methods used for the detection of slag below the surface of single-pass or multipass welds are magnetic-particle, radiographic, and ultrasonic inspection.

Tungsten Inclusions

Tungsten inclusions are particles found in the weld metal from the nonconsumable tungsten electrode used in gas tungsten arc welding. These inclusions are the result of:

- Exceeding the maximum current for a given electrode size or type
- Letting the tip of the electrode make contact with the molten weld pool
- Using an excessive electrode extension
- Inadequate gas shielding or excessive wind drafts, which result in oxidation
- Using improper shielding gases such as argon-oxygen or argon-CO_2 mixtures, which are used for gas metal arc welding

Tungsten inclusions, which are not acceptable for high-quality work, can only be found by internal inspection techniques, particularly radiographic testing. In general, they must be ground out and repair welded.

Lack of Fusion and Lack of Penetration

Lack of fusion and lack of penetration result from improper electrode manipulation and the use of incorrect welding conditions. Fusion refers to the degree to which the original base-metal surfaces to be welded have been fused to the filler metal. Penetration refers to the degree to which the base metal has been melted and resolidified to result in a deeper throat than what was present in the joint before welding. In effect, a joint can be completely fused but have incomplete root penetration to obtain the throat size specified. Based on these definitions, lack-of-fusion discontinuities are located on the sidewalls of a joint, and lack-of-penetration discontinuities are located near the root (Fig. 18). With some joint configurations, such as butt joints, the two terms can be used interchangeably. The causes of lack of fusion include excessive travel speed, bridging, excessive electrode size, insufficient current, poor joint preparation, overly acute joint angle, improper electrode manipulation, and excessive arc blow. Lack of penetration can be the result of low welding current, excessive travel speed, improper electrode manipulation, or surface contaminants such as oxide, oil, or dirt that prevent full melting of the underlying metal.

Radiographic methods may be unable to detect these discontinuities in certain cases, because of the small effect they have on x-ray absorption. However, lack of sidewall fusion is readily detected by radiography. Ultrasonically, both types of discontinuities often appear as severe, almost continuous, linear porosity because of the nature of the unbonded areas of the joint. Except in thin sheet or plate, these discontinuities may be too deep to be detected by magnetic-particle inspection.

Geometric Weld Discontinuities

Geometric weld discontinuities are those associated with imperfect shape or unacceptable weld contour. Undercut, underfill, overlap, excessive reinforcement, fillet shape, and melt-through are included in this grouping. Geometric discontinuities are shown schematically in Fig. 19. Visual inspection and radiography are most often used to detect these flaws.

Cracking Associated with Welding

Four types of cracking are a concern in welded structures: hot cracks, heat-affected zone (HAZ) microfissures, cold cracks, and lamellar tearing.

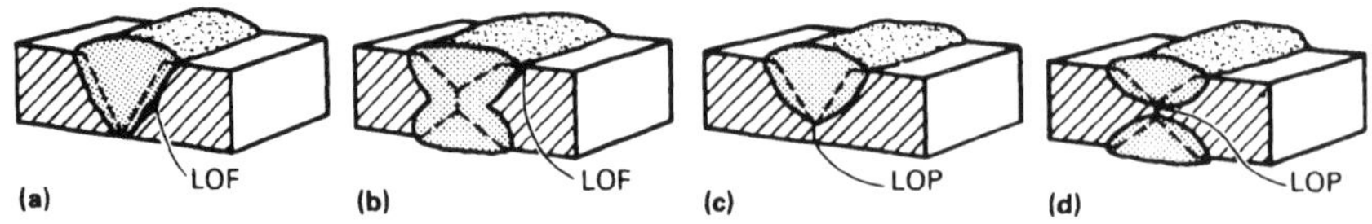

Fig. 18 Lack of fusion (LOF) in (a) a single-V-groove weld, and (b) double-V groove weld. Lack of penetration (LOP) in (c) a single-V-groove weld, and (d) a double-V-groove weld. Source: Ref 14

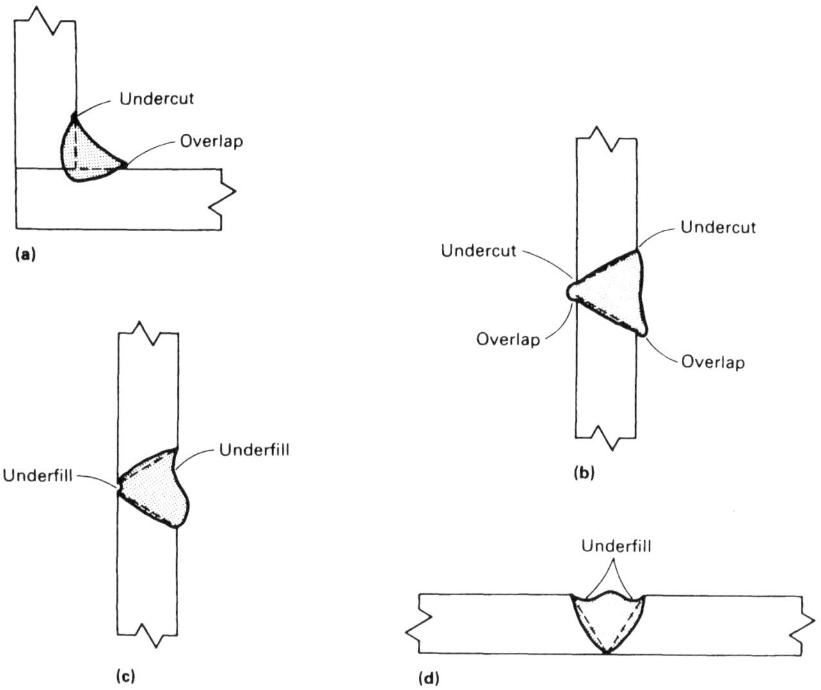

Fig. 19 Weld discontinuities affect weld shape and contour. (a) Undercut and overlapping a fillet weld. (b) Undercut and overlapping in a groove weld. (c) and (d) Underfill in groove welds. Source: Ref 14

Solidification Cracking (Hot Cracking). Hot cracks are solidification cracks that occur in the fusion zone near the end of solidification. They result from the inability of the semisolid material to accommodate the thermal shrinkage strains associated with weld solidification and cooling. Cracks then form at susceptible sites to relieve the accumulating strain. Susceptible sites are interfaces, such as solidification grain boundaries and interdendritic regions, that are at least partially wetted.

Solidification cracking requires both a sufficient amount of mechanical restraint (strain) and a susceptible microstructure. Under conditions of rapid solidification and cooling, the rate of strain accumulation is rapid, leading to an increased cracking susceptibility. Strains for solidification cracking are more likely to be experienced with welding processes that promote rapid solidification and cooling. The use of preheating and controlled heating and cooldown rates help to prevent hot cracking.

The approach used to minimize the mechanical factor is to reduce the overall weld restraint through joint design and appropriate choice of welding parameters. A simple way to minimize the restraint on a solidifying weld joint is to keep the joint gap to a minimum by designing hardware with good fitup. Welding parameters can also have a profound influence on the occurrence of solidification cracking. The natural tendency to use high-speed welding to improve productivity can have detrimental effects. Formation of a teardrop-shaped weld pool, which may occur as the weld travel speed increases, can result in centerline solidification cracks. The solidification pattern associated with this type of weld pool is such that solidifying grains meet at the weld centerline, forming a particularly susceptible site for solidification crack initiation.

Alloys with a wide solidification temperature range are more susceptible to solidification cracking than alloys that solidify over a narrow temperature range. This occurs because accumulated thermal strain is proportional to the temperature range over which a material solidifies. Composition effects in steel alloys that can cause hot cracking include high carbon contents, high alloy contents, and high sulfur contents. Manganese additions are frequently used to tie up sulfur in the form of harmless globular MnS particles.

Heat-Affected Zone Cracks. Microfissures are cracks that occur in the area of partial melting and the HAZ adjacent to the fusion line. Microfissures often occur as a result of local chemical variations in the metal that result in local variations in the melting point. Segregation of specific alloying elements to grain boundaries may cause a reduction in the melting temperatures of these areas. During fusion welding, this manifests itself in the formation of a zone of partial melting at temperatures below which the alloy liquidus exists.

Many microfissures form in a manner somewhat analogous to the formation of solidification cracks, in that susceptible sites, generally grain boundaries that intersect the fusion zone, are wetted with liquid from one of various sources. Shrinkage strains accumulating as the weld pool advances past the liquated boundary can develop to a level sufficient to cause boundary separation (i.e., cracking). Figure 20 is a micrograph of an HAZ microfissure in a cast austenitic stainless steel (this crack also extends into the fusion zone as a solidification crack). Cast microstructures

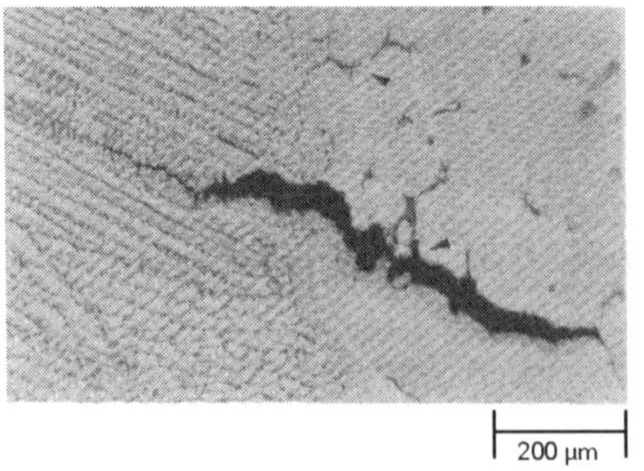

Fig. 20 Heat-affected zone microfissure in a cast stainless steel. Note extensive region of partial melting. Source: Ref 15

are highly inhomogeneous, raising the possibility of extensive partial melting. Note the melting associated with interdendritic constituents in Fig. 20.

The presence of a liquid is not essential to the formation of HAZ cracks, because they may occur at temperatures well below the solidus. Poor ductility inherent in certain materials, such as intermetallics, can lead to solid-state failure during welding if the thermal stresses associated with the weld thermal cycle exceed the local tensile strength. Cracks can also occur in alloys that undergo phase transformations during the weld thermal cycle. The act of fusion welding subjects the fusion zone and the HAZ to temperatures above those normally encountered during prior deformation processing. The resultant HAZ microstructure may be substantially different from that of the base material. If the newly transformation microstructure has limited ductility, which is not uncommon in shear transformations or in transformations that result in ordered structures, the likelihood of cracks forming at susceptible sites, such as grain boundaries, matrix/minor-phase interfaces, and slip band intersections, is increased.

Reheat cracking, also referred to as stress-relief or strain age cracking, is another defect type observed in certain alloys that undergo precipitation reactions. Historically observed in several nickel-base superalloys and creep-resistant steels, this type of defect manifests itself during postweld heat treatment, generally as intergranular HAZ cracks. Postweld heat treatment is often a recommended practice for high-strength, thick-section welded steels, where a reduction of residual stresses developed as a result of welding is desired. In the case of welded nickel-base superalloys, postweld heat treatment is used both to relieve residual stresses and to achieve optimum mechanical properties through precipitation-hardening reactions.

In alloys that undergo precipitation reactions, the rate at which the alloy strengthens may greatly exceed the rate at which residual stresses are thermally diminished. This can be especially true in the case of heavy-section welds or in alloys with relatively poor thermal conductivity. That is, before the alloy can reach the temperature at which residual stresses begin to be eliminated, it has aged and lost ductility relative to its as-welded state. At this new lower level of ductility, the residual stresses present may induce cracks to eliminate the accumulated strain energy. In principle, the problem of reheat cracking could be largely eliminated if the rate of heating through the precipitation temperature range was rapid enough to prevent precipitate formation. However, extremely rapid rates of heating in welded hardware would likely lead to problems such as excessive distortion; therefore, improved alloys (e.g., alloy 718 among the nickel-base superalloys) have been developed in which the kinetics of precipitate formation have been sufficiently retarded to allow for successful postweld heat treatment.

Hydrogen-Induced Cracking (Cold Cracking). Cold cracks are defects that form as the result of the contamination of the weld microstructure by hydrogen. Whereas solidification cracking and HAZ cracking occur during or soon after the actual welding process, hydrogen-induced cracking is usually a delayed phenomenon, occurring possibly weeks or even months after welding. The temperature at which these defects tend to form ranges from −50 to 150 °C (−60 to 300 °F) in steels. The fracture can either be intergranular or transgranular cleavage.

As with other forms of cracking, hydrogen-induced cracking involves both a requisite microstructure and a threshold level of stress. It also involves a critical level of hydrogen, which is alloy and microstructure dependent. In the case of ideal weld processing, hydrogen-induced cracking would be, at most, a minor welding engineering concern. However, excluding hydrogen from structures during welding is exceedingly difficult. Although the primary source of hydrogen in weld metal is considered to be the disassociation of water vapor in the arc and absorption of gaseous or ionized hydrogen into the liquid, other sources are also available. All organic compounds contain hydrogen in their molecular structure, and all may be broken down in the intense thermal environment of a welding heat source. Organic compounds are ubiquitous in the welding environment, from lubricants in assembly areas to body oils on the hands of welding operators. Plated hardware may also contain high levels of residual hydrogen.

The mechanism of hydrogen-induced crack formation is still being investigated. The most widely accepted model involves the presence of preexisting defect sites in the material—small cracks or discontinuities caused by minor phase particles or inclusions. In the presence of existing stress, these sites may develop high local areas of biaxial or triaxial tensile stress. Hydrogen diffuses preferentially to these sites of the dilated lattice structure. As the local hydrogen concentration increases, the cohesive energy and stress of the lattice decreases. When the cohesive stress falls below the local intensified stress level, fracture occurs spontaneously. Hydrogen then evolves in the crack volume, and the process is repeated. This model of hydrogen-induced cracking is consistent with the relatively slow and discontinuous nature of the process.

In steels, where the problem of hydrogen-induced cracking is significant, cracking susceptibility has been correlated both with material hardness and strength and with specific microstructure. Higher-strength steels are more susceptible to hydrogen-induced cracking than low-strength steels. Steels that transform martensitically are particularly susceptible, especially the higher-carbon alloys with twinned martensitic structures. The desire to avoid martensite formation has driven the development of high-strength structural steels for welded applications. Production of the newer high-strength, low-alloy steels uses a variety of precisely controlled alloying additions (for example, aluminum, titanium, vanadium, and niobium) along with meticulous thermomechanical processing to develop a very fine-grained ferrite microstructure possessing substantial strength and fracture toughness with a high degree of resistance to hydrogen-induced cracking.

A useful concept for understanding the susceptibility of carbon and alloy steels to hydrogen-induced cracking is the carbon equivalent (CE), an empirical relationship that attempts to reduce the number of significant compositional variables affecting the weldability of steels into a single quantity. Several carbon equivalent relationships have been developed for different classes of steels. An example is:

$$CE = \%C + \frac{\%Mn}{6} + \frac{\%Cr + \%Mo + \%V}{5} + \frac{\%Si + \%Ni + \%Cu}{15} \quad \text{(Eq 2)}$$

From a metallurgical perspective, the carbon equivalent can be related to the development of hydrogen-sensitive microstructures. That is, as the carbon equivalent increases, microstructures are evolved during cooling through the transformation temperature range that are increasingly more susceptible to hydrogen-induced cracking. At high carbon equivalent values, martensitic structures can be expected. In the example (Eq 2), when the carbon equivalent exceeds 0.35%, preheats are recommended to minimize susceptibility to hydrogen cracking. At higher levels of carbon equivalent, both preheats and postheats may be required.

The subsequent guidelines should be followed to minimize the occurrence of hydrogen-induced cracking. For a given level of required strength, the steel with the lowest carbon equivalent should be considered. Low-hydrogen welding practice should be followed. This involves elimination of possible sources of hydrogen by using ultrahigh-purity gases and moisture-free gas lines and by baking coated electrodes, following the manufacturer's recommendations, to ensure removal of nascent water. Finally, preheating and postheating requirements should be used.

Lamellar tearing is cracking that occurs beneath welds. It is found in rolled steel plate weldments. The tearing always lies within the base metal, generally outside the HAZ and parallel to the weld fusion boundary. The problem is caused by welds that subject the base metal to tensile loads in the *z*- or through-direction of the rolled steel. Occasionally, the tearing comes to the surface of the metal but more commonly remains under the weld (Fig. 21) and is detectable only by ultrasonic testing. Lamellar tearing occurs when three conditions are simultaneously present:

- Strains develop in the through-direction of the plate. They are caused by weld metal shrinkage in the joint and can be increased by residual stresses and by loading.
- The weld orientation is such that the stress acts through the joint across the plate thickness (the *z*-direction). The fusion line beneath the weld is roughly parallel to the laminar separation.
- The material has poor ductility in the *z*-direction.

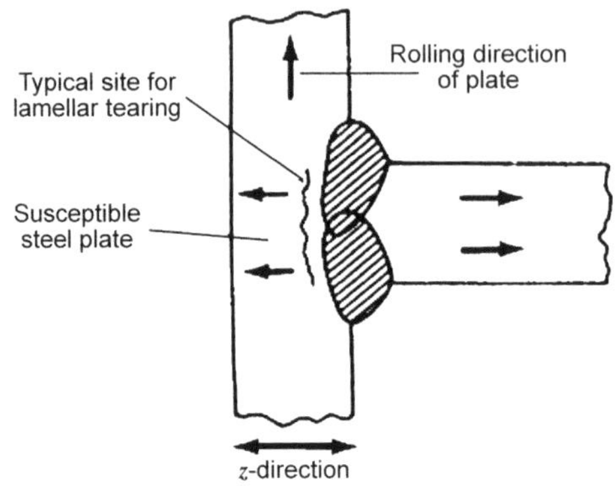

Fig. 21 Typical location for lamellar tearing in a T-joint. Source: Ref 15

The problem can be avoided by proper attention to joint details. In T-joints, double-fillet welds appear to be less susceptible than full-penetration welds. Also, balanced welds on both sides of the joint appear to present less risk than large, single-sided welds.

Service-Life Anomalies

The life of a component or system is heavily dependent on the conditions under which the product operates in service. The service life of a product includes its operation, maintenance, inspection, repair, and modification. Failures due to anomalies in any one of these aspects of service life are unique from those created during the design, procurement of materials, and manufacture of products, as described previously. Examples of the types of root causes of failures that result from unanticipated service conditions are summarized in the following paragraphs.

Operation of the equipment outside of the manufacturer's design parameters would include an example such as a military fighter aircraft in a turn that causes "g" forces that are outside of the operating envelope of the aircraft. Another example is inlet-flow blockage on a high-performance air compressor, resulting in excessive cyclic loads applied to the blades and causing blade (Fig. 22) and drive shaft (Fig. 23) failures. Failure analysis revealed both the compressor rotor and the shaft sustained fatigue failures.

Careful fracture analysis revealed fatigue cracks initiated on the low-pressure side of the blades, which are in compression during normal compressor operation. However, when the inlet flow is blocked, particularly when the blockage is only partial, the blades sustain alternating tensile forces, one load cycle per revolution on the low-pressure side of the blades, resulting in the observed blade fractures. The shaft failed subsequently, due to the severe imbalance and rubbing caused by the blade failures.

Fig. 22 Compressor blade fracture surface showing fatigue origins on low-pressure (i.e., right) side of blade, as indicated by the arrows. Original magnification: 13×. Source: Ref 1

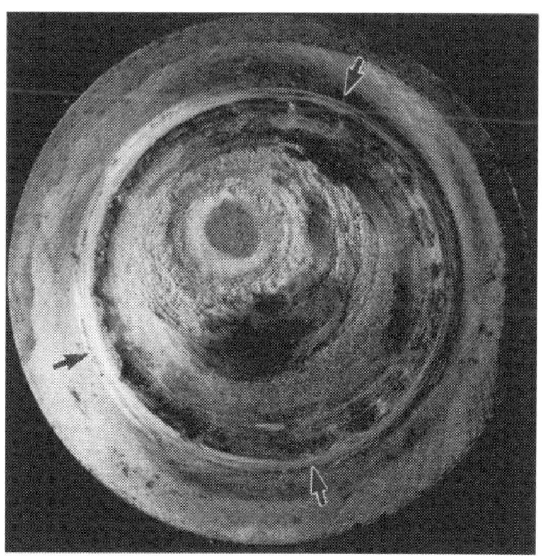

Fig. 23 Failed compressor rotor shaft. Fracture occurred at radius between large and small diameters. Arrows indicate some of fatigue origins. Original magnification: 1×. Source: Ref 1

The application-life diagram is useful in exploring the effects of service-life anomalies on the lives of products. For the compressor inlet blockage case described previously, Fig. 24 depicts the significant loss of service life when the rotor blades sustain the unintended cyclic stresses that occur during an inlet blockage event.

Exposure of the product or system to environments more aggressive than forethought would include examples such as:

- Microbiologically influenced corrosion in a cooling-water system using river water in which the ecosystem has changed
- Organic chloride-containing environment exposing a titanium centrifuge bowl, resulting in stress-corrosion cracking
- Faulty sensor cable resulting in an overtemperature condition in a jet engine, which consumes the high-pressure turbine blade life

Improper maintenance would include examples such as:

- Installing a metallic fuel line onto the mating fitting by forcing the tube to align with the mating fitting. Adding the installation stress to the normal cyclic stresses results in a leak due to fatigue cracking.
- Weld repair of a material that is sensitive to high heat cycles, causing brittle cracking and subsequent fatigue failure
- Misalignment of a bearing during rebuild, causing bending loads on the shaft and resulting in failure by rotating-bending fatigue

Inappropriate Modifications. An example of this would be part-through drill holes in a bicycle handlebar stem, resulting in fatigue initiation at the holes and subsequent fracture (Fig. 25).

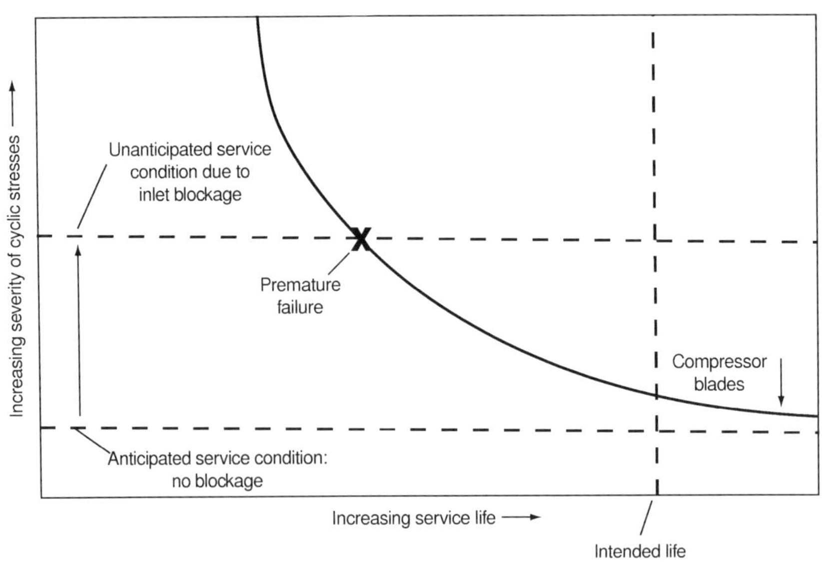

Fig. 24 Application-life diagram showing effects of increasing the severity of the service condition. Source: Ref 1

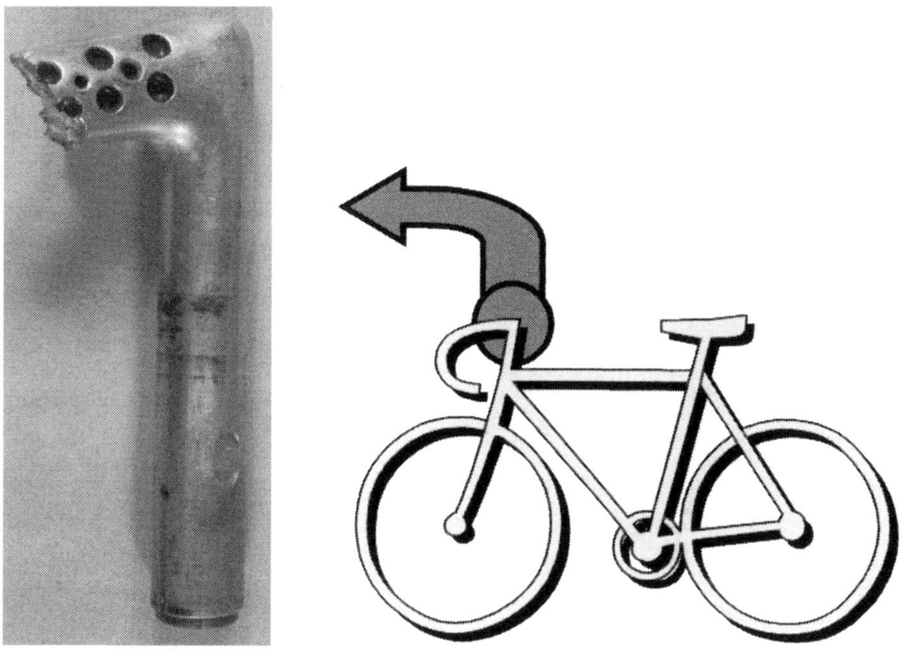

(a)

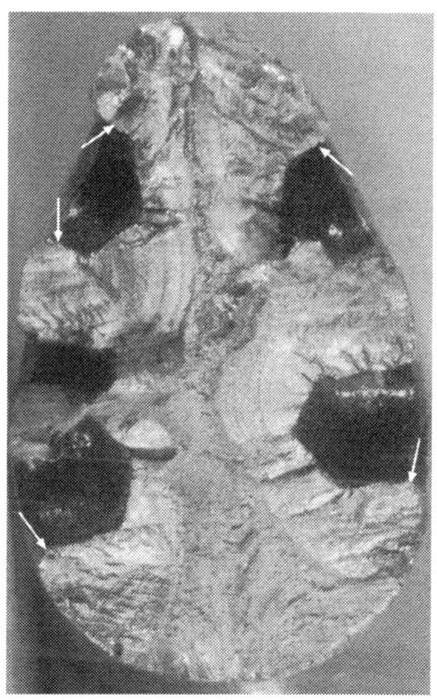

(b)

Fig. 25 (a) User-modified bicycle handlebar stem that failed in service. (b) Multiple fatigue initiations at part-through drill holes. Original magnification: 3×. Source: Ref 1

Materials Selection for Failure Prevention

The use of inappropriate materials and processes accounts for a significant number of failed parts. Table 3, for example, itemizes the general causes of failure, with the frequency of occurrence determined from a sur-

Table 3 Frequency of causes for failure

Cause	Percentage of failures: Engineering components	Percentage of failures: Aircraft components
Improper materials selection	38	...
Fabrication imperfections	15	17
Faulty heat treatment	15	...
Design errors	11	16
Unanticipated service conditions	8	10
Uncontrolled environmental conditions	6	...
Inadequate inspection/quality control	5	...
Material mix	2	...
Inadequate maintenance	...	44
Defective material	...	7
Unknown	...	6

Source: Ref 16

vey. In this survey, materials selection is the most frequent cause of failure for engineered components. However, in the case of aircraft components, the survey did not identify any failures caused by improper materials. This difference illustrates the important point of how different design methods may influence the process of materials selection. In aerospace, for example, design methods should involve more critical evaluations of material alternatives, because the hazards of failure can be severe.

The selection of materials to prevent failure is typically a structured approach including thorough and diligent research into suitable materials. There is no universal guide that will automatically identify the best material for any service, because the number of interrelated input variables can be difficult to manage by a formal decision structure. Executive decision trees and computer expert systems have been developed to simplify materials selection, and these systems can identify candidate materials from very large databases of engineering materials, with cross-referenced mechanical and physical properties. However, these systems are narrowly applicable and are usually industry or company specific.

Effective materials selection is aided by access to materials property information and acquired engineering knowledge of all engineers participating on the design. No generalizations can be made that will be valid for all materials selection problems, because each design problem must be considered individually or on the basis of closely related experience. However, Table 4 provides some general guidance to the criteria that are typically significant in selecting materials in relation to possible failure mechanisms, types of stress, and operating temperatures.

Perhaps one of the most troublesome areas of materials selection relates to the change (or variation) in properties and performance. Property variations can occur within the part geometry from processing and fabrication, or changes in properties can occur over time from factors such as:

- Wear
- Temperature extremes or changes

- Corrosion
- Fatigue

These application conditions require a great deal of judgment in interpreting laboratory test data into design and extrapolating properties and performance over extended periods of time. Often, simulated service testing may be required. An important role of the materials engineer is to assist the designer in making meaningful connections between materials properties and the performance of the part or system being designed. For most mechanical systems, performance is limited not by a single property but by a combination of them. For example, the materials with the best thermal shock resistance are those with the largest values of $\sigma_f/E\alpha$, where σ_f is the failure stress, E is Young's modulus, and α is the coefficient of thermal expansion. These types of performance indices (i.e., groupings of material properties that, when maximized, maximize some aspect of performance) can be useful for comparing materials.

Understanding the connection between properties and the failure modes is also important. Figure 26 is a chart of relationships between common failure modes and material properties. For most modes of failure, two or more material properties act to control the material behavior. However, it is also important to understand how property data should be interpreted. For example, even though most standard specifications require tensile test data, these data are only partially indicative of mechanical performance in

Table 4 Guide to criteria generally useful for selection of material in relation to possible failure mechanisms, types of loading, types of stress, and intended operating temperatures

	Types of loading			Types of stress			Operating temperatures			
Failure mechanisms	Static	Repeated	Impact	Tension	Compression	Shear	Low	Room	High	Criteria generally useful for selection of material
Brittle fracture	X	X	X	X	...	...	X	X	...	Charpy V-notch transition temperature. Notch toughness. K_{Ic} toughness measurements
Ductile fracture(a)	X	...	...	X	...	X	...	X	X	Tensile strength. Shearing yield strength
High-cycle fatigue(b)	...	X	...	X	...	X	X	X	X	Fatigue strength for expected life, with typical stress raisers present
Low-cycle fatigue	...	X	...	X	...	X	X	X	X	Static ductility available and the peak cyclic plastic strain expected at stress raisers during prescribed life
Corrosion fatigue	...	X	...	X	...	X	...	X	X	Corrotion-fatigue strength for the metal and contaminant and for similar time(c)
Buckling	X	...	X	...	X	...	X	X	X	Modulus of elasticity and compressive yield strength
Gross yielding(a)	X	...	...	X	X	X	X	X	X	Yield strength
Creep	X	...	...	X	X	X	...	...	X	Creep rate or sustained stress-rupture strength for the temperature and expected life(c)
Caustic or hydrogen embrittlement	X	...	...	X	...	...	...	X	X	Stability under simultaneous stress and hydrogen or other chemical environment(c)
Stress-corrosion cracking	X	...	...	X	...	X	...	X	X	Residual or imposed stress and corrosion resistance to the environment. K_{ISCC} measurements(c)

K_{Ic}, plane-strain fracture toughness; K_{ISCC}, threshold stress intensity to produce stress-corrosion cracking. (a) Applies to ductile metals only. (b) Millions of cycles. (c) Items strongly dependent on elapsed time. Source: Ref 17

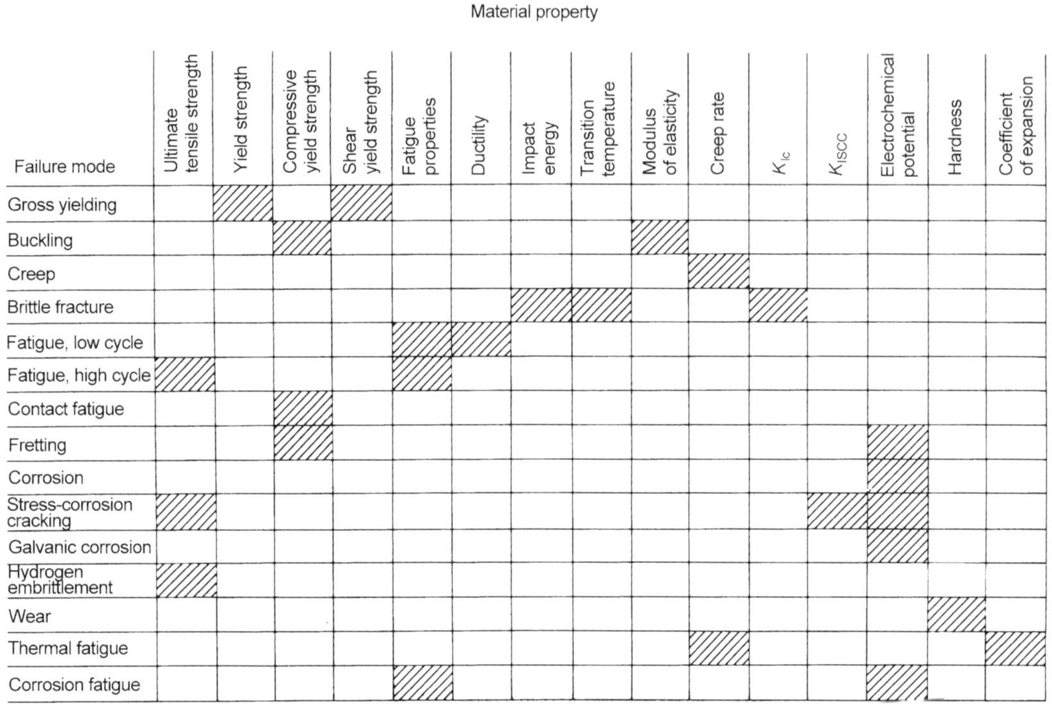

Fig. 26 General relationships between failure modes and material properties. Shaded blocks indicate properties that are influential in controlling a particular failure mode. K_{Ic}, plane-strain fracture toughness; K_{ISCC}, threshold stress intensity for stress-corrosion cracking. Source: Ref 18

specific conditions. The purpose of tensile testing is often to monitor relative quality of different lots, not necessarily for design. Moreover, except in those conditions where ductile fracture or gross yielding may be the limiting condition for failure, tensile strength and yield strength may be inadequate criteria for avoiding failure. A high tensile strength, for example, may be indicative of lower ductility and toughness, and thus, a part with severe stress raisers may be prone to failure.

ACKNOWLEDGMENTS

Portions of this chapter came from "Introduction to Failure Analysis and Prevention" by J.J. Scutti and W.J. McBrine and "Failures Related to Metalworking," both in *Failure Analysis and Prevention*, Vol 11, *ASM Handbook*, ASM International, 2002; "Defects and Distortion in Heat-Treated Parts" by A.K. Sinha, *Heat Treating*, Vol 4, *ASM Handbook*, ASM International, 1991; "Classification of Weld Discontinuities" and "Cracking Phenomena Associated with Welding" by M.J. Cieslak, both in *Welding, Brazing, and Soldering*, Vol 6, *ASM Handbook*, ASM International, 1993.

REFERENCES

1. J.J. Scutti and W.J. McBrine, Introduction to Failure Analysis and Prevention, *Failure Analysis and Prevention,* Vol 11, *ASM Handbook,* ASM International, 2002
2. Failures Related to Metalworking, *Failure Analysis and Prevention,* Vol 11, *ASM Handbook,* ASM International, 2002
3. A.D. Wilson, Calcium Treatment of Plate Steels and Its Effect on Fatigue and Toughness Properties, *Proc. 11th Annual Offshore Technology Conference,* OTC 3465, May 1979, p 939–948
4. A.D. Wilson, Effect of Calcium Treatment on Inclusions in Constructional Steels, *Met. Prog.,* Vol 121 (No. 5), April 1982, p 41–46
5. R. DasGupta, Common Defects in Various Casting Processes, *Casting,* Vol 15, *ASM Handbook,* ASM International, 2008
6. T.S. Piwonka, Design for Casting, *Materials Selection and Design,* Vol 20, *ASM Handbook,* ASM International, 1997
7. Permanent Mold Casting, *Forging and Casting,* Vol 5, *Metals Handbook,* 8th ed., American Society for Metals, 1970, p 279
8. R.W. Bohl, Difficulties and Imperfections Associated with Heat Treated Steel, Lesson 13, *Heat Treatment of Steel,* Materials Engineering Institute Course 10, ASM International, 1978
9. R.F. Kern and M.E. Suess, *Steel Selection,* Wiley-Interscience, 1979
10. R.F. Kern, *Selecting Steels and Designing Parts for Heat Treatment,* American Society for Metals, 1969
11. E.B. Evans, in *Encyclopedia of Materials Science and Engineering,* Pergamon Press, 1986, p 4183–4188
12. A. Rose and H.P. Hougardy, in *Proceedings of the Transformation and Hardenability in Steels Symposium,* Climax Molybdenum Company, 1967, p 155–167
13. R.R. Blackwood, L.M. Jarvis, D.G. Hoffman, and G.E. Totten, Conditions Leading to Quench Cracking Other than Severity of Quench, *Heat Treating, Including the Liu Dai Memorial Symposium, Proc. of the 18th Conf.,* R.A. Wallis and H.W. Walton, Ed., ASM International, 1998, p 575–585
14. Overview of Weld Discontinuities, *Welding, Brazing, and Soldering,* Vol 6, *ASM Handbook,* ASM International, 1993
15. M.J. Cieslak, Cracking Phenomena Associated with Welding, *Welding, Brazing, and Soldering,* Vol 6, *ASM Handbook,* ASM International, 1993
16. G.J. Davies, *Performance in Service: Essential Metallurgy for Engineers,* E.J. Bradbury, Ed., Van Nostrand Reinhold, London, 1985, p 126–155
17. Fundamental Sources of Failure, *Failure Analysis and Prevention,* Vol 10, *Metals Handbook,* 8th ed., American Society for Metals, 1975, p 4

18. C.O. Smith and B.E. Boardman, Concepts and Criteria in Materials Engineering, *Properties and Selection: Stainless Steels, Tool Materials, and Special-Purpose Metals,* Vol 3, *Metals Handbook,* 9th ed., American Society for Metals, 1980, p 825–834

Index

A

B

C

D

E

F

J

K

L

M

N

O

P

S

T

U

X

Y

Z

FSC
www.fsc.org
MIX
Papier | Fördert gute Waldnutzung
FSC® C083411

Kolibri 360 GmbH
Ferdinand-Jühlke-Straße 7
99095 Erfurt, Deutschland
produktsicherheit@kolibri360.de